Robert Qiu • Michael Wicks

Cognitive Networked Sensing and Big Data

Springer

Robert Qiu
Tennessee Technological University
Cookeville, Tennessee, USA

Michael Wicks
Utica, NY, USA

ISBN 978-1-4614-4543-2 ISBN 978-1-4614-4544-9 (eBook)
DOI 10.1007/978-1-4614-4544-9
Springer New York Heidelberg Dordrecht London

Library of Congress Control Number: 2013942391

Printed on acid-free paper

Springer is part of Springer Science+Business Media (www.springer.com)

To
Lily Liman Li

Preface

The idea of writing this book entitled "Cognitive Networked Sensing and Big Data" started with the plan to write a briefing book on wireless distributed computing and cognitive sensing. During our research on large-scale cognitive radio network (and its experimental testbed), we realized that big data played a central role. As a result, the book project reflects this paradigm shift. In the context, sensing roughly is equivalent to "measurement."

We attempt to answer the following basic questions. How do we sense the radio environment using a large-scale network? What is unique to cognitive radio? What do we do with the big data? How does the sample size affect the sensing accuracy?

To address these questions, we are naturally led to ask ourselves: What mathematical tools are required? What are the state-of-the-art for the analytical tools? How these tools are used?

Our prerequisite is the graduate-level course on random variables and processes. Some familiarity with wireless communication and signal processing is useful. This book is complementary with our previous book entitled "Cognitive Radio Communications and Networking: Principles and Practice" (John Wiley and Sons 2012). This book is also complementary with another book of the first author "Introduction to Smart Grid" (John Wiley and Sons 2014). This current book can be viewed as the mathematical tools for the two Wiley books.

Chapter 1 provides the necessary background to support the rest of the book. No attempt has been made to make this book really self-contained. The book will survey many latest results in the literature. We often include preliminary tools from publications. These preliminary tools may be still too difficult for many of the audience. Roughly, our prerequisite is the graduate-level course on random variables and processes.

Chapters 2–5 (Part I) are the core of this book. The contents of these chapters should be new to most graduate students in electrical and computer engineering.

Chapter 2 deals with the sum of matrix-valued random variables. One basic question is "how does the sample size affect the accuracy." The basic building block of the data is the sample covariance matrix, which is a random matrix. Bernstein-type concentration inequalities are of special interest.

Chapter 3 collects together the deepest mathematical theorems in this book. This chapter is really the departure point of this whole book. Chapter 2 is put before this chapter since we want the audience to understand how to deal with the basic linear functions of matrices. The theory of concentration inequality tries to answer the following question: Given a random vector $\mathbf{x}$ taking value in some measurable space $\mathcal{X}$ (which is usually some high dimensional Euclidean space), and a measurable map $f : \mathcal{X} \to \mathbb{R}$, what is a good explicit bound on $\mathbb{P}\left(\left|f\left(\mathbf{x}\right) - \mathbb{E}f\left(\mathbf{x}\right)\right| \geqslant t\right)$? Exact evaluation or accurate approximation is, of course, the **central** purpose of probability theory itself. In situations where exact evaluation or accurate approximation is not possible, which is the case for many practical problems, concentration inequalities aim to do the **next best job** by providing rapidly decaying tail bounds. It is our goal of this book is to systemically deal with the "next best job," when the classical probability theory fails to be valid in these situations.

The sum of random matrices is a sum of linear matrix functions. Non-linear matrix functions are encountered in practice. This motivates us to study, in Chap. 4, the concentration of measure phenomenon, unique to high dimensions. The so-called Lipschitz functions of matrices such as eigenvalues are the mathematical objects.

Chapter 5 culminates for the theoretical development of the random matrix theory. The goal of this chapter is to survey the latest results in the mathematical literature. We tried to be exhaustive in recent results. To our best knowledge, these results are never used in the engineering applications. Although the prerequisites for this chapter are highly demanding, it is out belief that the pay-off will be significant to engineering graduates if they can manage to understand the chapter.

Chapter 6 is included for completion, with the major goal for the readers to compare the parallel results with Chap. 5. Our book "Cognitive Radio Communications and Networking: Principles and Practice" (John Wiley and Sons 2012) contains complementary materials of 230 pages on this subject.

In Part II, we attempt to apply these mathematical tools to different applications. The emphasis is on the connection between the theory and the diverse applications. No attempt is made to collect all the scattered results in one place.

Chapter 7 deals with compressed sensing and recovery of sparse vectors. Concentration inequalities play the central role in the sparse recovery. The so-called restricted isometry property for sensing matrices is another aspect of stating concentration of measure.

A matrix is decomposed into the eigenvalues and the corresponding eigenvectors. When the matrix is of low rank, we can equivalently say the vector of eigenvalues are sparse. Chapter 8 deals with this aspect in the context of concentration of measure.

Statistics starts with covariance matrix estimation. In Chap. 9, we deals with this problem in high dimensions. We think that compressed sensing and low-rank matrix recovery are more basic than covariance matrix estimation.

Once the covariance matrix is estimated, we can apply the statistical information to different applications. In Chap. 10, we apply the covariance matrix to hypothesis detection in high dimensions. During the study of information plus noise model, the low-rank structure is explicitly exploited. This is one justification for putting low-rank matrix recovery (Chap. 9) before this chapter. A modern trend is to exploit the structure of the data (sparsity and low rank) during the detection theory. The research in this direction is growing rapidly. Indeed, we surveyed some latest results in this chapter.

An unexpected chapter is Chap. 11 on probability constrained optimization. Due to the recent progress (as late as 2003 by Nemirovski), optimization with probabilistic constraints, often regarded as computationally intractable in the past, may be formulated in terms of deterministic convex problems that can be solved using modern convex optimization solvers. The "closed-form" Bernstein concentration inequalities play a central role in this formulation.

In Chap. 12, we show how concentration inequalities play a central role in data friendly data processing such as low rank matrix approximation. We only want to point out the connection.

Chapter 13 is designed to put all pieces together. This chapter may be put as Chap. 1. We can see that so many problems are open. We only touched the tip of the iceberg of the big data. Chapter 1 also gives us motivations of other chapters of this book.

The first author wants to thank his students for the course in the Fall semester of 2012: ECE 7970 Random Matrices, Concentration and Networking. Their comments greatly improved this book. We also want to thank PhD students at TTU for their help in proof-reading: Jason Bonior, Shujie Hou, Xia Li, Feng Lin, and Changchun Zhang. The simulations made by Feng Lin, indeed, shaped the conceptions and formulations of many places of this book, in particular on hypothesis detection. Dr. Zhen Hu and Dr. Nan Guo at TTU are also of help for their discussions. The first author's research collaborator Professor Husheng Li (University of Tennessee at Knoxville) is acknowledged for many inspired discussions.

The first author's work has been supported, for many years, by Office of Naval Research (ONR) through the program manager Dr. Santanu K. Das (code 31). Our friend Paul James Browning is instrumental in making this book possible. This work is partly funded by National Science Foundation (NSF) through two grants (ECCS-0901420 and CNS-1247778), Office of Naval Research through two grants (N00010-10-10810 and N00014-11-1-0006), and Air Force Office of Scientific Research, via a local contractor (prime contract number FA8650-10-D-1750-Task 4). Some parts of this book were finished while the first author was a visiting professor during the summer of 2012 at the Centre for Quantifiable Quality of Service in Communication Systems (Q2S), the Norwegian University of Science and Technology (NTNU), Trondheim, Norway. Many discussions with his host Professor Yuming Jiang are acknowledged here.

The authors want to thank our editor Brett Kurzman at Springer (US) for his interest in this book. We acknowledge Rebecca Hytowitz at Springer (US) for her help.

The first author wants to thank his mentors who, for the first time, exposed him to many subjects of this book: Weigan Lin (UESTC, China) on remote sensing,

Zhengde Wu (UESTC, China) on electromagnetics, Shuzhang Liu (UESTC, China) on electromagnetic materials, I-Tai Lu (Polytechnic Institute of NYU) on radio propagation, Lawrence Carin (Duke) on physics-based signal processing, Leopold Felsen on scattering, and Henry Bertoni on radio propagation in wireless channels. His industrial colleagues at GTE Labs (now Verizon Wireless) and Bell Labs (Alcatel-Lucent) greatly shaped his view.

Last, but not the least, the first author wants to thank his wife Lily Liman Li for her love, encouragement, and support that sustains him during the lonely (but exciting) book writing journey—she is always there; and his children Michelle, David and Jackie light his life. This is, indeed, a special moment to record a personal note: This summer his eldest daughter Michelle is going to drive on her own, David is happy in his high school, and Jackie smiles again. Also this summer will mark the time of one decade after his moving back to academia—after spending 8 years in industry. Wring books like this was among his deepest wishes that drove him to read mathematical papers and books while watching his TV for the last two decades. Finally, he is in memory of her mother Suxian Li who passed away in 2006. His father Dafu Qiu lives in China. His parents-in-laws Lumei Li and Jinxue Chen live with him for many years. Their love and daily encouragement really make a difference in his life.

Contents

Part I Theory

Introduction

This book deals with the data that is collected from a cognitive radio network. Although this is the motivation, the contents really treat the general mathematical models and the latest results in the literature.

Big data, only at its dawn, refers to things one can do at a large scale that cannot be done at a smaller one [4]. Mankind's constraints are really functions of the *scale* in which we operate. Scale really matters. Big data see a shift from causation to correlation, to infer probabilities. Big data is messy. The data is huge, and can tolerate inexactitude. Mathematical models crunch mountains of data to predict gains, while trying to reduce risks.

At this writing, big data is viewed as a paradigm shift in science and engineering, as illustrated in Fig. 1. In November 2011 when we gave the final touch to our book [5] on cognitive radio network. The authors of this book recognized the fundamental significance of big data. So at the first page and first section, our section title (Sect. 1.1 of [5]) was called "big data." Our understanding was that due to the spectrum sensing, the cognitive radio network leads us naturally towards big data. In the last 18 months, as a result of this book writing, this understanding went even further: we swam in the mathematical domains, understanding the beauty of the consequence of big data—high dimensions. Book writing is truly a journey, and helps one to understand the subject much deeper than otherwise. It is believed that smart grid [6] will use many big data concepts and hopefully some mathematical tools that are covered in this book. Many mathematical insights could not be explicit, if the high dimensions were not assumed to be large. As a result, concentration inequalities are natural tools to capture this insight in a non-asymptotic manner.

Figure 13.1 illustrates the vision of big data that will be the foundation to understand cognitive networked sensing, cognitive radio network, cognitive radar and even smart grid. We will further develop this vision in the book on smart grid [6]. High dimensional statistics is the driver behind these subjects. Random matrices are natural building blocks to model big data. Concentration of measure phenomenon is of fundamental significance to modeling a large number of random matrices. Concentration of measure phenomenon is a phenomenon unique to high-dimensional spaces.

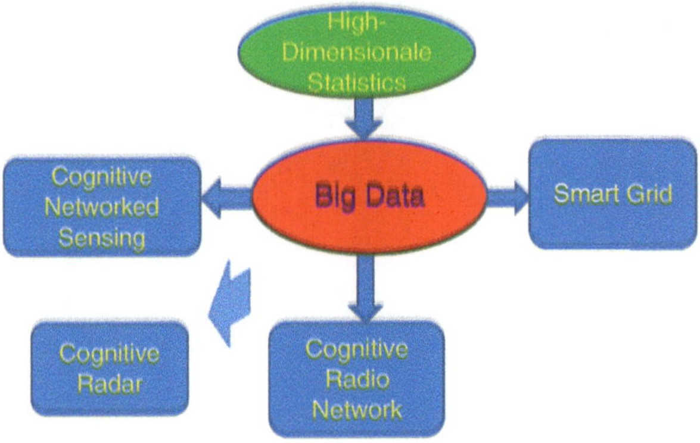

Fig. 1 Big data vision

To get a feel for the book, let us consider one basic problem. The large data sets are conveniently expressed as a matrix

$$\mathbf{X} = \begin{bmatrix} X_{11} & X_{12} & \cdots & X_{1n} \\ X_{21} & X_{22} & \cdots & X_{2n} \\ \vdots & \vdots & \cdots & \vdots \\ X_{m1} & X_{m2} & \cdots & X_{mn} \end{bmatrix} \in \mathbb{C}^{m \times n}$$

where X_{ij} are random variables, e.g., sub-Gaussian random variables. Here m, n are finite and large. For example, $m = 100, n = 100$. The spectrum of a random matrix X tends to stabilize as the dimensions of $\mathbf{X}$ grows to infinity. In the last few years, local and non-asymptotic regimes, the dimensions of $\mathbf{X}$ are fixed rather than grow to infinity. Concentration of measure phenomenon naturally occurs. The eigenvalues $\lambda_i\left(\mathbf{X}^T\mathbf{X}\right), i = 1, \ldots, n$ are natural mathematical objects to study. The eigenvalues can be viewed as Lipschitz functions that can be handled by Talagrand's concentration inequality. It expresses the insight: The sum of a large number of random variables is a constant with *high probability*. We can often treat both standard Gaussian and Bernoulli random variables in the unified framework of the sub-Gaussian family.

Theorem 1 (Talagrand's Concentration Inequality). *For every product probability* $\mathbb{P}$ *on* $\{-1, 1\}^n$, *consider a convex and Lipschitz function* $f : \mathbb{R}^n \to \mathbb{R}$ *with Lipschitz constant L. Let* $X_1, \ldots, X_n$ *be independent random variables taking values* $\{-1, 1\}$. *Let* $Y = f(X_1, \ldots, X_n)$ *and let* $\mathbb{M}Y$ *be a median of* Y. *Then For every* $t > 0$, *we have*

$$\mathbb{P}\left(|Y - \mathbb{M}Y| \geqslant t\right) \leqslant 4e^{-t^2/16L^2}. \tag{1}$$

The random variable Y has the following property

$$\mathrm{Var}\left(Y\right)\leqslant 16L^2,\qquad \mathbb{E}\left[Y\right]-16L\leqslant \mathbb{M}\left[Y\right]\leqslant \mathbb{E}\left[Y\right]+16L. \tag{2}$$

For a random matrix $\mathbf{X}\in\mathbb{R}^{n\times n}$, the following functions are Lipschitz functions:

$$(1)\lambda_{\max}\left(\mathbf{X}\right);(2)\lambda_{min}\left(\mathbf{X}\right);(3)\,\mathrm{Tr}\left(\mathbf{X}\right);(4)\sum_{i=1}^{k}\lambda_i\left(\mathbf{X}\right);(5)\sum_{i=1}^{k}\lambda_{n-i+1}\left(\mathbf{X}\right)$$

where $\mathrm{Tr}\left(\mathbf{X}\right)$ has a Lipschitz constant of $L=1/n$, and $\lambda_i\left(\mathbf{X}\right), i=1,\ldots,n$ has a Lipschitz constant of $L=1/\sqrt{n}$. So the variance of $\mathrm{Tr}\left(\mathbf{X}\right)$ is upper bounded by $16/n^2$, while the variance of $\lambda_i\left(\mathbf{X}\right), i=1,\ldots,n$ by $16/n$. The variance of $\mathrm{Tr}\left(\mathbf{X}\right)$ is $1/n$ smaller than that of $\lambda_i\left(\mathbf{X}\right), i=1,\ldots,n$. For example, $n=100$, their ratio is 20 dB. The variance has a fundamental control over the hypothesis detection.

Part I
Theory

Chapter 1
Mathematical Foundation

This chapter provides the necessary background to support the rest of the book. No attempt has been made to make this book really self-contained. The book will survey many recent results in the literature. We often include preliminary tools from publications. These preliminary tools may be still too difficult for many of the audience. Roughly, our prerequisite is the graduate-level course on random variables and processes.

1.1 Basic Probability

The probability of an event is expressed as $(\cdot)$, and we use $\mathbb{E}$ for the expectation operator. For conditional expectation, we use the notation $\mathbb{E}_X Z$, which represents integration with respect to X, holding all other variables fixed. We sometimes omit the parentheses when there is no possibility of confusion. Finally, we remind the reader of the analysts convention that roman letters c, C, etc. denote universal constants that may change at every appearance.

1.1.1 Union Bound

Let $(\Omega, \mathcal{F}, \mathbb{P})$ be a probability space, $\mathcal{F}$ denotes a σ-algebra on the sample space Ω and $\mathbb{P}$ a probability measure on $(\Omega, \mathcal{F})$. The probability of an event $A \in \mathcal{F}$ is denoted by

$$\mathbb{P}(A) = \int_A d\mathbb{P}(\omega) = \int_\Omega I_A(\omega)\, d\mathbb{P}(\omega),$$

R. Qiu and M. Wicks, *Cognitive Networked Sensing and Big Data*,
DOI 10.1007/978-1-4614-4544-9_1,

where the indicator function $I_A(\omega)$ takes the value of 1 if $\omega \in A$ and 0 otherwise. The union bound (or Bonferroni's inequality, or Boole's inequality) states that for a collection of events $A_i \in \mathcal{F}, i = 1, \ldots, n$, we have

$$\mathbb{P}\left(\bigcup_{i=1}^{n} A_i\right) \leqslant \sum_{i=1}^{n} \mathbb{P}(A_i). \tag{1.1}$$

1.1.2 Independence

In general, it can be shown that the random variable X and Y are *independent* if and only if their joint cdf is equal to the product of its marginal cdf's:

$$F_{X,Y}(x,y) = F_X(x)F_Y(y) \text{for all } x \text{ and } y. \tag{1.2}$$

Similarly, if X and Y are jointly continuous, then X and Y are independent if and only if their joint pdf is equal to the product of its marginal pdf's:

$$f_{X,Y}(x,y) = f_X(x)f_Y(y) \text{for all } x \text{ and } y. \tag{1.3}$$

Equation (1.3) is obtained from (1.2) by differentiation. Conversely, (1.2) is obtained from (1.3) by integration.

1.1.3 Pairs of Random Variables

Suppose that X and Y are independent random variables, and let $g(X,Y) = g_1(X)g_2(Y)$. Find $\mathbb{E}[g(X,Y)] = \mathbb{E}[g_1(X)g_2(Y)]$. It follows that

$$\begin{aligned}\mathbb{E}[g(X,Y)] &= \int_{-\infty}^{\infty} g_1(x')g_2(y') \int_{-\infty}^{\infty} f_X(x')f_Y(y')\,dx'dy' \\ &= \left\{\int_{-\infty}^{\infty} g_1(x')f_X(x')\,dx'\right\}\left\{\int_{-\infty}^{\infty} g_2(y')f_Y(y')\,dy'\right\} \\ &= \mathbb{E}[g_1(X)\,g_2(Y)]\,. \end{aligned} \tag{1.4}$$

Let us consider the sum of two random variables. $Z = X + Y$. Find $F_Z(z)$ and $f_Z(z)$ in terms of the joint pdf of X and Y.

The cdf of Z is found by integrating the joint pdf of X and Y over the region of the plane corresponding to the event $\mathbb{P}(Z \leq z) = \mathbb{P}(X + Y \leq z)$.

$$F_Z(z) = \int_{-\infty}^{\infty}\int_{-\infty}^{z-x'} f_{X,Y}(x',y')\,dx'dy'.$$

The pdf of Z is

$$f_Z(z) = \frac{d}{dz} F_Z(z) = \int_{-\infty}^{\infty} f_{X,Y}(x', z - x')\, dx' dy'.$$

Thus the pdf for the sum of two random variables is given by a superposition integral.

If X and Y are independent random variables, then by (1.3), the pdf is given by the convolution integral of the marginal pdf's of X and Y:

$$f_Z(z) = \int_{-\infty}^{\infty} f_X(x') f_Y(z - x')\, dx'.$$

Thus, the pdf of $Z = X + Y$ is given by the convolution of the pdf's of X and Y:

$$\mathrm{f}_Z(z) = f_X(x) * f_Y(y). \tag{1.5}$$

1.1.4 The Markov and Chebyshev Inequalities and Chernoff Bound

These inequalities in this subsection will be generalized to the matrix setting, replacing the scalar-valued random variable X with the matrix-valued random variables $\mathbf{X}$—random matrices.

In general, the mean and variance of a random variable do not provide enough information to determine the cdf/pdf. However, the mean and variance do allow us to obtain bounds for probabilities of the form $\mathbb{P}(|X| \geqslant t)$. Suppose that X is a nonnegative random variable with mean $\mathbb{E}[X]$. The **Markov inequality** then states that

$$\mathbb{P}(X \geqslant a) \leqslant \frac{\mathbb{E}[X]}{a} \quad \text{for } X \text{ nonnegative.} \tag{1.6}$$

It follows from Markov's inequality that if φ is a strictly monotonically increasing nonnegative-valued function, then for any random variable X and real number, we have [7]

$$\mathbb{P}(X \geqslant t) = \mathbb{P}\{\phi(X) \geqslant \phi(t)\} \leqslant \frac{\mathbb{E}\phi(X)}{\phi(t)}. \tag{1.7}$$

An application of (1.7) with $\varphi(x) = x^2$ is Chebyshev's inequality: if X is an arbitrary random variable and $t > 0$, then

$$\mathbb{P}(|X - \mathbb{E}X| \geqslant t) = \mathbb{P}\left(|X - \mathbb{E}X|^2 \geqslant t^2\right) \leqslant \frac{\mathbb{E}|X - \mathbb{E}X|^2}{t^2} = \frac{\mathrm{Var}[X]}{t^2}.$$

Now suppose that the mean $\mathbb{E}[X] = m$ and the variance $\mathrm{Var}[X] = \sigma^2$ of a random variable are known, and that we are interested in bounding $\mathbb{P}(|X - m| \geqslant a)$. The *Chebyshev inequality* states that

$$\mathbb{P}(|X - m| \geqslant a) \leqslant \frac{\sigma^2}{a^2}. \tag{1.8}$$

The Chebyshev inequality is a consequence of the Markov inequality. More generally, taking $\phi(x) = x^q\ (x \geqslant 0)$, for any $q > 0$ we have the moment bound

$$\mathbb{P}(|X - \mathbb{E}X| \geqslant t) \leqslant \frac{\mathbb{E}|X - \mathbb{E}X|^q}{t^q}. \tag{1.9}$$

In specific examples, we may choose the value of q to optimize the obtained upper bound. Such moment bounds often provide with very sharp estimates of the tail probabilities. A related idea is at the basis of *Chernoff's bounding method.* Taking $\varphi(x) = e^{sx}$ where x is an arbitrary positive number, for any random variable X, and any $t \in \mathbb{R}$, we have

$$\mathbb{P}(X \geqslant t) = \mathbb{P}\left\{e^{sX} \geqslant e^{st}\right\} \leqslant \frac{\mathbb{E}e^{sX}}{e^{st}}. \tag{1.10}$$

If more information is available than just the mean and variance, then it is possible to obtain bounds that are tighter than the Markov and Chebyshev inequalities. The region of interest is $A = \{t \geqslant a\}$, so let $I_A(t)$ be the indicator function, that is $I_A(t) = 1, t \in A$ and $I_A(t) = 0$ otherwise. Consider the bound $I_A(t) \leqslant e^{s(t-a)}$, $s > 0$. The resulting bound is

$$\begin{aligned}\mathbb{P}(X \geqslant a) &= \int_0^\infty I_A(t) f_X(t) dt \leqslant \int_0^\infty e^{s(t-a)} f_X(t) dt \\ &= e^{-sa} \int_0^\infty e^{st} f_X(t) dt = e^{-sa}\mathbb{E}\left[e^{sX}\right].\end{aligned} \tag{1.11}$$

This bound is the *Chernoff bound,* which can be seen to depend on the expected value of an exponential function of X. This function is called the moment generating function and is related to the Fourier and Laplace transforms in the following subsections. In Chernoff's method, we find an $s \geq 0$ that minimizes the upper bound or makes the upper bound small. Even though Chernoff bounds are never as good as the best moment bound, in many cases they are easier to handle [7].

1.1.5 Characteristic Function and Fourier Transform

Transform methods are extremely useful. In many applications, the solution is given by the convolution of two functions $f_1(x) * f_2(x)$. The Fourier transform will convert this convolution integral into a product of two functions in the transformed domains. This is a result of a linear system, which is most fundamental.

The characteristic function of a random variable is defined by

$$\Phi_X(\omega) = \mathbb{E}\left[e^{j\omega X}\right] = \int_{-\infty}^{\infty} f_X(x)e^{j\omega x}dx,$$

where $j = \sqrt{-1}$ is the imaginary unit number. If we view $\Phi_X(\omega)$ as a Fourier transform, then we have from the Fourier transform inversion formula that the pdf of X is given by

$$f_X(x) = \frac{1}{2\pi}\int_{-\infty}^{\infty} \Phi_X(\omega)e^{-j\omega x}d\omega.$$

If X is a discrete random variable, it follows that

$$\Phi_X(\omega) = \sum_k p_X(x_k)e^{j\omega x_k} \text{discrete random variables.}$$

Most of time we deal with discrete random variables that are integer-valued. The characteristic function is then defined as

$$\Phi_X(\omega) = \sum_k p_X(k)e^{j\omega k} \text{ integer-valued random variables.} \quad (1.12)$$

Equation (1.12) is the Fourier transform of the sequence $p_X(k)$. The following inverse formula allows us to recover the probabilities $p_X(k)$ from $\Phi_X(\omega)$

$$p_X(k) = \frac{1}{2\pi}\int_0^{2\pi} \Phi_X(\omega)\, e^{-j\omega k}d\omega \quad \text{k} = 0, \pm 1, \pm 2, \ldots$$

$p_X(k)$ are simply the coefficients of the Fourier series of the periodic function $\Phi_X(\omega)$. The moments of X can be defined by $\Phi_X(\omega)$—a very basic idea. The power series (Taylor series) can be used to expand the complex exponential $e^{-j\omega x}$ in the definition of $\Phi_X(\omega)$:

$$\Phi_X(\omega) = \int_{-\infty}^{\infty} f_X(x)\left\{1 + j\omega X + \frac{1}{2!}(j\omega X)^2 + \cdots\right\} e^{j\omega x}dx.$$

Assuming that all the moments of X are finite and the series can be integrated term by term, we have

$$\Phi_X(\omega) = 1 + j\omega\mathbb{E}[X] + \frac{1}{2!}(j\omega)^2\mathbb{E}\left[X^2\right] + \cdots + \frac{1}{n!}(j\omega)^n\mathbb{E}[X^n] + \cdots.$$

If we differentiate the above expression once and evaluate the result at $\omega = 0$, we obtain

$$\left(\frac{d}{d\omega}\right)^n \Phi_X(\omega)\bigg|_{\omega=0} = j^n\mathbb{E}[X^n],$$

which yields the final result

$$\mathbb{E}\left[X^n\right] = \frac{1}{j^n}\left(\frac{d}{d\omega}\right)^n \Phi_X\left(\omega\right)\bigg|_{\omega=0}.$$

Example 1.1.1 (Chernoff Bound for Gaussian Random Variable). Let X be a Gaussian random variable with mean m and variance σ^2. Find the Chernoff bound for X. □

1.1.6 Laplace Transform of the pdf

When we deal with nonnegative continuous random variables, it is customary to work with the **Laplace transform** of the pdf,

$$X^*(s) = \int_0^{\infty} f_X(x)e^{-sx}dx = \mathbb{E}\left[e^{-sX}\right]. \tag{1.13}$$

Note that $X^*(s)$ can be regarded as a Laplace transform of the pdf or as an expected value of a function of X, e^{-sX}. When X is replaced with a matrix-valued random variable $\mathbf{X}$, we are motivated to study

$$\mathbf{X}^*(s) = \int_0^{\infty} f_{\mathbf{X}}(x)e^{-sx}dx = \mathbb{E}\left[e^{-s\mathbf{X}}\right]. \tag{1.14}$$

Through the spectral mapping theorem defined in Theorem 1.4.4, $f(\mathbf{A})$ is defined simply by applying the function f to the eigenvalues of $\mathbf{A}$, where $f(x)$ is an arbitrary function. Here we have $f(x) = e^{-sx}$. The eigenvalues of the matrix-valued random variable $e^{-s\mathbf{X}}$ are scalar-valued random variables.

The moment theorem also holds for $X^*(s)$:

$$\mathbb{E}\left[X^n\right] = (-1)^n \frac{d^n}{ds^n} X^*(s)\bigg|_{s=0}.$$

1.1.7 Probability Generating Function

In problems where random variables are nonnegative, it is usually more convenient to use the z-transform or the Laplace transform. The **probability generating function** $G_N(z)$ of a nonnegative integer-valued random variables N is defined by

$$G_N(z) = \mathbb{E}\left[z^N\right] = \sum_{k=0}^{\infty} p_N(k)z^k. \tag{1.15}$$

The first expression is the expected value of the function of N, z^N. The second expression is the z-transform the probability mass function (with a sign change in the exponent). Table 3.1 of [8, p. 175] shows the probability generating function for some discrete random variables. Note that the characteristic function of N is given by $G_N(z) = \mathbb{E}\left[z^N\right] = G_N(e^{j\omega})$.

1.2 Sums of Independent (Scalar-Valued) Random Variables and Central Limit Theorem

We follow [8] on the standard Fourier-analytic proof of the central limit theorem for scalar-valued random variables. This material allows us to warm up, and set the stage for a parallel development of a theory for studying the sums of the matrix-valued random variables. The Fourier-analytical proof of the central limit theorem is one of the quickest (and slickest) proofs available for this theorem, and accordingly the "standard" proof given in probability textbooks [9].

Let $X_1, X_2, \ldots, X_n$ be n *independent* random variables. In this section, we show how the standard Fourier transform methods can be used to find the pdf of $S_n = X_1 + X_2 + \ldots + X_n$.

First, consider the $n = 2$ case, $Z = X + Y$, where X and Y are independent random variables. The characteristic function of Z is given by

$$\begin{aligned}\Phi_Z(\omega) &= \mathbb{E}\left[e^{j\omega Z}\right] = \mathbb{E}\left[e^{j\omega(X+Y)}\right] = \mathbb{E}\left[e^{j\omega X}e^{j\omega Y}\right] \\ &= \mathbb{E}\left[e^{j\omega X}\right]\mathbb{E}\left[e^{j\omega Y}\right] = \Phi_X(\omega)\,\Phi_Y(\omega),\end{aligned}$$

where the second line follows from the fact that functions of independent random variables (i.e., $e^{j\omega X}$ and $e^{j\omega Y}$) are also independent random variables, as discussed in (1.4). Thus the characteristic function of Z is the product of the individual characteristic functions of X and Y.

Recall that $\Phi_Z(\omega)$ can be also viewed as the Fourier transform of the pdf of Z:

$$\Phi_Z(\omega) = \mathcal{F}\left\{\mathrm{f}_Z(z)\right\}.$$

According to (1.5), we obtain

$$\Phi_Z(\omega) = \mathcal{F}\left\{\mathrm{f}_Z(z)\right\} = \mathcal{F}\left\{f_X(x) * f_Y(y)\right\} = \Phi_X(\omega)\,\Phi_Y(\omega). \tag{1.16}$$

Equation (1.16) states the well-known result that the Fourier transform of a convolution of two functions is equal to the product of the individual Fourier transforms. Now consider the sum of n independent random variables:

$$S_n = X_1 + X_2 + \cdots + X_n.$$

The characteristic function of S_n is

$$\begin{aligned}\Phi_{S_n}(\omega) &= \mathbb{E}\left[e^{j\omega S_n}\right] = \mathbb{E}\left[e^{j\omega(X_1+X_2+\cdots+X_n)}\right] = \mathbb{E}\left[e^{j\omega X_1}e^{j\omega X_2}\cdots e^{j\omega X_n}\right]\\ &= \mathbb{E}\left[e^{j\omega X_1}\right]\mathbb{E}\left[e^{j\omega X_2}\right]\cdots\mathbb{E}\left[e^{j\omega X_n}\right]\\ &= \Phi_{X_1}(\omega)\,\Phi_{X_2}(\omega)\cdots\Phi_{X_n}(\omega)\,. \end{aligned} \tag{1.17}$$

Thus the pdf of S_n can then be found by finding the inverse Fourier transform of the product of the individual characteristic functions of the X_i's:

$$f_{s_n}(x) = \mathcal{F}^{-1}\left\{\Phi_{X_1}(\omega)\,\Phi_{X_1}(\omega)\cdots\Phi_{X_n}(\omega)\right\}. \tag{1.18}$$

Example 1.2.1 (Sum of Independent Gaussian Random Variables). Let S_n be the sum of n independent Gaussian random variables with respective means and variances, $m_1, m_2, \ldots, m_n$ and ${\sigma^2}_1, {\sigma^2}_2, \ldots, {\sigma^2}_n$. Find the pdf of S_n. The characteristic function of X_k is

$$\Phi_{X_k}(\omega) = e^{+j\omega m_k - \omega^2\sigma_k^2/2}$$

so by (1.17),

$$\begin{aligned}\Phi_{S_n}(\omega) &= \prod_{k=1}^{n} e^{+j\omega m_k - \omega^2\sigma_k^2/2}\\ &= \exp\left\{+j\omega\left(m_1+m_2+\cdots+m_n\right) - \omega^2\left(\sigma_1^2+\sigma_2^2+\cdots+\right)\sigma_n^2\right\}.\end{aligned}$$

This is the characteristic function of a Gaussian random variable. Thus S_n is a Gaussian random variable with mean $m_1 + m_2 + \cdots + m_n$ and variance $\sigma_1^2 + \sigma_2^2 + \cdots + \sigma_n^2$. □

Example 1.2.2 (Sum of i.i.d. Random Variables). Find the pdf of a sum of n independent, identically distributed random variables with characteristic functions

$$\Phi_{X_k}(\omega) = \Phi_X(\omega) \text{ for } k = 1, 2, \ldots, n.$$

Equation (1.17) immediately implies that the characteristic function S_n is

$$\Phi_{S_n}(\omega) = \mathbb{E}\left[e^{j\omega X_1}\right]\mathbb{E}\left[e^{j\omega X_2}\right]\cdots\mathbb{E}\left[e^{j\omega X_n}\right] = \left\{\Phi_X(\omega)\right\}^n.$$

The pdf of S_n is found by taking the inverse transform of this expression. □

Example 1.2.3 (Sum of i.i.d. Exponential Random Variables). Find the pdf of a sum of n independent exponentially distributed random variables, all with parameter α. The characteristic function of a single exponential random variable is

$$\Phi_X(\omega) = \frac{\alpha}{\alpha - j\omega}.$$

From the previous example we then have that

$$\Phi_{S_n}(\omega) = \{\Phi_X(\omega)\}^n = \left\{\frac{\alpha}{\alpha - j\omega}\right\}^n.$$

From Table 4.1 of [8], we see that S_n is an m-Erlang random variables. □

When dealing with integer-valued random variables it is usually preferable to work with the probability generating function defined in (1.15)

$$G_N(z) = \mathbb{E}\left[z^N\right].$$

The generating function for a sum of independent discrete random variables, $N = X_1 + \cdots + X_n$, is

$$G_N(z) = \mathbb{E}\left[z^{X_1+\cdots+X_n}\right] = \mathbb{E}\left[z^{X_1}\right]\cdots\mathbb{E}\left[z^{X_n}\right] = G_{X_1}(z)\cdots G_{X_n}(z). \quad (1.19)$$

Example 1.2.4. Find the generating function for a sum of n independent, identically distributed random variables.

The generating function for a single geometric random variable is given by

$$G_X(z) = \frac{pz}{1 - qz}.$$

Therefore, the generating function for a sum of n such independent random variables is

$$G_N(z) = \left\{\frac{pz}{1 - qz}\right\}^n.$$

From Table 3.1 of [8], we see that this is the generating function of a negative binomial random variable with parameter p and n. □

1.3 Sums of Independent (Scalar-Valued) Random Variables and Classical Deviation Inequalities: Hoeffding, Bernstein, and Efron-Stein

We are mainly concerned with upper bounds for the probabilities of deviations from the mean, that is, to obtain $\mathbb{P}(S_n - \mathbb{E}S_n \geqslant t)$, with $S_n = \sum_{i=1}^{n} X_i$, where $X_1, \ldots, X_n$ are independent real-valued random variables.

Chebyshev's inequality and independence imply

$$\mathbb{P}\left(S_n - \mathbb{E}S_n \geqslant t\right) \leqslant \frac{\mathrm{Var}\left[S_n\right]}{t^2} = \frac{\sum\limits_{i=1}^{n} \mathrm{Var}\left[X_n\right]}{t^2}.$$

In other words, writing $\sigma^2 = \frac{1}{n}\sum\limits_{i=1}^{n} \mathrm{Var}\left[X_n\right]$, we have

$$\mathbb{P}\left(S_n - \mathbb{E}S_n \geqslant t\right) \leqslant \frac{n\sigma^2}{t^2}.$$

This simple inequality is at the basis of the *weak law of large numbers.*

1.3.1 Transform of Probability Bounds to Expectation Bounds

We often need to convert a probability bound for the random variable X to an expectation bound for X^p, for all $p \geq 1$. This conversion is of independent interest. It may be more convenient to apply expectation bounds since expectation can be approximated by average. Our result here is due to [10]. Let X be a random variable assuming non-negative values. Let a, b, t and h be non-negative parameters. If we have exponential-like tail

$$\mathbb{P}\left(X \geqslant a + tb\right) \leqslant e^{-t+h},$$

then, for all $p \geq 1$,

$$\mathbb{E}X^p \leqslant 2(a + bh + bp)^p. \tag{1.20}$$

On the other hand, if we have Gaussian-like tail

$$\mathbb{P}\left(X \geqslant a + tb\right) \leqslant e^{-t^2+h},$$

then, for all $p \geq 1$,

$$\mathbb{E}X^p \leqslant 3 * \sqrt{p}\left(a + b\sqrt{h} + b\sqrt{p/2}\right)^p. \tag{1.21}$$

1.3.2 Hoeffding's Inequality

Chernoff's bounding method, described in Sect. 1.1.4, is especially convenient for bounding tail probabilities of sums of independent random variables. The reason is that since the expected value of a product of independent random variables equals the product of the expected variables—this is not true for matrix-valued random variables, Chernoff's bound becomes

$$\begin{aligned}\mathbb{P}\left(S_n - \mathbb{E}S_n \geqslant t\right) &\leqslant e^{-st}\mathbb{E}\left[\exp\left(s\sum_{i=1}^{n}\left(X_i - \mathbb{E}X_i\right)\right)\right] \\ &= e^{-st}\prod_{i=1}^{n}\mathbb{E}\left[e^{s(X_i - \mathbb{E}X_i)}\right] \qquad \text{by independence.}\end{aligned} \tag{1.22}$$

Now the problem of finding tight bounds comes down to finding a good upper bound for the moment generating function of the random variables $X_i - \mathbb{E}X_i$. There are many ways of doing this. In the case of bounded random variables, perhaps the most elegant version is due to Hoeffding [11]:

Lemma 1.3.1 (Hoeffding's Inequality). *Let X be a random variable with $\mathbb{E}X = 0$, $a \le X \le b$. Then for $s \ge 0$,*

$$\mathbb{E}\left[e^{sX}\right] \leqslant e^{s^2(b-a)^2/8}. \tag{1.23}$$

Proof. The convexity of the exponential function implies

$$e^{sx} \leqslant \frac{x-a}{b-a}e^{sb} + \frac{b-x}{b-a}e^{sa} \quad \text{for} \quad a \leqslant x \leqslant b.$$

Expectation is linear. Exploiting $\mathbb{E}X = 0$, and using the notation $p = \frac{-a}{b-a}$ we have

$$\begin{aligned}\mathbb{E}e^{sX} &\leqslant \tfrac{b}{b-a}e^{sa} - \tfrac{a}{b-a}e^{sb} \\ &= \left(1 - p + pe^{s(b-a)}\right)e^{-ps(b-a)} \\ &\triangleq e^{\phi(u)},\end{aligned}$$

where $u = s(b-a)$, and $\phi(u) = -pu + \log\left(1 - p + pe^u\right)$. But by direct calculation, we can see that the derivative of ϕ is

$$\phi'(u) = -p + \frac{p}{p + (1-p)e^{-u}},$$

thus $\phi(u) = \phi'(0) = 0$. Besides,

$$\phi''(u) = \frac{p(1-p)e^{-u}}{\left(p + (1-p)e^{-u}\right)^2} \leqslant \frac{1}{4}.$$

Therefore, by Taylor's theorem, for some $\theta \in [0, u]$,

$$\phi(u) = \phi(0) + u\phi'(0) + \frac{u^2}{2}\phi''(\theta) \leqslant \frac{u^2}{8} = \frac{s^2(b-a)^2}{8}. \qquad \square$$

Now we directly plug this lemma into (1.22):

$$\begin{aligned}
&\mathbb{P}(S_n - \mathbb{E}S_n \geqslant t) \\
&\leqslant e^{-st}\prod_{i=1}^{n} e^{s^2(b_i-a_i)^2/8} \text{(by Lemma 1.3.1)} \\
&= e^{-st} e^{s^2 \sum_{i=1}^{n}(b_i-a_i)^2/8} \\
&= e^{-2t^2/\sum_{i=1}^{n}(b_i-a_i)^2} \qquad \left(\text{by choosing } \; s = 4t/\sum_{i=1}^{n}(b_i-a_i)^2\right).
\end{aligned}$$

Theorem 1.3.2 (Hoeffding's Tail Inequality [11]). *Let $X_1, \ldots, X_n$ be independent bounded random variables such that X_i falls in the interval $[a_i, b_i]$ with probability one. Then, for any $t > 0$, we have*

$$\mathbb{P}(S_n - \mathbb{E}S_n \geqslant t) \leqslant e^{-2t^2/\sum_{i=1}^{n}(b_i-a_i)^2} \quad \textit{and}$$

$$\mathbb{P}(S_n - \mathbb{E}S_n \leqslant -t) \leqslant e^{-2t^2/\sum_{i=1}^{n}(b_i-a_i)^2}.$$

1.3.3 Bernstein's Inequality

Assume, without loss of generality, that $\mathbb{E}X_i = 0$ for all $i = 1, \ldots, n$. Our starting point is again (1.22), that is, we need bounds for $\mathbb{E}e^{sX_i}$. Introduce $\sigma_i^2 = \mathbb{E}\left[X_i^2\right]$, and

$$F_i = \sum_{k=2}^{\infty} \frac{s^{k-2}\mathbb{E}\left[X_i^k\right]}{k!\sigma_i^2}.$$

Since $e^{sx} = 1 + sx + \sum_{k=2}^{\infty} s^k x^k/k!$, and the expectation is linear, we may write

$$\begin{aligned}
\mathbb{E}e^{sX_i} &= 1 + s\mathbb{E}[X_i] + \sum_{k=2}^{\infty} \frac{s^k\mathbb{E}[X_i^k]}{k!} \\
&= 1 + s^2\sigma_i^2 F_i \qquad (\text{since} \quad \mathbb{E}[X_i] = 0) \\
&\leqslant e^{s^2\sigma_i^2 F_i}.
\end{aligned}$$

Now assume that X_i's are bounded such that $|X_i| \leq c$. Then for each $k \geq 2$,

$$\mathbb{E}\left[X_i^k\right] \leqslant c^{k-2}\sigma_i^2.$$

Thus,

$$F_i \leqslant \sum_{k=2}^{\infty} \frac{s^{k-2} c^{k-2} \sigma_i^2}{k! \sigma_i^2} = \frac{1}{(sc)^2} \sum_{k=2}^{\infty} \frac{(sc)^k}{k!} = \frac{e^{sc} - 1 - sc}{(sc)^2}.$$

Thus we have obtained

$$\mathbb{E}\left[e^{sX_i}\right] \leqslant e^{s^2 \sigma_i^2 \frac{e^{sc} - 1 - sc}{(sc)^2}}.$$

Returning to (1.22) and using the notation $\sigma^2 = (1/n) \sum_{i=1}^{n} \sigma_i^2$, we have

$$\mathbb{P}\left(\sum_{i=1}^{n} X_i \geqslant t\right) \leqslant e^{n\sigma^2 (e^{sc} - 1 - sc)/c^2 - st}.$$

Now we are free to choose s. The upper bound is minimized for

$$s = \frac{1}{c} \log\left(1 + \frac{tc}{n\sigma^2}\right).$$

Resubstituting this value, we obtain Bennett's inequality [12]:

Theorem 1.3.3 (Bennett's Inequality). *Let $X_1, \ldots, X_n$ be independent real-valued random variables with zero mean, and assume with zero mean, and assume that $|X_i| \le c$ with probability one. Let*

$$\sigma^2 = \frac{1}{n} \sum_{i=1}^{n} \mathrm{Var}\{X_i\}.$$

Then, for any $t > 0$,

$$\mathbb{P}\left(\sum_{i=1}^{n} X_i \geqslant t\right) \leqslant \exp\left(-\frac{n\sigma^2}{c^2} h\left(\frac{ct}{n\sigma^2}\right)\right),$$

where $h(u) = (1+u)\log(1+u) - u$ for $u \ge 0$.

The following inequality is due to Bennett (also referred to as Bernstein's inequality) [12, Eq. (7)] and [13, Lemma 2.2.11].

Theorem 1.3.4 (Bennett [12]). *Let $X_1, \ldots, X_n$ be independent random variables with zero mean such that*

$$\mathbb{E}|X_i|^p \leqslant p! M^{p-2} \sigma_i^2 / 2 \tag{1.24}$$

for some $p \geq 2$ *and some constants* M *and* σ_i, $i = 1, \ldots, n$. *Then for* $t > 0$

$$\mathbb{P}\left(\left|\sum_{i=1}^{n} X_i\right| \geqslant t\right) \leqslant 2e^{-t^2/2(\sigma^2+Mt)} \tag{1.25}$$

with $\sigma^2 = \sum_{i=1}^{n} \sigma_i^2$.

See Example 7.5.5 for its application.

Applying the elementary inequality $h(u) \geqslant u^2/(2 + 2u/3)$, $u \geq 0$ which can be seen by comparing the derivatives of both sides, we obtain a classical inequality of Bernstein [7]:

Theorem 1.3.5 (Bernstein Inequality). *Under the conditions of the previous theorem, for any* $t > 0$,

$$\mathbb{P}\left(\frac{1}{n}\sum_{i=1}^{n} X_i \geqslant t\right) \leqslant \exp\left(-\frac{nt^2}{2\sigma^2 + 2ct/3}\right). \tag{1.26}$$

We see that, except for the term $2ct/3$, Bernstein's inequality is quantitatively right when compared with the central limit theorem: the central limit theorem states that

$$\mathbb{P}\left(\sqrt{\frac{n}{\sigma^2}}\left(\frac{1}{n}\sum_{i=1}^{n} X_i - \mathbb{E}X_i\right) \geqslant y\right) \to 1 - \Phi(y) \leqslant \frac{1}{\sqrt{2\pi}}\frac{e^{-y^2/2}}{y},$$

from which we would expect, at least in a certain range of the parameters, something like

$$\mathbb{P}\left(\frac{1}{n}\sum_{i=1}^{n} X_i - \mathbb{E}X_i \geqslant t\right) \approx e^{-nt^2/(2\sigma^2)},$$

which is comparable with (1.26).

Exercise 1.3.6 (Sampling Without Replacement). Let $\mathcal{X}$ be a finite set with N elements, and let $X_1, \ldots, X_n$ be a random sample without replacement from $\mathcal{X}$ and $Y_1, \ldots, Y_n$ a random sample with replacement from $\mathcal{X}$. Show that for any convex real-valued function f,

$$\mathbb{E}f\left(\sum_{i=1}^{n} X_i\right) \leqslant \mathbb{E}f\left(\sum_{i=1}^{n} Y_i\right).$$

In particular, by taking $f(x) = e^{sx}$, we see that all inequalities derived from the sums of independent random variables Y_i using Chernoff's bounding remain true for the sums of the X_i's. (This result is due to Hoeffding [11].)

1.3.4 Efron-Stein Inequality

The main purpose of these notes [7] is to show how many of the tail inequalities of the sums of independent random variables can be extended to general functions of independent random variables. The simplest, yet surprisingly powerful inequality of this kind is known as the *Efron-Stein inequality*.

1.4 Probability and Matrix Analysis

1.4.1 Eigenvalues, Trace and Sums of Hermitian Matrices

Let $\mathbf{A}$ is a Hermitian $n \times n$ matrix. By the spectral theorem for Hermitian matrices, one diagonalize $\mathbf{A}$ using a sequence

$$\lambda_1 (\mathbf{A}) \geqslant \cdots \geqslant \lambda_n (\mathbf{A})$$

of n real eigenvalues, together with an orthonormal basis of eigenvectors

$$\mathbf{u}_1 (\mathbf{A}), \ldots, \mathbf{u}_n (\mathbf{A}) \in \mathbf{C}^n.$$

The set of the eigenvalues $\{\lambda_1 (\mathbf{A}), \ldots, \lambda_n (\mathbf{A})\}$ is known as the spectrum of $\mathbf{A}$. The eigenvalues are sorted in a non-increasing manner. The trace of a $n \times n$ matrix is equal to the sum of the its eigenvalues

$$\mathrm{Tr}(\mathbf{A}) = \sum_{i=1}^{n} \lambda_n.$$

The linearity of trace

$$\mathrm{Tr}(\mathbf{A}+\mathbf{B}) = \mathrm{Tr}(\mathbf{A}) + \mathrm{Tr}(\mathbf{B}).$$

The first eigenvalue is defined as

$$\lambda_1 (\mathbf{A}) = \sup_{|\mathbf{v}|=1} \mathbf{v}^H \mathbf{A} \mathbf{v}.$$

We have

$$\lambda_1 (\mathbf{A}+\mathbf{B}) \leqslant \lambda_1 (\mathbf{A}) + \lambda_1 (\mathbf{B}). \tag{1.27}$$

The Weyl inequalities are

$$\lambda_{i+j-1} (\mathbf{A}+\mathbf{B}) \leqslant \lambda_i (\mathbf{A}) + \lambda_j (\mathbf{B}),$$

and the Ky Fan inequality

$$\lambda_1(\mathbf{A}+\mathbf{B})+\cdots+\lambda_k(\mathbf{A}+\mathbf{B}) \leqslant \lambda_1(\mathbf{A})+\cdots+\lambda_k(\mathbf{A})+\lambda_1(\mathbf{B})+\cdots+\lambda_k(\mathbf{B}) \tag{1.28}$$

In particular, we have

$$\mathrm{Tr}(\mathbf{A}+\mathbf{B}) \leqslant \mathrm{Tr}(\mathbf{A}) + \mathrm{Tr}(\mathbf{B}). \tag{1.29}$$

One consequence of these inequalities is that the spectrum of a Hermitian matrix is *stable* with respect to small perturbations. This is very important when we deal with an extremely weak signal that can be viewed as small perturbations within the noise [14].

An $n \times n$ density matrix ρ is a positive definite matrix with $\mathrm{Tr}(\rho) = 1$. Let $\mathbb{S}_n$ denote the set of density matrices on $\mathbb{C}^n$. This is a convex set [15].

$\mathbf{A} \geq \mathbf{0}$ is equivalent to saying that all eigenvalues of $\mathbf{A}$ are nonnegative, i.e., $\lambda_i(\mathbf{A}) \geqslant 0$.

1.4.2 Positive Semidefinite Matrices

Inequality is one of the main topics in modern matrix theory [16]. An arbitrary complex matrix $\mathbf{A}$ is Hermitian, if $\mathbf{A} = \mathbf{A}^H$, where H stands for conjugate and transpose of a matrix. If a Hermitian matrix $\mathbf{A}$ is positive semidefinite, we say

$$\mathbf{A} \geq \mathbf{0}. \tag{1.30}$$

Matrix $\mathbf{A}$ is positive semidefinite, i.e., $\mathbf{A} \geq \mathbf{0}$, if all the eigenvalues $\lambda_i(\mathbf{A})$ are nonnegative [16, p. 166]. In addition,

$$\mathbf{A} \geq \mathbf{0} \Rightarrow \det \mathbf{A} \geq \mathbf{0} \;\text{ and }\; \mathbf{A} > \mathbf{0} \Rightarrow \det \mathbf{A} > \mathbf{0}, \tag{1.31}$$

where $\Rightarrow$ has the meaning of "implies." When $\mathbf{A}$ is a random matrix, its determinant $\det \mathbf{A}$ and its trace $\mathrm{Tr}\,\mathbf{A}$ are scalar random variables. Trace is a linear operator [17, p. 30].

For every complex matrix $\mathbf{A}$, the Gram matrix $\mathbf{A}\mathbf{A}^H$ is positive semidefinite [16, p. 163]:

$$\mathbf{A}\mathbf{A}^H \geq 0. \tag{1.32}$$

The eigenvalues of $\left(\mathbf{A}\mathbf{A}^H\right)^{\frac{1}{2}}$ are the singular values of $\mathbf{A}$.

It follows from [17, p. 189] that

$$\mathrm{Tr}\,\mathbf{A} = \sum_{i=1}^{n} \lambda_i, \quad \det \mathbf{A} = \prod_{i=1}^{n} \lambda_i,$$

$$\operatorname{Tr} \mathbf{A}^k = \sum_{i=1}^{n} \lambda_i^k, k = 1, 2, \ldots \tag{1.33}$$

where λ_i are the eigenvalues of the matrix $\mathbf{A} \in \mathbb{C}^{n \times n}$.

It follows from [17, p. 392] that

$$(\det \mathbf{A})^{1/n} \leqslant \frac{1}{n} \operatorname{Tr} \mathbf{A}, \tag{1.34}$$

for every positive semidefinite matrix $\mathbf{A} \in \mathbb{C}^{n \times n}$.

It follows from [17, p. 393] that

$$\frac{1}{n} \operatorname{Tr} \mathbf{AX} \geqslant (\det \mathbf{A})^{1/n} \tag{1.35}$$

for every positive semidefinite matrix $\mathbf{A}, \mathbf{X} \in \mathbb{C}^{n \times n}$ and further $\det \mathbf{X} = 1$.

1.4.3 Partial Ordering of Positive Semidefinite Matrices

By the Cauchy-Schwarz inequality, it is immediate that

$$|\operatorname{Tr}(\mathbf{AB})| \leqslant \operatorname{Tr}\left(\mathbf{AA}^H\right) \operatorname{Tr}\left(\mathbf{BB}^H\right) \tag{1.36}$$

and

$$\operatorname{Tr}\left(\mathbf{AA}^H\right) = 0 \text{ if and only if } \mathbf{A} = 0. \tag{1.37}$$

For a pair of positive semidefinite matrices $\mathbf{A}$ and $\mathbf{B}$, we say

$$\mathbf{B} \geq \mathbf{A} \text{ if } \mathbf{B} - \mathbf{A} \geq 0. \tag{1.38}$$

A partial order may be defined using (1.38). We hold the intuition that matrix $\mathbf{B}$ is somehow "greater" than matrix $\mathbf{A}$. If $\mathbf{B} \geq \mathbf{A} \geq \mathbf{0}$, then [16, p. 169]

$$\operatorname{Tr} \mathbf{B} \geqslant \operatorname{Tr} \mathbf{A}, \det \mathbf{B} \geqslant \det \mathbf{A}, \mathbf{B}^{-1} \leqslant \mathbf{A}^{-1}. \tag{1.39}$$

If $\mathbf{A} \geqslant 0$ and $\mathbf{B} \geqslant 0$ be of the same size. Then [16, p. 166]

1.

$$\mathbf{A} + \mathbf{B} \geq \mathbf{B}, \tag{1.40}$$

2.

$$\mathbf{A}^{1/2}\mathbf{B}\mathbf{A}^{1/2} \geqslant 0, \tag{1.41}$$

3.

$$\operatorname{Tr}(\mathbf{AB}) \le \operatorname{Tr}(\mathbf{A}) \operatorname{Tr}(\mathbf{B}), \tag{1.42}$$

4. The eigenvalues of $\mathbf{AB}$ are all nonnegative. Furthermore, $\mathbf{AB}$ is positive semidefinite if and only if $\mathbf{AB} = \mathbf{BA}$.

A sample covariance matrix is a random matrix. For a pair of random matrices X and Y, in analogy with their scalar counterparts, their expectations are of particular interest:

$$\mathbf{Y} \ge \mathbf{X} \ge 0 \;\Rightarrow\; \mathbb{E}\mathbf{Y} \ge \mathbb{E}\mathbf{X}. \tag{1.43}$$

Proof. Since the expectation of a random matrix can be viewed as a convex combination, and also the positive semidefinite (PSD) cone is convex [18, p. 459], expectation preserves the semidefinite order [19]. □

1.4.4 Definitions of $f(\mathrm{A})$ for Arbitrary f

The definitions of $f(\mathbf{A})$ of matrix $\mathbf{A}$ for general function f were posed by Sylvester and others. A eleglant treatment is given in [20]. A special function called spectrum is studied [21]. Most often, we deal with the PSD matrix, $\mathbf{A} \ge 0$. References [17, 18, 22–24].

When $f(t)$ is a polynomial or rational function with scalar coefficients and a scalar argument, t, it is natural to define $f(\mathbf{A})$ by substituting $\mathbf{A}$ for t, replacing division by matrix inverse, and replacing 1 by the identity matrix. Then, for example,

$$f(t) = \frac{1+t^2}{1-t} \Rightarrow f(\mathbf{A}) = (\mathbf{I} - \mathbf{A})^{-1}\left(\mathbf{I} + \mathbf{A}^2\right) \text{ if } 1 \notin \Lambda(\mathbf{A}). \tag{1.44}$$

Here, $\Lambda(\mathbf{A})$ denotes the set of eigenvalues of $\mathbf{A}$ (the spectrum of $\mathbf{A}$). For a general theory, we need a way of defining $f(\mathbf{A})$ that is applicable to arbitrary functions f.

Any matrix $\mathbf{A} \in \mathbb{C}^{n \times n}$ can be expressed in the Jordan canonical form

$$\mathbf{Z}^{-1}\mathbf{AZ} = \mathbf{J} = \operatorname{diag}(\mathbf{J}_1, \mathbf{J}_2, \ldots, \mathbf{J}_p) \tag{1.45}$$

where

$$\mathbf{J}_k = \mathbf{J}_k(\lambda_k) = \begin{pmatrix} \lambda_1 & 1 & 0 \\ & \ddots & 1 \\ 0 & & \lambda_k \end{pmatrix} \in \mathbb{C}^{m_k \times m_k}, \tag{1.46}$$

where $\mathbf{Z}$ is nonsingular and $m_1 + m_2 + \cdots + m_p = n$.

Let f be defined on the spectrum of $\mathbf{A} \in \mathbb{C}^{n \times n}$ and let $\mathbf{A}$ have the Jordan canonical form (1.45). Then,

$$f(\mathbf{A}) \triangleq \mathbf{Z} f(\mathbf{J}) \mathbf{Z}^{-1} = \mathbf{Z}\,\mathrm{diag}\left(f(\mathbf{J}_k)\right) \mathbf{Z}^{-1}, \tag{1.47}$$

where

$$f(\mathbf{J}_k) \triangleq \begin{pmatrix} f(\lambda_k) & f'(\lambda_k) & \cdots & \frac{f^{(m_k-1)}(\lambda_k)}{(m_k-1)!} \\ & f(\lambda_k) & \ddots & \vdots \\ & & \ddots & f'(\lambda_k) \\ & & & f(\lambda_k) \end{pmatrix}. \tag{1.48}$$

Several remarks are in order. First, the definition yields an $f(\mathbf{A})$ that can be shown to be independent of the particular Jordan canonical form that is used. Second, if $\mathbf{A}$ is diagonalizable, then the Jordan canonical form reduces to an eigen-decomposition $\mathbf{A} = \mathbf{Z}^{-1}\mathbf{D}\mathbf{Z}$, with $\mathbf{D} = \mathrm{diag}(\lambda_i)$ and the columns of $\mathbf{Z}$ (renamed $\mathbf{U}$) eigenvectors of $\mathbf{A}$, the above definition yields

$$f(\mathbf{A}) = \mathbf{Z} f(\mathbf{J}) \mathbf{Z}^{-1} = \mathbf{U} f(\mathbf{D}) \mathbf{U}^{-1} = \mathbf{U} f(\lambda_i) \mathbf{U}^{-1}. \tag{1.49}$$

Therefore, for diagonalizable matrices, $f(\mathbf{A})$ has the same eigenvectors as $\mathbf{A}$ and its eigenvalues are obtained by applying f to those of $\mathbf{A}$.

1.4.5 Norms of Matrices and Vectors

See [25] for matrix norms. The matrix p-norm is defined, for $1 \le p \le \infty$, as

$$\|\mathbf{A}\|_p = \max_{\mathbf{x}\neq 0} \frac{\|\mathbf{A}\mathbf{x}\|_p}{\|\mathbf{x}\|_p},$$

where $\|\mathbf{x}\|_p = \left(\sum_{i=1}^{n} |x_i|^p\right)^{1/p}$. When $p = 2$, it is called spectral norm $\|\mathbf{A}\|_2 = \|\mathbf{A}\|$. The Frobenius norm is defined as

$$\|\mathbf{A}\|_F = \left(\sum_{i=1}^{n}\sum_{j=1}^{n} |a_{ij}|^2\right)^{1/2},$$

which can be computed element-wise. It is the same as the Euclidean norm on vectors. Let $\mathbf{C} = \mathbf{A}\mathbf{B}$. Then $c_{ij} = \sum_{k=1}^{n} a_{ik} b_{kj}$. Thus

$$\|\mathbf{A}\mathbf{B}\|_F^2 = \|\mathbf{C}\|_F^2 = \sum_{i=1}^{n}\sum_{j=1}^{n} |c_{ij}|^2 = \sum_{i=1}^{n}\sum_{j=1}^{n}\sum_{k=1}^{n} |a_{ik} b_{kj}|^2.$$

Applying the Cauchy-Schwarz inequality to the expression $\sum_{k=1}^{n} a_{ik}b_{kj}$, we find that

$$\begin{aligned}\|\mathbf{AB}\|_F^2 &\leqslant \sum_{i=1}^{n}\sum_{j=1}^{n}\left(\sum_{k=1}^{n}\left|a_{ik}^2\right|\sum_{k=1}^{n}\left|b_{kj}^2\right|\right)\\ &= \left(\sum_{i=1}^{n}\sum_{k=1}^{n}\left|a_{ik}^2\right|\right)\left(\sum_{j=1}^{n}\sum_{k=1}^{n}\left|b_{kj}^2\right|\right)\\ &= \|\mathbf{A}\|_F\|\mathbf{B}\|_F.\end{aligned}$$

Let $\sigma_i, i = 1, \ldots, n$ be singular values that are sorted in decreasing magnitude. The Schatten-p norm is defined as

$$\|\mathbf{A}\|_{S_p} = \left(\sum_{i=1}^{n}\sigma_i^p\right)^{1/p} \quad \text{for} \quad 1 \leqslant p \leqslant \infty, \quad \text{and} \quad \|\mathbf{A}\|_\infty = \|\mathbf{A}\|_{\text{op}} = \sigma_1,$$

for a matrix $\mathbf{A} \in \mathbb{R}^{n\times n}$. When $p = \infty$, we obtain the operator norm (or spectral norm) $\|\mathbf{A}\|_{\text{op}}$, which is the largest singular value. It is so commonly used that we sometimes use $\|\mathbf{A}\|$ to represent it. When $p = 2$, we obtain the commonly called Hilbert-Schmidt norm or Frobenius norm $\|\mathbf{A}\|_{S_2} = \|\mathbf{A}\|_F$. When $p = 1$, $\|\mathbf{A}\|_{S_1}$ denotes the nuclear norm. Note that $\|\mathbf{A}\|$ is the spectral norm, while $\|\mathbf{A}\|_F$ is the Frobenius norm. The drawback of the spectral norm it that it is expensive to compute; it is *not* the Frobenius norm. We have the following properties of Schatten p-norm

1. When $p < q$, the inequality occurs: $\|\mathbf{A}\|_{S_q} \leqslant \|\mathbf{A}\|_{S_p}$.
2. If r is a rank of $\mathbf{A}$, then with $p > \log(r)$, it holds that $\|\mathbf{A}\| \leqslant \|\mathbf{A}\|_{S_p} \leqslant e\,\|\mathbf{A}\|$.

Let $\langle \mathbf{X}, \mathbf{Y}\rangle = \text{Tr}\left(\mathbf{X}^T\mathbf{Y}\right)$ represent the Euclidean inner product between two matrices and $\|\mathbf{X}\|_F = \langle \mathbf{X}, \mathbf{X}\rangle$. It can be easily shown that

$$\|\mathbf{X}\|_F = \sup_{\|\mathbf{G}\|_F=1} \text{Tr}\left(\mathbf{X}^T\mathbf{G}\right) = \sup_{\|\mathbf{G}\|_F=1} \langle \mathbf{X}, \mathbf{G}\rangle\,.$$

Note that trace and inner product are both linear.

For vectors, the only norm we consider is the l_2-norm, so we simply denote the l_2-norm of a vector by $\|\mathbf{x}\|$ which is equal to $\sqrt{\langle \mathbf{x}, \mathbf{x}\rangle}$, where $\langle \mathbf{x}, \mathbf{y}\rangle$ is the Euclidean inner product between two vectors. Like matrices, it is easy to show

$$\|\mathbf{x}\| = \sup_{\|\mathbf{y}\|=1} \langle \mathbf{x}, \mathbf{y}\rangle\,.$$

1.4.6 Expectation

We follow closely [9] for this basic background knowledge to set the stage for future applications. Given an unsigned random variable X (i.e., a random variable taking values in $[0, +\infty]$,) one can define the expectation or mean $\mathbb{E}X$ as the unsigned integral

$$\mathbb{E}X = \int_0^\infty x d\mu_X(x),$$

which by the Fubini-Tonelli theorem [9, p. 13] can also be rewritten as

$$\mathbb{E}X = \int_0^\infty \mathbb{P}(X \geqslant \lambda) d\lambda.$$

The expectation of an unsigned variable lies in $[0, +\infty]$. If X is a scalar random variable (which is allowed to take the value ∞), for which $\mathbb{E}|X| < \infty$, we find X is absolutely integrable, in which case we can define its expectation as

$$\mathbb{E}X = \int_{\mathbb{R}} x d\mu_X(x)$$

in the real case, or

$$\mathbb{E}X = \int_{\mathbb{C}} x d\mu_X(x)$$

in the complex case. Similarly, for a vector-valued random variable (note in finite dimensions, all norms are equivalent, so the precise choice of norm used to define $|X|$ is not relevant here). If $\mathbf{x} = (X_1, \ldots, X_n)$ is a vector-valued random variable, then X is absolutely integrable if and only if the components X_i are all absolutely integrable, in which case one has

$$\mathbb{E}\mathbf{x} = (\mathbb{E}X_1, \ldots, \mathbb{E}X_n).$$

A fundamentally important property of expectation is that it is linear: if $X_1, \ldots, X_n$ are absolutely integrable scalar random variables and $c_1, \ldots, c_k$ are finite scalars, then $c_1 X_1, \ldots, c_k X_k$ is also absolutely integrable and

$$\mathbb{E}(c_1 X_1 + \cdots + c_k X_k) = c_1 \mathbb{E}X_1 + \cdots + c_k \mathbb{E}X_k. \tag{1.50}$$

By the Fubini-Tonelli theorem, the same result also applies to infinite sums $\sum_{i=1}^{\infty} c_i X_i$, provided that $\sum_{i=1}^{\infty} |c_i| \mathbb{E}|X_i|$ is finite.

Linearity of expectation requires no assumption of independence or dependence amongst the individual random variables X_i; that is what makes this property of expectation so powerful [9, p. 14]. We use this linearity of expectation so often that we typically omit an explicit reference to it when it is being used.

Expectation is also monotone: if $X \le Y$ is true for some unsigned or real absolutely integrable X, Y, then

$$\mathbb{E}X \leqslant \mathbb{E}Y.$$

For an unsigned random variable, we have the obvious but very useful Markov inequality

$$\mathbb{P}\left(X \geqslant \lambda\right) \leqslant \frac{1}{\lambda}\mathbb{E}X$$

for some $\lambda > 0$. For the signed random variables, Markov's inequality becomes

$$\mathbb{P}\left(|X| \geqslant \lambda\right) \leqslant \frac{1}{\lambda}\mathbb{E}\left|X\right|.$$

If X is an absolutely integrable or unsigned scalar random variable, and F is a measurable function from the scalars to the unsigned extended reals $[0, +\infty]$, then one has the change of variables formula

$$\mathbb{E}F(X) = \int_{\mathbb{R}} F(x) d\mu_X(x)$$

when X is real-valued and

$$\mathbb{E}F(X) = \int_{\mathbb{C}} F(x) d\mu_X(x)$$

when X is complex-valued. The same formula applies to signed or complex F if it is known that $|F(x)|$ is absolutely integrable. Important examples of expressions such as $\mathbb{E}F(X)$ are moments

$$\mathbb{E}|X|^k$$

for $k \ge 1$ (particularly k$= 1, 2, 4$), exponential moments

$$\mathbb{E}e^{tX}$$

for real t, X, and Fourier moments (or the characteristic function)

$$\mathbb{E}e^{jtX}$$

for real t, X, or

$$e^{j\mathbf{t}\cdot\mathbf{x}}$$

for complex or vector-valued $\mathbf{t}, \mathbf{x}$, where $\cdot$ denotes a real inner product. We shall encounter the resolvents

$$\mathbb{E}\frac{1}{X - z}$$

for complex z.

The reason for developing the scalar and vector cases is because we are motivated to study

$$\mathbb{E}e^{\mathbf{X}}$$

for matrix-valued $\mathbf{X}$, where the entries of X may be deterministic or random variables. A random matrix $\mathbf{X}$ and its functional $f(\mathbf{X})$ can be studied using this framework ($f(t)$ is an arbitrary function of r). For example,

$$\mathbb{E}e^{\mathbf{X}_1 + \cdots + \mathbf{X}_n}$$

is of special interest when $\mathbf{X}_i$ are independent random matrices.

Once the second moment of a scalar random variable is finite, one can define the variance

$$\text{var}\,(X) = \mathbb{E}|X - \mathbb{E}X|^2.$$

From Markov's inequality we thus have Chebyshev's inequality

$$\mathbb{P}\left(|X - \mathbb{E}X| \geqslant \lambda\right) \leqslant \frac{\text{var}\,(X)}{\lambda^2}.$$

A real-valued random variable X is sub-Gaussian if and only if there exists $C > 0$ such that

$$\mathbb{E}e^{tX} \leqslant C\exp\left(Ct^2\right)$$

for all real t, and if and only if there exists $C > 0$ such that

$$\mathbb{E}|X|^k \leqslant (Ck)^{k/2}$$

for all integers $k \geq 1$.

A real-valued random variable X has sub-exponential tails if and only if there exists $C > 0$ such that

$$\mathbb{E}|X|^k \leqslant \exp\left(Ck^C\right)$$

for all positive integers k.

If X is sub-Gaussian (or has sub-exponential tails with exponent $a > 1$), then from dominated convergence we have the Taylor expansion

$$\mathbb{E}e^{tX} = 1 + \sum_{k=1}^{\infty} \frac{t^k}{k!} \mathbb{E}X^k$$

for any real or complex t, thus relating the exponential and Fourier moments with the kth moments.

1.4.7 Moments and Tails

The quantities $\mathbb{E}|X|^p, 0 < p < \infty$ are called absolute moments. The absolute moments of a random variable X can be expressed as

$$\mathbb{E}|X|^p = p \int_0^\infty \mathbb{P}(|X| > t) t^{p-1} dt, \quad p > 0. \tag{1.51}$$

The proof of this is as follows. Let $I_{\{|X|^p \geqslant x\}}$ is the indicator random variable: takes 1 on the event $|X|^p \geqslant x$ and 0 otherwise. Using Fubini's theorem, we derive

$$\begin{aligned}
\mathbb{E}|X|^p &= \int_\Omega |X|^p d\mathbb{P} = \int_\Omega \int_0^{|X|^p} dx d\mathbb{P} = \int_\Omega \int_0^\infty I_{\{|X|^p \geqslant x\}} dx d\mathbb{P} \\
&= \int_0^\infty \int_\Omega I_{\{|X|^p \geqslant x\}} d\mathbb{P} dx = \int_0^\infty \mathbb{P}(|X|^p \geqslant x)\, dx \\
&= p \int_0^\infty \mathbb{P}(|X| \geqslant t^p) t^{p-1} dt = p \int_0^\infty \mathbb{P}(|X| \geqslant t) t^{p-1} dt,
\end{aligned}$$

where we also used a change of variables.

For $1 \leqslant p < \infty, (\mathbb{E}|X|^p)^{1/p}$ defines a norm on the $L^p(\Omega, \mathbb{P})$-space of all p-integrable random variables, in particular, the triangular inequality

$$(\mathbb{E}|X+Y|^p)^{1/p} \leqslant (\mathbb{E}|X|^p)^{1/p} + (\mathbb{E}|Y|^p)^{1/p} \tag{1.52}$$

holds for $X, Y \in L^p(\Omega, \mathbb{P}) = \{X \text{ measurable}, \mathbb{E}|X|^p < \infty\}$.

Let $p, q \geq 1$ with $1/p + 1/q = 1$, Hölder's inequality states that

$$\mathbb{E}|XY| \leqslant (\mathbb{E}|X|^p)^{1/p} (\mathbb{E}|Y|^q)^{1/q}$$

for random variables X, Y. The space case $p = q = 2$ is the Cauchy-Schwartz inequality. It follows from Hölder's inequality that for $0 < p \leqslant q < \infty$,

$$(\mathbb{E}|X|^p)^{1/p} \leqslant (\mathbb{E}|Y|^q)^{1/q}.$$

The function $\mathbb{P}(|X| > t)$ is called the tail of X. The Markov inequality is a simple way of estimating the tail. We can also use the moments to estimate the tails. The next statement due to Tropp [26] is simple but powerful. Suppose X is a random variable satisfying

$$(\mathbb{E}|X|^p)^{1/p} \leqslant \alpha\beta^{1/p}p^{1/\gamma}, \qquad \text{for all } p \geqslant p_0$$

for some constants $\alpha, \beta, \gamma, p_0 > 0$. Then

$$\mathbb{P}\left(|X| \geqslant e^{1/\gamma}\alpha t\right) \leqslant \beta e^{-t^\gamma/\gamma}$$

The proof of the claim is short. By Markov's inequality, we obtain for an arbitrary $\kappa > 0$

$$\mathbb{P}\left(|X| \geqslant e^{1/\gamma}\alpha t\right) \leqslant \frac{\mathbb{E}|X|^p}{(e^\kappa \alpha t)^p} \leqslant \beta \left(\frac{\alpha p^{1/\gamma}}{e^\kappa \alpha t}\right)^p.$$

Choose $p = t^\gamma$ and the optimal value $\kappa = 1/\gamma$ to obtain the claim.

Also the converse of the above claim can be shown [27]. Important special cases are $\gamma = 1, 2$. In particular, if $(\mathbb{E}|X|^p)^{1/p} \leqslant \alpha\beta^{1/p}\sqrt{p}$, for all $p \geq 2$, then X satisfies the subGaussian tail estimate

$$\mathbb{P}\left(|X| \geqslant e^{1/2}\alpha t\right) \leqslant \beta e^{-t^2/2} \qquad \text{for all } t \geqslant \sqrt{2}. \tag{1.53}$$

For a random variable Z, we define its L_p norm

$$\mathbb{E}^p(Z) = (\mathbb{E}|Z|^p)^{1/p}.$$

We use a simple technique for bounding the moments of a maximum. Consider an arbitrary set $\{Z_1, \ldots, Z_N\}$ of random variables. We have that

$$\mathbb{E}^p(\max_i Z_i) = N^{1/p}\max_i \mathbb{E}^p(Z_i).$$

To check this claim, simply note that [28]

$$(\mathbb{E}\max_i |Z_i|^p)^{1/p} \leqslant \left(\mathbb{E}\sum_i |Z_i|^p\right)^{1/p} \leqslant \left(\sum_i \mathbb{E}|Z_i|^p\right)^{1/p} \leqslant (N \cdot \max_i \mathbb{E}|Z_i|^p)^{1/p}.$$

In many cases, this inequality yields essentially sharp results for the appropriate choice of p.

If X, Y are independent with $\mathbb{E}[Y] = 0$ and $k \geq 2$, then [29] $\mathbb{E}\left[|X|^k\right] \leqslant \mathbb{E}\left[|X - Y|^k\right]$.

For random variables satisfying a subGaussian tail estimate, we have the following useful lemma [27]. See also [30].

Lemma 1.4.1. *Let* $X_1, \ldots, X_N$ *be random variables satisfying*

$$\mathbb{P}\left(|X_i| \geqslant t\right) \leqslant \beta e^{-t^2/2} \qquad \text{for all } t \geqslant \sqrt{2}, \qquad i = 1, \ldots, N,$$

for some $\beta \geq 1$. *Then*

$$\mathbb{E} \max_{i=1,\ldots,N} |X_i| \leqslant C_\beta \sqrt{\ln\left(4\beta N\right)}$$

with $C_\beta \leqslant \sqrt{2} + \frac{1}{4\sqrt{2(4\beta)}}$.

Proof. According to (1.51), we have, for some $\alpha \geqslant \sqrt{2}$

$$\begin{aligned}
\mathbb{E} \max_{i=1,\ldots,N} |X_i| &= \textstyle\int_0^\infty \mathbb{P}\left(\max_{i=1,\ldots,N} |X_i| > t\right) dt \\
&\leqslant \textstyle\int_0^\alpha 1 dt + \int_\alpha^\infty \mathbb{P}\left(\max_{i=1,\ldots,N} |X_i| > t\right) dt \\
&\leqslant \alpha + \textstyle\int_\alpha^\infty \sum_{i=1}^N \mathbb{P}\left(|X_i| > t\right) dt \\
&\leqslant \alpha + N\beta \textstyle\int_\alpha^\infty e^{-t^2/2} dt.
\end{aligned}$$

In the second line, we used the union bound.

A change of variable gives

$$\int_u^\infty e^{-t^2/2} dt = \int_0^\infty e^{-(t+u)^2/2} dt = e^{-u^2/2} \int_0^\infty e^{-tu} e^{-t^2/2} dt.$$

On the right hand side, using $e^{-tu} \leqslant 1$ for $t, u \geq 0$ gives

$$\int_u^\infty e^{-t^2/2} dt \leqslant e^{-u^2/2} \int_0^\infty e^{-t^2/2} dt = \sqrt{\frac{\pi}{2}} e^{-u^2/2}.$$

On the other hand, using $e^{-t^2/2} \leqslant 1$ for $t \geq 0$ gives

$$\int_u^\infty e^{-t^2/2} dt \leqslant e^{-u^2/2} \int_0^\infty e^{-tu} dt = \frac{1}{u} e^{-u^2/2}.$$

Combining the two results, we have

$$\int_u^\infty e^{-t^2/2} dt \leqslant \min\left\{\sqrt{\frac{\pi}{2}}, \frac{1}{u}\right\} e^{-u^2/2}.$$

Thus we have

$$\mathbb{E}\max_{i=1,\ldots,N}|X_i| \leqslant \alpha + N\beta\int_\alpha^\infty e^{-t^2/2}dt \leqslant \alpha + N\beta\frac{1}{\alpha}e^{-\alpha^2/2}.$$

Now we choose $\alpha = \sqrt{2\ln(4\beta N)} \geqslant \sqrt{2\ln(4)} \geqslant \sqrt{2}$. This gives

$$\begin{aligned}\mathbb{E}\max_{i=1,\ldots,N}|X_i| &\leqslant \sqrt{2\ln(4\beta N)} + \tfrac{1}{4\sqrt{2\ln(4\beta N)}} \\ &= \left(\sqrt{2} + \tfrac{1}{4\sqrt{2}\ln(4\beta N)}\right)\sqrt{\ln(4\beta N)} \leqslant C_\beta\sqrt{\ln(4\beta N)}.\end{aligned}$$

□

Some results are formulated in terms of moments; the transition to a tail bound can be established by the following standard result, which easily follows from Markov's inequality.

Theorem 1.4.2 ([31]). *Suppose X is a random variable satisfying*

$$(\mathbb{E}|X|^p)^{1/p} \leqslant \alpha + \beta\sqrt{p} + \gamma p \qquad \textit{for all } p \geqslant p_0$$

for some $\alpha, \beta, \gamma, p_0 > 0$. Then, for $t \geq p_0$,

$$\mathbb{P}\left(\mathbb{E}|X| \geqslant e\left(\alpha + \beta\sqrt{t} + \gamma t\right)\right) \leqslant e^{-t}.$$

If a Bernoulli vector $\mathbf{y}$ weakly dominates random vector $\mathbf{x}$ then $\mathbf{y}$ strongly dominates $\mathbf{x}$. See also [32].

Theorem 1.4.3 (Bednorz and Latala [33]). *Let $\mathbf{x}, y$ be random vectors in a separate Banach space $(F, ||\cdot||)$ such that $\mathbf{y} = \sum_{i\geqslant 1}\mathbf{u}_i\varepsilon_i$ for some vectors $\mathbf{u}_i \in F$ and*

$$\mathbb{P}(|\varphi(\mathbf{x})| \geqslant t) \leqslant \mathbb{P}(|\varphi(\mathbf{y})| \geqslant t) \textit{ for all } \varphi \in F^*, t > 0.$$

Then there exists universal constant L such that:

$$\mathbb{P}(\|\mathbf{x}\| \geqslant t) \leqslant L\mathbb{P}(\|\mathbf{y}\| \geqslant t/L) \textit{ for all } t > 0.$$

1.4.8 Random Vector and Jensen's Inequality

A random vector $\mathbf{x} = (X_1, \ldots, X_n)^T \in \mathbb{R}^n$ is a collection of n random variables X_i on a common probability space. Its expectation is the vector

$$\mathbb{E}\mathbf{x} = (\mathbb{E}X_1, \ldots, \mathbb{E}X_n)^T \in \mathbb{R}^n.$$

A complex random vector $\mathbf{z} = \mathbf{x} + j\mathbf{y} \in \mathbb{C}^n$ is a special case of a $2n$-dimensional real random vector $(\mathbf{x}, \mathbf{y}) \in \mathbb{R}^{2n}$.

A collection of random vectors $\mathbf{x}_1, \ldots, \mathbf{x}_N \in \mathbb{C}^n$ is called (stochastic ally) independent if for all measurable subsets $A_1, \ldots, A_N \subset \mathbb{C}^n$,

$$\mathbb{P}(\mathbf{x}_1 \in A_1, \ldots, \mathbf{x}_N \in A_N) = \mathbb{P}(\mathbf{x}_1 \in A_1) \cdots \mathbb{P}(\mathbf{x}_N \in A_N).$$

Functions of independent random vectors are again independent. A random vector $\mathbf{x}'$ in $\mathbb{C}^n$ will be called an independent copy of $\mathbf{x}$ if $\mathbf{x}$ and $\mathbf{x}'$ are independent and have the same distribution, that is, $\mathbb{P}(\mathbf{x} \in A) = \mathbb{P}(\mathbf{x}' \in A)$ for all $A \in \mathbb{C}^n$.

Jensen's inequality says that: let $f : \mathbb{C}^n \to \mathbb{R}$ be a convex function, and let $\mathbf{x} \in \mathbb{C}^n$ be a random vector. Then

$$f(\mathbb{E}\mathbf{x}) \leqslant \mathbb{E}f(\mathbf{x}). \tag{1.54}$$

1.4.9 Convergence

For a sequence x_n of scalars to converge to a limit x, for every $\varepsilon > 0$, we have $|x - x_n| \leq \varepsilon$ for all sufficiently large n. This notion of convergence is generalized to metric space convergence.

Let $R = (R, d)$ be a σ-compact metric space (with the σ-algebra, and let X_n be a sequence of random variables taking values in R. X_n *converges almost surely* to X if, for almost every $\omega \in \Omega$, $X_n(\omega)$ converges to $X(\omega)$, or equivalently

$$\mathbb{P}\left(\lim_{n\to\infty} d(X_n, X) \leqslant \varepsilon\right) = 1$$

for every $\varepsilon > 0$.

X_n *converges in probability* to X if, for every $\varepsilon > 0$, one has

$$\lim_{n\to\infty} \inf \mathbb{P}(d(X_n, X) \leqslant \varepsilon) = 1,$$

or equivalently if $d(X_n, X) \leq \varepsilon$ holds asymptotically almost surely for every $\varepsilon > 0$.

X_n *converges in distribution* to X if, for every bounded continuous function $F : R \to \mathbf{R}$, one has

$$\lim_{n\to\infty} \mathbb{E}F(X_n) = \mathbb{E}F(X).$$

1.4.10 Sums of Independent Scalar-Valued Random Variables: Chernoff's Inequality

Gnedenko and Kolmogorov [34] points out: "In the formal construction of a course in the theory of probability, limit theorems appear as a kind of superstructure over elementary chapters, in which all problems have *finite, purely arithmetic character*. In reality, however, the epistemological value of the theory of probability is revealed *only by limit theorems*. Moreover, without limit theorems it is impossible to understand the real content of the primary concept of all our sciences—the concept of probability. In fact, all epistemological value of the theory of probability is based on this: that large-scale random phenomena in their collective action create strict, nonrandom regularity. The very concept of mathematical probability would be fruitless if it did not find its realization in the frequency of occurrence of events under large-scale repetition of uniform conditions (a realization which is always approximate and not wholly reliable), but that becomes, in principle, arbitrarily precise and reliable as the number of repetitions increases."

The philosophy behind the above cited paragraph is especially relevant to the Big Data: The dimensions of the data are high but finite. We seek a theory of purely arithmetic character: a non-asymptotic theory.

We follow closely [35] in this development. Ashlwede and Winter [36] proposed a new approach to develop inequalities for sums of independent random matrices.

Ashlwede-Winter's method [36] is parallel to the classical approach to derivation inequalities for real valued random variables. Let $X_1, \ldots, X_n$ be independent mean zero random variables. We are interested in the magnitude

$$S_n = X_1 + \ldots + X_n = \sum_{i=1}^{n} X_i.$$

For simplicity, we shall assume that $|X_i| \le 1$ almost surely. This hypothesis can be relaxed to some control of the moments, precisely to having sub-exponential tail.

Fix $t > 0$ and let $\lambda > 0$ be a parameter to be determined later. Our task is to estimate

$$p \triangleq \mathbb{P}\left(S_n > t\right) = \mathbb{P}\left(e^{S_n} > e^t\right).$$

By Markov inequality and using independence, we have

$$p \leqslant e^{-\lambda t}\mathbb{E}e^{\lambda S_n} = e^{-\lambda t}\prod_i \mathbb{E}e^{\lambda X_i}.$$

Next, Taylor's expansion and the mean zero and boundedness hypotheses can be used to show that, for every i,

$$e^{\lambda X_i} \leqslant e^{\lambda^2 \operatorname{var} X_i}, \quad 0 \leqslant \lambda \leqslant 1.$$

This results in

$$p \leqslant e^{-\lambda t + \lambda^2\sigma^2}, \text{ where } \sigma^2 \triangleq \sum_{i=1}^{n} \operatorname{var} X_i.$$

The optimal choice of the parameter $\lambda \sim \min\left(\tau/2\sigma^2, 1\right)$ implies Chernoff's inequality

$$p \leqslant \max\left(e^{-t^2/\sigma^2}, e^{-t/2}\right).$$

1.4.11 Extensions of Expectation to Matrix-Valued Random Variables

If $\mathbf{X}$ is a matrix (or vector) with random variables, then

$$\mathbb{E}\mathbf{X} = [\mathbb{E}X_{ij}] .$$

In other words, the expectation of a matrix (or vector) is just the matrix of expectations of the individual elements.

The basic properties of expectation still hold in these extensions.

If $\mathbf{A}, \mathbf{B}, \mathbf{c}$ are nonrandom, then [37, p. 276]

$$\mathbb{E}\left(\mathbf{AX} + \mathbf{BX} + \mathbf{c}\right) = \mathbf{A}\mathbb{E}\mathbf{X} + \mathbf{B}\mathbb{E}\mathbf{X} + \mathbf{c}. \tag{1.55}$$

Define a weighted sum

$$S_n = a_1 X_1 + \cdots + a_n X_n = \sum_{i=1}^{n} \mathbf{a}^T \mathbf{X}.$$

1.4.12 Eigenvalues and Spectral Norm

$\mathbb{M}_d$ is the set of real symmetric $d \times d$ matrices. $\mathbb{C}_{\text{Herm}}^{d \times d}$ denote the set of complex Hermitian $d \times d$ matrices. The spectral theorem state that all $\mathbf{A} \in \mathbb{C}_{\text{Herm}}^{d \times d}$ have d real eigenvalues (possibly with repetitions) that correspond to an orthonormal set of eigenvectors. $\lambda_{max}(\mathbf{A})$ is the largest eigenvalue of $\mathbf{A}$. All the eigenvalues are sorted in non-increasing manner. The spectrum of $\mathbf{A}$, denoted by $\text{spec}(\mathbf{A})$, is the multiset of all eigenvalues, where each eigenvalue appears a number of times equal to its multiplicity. We let

$$\|\mathbf{C}\| \equiv \max_{\mathbf{v} \in \mathbb{C}^d |\mathbf{v}|=1} |\mathbf{C}\mathbf{v}| \tag{1.56}$$

denote the operator norm of $\mathbf{C} \in \mathbb{C}_{\text{Herm}}^{d \times d}$ ($|\cdot|$ is the Euclidean norm). By the spectral theorem,

$$\|\mathbf{A}\| = \max \left\{ \lambda_{\max}\left(\mathbf{A}\right), \lambda_{\max}\left(-\mathbf{A}\right) \right\}. \tag{1.57}$$

Using the spectral theorem for the identity matrix $\mathbf{I}$ gives: $\|\mathbf{I}\| = 1$. Moreover, the trace of $\mathbf{A}$, $\text{Tr}\left(\mathbf{A}\right)$ is defined as the sum of the sum of the diagonal entries of $\mathbf{A}$. The trace of a matrix is equal to the sum of the eigenvalues of $\mathbf{A}$, or

$$\text{Tr}\left(\mathbf{A}\right) = \sum_{i=1}^{d} \lambda_i \left(\mathbf{A}\right).$$

See Sect. 1.4.1 for more properties of trace and eigenvalues.

Given a matrix ensemble $\mathbf{X}$, there are many statistics of $\mathbf{X}$ that one may wish to consider, e.g., the eigenvalues or singular values of $\mathbf{X}$, the trace, and determinant, etc. Including basic statistics, namely the operator norm [9, p. 106]. This is a basic upper bound on many quantities. For example, $\|\mathbf{A}\|_{op}$ is also the largest singular value $\sigma_1(\mathbf{A})$ of $\mathbf{A}$, and thus dominates the other singular values; similarly, all eigenvalues $\lambda_i(\mathbf{A})$ of $\mathbf{A}$ clearly have magnitude at most $\|\mathbf{A}\|_{op}$ since $\lambda_i(\mathbf{A}) = \sigma_1(\mathbf{A})^2$. Because of this, it is particularly important to obtain good upper bounds,

$$\mathbb{P}\left(\|\mathbf{A}\|_{op} \geqslant \lambda\right) \leqslant \cdots,$$

on this quantity, for various thresholds λ. Lower tail bounds are also of interest; for instance, they give confidence that the upper tail bounds are sharp.

We denote $|\mathbf{A}|$ the positive operator (or matrix) $(\mathbf{A}^*\mathbf{A})^{1/2}$ and by $\mathbf{s}(\mathbf{A})$ the vector whose coordinates are the singular values of $\mathbf{A}$, arranged as $s_1(\mathbf{A}) \geqslant s_2(\mathbf{A}) \geqslant \cdots \geqslant s_n(\mathbf{A})$. We have [23]

$$\mathbf{A} = \|\ |\mathbf{A}|\ \| = s_1(\mathbf{A}).$$

Now, if $\mathbf{U}, \mathbf{V}$ are unitary operators on $\mathbb{C}^{n\times n}$, then $|\mathbf{UAV}| = \mathbf{V}^* |\mathbf{A}| \mathbf{V}$ and hence

$$\|\mathbf{A}\| = \|\mathbf{UAV}\|$$

for all unitary operators $\mathbf{U}, \mathbf{V}$. The last property is called *unitary invariance*. Several other norms have this property. We will use the symbol $|||\mathbf{A}|||$ to mean a norm on $n \times n$ matrices that satisfies

$$|||\mathbf{A}||| = |||\mathbf{UAV}||| \tag{1.58}$$

for all $\mathbf{A}$ and for unitary $\mathbf{U}, \mathbf{V}$. We will call such a norm a *unitarily invariant norm*. We will normalize such norms such that they take the value 1 on the matrix $\text{diag}(1, 0, \ldots, 0)$.

1.4.13 Spectral Mapping

The multiset of all the eigenvalues of $\mathbf{A}$ is called the spectrum of $\mathbf{A}$, denoted $\text{spec}(\mathbf{A})$, where each eigenvalue appears a number of times equal to its multiplicity.

When $f(t)$ is a polynomial or rational function with scalar coefficients and a scalar argument, t, it is natural to define $f(\mathbf{A})$ by substituting $\mathbf{A}$ for t, replacing division by matrix inversion, and replacing 1 by the identity matrix [1, 20]. For example,

$$f(t) = \frac{1+t^2}{1-t} \Rightarrow f(\mathbf{A}) = (\mathbf{I}-\mathbf{A})^{-1}(\mathbf{I}+\mathbf{A})^2$$

if $1 \notin \text{spec}(\mathbf{A})$.

If $f(t)$ has a convergent power series representation, such as

$$\log(1+t) = t - \frac{t^2}{2} + \frac{t^3}{3} - \frac{t^4}{4}, |t| < 1,$$

we again can simply substitute $\mathbf{A}$ for t, to define

$$\log(\mathbf{I}+\mathbf{A}) = \mathbf{A} - \frac{\mathbf{A}^2}{2} + \frac{\mathbf{A}^3}{3} - \frac{\mathbf{A}^4}{4}, \rho(\mathbf{A}) < 1.$$

Here ρ denotes the spectral radius and the condition $\rho(\mathbf{A}) < 1$ ensures convergence of the matrix series. In this ad hoc fashion, a wide variety of matrix functions can be defined. This approach is certainly appealing to engineering communities, however, this approach has several drawbacks.

Theorem 1.4.4 (Spectral Mapping Theorem [38]). *Let $f : \mathbb{C} \to \mathbb{C}$ be an entire analytic function with a power-series representation $f(z) = \sum_{l \geqslant 0} c_l z^l, (z \in \mathbb{C})$. If all c_l are real, we define the mapping expression:*

$$f(\mathbf{A}) \equiv \sum_{l \geqslant 0} c_l \mathbf{A}^l, \quad \mathbf{A} \in \mathbb{C}_{Herm}^{d \times d}, \tag{1.59}$$

where $\mathbb{C}_{Herm}^{d \times d}$ is the set of Hermitian matrices of $d \times d$. The expression corresponds to a map from $\mathbb{C}_{Herm}^{d \times d}$ to itself. The so-called spectral mapping property is expressed as:

$$\text{spec}\, f(\mathbf{A}) = f(\text{spec}(\mathbf{A})). \tag{1.60}$$

By (1.60), we mean that the eigenvalues of $f(\mathbf{A})$ are the numbers $f(\lambda)$ with $\lambda \in \text{spec}(\mathbf{A})$. Moreover, the multiplicity of $\xi \in \text{spec}(\mathbf{A})$ is the sum of the multiplicity of all preimages of ξ under f that lie in spec$(\mathbf{A})$.

For any function $f : \mathbb{R} \to \mathbb{R}$, we extend f to a function on Hermitian matrices as follows. We define a map on diagonal matrices by applying the function to each diagonal entry. We extend f to a function on Hermitian matrices using the eigenvalue decomposition. Let $\mathbf{A} = \mathbf{U}\mathbf{D}\mathbf{U}^T$ be a spectral decomposition of $\mathbf{A}$. Then, we define

$$f(\mathbf{A}) = \mathbf{U} \begin{pmatrix} f(D_{1,1}) & & 0 \\ & \ddots & \\ 0 & & f(D_{d,d}) \end{pmatrix} \mathbf{U}^T. \tag{1.61}$$

In other words, $f(\mathbf{A})$ is defined simply by applying the function f to the eigenvalues of $\mathbf{A}$. The spectral mapping theorem states that each eigenvalue of $f(\mathbf{A})$ is equal to $f(\lambda)$ for some eigenvalue λ of $\mathbf{A}$. This point is obvious from our definition.

Standard inequalities for real functions typically *do not* have parallel versions that hold for the semidefinite ordering. Nevertheless, there is one type of relation (referred to as the transfer rule) for real functions that always extends to the semidefinite setting:

Claim 1.4.5 (Transfer Rule). *Let $f : \mathbb{R} \to \mathbb{R}$ and $g : \mathbb{R} \to \mathbb{R}$ satisfy $f(x) \leq g(x)$ for all $x \in [l, u] \subset \mathbb{R}$. Let $\mathbf{A}$ be a symmetric matrix for which all eigenvalues lie in $[l, u]$ (i.e., $l\mathbf{I} \leqslant \mathbf{A} \leqslant u\mathbf{I}$.) Then*

$$f(\mathbf{A}) \leqslant g(\mathbf{A}). \tag{1.62}$$

1.4.14 Operator Convexity and Monotonicity

We closely follow [15].

Definition 1.4.6 (Operator convexity and monotonicity). A function $f : (0, \infty) \to \mathbb{R}$ is said to be operator monotone in case whenever for all n, and all positive definite matrix $\mathbf{A}, \mathbf{B} > \mathbf{0}$,

$$\mathbf{A} > \mathbf{B} \Rightarrow f(\mathbf{A}) > f(\mathbf{B}). \tag{1.63}$$

A function $f : (0, \infty) \to \mathbb{R}$ is said to be operator convex in case whenever for all n, and all positive definite matrix$\mathbf{A}, \mathbf{B} > \mathbf{0}$, and $0 < t < 1$,

$$f((1-t)\mathbf{A} + t\mathbf{B}) \leqslant (1-t)f(\mathbf{A})\, tf(\mathbf{B}). \tag{1.64}$$

The square function $f(t) = t^2$ is monotone in the usual real-valued sense but not monotone in the operator monotone sense. It turns out that the square root function $f(t) = \sqrt{t}$ is also operator monotone.

The square function is, however, operator convex. The cube function is not operator convex. After seeing these examples, let us present the Lőwner-Hernz Theorem.

Theorem 1.4.7 (Lőwner-Hernz Theorem). *For $-1 \leq p \leq 0$, the function $f(t) = -t^p$ is operator monotone and operator concave. For $0 \leq p \leq 1$, the function $f(t) = t^p$ is operator monotone and operator concave. For $1 \leq p \leq 2$, the function $f(t) = -t^p$ is operator monotone and operator convex. Furthermore, $f(t) = \log(t)$ is operator monotone and operator concave., while $f(t) = t\log(t)$ is operator convex.*

$f(t) = t^{-1}$ is operator convex and $f(t) = -t^{-1}$ is operator monotone.

1.4.15 Convexity and Monotonicity for Trace Functions

Theorem 1.4.8 (Convexity and monotonicity for trace functions). *Let $f : \mathbb{R} \to \mathbb{R}$ be continuous, and let n be any natural number. Then if $t \mapsto f(t)$ is monotone increasing, so is $\mathbf{A} \mapsto \operatorname{Tr} f(\mathbf{A})$ on Hermitian matrices. Likewise, if $t \mapsto f(t)$ is convex, so is $\mathbf{A} \mapsto \operatorname{Tr} f(\mathbf{A})$ on Hermitian matrices, and strictly so if f is strictly convex.*

Much less is required of f in the context of the trace functions: f is *continuous and convex (or monotone increasing).*

Theorem 1.4.9 (Peierls-Bogoliubov Inequality). *For every natural number n, the map*

$$\mathbf{A} \mapsto \log \{\operatorname{Tr}[\exp(\mathbf{A})]\} \tag{1.65}$$

is convex on Hermitian matrices.

Indeed, for Hermitian matrices $\mathbf{A}, \mathbf{B}$ and $0 < t < 1$, let $\psi(t)$ be the function

$$\psi(t) : \mathbf{A} \mapsto \log \{\operatorname{Tr}[\exp(\mathbf{A})]\}.$$

By Theorem 1.4.9, this is convex, and hence

$$\psi(1) - \psi(0) \geqslant \frac{\psi(t) - \psi(0)}{t}$$

for all t. Taking the limit $t \to 0$, we obtain

$$\log\left(\frac{\operatorname{Tr}\left[e^{\mathbf{A}+\mathbf{B}}\right]}{\operatorname{Tr}\left[e^{\mathbf{A}}\right]}\right) \geqslant \frac{\operatorname{Tr}\left[\mathbf{B}e^{\mathbf{A}}\right]}{\operatorname{Tr}\left[e^{\mathbf{A}}\right]}. \tag{1.66}$$

Frequently, this consequence of Theorem 1.4.9 is referred to as the Peierls-Bogoliubov Inequality. Not only are both of the functions $\mathbf{H} \mapsto \log\left[\operatorname{Tr}\left(e^{\mathbf{H}}\right)\right]$ and $\rho \mapsto -S(\rho)$ are both convex, they are *Legendre Transforms* of one another. Here ρ is a density matrix. See [39] for a full mathematical treatment of the Legendre Transform.

Example 1.4.10 (A Novel Use of Peierls-Bogoliubov Inequality for Hypothesis Testing). To our best knowledge, this example provides a novel use of the Peierls-Bogoliubov Inequality. The hypothesis for signal plus noise model is

$$\mathcal{H}_0 : \mathbf{y} = \mathbf{w}$$
$$\mathcal{H}_1 : \mathbf{y} = \mathbf{x} + \mathbf{w}$$

where x, y, w are signal and noise vectors and y is the output vector. The covariance matrix relation is used to rewrite the problem as the matrix-valued hypothesis testing

$$\begin{aligned}\mathcal{H}_0 &: \mathbf{R}_{yy} = \mathbf{R}_{ww} := \mathbf{A} \\ \mathcal{H}_1 &: \mathbf{R}_{yy} = \mathbf{R}_{xx} + \mathbf{R}_{ww} := \mathbf{A} + \mathbf{B}\end{aligned}$$

when x, w are independent. The covariance matrices can be treated as density matrices with some normalizations: $\mathbf{C} \mapsto \frac{\mathbf{C}}{\mathrm{Tr}(\mathbf{C})}$. Our task is to decide between two alternative hypotheses: $\mathcal{H}_0$ and $\mathcal{H}_1$. This is the very nature, in analogy with the quantum testing of two alternative states. See [5]. The use of (1.66) gives

$$\log\left(\frac{\mathrm{Tr}\left[e^{\mathbf{R}_{xx}+\mathbf{R}_{ww}}\right]}{\mathrm{Tr}\left[e^{\mathbf{R}_{ww}}\right]}\right) \geqslant \frac{\mathrm{Tr}\left[\mathbf{R}_{xx}e^{\mathbf{R}_{ww}}\right]}{\mathrm{Tr}\left[e^{\mathbf{R}_{ww}}\right]}.$$

Let us consider a threshold detector:

$$\begin{aligned}&\mathcal{H}_0 : \text{otherwise} \\ &\mathcal{H}_1 : \log\left(\frac{\mathrm{Tr}\left[e^{\mathbf{R}_{xx}+\mathbf{R}_{ww}}\right]}{\mathrm{Tr}\left[e^{\mathbf{R}_{ww}}\right]}\right) \geqslant T_0, \text{ with } T_0 = \frac{\mathrm{Tr}\left[\mathbf{R}_{xx}e^{\mathbf{R}_{ww}}\right]}{\mathrm{Tr}\left[e^{\mathbf{R}_{ww}}\right]}.\end{aligned}$$

Thus, the a prior knowledge of $\mathbf{R}_{xx}, \mathbf{R}_{ww}$ can be used to set the threshold of T_0.

In real world, an estimated covariance matrix must be used to replace the above covariance matrix. We often consider a number of estimated covariance matrices, which are random matrices. We naturally want to consider a sum of these estimated covariance matrices. Thus, we obtain

$$\begin{aligned}&\mathcal{H}_0 : \text{otherwise} \\ &\mathcal{H}_1 : \log\left(\frac{\mathrm{Tr}\left[e^{\left(\hat{\mathbf{R}}_{xx,1}+\cdots+\hat{\mathbf{R}}_{xx,n}\right)+\left(\hat{\mathbf{R}}_{ww,1}+\cdots+\hat{\mathbf{R}}_{ww,n}\right)}\right]}{\mathrm{Tr}\left[e^{\hat{\mathbf{R}}_{ww,1}+\cdots+\hat{\mathbf{R}}_{ww,n}}\right]}\right) \geqslant T_0\end{aligned}$$

with

$$T_0 = \frac{\mathrm{Tr}\left[\left(\hat{\mathbf{R}}_{xx,1}+\cdots+\hat{\mathbf{R}}_{xx,n}\right)e^{\hat{\mathbf{R}}_{ww,1}+\cdots+\hat{\mathbf{R}}_{ww,n}}\right]}{\mathrm{Tr}\left[e^{\hat{\mathbf{R}}_{ww,1}+\cdots+\hat{\mathbf{R}}_{ww,n}}\right]}.$$

If the bounds of sums of random matrices can be used to bound the threshold T_0, the problem can be greatly simplified. This example provides one motivation for systematically studying the sums of random matrices in this book. □

1.4.16 *The Matrix Exponential*

The exponential of an Hermitian matrix $\mathbf{A}$ can be defined by applying (1.61) with the function $f(x) = e^x$. Alternatively, we may use the power series expansion (Taylor's series)

$$\exp(\mathbf{A}) = e^{\mathbf{A}} := \mathbf{I} + \sum_{k=1}^{\infty} \frac{\mathbf{A}^k}{k!}. \tag{1.67}$$

The exponential of an Hermitian matrix is ALWAYS positive definite because of the spectral mapping theorem (Theorem 1.4.4 and Eq. (1.62)): $e^{\lambda_i(\mathbf{A})} > 0$, where $\lambda_i(\mathbf{A})$ is the i-th eigenvalue of $\mathbf{A}$. On account of the transfer rule (1.62), the matrix exponential satisfies some simple semidefinite relations. For each Hermitian matrix $\mathbf{A}$, it holds that

$$\mathbf{I} + \mathbf{A} \leqslant e^{\mathbf{A}}, \text{ and} \tag{1.68}$$

$$\cosh(\mathbf{A}) \leqslant e^{\mathbf{A}^2/2}. \tag{1.69}$$

We often work with the trace of the matrix exponential, $\mathrm{Tr}\exp : \mathbf{A} \mapsto \mathrm{Tr}\, e^{\mathbf{A}}$. The trace exponential function is convex [22]. It is also monotone [22] with respect to the semidefinite order:

$$\mathbf{A} \leqslant \mathbf{H} \Rightarrow \mathrm{Tr}\, e^{\mathbf{A}} \leqslant \mathrm{Tr}\, e^{\mathbf{H}}. \tag{1.70}$$

See [40] for short proofs of these facts.

1.4.17 *Golden-Thompson Inequality*

The matrix exponential *doe not* convert sums into products, but the trace exponential has a related property that serves as a limited substitute.

For $n \times n$ complex matrices, the matrix exponential is defined by the Taylor series (of course a power series representation) as

$$\exp(\mathbf{A}) = e^{\mathbf{A}} = \sum_{k=0}^{\infty} \frac{1}{k!}\mathbf{A}^k.$$

For commutative matrices $\mathbf{A}$ and $\mathbf{B}$: $\mathbf{AB} = \mathbf{BA}$, we see that $e^{\mathbf{A}+\mathbf{B}} = e^{\mathbf{A}}e^{\mathbf{B}}$ by multiplying the Taylor series. This identity is not true for general non-commutative matrices. In fact, it always fails if $\mathbf{A}$ and $\mathbf{B}$ do not commute, see [40].

The matrix exponential is convergent for all square matrices. Furthermore, it is not hard to see for $\mathbf{A} \in \mathbb{M}_d$ that an eigenbasis of $\mathbf{A}$ is also an eigenbasis of $\exp(\mathbf{A})$ and that $\lambda_i\left(e^{\mathbf{A}}\right) = e^{\lambda_i(\mathbf{A})}$ for all $1 \leq i \leq d$. Also, for all $\mathbf{A} \in \mathbb{M}_d$, it holds that $e^{\mathbf{A}} \geq 0$.

We will be interested in the case of the exponential function $f(x) = e^x$. For any Hermitian matrix $\mathbf{A}$, note that $e^{\mathbf{A}}$ is positive semi-definite. Whereas in the scalar case $e^{a+b} = e^a e^b$ holds, it is not necessarily true in the matrix case that $e^{\mathbf{A}+\mathbf{B}} = e^{\mathbf{A}} \cdot e^B$. However, the following useful inequality does hold:

Theorem 1.4.11 (Golden-Thompson Inequality). *Let* $\mathbf{A}$ *and* $\mathbf{B}$ *be arbitrary Hermitian* $d \times d$ *matrices. Then*

$$\operatorname{Tr}\left(e^{\mathbf{A}+\mathbf{B}}\right) \leqslant \operatorname{Tr}\left(e^{\mathbf{A}} \cdot e^{\mathbf{B}}\right). \tag{1.71}$$

For a proof, we refer to [23, 41]. For a survey of Golden-Thompson and other trace inequalities, see [40]. Golden-Thompson inequality holds for arbitrary unitary-invariant norm replacing the trace, see [23] Theorem 9.3.7. A version of Golden-Thompson inequality for three matrices fails:

$$\operatorname{Tr}\left(e^{\mathbf{A}+\mathbf{B}+\mathbf{C}}\right) \nleqslant \operatorname{Tr}\left(e^{\mathbf{A}} e^{\mathbf{B}} e^{\mathbf{C}}\right).$$

1.4.18 The Matrix Logarithm

We define the matrix logarithm as the functional inverse of the matrix exponential:

$$\log\left(e^{\mathbf{A}}\right) := \mathbf{A} \text{ for all Hermitian matrix } \mathbf{A}. \tag{1.72}$$

This formula determines the logarithm on the positive definite cone.

The matrix logarithm interfaces beautifully with the semidefinite order [23, Exercise 4.2.5]. In fact, the logarithm is operator monotone:

$$\mathbf{0} \leqslant \mathbf{A} \leqslant \mathbf{H} \Rightarrow \log \mathbf{A} \leqslant \log \mathbf{H}, \tag{1.73}$$

where $\mathbf{0}$ denotes the zero matrix whose entries are all zeros. The logarithm is also operator concave:

$$\tau \log \mathbf{A} + (1-\tau) \log \mathbf{H} \leqslant \log\left(\tau \mathbf{A} + (1-\tau)\mathbf{H}\right) \tag{1.74}$$

for all positive definite $\mathbf{A}, \mathbf{H}$ and $\tau \in [0,1]$. Operator monotone functions and operator convex functions are depressingly rare. In particular, the matrix exponential does not belong to either class [23, Chap. V]. Fortunately, the trace inequalities of a matrix-valued function can be used as limited substitute. For a survey, see [40]which is very accessible. Carlen [15] is also ideal for a beginner.

1.4.19 Quantum Relative Entropy and Bregman Divergence

Quantum relative entropy can be interpreted as a measure of dissimilarity between two positive-definite matrices.

Definition 1.4.12 (Quantum relative entropy). Let $\mathbf{X}, \mathbf{Y}$ be positive-definite matrices. The quantum relative entropy of $\mathbf{X}$ with respect to $\mathbf{Y}$ is defined as

$$\begin{aligned} D(\mathbf{X}; \mathbf{Y}) &= \mathrm{Tr}\left[\mathbf{X}\log\mathbf{X} - \mathbf{X}\log\mathbf{Y} - (\mathbf{X} - \mathbf{Y})\right] \\ &= \mathrm{Tr}\left[\mathbf{X}(\log\mathbf{X} - \log\mathbf{Y}) - (\mathbf{X} - \mathbf{Y})\right]. \end{aligned} \tag{1.75}$$

Quantum relative entropy is also called quantum information divergence and von Neumann divergence. It has a nice geometric interpretation [42].

A new class of matrix nearness problems uses a directed distance measure called a Bregman divergence. We define the Bregman divergence of the matrix $\mathbf{X}$ from the matrix $\mathbf{Y}$ as

$$D_{\varphi}(\mathbf{X}; \mathbf{Y}) \triangleq \varphi(\mathbf{X}) - \varphi(\mathbf{Y}) - \langle \nabla\varphi(\mathbf{X}), \mathbf{X} - \mathbf{Y} \rangle, \tag{1.76}$$

where the matrix inner product $\langle \mathbf{X}, \mathbf{Y} \rangle = \mathrm{Re}\,\mathrm{Tr}\,\mathbf{X}\mathbf{Y}^*$. Two principal examples of Bregman divergences are the following. When $\varphi(\mathbf{X}) = \frac{1}{2}\|\mathbf{X}\|_F^2$, the associated divergence is the squared Frobenius norm $\frac{1}{2}\|\mathbf{X} - \mathbf{Y}\|_F^2$. When $\varphi(\mathbf{X})$ is the negative Shannon entropy, we obtain the Kullback-Leibler divergence, which is also known as relative entropy. But these two cases are just the tip of the iceberg [42]. In general, Bregman divergences provide a powerful way to measure the distance between matrices. The problem can be formulated in terms of convex optimization

$$\underset{\mathbf{X}}{\text{minimize}}\ D_{\varphi}(\mathbf{X}; \mathbf{Y}) \text{ subject to } \mathbf{X} \in \bigcap\nolimits_k C_k,$$

where C_k is a finite collection of closed, convex sets whose intersection is nonempty. For example, we can apply it to the problem of learning a divergence from data.

Define the quantum entropy function

$$\varphi(\mathbf{X}) = \mathrm{Tr}(\mathbf{X}\log\mathbf{X}) \tag{1.77}$$

for a positive-definite matrix. Note that the trace function is linear. The divergence $D(\mathbf{X}; \mathbf{Y})$ can be viewed as the difference between $\varphi(\mathbf{X})$ and the best affine approximation of the entropy at the matrix $\mathbf{Y}$. In other words, (1.75) is the special case of (1.76) when $\varphi(\mathbf{X})$ is given by (1.77). The entropy function φ given in (1.77) is a strictly convex function, which implies that the affine approximation strictly underestimates this φ. This observation gives us the following fact.

Fact 1.4.13 (Klein's inequality). *The quantum relative entropy is nonnegative*

$$D(\mathbf{X}; \mathbf{Y}) \geqslant 0.$$

Equality holds if and only if $\mathbf{X} = \mathbf{Y}$.

Introducing the definition of the quantum relative entropy into Fact (1.4.13), and rearranging, we obtain

$$\operatorname{Tr}\mathbf{Y} \geqslant \operatorname{Tr}\left(\mathbf{X}\log\mathbf{Y} - \mathbf{X}\log\mathbf{X} + \mathbf{X}\right).$$

When $\mathbf{X} = \mathbf{Y}$, both sides are equal. We can summarize this observation in a lemma for convenience.

Lemma 1.4.14 (Variation formula for trace [43]). *Let* $\mathbf{Y}$ *be a positive-definite matrix. Then,*

$$\operatorname{Tr}\mathbf{Y} = \max_{\mathbf{X}>0}\operatorname{Tr}\left(\mathbf{X}\log\mathbf{Y} - \mathbf{X}\log\mathbf{X} + \mathbf{X}\right).$$

This lemma is a restatement of the fact that quantum relative entropy is nonnegative.

The convexity of quantum relative entropy has paramount importance.

Fact 1.4.15 (Lindblad [44]). *The quantum relative entropy defined in* (1.75) *is a jointly convex function. That is,*

$$D\left(t\mathbf{X}_1 + (1-t)\,\mathbf{X}_2; t\mathbf{Y}_1 + (1-t)\,\mathbf{Y}_2\right) \leqslant tD\left(\mathbf{X}_1; \mathbf{Y}_1\right) + (1-t)\,D\left(\mathbf{X}_2; \mathbf{Y}_2\right), \quad t \in [0,1],$$

where $\mathbf{X}_i$ *and* $\mathbf{Y}_i$ *are positive definite for* $i = 1, 2$.

Bhatia's book [23, IX.6 and Problem IX.8.17] gives a clear account of this approach. A very accessible work is [45].

A final useful tool is a basic result in matrix theory and convex analysis [46, Lemma 2.3]. Following [43], a short proof originally from [47] is included here for convenience.

Proposition 1.4.16. *Let* $f(\cdot;\cdot)$ *be a jointly concave function. Then, the function* $y \mapsto \max_x f(x;y)$ *obtained by partial maximization is concave, assuming the maximization is always attained.*

Proof. For a pair of points y_1 and y_2, there are points x_1 and x_2 that meet

$$f\left(x_1; y_1\right) = \max_x f\left(x; y_1\right) \quad \text{and} \quad f\left(x_2; y_2\right) = \max_x f\left(x; y_2\right).$$

For each $t \in [0,1]$, the joint concavity of f says that

$$\begin{aligned}
\max\nolimits_x f\left(x; ty_1 + (1-t)\,y_2\right) &\geqslant f\left(tx_1 + (1-t)\,x_2; ty_1 + (1-t)\,y_2\right) \\
&\geqslant t \cdot f\left(x_1; y_1\right) + (1-t)\,f\left(x_2; y_2\right) \\
&= t \cdot \max\nolimits_x f\left(x; y_1\right) + (1-t)\,f\left(x_2; y_2\right).
\end{aligned}$$

The second line follows from the assumption that $f(\cdot;\cdot)$ be a jointly concave function. In words, the partial maximum is a concave function. □

If f is a convex function and $\alpha > 0$, then the function αf is convex. If f_1 and f_2 are both convex, then so is their sum $f_1 + f_2$. Combining nonnegative scaling and addition [48, p. 79], we see that the set of convex functions is itself a convex cone: a nonnegative weighted sum of convex functions

$$f = w_1 f_1 + \cdots + w_m f_m \tag{1.78}$$

is convex. Similarly, a nonnegative weighted sum of concave functions is concave. A linear function is of course convex. Let $\mathbb{S}^{n\times n}$ stand for the set of the $n \times n$ symmetric matrix. Any linear function $f : \mathbb{S}^{n\times n} \mapsto \mathbb{R}$ can be represented in the form

$$f(\mathbf{X}) = \mathrm{Tr}(\mathbf{CX}), \quad \mathbf{C} \in \mathbb{S}^{n\times n}. \tag{1.79}$$

1.4.20 Lieb's Theorem

Lieb's Theorem is the foundation for studying the sum of random matrices in Chap. 2. We present a succinct proof this theorem, following the arguments of Tropp [49]. Although the main ideas of Tropp's presentation are drawn from [46], his proof provides a geometric intuition for Theorem 1.4.17 and connects it to another major result. Section 1.4.19 provides all the necessary tools for this proof.

Theorem 1.4.17 (Lieb [50]). *Fix a Hermitian matrix* $\mathbf{H}$. *The function*

$$\mathbf{A} \mapsto \mathrm{Tr}\exp(\mathbf{H} + \log(\mathbf{A}))$$

is concave on the positive-definite cone.

Proof. In the variational formula, Lemma 1.4.14, select

$$\mathbf{Y} = \exp(\mathbf{H} + \log \mathbf{A})$$

to obtain

$$\mathrm{Tr}\exp(\mathbf{H} + \log \mathbf{A}) = \max_{\mathbf{X}>0} \mathrm{Tr}(\mathbf{X}(\mathbf{H} + \log \mathbf{A}) - \mathbf{X}\log \mathbf{X} + \mathbf{X}).$$

Using the quantum relative entropy of (1.75), this expression can be rewritten as

$$\mathrm{Tr}\exp(\mathbf{H} + \log \mathbf{A}) = \max_{\mathbf{X}>0} [\mathrm{Tr}(\mathbf{XH}) - (D(\mathbf{X}; \mathbf{A}) - \mathrm{Tr}\,\mathbf{A})]. \tag{1.80}$$

Note that trace is a linear function.

For a Hermitian matrix $\mathbf{H}$, Fact 1.4.15 says that $D(\mathbf{X}; \mathbf{A})$ is a jointly convex function of the matrix variables $\mathbf{A}$ and $\mathbf{X}$. Due to the linearity of the trace function, the whole bracket on the right-hand-side of (1.80) is also a jointly convex function of the matrix variables $\mathbf{A}$ and $\mathbf{X}$. It follows from Proposition 1.4.16 that the right-hand-side of (1.80) defines a concave function of $\mathbf{A}$. This observation completes the proof. □

We require a simple but powerful corollary of Lieb's theorem. This result connects expectation with the trace exponential.

Corollary 1.4.18. *Let* $\mathbf{H}$ *be a fixed Hermitian matrix, and let* $\mathbf{X}$ *be a random Hermitian matrix. Then*

$$\mathbb{E}\,\mathrm{Tr}\exp\left(\mathbf{H}+\mathbf{X}\right) \leqslant \mathrm{Tr}\exp\left(\mathbf{H}+\log\left(\mathbb{E}e^{\mathbf{X}}\right)\right).$$

Proof. Define the random matrix $\mathbf{Y} = e\mathbf{X}$, and calculate that

$$\begin{aligned}\mathbb{E}\,\mathrm{Tr}\exp\left(\mathbf{H}+\mathbf{X}\right) &= \mathbb{E}\,\mathrm{Tr}\exp\left(\mathbf{H}+\log\mathbf{Y}\right)\\ &\leqslant \mathrm{Tr}\exp\left(\mathbf{H}+\log\left(\mathbb{E}\mathbf{Y}\right)\right)\\ &= \mathrm{Tr}\exp\left(\mathbf{H}+\log\left(\mathbb{E}e^{\mathbf{X}}\right)\right).\end{aligned}$$

The first relation follows from the definition (1.72) of the matrix logarithm because $\mathbf{Y}$ is always positive definite, $\mathbf{Y} > 0$. Lieb's result, Theorem 1.4.17, says that the trace function is concave in $\mathbf{Y}$, so in the second relation we may invoke Jensen's inequality to draw the expectation inside the logarithm. □

1.4.21 Dilations

An extraordinary fruitful idea from operator theory is to embed matrices within larger block matrices, called *dilations* [51]. The Hermitian dilation of a rectangular matrix $\mathbf{B}$ is

$$\varphi\left(\mathbf{B}\right) = \begin{bmatrix} \mathbf{0} & \mathbf{B} \\ \mathbf{B}^* & \mathbf{0} \end{bmatrix}. \tag{1.81}$$

Evidently, $\varphi\left(\mathbf{B}\right)$ is always Hermitian. A short calculation yields the important identity

$$\varphi(\mathbf{B})^2 = \begin{bmatrix} \mathbf{B}\mathbf{B}^* & \mathbf{0} \\ \mathbf{0} & \mathbf{B}^*\mathbf{B} \end{bmatrix}. \tag{1.82}$$

It is also be verified that the Hermitian dilation preserves spectral information:

$$\lambda_{\max}\left(\varphi\left(\mathbf{B}\right)\right) = \left\|\varphi\left(\mathbf{B}\right)\right\| = \left\|\mathbf{B}\right\|. \tag{1.83}$$

We use dilations to extend results for Hermitian matrices to rectangular matrices.

Consider a channel [52, p. 279] in which the additive Gaussian noise is a stochastic process with a finite-dimensional covariance matrix. Information theory tells us that the information (both quantum and classical) depends on only the eigenvalues of the covariance matrix. The dilations preserve the "information".

1.4.22 The Positive Semi-definite Matrices and Partial Order

The matrix $\mathbf{A}$ is called positive semi-definite if all of its eigenvalues are non-negative. This is denoted $\mathbf{A} \geqslant 0$. Furthermore, for any two Hermitian matrices $\mathbf{A}$ and $\mathbf{B}$, we write $\mathbf{A} \geq \mathbf{B}$ if $\mathbf{A} - \mathbf{B} \geq 0$. One can define a semidefinite order or partial order on all Hermitian matrices. See [22] for a treatment of this topic.

For any t, the eigenvalues of $\mathbf{A} - t\mathbf{I}$ are $\lambda_1 - t, \ldots, \lambda_d - t$. The spectral norm of A, denoted as $\|\mathbf{A}\|$, is defined to be $\max_i |\lambda_i|$. Thus $-\|\mathbf{A}\| \cdot \mathbf{I} \leqslant \mathbf{A} \leqslant \|\mathbf{A}\| \cdot \mathbf{I}$.

Claim 1.4.19. *Let* $\mathbf{A}$, $\mathbf{B}$ *and* $\mathbf{C}$ *be Hermitian* $d \times d$ *matrices satisfying* $\mathbf{A} \geq 0$ *and* $\mathbf{B} \leq \mathbf{C}$. *Then,* $\operatorname{Tr}(\mathbf{A} \cdot \mathbf{B}) \leqslant \operatorname{Tr}(\mathbf{A} \cdot \mathbf{C})$.

Notice that $\leq$ is a partial order and that

$$\mathbf{A}, \mathbf{B}, \mathbf{A}', \mathbf{B}' \in \mathbb{C}_{\text{Herm}}^{d \times d}, \mathbf{A} \leqslant \mathbf{B} \text{ and } \mathbf{A}' \leqslant \mathbf{B}' \Rightarrow \mathbf{A} + \mathbf{A}' \leqslant \mathbf{B} + \mathbf{B}'.$$

Moreover, spectral mapping (1.60) implies that

$$\mathbf{A} \in \mathbb{C}_{\text{Herm}}^{d \times d}, \mathbf{A}^2 \geqslant 0.$$

Corollary 1.4.20 (Trace-norm property). *If* $\mathbf{A} \geq 0$, *then*

$$\operatorname{Tr}(\mathbf{A} \cdot \mathbf{B}) \leqslant \|\mathbf{B}\| \operatorname{Tr}(\mathbf{A}), \tag{1.84}$$

where $\|\mathbf{B}\|$ *is the spectrum norm (largest singular value).*

Proof. Apply Claim 1.4.19 with $\mathbf{C} = \|\mathbf{B}\| \cdot \mathbf{I}$ and note that $\operatorname{Tr}(\alpha \mathbf{A}) = \alpha \operatorname{Tr}(\mathbf{A})$ for any scalar α. □

Suppose a real function f on an interval I has the following property [22, p. 60]: if $\mathbf{A}$ and $\mathbf{B}$ are two elements of $\mathbb{H}_n(I)$ and $\mathbf{A} \geq \mathbf{B}$, then $f(\mathbf{A}) \geqslant f(\mathbf{B})$. We say that such a function f is *matrix monotone* of order n on I. If f is matrix monotone of order n for $n = 1, 2, \ldots$, then we say f is operator monotone.

Matrix convexity of order n and operator convexity can be defined in a similar way. The function $f(t) = t^r$, on the interval $[0, \infty)$ is operator monotone for $0 \leq r \leq 1$, and is operator convex for $1 \leq r \leq 2$ and for $-1 \leq r \leq 0$.

1.4.23 Expectation and the Semidefinite Order

Since the expectation of a random matrix can be viewed as a convex combination and the positive semidefinite cone is convex, then expectation preserves the semidefinite order [53]:

$$\mathbf{X} \leqslant \mathbf{Y} \text{ almost surely } \Rightarrow \mathbb{E}\mathbf{X} \leqslant \mathbb{E}\mathbf{Y}. \tag{1.85}$$

Every operator convex function admits an operator Jensen's inequality [54]. In particular, the matrix square is operator convex, which implies that

$$(\mathbb{E}\mathbf{X})^2 \leqslant \mathbb{E}\left(\mathbf{X}^2\right). \tag{1.86}$$

The relation (1.86) is also a specific instance of Kadison's inequality [23, Theorem 2.3.3].

1.4.24 Probability with Matrices

Assume $(\Omega, \mathcal{F}, \mathbb{P})$ is a probability space and $\mathbf{Z} : \Omega \to \mathbb{C}_{\text{Herm}}^{d\times d}$ is measurable with respect to $\mathcal{F}$ and the Borel σ-field on $\mathbb{C}_{\text{Herm}}^{d\times d}$. This is equivalent to requiring that all entries of $\mathbf{Z}$ be complex-valued random variables. $\mathbb{C}_{\text{Herm}}^{d\times d}$ is a metrically complete vector space and one can naturally define an expected value $\mathbb{E}\left[\mathbf{Z}\right] \in \mathbb{C}_{\text{Herm}}^{d\times d}$. This turns out to be the matrix $\mathbb{E}\left[\mathbf{Z}\right] \in \mathbb{C}_{\text{Herm}}^{d\times d}$ whose (i,j)-entry is the expected value of the (i,j)-entry of $\mathbf{Z}$. Of course, $\mathbb{E}\left[\mathbf{Z}\right]$ is only defined if all entries of $\mathbf{Z}$ are integrable, but this will always be in the case in this section.

The definition of expectation implies that *trace and expectation commute*:

$$\operatorname{Tr}\left(\mathbb{E}\left[\mathbf{Z}\right]\right) = \mathbb{E}\left(\operatorname{Tr}\left[\mathbf{Z}\right]\right). \tag{1.87}$$

Moreover, one can check that the usual product rule is satisfied: If $\mathbf{Z}, \mathbf{W} : \Omega \to \mathbb{C}_{\text{Herm}}^{d\times d}$ are measurable and independent, then

$$\mathbb{E}\left[\mathbf{ZW}\right] = \mathbb{E}\left[\mathbf{Z}\right]\mathbb{E}\left[\mathbf{W}\right]. \tag{1.88}$$

Finally,

$$\text{if } \mathbf{Z} : \Omega \to \mathbb{C}_{\text{Herm}}^{d\times d} \text{ satisfies } \mathbf{Z} \geqslant 0 \text{ almost surely (a.s.), then } \mathbb{E}\left[\mathbf{Z}\right] \geq 0,$$

which is an easy consequence of another readily checked fact: $(\mathbf{v}, \mathbb{E}\left[\mathbf{Z}\right]\mathbf{v}) = \mathbb{E}\left[(\mathbf{v}, \mathbf{Z}\mathbf{v})\right], \mathbf{v} \in \mathbb{C}^d$, where $(\cdot,\cdot)$ is the standard Euclidean inner product.

1.4.25 Isometries

There is an analogy between numbers and transforms [55, p. 142]. A finite-dimensional transform is a finite-dimensional matrix.

Theorem 1.4.21 (Orthogonal or Unitary). *The following three conditions on a linear transform* $\mathbf{U}$ *on inner product space are equivalent to each other.*

1. $\mathbf{U}^*\mathbf{U} = \mathbf{I}$,
2. $(\mathbf{Ux}, \mathbf{Uy}) = (\mathbf{x}, \mathbf{y})$ *for all* $\mathbf{x}$ *and* $\mathbf{y}$,
3. $\|\mathbf{Ux}\| = \|\mathbf{x}\|$ *for all* $\mathbf{x}$.

Since condition 3 implies that

$$\|\mathbf{Ux} - \mathbf{Uy}\| = \|\mathbf{x} - \mathbf{y}\| \text{ for all } \mathbf{x} \text{ and } \mathbf{y},$$

we see that transforms of the type that the theorem deals with are characterized by the fact that *they preserve distances*. For this reason, we call such a transform an *isometry*. An isometry on a finite-dimensional space is necessarily orthogonal or unitary, use of this terminology will enable us to treat the real and the complex cases simultaneously. On a finite-dimensional space, we observe that an isometry is always invertible and that $\mathbf{U}^{-1}(= \mathbf{U}^*)$ is an isometry along with $\mathbf{U}$.

1.4.26 Courant-Fischer Characterization of Eigenvalues

The expectation of a random variable is $\mathbb{E}X$. We write $X \sim \text{Bern}(p)$ to indicate that X has a Bernoulli distribution with mean p. In Sect. 2.12, one of the central tools is the variational characterization of a Hermitian matrix given by the Courant-Fischer theorem. For integers d and n satisfying $1 \le d \le n$, the complex Stiefel manifold

$$\mathbb{V}_d^n = \{\mathbf{V} \in \mathbb{C}^{n \times d} : \mathbf{V}^*\mathbf{V} = \mathbf{I}\}$$

is the collection of orthonormal bases for the d-dimensional subspaces of $\mathbb{C}^n$, or, equivalently, the collection of all *isometric* embeddings of $\mathbb{C}^d$ into a subspace of $\mathbb{C}^n$. Then the matrix $\mathbf{V}^*\mathbf{AV}$ can be interpreted as the compression of $\mathbf{A}$ to the space spanned by $\mathbf{V}$.

Theorem 1.4.22 (Courant-Fischer). *Let* $\mathbf{A}$ *is a Hermitian matrix with dimension* n. *Then*

$$\lambda_k(\mathbf{A}) = \min_{\mathbf{V} \in \mathbb{V}_{n-k+1}^n} \lambda_{\max}(\mathbf{V}^*\mathbf{AV}) \quad \text{and} \tag{1.89}$$

$$\lambda_k(\mathbf{A}) = \max_{\mathbf{V} \in \mathbb{V}_k^n} \lambda_{\min}(\mathbf{V}^*\mathbf{AV}). \tag{1.90}$$

A matrix $\mathbf{V}_- \in \mathbb{V}_k^n$ *achieves equality in* (1.90) *if and only if its columns span a dominant* k*-dimensional invariant subspace* $\mathbf{A}$. *Likewise, a matrix* $\mathbf{V}_+ \in \mathbb{V}_{n-k+1}^n$ *achieves equality in* (1.89) *if and only if its columns span a bottom* $(n-k+1)$*-dimensional invariant subspace* $\mathbf{A}$.

The $\pm$ subscripts in Theorem 1.4.22 are chosen to reflect the fact that $\lambda_k(\mathbf{A})$ is the *minimum* eigenvalue of $\mathbf{V}_-^*\mathbf{A}\mathbf{V}_-$ and the *maximum* eigenvalue of $\mathbf{V}_+^*\mathbf{A}\mathbf{V}_+$. As a consequence of Theorem 1.4.22, when $\mathbf{A}$ is Hermitian,

$$\lambda_k(-\mathbf{A}) = -\lambda_{n-k+1}(\mathbf{A}). \tag{1.91}$$

In other words, for the minimum eigenvalue of a Hermitian matrix $\mathbf{A}$, we have [53, p. 13]

$$\lambda_{\min}(\mathbf{A}) = -\lambda_{\max}(-\mathbf{A}). \tag{1.92}$$

This fact (1.91) allows us to use the same techniques we develop for bounding the eigenvalues from above to bound them from below. The use of this fact is given in Sect. 2.13.

1.5 Decoupling from Dependance to Independence

Decoupling is a technique of replacing quadratic forms of random variables by bilinear forms. The monograph [56] gives a systematic study of decoupling and its applications. A simple decoupling inequality is given by Vershynin [57]. Both the result and its proof are well known but his short proof is not easy to find in the literature. In a more general form, for multilinear forms, this inequality can be found in [56, Theorem 3.1.1].

Theorem 1.5.1. *Let* $\mathbf{A}$ *be an* $n \times n$ $(n \geq 2)$ *matrix with zero diagonal. Let* $\mathbf{x} = (X_1, \ldots, X_n), n \geq 2$ *be a random vector with independent mean zero coefficients. Then, for every convex function* f, *one has*

$$\mathbb{E}f(\langle \mathbf{A}\mathbf{x}, \mathbf{x}\rangle) \leqslant \mathbb{E}f(4\langle \mathbf{A}\mathbf{x}, \mathbf{x}'\rangle) \tag{1.93}$$

where $\mathbf{x}'$ *is an independent copy of* $\mathbf{x}$.

The consequence of the theorem can be equivalently stated as

$$\mathbb{E}f\left(\sum_{i,j=1}^{n} a_{ij}X_iX_j\right) \leqslant \mathbb{E}f\left(4\sum_{i,j=1}^{n} a_{ij}X_iX_j'\right)$$

where $\mathbf{x}' = \left(X_1', \ldots, X_n'\right)$ is an independent copy of $\mathbf{x}$. In practice, the independent copy of a random vector is easily available but the true random vector is hardly available. The larger the dimension n, the bigger gap between both sides of (1.93). We see Fig. 1.2 for illustration. Jensen's inequality says that if f is convex,

$$f(\mathbb{E}x) \leqslant \mathbb{E}f(x), \tag{1.94}$$

provided that the expectations exist. Empirically, we found when n is large enough, say $n \geq 100$, we can write (1.93) as the form

$$\mathbb{E}f(\langle \mathbf{Ax}, \mathbf{x} \rangle) \leqslant \mathbb{E}f(C \langle \mathbf{Ax}, \mathbf{x}' \rangle) \tag{1.95}$$

where $C > 1$, say $C = 1.1$. In practice, the expectation is replaced by an average of K simulations. The size of K affects the tightness of the inequality. Empirical evidence shows that $K = 100$–1,000 is sufficient. For Gaussian random vectors, C is a function of K and n, or $C(K, n)$. We cannot rigorously prove (1.95), however.

Examples of the convex function $f(x)$ in (1.93) include [48, p. 71]

- *Exponential.* e^{ax} is convex on $\mathbb{R}$, for any $a \in \mathbb{R}$.
- *Powers.* x^a is convex on $\mathbb{R}_{++}$, the set of positive real numbers, when $a \geq 1$ or $a \leq 0$, and concave for $0 \leq a \leq 1$.
- *Powers of absolute value.* $|x|^p$, for $p \geq 1$, is convex on $\mathbb{R}$.
- *Logarithm.* $\log x$ is concave on $\mathbb{R}_{++}$.
- *Negative entropy.* $x \log x$ is convex, either on $\mathbb{R}_{++}$, or $\mathbb{R}_{+}$, (the set of nonnegative real numbers) defined as 0 for $x = 0$.

Example 1.5.2 (Quadratic form is bigger than bilinear form). The function $f(t) = |t|$ is convex on $\mathbb{R}$. We will use this simple function in our simulations below.

Let us illustrate Theorem 1.5.1 using MATLAB simulations (See the Box for the MATLAB code). Without loss of generality, we consider a fixed $n \times n$ matrix $\mathbf{A}$ whose entries are Gaussian random variables. The matrix $\mathbf{A}$ has zero diagonal. For the random vector $\mathbf{x}$, we also assume the Gaussian random variables as its n entries. An independent copy $\mathbf{x}'$ is assumed to be available to form the bilinear form. For the quadratic form, we assume that the true random vector $\mathbf{x}$ is available: this assumption is much stronger than the availability of an independent copy $\mathbf{x}'$ of the Gaussian random vector $\mathbf{x}$ (Fig. 1.1).

Using the MATLAB code in the Box, we obtain Fig. 1.2. It is seen that the right-hand side (bilinear form) of (1.93) is always greater than the left-hand side of (1.93).

Example 1.5.3 (Multiple-Input, Multiple-Output). Given a linear system, we have that

$$\mathbf{y} = \mathbf{Hx} + \mathbf{n}$$

where $\mathbf{x} = (X_1, \ldots, X_n)$ is the input random vector, $\mathbf{y} = (Y_1, \ldots, Y_n)$ is the output random vector and $\mathbf{n} = (N_1, \ldots, N_n)$ is the noise vector (often Gaussian). The $\mathbf{H}$ is an $n \times n$ matrix with zero diagonal. One is interested in the inner product (in the quadratic form)

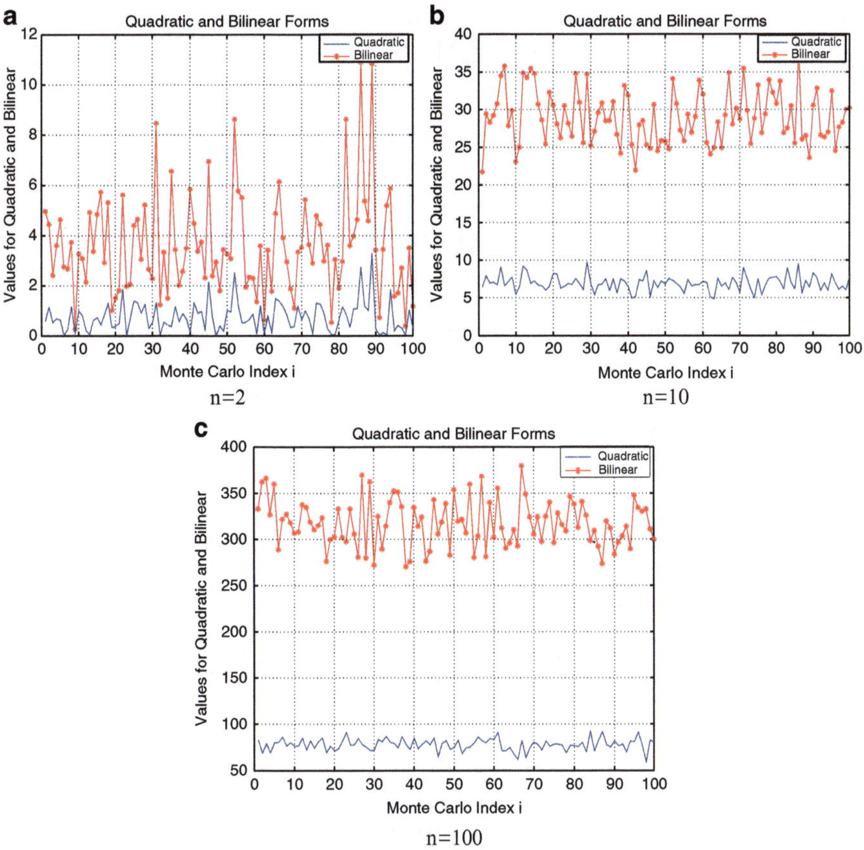

Fig. 1.1 Comparison of quadratic and bilinear forms as a function of dimension n. Expectation is approximated by an average of $K = 100$ Monte Carlo simulations. (**a**) n = 2. (**b**) n = 10. (**c**) n = 100

$$\langle \mathbf{y}, \mathbf{x} \rangle = \langle \mathbf{Hx}, \mathbf{x} \rangle + \langle \mathbf{n}, \mathbf{x} \rangle = \sum_{i,j=1}^{n} a_{ij} X_i X_j + \langle \mathbf{n}, \mathbf{x} \rangle .$$

if there is the complete knowledge of $\mathbf{x} = (X_1, \ldots, X_n)$. Unfortunately, sometimes we only know the *independent* copy $\mathbf{x}' = \left(X_1^{'}, \ldots, X_n^{'} \right)$, rather than the random vector $\mathbf{x} = (X_1, \ldots, X_n)$. Under this circumstance, we arrive at a bilinear form

$$\langle \mathbf{y}, \mathbf{x}' \rangle = \langle \mathbf{Hx}, \mathbf{x}' \rangle + \langle \mathbf{n}, \mathbf{x}' \rangle = \sum_{i,j=1}^{n} a_{ij} X_i X_j^{'} + \langle \mathbf{n}, \mathbf{x}' \rangle .$$

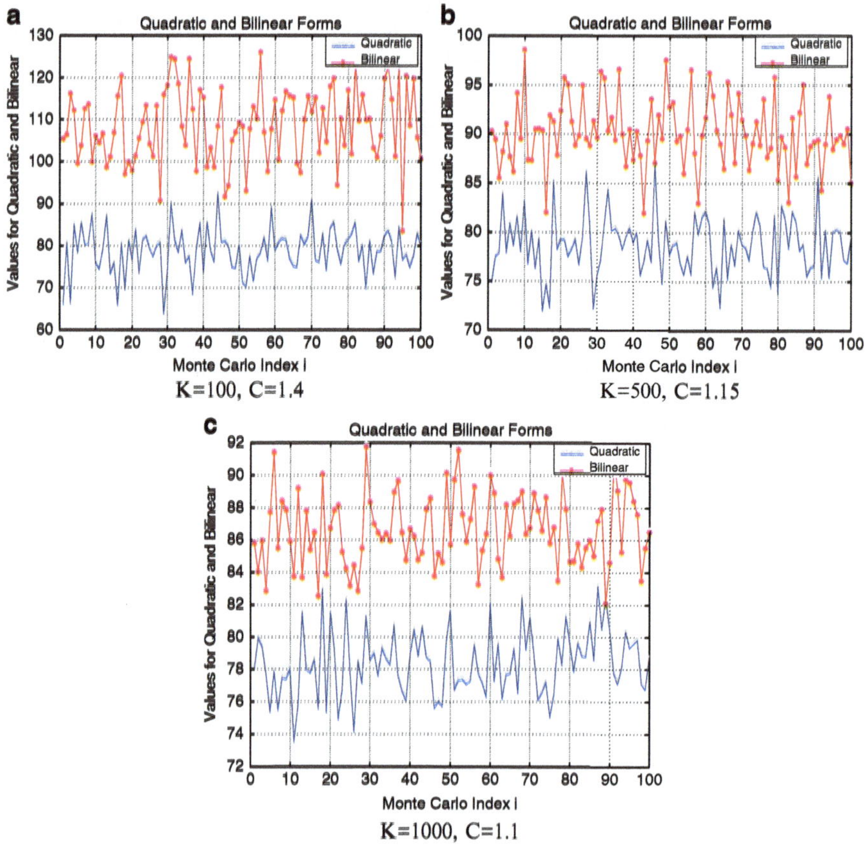

Fig. 1.2 Comparison of quadratic and bilinear forms as a function of K and C. Expectation is approximated by an average of K Monte Carlo simulations. $n = 100$. (**a**) K = 100, C = 1.4 (**b**) K = 500, C = 1.15 (**c**) K = 1,000, C = 1.1

Similarly, if we have the complete information of $\mathbf{n}$, we can form

$$\langle \mathbf{Ay}, \mathbf{n} \rangle = \langle \mathbf{AHx}, \mathbf{n} \rangle + \langle \mathbf{An}, \mathbf{n} \rangle \,,$$

where $\mathbf{A}$ is an $n \times n$ matrix with zero diagonal, and $\langle \mathbf{An}, \mathbf{n} \rangle$ is the quadratic form. On the other hand, if we have the independent copy $\mathbf{n}'$ of $\mathbf{n}$, we can form

$$\langle \mathbf{Ay}, \mathbf{n}' \rangle = \langle \mathbf{AHx}, \mathbf{n}' \rangle + \langle \mathbf{An}, \mathbf{n}' \rangle \,,$$

where $\langle \mathbf{An}, \mathbf{n}' \rangle$ is the bilinear form. Thus, (1.93) can be used to establish the inequality relation. □

MATLAB Code: Comparing Quadratic and Bilinear Forms

```
clear all;
for itest=1:100
K=100;n=10; Q=0; B=0; C=4; A1=randn(n,n); A=A1;
for itry=1:K        % We use K Monte Carlo simulations to approximate the expectation
for i=1:n
A(i,i)=0;           % A is an n x n matrix with zero diagonal
end
x=randn(n,1);       % random vector x
x1=randn(n,1);      % independent copy x' of random vector x
Q=Q+abs(x'*A*x);    % f(x) = |x|^p, for p ≥ 1, is convex on R . p=1 is chosen here.
B=B+abs(C*x1'*A*x); % The prime ' represents the Hermitian transpose in MATLAB
end
Q=Q/K; B=B/K;       % An average of K Monte Carlo simulations to approximate the expectation
Qtest(itest,1)=Q; Btest(itest,1)=B;
end
n=1:length(Qtest);
figure(1), plot(n,Qtest(n),'b–',n,Btest(n),'r-*'), xlabel('Monte Carlo Index i'), ylabel('Values for Quadratic and Bilinear'), title('Quadratic and Bilinear Forms'), legend('Quadratic', 'Bilinear'), grid
```

□

Proof of Theorem 1.5.1. We give a short proof due to [57]. Let $\mathbf{A} = (a_{ij})_{i,j=1}^{n}$, and let $\varepsilon_1, \ldots, \varepsilon_n$ be independent Bernoulli random variables with $\mathbb{P}(\varepsilon_i = 0) = \mathbb{P}(\varepsilon_i = 1) = \frac{1}{2}$. Let $\mathbb{E}_\varepsilon$ be the conditional expectation with respect to these random variables $\varepsilon_i, i = 1, \ldots, n.$ and similarly for the conditional expectation with respect to random vectors $\mathbf{x} = (X_1, \ldots, X_n)$ and $\mathbf{x}' = \left(X_1^{'}, \ldots, X_n^{'}\right)$. Let $[n] = \{1, 2, \ldots, n\}$ be the set. We have

$$\langle \mathbf{Ax}, \mathbf{x} \rangle = \sum_{i,j \in [n]} a_{ij} X_i X_j = 4\mathbb{E}_\varepsilon \sum_{i,j \in [n]} \varepsilon_i (1 - \varepsilon_i) a_{ij} X_i X_j.$$

By Jensen's inequality (1.94) and Fubini's inequality [58, 59],

$$\mathbb{E}f(\langle \mathbf{Ax}, \mathbf{x} \rangle) \leqslant \mathbb{E}_\varepsilon \mathbb{E}_{\mathbf{x}} f\left(4 \sum_{i,j \in [n]} \varepsilon_i (1 - \varepsilon_i) a_{ij} X_i X_j\right).$$

We fix a realization of $\varepsilon_1, \ldots, \varepsilon_n$ and consider the subset $I = \{i \in [n] : \varepsilon_i = 1\}$. Then we obtain

$$4 \sum_{i,j \in [n]} \varepsilon_i (1 - \varepsilon_i) a_{ij} X_i X_j = 4 \sum_{(i,j) \in I \times I^c} \varepsilon_i (1 - \varepsilon_i) a_{ij} X_i X_j.$$

where I^c is the complement the subset I. Since the $X_i, i \in I$ are independent of the $X_j, j \in I^c$, the distribution of the sum will not change if we replace $X_j, j \in I^c$ by their independent copies $X_j^{'}, j \in I^c$, the coordinates of the independent copy $\mathbf{x}^{'}$ of the $\mathbf{x}$. As a result, we have

$$\mathbb{E}f\left(\langle \mathbf{Ax}, \mathbf{x} \rangle\right) \leqslant \mathbb{E}_\varepsilon \mathbb{E}_{\mathbf{x},\mathbf{x}} f\left(4 \sum_{(i,j) \in I \times I^c} \varepsilon_i (1 - \varepsilon_i) a_{ij} X_i X_j^{'}\right).$$

We use a simple consequence of Jensen's inequality (1.94): If Y and Z are independent random variables and $\mathbb{E}Z = 0$ then

$$\mathbb{E}f(Y) = \mathbb{E}f(Y + \mathbb{E}Z) \leqslant \mathbb{E}(Y + Z).$$

Using this fact for

$$Y = 4 \sum_{(i,j) \in I \times I^c} a_{ij} X_i X_j^{'}, \quad Z = 4 \sum_{(i,j) \notin I \times I^c} a_{ij} X_i X_j^{'},$$

we arrive at

$$\mathbb{E}_{\mathbf{x},\mathbf{x}'} f(Y) = \mathbb{E}_{\mathbf{x},\mathbf{x}'} f(Y + Z) = f\left(4 \sum_{i,j=1}^{n} a_{ij} X_i X_j^{'}\right) = \mathbb{E}f\left(4 \langle \mathbf{Ax}, \mathbf{x}' \rangle\right).$$

Taking the expectation with respect to (ε_i), we complete the proof. □

The following result is valid for the Gaussian case.

Theorem 1.5.4 (Arcones and Giné [60]). *These exists an absolute constant C such that the following holds for all $p \geq 1$. Let $\mathbf{g} = (g_1, \ldots, g_n)$ be a sequence of independent standard Gaussian random variables. If $\mathcal{A}$ is a collection of Hermitian matrices and $\mathbf{g}'$ is an independent copy of $\mathbf{g}$, then*

$$\mathbb{E} \sup_{\mathbf{A} \in \mathcal{A}} \left| \sum_{i,j=1}^{n} g_i g_j \mathbf{A}_{i,j} + \sum_{i=1}^{n} \left(g_i^2 - 1\right) \mathbf{A}_{i,i} \right|^p \leqslant C^p \mathbb{E} \sup_{\mathbf{A} \in \mathcal{A}} \left| \sum_{i,j=1}^{n} g_i g_j^{'} \mathbf{A}_{i,j} \right|^p.$$

In other words, we have

$$\mathbb{E} \sup_{\mathbf{A}\in\mathcal{A}} \left|\langle \mathbf{Ag}, \mathbf{g}\rangle + \mathrm{Tr}\left(\mathbf{Agg}^T\right) - \mathrm{Tr}\left(\mathbf{A}\right)\right|^p \leqslant C^p \mathbb{E} \sup_{\mathbf{A}\in\mathcal{A}} \left|\langle \mathbf{Ag}, \mathbf{g}'\rangle\right|^p.$$

Theorem 1.5.5 dates back to [61], but appeared with explicit constants and with a much simplified proof in [62]. Let $\|\mathbf{X}\|$ be the operator norm and $\|\mathbf{X}\|_F$ be the Frobenius norm.

Theorem 1.5.5 (Theorem 17 of Boucheron et al. [62]). *Let* $\mathbf{X}$ *be the* $N \times N$ *matrix with extries* $x_{i,j}$ *and assume that* $x_{i,i} = 0$ *(zero diagonal) for all* $i \in \{1, \ldots, N\}$. *Let* $\boldsymbol{\xi} = \{\xi_i\}_{i=1}^n$ *be a Rademacher sequence. Then, for any* $t > 0$,

$$\mathbb{P}\left(\left|\sum_{i,j} \xi_i \xi_j x_{i,j}\right| > t\right) \leqslant 2\exp\left(-\frac{1}{64}\min\left\{\frac{\frac{96}{65}t}{\|\mathbf{X}\|}, \frac{t^2}{\|\mathbf{X}\|_F^2}\right\}\right). \quad (1.96)$$

Or

$$\mathbb{P}\left(\left|\boldsymbol{\xi}^T \mathbf{X} \boldsymbol{\xi}\right| > t\right) \leqslant 2\exp\left(-\frac{1}{64}\min\left\{\frac{\frac{96}{65}t}{\|\mathbf{X}\|}, \frac{t^2}{\|\mathbf{X}\|_F^2}\right\}\right).$$

Let $\mathcal{F}$ denote a collection of $n \times n$ symmetric matrices $\mathbf{X}$, and $\varepsilon_1, \ldots, \varepsilon_n$ are i.i.d. Rademacher variables. For convenience assume that the matrices $\mathbf{X}$ have zero diagonal, that is, $X_{ii} = 0$ for all $\mathbf{X} \in \mathcal{F}$ and $i = 1, \ldots, n$. Suppose the supremum of the L_2 operator norm of matrices $(\mathbf{X})_{\mathbf{X}\in\mathcal{F}}$ is finite, and without loss of generality we assume that this supremum equals one, that is,

$$\sup_{\mathbf{X}\in\mathcal{F}} \sup_{\|\mathbf{z}\|_2^2 \leqslant 1} \mathbf{z}^T \mathbf{X} \mathbf{z} = 1$$

for $\mathbf{z} \in \mathbb{R}^n$.

Theorem 1.5.6 (Theorem 17 of Boucheron et al. [62]). *For all* $t > 0$,

$$\mathbb{P}\left(Z \geqslant \mathbb{E}\left[Z\right] + t\right) \leqslant \exp\left(-\frac{t^2}{32\mathbb{E}\left[Y^2\right] + 65t/3}\right)$$

where the random variable Y *is defined as*

$$Y = \sup_{\mathbf{X}\in\mathcal{F}} \left(\sum_{i=1}^n \left(\sum_{j=1}^n \varepsilon_j X_{ij}\right)^2\right)^{1/2}.$$

1.6 Fundamentals of Random Matrices

Here, we highlight the fundamentals of random matrix theory that will be needed in Chap. 9 that deals with high-dimensional data processing motivated by the large-scale cognitive radio network testbed. As mentioned in Sect. 9.1, the basic building block for each node in the data processing is a random matrix (e.g., sample covariance matrix). A sum of random matrices arise naturally. Classical textbooks deal with a sum of scalar-valued random variables—and the central limit theorem. Here, we deal with a sum of random matrices—matrix-valued random variables. Many new challenges will be encountered due to this fundamental paradigm shift. For example, scalar-valued random variables are commutative, while matrix-valued random variables are non-commutative. See Tao [9].

1.6.1 *Fourier Method*

This method is standard method for the proof of the central limit theorem. Given any real random variable X, the *characteristic function* $F_X(t) : \mathbb{R} \to \mathbb{C}$ is defined as

$$F_X(t) = \mathbb{E}e^{jtX}.$$

Equivalently, F_X is the Fourier transform of the probability measure μ_X. The signed Bernoulli distribution has $F_X = \cos(t)$ and the normal distribution $\mathcal{N}\left(\mu, \sigma^2\right)$ has $F_X(t) = e^{jt\mu}e^{-\sigma^2 t^2/2}$.

For a random vector $\mathbf{X}$ taking values in $\mathbb{R}^n$, we define $F_{\mathbf{X}}(t) : \mathbb{R}^n \to \mathbb{C}$ as

$$F_{\mathbf{X}}(t) = \mathbb{E}e^{jt\cdot\mathbf{X}}$$

where $\cdot$ denotes the Euclidean inner product on $\mathbb{R}^n$. One can similarly define the characteristic function on complex vector spaces $\mathbb{C}^n$ by using the complex inner product

$$(z_1, \ldots, z_n) \cdot (w_1, \ldots, w_n) = \mathrm{Re}\left(z_1\bar{w}_1 + \cdots + z_n\bar{w}_n\right).$$

1.6.2 *The Moment Method*

The most elementary (but still remarkably effective) method is the moment method [63]. The method is to understand the distribution of a random variable X via its moments X^k. This method is equivalent to Fourier method. If we Taylor expand e^{jtX} and formally exchange the series and expectation, we arrive at the heuristic identity

$$F_X(t) = \sum_{k=0}^{\infty} \frac{(jk)^k}{k!} \mathbb{E}X^k$$

which connects the characteristic functions of a real variable X as a kind of generating functions for the moments. In practice, the moment method tends to look somewhat different from the Fourier methods, and it is more apparent how to modify them to non-independent or non-commutative settings.

The Fourier phases $x \to e^{itx}$ are bounded, but the moment function $x \to x^k$ becomes unbounded at infinity. One can deal with this issue, however, as long as one has sufficient decay:

Theorem 1.6.1 (Carleman continuity theorem). *Let X_n be a sequence of uniformly sub-Gaussian real random variables, and let X be another sub-Gaussian random variable. Then the following statements are equivalent:*

1. *For every $k = 0, 1, 2, \ldots$, $\mathbb{E}X_n^k$ converges to $\mathbb{E}X^k$.*
2. *X_n converges in distribution to X.*

See [63] for a proof.

1.6.3 Expected Moments of Random Matrices with Complex Gaussian Entries

Recall that for ξ in $\mathbb{R}$ and $\sigma^2 \in]0, \infty[$, $\mathcal{N}\left(\xi, \frac{1}{2}\sigma^2\right)$ denotes the Gaussian distribution with mean ξ and variance σ^2. The normalized trace is defined as

$$\mathrm{tr}_n = \frac{1}{n}\mathrm{Tr}.$$

The first class, denoted Hermitian Gaussian Random Matrices or HGRM(n,σ^2), is a class of Hermitian $n \times n$ random matrices $\mathbf{A} = (a_{ij})$, satisfying that the entries $\mathbf{A} = (a_{ij}), 1 \leqslant i \leqslant j \leqslant n$, forms a set of $\frac{1}{2}n(n+1)$ independent, Gaussian random variables, which are complex valued whenever $i < j$, and fulfill that

$$\mathbb{E}(a_{ij}) = 0, \text{ and } \mathbb{E}\left(|a_{ij}|^2\right) = \sigma^2, \text{ for all } i, j.$$

The case $\sigma^2 = \frac{1}{2}$ gives the normalization used by Wigner [64] and Mehta [65], while the case $\sigma^2 = \frac{1}{n}$ gives the normalization used by Voiculescu [66]. We say that $\mathbf{A}$ is a standard Hermitian Gaussian random $n \times n$ matrix with entries of variance σ^2, if the following conditions are satisfied:

1. The entries $a_{ij}, 1 \leqslant i \leqslant j \leqslant n$, form a set of $\frac{1}{2}n(n+1)$ independent, complex valued random variables.

2. For each i in $\{1, 2, \ldots, n\}$, a_{ii} is a real valued random variable with distribution $\mathcal{N}\left(0, \frac{1}{2}\sigma^2\right)$.
3. When $i \leq j$, the real and imaginary parts $\mathrm{Re}\left(a_{ij}\right), \mathrm{Im}\left(a_{ij}\right)$, of a_{ij} are independent, identically distributed random variables with distribution $\mathcal{N}\left(0, \frac{1}{2}\sigma^2\right)$.
4. When $j > i$, $a_{ij} = \bar{a}_{ji}$, where $\bar{d}$ is the complex conjugate of d.

We denote by HGRM(n,σ^2) the set of all such random matrices. If $\mathbf{A}$ is an element of HGRM(n,σ^2), then

$$\mathbb{E}\left(\left|a_{ij}\right|^2\right) = \sigma^2, \text{ for all } i, j.$$

The distribution of the real valued random variable a_{ii} has density

$$x \mapsto \frac{1}{\sqrt{2\pi\sigma^2}} \exp\left(-\frac{x^2}{2\sigma^2}\right), \quad x \in \mathbb{R},$$

with respect to Lebesgue measure on $\mathbb{R}$, whereas, if $i \leq j$, the distribution of the complex valued random variable a_{ij} has density

$$z \mapsto \frac{1}{\sqrt{2\pi\sigma^2}} \exp\left(-\frac{z^2}{2\sigma^2}\right), \quad z \in \mathbb{C},$$

with respect to (w.r.t.) Lebesgue measure on $\mathbb{C}$. For any complex matrix $\mathbf{H}$, we have that

$$\mathrm{Tr}\left(\mathbf{H}^2\right) = \sum_{i=1}^{n} h_{ii} + 2\sum_{i<j} \left|h_{ij}\right|^2.$$

The distribution of an element $\mathbf{A}$ of HGRM(n,σ^2) has the density

$$\mathbf{H} \mapsto \frac{1}{\sqrt{2\pi\sigma^2}} \exp\left(-\frac{1}{2\sigma^2}\mathrm{Tr}\left(\mathbf{H}^2\right)\right), \quad \mathbf{H} \in \mathbb{C}^{n\times n}.$$

The second class, denoted Gaussian Random Matrices or GRM(m,n,σ^2), is a class of $m \times n$ random matrices $\mathbf{B} = \left(b_{ij}\right), 1 \leqslant i \leqslant m, 1 \leqslant j \leqslant n$. This class forms a set of mn independent, complex-valued, Gaussian random variables, satisfying that

$$\mathbb{E}\left(b_{ij}\right) = 0, \text{ and } \mathbb{E}\left(\left|b_{ij}\right|^2\right) = \sigma^2, \text{ for all } i, j.$$

We say $\mathbf{B}$ is a standard Gaussian random matrix of $m\times n$ with entries of variance σ^2, if real valued random variables $\mathrm{Re}\left(b_{ij}\right), \mathrm{Im}\left(b_{ij}\right), 1 \leqslant i, j \leqslant n$, form a family of $2mn$ independent, identically distributed (i.i.d.) random variables, with distribution $\mathcal{N}\left(0, \frac{1}{2}\sigma^2\right)$. This class starts with Wishart [67] and Hsu [68].

We are interested in the explicit formulas for the mean values $\mathbb{E}\left(\mathrm{Tr}\left[\exp\left(s\mathbf{A}\right)\right]\right)$ in the Wigner case, and

$$\mathbb{E}\left(\mathrm{Tr}\left[\mathbf{B}^*\mathbf{B}\exp\left(s\mathbf{B}^*\mathbf{B}\right)\right]\right)$$

in the Wishart case, as functions of a complex parameter s. A new and entirely analytical treatment of these problems is given by the classical work of Haagerup and Thorbjørnsen [69] which we follow closely for this section.

1.6.4 Hermitian Gaussian Random Matrices HGRM(n, σ^2)

We need the confluent hyper-geometric function [70, Vol. 1, p. 248] $(a, c, x) \mapsto \Phi\left(a, c, x\right)$ that is defined as

$$\Phi\left(a,c,x\right) = \sum_{n=0}^{\infty} \frac{(a)_n x^n}{(c)_n n!} = 1 + \frac{a}{c}\frac{x}{1} + \frac{a(a+1)}{c(c+1)}\frac{x^2}{2} + \cdots,$$

for a, c, x in $\mathbb{C}$, such that $c \notin \mathbb{Z}\backslash\mathbb{N}$. In particular, if $a \in \mathbb{Z}\backslash\mathbb{N}$, then $(x) \mapsto \Phi\left(a, c, x\right)$ is a polynomial in x of degree $-a$, for any permitted c. For any non-negative integer n, and any complex number w, we apply the notation

$$(w)_n = \begin{cases} 1, & \text{if } n = 0, \\ w(w+1)(w+2)\cdots(w+n-1), & \text{if n} \in \mathbb{N}. \end{cases}$$

For any element $\mathbf{A}$ of HGRM(n, σ^2) and any $s \in \mathbb{C}$, we have that

$$\mathbb{E}\left(\mathrm{Tr}\left[\exp\left(s\mathbf{A}\right)\right]\right) = n \cdot \exp\left(\frac{\sigma^2 s^2}{2}\right) \cdot \Phi\left(1-n, 2; -\sigma^2 s^2\right).$$

If $(\mathbf{X}_n)$ is a sequence of random matrices, such that $\mathbf{X}_n \in$ HGRM $\left(n, \frac{1}{n}\right)$ for all n in $\mathbb{N}$. Then for any $s \in \mathbb{C}$, we have that

$$\lim_{n\to\infty} \mathbb{E}\left(\mathrm{Tr}\left[\exp\left(s\mathbf{X}_n\right)\right]\right) = \frac{1}{2\pi}\int_{-2}^{2} \exp\left(sx\right)\sqrt{4-x^2}dx,$$

and the convergence is uniform on compact subsets of $\mathbb{C}$. Further we have that for the k-th moment of $\mathbf{X}_n$

$$\lim_{n\to\infty} \mathbb{E}\left(\mathrm{tr}_n\left[\mathbf{X}_n^k\right]\right) = \frac{1}{2\pi}\int_{-2}^{2} x^k\sqrt{4-x^2}dx,$$

and in general, for every continuous bounded function $f : \mathbb{R} \mapsto \mathbb{C}$,

$$\lim_{n\to\infty} \mathbb{E}\left(\mathrm{tr}_n\left[f\left(\mathbf{X}_n\right)\right]\right) = \frac{1}{2\pi}\int_{-2}^{2} f\left(x\right)\sqrt{4-x^2}dx.$$

Often, a recursion formula is efficient in calculation. Let $\mathbf{A}$ is an element of HGRM$(n, 1)$, and for integer k define

$$C(k,n) = \mathbb{E}\left(\mathrm{Tr}\left[\mathbf{A}^{2k}\right]\right).$$

Then the initial values are $C(0,n) = n, C(1,n) = n^2$, and for fixed n in $\mathbb{N}$, the numbers $C(k,n)$ satisfy the recursion formula:

$$C(k+1,n) = n \cdot \frac{4k+2}{k+2} \cdot C(k,n) + \frac{k\left(4k^2-1\right)}{k+2} \cdot C(k-1,n), \qquad k \geqslant 1. \tag{1.97}$$

We can further show that $C(k,n)$ has the following form [71]

$$C(k,n) = \sum_{i=0}^{\left[\frac{k}{2}\right]} a_i(k)\, n^{k+1-2i}, \qquad k \in \mathbb{N}_0, n \in \mathbb{N}. \tag{1.98}$$

Here the notation $\mathbb{N}_0$ denotes the integer that does not include zero, in contrast with $\mathbb{N}$. The coefficients $a_i(k), i, k \in \mathbb{N}_0$ are determined by the following recursive formula

$$a_i(k) = 0, \quad i \geqslant \left[\tfrac{k}{2}\right] + 1,$$

$$a_0(k) = \frac{1}{k+1}\binom{2k}{k}, \quad k \in \mathbb{N}_0$$

$$a_i(k+1) = \frac{4k+2}{k+2} \cdot a_i(k) + \frac{k\left(4k^2-1\right)}{k+2} \cdot a_{i-1}(k-1), \quad k, i \in \mathbb{N}.$$

A list of the numbers of $a_i(k)$ is given in [71, p. 459].

From (1.97) and (1.98), we can get [69] for any $\mathbf{A}$ in HGRM$(n, 1)$,

$$\begin{aligned}
\mathbb{E}\left(\mathrm{Tr}\left[\mathbf{A}^2\right]\right) &= n^2,\\
\mathbb{E}\left(\mathrm{Tr}\left[\mathbf{A}^4\right]\right) &= 2n^3 + n,\\
\mathbb{E}\left(\mathrm{Tr}\left[\mathbf{A}^6\right]\right) &= 5n^4 + 10n^2,\\
\mathbb{E}\left(\mathrm{Tr}\left[\mathbf{A}^8\right]\right) &= 14n^5 + 70n^3 + 21n,\\
\mathbb{E}\left(\mathrm{Tr}\left[\mathbf{A}^{10}\right]\right) &= 42n^6 + 420n^4 + 483n^2,
\end{aligned}$$

etc. If we replace the $\mathbf{A}$ above by an element $\mathbf{X}$ of HGRM$(n, \frac{1}{n})$, and Tr by tr_n, then we have to divide the above numbers by n^{k+1}. Finally, for $\mathbf{X}$ of HGRM$(n, \frac{1}{n})$, we have

$$\begin{aligned}
\mathbb{E}\left(\operatorname{tr}_n\left[\mathbf{A}^2\right]\right) &= 1,\\
\mathbb{E}\left(\operatorname{tr}_n\left[\mathbf{A}^4\right]\right) &= 2+\tfrac{1}{n^2},\\
\mathbb{E}\left(\operatorname{tr}_n\left[\mathbf{A}^6\right]\right) &= 5+\tfrac{10}{n^2},\\
\mathbb{E}\left(\operatorname{tr}_n\left[\mathbf{A}^8\right]\right) &= 14+\tfrac{70}{n^2}+\tfrac{21}{n^4},\\
\mathbb{E}\left(\operatorname{tr}_n\left[\mathbf{A}^{10}\right]\right) &= 42+\tfrac{420}{n^2}+\tfrac{483}{n^4},
\end{aligned}$$

etc. The constant term in $\mathbb{E}\left(\operatorname{tr}_n\left[\mathbf{A}^{2k}\right]\right)$ is

$$a_0\left(k\right)=\frac{1}{k+1}\binom{2k}{k}=\frac{1}{2\pi}\int_{-2}^{2}x^{2k}\sqrt{4-x^2}dx,$$

in concordance with Wigner's semi-circle law.

1.6.5 Hermitian Gaussian Random Matrices GRM(m, n, σ^2)

We first define the function $\varphi_k^{\alpha}\left(x\right)$ as

$$\varphi_k^{\alpha}\left(x\right)=\left[\frac{k!}{\Gamma\left(k+\alpha+1\right)}x^{\alpha}\exp\left(-x\right)\right]^{1/2}\cdot L_k^{\alpha}\left(x\right),\quad k\in\mathbb{N}_0, \tag{1.99}$$

where $L_k^{\alpha}(x)_{k\in\mathbb{N}_0}$ is the sequence of generalized Laguerre polynomials of order α, i.e.,

$$L_k^{\alpha}\left(x\right)=\left(k!\right)^{-1}x^{-\alpha}\exp\left(x\right)\cdot\frac{d^k}{dx^k}\left(x^{k+\alpha}\exp\left(-x\right)\right),\quad k\in\mathbb{N}_0.$$

Here $\Gamma\left(x\right)$ is the Gamma function.

Now we can state a corollary from [69]. Let $\mathbf{B}$ be an element of GRM$(m, n, 1)$, let $\varphi_k^{\alpha}\left(x\right), \alpha\in\left]0,\infty\right[, k\in\mathbb{N}_0$, be the functions introduced in (1.99), and let $f:\left]0,\infty\right[\mapsto\mathbb{R}$ a Borel function.[1] If $m\geq n$, we have that

$$\mathbb{E}\left(\operatorname{Tr}\left[f\left(\mathbf{B}^*\mathbf{B}\right)\right]\right)=\int_0^{\infty}f(x)\left[\sum_{i=0}^{n-1}\left(\varphi_k^{m-n}\left(x\right)\right)^2\right]dx.$$

If $m\leq n$, we have that

[1] A map $f: X\mapsto Y$ between two topological spaces is called Borel (or Borel measurable) if $f^{-1}(A)$ is a Borel set for any open set A.

$$\mathbb{E}\left(\text{Tr}\left[f\left(\mathbf{B}^*\mathbf{B}\right)\right]\right) = (n-m)f(0) + \int_0^\infty f(x)\left[\sum_{i=0}^{n-1}\left(\varphi_k^{n-m}(x)\right)^2\right]dx.$$

We need to define the hyper-geometric function F

Q of $m \times n$ is an element from GRM($m, n, \frac{1}{n}$). Denote c = $\frac{m}{n}$. We have

$$\mathbb{E}\left(\text{tr}_n\left[\mathbf{Q}^*\mathbf{Q}\right]\right) = c$$

$$\mathbb{E}\left(\text{tr}_n(\mathbf{Q}^*\mathbf{Q})^2\right) = c^2 + c,$$

$$\mathbb{E}\left(\text{tr}_n(\mathbf{Q}^*\mathbf{Q})^3\right) = \left(c^3 + 3c^2 + c\right) + cn^{-2}$$

$$\mathbb{E}\left(\text{tr}_n(\mathbf{Q}^*\mathbf{Q})^4\right) = \left(c^4 + 6c^3 + 6c^2 + c\right) + \left(5c^2 + 5c\right)n^{-2}$$

$$\mathbb{E}\left(\text{tr}_n(\mathbf{Q}^*\mathbf{Q})^5\right) = \left(c^5+10c^4+20c^3+10c^2+c\right) + \left(15c^3+40c^2+15c\right)n^{-2}+8cn^{-4}. \tag{1.100}$$

In general, $\mathbb{E}\left(\text{tr}_n(\mathbf{Q}^*\mathbf{Q})^k\right)$ is a polynomial of degree $[\frac{k-1}{2}]$ in n^{-2}, for fixed c.

1.7 Sub-Gaussian Random Variables

The material here is taken from [72–74]. Buldygin and Solntsev [74] develops and uses this tool systemically. If $S_N = \sum_{i=1}^{N} a_i X_i$, where X_i are the Bernoulli random variables, then its generating moment function $\mathbb{E}\left(e^{tX}\right)$ satisfies $\mathbb{E}\left(e^{tX}\right) \leqslant e^{\sigma^2 t^2/2}$. On the other hand, if X is a Gaussian random variable with mean zero and variance $\mathbb{E}\left(X^2\right) = \sigma^2$, its moment generating function $\mathbb{E}\left(e^{tX}\right)$ is $e^{\sigma^2 t^2/2}$. This led Kahane [75] to make the following definition. A random variable X is *sub-Gaussian*, with exponent b, if

$$\mathbb{E}\left(e^{tX}\right) \leqslant e^{b^2 t^2/2} \tag{1.101}$$

for all $-\infty < t < \infty$.

Lemma 1.7.1 (Equivalence of sub-Gaussian properties [72]). *Let X be a random variable. Then the following properties are equivalent with parameters $K_i > 0$ that are different from each other by at most an absolute constant.*[2]

1. Tails: $\mathbb{P}\left(|X| > t\right) \leqslant \exp\left(1 - t^2/K_1^2\right)$, *for all* $t \geq 0$.

[2]The precise meaning of this equivalence is the following: There is an absolute constant C such that property i implies property j with parameter $K_j \leqslant CK_i$ for any two properties $i, j = 1, 2, 3$.

2. *Moments:* $(\mathbb{E}|X|^p)^{1/p} \leqslant K_2\sqrt{p}$, *for all* $p \geq 1$.
3. *Super-exponential moment:* $\mathbb{E}\left[\exp\left(X^2/K_3^2\right)\right] \leqslant e$.

   Moreover, if $\mathbb{E}X = 0$, *then properties 1–3 are also equivalent to the following one:*
4. *Laplace transform condition:* $\mathbb{E}\left[\exp\left(tX\right)\right] \leqslant \exp\left(t^2K_4^2\right)$ *for all* $t \in \mathbb{R}$.

If $X_1, \ldots, X_N$ are independent sub-Gaussian random variables with exponents $b_1, \ldots, b_N$ respectively and $a_1, \ldots, a_N$ are real numbers, then [73, p. 109]

$$\mathbb{E}\left(e^{t(a_1X_1+\cdots+a_NX_N)}\right) = \prod_{i=1}^{N}\mathbb{E}\left(e^{ta_iX_i}\right) \leqslant \prod_{i=1}^{N} e^{a_i^2b_i^2/2},$$

so that $a_1X_1+\cdots+a_NX_N$ is sub-Gaussian, with exponent $\left(a_1^2b_1^2+\cdots+a_N^2b_N^2\right)^{1/2}$.

We say that a random variable Y *majorizes in distribution* another random variable X if there exists a number $\alpha \in (0,1]$ such that, for all $t > 0$, one has [74]

$$\alpha\mathbb{P}\left(|X| > t\right) \leqslant \mathbb{P}\left(|Y| > t\right).$$

In a similar manner, we say that a sequence of random variables $\{Y_i, i \geqslant 1\}$ *uniformly majorizes in distribution* another sequence of random variables $\{X_i, i \geqslant 1\}$ if there exists a number $\alpha \in (0,1]$ such that, for all $t > 0$, and $i \geq 1$, one has

$$\alpha\mathbb{P}\left(|X_i| > t\right) \leqslant \mathbb{P}\left(|Y_i| > t\right).$$

Consider the quantity

$$\tau\left(X\right) = \sup_{|t|>0}\left[\frac{2\ln\mathbb{E}\exp\left(Xt\right)}{t^2}\right]^{1/2}. \tag{1.102}$$

We have that

$$\tau\left(X\right) = \inf\left\{t \geqslant 0 : \mathbb{E}\exp\left(Xt\right) \leqslant \exp\left(X^2t^2/2\right), \quad t \in \mathbb{R}\right\}.$$

Further if X is sub-Gaussian if and only if $\tau\left(X\right) < \infty$.

The quantity $\tau\left(X\right)$ will be called the sub-Gaussian standard. Since one has

$$\mathbb{E}\exp\left(Xt\right) = 1 + t\mathbb{E}X + \frac{1}{2}t^2\mathbb{E}X^2 + o(t^2),$$

$$\exp\left(a^2t^2/2\right) = 1 + \frac{1}{2}t^2a^2 + o(t^2),$$

as $t \to 0$, then the inequality

$$\mathbb{E}\exp\left(Xt\right) \leqslant \exp\left(X^2t^2/2\right), \quad t \in \mathbb{R}$$

may only hold if

$$\mathbb{E}X = 0, \quad \mathbb{E}X^2 \leqslant a^2.$$

This is why each sub-Gaussian random variable X has zero mean and satisfies the condition

$$\mathbb{E}X^2 \leqslant \tau^2 (X).$$

We say the sub-Gaussian random variable X has parameters $(0, \tau^2)$.

If X is a sub-Gaussian random variable X has parameters $(0, \tau^2)$, then

$$\mathbb{P}(X > t) \leqslant \exp\left(-t^2/\tau^2\right),$$
$$\mathbb{P}(-X > t) \leqslant \exp\left(-t^2/\tau^2\right),$$
$$\mathbb{P}(|X| > t) \leqslant 2\exp\left(-t^2/\tau^2\right).$$

Assume that $X_1, \ldots, X_N$ are independent sub-Gaussian random variables. Then one has

$$\tau^2\left(\sum_{i=1}^{n} X_i\right) \leqslant \sum_{i=1}^{n} \tau^2 (X_i),$$
$$\max_{1 \leqslant i \leqslant m \leqslant n} \tau^2\left(\sum_{k=i}^{m} X_i\right) \leqslant \tau^2\left(\sum_{k=1}^{n} X_k\right).$$

Assume that X is a zero-mean random variable. Then the following inequality holds

$$\tau(X) \leqslant \sqrt{2}\theta(X),$$

where

$$\theta(X) = \sup_{n \geqslant 1}\left[\frac{2^n \cdot n!}{(2n)!}\mathbb{E}X^{2n}\right].$$

We say that a random variable X is *strictly sub-Gaussian* if X is sub-Gaussian and $\mathbb{E}X^2 = \tau^2(X)$. If $a \in \mathbb{R}$ and $\mathbf{X}$ is strictly sub-Gaussian then we have

$$\tau^2(aX) = a^2\tau^2(X) = a^2\mathbb{E}X^2 = \mathbb{E}(aX)^2.$$

In such a way, the class of sub-Gaussian random variable is closed with respect to multiplication by scalars. This class, however, is not closed with respect to addition of random variables. The next statement motivates us to set this class out.

Lemma 1.7.2 ([74]). *Let $X_1, \ldots, X_N$ be independently sub-Gaussian random variables, and $\{c_1, \ldots, c_N\} \in \mathbb{R}$. Then $\sum\limits_{i=1}^{n} c_i X_i$ is strictly sub-Gaussian random variable.*

Theorem 1.7.3 ([74]). *Assume that Y is a Gaussian random variable with zero mean and variance σ^2 (or with parameters $(0, \sigma^2)$). Then*

1. *If X is a sub-Gaussian random variable with parameters $(0, \tau^2)$ and $\sigma^2 > \tau^2$, then Y majorizes X in distribution.*
2. *If Y majorizes in distribution some zero-mean random variable X, then the random variable$X$ is a sub-Gaussian random variable.*

Assume that $\{X_i, i \geqslant 1\}$ is a sequence of sub-Gaussian random variable with parameters $(0, \tau_i^2), i \geqslant 1$ while $\{Y_i, i \geqslant 1\}$ is a sequence of Gaussian random variable with parameters $(0, \alpha\sigma_i^2), i \geqslant 1, \alpha > 1$. Then, the sequence $\{Y_i, i \geqslant 1\}$ uniformly majorizes in distribution $\{X_i, i \geqslant 1\}$.

A random n-dimensional vector $\mathbf{x}$ will be called standard sub-Gaussian vector [74, p. 209] if, in some orthogonal basis of the space $\mathbb{R}^n$, its components $X^{(1)}, \ldots, X^{(n)}$ are jointly independently sub-Gaussian random variables. Then we set

$$\tau_n(\mathbf{x}) = \max_{1 \leqslant i \leqslant n} \tau\left(X^{(i)}\right),$$

where $\tau(X)$ is defined in (1.102). The simplest example of standard sub-Gaussian vector is the standard n-dimensional Gaussian random vector $\mathbf{y}$.

A linear combination of independent subGaussian random variables is subGaussian. As a special case, a linear combination of independent Gaussian random variables is Gaussian.

Theorem 1.7.4. *Let $X_1, \ldots, X_n$ be independent centered subGaussian random variables. Then for any $a_1, \ldots, a_n \in \mathbb{R}$*

$$\mathbb{P}\left(\left|\sum_{i=1}^{n} a_i X_i\right| > t\right) \leqslant 2\exp\left(-\frac{ct^2}{\sum\limits_{i=1}^{n} a_i^2}\right).$$

Proof. We follow [76] for the short proof. Set $v_i = a_i / \left(\sum\limits_{i=1}^{n} a_i^2\right)^{1/2}$. We have to show that the random variable $Y = \sum\limits_{i=1}^{n} v_i X_i$ is subGaussian. Let us check the Laplace transform condition (4) of the definition of a subGaussian random variable. For any $t \in \mathbb{R}$

$$\mathbb{E}\exp\left(t\sum_{i=1}^{n} v_i X_i\right) = \prod_{i=1}^{n} \mathbb{E}\exp\left(t v_i X_i\right)$$
$$\leqslant \prod_{i=1}^{n} \exp\left(t^2 v_i^2 K_4^2\right) = \exp\left(t^2 K_4^2 \sum_{i=1}^{n} v_i^2\right) = e^{t^2 K_4^2}.$$

The inequality here follows from Laplace transform condition (4). The constant in front of the exponent in Laplace transform condition (4) is 1, this fact plays the crucial role here. □

Theorem 1.7.4 can be used to give a very short proof of a classical inequality due to Khinchin.

Theorem 1.7.5 (Khinchin's inequality). *Let $X_1, \ldots, X_n$ be independent centered subGaussian random variables. Then for any $p \geq 1$, there exists constants $A_p, B_p > 0$ such that the inequality*

$$A_p\left(\sum_{i=1}^{n} a_i^2\right)^{1/2} \leqslant \left(\mathbb{E}\left|\sum_{i=1}^{n} a_i X_i\right|^p\right)^{1/p} \leqslant B_p\left(\sum_{i=1}^{n} a_i^2\right)^{1/2}$$

holds for all $a_1, \ldots, a_n \in \mathbb{R}$.

Proof. We follow [76] for the proof. Without loss of generality, we assume that $\left(\sum_{i=1}^{n} a_i^2\right)^{1/2} = 1$. Let $p \geq 2$. Then by Hölder's inequality

$$\left(\sum_{i=1}^{n} a_i^2\right)^{1/2} = \left(\mathbb{E}\left|\sum_{i=1}^{n} a_i X_i\right|^2\right)^{1/2} \leqslant \left(\left|\mathbb{E}\sum_{i=1}^{n} a_i X_i\right|^p\right)^{1/p},$$

so $A_p = 1$. Using Theorem 1.7.4, we know that the linear combination $Y = \sum_{i=1}^{n} a_i X_i$ is a subGaussian random variable. Hence,

$$\left(\mathbb{E}|Y|^p\right)^{1/p} \leqslant C\sqrt{p} : B_p.$$

This is the right asymptotic as $p \to \infty$.

In the case $1 \leq p \leq 2$ it is enough to prove the inequality for $p = 1$. Again, using Hölder's inequality, we can choose $B_p = 1$. Applying Khinchin's inequality with $p = 3$, we have

$$\mathbb{E}|Y|^2 = \mathbb{E}\left(|Y|^{1/2} \cdot |Y|^{3/2}\right) \leqslant \left(\mathbb{E}\,|Y|\right)^{1/2} \cdot \left(\mathbb{E}|Y|^3\right)^{1/2} \leqslant \left(\mathbb{E}\,|Y|\right)^{1/2} \cdot B_3^{3/2}\left(\mathbb{E}|Y|^2\right)^{3/4}.$$

Thus,

$$B_3^{-3}\left(\mathbb{E}|Y|^2\right)^{1/2} \leqslant \mathbb{E}\,|Y|.$$

□

1.8 Sub-Gaussian Random Vectors

Let S^{n-1} denote the unit sphere in $\mathbb{R}^n$ (resp. in $\mathbb{C}^n$). For two complex vectors $\mathbf{a}, \mathbf{b} \in \mathbb{C}^n$, the inner product is $\langle \mathbf{a}, \mathbf{b} \rangle = \sum_{i=1}^{n} a_i \bar{b}_i$, where the bar standards for the complex conjugate. A mean-zero random vector $\mathbf{x}$ on $\mathbb{C}^n$ is called *isotropic* if for every $\boldsymbol{\theta} \in S^{n-1}$,

$$\mathbb{E}|\langle \mathbf{x}, \boldsymbol{\theta} \rangle|^2 = 1.$$

A random vector $\mathbf{x}$ is called *L-sub-Gaussian* if it is isotropic and

$$\mathbb{P}\left(|\langle \mathbf{x}, \boldsymbol{\theta} \rangle| \geqslant t\right) \leqslant 2 \exp\left(-t^2/2L^2\right)$$

for every $\boldsymbol{\theta} \in S^{n-1}$, and any $t > 0$. It is well known that, up to an absolute constant, the tail estimates in the definition of a sug-Gaussian random vector are equivalent to the moment characterization

$$\sup_{\theta \in S^{n-1}} \left(|\langle \mathbf{x}, \theta \rangle|^p\right)^{1/p} \leqslant \sqrt{p} L.$$

Assume that a random vector $\boldsymbol{\xi}$ has *independent* coordinates ξ_i, each of which is an L-sub-Gaussian random variable of mean zero and variance one. One may verify by direct computation that $\boldsymbol{\xi}$ is L-sub-Gaussian. Rademacher vectors, standard Gaussian vectors, (that is, random vectors with independent normally distributed entries of mean zero and variance one), as well as Steinhaus vectors (that is, random vectors with independent entries that are uniformly distributed on $\{z \in \mathbb{C} : |z| = 1\}$), are examples of isotropic, L-subGaussian random vectors for an absolute constant L. Bernoulli random vectors ($X = \pm 1$ with equal probability $1/2$ for $X = +1$ and $X = -1$) are special cases of Steinhaus vectors.

The following well-known bound is relating strong and weak moments. A proof based on chaining and the majorizing measures theorem is given in [31].

Theorem 1.8.1 (Theorem 2.3 of Krahmer and Mendelson and Rauhut [31]). *Let* $\mathbf{x}_1, \ldots, \mathbf{x}_n \in \mathbb{C}^N$ *and* $T \in \mathbb{C}^N$. *If* $\boldsymbol{\xi}$ *is an isotropic, L-sub-Gaussian random vector and* $\mathbf{Y} = \sum_{i=1}^{n} \xi_i \mathbf{x}_i$, *then for every* $p \geq 1$,

$$\left(\mathbb{E} \sup_{\mathbf{t} \in T} |\langle \mathbf{t}, \mathbf{Y} \rangle|^p\right)^{1/p} \leqslant c \left(\mathbb{E} \sup_{\mathbf{t} \in T} |\langle \mathbf{t}, \mathbf{G} \rangle| + \sup_{\mathbf{t} \in T} \left(\mathbb{E} |\langle \mathbf{t}, \mathbf{Y} \rangle|^p\right)^{1/p}\right),$$

where c is a constant which depends only on L and $\mathbf{G} = \sum_{i=1}^{N} g_i \mathbf{x}_i$ *for* $g_1, \ldots, g_N$ *independent standard Gaussian random variables.*

If $p, q \in [1, \infty)$ satisfy $1/p + 1/q = 1$, then the ℓ_p and ℓ_q norms are dual to each other. In particular, the Euclidean norm is self-dual $p = q = 2$. Similarly, the Schatten p-norm is dual to the Schatten q-norm. If $\|\cdot\|$ is some norm on $\mathbb{C}^N$ and $\mathcal{B}_*$ is the unit ball in the dual norm of $\|\cdot\|$, then the above theorem implies that

$$\left(\mathbb{E}\|\mathbf{Y}\|^p\right)^{1/p} \leqslant c\left(\mathbb{E}\|\mathbf{G}\| + \sup_{\mathbf{t}\in\mathcal{B}_*}\left(\mathbb{E}|\langle \mathbf{t}, \mathbf{Y}\rangle|^p\right)^{1/p}\right).$$

1.9 Sub-exponential Random Variables

Some random variables have tails heavier than Gaussian. The following properties are equivalent for $K_i > 0$

$$\begin{aligned} \mathbb{P}(|X| > t) \leqslant \exp(1 - t/K_1) \text{ for all } t \geqslant 0; \\ (\mathbb{E}|Y|^p)^{1/p} \leqslant K_2 p \text{ for all } p \geqslant 1; \\ \mathbb{E}\exp(X/K_3) \leqslant e. \end{aligned} \tag{1.103}$$

A random variable X that satisfies one of the equivalent properties of (1.103) is called a *sub-exponential* random variable. The sub-exponential norm, denoted $\|X\|_{\psi_1}$, is defined to be the smallest parameter K_2. In other words,

$$\|X\|_{\psi_1} = \sup_{p\geqslant 1}\frac{1}{p}\left(\mathbb{E}|X|^p\right)^{1/p}.$$

Lemma 1.9.1 (Sub-exponential is sub-Gaussian squared [72]). *A (scalar valued) random variable X is sub-Gaussian if and only if X^2 is sub-exponential. Moreover,*

$$\|X\|_{\psi_2}^2 \leqslant \|X\|_{\psi_1}^2 \leqslant 2\|X\|_{\psi_2}^2.$$

Lemma 1.9.2 (Moment generating function [72]). *Let X be a centered sub-exponential random variable. Then, for t such that $|t| \leqslant c/\|X\|_{\psi_1}$, one has*

$$\mathbb{E}\exp(tX) \leqslant \exp\left(Ct^2\|X\|_{\psi_1}^2\right)$$

where C, c are absolute constants.

Corollary 1.9.3 (Bernstein-type inequality [72]). *Let $X_1, \ldots, X_N$ be independent centered sub-exponential random variables, and let $K = \max_i \|X_i\|_{\psi_1}^2$. Then for every $\mathbf{a} = (a_1, \ldots, a_N) \in \mathbb{R}^N$ and every $t \geq 0$, we have*

$$\mathbb{P}\left(\left|\sum_{i=1}^{N} a_i X_i\right| \geqslant t\right) \leqslant 2\exp\left[-c\min\left(\frac{t^2}{K^2\|\mathbf{a}\|_2^2}, \frac{t}{K\|\mathbf{a}\|_\infty}\right)\right] \qquad (1.104)$$

where $c > 0$ is an absolute constant.

Corollary 1.9.4 (Corollary 17 of Vershynin [72]). *Let $X_1, \ldots, X_N$ be independent centered sub-exponential random variables, and let $K = \max_i \|X_i\|_{\psi_1}^2$. Then, for every $t \geq 0$, we have*

$$\mathbb{P}\left(\left|\sum_{i=1}^{N} a_i X_i\right| \geqslant tN\right) \leqslant 2\exp\left[-c\min\left(\frac{t^2}{K^2}, \frac{t}{K}\right)N\right]$$

where $c > 0$ is an absolute constant.

Remark 1.9.5 (Centering).

The definition of sub-Gaussian and sub-exponential random variables X does not require them to be centered. In any case, one can always center X using the simple fact that if X is sub-Gaussian (or sub-exponential), then so is $X - \mathbb{E}X$. Also,

$$\|X - \mathbb{E}X\|_{\psi_2}^2 \leqslant 2\|X\|_{\psi_2}^2, \|X - \mathbb{E}X\|_{\psi_1}^2 \leqslant 2\|X\|_{\psi_1}^2.$$

This follows from triangle inequality $\|X - \mathbb{E}X\|_{\psi_2}^2 \leqslant \|X\|_{\psi_2}^2 + \|\mathbb{E}X\|_{\psi_2}^2$ along with $\|\mathbb{E}X\|_{\psi_2}^2 = |\mathbb{E}X| \leqslant \|X\|_{\psi_2}^2$, and similarly for the sub-exponential norm.

1.10 ε-Nets Arguments

Let (T, d) be a metric space. Let $K \subset T$. A set $\mathcal{N} \subset T$ is called an ε-net for K if

$$\forall \mathbf{x} \in K, \quad \exists \mathbf{y} \in \mathcal{N} \quad d(\mathbf{x}, \mathbf{y}) < \varepsilon.$$

A set $\mathcal{S} \subset K$ is called ε-separated if

$$\forall \mathbf{x} \in K, \quad \exists \mathbf{y} \in \mathcal{S} \quad d(\mathbf{x}, \mathbf{y}) \geqslant \varepsilon.$$

These two notions are closely related. Namely, we have the following elementary Lemma.

Lemma 1.10.1. *Let K be a subset of a metric space (T, d), and let set $\mathcal{N} \subset T$ be an ε-net for K. Then*

1. *There exists a 2ε-net $\mathcal{N}' \subset K$ such that $\mathcal{N}' \leq \mathcal{N}$;*
2. *Any 2ε-separated set $\mathcal{S} \subset K$ satisfies $\mathcal{S} \leq \mathcal{N}$;*
3. *From the other side, any maximal ε-separated set $\mathcal{S}' \subset K$ is an ε-net for K.*

Let $N = |\mathcal{N}|$ be the minimum cardinality of an ε-net of T, also called the covering number of T at scale ε.

Lemma 1.10.2 (Covering numbers of the sphere). *The unit Euclidean sphere S^{n-1} equipped with the Euclidean metric satisfies for every $\varepsilon > 0$ that*

$$N\left(S^{n-1}, \varepsilon\right) \leqslant \left(1 + \frac{2}{\varepsilon}\right)^n.$$

Lemma 1.10.3 (Volumetric estimate). *For any $\varepsilon < 1$ there exists an ε-net such that $\mathcal{N} \subset S^{n-1}$ such that*

$$|\mathcal{N}| \leqslant \left(1 + \frac{2}{\varepsilon}\right)^n \leqslant \left(\frac{3}{\varepsilon}\right)^n.$$

Proof. Let $\mathcal{N}$ be a maximal ε-separated subset of sphere S^{n-1}. Let B_2^n be Euclidean ball. Then for any distinct points $\mathbf{x}, \mathbf{y} \in \mathcal{N}$

$$\left(\mathbf{x} + \frac{\varepsilon}{2} B_2^n\right) \cap \left(\mathbf{y} + \frac{\varepsilon}{2} B_2^n\right) = \emptyset.$$

So,

$$|\mathcal{N}| \cdot \operatorname{vol}\left(\frac{\varepsilon}{2} B_2^n\right) = \operatorname{vol}\left(\bigcup_{\mathbf{x} \in \mathcal{N}} \left(\mathbf{x} + \frac{\varepsilon}{2} B_2^n\right)\right) \leqslant \operatorname{vol}\left(\left(1 + \frac{\varepsilon}{2}\right) B_2^n\right),$$

which implies

$$|\mathcal{N}| \leqslant \left(1 + \frac{2}{\varepsilon}\right)^n \leqslant \left(\frac{3}{\varepsilon}\right)^n. \qquad \square$$

Using ε-nets, we prove a basic bound on the first singular value of a random subGaussian matrix: Let $\mathbf{A}$ be an $m \times n$ random matrix, $m \geq n$, whose entries are independent copies of a subGaussian random variable. Then

$$\mathbb{P}\left(s_1 > t\sqrt{m}\right) \leqslant e^{-c_1 t^2 m} \quad \text{for } t \geqslant C_0.$$

See [76] for a proof.

Lemma 1.10.4 (Computing the spectral norm on a net). *Let $\mathbf{A}$ be a symmetric $n \times n$ matrix, and let $\mathcal{N}_\varepsilon$ be an ε-net of S^{n-1} for some $\varepsilon \in (0, 1)$. Then*

$$\|\mathbf{A}\| = \sup_{\mathbf{x} \in S^{n-1}} |\langle \mathbf{A}\mathbf{x}, \mathbf{x} \rangle| \leqslant (1 - 2\varepsilon)^{-1} \sup_{\mathbf{x} \in \mathcal{N}_\varepsilon} |\langle \mathbf{A}\mathbf{x}, \mathbf{x} \rangle|.$$

See [72] for a proof.

1.11 Rademacher Averages and Symmetrization

One simple but basic idea in the study of sums of independent random variables is the concept of symmetrization [27]. The simplest probabilistic object is the Rademacher random variable ε, which takes the two values ± 1 with equal probability $1/2$. A random vector is *symmetric* (or has symmetric distribution) if $\mathbf{x}$ and $-\mathbf{x}$ have the same distribution [77]. In this case $\mathbf{x}$ and $\varepsilon\mathbf{x}$, where ε is a Rademacher random variable independent of $\mathbf{x}$, have the same distribution. Let $\mathbf{x}_i, i = 1, \ldots, n$ be independent symmetric random vectors. The joint distribution of $\varepsilon_i \mathbf{x}_i, i = 1, \ldots, n$ is that of the original sequence if the coefficients ε_i are either non-random with values ± 1, or they are random and independent from each other and all $\mathbf{x}_i$ with $\mathbb{P}(\varepsilon_i = \pm 1) = \frac{1}{2}$.

The technique of symmetrization leads to so-called Rademacher sums $\sum_{i=1}^{N} \varepsilon_i x_i$, where x_i are scalars, $\sum_{i=1}^{N} \varepsilon_i \mathbf{x}_i$, where $\mathbf{x}_i$ are vectors and $\sum_{i=1}^{N} \varepsilon_i \mathbf{X}_i$, where $\mathbf{X}_i$ are matrices. Although quite simple, symmetrization is very powerful since there are nice estimates for Rademacher sums available—the so-called Khintchine inequalities.

A sequence of independent Rademacher variables is referred to as a *Rademacher sequence*. A Rademacher series in a Banach space $\mathcal{X}$ is a sum of the form

$$\sum_{i=1}^{\infty} \varepsilon_i \mathbf{x}_i$$

where $\mathbf{x}$ is a sequence of points in $\mathcal{X}$ and ε_i is an (independent) Rademacher sequence.

For $0 < p \leq \infty$, the l_p-norm is defined as

$$\|\mathbf{x}\|_p = \left(\sum_i |x_i|^p \right)^{1/p} < \infty.$$

$\|\cdot\|_2$ denotes the Euclidean norm. For $p = \infty$,

$$\|\mathbf{x}\|_p = \|\mathbf{x}\| = \sup_i |x_i|, p = \infty.$$

We use $\|\cdot\|$ to represent the case of $p = \infty$.

For Rademacher series with scalar coefficients, the most important result is the inequality of Khintchine. The following sharp version is due to Haagerup [78].

Proposition 1.11.1 (Khintchine). *Let $p \geq 2$. For every sequence $\{a_i\}$ of complex scalars,*

$$\mathbb{E}^p \left| \sum_i \varepsilon_i a_i \right| \leqslant C_p \left[\sum_i |a_i|^2 \right]^{1/2},$$

where the optimal constant

$$C_p = \left[\frac{p!}{2^{p/2}\,(p/2)!}\right]^{1/p} \leqslant \left(\sqrt{2}\right)^{1/p} e^{-0.5}\sqrt{p}.$$

This inequality is typically established only for real scalars, but the real case implies that the complex case holds with the same constant.

Lemma 1.11.2 (Khintchine inequality [29]). *For* $\mathbf{a} \in \mathbb{R}^n$, $\mathbf{x} \in \{-1,1\}^n$ *uniform, and* $k \geq 2$ *an even integer,* $\mathbb{E}\left[\left(\mathbf{a}^T\mathbf{x}\right)^k\right] \leqslant \|\mathbf{a}\|_2^k \cdot k^{k/2}$.

For a family of random variables, it is often useful to consider a number of its independent copies, viz., independent families of random vectors having the same distributions.

A *symmetrization* of the sequence of random vectors is the difference of two independent copies of this sequence

$$\tilde{\mathbf{x}} = \mathbf{x}^{(1)} - \mathbf{x}^{(2)}. \tag{1.105}$$

If the original sequences consist of independent random vectors, all the random vectors $\mathbf{x}^{(1)}$ and $\mathbf{x}^{(2)}$ used to construct symmetrization are independent. Random vectors defined in (1.105) are also independent and symmetric.

A Banach space $\mathcal{B}$ is a *vector space* over the field of the real or complex numbers equipped with a norm $\|\cdot\|$ for which it is complete. We consider Rademacher averages $\sum_i \varepsilon_i \mathbf{x}_i$ with vector valued coefficients as a natural analog of the Gaussian averages $\sum_i g_i \mathbf{x}_i$. A sequence (ε_i) of independent random variables taking the values $+1$ and -1 with equal probability 1/2, that is symmetric Bernoulli or Rademacher random variables. We usually call (ε_i) a Rademacher sequence or Bernoulli sequence. We often investigate finite or convergent sums $\sum_i \varepsilon_i \mathbf{x}_i$ with vector valued coefficients $\mathbf{x}_i$.

For arbitrary $m \times n$ matrices, $\|\mathbf{A}\|_{S_p}$ denotes the Schatten p-norm of an $m \times n$ matrix $\mathbf{A}$, i.e.,

$$\|\mathbf{A}\|_{S_p} = \|\boldsymbol{\sigma}(\mathbf{A})\|_p,$$

where $\boldsymbol{\sigma} \in \mathbb{R}^{\min\{m,n\}}$ is the vector of singular values of $\mathbf{A}$, and $\|\cdot\|_p$ is the usual l_p-norm defined above. When $p = \infty$, it is also called the spectrum (or matrix) norm. The Rademacher average is given by

$$\left\|\sum_i \varepsilon_i \mathbf{A}_i\right\|_{S_p} = \left\|\boldsymbol{\sigma}\left(\sum_i \varepsilon_i \mathbf{A}_i\right)\right\|_p.$$

The infinite dimensional setting is characterized by the *lack* of the orthogonal property

$$\mathbb{E}\left\|\sum_i \mathbf{x}_i\right\|_2^2 = \sum_i \mathbb{E}\|\mathbf{x}_i\|^2,$$

where (X_i) is a finite sequence of independent mean zero real valued random vectors. This type of identity extends to Hilbert space valued random variables, but *does not* in general hold for arbitrary Banach space valued random variables. The classical theory is developed under this orthogonal property.

Lemma 1.11.3 (Ledoux and Talagrand [27]). *Let $F : \mathbb{R}_+ \to \mathbb{R}_+$ be convex. Then, for any finite sequence (X_i) of independent zero mean random variables in a Banach space $\mathcal{B}$ such that $\mathbb{E}F(\|X_i\|) < \infty$ for every i,*

$$\mathbb{E}F\left(\frac{1}{2}\left\|\sum_i \varepsilon_i X_i\right\|\right) \leqslant \mathbb{E}F\left(\left\|\sum_i X_i\right\|\right) \leqslant \mathbb{E}F\left(2\left\|\sum_i \varepsilon_i X_i\right\|\right).$$

Rademacher series appear as a basic tool for studying sums of independent random variables in a Banach space [27, Lemma 6.3].

Proposition 1.11.4 (Symmetrization [27]). *Let $\{X_i\}$ be a finite sequence of independent, zero-mean random variables taking values in a Banach space $\mathcal{B}$. Then*

$$\mathbb{E}^p\left\|\sum_i X_i\right\|_{\mathcal{B}} \leqslant 2\mathbb{E}^p\left\|\sum_i \varepsilon_i X_i\right\|_{\mathcal{B}},$$

where ε_i is a Rademacher sequence independent of $\{X_i\}$.

In other words [28], the moments of the sum are controlled by the moments of the associated Rademacher series. The advantage of this approach is that we can condition on the choice of X_i and apply sophisticated methods to estimate the moments of the Rademacher series.

We need some facts about symmetrized random variables. Suppose that Z is a zero-mean random variable that takes values in a Banach space $\mathcal{B}$. We may define the symmetrized variable $Y = Z - Z_0$, where Z_0 is an independent copy of Z. The tail of the symmetrized variable Y is closely related to the tail of Z. Indeed, we have [28]

$$\mathbb{P}\left(\|Z\|_{\mathcal{B}} > 2\mathbb{E}\|Z\|_{\mathcal{B}} + t\right) \leqslant \mathbb{P}\left(\|Y\|_{\mathcal{B}} > t\right). \quad (1.106)$$

The relation follows from [27, Eq. (6.2)] and the fact that $\mathbb{M}(Y) \leqslant 2\mathbb{E}Y$ for every nonnegative random variable Y. Here $\mathbb{M}$ denotes the median.

1.12 Operators Acting on Sub-Gaussian Random Vectors

See Sect. 5.7 for the applications of the results here. The proofs below are taken from [79]. We also follow [79] for the exposition and the notation here. By g, g_i, we denote independent $\mathcal{N}(0,1)$ Gaussian random variables. For a random variable ξ and $p > 0$, we put $\|\xi\|_p = \left(\mathbb{E}|\xi|^p\right)^{1/p}$. Let $\|\cdot\|_F$ be the Frobenius norm and $\|\cdot\|_{\rm op}$ the operator norm. By $|\cdot|$ and $< \cdot, \cdot >$ we denote the standard Euclidean norm and the inner product on $\mathbb{R}^n$.

A random variable ξ is called sub-Gaussian if there is a constant $\beta < \infty$ such that

$$\|\xi\|_{2k} \leqslant \beta \|g\|_{2k} \qquad k = 1, 2, \ldots \tag{1.107}$$

We refer to the infimum overall all β satisfying (1.107) as the sub-Gaussian constant of ξ. An equivalent definition is often given in terms of the ψ_2-norm. Denoting the Orlicz function $\psi_2(x) = \exp(x^2) - 1$ by ψ_2, ξ is sub-Gaussian if and only if

$$\|\xi\|_{\psi_2} := \inf\left\{t > 0 \,|\psi_2(\xi/t) \leqslant 1\right\} < \infty. \tag{1.108}$$

Denting the sub-Gaussian constant of ξ by $\bar{\beta}$, a direct calculation will show the following common (and not optimal) estimate

$$\tilde{\beta} \leqslant \|\xi\|_{\psi_2} \leqslant \tilde{\beta}\|g\|_{\psi_2} = \tilde{\beta}\sqrt{8/3}.$$

The lower estimate follows since $\mathbb{E}\psi_2(X) \geqslant \mathbb{E}X^{2k}/k!$ for $k = 1, 2, \ldots$. The upper one is using the fact that $\mathbb{E}\exp(tg^2) = 1/\sqrt{1-2t}$ for $t < 1/2$.

Apart from the Gaussian random variables, the prime example of sub-Gaussian random variables are Bernoulli random variables, taking values $+1$ and -1 with equal probability $\mathbb{P}\left(\xi = +1\right) = \mathbb{P}\left(\xi = -1\right) = 1/2$.

Very often, we work with random vectors in $\mathbb{R}^n$ of the form $\boldsymbol{\xi} = (\xi_1, \xi_2, \ldots, \xi_n)$, where ξ_i are independent sub-Gaussian random variables, and we refer to such vectors as sub-Gaussian random vectors. We require that $\mathrm{Var}\left(\xi_i\right) \geqslant 1$ and sub-Gaussian constants are at most β. Under these assumptions, we have

$$\mathbb{E}\xi_i^2 \geqslant \mathrm{Var}\left(\xi_i\right) \geqslant 1 = \mathbb{E}g^2,$$

hence $\beta \geq 1$.

We have the following fact: for any $t > 0$,

$$\mathbb{P}\left(|\boldsymbol{\xi}| \geqslant t\sqrt{n}\right) \leqslant \exp\left(n\left(\ln 2 - t^2/\left(3\beta^2\right)\right)\right). \tag{1.109}$$

In particular, $\mathbb{P}\left(|\xi| \geqslant 3\beta\sqrt{n}\right) \leqslant e^{-2n}$. Let us prove the result. For an arbitrary $s > 0$ and $1 \leq i \leq n$ we have

$$\mathbb{E}\exp\left(\frac{\xi_i^2}{s^2}\right) = \sum_{k=0}^{\infty} \frac{1}{k! \cdot s^{2k}} \mathbb{E}\xi_i^{2k} \leqslant \sum_{k=0}^{\infty} \frac{\beta^{2k}}{k! \cdot s^{2k}} \mathbb{E}g^{2k} = \mathbb{E}\exp\left(\frac{(\beta g)^2}{s^2}\right).$$

This last quantity is less than or equal to 2 since, e.g., $s = \sqrt{3}\beta$. For this choice of s,

$$\begin{aligned} \mathbb{P}\left(\sum_{i=1}^{n} \xi_i^2 \geqslant t^2 n\right) &\leqslant \mathbb{E}\exp\left(\tfrac{1}{s^2}\left(\sum_{i=1}^{n} \xi_i^2 - t^2 n\right)\right) \\ &\leqslant \exp\left(-\tfrac{t^2 n}{s^2}\right) \prod_{i=1}^{n} \mathbb{E}\exp\left(\tfrac{\xi_i^2}{s^2}\right) \leqslant \exp\left(-\tfrac{t^2 n}{3\beta^2}\right) \cdot 2^n, \end{aligned}$$

which is the desired result.

Theorem 1.12.1 (Lemma 3.1 of Latala [79]). *Let $\xi_1, \xi_2, \ldots, \xi_n$ be a sequence of independent symmetric sub-Gaussian random variables satisfying* (1.107) *and let* $\mathbf{A} = (a_{ij})$ *be a symmetic matrix with zero diagonal. Then, for any $t > 1$,*

$$\mathbb{P}\left(\left|\sum_{i<j} a_{ij}\xi_i\xi_j\right| \geqslant C\beta^2\left(\sqrt{t}\|\mathbf{A}\|_F + t\|\mathbf{A}\|_{\text{op}}\right)\right) \leqslant e^{-t},$$

where C is a universal constant.

Proof. By (1.107) and by the symmetry of ξ_i, we immediately get $\|a + b\xi_i\|_{2k} \leqslant \|a + b\beta g_i\|_{2k}$ for any real numbers a, b and a positive integer k. So we have

$$\left\|\sum_{i<j} a_{ij}\xi_i\xi_j\right\|_{2k} \leqslant \left\|\sum_{i<j} a_{ij}\beta g_i \beta g_j\right\|_{2k} = \beta^2 \left\|\sum_{i<j} a_{ij} g_i \times g_j\right\|_{2k}.$$

Using the Hanson-Wright estimate [61], we get

$$\left\|\sum_{i<j} a_{ij} g_i \times g_j\right\|_{2k} \leqslant C'\beta^2\left(\sqrt{2k}\|\mathbf{A}\|_F + 2k\|\mathbf{A}\|_{\text{op}}\right)$$

with some universal constant C'. Taking $k = \lceil t/2 \rceil$, then by Chebyshev's inequality

$$\mathbb{P}\left(\left|\sum_{i<j} a_{ij}\xi_i\xi_j\right| \geqslant eC'\beta^2\left(\sqrt{2k}\|\mathbf{A}\|_F + 2k\|\mathbf{A}\|_{\text{op}}\right)\right) \leqslant e^{-2k} \leqslant e^{-t}.$$

Statement follows, since $k \leq t$. □

Theorem 1.12.2 (Lemma 3.2 of Latala [79]). *Let $\xi_1, \xi_2, \ldots, \xi_n$ be a sequence of independent random variables with finite fourth moments. Then for any nonnegative coefficients b_i and $t > 0$,*

$$\mathbb{P}\left(\left|\sum_{i=1}^{n} b_i\left(\mathbb{E}\xi_i^2-\xi_i^2\right)\right|>\left(2t\sum_{i=1}^{n} b_i^2\mathbb{E}\xi_i^4\right)^{\frac{1}{2}}\right)\leqslant e^{-t}.$$

Proof. We may obviously assume that $\sum_{i=1}^{n} b_i^2\mathbb{E}\xi_i^4>0$. For $x>0$, we have $e^{-x}=(e^x)^{-1}\leqslant\left(1+x+x^2/2\right)^{-1}\leqslant 1-x+x^2/2$. Thus for $\lambda\geq 0$,

$$\mathbb{E}\exp\left(-\lambda\xi_i^2\right)\leqslant 1-\mathbb{E}\xi_i^2+\tfrac{1}{2}\lambda^2\mathbb{E}\xi_i^4\leqslant\exp\left(-\lambda\mathbb{E}\xi_i^2+\tfrac{1}{2}\lambda^2\mathbb{E}\xi_i^4\right).$$

Letting $S=\sum_{i=1}^{n} b_i\left(\mathbb{E}\xi_i^2-\xi_i^2\right)$, we get $\mathbb{E}\exp(\lambda S)\leqslant\exp\left(\frac{1}{2}\lambda^2\sum_{i=1}^{n} b_i^2\mathbb{E}\xi_i^4\right)$, and for any $u\geq 0$,

$$\mathbb{P}(S\geqslant u)\leqslant\inf_{\lambda\geqslant 0}\mathbb{E}\exp(\lambda S-\lambda u)\leqslant\exp\left(-\frac{u^2}{2\sum_{i=1}^{n} b_i^2\mathbb{E}\xi_i^4}\right).$$

□

Lemma 1.12.2 is somewhat special since we assume that coefficients are non-negative. In the general case one has for any sequence of independent random variables ξ_i with sub-Gaussian constant at most β and $t>1$,

$$\mathbb{P}\left(\left|\sum_{i=1}^{n} a_i\left(\mathbb{E}\xi_i^2-\xi_i^2\right)\right|>C\beta^2\left(t\|(a_i)\|_\infty+\sqrt{t}\,|(a_i)|\right)\right)\leqslant e^{-t}. \tag{1.110}$$

We provide a sketch of the proof for the sake of completeness. Let $\tilde{\xi}_i$ be an independent copy of ξ_i. We have by Jensen's inequality for $p\geq 1$,

$$\left\|\sum_{i=1}^{n} a_i\left(\mathbb{E}\xi_i^2-\xi_i^2\right)\right\|_p\leqslant\left\|\sum_{i=1}^{n} a_i\left(\xi_i^2-\tilde{\xi}_i^2\right)\right\|_p.$$

Random variables $\xi_i^2-\tilde{\xi}_i^2$ are independent, symmetric. For $k\geq 1$,

$$\left\|\xi_i^2-\tilde{\xi}_i^2\right\|_{2k}\leqslant 2\beta^2\left\|g_i^2\right\|_{2k}\leqslant 4\beta^2\left\|\eta_i^2\right\|_{2k}$$

where η_i are i.i.d. symmetric exponential random variables with variance 1. So, for positive integer k,

$$\left\|\sum_{i=1}^{n} a_i\left(\xi_i^2-\tilde{\xi}_i^2\right)\right\|_{2k} \leqslant 4\beta^2\left\|\sum_{i=1}^{n} a_i\eta_i\right\|_{2k} \leqslant C_1\beta^2\left(k\|(a_i)\|_\infty+\sqrt{k}\|(a_i)\|_2\right),$$

where the last inequality follows by the Gluskin-Kwapien estimate [80]. So by Chebyshev's inequality

$$\mathbb{P}\left(\left|\sum_{i=1}^{n} a_i\left(\mathbb{E}\xi_i^2-\xi_i^2\right)\right| > eC_1\beta^2\left(k\|(a_i)\|_\infty+\sqrt{k}\|(a_i)\|_2\right)\right) \leqslant e^{-2t}$$

and the assertion easily follows.

1.13 Supremum of Stochastic Processes

We follow [33] for this introduction. The standard reference is [81]. See also [27,82]. One of the *fundamental* issues of the probability theory is the study of suprema of stochastic processes. In particular, in many situations one needs to estimate the quantity $\mathbb{E}\sup_{t\in T} X_t$, where $\sup_{t\in T} X_t$ is a stochastic process. T in order to avoid measurability problems, one may assume that T is countable. The modern approach to this problem is based on chaining techniques. The most important case of centered Gaussian process is well understood. In this case, the boundedness of the process is related to the geometry of the metric space (T, d), where

$$d(t,s) = \left(\mathbb{E}(X_t - X_s)^2\right)^{1/2}.$$

In 1967, R. Dudley [83] obtained an upper bound for $\mathbb{E}\sup_{t\in T} X_t$ in terms of entry numbers and in 1975 X. Fernique [84] improved Dudley's bound using so-called majorizing measures. In 1987, Talagrand [85] showed that Fernique's bound may be reversed and that for centered Gaussian processes (X_t),

$$\frac{1}{L}\gamma_2(T,d) \leqslant \mathbb{E}\sup_{t\in T} X_t \leqslant L\gamma_2(T,d),$$

where L is a universal constant. There are many equivalent definitions of the Talagrand's gamma function γ_2, for example one may define

$$\gamma_\alpha(T,d) = \inf\sup_{t\in T}\sum_{i=0}^{\infty} 2^{n/\alpha} d(t, T_i),$$

where the infimum runs over all sequences T_i of subsets of T such that $|T_0| = 1$ and $|T_i| = 2^{2^i}$.

1.14 Bernoulli Sequence

Another *fundamental* class of processes is based on the Bernoulli sequence [33], i.e. the sequence $(\varepsilon_i)_{i\geqslant 1}$ of i.i.d. symmetric random variables taking values ± 1. For $t = \ell_2$, the series $X_t := \sum\limits_{i\geqslant 1} t_i\varepsilon_i$ converges almost surely and for $T \in \ell_2$, we may define a Bernoulli process $(X_t)_{t\in T}$ and try to estimate $b(T) = \mathbb{E}\sup\limits_{t\in T} X_t$. There are two ways to bound $b(T)$. The first one is a consequence of the uniform bound $|X_t| \leqslant \|t\|_1 = \sum\limits_{i\geqslant 1} |t_i|$, so that $b(T) \leqslant \sup\limits_{t\in T} \|t\|_1$. Another is based on the domination by the canonical Gaussian process $G_t := \sum\limits_{i\geqslant 1} t_i g_i$, where g_i are i.d.d. $\mathcal{N}(0,1)$ random variables. Assuming the independence of (g_i) and (ε_i), Jensen's inequality implies:

$$g(T) = \mathbb{E}\sup_{t\in T}\sum_{i\geqslant 1} t_i g_i = \mathbb{E}\sup_{t\in T}\sum_{i\geqslant 1} t_i\varepsilon_i\,|g_i| \geqslant \mathbb{E}\sup_{t\in T}\sum_{i\geqslant 1} t_i\varepsilon_i \mathbb{E}\,|g_i| = \sqrt{\frac{2}{\pi}}b(T).$$

1.15 Converting Sums of Random Matrices into Sums of Random Vectors

Sums of independent random vectors [27, 74, 86] are classical topics nowadays. It is natural to convert sums of random matrices into sums of independent random vectors that can be handled using the classical machinery. Sums of dependent random vectors are much less understood: Stein's method is very powerful [87].

Often we are interested in the sample covariance matrix $\frac{1}{N}\sum\limits_{i=1}^{N} \mathbf{x}_i \otimes \mathbf{x}_i$ the sums of N rank-one matrices where $\mathbf{x}_1, \ldots, \mathbf{x}_N$ are N independent random vectors. More generally, we consider $\frac{1}{N}\sum\limits_{i=1}^{N} \mathbf{X}_i$ where $\mathbf{X}_i$ are independent random Hermitian matrices. Let

$$\mathbf{\Lambda} = \begin{bmatrix} \lambda_1 \\ \lambda_2 \\ \vdots \\ \lambda_n \end{bmatrix} \in \mathbb{R}^n$$

be a vector of eigenvalues of $\mathbf{Y}$. For each random matrix $\mathbf{X}_i$, we have one random vector $\mathbf{\Lambda}_i$ for all $i = 1, \ldots, N$. As a result, we obtain N independent random vectors consisting of eigenvalues. We are interested in the sums of these independent random vectors

$$\frac{1}{N}\sum_{i=1}^{N}\mathbf{\Lambda}_i.$$

Then we can use Theorem 1.15.1 to approximate characteristic functions with normal distribution. In [86], there are many other theorems that can be used in this context. Our technique is to convert a random matrix into a random vector that is much easier to handle using classical results [27, 74, 86], since very often, we are only interested in eigenvalues only. We shall pursue this connection more in the future research.

For example, we use the classical results [86]. Let $\mathbf{x}_1, \ldots, \mathbf{x}_N$ be N independent random vectors in $\mathbb{R}^n$, each having a zero mean and a finite s-th order absolute moment for some $s \geq 2$. Let $\hat{P}$ stand for Fourier transform of P, the characteristic function of a probability measure P. We here study $\hat{P}_{(\mathbf{x}_1+\ldots+\mathbf{x}_N)/\sqrt{N}}$, the rate of convergence of the characteristic function of a probability measure, to $\hat{\Phi}_{0,\mathbf{R}}$ where $\mathbf{R}$ is the average of the covariance matrices of $\mathbf{x}_1, \ldots, \mathbf{x}_N$. Here for normal distribution in $\mathbb{R}^n$,

$$\log \hat{\Phi}_{0,\mathbf{R}} = -\frac{1}{2}\left\langle \mathbf{y}, \mathbf{R}\mathbf{y} \right\rangle.$$

We assume that

$$\mathbf{R} = \frac{1}{N}\sum_{i=1}^{N}\mathrm{Cov}\left(\mathbf{x}_i\right) = \frac{1}{N}\sum_{i=1}^{N}\mathbf{R}_i$$

is nonsingular. Then, we define the Liapounov coefficient

$$l_{s,N} = \sup_{\|\mathbf{y}\|=1}\frac{\frac{1}{N}\sum\limits_{i=1}^{N}\mathbb{E}\left(\left|\left\langle \mathbf{y}, \mathbf{x}_i\right\rangle\right|^s\right)}{\left[\frac{1}{N}\sum\limits_{i=1}^{N}\mathbb{E}\left(\left\langle \mathbf{y}, \mathbf{x}_i\right\rangle^2\right)\right]^{s/2}}N^{-(s-2)/2} \quad (s \geqslant 2). \tag{1.111}$$

It is easy to check that $l_{s,N}$ is independent of scale. If $\mathbf{B}$ is a nonsingular $n \times n$ matrix, then $\mathbf{B}\mathbf{x}_1, \ldots, \mathbf{B}\mathbf{x}_N$ have the same Liapounov coefficient as $\mathbf{x}_1, \ldots, \mathbf{x}_N$. If we write

$$\begin{gathered}\rho_{r,i} = \mathbb{E}\left(\left\|\mathbf{x}_i\right\|^r\right), \quad 1 \leqslant i \leqslant N, \\ \rho_r = \frac{1}{N}\sum_{i=1}^{N}\rho_{r,i}, \quad r \geqslant 0.\end{gathered},$$

according to (1.111), we have that

$$l_{s,N} \leqslant N^{-(s-2)/2} \sup_{\|\mathbf{y}\|=1} \frac{\rho_s \|\mathbf{y}\|^s}{[\langle \mathbf{y}, \mathbf{R}\mathbf{y} \rangle]^{s/2}} = \frac{\rho_s}{\lambda_{\min}^{s/2}} N^{-(s-2)/2},$$

where $\lambda_{\min}$ is the smallest eigenvalue of the average covariance matrix $\mathbf{R}$. In one dimension (i.e., $n = 1$)

$$l_{s,N} = \frac{\rho_s}{\rho_2^{s/2}} N^{-(s-2)/2}, \quad s \geqslant 2.$$

If $\mathbf{R} = \mathbf{I}$, then

$$\mathbb{E}|\langle \mathbf{y}, \mathbf{x}_i \rangle|^s \leqslant \sum_{i=1}^{n} \mathbb{E}|\langle \mathbf{y}, \mathbf{x}_i \rangle|^s \leqslant N^{s/2} l_{s,N} \|\mathbf{y}\|^s.$$

Now we are ready to state a theorem.

Theorem 1.15.1 (Theorem 8.6 of [86]). *Let $\mathbf{x}_1, \ldots, \mathbf{x}_N$ be n independent random vectors in $\mathbb{R}^n$ having distribution $G_1, \ldots, G_N$, respectively. Suppose that each random vector $\mathbf{x}_i$ has zero mean and a finite fourth absolute moment. Assume that the average covariance matrix $\mathbf{R}$ is nonsingular. Also assume*

$$l_{4,N} \leqslant 1.$$

Then for all $\mathbf{t}$ *satisfying*

$$\|\mathbf{t}\| \leqslant \frac{1}{2} l_{4,N}^{-1/4},$$

One has

$$\begin{aligned}
&\left| \prod_{i=1}^{N} \hat{G}_i \left(\frac{1}{\sqrt{N}} \mathbf{B}\mathbf{t} \right) - \exp\left(-\tfrac{1}{2} \|\mathbf{t}\|^2 \right) \left(1 + \frac{j^3}{6\sqrt{N}} \mu_3(\mathbf{t}) \right) \right| \\
&\leqslant (0.175)\, l_{4,N} \|\mathbf{t}\|^4 \exp\left(-\tfrac{1}{2} \|\mathbf{t}\|^4 \right) \\
&+ \left[(0.018)\, l_{4,N}^2 \|\mathbf{t}\|^8 + \tfrac{1}{36} l_{3,N}^2 \|\mathbf{t}\|^6 \right] \exp\left\{ -(0.383) \|\mathbf{t}\|^2 \right\},
\end{aligned}$$

where $\mathbf{B}$ is the positive-definite symmetric matrix defined by $\mathbf{B}^2 = \mathbf{R}^{-1}$, and

$$\mu_3(\mathbf{t}) = \frac{1}{N} \sum_{i=1}^{N} \mathbb{E}\langle \mathbf{t}, \mathbf{x}_i \rangle^3.$$

1.16 Linear Bounded and Compact Operators

The material here is standard, taken from [88, 89]. Let X and Y always be normed spaces and $\mathcal{A} : X \to Y$ be a linear operator. The linear operator $\mathcal{A}$ is *bounded* if there exists $c > 0$ such that

$$\|\mathcal{A}x\| \leqslant c\,\|x\| \qquad \text{for all } x \in X.$$

The smallest of these constants is called the norm of $\mathcal{A}$, i.e.,

$$\|\mathcal{A}\| = \sup_{x \neq 0} \frac{\|\mathcal{A}x\|}{\|x\|}. \tag{1.112}$$

The following are equivalent

1. $\mathcal{A}$ is bounded.
2. $\mathcal{A}$ is continuous at $x = 0$, i.e. $x_i = 0$ implies that $\mathcal{A}x_i = 0$.
3. $\mathcal{A}$ is continuous for every $x \in X$.

The space $\mathcal{L}(X, Y)$ of all *linear bounded* mappings from X to Y with the operator norm is a normed space. Let $\mathcal{A} \in \mathcal{L}(X, Y)$, $\mathcal{B} \in \mathcal{L}(Y, Z)$; then $\mathcal{AB} \in \mathcal{L}(X, Z)$ and $\|\mathcal{AB}\| \leqslant \|\mathcal{A}\|\,\|\mathcal{B}\|$.

Let $k \in L^2((c, d) \times (a, b))$. The integral operator

$$(\mathcal{A}x)(t) := \int_a^b k(t, s)x(s)\,ds, \quad t \in (c, d), \quad x \in L^2(a, b), \tag{1.113}$$

is well-defined, linear, and bounded from $L^2(a, b)$ to $L^2(c, d)$. Furthermore,

$$\|\mathcal{A}\|_{L^2} \leqslant \int_c^d \int_a^b |k(t, s)|dsdt.$$

Let k be continuous on $[c, d] \times [a, b]$. Then $\mathcal{A}$ is also well-defined, linear, and bounded from $C[a, b]$ into $C[c, d]$ and

$$\|\mathcal{A}\|_\infty \leqslant \max_{t \in [c,d]} \int_a^b |k(s, t)|\,ds.$$

We can extend above results to integral operators with *weakly singular* kernels. A kernel is *weakly singular* on $[a, b] \times [a, b]$ if k is defined and continuous for all $t, s \in [a, b], t \neq s$, and there exists constants $c > 0$ and $\alpha \in [0, 1)$ such that

$$|k(s, t)| \leqslant c\frac{1}{|t - s|^\alpha} \qquad \text{for all t,s} \in [a, b], t \neq s.$$

Let k be weakly singular on $[a, b]$. Then the integral operator $\mathcal{A}$, defined in (1.113) for $[c, d] = [a, b]$, is well-defined and bounded as an operator in $L^2(a, b)$ as well as in $C[a, b]$.

Let $\mathcal{A} : X \to Y$ be a linear and bounded operator between Hilbert spaces. Then there exists one and only one linear bounded operator $\mathcal{A}^* : Y \to X$ with the property

$$\langle \mathcal{A}x, y \rangle = \langle x, \mathcal{A}^* y \rangle \quad \text{for all } x \in X, y \in Y.$$

This operator $\mathcal{A}^* : Y \to X$ is called the *adjoint operator* to $\mathcal{A}$. For $X = Y$, the operator $\mathcal{A}$ is called *self-adjoint* if $\mathcal{A}^* = \mathcal{A}$.

The operator $\mathcal{K} : X \to Y$ is called *compact* if it maps every bounded set S into a relatively compact set $\mathcal{K}(S)$. A set $M \subset Y$ is called *relatively compact* if every *bounded* sequence $(y_i) \subset M$ has an accumulation point in $\mathrm{cl}\,(M)$, i.e., the closure $\mathrm{cl}\,(M)$ is compact. The closure of a set M is defined as

$$\mathrm{cl}\,(M) := \left\{ x \in M : \text{there exists } (x_k)_k \subset M \text{ with } x = \lim_{k \to \infty} x_k \right\}.$$

The set of all compact operators from X to Y is a closed subspace of the vector space $\mathcal{L}^2\,(a, b)$ where

$$\mathcal{L}^2\,(a, b) = \left\{ x : (a, b) \to \mathbb{C} : x \text{ is measurable and } |x|^2 \text{ integrable} \right\}.$$

Let $k \in L^2\,((c, d) \times (a, b))$. The operator $\mathcal{K} : L^2\,(c, d) \to L^2\,(a, b)$, defined by

$$(\mathcal{K}x)\,(t) := \int_a^b k\,(t, s) x\,(s)\, ds, \quad t \in (c, d), \quad x \in L^2\,(a, b)\,, \tag{1.114}$$

is compact from (c, d) to (a, b). Let k be continuous on $(c, d) \times (a, b)$ or weakly singular on $(a, b) \times (a, b)$ (in this case $(c, d) = (a, b)$). Then $\mathcal{K}$ defined by (1.114) is also compact as an operator from $C[a, b]$ to $C[c, d]$.

1.17 Spectrum for Compact Self-Adjoint Operators

The material here is standard, taken from [88, 89]. The most important results in functional analysis are collected here. Define

$$\mathcal{N} = \left\{ x \in L^2\,(a, b) : x(t) = 0 \quad \text{almost everywhere on } [a, b] \right\}.$$

Let $\mathcal{K} : X \to X$ be *compact and self-adjoint* (and $\neq 0$). Then, the following holds:

1. The spectrum consists only of eigenvalues and possibly 0. Every eigenvalue of $\mathcal{K}$ is real-valued. $\mathcal{K}$ has at least one but at most a countable number of eigenvalues with 0 as the only possible accumulation point.

2. For every eigenvalue $\lambda \neq 0$, there exist only finitely, many linearly independent eigenvectors, i.e., the eigenvectors are finite-dimensional. Eigenvectors corresponding to different eigenvalues are orthonormal.
3. We order the eigenvalues in the form

$$|\lambda_1| \geqslant |\lambda_2| \geqslant |\lambda_3| \geqslant \cdots$$

and denote by $P_i : X \to \mathcal{N}(K - \lambda_i I)$ the orthogonal projection onto the eigenspace corresponding to λ_i. If there exist only a finite number $\lambda_1, \ldots, \lambda_m$ of eigenvalues, then

$$\mathcal{K} = \sum_{i=1}^{m} \lambda_i P_i.$$

If there exists an infinite sequence λ_i of eigenvalues, then

$$\mathcal{K} = \sum_{i=1}^{\infty} \lambda_i P_i,$$

where the series converges in the operator norm. Furthermore,

$$\left\| \mathcal{K} - \sum_{i=1}^{m} \lambda_i P_i \right\| = |\lambda_{m+1}| \, .$$

4. Let H be the linear span of all of the eigenvectors corresponding to the eigenvalues $\lambda_i \neq 0$ of $\mathcal{K}$. Then

$$X = \mathrm{cl}\,(H) \oplus \mathcal{N}\,(\mathcal{K})\,.$$

Let X and Y be Hilbert spaces and $\mathcal{K} : X \to Y$ is a compact operator with adjoint operator $\mathcal{K}^* : Y \to X$. Every eigenvalue λ of $\mathcal{K}^*\mathcal{K}$ is nonnegative because $\mathcal{K}^*\mathcal{K}x = \lambda x$ implies that $\lambda \langle x, x \rangle = \langle \mathcal{K}^*\mathcal{K}x, x \rangle = \langle \mathcal{K}x, \mathcal{K}x \rangle \geqslant 0$, i.e., $\lambda \geq 0$. The square roots $\sigma_i = \sqrt{\lambda_i}$ of the eigenvalues $\lambda_i, i \in J$ of the compact self-adjoint operator $\mathcal{K}^*\mathcal{K} : X \to X$ are called singular values of $\mathcal{K}$. Here again, $J \in \mathbb{N}$ could be either finite or $J = \mathbb{N}$.

A compact self-adjoint operator is *non-negative (positive)* if and only if all of the eigenvalues are non-negative (positive). The sum of two non-negative operators are non-negative and is positive if one of the summands is positive. If an operator is positive and bounded below, then it is invertible and its inverse is positive and bounded below.

Every non-negative compact operator $\mathcal{K}$ in a Hilbert space $\mathbb{H}$ has a unique non-negative square root $\mathcal{G}$; that is, if $\mathcal{K}$ is non-negative and compact, there is a unique non-negative bounded linear map $\mathcal{G}$ such that $\mathcal{G}^2 = \mathcal{K}$. $\mathcal{G}$ is compact and commutes with every bounded operator which commutes with $\mathcal{K}$.

An operator $\mathcal{A} : X \to Y$ is compact if and only if its adjoint $\mathcal{A}^* : Y \to X$ is compact. If $\mathcal{A}^*\mathcal{A}$ is compact, then $\mathcal{A}$ is also compact.

Example 1.17.1 (Integral). Let $\mathcal{K} : L^2(0,1) \to L^2(0,1)$ be defined by

$$(\mathcal{K}x)(t) := \int_0^t x(s)\,ds, \quad t \in (0,1), \quad x \in L^2(0,1).$$

Then

$$(\mathcal{K}^*x)(t) := \int_t^1 y(s)\,ds \quad \text{and} \quad (\mathcal{K}\mathcal{K}^*x)(t) := \int_t^1 \left(\int_0^t x(s)\,ds\right)dt.$$

The eigenvalue problem $\mathcal{K}\mathcal{K}^*x = \lambda x$ is equivalent to

$$\lambda x = \int_t^1 \left(\int_0^t x(s)\,ds\right)dt, \quad t \in (0,1).$$

Differentiating twice, we observe that for $\lambda \neq 0$ this is equivalent to the eigenvalue problem

$$\lambda x'' + x = 0 \text{ in } (0,1), \quad x(1) = x'(0) = 0.$$

Solving this gives

$$x_i(t) = \sqrt{\frac{2}{\pi}} \cos \frac{2i-1}{2}\pi t, \quad i \in \mathbb{N}, \quad \text{and } \lambda_i = \frac{4}{(2i-1)^2\pi^2}, \quad i \in \mathbb{N}. \quad \square$$

Example 1.17.2 (Porter and Stirling [89]). Let the operator $\mathcal{K}$ on $L_2(0,1)$ be defined by

$$(\mathcal{K}\phi)(x) = \int_0^1 \log\left|\frac{\sqrt{x}+\sqrt{t}}{\sqrt{x}-\sqrt{t}}\right|\phi(t)dt \quad (0 \leqslant x \leqslant 1).$$

To show that $\mathcal{K}$ is a positive operator, consider

$$(\mathcal{T}\phi)(x) = \int_0^x \frac{1}{\sqrt{x-t}}\phi(t)dt \quad (0 \leqslant x \leqslant 1).$$

Although the kernel is unbounded function, it is a Schur kernel and therefore $\mathcal{T}$ is a bounded operator on $L_2(0,1)$, with adjoint given by

$$(\mathcal{T}^*\phi)(x) = \int_x^1 \frac{1}{\sqrt{t-x}}\phi(t)dt \quad (0 \leqslant x \leqslant 1).$$

The kernel of $\mathcal{T}\mathcal{T}^*$ is, for $x \neq t$,

$$k(x,t) = \int_0^{\min(x,t)} \frac{1}{\sqrt{x-s}\sqrt{t-s}} ds = \log\left|\frac{\sqrt{x}+\sqrt{t}}{\sqrt{x}-\sqrt{t}}\right|,$$

the integration being most easily performed using the substitution $u = \sqrt{x-s} + \sqrt{t-s}$. Therefore, $\mathcal{K} = \mathcal{T}\mathcal{T}^*$ and $\mathcal{T}^*\phi = 0 \Rightarrow (\mathcal{T}^*)^2\phi = 0 \Rightarrow \phi = 0$ gives the positivity. □

Chapter 2
Sums of Matrix-Valued Random Variables

This chapter gives an exhaustive treatment of the line of research for sums of matrix-valued random matrices. We will present eight different derivation methods in this context of matrix Laplace transform method. The emphasis is placed on the methods that will be hopefully useful to some engineering applications. Although powerful, the methods are elementary in nature. It is remarkable that some modern results on matrix completion can be simply derived, by using the framework of sums of matrix-valued random matrices. The treatment here is self-contained. All the necessary tools are developed in Chap. 1. The contents of this book are complementary to our book [5]. We have a small overlapping on the results of [36].

In this chapter, the classical, commutative theory of probability is generalized to the more general theory of non-communicative probability. Non-communicative algebras of random variables ("observations") and their expectations (or "trace") are built. *Matrices or operators takes the role of scalar random variables and the trace takes the role of expectation*. This is very similar to free probability [9].

2.1 Methodology for Sums of Random Matrices

The theory of real random variables provides the framework of much of modern probability theory [8], such as laws of large numbers, limit theorems, and probability estimates for "deviations", when sums of independent random variables are involved. However, some authors have started to develop analogous theories for the case that the algebraic structure of the reals is substituted by more general structures such as groups, vector spaces, etc., see for example [90].

In a remarkable work [36], Ahlswede and Winter has laid the ground for the fundamentals of a theory of (self-adjoint) operator valued random variables. There, the large deviation bounds are derived. A self-adjoint operator includes finite dimensions (often called Hermitian matrix) and infinite dimensions. For the purpose of this book, finite dimensions are sufficient. We will prefer Hermitian matrix.

R. Qiu and M. Wicks, *Cognitive Networked Sensing and Big Data*,
DOI 10.1007/978-1-4614-4544-9_2,

To extend the theory from scalars to the matrices, the fundamental difficulty arises from that fact, in general, two matrices are not commutative. For example, $\mathbf{AB} \neq \mathbf{BA}$. The functions of a matrix can be defined; for example, the matrix exponential is defined [20] as $e^{\mathbf{A}}$. As expected, $e^{\mathbf{AB}} \neq e^{\mathbf{BA}}$, although a scalar exponential has the elementary property $e^{ab} = e^{ba}$, for two scalars a, b. Fortunately, we have the Golden-Thompson inequality that has the limited replacement for the above elementary property of the scalar exponential. The Golden-Thompson inequality

$$\mathrm{Tr}\left(e^{\mathbf{A}+\mathbf{B}}\right) \leq \mathrm{Tr}\left(e^{\mathbf{A}}e^{\mathbf{B}}\right),$$

for Hermitian matrices $\mathbf{A}, \mathbf{B}$, is the most complicate result that we will use.

Through the spectral mapping theorem, the eigenvalues of arbitrary matrix function $f(\mathbf{A})$, are $f(\lambda_i)$ where λ_i is the i-th eigenvalue of $\mathbf{A}$. In particular, for $f(x) = e^x$ for a scalar x; the eigenvalues of $e^{\mathbf{A}}$ are e^{λ_i}, which is, of course, positive (i.e., $e^{\lambda_i} > 0$). In other words, the matrix exponential $e^{\mathbf{A}}$ is ALWAYS positive semidefinite for an arbitrary matrix $\mathbf{A}$. The positive real numbers have a lot of special structures to exploit, compared with arbitrary real numbers. The elementary fact motivates the wide use of positive semidefinite (PSD) matrices, for example, convex optimization and quantum information theory. Through the spectral mapping theorem, all the eigenvalues of positive semidefinite matrices are nonnegative.

For a sequence of scalar random variables (real or complex numbers), $x_1, \ldots, x_n$, we can study its convergence by studying the so-called partial sum $S_n = x_1 + \ldots + x_n = \sum_{i=1}^{n} x_i$. We say the sequence converges to a limit value $S = \mathbb{E}[x]$, if there exists a limit S as $n \to \infty$. In analogy with the scalar counterparts, we can similarly define

$$\mathbf{S}_n = \mathbf{X}_1 + \ldots + \mathbf{X}_n = \sum_{i=1}^{n} \mathbf{X}_i,$$

for a sequence of Hermitian matrices, $\mathbf{X}_1, \ldots, \mathbf{X}_n$. We say the sequence converges to a limit matrix $\mathbf{S} = \mathbb{E}[\mathbf{X}]$, if there exists a limit $\mathbf{S}$ as $n \to \infty$.

One nice thing about the positive number is the ordering. When $a = 0.4$ and $b = 0.5$, we can say $a < b$. In analogy, we say the partial order $\mathbf{A} \leq \mathbf{B}$ if all the eigenvalues of $\mathbf{B} - \mathbf{A}$ are nonnegative, which is equivalent to say $\mathbf{B} - \mathbf{A}$ is positive semidefinite matrix. Since a matrix exponential is $e^{\mathbf{A}}$ is always positive semidefinite for an arbitrary matrix $\mathbf{A}$, we can instead study $e^{\mathbf{A}} \leq e^{\mathbf{B}}$, to infer about the partial order $\mathbf{A} \leq \mathbf{B}$. The function $x \to e^{sx}$ is monotone, non-decreasing and positive for all $s \geq 0$. we can, by the spectral mapping theorem to study their eigenvalues which are scalar random variables. Thus a matrix-valued random variable is converted into a scalar-valued random variable, by using the bridge of the spectral mapping theorem. For our interest, what matters is the spectrum (spec$(\mathbf{A})$, the set of all eigenvalues).

In summary, the sums of random matrices are of elementary nature. We emphasize the fundamental contribution of Ahlswede and Winter [36] since their work has triggered a snow ball of this line of research.

2.2 Matrix Laplace Transform Method

Due to the basic nature of sums of random matrices, we give several versions of the theorems and their derivations. Although essentially their techniques are equivalent, the assumptions and arguments are sufficiently different to justify the space. The techniques for handling matrix-valued random variables are very subtle; it is our intention to give an exhaustive survey of these techniques. Even a seemingly small twist of the problems can cause a lot of technical difficulties. These presentations serve as examples to illustrate the key steps. Repetition is the best teacher—practice makes it perfect. This is the rationale behind this chapter. It is hoped that the audience pays attention to the methods, not the particular derived inequalities.

The Laplace transform method is the standard technique for the scalar-valued random variables; it is remarkable that this method can be extended to the matrix setting. We argue that this is a break-through in studying the matrices concentration. This method is used as a thread to tie together all the surveyed literature. For completion, we run the risk of "borrowing" too much from the cited references. Here we give credit to those cited authors. We try our best to add more details about their arguments with the hope of being more accessible.

2.2.1 Method 1—Harvey's Derivation

The presentation here is essentially the same as [91,92] whose style is very friendly and accessible. We present Harvey's version first.

2.2.1.1 The Ahlswede-Winter Inequality

Let $\mathbf{X}$ be a random $d \times d$ matrix, i.e., a matrix whose entries are all random variables. We define $\mathbb{E}\mathbf{X}$ to be the matrix whose entries are the expectation of the entries of $\mathbf{X}$. Since expectation and trace are both linear, they commute:

$$\begin{aligned}\mathbb{E}\left[\operatorname{Tr}\mathbf{X}\right] &\triangleq \sum_{\mathbf{A}} \mathbb{P}\left(\mathbf{X}=\mathbf{A}\right) \cdot \sum_i A_{i,i} = \sum_i \sum_{\mathbf{A}} \mathbb{P}\left(\mathbf{X}=\mathbf{A}\right) \cdot A_{i,i} \\ &= \sum_i \sum_a \mathbb{P}\left(X_{i,i}=a\right) \cdot a = \sum_i \mathbb{E}\left(X_{i,i}\right) = \operatorname{Tr}\left(\mathbb{E}\mathbf{X}\right).\end{aligned}$$

Let $\mathbf{X}_1, \cdots, \mathbf{X}_n$ be random, *symmetric*[1] matrices of size $d \times d$. Define the partial sums

$$\mathbf{S}_n = \mathbf{X}_1 + \cdots + \mathbf{X}_n = \sum_{i=1}^{n} \mathbf{X}_i.$$

$\mathbf{A} \geq 0$ is equivalent to saying that all eigenvalues of $\mathbf{A}$ are nonnegative, i.e., $\lambda_i(\mathbf{A}) \geqslant 0$. We would like to analyze the probability that eigenvalues of $\mathbf{S}_n$ are at most t, i.e., $\mathbf{S}_n \leq t\mathbf{I}$. This is equivalent to the event that all eigenvalues of $e^{\mathbf{S}_n}$ are at most $e^{\lambda t}$, i.e., $e^{\mathbf{S}_n \lambda} \leqslant e^{\lambda t \mathbf{I}}$. If this event fails to hold, then certainly $\operatorname{Tr} e^{\mathbf{S}_n \lambda} > \operatorname{Tr} e^{\lambda t \mathbf{I}}$, since all eigenvalues of $e^{\mathbf{S}_n}$ are non-negative. Thus, we have argued that

$$\begin{aligned}
&\Pr\left[\text{some eigenvalues of matrix } \mathbf{S}_n \text{ is greater than t}\right] \\
&\leqslant \mathbb{P}\left(\operatorname{Tr} e^{\mathbf{S}_n t} > \operatorname{Tr} e^{\lambda t \mathbf{I}}\right) \qquad (2.1)\\
&\leqslant \mathbb{E}\left(\operatorname{Tr} e^{\mathbf{S}_n t}\right)/e^{\lambda t},
\end{aligned}$$

by Markov's inequality. Now, as in the proof of the Chernoff bound, we want to bound this expectation by a product of expectations, which will lead to an exponentially decreasing tail bound. This is where the Golden-Thompson inequality is needed.

$$\begin{aligned}
&\mathbb{E}\left(\operatorname{Tr} e^{\mathbf{S}_n \lambda}\right) = \mathbb{E}\left(\operatorname{Tr} e^{\lambda \mathbf{X}_n + \lambda \mathbf{S}_{n-1}}\right) \text{ (since } \mathbf{S}_n = \mathbf{X}_n + \mathbf{S}_{n-1}\text{)}\\
&\leqslant \mathbb{E}\left[\operatorname{Tr}\left(e^{\lambda \mathbf{X}_n} \cdot e^{\lambda \mathbf{S}_{n-1}}\right)\right] \text{ (by Golden - Thompson inequality)}\\
&= \mathbb{E}_{\mathbf{X}_1, \cdots, \mathbf{X}_{n-1}}\left\{\mathbb{E}_{\mathbf{X}_n}\left[\operatorname{Tr}\left(e^{\lambda \mathbf{X}_n} \cdot e^{\lambda \mathbf{S}_{n-1}}\right)\right]\right\} \text{ (since the } \mathbf{X}_i\text{'s are mutually independent)}\\
&= \mathbb{E}_{\mathbf{X}_1, \cdots, \mathbf{X}_{n-1}}\left\{\operatorname{Tr}\left[\mathbb{E}_{\mathbf{X}_n}\left(e^{\lambda \mathbf{X}_n} \cdot e^{\lambda \mathbf{S}_{n-1}}\right)\right]\right\} \text{ (since trace and expectation commute)}\\
&= \mathbb{E}_{\mathbf{X}_1, \cdots, \mathbf{X}_{n-1}}\left\{\operatorname{Tr}\left[\mathbb{E}_{\mathbf{X}_n}\left(e^{\lambda \mathbf{X}_n}\right) \cdot e^{\lambda \mathbf{S}_{n-1}}\right]\right\} \text{ (since } \mathbf{X}_n \text{ and } \mathbf{S}_{n-1} \text{are independent)}\\
&= \mathbb{E}_{\mathbf{X}_1, \cdots, \mathbf{X}_{n-1}}\left[\left\|\mathbb{E}_{\mathbf{X}_n}\left(e^{\lambda \mathbf{X}_n}\right)\right\| \cdot \operatorname{Tr}\left(e^{\lambda \mathbf{S}_{n-1}}\right)\right] \text{ (by Corollary of trace-norm property)}\\
&= \left\|\mathbb{E}_{\mathbf{X}_n}\left(e^{\lambda \mathbf{X}_n}\right)\right\| \cdot \mathbb{E}_{\mathbf{X}_1, \cdots, \mathbf{X}_{n-1}}\left[\operatorname{Tr}\left(e^{\lambda \mathbf{S}_{n-1}}\right)\right]
\end{aligned} \qquad (2.2)$$

Applying this inequality inductively, we get

$$\mathbb{E}\left(\operatorname{Tr} e^{\mathbf{S}_n \lambda}\right) \leqslant \prod_{i=1}^{n} \left\|\mathbb{E}_{\mathbf{X}_i}\left(e^{\lambda \mathbf{X}_i}\right)\right\| \cdot \operatorname{Tr}\left(e^{\lambda \mathbf{0}}\right) = \prod_{i=1}^{n} \left\|\mathbb{E}\left(e^{\lambda \mathbf{X}_i}\right)\right\| \cdot \operatorname{Tr}\left(e^{\lambda \mathbf{0}}\right),$$

[1] The assumption symmetric matrix is too strong for many applications. Since we often deal with complex entries, the assumption of Hermitian matrix is reasonable. This is the fatal flaw of this version. Otherwise, it is very useful.

where $\mathbf{0}$ is the zero matrix of size $d \times d$. So $e^{\lambda \mathbf{0}} = \mathbf{I}$ and $\mathrm{Tr}\,(\mathbf{I}) = d$, where $\mathbf{I}$ is the identity matrix whose diagonal are all 1. Therefore,

$$\mathbb{E}\left(\mathrm{Tr}\, e^{\mathbf{S}_n \lambda}\right) \leqslant d \cdot \prod_{i=1}^{n} \left\| \mathbb{E}\left(e^{\lambda \mathbf{X}_i}\right) \right\|.$$

Combining this with (2.1), we obtain

$$\Pr\left[\text{some eigenvalues of matrix } \mathbf{S}_n \text{ is greater than t}\right] \leqslant d e^{-\lambda t} \prod_{i=1}^{n} \left\| \mathbb{E}\left(e^{\lambda \mathbf{X}_i}\right) \right\|.$$

We can also bound the probability that any eigenvalue of $\mathbf{S}_n$ is less than $-t$ by applying the same argument to $-\mathbf{S}_n$. This shows that the probability that any eigenvalue of $\mathbf{S}_n$ lies outside $[-t, t]$ is

$$\mathbb{P}\left(\|\mathbf{S}_n\| > t\right) \leqslant d e^{-\lambda t} \left\{ \prod_{i=1}^{n} \left\| \mathbb{E}\left(e^{\lambda \mathbf{X}_i}\right) \right\| + \prod_{i=1}^{n} \left\| \mathbb{E}\left(e^{-\lambda \mathbf{X}_i}\right) \right\| \right\}. \tag{2.3}$$

This is the basis inequality. Much like the Chernoff bound, numerous variations and generalizations are possible. Two useful versions are stated here without proof.

Theorem 2.2.1. *Let* $\mathbf{Y}$ *be a random, symmetric, positive semi-definite* $d \times d$ *matrix such that* $\mathbb{E}[\mathbf{Y}] = \mathbf{I}$. *Suppose* $\|Y\| \leq R$ *for some fixed scalar* $R \geq 1$. *Let* $\mathbf{Y}_1, \ldots, \mathbf{Y}_k$ *be independent copies of* $\mathbf{Y}$ *(i.e., independently sampled matrices with the same distribution as* $\mathbf{Y}$*). For any* $\varepsilon \in (0, 1)$, *we have*

$$\mathbb{P}\left[(1 - \varepsilon)\,\mathbf{I} \leqslant \frac{1}{k}\sum_{i=1}^{k} \mathbf{Y}_i \leqslant (1 + \varepsilon)\,\mathbf{I}\right] \geqslant 1 - 2d \cdot \exp\left(-\varepsilon^2 k / 4R\right).$$

This event is equivalent to the sample average $\frac{1}{k}\sum_{i=1}^{k} \mathbf{Y}_i$ *having minimum eigenvalue at least* $1 - \varepsilon$ *and maximum eigenvalue at most* $1 + \varepsilon$.

Proof. See [92]. □

Corollary 2.2.2. *Let* $\mathbf{Z}$ *be a random, symmetric, positive semi-definite* $d \times d$ *matrix. Define* $\mathbf{U} = \mathbb{E}[\mathbf{Z}]$ *and suppose* $\mathbf{Z} \leq R \cdot \mathbf{U}$ *for some scalar* $R \geq 1$. *Let* $\mathbf{Z}_1, \ldots, \mathbf{Z}_k$ *be independent copies of* $\mathbf{Z}$ *(i.e., independently sampled matrices with the same distribution as* $\mathbf{Z}$*). For any* $\varepsilon \in (0, 1)$, *we have*

$$\mathbb{P}\left[(1 - \varepsilon)\,\mathbf{U} \leqslant \frac{1}{k}\sum_{i=1}^{k} \mathbf{Z}_i \leqslant (1 + \varepsilon)\,\mathbf{U}\right] \geqslant 1 - 2d \cdot \exp\left(-\varepsilon^2 k / 4R\right).$$

Proof. See [92]. □

2.2.1.2 Rudelson's Theorem

In this section, we use how the Ahlswede-Winter inequality is used to prove a concentration inequality for random vectors due to Rudelson. His original proof was quite different [93].

The motivation for Rudelson's inequality comes from the problem of approximately computing the volume of a convex body. When solving this problem, a convenient first step is to transform the body into the "isotropic position", which is a technical way of saying "roughly like the unit sphere." To perform this first step, one requires a concentration inequality for randomly sampled vectors, which is provided by Rudelson's theorem.

Theorem 2.2.3 (Rudelson's Theorem [93]). *Let* $\mathbf{x} \in \mathbb{R}^d$ *be a random vector such that* $\mathbb{E}\left(\mathbf{x}\mathbf{x}^T\right) = \mathbf{I}$*. Suppose* $\|\mathbf{x}\| \leqslant R$*. Let* $\mathbf{x}_1, \ldots, \mathbf{x}_n$ *be independent copies of* $\mathbf{x}$*. For any* $\varepsilon \in (0,1)$*, we have*

$$\mathbb{P}\left(\left\|\frac{1}{n}\sum_{i=1}^{n}\mathbf{x}_i\mathbf{x}_i^T - \mathbf{I}\right\| > \varepsilon\right) \leqslant 2d \cdot \exp\left(-\varepsilon^2 n/4R^2\right).$$

Note that $R \geqslant \sqrt{d}$ because

$$d = \mathrm{Tr}\,\mathbf{I} = \mathrm{Tr}\left[\mathbb{E}\left(\mathbf{x}\mathbf{x}^T\right)\right] = \mathbb{E}\left[\mathrm{Tr}\left(\mathbf{x}\mathbf{x}^T\right)\right] = \mathbb{E}\left[\mathbf{x}^T\mathbf{x}\right],$$

since $\mathrm{Tr}\left(\mathbf{AB}\right) = \mathrm{Tr}\left(\mathbf{BA}\right)$.

Proof. We apply the Ahlswede-Winter inequality with the rank-1 matrix $\mathbf{X}_i$

$$\mathbf{X}_i = \frac{1}{2R^2}\left(\mathbf{x}_i\mathbf{x}_i^T - \mathbb{E}\left[\mathbf{x}_i\mathbf{x}_i^T\right]\right) = \frac{1}{2R^2}\left(\mathbf{x}_i\mathbf{x}_i^T - \mathbf{I}\right).$$

Note that $\mathbb{E}\mathbf{X}_i = \mathbf{0}$, $\|\mathbf{X}_i\| \leqslant 1$, and

$$\begin{aligned}
\mathbb{E}\left[\mathbf{X}_i^2\right] &= \frac{1}{4R^4}\mathbb{E}\left[\left(\mathbf{x}_i\mathbf{x}_i^T - \mathbf{I}\right)^2\right] \\
&= \frac{1}{4R^4}\left\{\mathbb{E}\left[\left(\mathbf{x}_i\mathbf{x}_i^T\right)^2\right] - \mathbf{I}\right\} \text{ (since } \mathbb{E}\left[\mathbf{x}_i\mathbf{x}_i^T\right] = \mathbf{I}) \\
&\leqslant \frac{1}{4R^4}\mathbb{E}\left[\left(\mathbf{x}_i^T\mathbf{x}_i\right)\left(\mathbf{x}_i\mathbf{x}_i^T\right)\right] \\
&\leqslant \frac{R^2}{4R^4}\mathbb{E}\left[\mathbf{x}_i\mathbf{x}_i^T\right] \text{ (since } \|\mathbf{x}_i\| \leqslant R) \\
&= \frac{\mathbf{I}}{4R^2}.
\end{aligned} \tag{2.4}$$

Now using Claim 1.4.5 together with the inequalities

$$1 + y \leqslant e^y, \forall y \in \mathbb{R}$$
$$e^y \leqslant 1 + y + y^2, \forall y \in [-1, 1].$$

Since $\|\mathbf{X}_i\| \leqslant 1$, for any $\lambda \in [0,1]$, we have $e^{\lambda \mathbf{X}_i} \leqslant \mathbf{I} + \lambda \mathbf{X}_i + \lambda^2 \mathbf{X}_i^2$, and so

$$\mathbb{E}\left[e^{\lambda \mathbf{X}_i}\right] \leqslant \mathbb{E}\left[\mathbf{I} + \lambda \mathbf{X}_i + \lambda^2 \mathbf{X}_i^2\right] \leqslant \mathbf{I} + \lambda^2 \mathbb{E}\left[\mathbf{X}_i^2\right]$$
$$\leqslant e^{\lambda^2 \mathbb{E}\left[\mathbf{X}_i^2\right]} \leqslant e^{\lambda^2/4R^2 \mathbf{I}},$$

by Eq. (2.4). Thus, $\left\|\mathbb{E}\left[e^{\lambda \mathbf{X}_i}\right]\right\| \leqslant e^{\lambda^2/4R^2}$. The same analysis also shows that $\left\|\mathbb{E}\left[e^{-\lambda \mathbf{X}_i}\right]\right\| \leqslant e^{\lambda^2/4R^2}$. Substituting this into Eq. (2.3), we obtain

$$\mathbb{P}\left(\left\|\sum_{i=1}^{n} \frac{1}{2R^2}\left(\mathbf{x}_i \mathbf{x}_i^T - \mathbf{I}\right)\right\| > t\right) \leqslant 2d \cdot e^{-\lambda t} \prod_{i=1}^{n} e^{\lambda^2/4R^2} = 2d \cdot \exp\left(-\lambda t + n\lambda^2/4R^2\right).$$

Substituting $t = n\varepsilon/2R^2$ and $\lambda = \varepsilon$ proves the theorem. □

2.2.2 *Method 2—Vershynin's Derivation*

We give the derivation method, taken from [35], by Vershynin.

Let $\mathbf{X}_1, \ldots, \mathbf{X}_n$ be independent random $d \times d$ real matrices, and let

$$\mathbf{S} = \mathbf{X}_1 + \cdots + \mathbf{X}_n.$$

We will be interested in the magnitude of the derivation $\|\mathbf{S}_n - \mathbb{E}\mathbf{S}_n\|$ in the operator norm $\|\cdot\|$.

Now we try to generalize the method of Sect. 1.4.10 when $\mathbf{X}_i \in \mathbb{M}_d$ are independent random matrices of mean zero, where $\mathbb{M}_d$ denotes the class of symmetric $d \times d$ matrices.

For, $\mathbf{A} \in \mathbb{M}_d$, the matrix exponential $e^{\mathbf{A}}$ is defined as usual by Taylor series. $e^{\mathbf{A}}$ has the same eigenvectors as $\mathbf{A}$, and eigenvalues $e^{\lambda_i(\mathbf{A})} > 0$. The partial order $\mathbf{A} \geq \mathbf{B}$ means $\mathbf{A} - \mathbf{B} \geqq 0$, i.e., $\mathbf{A} - \mathbf{B}$ is positive semi-definite (their eigenvalues are non-negative). By using the exponential function of $\mathbf{A}$, we deal with the positive semi-definite matrix which has a fundamental structure to exploit.

The non-trivial part is that, in general,

$$e^{\mathbf{A}+\mathbf{B}} \neq e^{\mathbf{A}} e^{\mathbf{B}}.$$

However, the famous Golden-Thompson's inequality [94] sates that

$$\operatorname{Tr}\left(e^{\mathbf{A}+\mathbf{B}}\right) \leqslant \operatorname{Tr}\left(e^{\mathbf{A}} e^{\mathbf{B}}\right)$$

holds for arbitrary $\mathbf{A}, \mathbf{B} \in \mathbb{M}_d$ (and in fact for arbitrary unitary-invariant norm replacing the trace [23]). Therefore, for $\mathbf{S}_n = \mathbf{X}_1 + \cdots + \mathbf{X}_n = \sum_{i=1}^{n} \mathbf{X}_i$ and for $\mathbf{I}$ being the identify matrix on $\mathbb{M}_d$, we have

$$p \triangleq \mathbb{P}\left(\mathbf{S}_n \nleqslant t\mathbf{I}\right) = \mathbb{P}\left(e^{\lambda \mathbf{S}_n} \nleqslant e^{\lambda t\mathbf{I}}\right) \leqslant \mathbb{P}\left(\operatorname{Tr} e^{\lambda \mathbf{S}_n} > e^{\lambda t}\right) \leqslant e^{-\lambda t}\mathbb{E}\operatorname{Tr}\left(e^{\lambda \mathbf{S}_n}\right).$$

This estimate is not sharp: $e^{\lambda \mathbf{S}_n} \nleqslant e^{t\mathbf{I}}$ means the biggest eigenvalue of $e^{\lambda \mathbf{S}_n}$ exceeds $e^{\lambda t}$, while $\operatorname{Tr} e^{\lambda \mathbf{S}_n} > e^{\lambda t}$ means that the sum of all d eigenvalues exceeds the same.

Since $\mathbf{S}_n = \mathbf{X}_n + \mathbf{S}_{n-1}$, we use the Golden-Thomas's inequality to separate the last term from the sum:

$$\mathbb{E}\operatorname{Tr}\left(e^{\lambda \mathbf{S}_n}\right) \leqslant \mathbb{E}\operatorname{Tr}\left(e^{\lambda \mathbf{X}_n} e^{\lambda \mathbf{S}_{n-1}}\right).$$

Now, using independence and that $\mathbb{E}$ and trace commute, we continue to write

$$= \mathbb{E}_{n-1}\operatorname{Tr}\left[\left(\mathbb{E}_n e^{\lambda \mathbf{X}_n}\right) \cdot e^{\lambda \mathbf{S}_{n-1}}\right] \leqslant \left\|\mathbb{E}_n e^{\lambda \mathbf{X}_n}\right\| \cdot \mathbb{E}_{n-1}\operatorname{Tr}\left(e^{\lambda \mathbf{S}_{n-1}}\right),$$

since

$$\operatorname{Tr}(\mathbf{AB}) \leqslant \|\mathbf{A}\| \operatorname{Tr}(\mathbf{B}),$$

for $\mathbf{A}, \mathbf{B} \in \mathbb{M}_d$.

Continuing by induction, we reach (since $\operatorname{Tr}\mathbf{I} = d$) to

$$\mathbb{E}\operatorname{Tr}\left(e^{\lambda \mathbf{S}_n}\right) \leqslant d \cdot \prod_{i=1}^{n} \mathbb{E} e^{\lambda \mathbf{X}_i}.$$

We have proved that

$$\mathbb{P}\left(\mathbf{S}_n \nleqslant t\mathbf{I}\right) \leqslant d \cdot \prod_{i=1}^{n} \mathbb{E} e^{\lambda \mathbf{X}_i}.$$

Repeating for $-\mathbf{S}_n$ and using that $-t\mathbf{I} \leqslant \mathbf{S}_n \leqslant t\mathbf{I}$ is equivalent to $\|\mathbf{S}_n\| \leqslant t$ we have shown that

$$\mathbb{P}\left(\|\mathbf{S}_n\| > t\right) \leqslant 2de^{-\lambda t} \cdot \prod_{i=1}^{n} \left\|\mathbb{E} e^{\lambda \mathbf{X}_i}\right\|. \tag{2.5}$$

As in the real valued case, full independence is never needed in the above argument. It works out well for martingales.

Theorem 2.2.4 (Chernoff-type inequality). *Let $\mathbf{X}_i \in \mathbb{M}_d$ be independent mean zero random matrices, $\|\mathbf{X}_i\| \leqslant 1$ for all i almost surely. Let*

$$\mathbf{S}_n = \mathbf{X}_1 + \cdots + \mathbf{X}_n = \sum_{i=1}^{n} \mathbf{X}_i,$$

$$\sigma^2 = \sum_{i=1}^{n} \|\mathrm{var}\, \mathbf{X}_i\|.$$

Then, for every $t > 0$, we have

$$\mathbb{P}\left(\|\mathbf{S}_n\| > t\right) \leqslant d \cdot \max\left(e^{-t^2/4\sigma^2}, e^{-t/2}\right).$$

To prove this theorem, we have to estimate (2.5). The standard estimate

$$1 + y \leqslant e^y \leqslant 1 + y + y^2$$

is valid for real number $y \in [-1, 1]$ (actually a bit beyond) [95]. From the two bounds, we get (replacing y with $\mathbf{Y}$)

$$\mathbf{I} + \mathbf{Y} \leqslant e^{\mathbf{Y}} \leqslant \mathbf{I} + \mathbf{Y} + \mathbf{Y}^2$$

Using the bounds twice (first the upper bound and then the lower bound), we have

$$\mathbb{E}e^{\mathbf{Y}} \leqslant \mathbb{E}\left(\mathbf{I} + \mathbf{Y} + \mathbf{Y}^2\right) = \mathbf{I} + \mathbb{E}\left(\mathbf{Y}^2\right) \leqslant e^{\mathbb{E}\left(\mathbf{Y}^2\right)}.$$

Let $0 < \lambda \leq 1$. Therefore, by the Theorem's hypothesis,

$$\left\|\mathbb{E}e^{\lambda \mathbf{X}_i}\right\| \leqslant \left\|e^{\lambda^2 \mathbb{E}\left(\mathbf{X}_i^2\right)}\right\| = e^{\lambda^2 \left\|\mathbb{E}\left(\mathbf{X}_i^2\right)\right\|}.$$

Hence by (2.5),

$$\mathbb{P}\left(\|\mathbf{S}_n\| > t\right) \leqslant 2d \cdot e^{-\lambda t + \lambda^2 \sigma^2}.$$

With the optimal choice of $\lambda = \min\left(t/2\sigma^2, 1\right)$, the conclusion of the Theorem follows.

Does the Theorem hold for σ^2 replaced by $\sum_{i=1}^{n} \left\|\mathbb{E}\left(\mathbf{X}_i^2\right)\right\|$?

Corollary 2.2.5. *Let $\mathbf{X}_i \in \mathbb{M}_d$ be independent mean zero random matrices, $\|\mathbf{X}_i\| \leqslant 1$ for all i almost surely. Let*

$$\mathbf{S}_n = \mathbf{X}_1 + \cdots + \mathbf{X}_n = \sum_{i=1}^{n} \mathbf{X}_i, \quad E = \sum_{i=1}^{n} \|\mathbb{E}\mathbf{X}_i\|.$$

Then, for every $\varepsilon \in (0,1)$, we have

$$\mathbb{P}\left(\|\mathbf{S}_n - \mathbb{E}\mathbf{S}_n\| > \varepsilon E\right) \leqslant d \cdot e^{-\varepsilon^2 E/4}.$$

2.2.3 Method 3—Oliveria's Derivation

Consider the random matrix $\mathbf{Z}_n$. We closely follow Oliveria [38] whose exposition is highly accessible. In particular, he reviews all the needed theorems, all of those are collected in Chap. 1 for easy reference. In this subsection, the matrices are assumed to be $d \times d$ Hermitian matrices, that is, $\mathbf{A} \in \mathbb{C}^{d\times d}_{\text{Herm}}$, where $\mathbb{C}^{d\times d}_{\text{Herm}}$ is the set of $d \times d$ Hermitian matrices.

2.2.3.1 Bernstein Trick

The usual Bernstein trick implies that for all $t \geq 0$,

$$\forall t \geqslant 0, \mathbb{P}\left(\lambda_{max}\left(\mathbf{Z}_n\right) > t\right) \leqslant \inf_{s>0} e^{-st}\mathbb{E}\left[e^{s\lambda_{max}(\mathbf{Z}_n)}\right]. \tag{2.6}$$

Notice that

$$\mathbb{E}\left[e^{s\|\mathbf{Z}_n\|}\right] \leqslant \mathbb{E}\left[e^{s\lambda_{\max}(\mathbf{Z}_n)}\right] + \mathbb{E}\left[e^{s\lambda_{\max}(-\mathbf{Z}_n)}\right] = 2\mathbb{E}\left[e^{s\lambda_{\max}(\mathbf{Z}_n)}\right] \tag{2.7}$$

since $\|\mathbf{Z}_n\| = \max\left\{\lambda_{\max}\left(\mathbf{Z}_n\right), \lambda_{\max}\left(-\mathbf{Z}_n\right)\right\}$ and $\mathbf{Z}_n$ has the same law as $-\mathbf{Z}_n$.

2.2.3.2 Spectral Mapping

The function $x \mapsto e^{sx}$ is monotone, non-decreasing and positive for all $s \geq 0$. It follows from the spectral mapping property (1.60) that for all $s \geq 0$, the largest eigenvalue of $e^{s\mathbf{Z}_n}$ is $e^{s\lambda_{\max}(\mathbf{Z}_n)}$ and all eigenvalues of $e^{s\mathbf{Z}_n}$ are nonnegative. Using the equality "trace = sum of eigenvalues" implies that for all $s \geq 0$,

$$\mathbb{E}\left[e^{s\lambda_{\max}(\mathbf{Z}_n)}\right] = \mathbb{E}\left[\lambda_{\max}\left(e^{s\mathbf{Z}_n}\right)\right] \leqslant \mathbb{E}\left[\operatorname{Tr}\left(e^{s\mathbf{Z}_n}\right)\right]. \tag{2.8}$$

Combining (2.6), (2.7) with (2.8) gives

$$\forall t \geq 0, \mathbb{P}\left(\|\mathbf{Z}_n\| > t\right) \leqslant 2\inf_{s\geq 0} e^{-st}\mathbb{E}\left[\operatorname{Tr}\left(e^{s\mathbf{Z}_n}\right)\right]. \tag{2.9}$$

Up to now, the Oliveira's proof in [38] has followed Ahlswede and Winter's argument in [36]. The next lemma is originally due to Oliveira [38]. Now Oliveira considers the special case

$$\mathbf{Z}_n = \sum_{i=1}^{n} \varepsilon_i \mathbf{A}_i, \tag{2.10}$$

where ε_i are random coefficients and $\mathbf{A}_1, \ldots, \mathbf{A}_n$ are **deterministic** Hermitian matrices. Recall that a Rademacher sequence is a sequence of $\varepsilon_{i}{}_{i=1}^{n}$ of i.i.d. random variables with $\varepsilon_i = \varepsilon_1$ uniform over $\{-1, 1\}$. A standard Gaussian sequence is a sequence i.i.d. standard Gaussian random variables.

Lemma 2.2.6 (Oliveira [38]). *For all $s \in \mathbb{R}$,*

$$\mathbb{E}\left[\mathrm{Tr}\left(e^{s\mathbf{Z}_n}\right)\right] = \mathrm{Tr}\left[\mathbb{E}\left(e^{s\mathbf{Z}_n}\right)\right] \leqslant \mathrm{Tr}\left[\exp\left(\frac{s^2 \sum_{i=1}^{n} \mathbf{A}_i^2}{2}\right)\right]. \tag{2.11}$$

Proof. In (2.11), we have used the fact that trace and expectation commute, according to (1.87). The key proof steps have been followed by Rudelson [93], Harvey [91,92], and Wigderson and Xiao [94]. □

2.2.4 Method 4—Ahlswede-Winter's Derivation

Ahlswede and Winter [36] were the first who used the matrix Laplace transform method. Ahlswede-Winter's derivation, taken from [36], is presented in detail below. We postpone their original version until now, for easy understanding. Their paper and Tropp's long paper [53] are two of the most important sources on this topic. We first digress to study the problem of hypothesis for motivation.

Consider a hypothesis testing problem for a motivation

$$\mathcal{H}_0 : \mathbf{A}_1, \ldots, \mathbf{A}_K$$
$$\mathcal{H}_1 : \mathbf{B}_1, \ldots, \mathbf{B}_K$$

where a sequence of positive, random matrices $\mathbf{A}_i, i = 1, \ldots, K$ and $\mathbf{B}_i, i = 1, \ldots, K$ are considered.

Algorithm 2.2.7 (Detection Using Traces of Sums of Covariance Matrices).

1. Claim $\mathcal{H}_1$ if

$$\mathrm{Tr} \sum_{k=1}^{K} \mathbf{A}_k = \xi \leq \mathrm{Tr} \sum_{k=1}^{K} \mathbf{B}_k,$$

2. Otherwise, claim $\mathcal{H}_0$.

Only diagonal elements are used in Algorithm 2.2.7; However, non-diagonal elements contain information of use to detection. The exponential of a matrix provides one tool. See Example 2.2.9. In particular, we have

$$\mathrm{Tr}e^{\mathbf{A}+\mathbf{B}} \leq \mathrm{Tr}e^{\mathbf{A}}e^{\mathbf{B}}.$$

The following matrix inequality

$$\mathrm{Tr}e^{\mathbf{A}+\mathbf{B}+\mathbf{C}} \leq \mathrm{Tr}e^{\mathbf{A}}e^{\mathbf{B}}e^{\mathbf{C}}$$

is known to be false.

Let $\mathbf{A}$ and $\mathbf{B}$ be two Hermitian matrices of the same size. If $\mathbf{A} - \mathbf{B}$ is positive semidefinite, we write [16]

$$\mathbf{A} \geq \mathbf{B} \quad \text{or} \quad \mathbf{B} \leq \mathbf{A}. \tag{2.12}$$

$\geq$ is a partial ordering, referred to as Löwner partial ordering, on the set of Hermitian matrices, that is,

1. $\mathbf{A} \geq \mathbf{A}$ for every Hermitian matrix $\mathbf{A}$,
2. If $\mathbf{A} \geq \mathbf{B}$ and $\mathbf{B} \geq \mathbf{A}$, then $\mathbf{A} = \mathbf{B}$, and
3. If $\mathbf{A} \geq \mathbf{B}$ and $\mathbf{B} \geq \mathbf{C}$, then $\mathbf{A} \geq \mathbf{C}$.

The statement $\mathbf{A} \geq 0 \Leftrightarrow \mathbf{X}^*\mathbf{A}\mathbf{X} \geq 0$ is generalized as follows:

$$\mathbf{A} \geq \mathbf{B} \Leftrightarrow \mathbf{X}^*\mathbf{A}\mathbf{X} \geq \mathbf{X}^*\mathbf{B}\mathbf{X} \tag{2.13}$$

for every complex matrix $\mathbf{X}$.

A hypothesis detection problem can be viewed as a problem of partially ordering the measured matrices for individual hypotheses. If many (K) copies of the measured matrices $\mathbf{A}_k$ and $\mathbf{B}_k$ are at our disposal, it is natural to ask this fundamental question:

Is $\mathbf{B}_1 + \mathbf{B}_2 + \cdots + \mathbf{B}_K$ **(statistically) different than** $\mathbf{A}_1 + \mathbf{A}_2 + \cdots + \mathbf{A}_K$?

To answer this question motivates this whole section. It turns out that a new theory is needed. We freely use [36] that contains a relatively complete appendix for this topic.

The theory of real random variables provides the framework of much of modern probability theory, such as laws of large numbers, limit theorems, and probability estimates for large deviations, when sums of independent random variables are involved. Researchers develop analogous theories for the case that the algebraic structure of the reals is substituted by more general structures such as groups, vector spaces, etc.

At the hands of our current problem of hypothesis detection, we focus on a structure that has vital interest in quantum probability theory and names the algebra of operators[2] on a (complex) Hilbert space. In particular, the real vector space of self-adjoint operators (Hermitian matrices) can be regarded as a partially ordered generalization of the reals, as reals are embedded in the complex numbers.

2.2.4.1 Fundamentals of Matrix-Valued Random Variables

In the ground-breaking work of [36], they focus on a structure that has vital interest in the algebra of operators on a (complex) Hilbert space, and in particular, the real vector space of self-adjoint operators. Through the spectral mapping theorem, these self-adjoint operators can be regarded as a partially ordered generalization of the reals, as reals are embedded in the complex numbers. To study the convergence of sums of matrix-valued random variables, this partial order is necessary. It will be clear later.

One can generalize the exponentially good estimate for large deviations by the so-called Bernstein trick that gives the famous Chernoff bound [96, 97].

A matrix-valued random variable $\mathbf{X}:\boldsymbol{\Omega}\to\mathcal{A}_s$, where

$$\mathcal{A}_s=\{\mathbf{A}\in\mathcal{A}:\mathbf{A}=\mathbf{A}^*\} \tag{2.14}$$

is the self-adjoint part of the $\mathbf{C}^*$-algebra $\mathcal{A}$ [98], which is a real vector space. Let $\mathcal{L}(\mathcal{H})$ be the full operator algebra of the complex Hilbert space $\mathcal{H}$. We denote $d=\dim(\mathcal{H})$, which is assumed to be finite. Here dim means the dimensionality of the vector space. In the general case, $d=\mathrm{Tr}\mathbf{I}$, and $\mathcal{A}$ can be embedded into $\mathcal{L}(C^d)$ as an algebra, *preserving the trace*. Note the trace (often regarded as expectation) has the property $\mathrm{Tr}\,(\mathbf{AB})=\mathrm{Tr}\,(\mathbf{BA})$, for any two matrices (or operators) of $\mathbf{A},\mathbf{B}$. In free probability[3] [99], this is a (optional) axiom as very weak form of commutativity in the trace [9, p. 169].

The real cone

$$\mathcal{A}_+=\{\mathbf{A}\in\mathcal{A}:\mathbf{A}=\mathbf{A}^*\geq 0\} \tag{2.15}$$

induces a *partial order* $\leq$ in $\mathcal{A}_s$. This partial order is in analogy with the order of two real numbers $a\leq b$. The partial order is the main interest in what follows. We can introduce some convenient notation: for $\mathbf{A},\mathbf{B}\in\mathcal{A}_s$ the closed interval $[\mathbf{A},\mathbf{B}]$ is defined as

$$[\mathbf{A},\mathbf{B}]=\{\mathbf{X}\in\mathcal{A}_s:\mathbf{A}\leq\mathbf{X}\leq\mathbf{B}\}. \tag{2.16}$$

---

[2]The finite-dimensional operators and matrices are used interchangeably.

[3]The idea of free probability is to make algebra (such as operator algebras C^*-algebra, von Neumann algebras) the foundation of the theory, as opposed to other possible choices of foundations such as sets, measures, categories, etc.

This is an analogy with the interval $x \in [a, b]$ when $a \leq x \leq b$ for $a, b \in \mathbb{R}$. Similarly, open and half-open intervals $(\mathbf{A}, \mathbf{B})$, $[\mathbf{A}, \mathbf{B})$, etc.

For simplicity, the space $\boldsymbol{\Omega}$ on which the random variable lives is discrete. Some remarks on the matrix (or operator) order is as follows.

1. The notation "$\leq$" when used for the matrices is *not a total order* unless $\mathcal{A}$, the set of $\mathbf{A}$, spans the entire complex space, i.e., $\mathcal{A} = \mathbb{C}$, in which case the set of self-adjoint operators is the real number space, i.e., $\mathcal{A}_s = \mathbb{R}$. Thus in this case (classical case), the theory developed below reduces to the study of the real random variables.
2. $\mathbf{A} \geq 0$ is equivalent to saying that all eigenvalues of $\mathbf{A}$ are nonnegative. These are d *nonlinear inequalities*. However, we can have the alternative characterization:

$$\begin{aligned} \mathbf{A} \geq 0 &\Leftrightarrow \forall \rho \text{ density operator } \mathrm{Tr}(\rho \mathbf{A}) \geq 0 \\ &\Leftrightarrow \forall \pi \text{ one} - \text{dimensional projector } \mathrm{Tr}(\pi \mathbf{A}) \geq 0 \end{aligned} \tag{2.17}$$

   From which, we see that these nonlinear inequalities are equivalent to infinitely many *linear* inequalities, which is better adapted to the vector space structure of $\mathcal{A}_s$.
3. The operator mapping $\mathbf{A} \mapsto \mathbf{A}^s$, for $s \in [0, 1]$ and $\mathbf{A} \mapsto \log \mathbf{A}$ are defined on $\mathcal{A}_+$, and both are operator monotone and operator concave. In contrast, $\mathbf{A} \mapsto \mathbf{A}^s$, for $s > 2$ and $\mathbf{A} \mapsto \exp \mathbf{A}$ are neither operator monotone nor operator convex. Remarkably, $\mathbf{A} \mapsto \mathbf{A}^s$, for $s \in [1, 2]$ is operator convex (though not operator monotone). See Sect. 1.4.22 for definitions.
4. The mapping $\mathbf{A} \mapsto \mathrm{Tr} \exp \mathbf{A}$ is monotone and convex. See [50].
5. Golden-Thompson-inequality [23]: for $\mathbf{A}, \mathbf{B} \in \mathcal{A}_s$

$$\mathrm{Tr} \exp(\mathbf{A} + \mathbf{B}) \leq \mathrm{Tr}((\exp \mathbf{A})(\exp \mathbf{B})). \tag{2.18}$$

Items 1–3 follows from Loewner's theorem. A good account of the partial order is [18, 22, 23]. Note that a rarely few of mappings (functions) are operator convex (concave) or operator monotone. Fortunately, we are interested in the trace functions that have much bigger sets [18].

Take a look at (2.19) for example. Since $\mathcal{H}_0 : \mathbf{A} = \mathbf{I} + \mathbf{X}$, and $\mathbf{A} \in \mathcal{A}_s$ (even stronger $\mathbf{A} \in \mathcal{A}_+$), it follows from (2.18) that

$$\mathcal{H}_0 : \mathrm{Tr} \exp(\mathbf{A}) = \mathrm{Tr} \exp(\mathbf{I} + \mathbf{X}) \leq \mathrm{Tr}((\exp \mathbf{I})(\exp \mathbf{X})). \tag{2.19}$$

The use of (2.19) allows us to separately study the diagonal part and the non-diagonal part of the covariance matrix of the noise, since all the diagonal elements are equal for a WSS random process. At low SNR, the goal is to find some ratio or threshold that is statistically stable over a large number of Monte Carlo trials.

Algorithm 2.2.8 (Ratio detection algorithm using the trace exponentials).
1. Claim $\mathcal{H}_1$*, if* $\xi = \frac{\mathrm{Tr}\exp \mathbf{A}}{\mathrm{Tr}((\exp \mathbf{I})(\exp \mathbf{X}))} \geq 1$*, where* $\mathbf{A}$ *is the measured covariance matrix with or without signals and* $\mathbf{X} = \frac{\mathbf{R}_w}{\sigma_w^2} - \mathbf{I}$.
2. Otherwise, claim $\mathcal{H}_0$.

Example 2.2.9 (Exponential of the 2×2 *matrix).* The 2×2 covariance matrix for L sinusoidal signals has symmetric structure with identical diagonal elements

$$\mathbf{R}_s = \mathrm{Tr}\mathbf{R}_s(\mathbf{I} + b\sigma_1)$$

where

$$\sigma_1 = \begin{pmatrix} 0 & 1 \\ 1 & 0 \end{pmatrix}$$

and b is a positive number. Obviously, $\mathrm{Tr}\sigma_1 = 0$. We can study the diagonal elements and non-diagonal elements separately. The two eigenvalues of the 2×2 matrix [100]

$$A = \begin{pmatrix} a & b \\ c & d \end{pmatrix}$$

are

$$\lambda_{1,2} = \tfrac{1}{2}\mathrm{Tr}\mathbf{A} \pm \tfrac{1}{2}\sqrt{\mathrm{Tr}^2\mathbf{A} - 4\det \mathbf{A}}$$

and the corresponding eigenvectors are, respectively,

$$u_1 = \frac{1}{||u_1||}\begin{pmatrix} b \\ \lambda_1 - a \end{pmatrix}; \quad u_2 = \frac{1}{||u_2||}\begin{pmatrix} b \\ \lambda_2 - a \end{pmatrix}.$$

To study how the zero-trace 2×2 matrix σ_1 affects the exponential, consider

$$\mathbf{X} = \begin{pmatrix} 0 & b \\ a^{-1} & 0 \end{pmatrix}.$$

The exponential of the matrix $\mathbf{X}$, $e^{\mathbf{X}}$, has positive entries, and in fact [101]

$$e^{\mathbf{X}} = \begin{pmatrix} \cosh\sqrt{\frac{b}{a}} & \sqrt{ab}\sinh\sqrt{\frac{b}{a}} \\ \frac{1}{\sqrt{ab}}\sinh\sqrt{\frac{b}{a}} & \cosh\sqrt{\frac{b}{a}} \end{pmatrix}$$

□

2.2.4.2 Matrix-Valued Concentration Inequalities

In analogy with the scalar-valued random variables, we can develop a matrix-valued Markov inequality. Suppose that X is a nonnegative random variable with mean $\mathbb{E}[X]$. The scalar-valued **Markov inequality** of (1.6) states that

$$\mathbb{P}(X \geqslant a) \leqslant \frac{\mathbb{E}[X]}{a} \quad \text{for } X \text{ nonnegative.} \tag{2.20}$$

Theorem 2.2.10 (Markov inequality). *Let* $\mathbf{X}$ *a matrix-valued random variable with values in* $\mathcal{A}_+$ *and expectation*

$$\mathbf{M} = \mathbb{E}\mathbf{X} = \sum_{\mathcal{X}} \Pr\{\mathbf{X} = \mathcal{X}\}\mathcal{X}, \tag{2.21}$$

and $\mathbf{A} \geq 0$ *is a fixed positive semidefinite matrix. Then*

$$\Pr\{\mathbf{X} \nleq \mathbf{A}\} \leq \mathrm{Tr}\left(\mathbf{M}\mathbf{A}^{-1}\right). \tag{2.22}$$

Proof. The support of $\mathbf{A}$ is assumed to contain the support of $\mathbf{M}$, otherwise, the theorem is trivial. Let us consider the positive matrix-valued random variable $\mathbf{Y} = \mathbf{A}^{-1/2}\mathbf{X}\mathbf{A}^{-1/2}$ which has expectation $\mathbb{E}[\mathbf{Y}] = \mathbf{A}^{-1/2}\mathbb{E}[\mathbf{X}]\mathbf{A}^{-1/2}$, using the product rule of (1.88): $\mathbb{E}[\mathbf{XY}] = \mathbb{E}[\mathbf{X}]\mathbb{E}[\mathbf{Y}]$. we have used the fact that the expectation of a constant matrix is itself: $\mathbb{E}[\mathbf{A}] = \mathbf{A}$.

Since the events $\{\mathbf{X} \leqslant \mathbf{A}\}$ and $\{\mathbf{Y} \leqslant \mathbf{I}\}$ coincide, we have to show that

$$\mathbb{P}(\mathbf{Y} \leqslant \mathbf{I}) \nleqslant \mathrm{Tr}(\mathbb{E}[\mathbf{X}])$$

Note from (1.87), the trace and expectation commute! This is seen as follows:

$$\mathbb{E}[\mathbf{Y}] = \sum_{\mathcal{Y}} \mathbb{P}(\mathbf{Y} = \mathcal{Y})\,\mathcal{Y} \geqslant \sum_{\mathcal{Y} \nleqslant \mathbf{I}} \mathbb{P}(\mathbf{Y} = \mathcal{Y})\,\mathcal{Y}.$$

The second inequality follows from the fact that $\mathbf{Y}$ is positive and $\mathcal{Y} = \{\mathcal{Y} \nleqslant \mathbf{I}\} \cup \{\mathcal{Y} > \mathbf{I}\}$. All eigenvalues of $\mathbf{Y}$ are positive. Ignoring the event $\{\mathcal{Y} > \mathbf{I}\}$ is equivalent to remove some positive eigenvalues from the spectrum of the $\mathbf{Y}$, spec$(\mathbf{Y})$.

Taking traces, and observing that a positive operator (or matrix) which is not less than or equal to $\mathbf{I}$ must have trace at least 1, we find

$$\mathrm{Tr}(\mathbb{E}[\mathbf{Y}]) \geqslant \sum_{\mathcal{Y} \nleqslant \mathbf{I}} \mathbb{P}(\mathbf{Y} = \mathcal{Y})\,\mathrm{Tr}(\mathcal{Y}) \geqslant \sum_{\mathcal{Y} \nleqslant \mathbf{I}} \mathbb{P}(\mathbf{Y} = \mathcal{Y}) = \mathbb{P}(\mathbf{Y} \nleqslant \mathbf{I}),$$

which is what we wanted. □

In the case of $\mathcal{H} = \mathbb{C}$ the theorem reduces to the well-known Markov inequality for nonnegative real random variables. One easily see that, as in the classical case, the inequality is optimal in the sense that there are examples when the inequality is assumed with equality.

Suppose that the mean m and the variance σ^2 of a scalar-valued random variable X are known. The *Chebyshev inequality* of (1.8) states that

$$\mathbb{P}\left(|X - m| \geqslant a\right) \leqslant \frac{\sigma^2}{a^2}. \tag{2.23}$$

The Chebyshev inequality is a consequence of the Markov inequality.

In analogy with the scalar case, if we assume knowledge about the matrix-valued expectation and the matrix-valued variance, we can prove the matrix-valued Chebyshev inequality.

Theorem 2.2.11 (Chebyshev inequality). *Let* $\mathbf{X}$ *a matrix-valued random variable with values in* $\mathcal{A}_s$*, expectation* $\mathbf{M} = \mathbb{E}\mathbf{X}$*, and variance*

$$\mathrm{Var}\mathbf{X} = \mathbf{S}^2 = \mathbb{E}\left((\mathbf{X} - \mathbf{M})^2\right) = \mathbb{E}(\mathbf{X}^2) - \mathbf{M}^2. \tag{2.24}$$

For $\mathbf{\Delta} \geq 0$,

$$\mathbb{P}\{|\mathbf{X} - \mathbf{M}| \not\leq \mathbf{\Delta}\} \leq \mathrm{Tr}\left(\mathbf{S}^2\mathbf{\Delta}^{-2}\right). \tag{2.25}$$

Proof. Observe that

$$|\mathbf{X} - \mathbf{M}| \leq \mathbf{\Delta} \Leftarrow (\mathbf{X} - \mathbf{M})^2 \leq \mathbf{\Delta}^2$$

since $\sqrt{(\cdot)}$ is operator monotone. See Sect. 1.4.22. We find that

$$\mathbb{P}\left(|\mathbf{X} - \mathbf{M}| \not\leqslant \mathbf{\Delta}\right) \leqslant \mathbb{P}\left((\mathbf{X} - \mathbf{M})^2 \not\leqslant \mathbf{\Delta}^2\right) \leqslant \mathrm{Tr}\left(\mathbf{S}^2\mathbf{\Delta}^{-2}\right).$$

The last step follows from Theorem 2.2.10. □

If $\mathbf{X}, \mathbf{Y}$ are independent, then $\mathrm{Var}(\mathbf{X} + \mathbf{Y}) = \mathrm{Var}\mathbf{X} + \mathrm{Var}\mathbf{Y}$. This is the same as in the classical case but one has to pay attention to the noncommunicativity that causes technical difficulty.

Corollary 2.2.12 (Weak law of large numbers). *Let* $\mathbf{X}, \mathbf{X}_1, \mathbf{X}_2, \ldots, \mathbf{X}_n$ *be identically, independently, distributed (i.i.d.) matrix-valued random variables with values in* $\mathcal{A}_s$*, expectation* $\mathbf{M} = \mathbb{E}\mathbf{X}$*, and variance* $\mathrm{Var}\mathbf{X} = \mathbf{S}^2$*. For* $\mathbf{\Delta} \geq 0$*, then*

$$\begin{aligned}
&\mathbb{P}\left\{\frac{1}{n}\sum_{n=1}^{n}\mathbf{X}_i \notin [\mathbf{M}-\boldsymbol{\Delta},\mathbf{M}+\boldsymbol{\Delta}]\right\} \le \frac{1}{n}\mathrm{Tr}\left(\mathbf{S}^2\boldsymbol{\Delta}^{-2}\right),\\
&\mathbb{P}\left\{\sum_{n=1}^{n}\mathbf{X}_i \notin [n\mathbf{M}-\sqrt{n}\boldsymbol{\Delta},n\mathbf{M}-\sqrt{n}\boldsymbol{\Delta}]\right\} \le \frac{1}{n}\mathrm{Tr}\left(\mathbf{S}^2\boldsymbol{\Delta}^{-2}\right).
\end{aligned} \tag{2.26}$$

Proof. Observe that $\mathbf{Y} \notin [\mathbf{M}-\boldsymbol{\Delta},\mathbf{M}+\boldsymbol{\Delta}]$ is equivalent to $|\mathbf{Y}-\mathbf{M}| \nleq \boldsymbol{\Delta}$, and apply the previous theorem. The event $|\mathbf{Y}-\mathbf{M}| \nleq \boldsymbol{\Delta}$ says that the absolute values of the eigenvalues of the matrix $\mathbf{Y}-\mathbf{M}$ is bounded by the eigenvalues of $\boldsymbol{\Delta}$ which are of course nonnegative. The matrix $\mathbf{Y}-\mathbf{M}$ is Hermitian (but not necessarily nonnegative or nonpositive). That is why the absolute value operation is needed. □

When we see these functions of matrix-valued inequalities, we see the functions of their eigenvalues. The spectral mapping theorem must be used all the time.

Lemma 2.2.13 (Large deviations and Bernstein trick). *For a matrix-valued random variable* $\mathbf{Y}$, $\mathbf{B} \in \mathcal{A}_s$, *and* $\mathbf{T} \in \mathcal{A}$ *such that* $\mathbf{T}^*\mathbf{T} > 0$

$$\mathbb{P}\{\mathbf{Y} \nleq \mathbf{B}\} \le \mathrm{Tr}\left[\mathbb{E}\mathrm{e}^{\mathbf{TYT}^*-\mathbf{TBT}^*}\right]. \tag{2.27}$$

Proof. We directly calculate

$$\begin{aligned}
\mathbb{P}\left(\mathbf{Y} \nleqq \mathbf{B}\right) &= \mathbb{P}\left(\mathbf{Y}-\mathbf{B} \nleqq 0\right)\\
&= \mathbb{P}\left(\mathbf{TYT}^* - \mathbf{TBT}^* \nleqq 0\right)\\
&= \mathbb{P}\left[\mathrm{e}^{\mathbf{TYT}^*-\mathbf{TBT}^*} \nleqq \mathbf{I}\right]\\
&\leqslant \mathrm{Tr}\left[\mathbb{E}\mathrm{e}^{\mathbf{TYT}^*-\mathbf{TBT}^*}\right].
\end{aligned}$$

Here, the second line is because the mapping $\mathbf{X} \mapsto \mathbf{TXT}^*$ is bijective and preserve the order. The $\mathbf{TYT}^*$ and $\mathbf{TBT}^*$ are two *commutative* matrices. For commutative matrices $\mathbf{A},\mathbf{B}$, $\mathbf{A} \le \mathbf{B}$ is equivalent to $e^{\mathbf{A}} \le e^{\mathbf{B}}$, from which the third line follows. The last line follows from Theorem 2.2.10. □

The Bernstein trick is a crucial step. The problem is reduced to the form of $\mathrm{Tr}\left[\mathbb{E}\mathrm{e}^{\mathbf{Z}}\right]$ where $\mathbf{Z} = \mathbf{TYT}^* - \mathbf{TBT}^*$ is Hermitian. We really do not know if $\mathbf{Z}$ is nonnegative or positive. But we do not care since the matrix exponential of any Hermitian $\mathbf{A}$ is always nonnegative. As a consequence of using the Bernstein trick, we only need to deal with nonnegative matrices.

But we need another key ingredient—Golden-Thompson inequality—since for Hermitian $\mathbf{AB}$, we have $e^{\mathbf{A}+\mathbf{B}} \neq e^{\mathbf{A}} \cdot e^{\mathbf{B}}$, unlike $e^{a+b} = e^a \cdot e^b$, for two scalars a, b. For two Hermitian matrices $\mathbf{A},\mathbf{B}$, we have the Golden-Thompson inequality

$$\mathrm{Tr}\left(e^{\mathbf{A}+\mathbf{B}}\right) \leqslant \mathrm{Tr}\left(e^{\mathbf{A}} \cdot e^{\mathbf{B}}\right).$$

Theorem 2.2.14 (Concentration for i.i.d matrix-valued random variables). *Let* $\mathbf{X}, \mathbf{X}_1, \ldots, \mathbf{X}_n$ *be i.i.d. matrix-valued random variables with values in* $\mathcal{A}_s$, $\mathbf{A} \in \mathcal{A}_s$. *Then for* $\mathbf{T} \in \mathcal{A}$, $\mathbf{T}^*\mathbf{T} > 0$

$$\mathbb{P}\left\{\sum_{n=1}^{n} \mathbf{X}_i \nleq n\mathbf{A}\right\} \leqslant d \cdot \mathbb{E}\|\exp\left(\mathbf{TXT}^* - \mathbf{TAT}^*\right)\|^n. \tag{2.28}$$

Define the binary I-divergence as

$$D(u\|v) = u(\log u - \log v) + (1-u)(\log(1-u) - \log(1-v)). \tag{2.29}$$

Proof. Using previous lemma (obtained from the Bernstein trick) with $\mathbf{Y} = \sum_{i=1}^{n} \mathbf{X}_i$ and $\mathbf{B} = n\mathbf{A}$, we find

$$\begin{aligned}
\mathbb{P}\left\{\sum_{i=1}^{n} \mathbf{X}_i \nleq n\mathbf{A}\right\} &\leqslant \mathrm{Tr}\left\{\mathbb{E}\exp\left[\sum_{i=1}^{n} \mathbf{T}\left(\mathbf{X}_i - \mathbf{A}\right)\mathbf{T}^*\right]\right\} \\
&= \mathbb{E}\left\{\mathrm{Tr}\exp\left[\sum_{i=1}^{n} \mathbf{T}\left(\mathbf{X}_i - \mathbf{A}\right)\mathbf{T}^*\right]\right\} \\
&\leqslant \mathbb{E}\,\mathrm{Tr}\left\{\exp\left[\sum_{i=1}^{n-1} \mathbf{T}\left(\mathbf{X}_i - \mathbf{A}\right)\mathbf{T}^*\right]\exp\left[\mathbf{T}\left(\mathbf{X}_n - \mathbf{A}\right)\mathbf{T}^*\right]\right\} \\
&= \mathbb{E}_{\mathbf{X}_1,\ldots,\mathbf{X}_{n-1}}\mathrm{Tr}\left\{\exp\left[\sum_{i=1}^{n-1} \mathbf{T}\left(\mathbf{X}_i - \mathbf{A}\right)\mathbf{T}^*\right]\mathbb{E}\exp\left[\mathbf{T}\left(\mathbf{X}_n - \mathbf{A}\right)\mathbf{T}^*\right]\right\} \\
&\leqslant \left\|\mathbb{E}\exp\left[\mathbf{T}\left(\mathbf{X}_n - \mathbf{A}\right)\mathbf{T}^*\right]\right\| \cdot \mathbb{E}_{\mathbf{X}_1,\ldots,\mathbf{X}_{n-1}}\mathrm{Tr}\left\{\exp\left[\sum_{i=1}^{n-1} \mathbf{T}\left(\mathbf{X}_i - \mathbf{A}\right)\mathbf{T}^*\right]\right\} \\
&\leqslant d \cdot \left\|\mathbb{E}\exp\left[\mathbf{T}\left(\mathbf{X}_n - \mathbf{A}\right)\mathbf{T}^*\right]\right\|^n
\end{aligned}$$

the first line follows from Lemma 2.2.13. The second line is from the fact that the trace and expectation commute, according to (1.87). In the third line, we use the famous Golden-Thompson inequality (1.71). In the fourth line, we take the expectation on the $\mathbf{X}_n$. The fifth line is due to the norm property (1.84). The sixth line is using the fifth step by induction for n times. d comes from the fact $\mathrm{Tr}\exp(\mathbf{0}) = \mathrm{Tr}\mathbf{I} = d$, where $\mathbf{0}$ is a zero matrix whose entries are all zero. □

The problem is now to minimize $\left\|\mathbb{E}\exp\left[\mathbf{T}\left(\mathbf{X}_n - \mathbf{A}\right)\mathbf{T}^*\right]\right\|$ with respect to $\mathbf{T}$. For a Hermitian matrix T, we have the polar decomposition $\mathbf{T} = |\mathbf{T}| \cdot \mathbf{U}$, where $\mathbf{Y}$ is a unitary matrix; so, without loss of generality, we may assume that $\mathbf{T}$ is Hermitian. Let us focus pursue the special case of a bounded matrix-valued random variable. Defining

$$D\left(u\|v\right) = u\left(\log u - \log v\right) + (1-u)\left[\log\left(1-u\right) - \log\left(1-v\right)\right]$$

we find the following matrix-valued Chernoff bound.

Theorem 2.2.15 (Matrix-valued Chernoff Bound). *Let* $\mathbf{X}, \mathbf{X}_1, \ldots, \mathbf{X}_n$ *be i.i.d. matrix-valued random variables with values in* $[\mathbf{0}, \mathbf{I}] \in \mathcal{A}_s$, $\mathbb{E}\mathbf{X} \leq m\mathbf{I}$, $\mathbf{A} \geq a\mathbf{I}$, $1 \geq a \geq m \geq 0$. *Then*

$$\mathbb{P}\left\{\sum_{n=1}^{n} \mathbf{X}_i \nleq n\mathbf{A}\right\} \leq d \cdot \exp\left(-nD\left(a||m\right)\right), \tag{2.30}$$

Similarly, $\mathbb{E}\mathbf{X} \geq m\mathbf{I}$, $\mathbf{A} \leq a\mathbf{I}$, $0 \leq a \leq m \leq 1$. *Then*

$$\mathbb{P}\left(\sum_{n=1}^{n} \mathbf{X}_i \ngeq n\mathbf{A}\right) \leqslant d \cdot \exp\left(-nD\left(a||m\right)\right), \tag{2.31}$$

As a consequence, we get, for $\mathbb{E}\mathbf{X} = \mathbf{M} \geq \mu\mathbf{I}$ *and* $0 \leq \epsilon \leq \frac{1}{2}$, *then*

$$\mathbb{P}\left\{\frac{1}{n}\sum_{n=1}^{n} \mathbf{X}_i \notin \left[\left(1-\varepsilon\right)\mathbf{M}, \left(1+\varepsilon\right)\mathbf{M}\right]\right\} \leq 2d \cdot \exp\left(-n \cdot \frac{\varepsilon^2 \mu}{2\ln 2}\right). \tag{2.32}$$

Proof. The second part follows from the first by considering $\mathbf{Y}_i = \mathbf{X}_i$, and the observation that $D\left(a||m\right) = D\left(1-a||1-m\right)$. To prove it, we apply Theorem 2.2.14 with a special case of $\mathbf{T} = \sqrt{t}\mathbf{I}$ to obtain

$$\begin{aligned}
\mathbb{P}\left\{\sum_{n=1}^{n} \mathbf{X}_i \nleq n\mathbf{A}\right\} &\leqslant \mathbb{P}\left\{\sum_{n=1}^{n} \mathbf{X}_i \nleq na\mathbf{I}\right\} \\
&\leqslant d \cdot ||\mathbb{E}\exp(t\mathbf{X})\exp(-ta\mathbf{I})||^n \\
&= d \cdot ||\mathbb{E}\exp(t\mathbf{X})\exp(-ta)||^n.
\end{aligned}$$

Note $\exp(-ta\mathbf{I}) = \exp(-ta)\mathbf{I}$ and $\mathbf{AI} = \mathbf{A}$. Now the convexity of the exponential function $\exp(x)$ implies that

$$\frac{\exp\left(tx\right)-1}{x} \leqslant \frac{\exp\left(t\right)-1}{1}, 0 \leqslant x \leqslant 1, x \in \mathbb{R},$$

which, by replacing x with matrix $\mathbf{X} \in \mathcal{A}_\mathbf{s}$ and 1 with the identify matrix $\mathbf{I}$ (see Sect. 1.4.13 for this rule), yields

$$\exp(t\mathbf{X}) - 1 \leqslant \mathbf{X}\left(\exp\left(t\right)-1\right).$$

As a consequence, we have

$$\mathbb{E}\exp(t\mathbf{X}) \leqslant \mathbf{I} + (\exp(t) - 1)\,\mathbb{E}\mathbf{X}. \tag{2.33}$$

hence, we have

$$\begin{aligned} \|\mathbb{E}\exp(t\mathbf{X})\exp(-ta)\| &\leqslant \exp(-ta)\,\|\mathbf{I} + (\exp(t) - 1)\,\mathbb{E}\mathbf{X}\| \\ &\leq \exp(-ta)\,\|\mathbf{I} + (\exp(t) - 1)\,m\mathbf{I}\| \\ &= \exp(-ta)\,[1 + (\exp(t) - 1)\,m]\,. \end{aligned}$$

The first line follows from using (2.33). The second line follows from the hypothesis of $\mathbb{E}\mathbf{X} \leqslant m\mathbf{I}$. The third line follows from using the spectral norm property of (1.57) for the identity matrix $\mathbf{I}$:$\|\mathbf{I}\| = 1$. Choosing

$$t = \log\left(\frac{a}{m}\cdot\frac{1-m}{1-a}\right) > 0$$

the right-hand side becomes exactly $\exp\left(-D\left(a\|m\right)\right)$.

To prove the last claim of the theorem, consider the variables $\mathbf{Y}_i = \mu\mathbf{M}^{-1/2}\mathbf{X}_i\mathbf{M}^{-1/2}$ with expectation $\mathbb{E}\mathbf{Y}_i = \mu\mathbf{I}$ and $\mathbf{Y}_i \in [\mathbf{0}, \mathbf{I}]$, by hypothesis. Since

$$\frac{1}{n}\sum_{n=1}^{n}\mathbf{X}_i \in [(1-\varepsilon)\,\mathbf{M}, (1+\varepsilon)\,\mathbf{M}] \Leftrightarrow \frac{1}{n}\sum_{n=1}^{n}\mathbf{Y}_i \in [(1-\varepsilon)\,\mu\mathbf{I}, (1+\varepsilon)\,\mu\mathbf{I}]$$

we can apply what we just proved to obtain

$$\begin{aligned} &\mathbb{P}\left\{\frac{1}{n}\sum_{n=1}^{n}\mathbf{X}_i \notin [(1-\varepsilon)\,\mathbf{M}, (1+\varepsilon)\,\mathbf{M}]\right\} \\ &= \mathbb{P}\left\{\frac{1}{n}\sum_{n=1}^{n}\mathbf{X}_i \geqslant (1+\varepsilon)\,\mathbf{M}\right\} + \mathbb{P}\left\{\frac{1}{n}\sum_{n=1}^{n}\mathbf{X}_i \leqslant (1-\varepsilon)\,\mathbf{M}\right\} \\ &\leqslant d\left\{\exp\left[-nD\left((1-\varepsilon)\,\mu\|\mu\right)\right] + \exp\left[-nD\left((1+\varepsilon)\,\mu\|\mu\right)\right]\right\} \\ &\leqslant 2d\cdot\exp\left(-n\frac{\varepsilon^2\mu}{2\ln 2}\right). \end{aligned}$$

The last line follows from the already used inequality

$$D\left((1+x)\mu\|\mu\right) \geqslant \frac{x^2\mu}{2\ln 2}.$$

□

2.2.5 Derivation Method 5—Gross, Liu, Flammia, Becker, and Eisert

Let $||\mathbf{A}||$ be the operator norm of matrix $\mathbf{A}$.

Theorem 2.2.16 (Matrix-Bernstein inequality—Gross [102]). *let* $\mathbf{X}_i, i = 1, \ldots, N$ *be i.i.d. zero mean, Hermitian matrix-valued random variables of size* $n \times n$. *Assume* $\sigma_0, c \in \mathbb{R}$ *are such that* $\left\|\mathbf{X}_i^2\right\| \leqslant \sigma_0^2$ *and* $\left\|\mathbf{X}_i\right\| \leqslant \mu$. *Set* $\mathbf{S} = \sum_{i=1}^{N} \mathbf{X}_i$ *and let* $\sigma^2 = N\sigma_0^2$, *an upper bound to the variance of* $\mathbf{S}$. *Then*

$$\begin{aligned} \mathbb{P}\left(\|\mathbf{S}\| > t\right) &\leqslant 2n \exp\left(-\frac{t^2}{4\sigma^2}\right), \quad t \leqslant \frac{2\sigma}{\mu}, \\ \mathbb{P}\left(\|\mathbf{S}\| < t\right) &\leqslant 2n \exp\left(-\frac{t}{2\mu}\right), \quad t > \frac{2\sigma}{\mu}. \end{aligned} \tag{2.34}$$

We refer to [102] for a proof. His proof directly follows from Ahlswede-Winter [36] with some revisions.

2.2.6 Method 6—Recht's Derivation

The version of derivation, taken from [103], is more general in that the random matrices need not be identically distributed. A symmetric matrix is assumed. It is our conjecture that results of [103] may be easily extended to a Hermitian matrix.

Theorem 2.2.17 (Noncommutative Bernstein Inequality [104]). *Let* $\mathbf{X}_1, \ldots, \mathbf{X}_L$ *be independent zero-mean random matrices of dimension* $d_1 \times d_2$. *Suppose* $\rho_k^2 = \max\left\{\left\|\mathbb{E}\left(\mathbf{X}_k\mathbf{X}_k^*\right)\right\|, \left\|\mathbb{E}\left(\mathbf{X}_k^*\mathbf{X}_k\right)\right\|\right\}$ *and* $\left\|\mathbf{X}_k\right\| \leqslant M$ *almost surely for all* k. *Then, for any* $\tau > 0$,

$$\mathbb{P}\left[\left\|\sum_{k=1}^{L} \mathbf{X}_k\right\| > \tau\right] \leqslant (d_1 + d_2) \exp\left(\frac{-\frac{\tau^2}{2}}{\sum_{k=1}^{L} \rho_k^2 + M\tau/3}\right). \tag{2.35}$$

Note that in the case that $d_1 = d_2 = 1$, this is precisely the two sided version of the standard Bernstein Inequality. When the $\mathbf{X}_k$ are diagonal, this bound is the same as applying the standard Bernstein Inequality and a union bound to the diagonal of the matrix summation. Besides, observe that the right hand side is less than $(d_1 + d_2) \exp\left(\frac{-\frac{3}{8}\tau^2}{\sum_{k=1}^{L} \rho_k^2}\right)$ as long as $\tau \leqslant \frac{1}{M} \sum_{k=1}^{L} \rho_k^2$. This condensed form of the inequality is used exclusively throughout in [103]. Theorem 2.2.17 is a corollary of an Chernoff bound for finite dimensional operators developed by Ahlswede and Winter [36]. A similar inequality for symmetric i.i.d. matrices is proposed in [95] $\|\cdot\|$ denotes the spectral norm (the top singular value) of an operator.

2.2.7 Method 7—Derivation by Wigderson and Xiao

Chernoff bounds are extremely useful in probability. Intuitively, they say that a random sample approximates the average, with a probability of deviation that goes down exponentially with the number of samples. Typically we are concerned about real-valued random variables, but recently several applications have called for large-deviations bounds for matrix-valued random variables. Such a bound was given by Ahlswede and Winter [36, 105].

All of Wigderson and Xiao's results [94] are extended to complex Hermitian matrices, or abstractly to self-adjoint operators over any Hilbert spaces where the operations of addition, multiplication, trace exponential, and norm are efficiently computable. Wigderson and Xiao [94] essentially follows the original style of Ahlswede and Winter [36] in the validity of their method.

2.2.8 Method 8—Tropp's Derivation

The derivation follows [53].

$$\begin{aligned}
&\mathbb{P}\left(\left\|\lambda_{\max}\left(\sum_{i=1}^{n}\mathbf{X}_i\right)\right\| \geqslant t\right)\\
&= \mathbb{P}\left(\left\|\lambda_{\max}\left(\sum_{i=1}^{n}\theta\mathbf{X}_i\right)\right\| \geqslant e^{\theta t}\right) \text{ (the positive homogeneity of the eigenvalue map)}\\
&\leqslant e^{-\theta t}\cdot\mathbb{E}\exp\left\{\lambda_{\max}\left(\sum_{i=1}^{n}\theta\mathbf{X}_i\right)\right\} \text{ (Markov's inequality)}\\
&= e^{-\theta t}\cdot\mathbb{E}\lambda_{\max}\left(\exp\left\{\sum_{i=1}^{n}\theta\mathbf{X}_i\right\}\right) \text{ (the spectral mapping theorem)}\\
&< e^{-\theta t}\cdot\mathbb{E}\,\mathrm{Tr}\left(\exp\left\{\sum_{i=1}^{n}\theta\mathbf{X}_i\right\}\right) \text{ (the exponential of a Hermitian matrix is positive definite)}
\end{aligned} \tag{2.36}$$

2.3 Cumulate-Based Matrix-Valued Laplace Transform Method

This section develops some general probability inequalities for the maximum eigenvalue of a sum of independent random matrices. The main ingredient is a matrix extension of the scalar-valued Laplace transform method for sums of independent real random variables, see Sect. 1.1.6.

Before introducing the matrix-valued Laplace transform, we need to define matrix and cumulants, in analogy with Sect. 1.1.6 for the scalar setting. At this point, a quick review of Sect. 1.1 will illuminate the contrast between the scalar and matrix settings. The central idea of Ahswede and Winter [36] is to extend the textbook idea of the Laplace Transform Method from the scalar setting to the matrix setting.

Consider a Hermitian matrix $\mathbf{X}$ that has moments of all orders. By analogy with the classical scalar definitions (Sect. 1.1.7), we may construct matrix extensions of the moment generating function and the cumulant generating function:

$$\mathbf{M}_{\mathbf{X}}\left(\theta\right) := \mathbb{E}e^{\theta\mathbf{X}} \text{ and } \Xi_{\mathbf{X}}\left(\theta\right) := \log \mathbb{E}e^{\theta\mathbf{X}} \text{ for } \theta \in \mathbb{R}. \tag{2.37}$$

We have the formal power series expansions:

$$\mathbf{M}_{\mathbf{X}}\left(\theta\right) = \mathbf{I} + \sum_{k=1}^{\infty} \frac{\theta^k}{k!} \cdot \left(\mathbf{X}^k\right) \text{ and } \Xi_{\mathbf{X}}\left(\theta\right) = \sum_{k=1}^{\infty} \frac{\theta^k}{k!} \cdot \Xi_k.$$

The coefficients $\left(\mathbb{E}\mathbf{X}^k\right)$ are called matrix moments and Ξ_k are called matrix cumulants. The matrix cumulant Ξ_k has a formal expression as a noncommutative polynomial in the matrix moments up to order k. In particular, the first cumulant is the mean and the second cumulant is the variance:

$$\Xi_1 = \mathbb{E}\left(\mathbf{X}\right) \text{ and } \Xi_2 = \mathbb{E}\left(\mathbf{X}^2\right) - \mathbb{E}\left(\mathbf{X}\right)^2.$$

Higher-order cumulants are harder to write down and interpret.

Proposition 2.3.1 (The Lapalce Transform Method). *Let $\mathbf{Y}$ be a random Hermitian matrix. For all $t \in \mathbb{R}$,*

$$\mathbb{P}\left(\lambda_{\max}\left(\mathbf{Y}\right) \geqslant t\right) \leqslant \inf_{\theta > 0} \left\{e^{-\theta t} \cdot \mathbb{E}\operatorname{Tr} e^{\theta\mathbf{Y}}\right\}.$$

In words, we can control tail probabilities for the maximum eigenvalue of a random matrix by producing a bound for the trace of the matrix moment generating function defined in (2.37). Let us show how Bernstein's Laplace transform technique extends to the matrix setting. The basic idea is due to Ahswede-Winter [36], but we follow Oliveria [38] in this presentation.

Proof. Fix a positive number θ. Observe that

$$\mathbb{P}(\lambda_{\max}(\mathbf{Y}) \geqslant t) = \mathbb{P}(\lambda_{\max}(\theta\mathbf{Y}) \geqslant \theta t) = \mathbb{P}\left(e^{\lambda_{\max}(\theta\mathbf{Y})} \geqslant e^{\theta t}\right) \leqslant e^{-\theta t} \cdot \mathbb{E} e^{\lambda_{\max}(\theta\mathbf{Y})}.$$

The first identity uses the homogeneity of the maximum eigenvalue map. The second relies on the monotonicity of the scalar exponential functions; the third relation is Markov's inequality. To bound the exponential, note that

$$e^{\lambda_{\max}(\theta\mathbf{Y})} = \lambda_{\max}\left(e^{\theta\mathbf{Y}}\right) \leqslant \operatorname{Tr} e^{\theta\mathbf{Y}}.$$

The first relation is the spectral mapping theorem (Sect. 1.4.13). The second relation holds because the exponential of an Hermitian matrix is always positive definite—the eigenvalues of the matrix exponential are always positive (see Sect. 1.4.16 for the matrix exponential); thus, the maximum eigenvalue of a positive definite matrix is dominated by the trace. Combine the latter two relations, we reach

$$\mathbb{P}(\lambda_{\max}(\mathbf{Y}) \geqslant t) \leqslant \inf_{\theta > 0} \left\{ e^{-\theta t} \cdot \mathbb{E} \operatorname{Tr} e^{\theta\mathbf{Y}} \right\}.$$

This inequality holds for any positive θ, so we may take an infimum[4] to complete the proof. □

2.4 The Failure of the Matrix Generating Function

In the scalar setting of Sect. 1.2, the Laplace transform method is very effective for studying sums of independent (scalar-valued) random variables, because the matrix generating function decomposes. Consider an independent sequence X_k of real random variables. Operating formally, we see that the scalar matrix generating function of the sum satisfies a multiplicative rule:

$$M_{\left(\sum_k X_k\right)}(\theta) = \mathbb{E}\exp\left(\sum_k \theta X_k\right) = \mathbb{E}\prod_k e^{\theta X_k} = \prod_k \mathbb{E} e^{\theta X_k} = \prod_k M_{X_k}(\theta). \tag{2.38}$$

This calculation relies on the fact that the scalar exponential function converts sums to products, a property the matrix exponential does not share, see Sect. 1.4.16. Thus, there is no immediate analog of (2.38) in the matrix setting.

[4]In analysis the infimum or greatest lower bound of a subset S of real numbers is denoted by $\inf(S)$ and is defined to be the biggest real number that is smaller than or equal to every number in S. An important property of the real numbers is that every set of real numbers has an infimum (any bounded nonempty subset of the real numbers has an infimum in the non-extended real numbers). For example, $\inf\{1,2,3\} = 1, \inf\{x \in \mathbb{R}, 0 < x < 1\} = 0$.

Ahlswede and Winter attempts to imitate the multiplicative rule of (2.38) using the following observation. When $\mathbf{X}_1$ and $\mathbf{X}_2$ are independent random matrices,

$$\operatorname{Tr} \mathbf{M}_{\mathbf{X}_1+\mathbf{X}_2}(\theta) \leqslant \mathbb{E} \operatorname{Tr}\left[e^{\theta \mathbf{X}_1} e^{\theta \mathbf{X}_2}\right] = \operatorname{Tr}\left[\left(\mathbb{E} e^{\theta \mathbf{X}_1}\right)\left(\mathbb{E} e^{\theta \mathbf{X}_2}\right)\right] = \operatorname{Tr}\left[\mathbf{M}_{\mathbf{X}_1}(\theta) \cdot \mathbf{M}_{\mathbf{X}_2}(\theta)\right]. \tag{2.39}$$

The first relation is the Golden-Thompson trace inequality (1.71). Unfortunately, we cannot extend the bound (2.39) to include additional matrices. This cold fact may suggest that the Golden-Thompson inequality may not be the natural way to proceed. In Sect. 2.2.4, we have given a full exposition of the Ahlswede-Winter Method. Here, we follow a different path due to [53].

2.5 Subadditivity of the Matrix Cumulant Generating Function

Let us return to the problem of bounding the matrix moment generating function of an independent sum. Although the multiplicative rule (2.38) for the matrix case is a dead end, the scalar cumulant generating function has a related property that can be extended. For an independent sequence X_k of real random variables, the scalar cumulant generating function is additive:

$$\Xi_{\left(\sum_k X_k\right)}(\theta) = \log \mathbb{E} \exp\left(\sum_k \theta X_k\right) = \sum_k \log \mathbb{E} e^{\theta X_k} = \sum_k \Xi_{X_k}(\theta), \tag{2.40}$$

where the second relation follows from (2.38) when we take logarithms.

The key insight of Tropp's approach is that Corollary 1.4.18 offers a completely way to extend the addition rule (2.40) for the scalar setting to the matrix setting. Indeed, this is a remarkable breakthrough. Much better results have been obtained due to this breakthrough. This justifies the parallel development of Tropp's method with the Ahlswede-Winter method of Sect. 2.2.4.

Lemma 2.5.1 (Subadditivity of Matrix Cumulant Generating Functions). *Consider a finite sequence $\{\mathbf{X}_k\}$ of independent, random, Hermitian matrices. Then*

$$\mathbb{E} \operatorname{Tr} \exp\left(\sum_k \theta \mathbf{X}_k\right) \leqslant \operatorname{Tr} \exp\left(\sum_k \log \mathbb{E} e^{\theta \mathbf{X}_k}\right) \text{ for } \theta \in \mathbb{R}. \tag{2.41}$$

Proof. It does not harm to assume $\theta = 1$. Let $\mathbb{E}_k$ denote the expectation, conditioned on $\mathbf{X}_1, \ldots, \mathbf{X}_k$. Abbreviate

$$\mathbf{\Xi}_k := \log\left(\mathbb{E}_{k-1} e^{\mathbf{X}_k}\right) = \log\left(\mathbb{E} e^{\mathbf{X}_k}\right),$$

where the equality holds because the sequence $\{\mathbf{X}_k\}$ is independent.

$$
\begin{aligned}
\mathbb{E}\,\mathrm{Tr}\exp\left(\sum_{k=1}^{n}\mathbf{X}_k\right) &= \mathbb{E}_0\cdots\mathbb{E}_{n-1}\,\mathrm{Tr}\exp\left(\sum_{k=1}^{n-1}\mathbf{X}_k+\mathbf{X}_n\right)\\
&\leqslant \mathbb{E}_0\cdots\mathbb{E}_{n-2}\,\mathrm{Tr}\exp\left(\sum_{k=1}^{n-1}\mathbf{X}_k+\log\left(\mathbb{E}_{n-1}e^{\mathbf{X}_n}\right)\right)\\
&= \mathbb{E}_0\cdots\mathbb{E}_{n-2}\,\mathrm{Tr}\exp\left(\sum_{k=1}^{n-2}\mathbf{X}_k+\mathbf{X}_{n-1}+\mathbf{\Xi}_n\right)\\
&\leqslant \mathbb{E}_0\cdots\mathbb{E}_{n-3}\,\mathrm{Tr}\exp\left(\sum_{k=1}^{n-2}\mathbf{X}_k+\mathbf{\Xi}_{n-1}+\mathbf{\Xi}_n\right)\\
&\cdots\leqslant \mathrm{Tr}\exp\left(\sum_{k=1}^{n}\mathbf{\Xi}_k\right).
\end{aligned}
$$

The first line follows from the tower property of conditional expectation. At each step, $m = 1, 2, \ldots, n$, we use Corollary 1.4.18 with the fixed matrix $\mathbf{H}$ equal to

$$\mathbf{H}_m = \sum_{k=1}^{m-1}\mathbf{X}_k + \sum_{k=m+1}^{n}\mathbf{\Xi}_k.$$

This act is legal because $\mathbf{H}_m$ does not depend on $\mathbf{X}_m$. □

To be in contrast with the additive rule (2.40), we rewrite (2.41) in the form

$$\mathbb{E}\,\mathrm{Tr}\exp\left(\mathbf{\Xi}_{\left(\sum_k \mathbf{X}_k\right)}(\theta)\right) \leqslant \mathrm{Tr}\exp\left(\sum_k \mathbf{\Xi}_{\mathbf{X}_k}(\theta)\right) \text{ for } \theta\in\mathbb{R}$$

by using definition (2.37).

2.6 Tail Bounds for Independent Sums

This section contains abstract tail bounds for the sums of random matrices.

Theorem 2.6.1 (Master Tail Bound for Independent Sums—Tropp [53]). *Consider a finite sequence $\{\mathbf{X}_k\}$ of independent, random, Hermitian matrices. For all $t \in \mathbb{R}$,*

$$\mathbb{P}\left(\lambda_{\max}\left(\sum_k \mathbf{X}_k\right)\geqslant t\right)\leqslant \inf_{\theta>0}\left\{e^{-\theta t}\cdot \mathrm{Tr}\exp\left(\sum_k \log \mathbb{E}e^{\theta\mathbf{X}_k}\right)\right\}. \quad (2.42)$$

Proof. Substitute the subadditivity rule for matrix cumulant generating functions, Eq.2.41, into the Lapalace transform bound, Proposition 2.3.1. □

Now we are in a position to apply the very general inequality of (2.42) to some specific situations. The first corollary adapts Theorem 2.6.1 to the case that arises most often in practice.

Corollary 2.6.2 (Tropp [53]). *Consider a finite sequence* $\{\mathbf{X}_k\}$ *of independent, random, Hermitian matrices with dimension d. Assume that there is a function* $g : (0,\infty) \to [0,\infty]$ *and a sequence of* $\{\mathbf{A}_k\}$ *of fixed Hermitian matrices that satisfy the relations*

$$\mathbb{E}e^{\theta \mathbf{X}_k} \leqslant \mathbb{E}e^{g(\theta)\cdot \mathbf{A}_k} \ \textit{for} \ \theta > 0. \tag{2.43}$$

Define the scale parameter

$$\rho := \lambda_{max}\left(\sum_k \mathbf{A}_k\right).$$

Then, For all $t \in \mathbb{R}$,

$$\mathbb{P}\left(\lambda_{\max}\left(\sum_k \mathbf{X}_k\right) \geqslant t\right) \leqslant d \cdot \inf_{\theta>0}\left\{e^{-\theta t + g(\theta)\cdot\rho}\right\}. \tag{2.44}$$

Proof. The hypothesis (2.44) implies that

$$\log \mathbb{E}e^{\theta \mathbf{X}_k} \leqslant g(\theta)\cdot \mathbf{A}_k \ \text{ for } \theta > 0. \tag{2.45}$$

because of the property (1.73) that the matrix logarithm is operator monotone. Recall the fact (1.70) that the trace exponential is monotone with respect to the semidefinite order. As a result, we can introduce each relation from the sequence (2.45) into the master inequality (2.42). For $\theta > 0$, it follows that

$$\begin{aligned}
\mathbb{P}\left(\lambda_{\max}\left(\sum_k \mathbf{X}_k\right) \geqslant t\right) &\leqslant e^{-\theta t}\cdot \operatorname{Tr}\exp\left(g(\theta)\cdot\sum_k \mathbf{A}_k\right)\\
&\leqslant e^{-\theta t}\cdot d\cdot \lambda_{\max}\left[\exp\left(g(\theta)\cdot\sum_k \mathbf{A}_k\right)\right]\\
&= e^{-\theta t}\cdot d\cdot \exp\left(g(\theta)\cdot\lambda_{\max}\left(\sum_k \mathbf{A}_k\right)\right).
\end{aligned}$$

The second inequality holds because the trace of a positive definite matrix, such as the exponential, is bounded by the dimension d times the maximum eigenvalue. The last line depends on the spectral mapping Theorem 1.4.4 and the fact that the function g is nonnegative. Identify the quantity ρ, and take the infimum over positive θ to reach the conclusion (2.44). □

Let us state another consequence of Theorem 2.6.1. This bound is sometimes more convenient than Corollary 2.6.2, since it combines the matrix generating functions of the random matrices together under a single logarithm.

Corollary 2.6.3. *Consider a sequence* $\{\mathbf{X}_k, k = 1, 2, \ldots, n\}$ *of independent, random, Hermitian matrices with dimension d. For all* $t \in \mathbb{R}$,

$$\mathbb{P}\left(\lambda_{\max}\left(\sum_{k=1}^{n}\mathbf{X}_k\right) \geqslant t\right) \leqslant d \cdot \inf_{\theta>0} \exp\left\{-\theta t + n \cdot \log \lambda_{max}\left(\frac{1}{n}\sum_{k=1}^{n}\mathbb{E}e^{\theta \mathbf{X}_k}\right)\right\}. \tag{2.46}$$

Proof. Recall the fact (1.74) that the matrix logarithm is operator concave. For $\theta > 0$, it follows that

$$\sum_{k=1}^{n}\log \mathbb{E}e^{\theta \mathbf{X}_k} = n \cdot \frac{1}{n}\sum_{k=1}^{n}\log \mathbb{E}e^{\theta \mathbf{X}_k} \leqslant n \cdot \log\left(\frac{1}{n}\sum_{k=1}^{n}\mathbb{E}e^{\theta \mathbf{X}_k}\right).$$

The property (1.70) that the trace exponential is monotone allows us to introduce the latter relation into the master inequality (2.42) to obtain

$$\mathbb{P}\left(\lambda_{\max}\left(\sum_{k=1}^{n}\mathbf{X}_k\right) \geqslant t\right) \leqslant e^{-\theta t} \cdot \operatorname{Tr}\exp\left(n \cdot \log\left(\frac{1}{n}\sum_{k=1}^{n}\mathbb{E}e^{\theta \mathbf{X}_k}\right)\right).$$

To bound the proof, we bound the trace by d times the maximum eigenvalue, and we invoke the spectral mapping Theorem (twice) 1.4.4 to draw the maximum eigenvalue map inside the logarithm. Take the infimum over positive θ to reach (2.46).

□

We can study the minimum eigenvalue of a sum of random Hermitian matrices because

$$\lambda_{\min}(\mathbf{X}) = -\lambda_{\max}(-\mathbf{X}).$$

As a result,

$$\mathbb{P}\left(\lambda_{\min}\left(\sum_{k=1}^{n}\mathbf{X}_k\right) \leqslant t\right) = \mathbb{P}\left(\lambda_{\max}\left(\sum_{k=1}^{n}-\mathbf{X}_k\right) \geqslant -t\right).$$

We can also analyze the maximum singular value of a sum of random *rectangular* matrices by applying the results to the Hermitian dilation (1.81). For a finite sequence $\{\mathbf{Z}_k\}$ of independent, random, rectangular matrices, we have

$$\mathbb{P}\left(\left\|\sum_{k}\mathbf{X}_k\right\| \geqslant t\right) = \mathbb{P}\left(\lambda_{\max}\left(\sum_{k}\varphi(\mathbf{Z}_k)\right) \geqslant t\right)$$

on account of (1.83) and the property that the dilation is real-linear. φ means dilation. This device allows us to extend most of the tail bounds developed in this book to rectangular matrices.

2.6.1 Comparison Between Tropp's Method and Ahlswede–Winter Method

Ahlswede and Winter uses a different approach to bound the matrix moment generating function, which uses the multiplication bound (2.39) for the trace exponential of a sum of two independent, random, Hermitian matrices.

Consider a sequence $\{\mathbf{X}_k, k = 1, 2, \ldots, n\}$ of independent, random, Hermitian matrices with dimension d, and let $\mathbf{Y} = \sum_k \mathbf{X}_k$. The trace inequality (2.39) implies that

$$\begin{aligned}
\operatorname{Tr}\mathbf{M}_{\mathbf{Y}}(\theta) &\leqslant \mathbb{E}\operatorname{Tr}\left[e^{\sum_{k=1}^{n-1}\theta\mathbf{X}_k}e^{\theta\mathbf{X}_n}\right] = \operatorname{Tr}\mathbb{E}\left[e^{\sum_{k=1}^{n-1}\theta\mathbf{X}_k}e^{\theta\mathbf{X}_n}\right] \\
&= \operatorname{Tr}\left[\left(\mathbb{E}e^{\sum_{k=1}^{n-1}\theta\mathbf{X}_k}\right)\left(\mathbb{E}e^{\theta\mathbf{X}_n}\right)\right] \\
&\leqslant \operatorname{Tr}\left(\mathbb{E}e^{\sum_{k=1}^{n-1}\theta\mathbf{X}_k}\right)\cdot\lambda_{\max}\left(\mathbb{E}e^{\theta\mathbf{X}_n}\right).
\end{aligned}$$

These steps are carefully spelled out in previous sections, for example Sect. 2.2.4.

Iterating this procedure leads to the relation

$$\operatorname{Tr}\mathbf{M}_{\mathbf{Y}}(\theta) \leqslant (\operatorname{Tr}\mathbf{I})\prod_k\lambda_{\max}\left(\mathbb{E}e^{\theta\mathbf{X}_k}\right) = d\cdot\exp\left(\sum_k\lambda_{\max}\left(\log\mathbb{E}e^{\theta\mathbf{X}_k}\right)\right). \tag{2.47}$$

This bound (2.47) is the key to the Ahlswede–Winter method. As a consequence, their approach generally leads to tail bounds that depend on a scale parameter involving "the sum of eigenvalues." In contrast, the Tropp's approach is based on the subadditivity of cumulants, Eq. 2.41, which implies that

$$\operatorname{Tr}\mathbf{M}_{\mathbf{Y}}(\theta) \leqslant d\cdot\exp\left(\lambda_{\max}\left(\sum_k\log\mathbb{E}e^{\theta\mathbf{X}_k}\right)\right). \tag{2.48}$$

(2.48) contains a scale parameter that involves the "eigenvalues of a sum."

2.7 Matrix Gaussian Series—Case Study

A matrix Gaussian series stands among the simplest instances of a sum of independent random matrices. We study this fundamental problem to gain insights.

Consider a finite sequence a_k of real numbers and finite sequence $\{\gamma_k\}$ of independent, standard Gaussian variables. We have

$$\mathbb{P}\left(\sum_k \gamma_k a_k \geqslant t\right) \leqslant e^{-t^2/2\sigma^2} \quad \text{where} \quad \sigma^2 := \sum_k a_k^2. \tag{2.49}$$

This result justifies that a Gaussian series with real coefficients satisfies a normal-type tail bound where the variance is controlled by the sum of the sequence coefficients. The relation (2.49) follows easily from the scalar Laplace transform method. See Example 1.2.1 for the derivation of the characteristic function; A Fourier inverse transform of this derived characteristic function will lead to (2.49). So far, our exposition in this section is based on the standard textbook.

The inequality (2.49) can be generalized directly to the noncommutative setting. The matrix Laplace transform method, Proposition 2.3.1, delivers the following result.

Theorem 2.7.1 (Matrix Gaussian and Rademacher Series—Tropp [53]). *Consider a finite sequence $\{\mathbf{A}_k\}$ of fixed Hermitian matrices with dimension d, and let γ_k be a finite sequence of independent standard normal variables. Compute the variance parameter*

$$\sigma^2 := \left\|\sum_k \mathbf{A}_k^2\right\|. \tag{2.50}$$

Then, for all $t > 0$,

$$\mathbb{P}\left(\lambda_{\max}\left(\sum_k \gamma_k \mathbf{A}_k\right) \geqslant t\right) \leqslant d \cdot e^{-t^2/2\sigma^2}. \tag{2.51}$$

In particular,

$$\mathbb{P}\left(\left\|\sum_k \gamma_k \mathbf{A}_k\right\| \geqslant t\right) \leqslant 2d \cdot e^{-t^2/2\sigma^2}. \tag{2.52}$$

The same bounds hold when we replace γ_k by a finite sequence of independent Rademacher random variables.

Observe that the bound (2.51) reduces to the scalar result (2.49) when the dimension $d = 1$. The generalization of (2.50) has been proven by Tropp [53] to be sharp and is also demonstrated that Theorem 2.7.1 cannot be improved without changing its form.

Most of the inequalities in this book have variants that concern the maximum singular value of a sum of *rectangular* random matrices. These extensions follow immediately, as mentioned above, when we apply the Hermitian matrices to the Hermitian dilation of the sums of rectangular matrices. Here is a general version of Theorem 2.7.1.

Corollary 2.7.2 (Rectangular Matrix Gaussian and Radamacher Series—Tropp [53]). *Consider a finite sequence $\{\mathbf{B}_k\}$ of fixed matrices with dimension $d_1 \times d_2$, and let γ_k be a finite sequence of independent standard normal variables. Compute the variance parameter*

$$\sigma^2 := \max\left\{ \left\| \sum_k \mathbf{B}_k \mathbf{B}_k^* \right\|, \left\| \sum_k \mathbf{B}_k^* \mathbf{B}_k \right\| \right\}.$$

Then, for all $t > 0$,

$$\mathbb{P}\left(\left\| \sum_k \gamma_k \mathbf{B}_k \right\| \geqslant t \right) \leqslant (d_1 + d_2) \cdot e^{-t^2/2\sigma^2}.$$

The same bounds hold when we replace γ_k by a finite sequence of independent Rademacher random variables.

To prove Theorem 2.7.1 and Corollary 2.7.2, we need a lemma first.

Lemma 2.7.3 (Rademacher and Gaussian moment generating functions). *Suppose that $\mathbf{A}$ is an Hermitian matrix. Let ε be a Rademacher random variable, and let γ be a standard normal random variable. Then*

$$\mathbb{E}e^{\varepsilon\theta\mathbf{A}} \leqslant e^{\theta^2\mathbf{A}^2/2} \text{ and } \mathbb{E}e^{\gamma\theta\mathbf{A}} = e^{\theta^2\mathbf{A}^2/2} \text{ for } \theta \in \mathbb{R}.$$

Proof. Absorbing θ into $\mathbf{A}$, we may assume $\theta = 1$ in each case. By direct calculation,

$$\mathbb{E}e^{\varepsilon\mathbf{A}} = \cosh(\mathbf{A}) \leqslant e^{\mathbf{A}^2/2},$$

where the second relation is (1.69).

For the Gaussian case, recall that the moments of a standard normal variable satisfy

$$\mathbb{E}\gamma^{2k+1} = 0 \text{ and } \mathbb{E}\gamma^{2k} = \frac{(2k)!}{k!2^k} \text{ for } k = 0, 1, 2, \ldots.$$

Therefore,

$$\mathbb{E}e^{\gamma \mathbf{A}} = \mathbf{I} + \sum_{k=1}^{\infty} \frac{\mathbb{E}\left(\gamma^{2k}\right) \mathbf{A}^{2k}}{(2k)!} = \mathbf{I} + \sum_{k=1}^{\infty} \frac{\left(\mathbf{A}^2/2\right)^k}{k!} = e^{\mathbf{A}^2/2}.$$

The first relation holds since the odd terms in the series vanish. With this lemma, the tail bounds for Hermitian matrix Gaussian and Rademacher series follow easily. □

Proof of Theorem 2.7.1. Let $\{\xi_k\}$ be a finite sequence of independent standard normal variables or independent Rademacher variables. Invoke Lemma 2.7.3 to obtain

$$\mathbb{E}e^{\xi_k \theta \mathbf{A}} \leqslant e^{g(\theta)\cdot \mathbf{A}_k^2} \text{ where } g(\theta) := \theta^2/2 \text{ for } \theta > 0.$$

Recall that

$$\sigma^2 = \left\| \sum_k \mathbf{A}_k^2 \right\| = \lambda_{\max}\left(\sum_k \mathbf{A}_k^2\right).$$

Corollary 2.6.2 gives

$$\mathbb{P}\left(\lambda_{\max}\left(\sum_k \xi_k \mathbf{A}_k\right) \geqslant t\right) \leqslant d \cdot \inf_{\theta>0} \left\{e^{-\theta t + g(\theta)\cdot\sigma^2}\right\} = d \cdot e^{-t^2/2\sigma^2}. \quad (2.53)$$

For the record, the infimum is attained when $\theta = t/\sigma^2$.

To obtain the norm bound (2.52), recall that

$$\|\mathbf{Y}\| = \max\left\{\lambda_{\max}(\mathbf{Y}), -\lambda_{\min}(\mathbf{Y})\right\}.$$

Since standard Gaussian and Rademacher variables are symmetric, the inequality (2.53) implies that

$$\mathbb{P}\left(-\lambda_{\min}\left(\sum_k \xi_k \mathbf{A}_k\right) \geqslant t\right) = \mathbb{P}\left(\lambda_{\max}\left(\sum_k (-\xi_k)\,\mathbf{A}_k\right) \geqslant t\right) \leqslant d \cdot e^{-t^2/2\sigma^2}.$$

Apply the union bound to the estimates for $\lambda_{\max}$ and $-\lambda_{\min}$ to complete the proof. We use the Hermitian dilation of the series. □

Proof of Corollary 2.7.2. Let $\{\xi_k\}$ be a finite sequence of independent standard normal variables or independent Rademacher variables. Consider the sequence $\{\xi_k \varphi(\mathbf{B}_k)\}$ of random Hermitian matrices with dimension $d_1 + d_2$. The spectral identity (1.83) ensures that

$$\left\|\sum_k \xi_k \mathbf{B}_k\right\| = \lambda_{\max}\left(\varphi\left(\sum_k \xi_k \mathbf{B}_k\right)\right) = \lambda_{\max}\left(\sum_k \xi_k \varphi\left(\mathbf{B}_k\right)\right).$$

Theorem 2.7.1 is used. Simply observe that the matrix variance parameter (2.50) satisfies the relation

$$\sigma^2 = \left\|\sum_k \varphi(\mathbf{B}_k)^2\right\| = \left\|\begin{bmatrix} \sum_k \mathbf{B}_k \mathbf{B}_k^* & 0 \\ 0 & \sum_k \mathbf{B}_k^* \mathbf{B}_k \end{bmatrix}\right\|$$
$$= \max\left\{\left\|\sum_k \mathbf{B}_k \mathbf{B}_k^*\right\|, \left\|\sum_k \mathbf{B}_k^* \mathbf{B}_k\right\|\right\}.$$

on account of the identity (1.82). □

2.8 Application: A Gaussian Matrix with Nonuniform Variances

Fix a $d_1 \times d_2$ matrix $\mathbf{B}$ and draw a random $d_1 \times d_2$ matrix $\mathbf{\Gamma}$ whose entries are independent, standard normal variables. Let $\odot$ denote the componentwise (i.e., Schur or Hadamard) product of matrices. Construct the random matrix $\mathbf{B} \odot \mathbf{\Gamma}$, and observe that its (i,j) component is a Gaussian variable with mean zero and variance $|b_{ij}|^2$. We claim that

$$\mathbb{P}\left\{\|\Gamma \odot \mathbf{B}\| \geqslant t\right\} \leqslant (d_1 + d_2) \cdot e^{-t^2/2\sigma^2}. \tag{2.54}$$

The symbols $\mathbf{b}_{i:}$ and $\mathbf{b}_{:j}$ represent the ith row and jth column of the matrix $\mathbf{B}$. An immediate sequence of (2.54) is that the median of the norm satisfies

$$\mathbb{M}\left(\|\Gamma \odot \mathbf{B}\|\right) \leqslant \sigma\sqrt{2\log\left(2\left(d_1 + d_2\right)\right)}. \tag{2.55}$$

These are nonuniform Gaussian matrices where the estimate (2.55) for the median has the correct order. We compare [106, Theorem 1] and [107, Theorem 3.1] although the results are not fully comparable. See Sect. 9.2.2 for extended work.

To establish (2.54), we first decompose the matrix of interest as a Gaussian series:

$$\mathbf{\Gamma} \odot \mathbf{B} = \sum_{ij} \gamma_{ij} \cdot b_{ij} \cdot \mathbf{C}_{ij}.$$

Now, let us determine the variance parameter σ^2. Note that

$$\sum_{ij}(b_{ij}\mathbf{C}_{ij})(b_{ij}\mathbf{C}_{ij})^* = \sum_{i}\left(\sum_{j}|b_{ij}|^2\right)\mathbf{C}_{ii} = \operatorname{diag}\left(\|\mathbf{b}_{1:}\|^2,\ldots,\|\mathbf{b}_{d_1:}\|^2\right).$$

Similarly,

$$\sum_{ij}(b_{ij}\mathbf{C}_{ij})^*(b_{ij}\mathbf{C}_{ij}) = \sum_{j}\left(\sum_{i}|b_{ij}|^2\right)\mathbf{C}_{jj} = \operatorname{diag}\left(\|\mathbf{b}_{:1}\|^2,\ldots,\|\mathbf{b}_{:d_2}\|^2\right).$$

Thus,

$$\begin{aligned}\sigma^2 &= \max\left\{\left\|\operatorname{diag}\left(\|\mathbf{b}_{1:}\|^2,\ldots,\|\mathbf{b}_{d_1:}\|^2\right)\right\|,\left\|\operatorname{diag}\left(\|\mathbf{b}_{:1}\|^2,\ldots,\|\mathbf{b}_{:d_2}\|^2\right)\right\|\right\}\\ &= \max\left\{\max_i\|\mathbf{b}_{i:}\|^2,\max_j\|\mathbf{b}_{:j}\|^2\right\}.\end{aligned}$$

An application of Corollary 2.7.2 gives the tail bound of (2.54).

2.9 Controlling the Expectation

The Hermitian Gaussian series

$$\mathbf{Y} = \sum_{k}\gamma_k\mathbf{A}_k \tag{2.56}$$

is used for many practical applications later in this book since it allows each sensor to be represented by the kth matrix.

Example 2.9.1 (NC-OFDM Radar and Communications). A subcarrier (or tone) has a frequency $f_k, k = 1,\ldots,N$. Typically, $N = 64$ or $N = 128$. A radio sinusoid $e^{j2\pi f_k t}$ is transmitted by the transmitter (cell phone tower or radar). This radio signal passes through the radio environment and "senses" the environment. Each sensor collects some length of data over the sensing time. The data vector $\mathbf{y}_k$ of length 10^6 is stored and processed for only one sensor. In other words, we receive typically $N = 128$ copies of measurements for using one sensor. Of course, we can use more sensors, say $M = 100$.

We can extract the data structure using a covariance matrix that is to be directly estimated from the data. For example, a sample covariance matrix can be used. We call the estimated covariance matrix $\hat{\mathbf{R}}_k, k = 1, 2, .., N$. We may desire to know the impact of N subcarriers on the sensing performance. Equation (2.56) is a natural model for this problem at hand. If we want to investigate the impact of $M = 100$ sensors on the sensing performance (via collaboration from a wireless network), we need a data fusion algorithm. Intuitively, we can simply consider the sum of these extracted covariance matrices (random matrices). So we have a total of $n = MN = 100 \times 128 = 12{,}800$ random matrices at our disposal. Formally, we have

$$\mathbf{Y} = \sum_k \gamma_k \mathbf{A}_k = \sum_{k=1}^{n=128{,}000} \gamma_k \hat{\mathbf{R}}_k.$$

Here, we are interested in the nonasymptotic view in statistics [108]: when the number of observations n is large, we fit large complex sets of data that one needs to deal with huge collections of models at different scales. Throughout the book, we promote this nonasymptotic view by solving practical problems in wireless sensing and communications. This is a problem with "Big Data". In this novel view, one takes the number of observations as it is and try to evaluate the effect of all the influential parameters. Here this parameter is n, the total number of measurements. Within one second, we have a total of $10^6 \times 128 \times 100 \approx 10^{10}$ points of data at our disposal. We need models at different scales to represent the data. □

A remarkable feature of Theorem 2.7.1 is that it always allows us to obtain reasonably accurate estimates for the expected norm of this Hermitian Gaussian series

$$\mathbf{Y} = \sum_k \gamma_k \mathbf{A}_k. \tag{2.57}$$

To establish this point, we first derive the upper and lower bounds for the second moment of $||\mathbf{Y}||$. Note $||\mathbf{Y}||$ is a scalar random variable. Using Theorem 2.7.1 gives

$$\begin{aligned}\mathbb{E}\left(\|\mathbf{Y}\|^2\right) &= \int_0^\infty \mathbb{P}\left(\|\mathbf{Y}\| > \sqrt{t}\right) dt \\ &= 2\sigma^2 \log(2d) + 2d \int_{2\sigma^2 \log(2d)}^\infty e^{-t/2\sigma^2} dt = 2\sigma^2 \log(2ed).\end{aligned}$$

Jensen's inequality furnishes the lower estimate:

$$\mathbb{E}\left(\|\mathbf{Y}\|^2\right) = \mathbb{E}\left(\|\mathbf{Y}^2\|\right) \geqslant \|\mathbb{E}\mathbf{Y}^2\| = \left\|\sum_k \mathbf{A}_k^2\right\| = \sigma^2.$$

The (homogeneous) first and second moments of the norm of a Gaussian series are equivalent up to a universal constant [109, Corollary 3.2], so we have

$$c\sigma \leqslant \mathbb{E}\left(\|\mathbf{Y}\|\right) \leqslant \sigma\sqrt{2\log(2ed)}. \tag{2.58}$$

According to (2.58), the matrix variance parameter σ^2 controls the expected norm $\mathbb{E}\left(\|\mathbf{Y}\|\right)$ up to a factor that depends very weakly on the dimension d. A similar remark goes to the median value $\mathbb{M}\left(\|\mathbf{Y}\|\right)$.

In (2.58), the dimensional dependence is a new feature of probability inequalities in the matrix setting. We cannot remove the factor d from the bound in Theorem 2.7.1.

2.10 Sums of Random Positive Semidefinite Matrices

The classical Chernoff bounds concern the sum of independent, nonnegative, and *uniformly bounded* random variables. In contrast, matrix Chernoff bounds deal with a sum of independent, positive semidefinite, random matrices whose maximum eigenvalues are subject to a uniform bound. For example, the sample covariance matrices satisfy the conditions of independent, positive semidefinite, random matrices. This connection plays a fundamental role when we deal with cognitive sensing in the network setting consisting of a number of sensors. Roughly, each sensor can be modeled by a sample covariance matrix.

The first result parallels with the strongest version of the scalar Chernoff inequality for the proportion of successes in a sequence of independent, (but not identical) Bernoulli trials [7, Excercise 7].

Theorem 2.10.1 (Matrix Chernoff I—Tropp [53]). *Consider a sequence* $\{\mathbf{X}_k : k = 1, \ldots, n\}$ *of independent, random, Hermitian matrices that satisfy*

$$\mathbf{X}_k \geqslant 0 \quad \textit{and} \quad \lambda_{\max}(\mathbf{X}_k) \leqslant 1 \quad \textit{almost surely.}$$

Compute the minimum and maximum eigenvalues of the average expectation,

$$\bar{\mu}_{\min} := \lambda_{\min}\left(\frac{1}{n}\sum_{k=1}^{n}\mathbb{E}\mathbf{X}_k\right) \quad \textit{and} \quad \bar{\mu}_{\max} := \lambda_{\max}\left(\frac{1}{n}\sum_{k=1}^{n}\mathbb{E}\mathbf{X}_k\right).$$

Then

$$\begin{aligned}
&\mathbb{P}\left\{\lambda_{\min}\left(\tfrac{1}{n}\sum_{k=1}^{n}\mathbf{X}_k\right) \leqslant \alpha\right\} \leqslant d\cdot e^{-n\cdot D(\alpha||\bar{\mu}_{\min})} \ \textit{for} \ \ 0 \leqslant \alpha \leqslant \bar{\mu}_{\min},\\
&\mathbb{P}\left\{\lambda_{\max}\left(\tfrac{1}{n}\sum_{k=1}^{n}\mathbf{X}_k\right) \geqslant \alpha\right\} \leqslant d\cdot e^{-n\cdot D(\alpha||\bar{\mu}_{\max})} \ \textit{for} \ \ 0 \leqslant \alpha \leqslant \bar{\mu}_{\max}.
\end{aligned}$$

the binary information divergence

$$D(a||u) = a(\log(a) - \log(u)) + (1-a)(\log(1-a) - \log(1-u))$$

for $a, u \in [0, 1]$.

Tropp [53] found the following weaker version of Theorem 2.10.1 produces excellent results but is simpler to apply. This corollary corresponds with the usual statement of the scalar Chernoff inequalities for sums of nonnegative random variables; see [7, Excercise 8] [110, Sect. 4.1]. Theorem 2.10.1 is a considerable strengthening of the version of Ahlswede-Winter [36, Theorem 19], in which case their result requires the assumption that the summands are identically distributed.

Corollary 2.10.2 (Matrix Chernoff II—Tropp [53]). *Consider a sequence* $\{\mathbf{X}_k : k = 1, \ldots, n\}$ *of independent, random, Hermitian matrices that satisfy*

$$\mathbf{X}_k \geqslant 0 \quad \textit{and} \quad \lambda_{\max}(\mathbf{X}_k) \leqslant R \quad \textit{almost surely.}$$

Compute the minimum and maximum eigenvalues of the average expectation,

$$\mu_{\min} := \lambda_{\min}\left(\sum_{k=1}^{n} \mathbb{E}\mathbf{X}_k\right) \quad \textit{and} \quad \mu_{\max} := \lambda_{\max}\left(\sum_{k=1}^{n} \mathbb{E}\mathbf{X}_k\right).$$

Then

$$\mathbb{P}\left\{\lambda_{\min}\left(\sum_{k=1}^{n} \mathbf{X}_k\right) \leqslant (1-\delta)\,\mu_{\min}\right\} \leqslant d \cdot \left[\frac{e^{-\delta}}{(1-\delta)^{1-\delta}}\right]^{\mu_{\min}/R} \textit{for } \delta \in [0,1],$$
$$\mathbb{P}\left\{\lambda_{\max}\left(\sum_{k=1}^{n} \mathbf{X}_k\right) \geqslant (1+\delta)\,\mu_{\max}\right\} \leqslant d \cdot \left[\frac{e^{\delta}}{(1+\delta)^{1+\delta}}\right]^{\mu_{\max}/R} \textit{for } \delta \geqslant 0.$$

The following standard simplification of Corollary 2.10.2 is useful:

$$\mathbb{P}\left\{\lambda_{\min}\left(\sum_{k=1}^{n} \mathbf{X}_k\right) \leqslant t\mu_{\min}\right\} \leqslant d \cdot e^{-(1-t)^2 \mu_{\min}/2R} \text{ for } t \in [0,1],$$
$$\mathbb{P}\left\{\lambda_{\max}\left(\sum_{k=1}^{n} \mathbf{X}_k\right) \geqslant t\mu_{\max}\right\} \leqslant d \cdot \left[\frac{e}{t}\right]^{t\mu_{\max}/R} \text{ for } t \geqslant e.$$

The minimum eigenvalues has norm-type behavior while the maximum eigenvalues exhibits Poisson-type decay.

Before giving the proofs, we consider applications.

Example 2.10.3 (Rectangular Random Matrix). Matrix Chernoff inequalities are very effective for studying random matrices with independent columns. Consider a rectangular random matrix

$$\mathbf{Z} = \left[\mathbf{z}_1\ \mathbf{z}_2\ \cdots\ \mathbf{z}_n\right]$$

where $\{\mathbf{z}_k\}$ is a family of independent random vector in $\mathbb{C}^m$. The sample covariance matrix is defined as

$$\hat{\mathbf{R}} = \frac{1}{n}\mathbf{Z}\mathbf{Z}^* = \frac{1}{n}\sum_{k=1}^{n} \mathbf{z}_k \mathbf{z}_k^*,$$

which is an estimate of the true covariance matrix $\mathbf{R}$. One is interested in the error $\left\|\hat{\mathbf{R}} - \mathbf{R}\right\|$ as a function of the number of sample vectors, n. The norm of $\mathbf{Z}$ satisfies

$$\|\mathbf{Z}\|^2 = \lambda_{\max}(\mathbf{ZZ}^*) = \lambda_{\max}\left(\sum_{k=1}^{n} \mathbf{z}_k\mathbf{z}_k^*\right). \tag{2.59}$$

Similarly, the minimum singular value s_m of the matrix satisfies

$$s_m(\mathbf{Z})^2 = \lambda_{\min}(\mathbf{ZZ}^*) = \lambda_{\min}\left(\sum_{k=1}^{n} \mathbf{z}_k\mathbf{z}_k^*\right).$$

In each case, the summands are stochastically independent and positive semidefinite (rank 1) matrices, so the matrix Chernoff bounds apply. □

Corollary 2.10.2 gives accurate estimates for the expectation of the maximum eigenvalue:

$$\mu_{\max} \leqslant \mathbb{E}\lambda_{\max}\left(\sum_{k=1}^{n} \mathbf{X}_k\right) \leqslant C \cdot \max\{\mu_{\max}, R\log d\}. \tag{2.60}$$

The lower bound is Jensen's inequality; the upper bound is from a standard calculation. The dimensional dependence vanishes, when the mean $\mu_{\max}$ is sufficiently large in comparison with the upper bound R! The a prior knowledge of knowing R accurately in $\lambda_{\max}(\mathbf{X}_k) \leqslant R$ converts into the tighter bound in (2.60).

Proof of Theorem 2.10.1. We start with a semidefinite bound for the matrix moment generating function of a random positive semidefinite contraction.

Lemma 2.10.4 (Chernoff moment generating function). *Suppose that* $\mathbf{X}$ *is a random positive semidefinite matrix that satisfies* $\lambda_{\max}(\mathbf{X}_k) \leqslant 1$. *Then*

$$\mathbb{E}\left(e^{\theta\mathbf{X}}\right) \leqslant \mathbf{I} + \left(e^{\theta} - 1\right)(\mathbb{E}\mathbf{X}) \quad \text{for} \quad \theta \in \mathbb{R}.$$

The proof of Lemma 2.10.4 parallels the classical argument; the matrix adaptation is due to Ashlwede and Winter [36], which is followed in the proof of Theorem 2.2.15.

Proof of Lemma 2.10.4. Consider the function

$$f(x) = e^{\theta x}.$$

Since f is convex, its graph has below the chord connecting two points. In particular,

$$f(x) \leqslant f(0) + [f(1) - f(0)] \cdot x \quad \text{for} \quad x \in [0, 1].$$

More explicitly,

$$e^{\theta x} \leqslant 1 + \left(e^{\theta} - 1\right) \cdot x \ \text{ for } \ x \in [0,1].$$

The eigenvalues of $\mathbf{X}$ lie in the interval of $[0,1]$, so the transfer rule (1.61) implies that

$$e^{\theta \mathbf{X}} \leqslant \mathbf{I} + \left(e^{\theta} - 1\right)\mathbf{X}.$$

Expectation respects the semidefinite order, so

$$\mathbb{E}e^{\theta \mathbf{X}} \leqslant \mathbf{I} + \left(e^{\theta} - 1\right)(\mathbb{E}\mathbf{X}).$$

This is the advertised result of Lemma 2.10.4. □

Proof Theorem 2.10.1, Upper Bound. The Chernoff moment generating function, Lemma 2.10.4, states that

$$\mathbb{E}e^{\theta \mathbf{X}_k} \leqslant \mathbf{I} + g(\theta)(\mathbb{E}\mathbf{X}_k) \ \text{ where } g(\theta) = \left(e^{\theta} - 1\right) \text{ for } \theta > 0.$$

As a result, Corollary 2.6.3 implies that

$$\begin{aligned}
\mathbb{P}\left\{\lambda_{\max}\left(\sum_{k=1}^{n}\mathbf{X}_k\right) \geqslant t\right\} &\leqslant d \cdot \exp\left(-\theta t + n \cdot \log \cdot \lambda_{\max}\left(\frac{1}{n}\sum_{k=1}^{n}(\mathbf{I} + g(\theta)(\mathbb{E}\mathbf{X}_k)\ )\right)\right) \\
&= d \cdot \exp\left(-\theta t + n \cdot \log \cdot \lambda_{\max}\left(\mathbf{I} + g(\theta) \cdot \frac{1}{n}\sum_{k=1}^{n}((\mathbb{E}\mathbf{X}_k)\ )\right)\right) \\
&= d \cdot \exp\left(-\theta t + n \cdot \log \cdot \left(1 + g(\theta) \cdot \bar{\mu}_{\max}\right)\right).
\end{aligned} \tag{2.61}$$

The third line follows from the spectral mapping Theorem 1.4.13 and the definition of $\bar{\mu}_{\max}$. Make the change of variable $t \mapsto n\alpha$. The right-hand side is smallest when

$$\theta = \log\left(\alpha/\left(1-\alpha\right)\right) - \log\left(\bar{\mu}_{\max}/\left(1-\bar{\mu}_{\max}\right)\right).$$

After substituting these quantiles into (2.61), we obtain the information divergence upper bound. □

Proof Corollary 2.10.2, Upper Bound. Assume that the summands satisfy the uniform eigenvalue bound with $R = 1$; the general result follows by re-scaling. The shortest route to the weaker Chernoff bound starts at (2.61). The numerical inequality $\log(1+x) \leq x$, valid for $x > -1$, implies that

$$\mathbb{P}\left\{\lambda_{\max}\left(\sum_{k=1}^{n}\mathbf{X}_k\right) \geqslant t\right\} \leqslant d \cdot \exp\left(-\theta t + g(\theta) \cdot n\bar{\mu}_{\max}\right) = d \cdot \exp\left(-\theta t + g(\theta) \cdot \mu_{\max}\right).$$

Make the change of variable $t \mapsto (1+\delta)\,\mu_{\max}$, and select the parameter $\theta = \log(1+\delta)$. Simplify the resulting tail bound to complete the proof. □

The lower bounds follow from a closely related arguments.

Proof Theorem 2.10.1, Lower Bound. Our starting point is Corollary 2.6.3, considering the sequence $\{-\mathbf{X}_k\}$. In this case, the Chernoff moment generating function, Lemma 2.10.4, states that

$$\mathbb{E}e^{\theta(-\mathbf{X}_k)} = \mathbb{E}e^{(-\theta)\mathbf{X}_k} \leqslant \mathbf{I} - g(\theta) \cdot (\mathbb{E}\mathbf{X}_k) \quad \text{where} \quad g(\theta) = 1 - e^{-\theta} \quad \text{for } \theta > 0.$$

Since $\lambda_{\min}(-\mathbf{A}) = -\lambda_{\max}(\mathbf{A})$, we can again use Corollary 2.6.3 as follows.

$$\begin{aligned}\mathbb{P}\left\{\lambda_{\min}\left(\sum_{k=1}^{n}\mathbf{X}_k\right) \leqslant t\right\} &= \mathbb{P}\left\{\lambda_{\max}\left(\sum_{k=1}^{n}(-\mathbf{X}_k)\right) \geqslant -t\right\} \\ &\leqslant d \cdot \exp\left(\theta t + n \cdot \log \lambda_{\max}\left(\frac{1}{n}\sum_{k=1}^{n}(\mathbf{I} - g(\theta)\cdot \mathbb{E}\mathbf{X}_k)\right)\right) \\ &= d \cdot \exp\left(\theta t + n \cdot \log\left(1 - g(\theta)\cdot \lambda_{\min}\left(\frac{1}{n}\sum_{k=1}^{n}\mathbb{E}\mathbf{X}_k\right)\right)\right) \\ &= d \cdot \exp\left(\theta t + n \cdot \log\left(1 - g(\theta)\cdot \bar{\mu}_{\min}\right)\right).\end{aligned} \tag{2.62}$$

Make the substitution $t \mapsto n\alpha$. The right-hand side is minimum when

$$\theta = \log\left(\bar{\mu}_{\min}/\left(1 - \bar{\mu}_{\min}\right)\right) - \log\left(\alpha/\left(1 - \alpha\right)\right).$$

These steps result in the information divergence lower bound. □

Proof Corollary 2.10.2, Lower Bound. As before, assume that the unform bound $R = 1$. We obtain the weaker lower bound as a consequence of (2.62). The numerical inequality $\log(1 + x) \leq x$, is valid for $x > -1$, so we have

$$\mathbb{P}\left\{\lambda_{\min}\left(\sum_{k=1}^{n}\mathbf{X}_k\right) \leqslant t\right\} \leqslant d \cdot \exp\left(\theta t - g(\theta)\cdot n\bar{\mu}_{\min}\right) = d \cdot \exp\left(\theta t - g(\theta)\cdot \mu_{\min}\right).$$

Make the substitution $t \mapsto (1 - \delta)\,\mu_{\min}$, and select the parameter $\theta = -\log(1 - \delta)$ to complete the proof. □

2.11 Matrix Bennett and Bernstein Inequalities

In the scalar setting, Bennett and Bernstein inequalities deal with a sum of independent, zero-mean random variables that are either bounded or subexponential. In the matrix setting, the analogous results concern a sum of *zero-mean random matrices*. Recall that the classical Chernoff bounds concern the sum of independent, nonnegative, and uniformly bounded random variables while, matrix Chernoff bounds deal with a sum of independent, *positive semidefinite*, random matrices whose maximum eigenvalues are subject to a uniform bound. Let us consider a motivating example first.

Example 2.11.1 (Signal plus Noise Model). For example, the sample covariance matrices of Gaussian noise, $\hat{\mathbf{R}}_{ww}$, satisfy the conditions of independent, zero-mean, random matrices. Formally

$$\hat{\mathbf{R}}_{yy} = \hat{\mathbf{R}}_{xx} + \hat{\mathbf{R}}_{ww},$$

$\hat{\mathbf{R}}_{yy}$ represent the sample covariance matrix of the received signal plus noise and $\hat{\mathbf{R}}_{xx}$ of the signal. Apparently, $\hat{\mathbf{R}}_{ww}$, is a zero-mean random matrix. All these matrices are independent, nonnegative, random matrices. □

Our first result considers the case where the maximum eigenvalue of each summand satisfies a uniform bound. Recall from Example 2.10.3 that the norm of a rectangular random matrix $\mathbf{Z}$ satisfies

$$\|\mathbf{Z}\|^2 = \lambda_{\max}\left(\mathbf{Z}\mathbf{Z}^*\right) = \lambda_{\max}\left(\sum_{k=1}^{n} \mathbf{z}_k \mathbf{z}_k^*\right). \tag{2.63}$$

Physically, we can call the norm as the power.

Example 2.11.2 (Transmitters with bounded power). Consider a practical application. Assume each transmitter is modeled as the random matrix $\{\mathbf{Z}_k\}, k = 1, \ldots, n$. We have the a prior knowledge that its transmission is bounded in some manner. A model is to consider

$$\mathbb{E}\mathbf{Z}_k = \mathbf{M}; \ \ \lambda_{\max}\left(\mathbf{Z}_k\right) \leqslant R_1, \ \ k = 1, 2, \ldots, n.$$

After the multi-path channel propagation with fading, the constraints become

$$\mathbb{E}\mathbf{X}_k = \mathbf{N}; \ \ \lambda_{\max}\left(\mathbf{X}_k\right) \leqslant R_2, \ \ k = 1, 2, \ldots, n.$$

Without loss of generality, we can always considered the centered matrix-valued random variable

$$\mathbb{E}\mathbf{X}_k = \mathbf{0}; \ \ \lambda_{\max}\left(\mathbf{X}_k\right) \leqslant R, \ \ k = 1, 2, \ldots, n.$$

When a number of transmitters, say n, are emitting at the same time, the total received signal is described by

$$\mathbf{Y} = \mathbf{X}_1 + \cdots, \mathbf{X}_n = \sum_{k=1}^{n} \mathbf{X}_k.$$

□

Theorem 2.11.3 (Matrix Bernstein:Bounded Case—Theorem 6.1 of Tropp [53]). *Consider a finite sequence $\{\mathbf{X}_k\}$ of independent, random, Hermitian matrices with dimension d. Assume that*

$$\mathbb{E}\mathbf{X}_k = \mathbf{0}; \lambda_{\max}(\mathbf{X}_k) \leqslant R, \quad \textit{almost surely}.$$

Compute the norm of the total variance,

$$\sigma^2 := \left\| \sum_{k=1}^{n} \mathbb{E}\left(\mathbf{X}_k^2\right) \right\|.$$

Then the following chain of inequalities holds for all $t \geq 0$.

$$\begin{aligned} \mathbb{P}\left\{ \lambda_{\max}\left(\sum_{k=1}^{n} \mathbf{X}_k \right) \geqslant t \right\} &\leqslant d \cdot \exp\left(-\tfrac{\sigma^2}{R^2} \cdot h\left(\tfrac{Rt}{\sigma^2} \right) \right) && (i) \\ &\leqslant d \cdot \exp\left(-\tfrac{t^2/2}{\sigma^2 + Rt/3} \right) && (ii) \\ &\leqslant \begin{cases} d \cdot \exp\left(-3t^2/8\sigma^2\right) & \textit{for } t \leqslant \sigma^2/R; \\ d \cdot \exp\left(-3t/8R\right) & \textit{for } t \geqslant \sigma^2/R. \end{cases} && (iii) \end{aligned} \qquad (2.64)$$

The function $h(x) := (1+x)\log(1+x) - x \;\; for \;\; x \geqslant 0$.

Theorem 2.11.4 (Matrix Bernstein: Subexponential Case—Theorem 6.2 of Tropp [53]). *Consider a finite sequence $\{\mathbf{X}_k\}$ of independent, random, Hermitian matrices with dimension d. Assume that*

$$\mathbb{E}\mathbf{X}_k = \mathbf{0}; \mathbb{E}\left(\mathbf{X}_k^p\right) \leqslant \frac{p!}{2!} \cdot R^{p-2}\mathbf{A}_k^2, \textit{for } p = 2,3,4,\ldots.$$

Compute the variance parameter

$$\sigma^2 := \left\| \sum_{k=1}^{n} \mathbf{A}_k^2 \right\|.$$

Then the following chain of inequalities holds for all $t \geq 0$.

$$\begin{aligned} \mathbb{P}\left\{ \lambda_{\max}\left(\sum_{k=1}^{n} \mathbf{X}_k \right) \geq t \right\} &\leqslant d \cdot \exp\left(-\tfrac{t^2/2}{\sigma^2 + Rt} \right) \\ &\leqslant \begin{cases} d \cdot \exp\left(-t^2/4\sigma^2\right) & \textit{for } t \leqslant \sigma^2/R; \\ d \cdot \exp\left(-t/4R\right) & \textit{for } t \geqslant \sigma^2/R. \end{cases} \end{aligned}$$

2.12 Minimax Matrix Laplace Method

This section, taking material from [43, 49, 53, 111], combines the matrix Laplace transform method of Sect. 2.3 with the Courant-Fischer characterization of eigenvalues (Theorem 1.4.22) to obtain nontrivial bounds on the interior eigenvalues of a sum of random Hermitian matrices. We will use this approach for estimates of the covariance matrix.

2.13 Tail Bounds for All Eigenvalues of a Sum of Random Matrices

In this section, closely following Tropp [111], we develop a generic bound on the tail probabilities of eigenvalues of sums of independent, random, Hermitian matrices. We establish this bound by supplementing the matrix Laplace transform methodology of Tropp [53], that is treated before in Sect. 2.3, with Theorem 1.4.22 and a new result, due to Lieb and Steiringer [112], on the concavity of a certain trace function on the cone of positive-definite matrices.

Theorem 1.4.22 allows us to relate the behavior of the kth eigenvalue of a matrix to the behavior of the largest eigenvalue of an appropriate compression of the matrix.

Theorem 2.13.1 (Tropp [111]). *Let* $\mathbf{X}$ *be a random, Hermitian matrix with dimension* n*, and let* $k \leq n$ *be an integer. Then, for all* $t \in \mathbb{R}$*,*

$$\mathbb{P}\left(\lambda_k\left(\mathbf{X}\right) \geqslant t\right) \leqslant \inf_{\theta>0} \min_{\mathbf{V} \in \mathbb{V}_{n-k+1}^{n}} \left\{e^{-\theta t} \cdot \mathbb{E}\operatorname{Tr} e^{\theta \mathbf{V}^{*}\mathbf{X}\mathbf{V}}\right\}. \tag{2.65}$$

Proof. Let θ be a fixed positive number. Then

$$\begin{aligned}\mathbb{P}\left(\lambda_k\left(\mathbf{X}\right) \geqslant t\right) = \mathbb{P}\left(\lambda_k\left(\theta\mathbf{X}\right) \geqslant \theta t\right) &= \mathbb{P}\left(e^{\lambda_k(\theta\mathbf{X})} \geqslant e^{\theta t}\right)\\ &\leqslant e^{-\theta t} \cdot \mathbb{E} e^{\lambda_k(\theta\mathbf{X})}\\ &= e^{-\theta t} \cdot \mathbb{E}\exp\left\{\min_{\mathbf{V} \in \mathbb{V}_{n-k+1}^{n}} \lambda_{\max}\left(\theta\mathbf{V}^{*}\mathbf{X}\mathbf{V}\right)\right\}.\end{aligned}$$

The first identify follows from the positive homogeneity of eigenvalue maps and the second from the monotonicity of the scalar exponential function. The final two steps are Markov's inequality and (1.89).

Let us bound the expectation. Interchange the order of the exponential and the minimum, due to the monotonicity of the scalar exponential function; then apply the spectral mapping Theorem 1.4.4 to see that

$$\begin{aligned}\mathbb{E}\exp\left\{\min_{\mathbf{V}\in\mathbb{V}_{n-k+1}^n}\lambda_{\max}\left(\theta\mathbf{V}^*\mathbf{X}\mathbf{V}\right)\right\} &= \mathbb{E}\min_{\mathbf{V}\in\mathbb{V}_{n-k+1}^n}\lambda_{\max}\left(\exp\left(\theta\mathbf{V}^*\mathbf{X}\mathbf{V}\right)\right)\\ &\leqslant \min_{\mathbf{V}\in\mathbb{V}_{n-k+1}^n}\mathbb{E}\lambda_{\max}\left(\exp\left(\theta\mathbf{V}^*\mathbf{X}\mathbf{V}\right)\right)\\ &\leqslant \min_{\mathbf{V}\in\mathbb{V}_{n-k+1}^n}\mathbb{E}\,\mathrm{Tr}\left(\exp\left(\theta\mathbf{V}^*\mathbf{X}\mathbf{V}\right)\right).\end{aligned}$$

The first step uses Jensen's inequality. The second inequality follows because the exponential of a Hermitian matrix is always positive definite—see Sect. 1.4.16, so its largest eigenvalue is smaller than its trace. The trace functional is linear, which is very critical. The expectation is also linear. Thus we can exchange the order of the expectation and the trace: trace and expectation commute—see (1.87).

Combine these observations and take the infimum over all positive θ to complete the argument. □

Now let apply Theorem 2.13.1 to the case that $\mathbf{X}$ can be expressed as a sum of independent, Hermitian, random matrices. In this case, we develop the right-hand side of the Laplace transform bound (2.65) by using the following result.

Theorem 2.13.2 (Tropp [111]). *Consider a finite sequence $\{\mathbf{X}_i\}$ of independent, Hermitian, random matrices with dimension n and a sequence $\{\mathbf{A}_i\}$ of* ***fixed*** *Hermitian matrices with dimension n that satisfy the relations*

$$\mathbb{E}\left(e^{\mathbf{X}_i}\right) \leqslant e^{\mathbf{A}_i}. \tag{2.66}$$

Let $\mathbf{V} \in \mathbb{V}_k^n$ be an isometric embedding of $\mathbb{C}^k$ into $\mathbb{C}^n$ for some $k \leq n$. Then

$$\mathbb{E}\,\mathrm{Tr}\exp\left\{\sum_i \mathbf{V}^*\mathbf{X}_i\mathbf{V}\right\} \leqslant \mathrm{Tr}\exp\left\{\sum_i \mathbf{V}^*\mathbf{A}_i\mathbf{V}\right\}. \tag{2.67}$$

In particular,

$$\mathbb{E}\,\mathrm{Tr}\exp\left\{\sum_i \mathbf{X}_i\right\} \leqslant \mathrm{Tr}\exp\left\{\sum_i \mathbf{A}_i\right\}. \tag{2.68}$$

Theorem 2.13.2 is an extension of Lemma 2.5.1, which establish the result of (2.68). The proof depends on a recent result of [112], which extends Lieb's earlier classical result [50, Theorem 6]. Here $\mathbb{M}_{\mathrm{H}}^n$ represents the set of Hermitian matrices of $n \times n$.

Proposition 2.13.3 (Lieb-Seiringer 2005). *Let $\mathbf{H}$ be a Hermitian matrix with dimension k. Let $\mathbf{V} \in \mathbb{V}_k^n$ be an isometric embedding of $\mathbb{C}^k$ into $\mathbb{C}^n$ for some $k \leq n$. Then the function*

$$\mathbf{A} \mapsto \mathrm{Tr}\exp\left\{\mathbf{H} + \mathbf{V}^*\left(\log \mathbf{A}\right)\mathbf{V}\right\}$$

is concave on the cone of positive-definite matrices in $\mathbb{M}_H^n$.

Proof of Theorem 2.13.2. First, combining the given condition (2.66) with the operator monotonicity of the matrix logarithm gives the following for each k:

$$\log \mathbb{E}e^{\mathbf{X}_k} \leqslant \mathbf{A}_k. \tag{2.69}$$

Let $\mathbb{E}_k$ denote the expectation conditioned on the first k summands, $\mathbf{X}_1$ through $\mathbf{X}_k$. Then

$$\begin{aligned}
\mathbb{E}\,\mathrm{Tr}\exp\left\{\sum_{i\leqslant j}\mathbf{V}^*\mathbf{X}_i\mathbf{V}\right\} &= \mathbb{E}\mathbb{E}_1\cdots\mathbb{E}_{j-1}\,\mathrm{Tr}\exp\left\{\sum_{i\leqslant j-1}\mathbf{V}^*\mathbf{X}_i\mathbf{V}+\mathbf{V}^*\left(\log e^{\mathbf{X}_j}\right)\mathbf{V}\right\}\\
&\leqslant \mathbb{E}\mathbb{E}_1\cdots\mathbb{E}_{j-2}\,\mathrm{Tr}\exp\left\{\sum_{i\leqslant j-1}\mathbf{V}^*\mathbf{X}_i\mathbf{V}+\mathbf{V}^*\left(\log \mathbb{E}e^{\mathbf{X}_j}\right)\mathbf{V}\right\}\\
&\leqslant \mathbb{E}\mathbb{E}_1\cdots\mathbb{E}_{j-2}\,\mathrm{Tr}\exp\left\{\sum_{i\leqslant j-1}\mathbf{V}^*\mathbf{X}_i\mathbf{V}+\mathbf{V}^*\left(\log e^{\mathbf{A}_j}\right)\mathbf{V}\right\}\\
&= \mathbb{E}\mathbb{E}_1\cdots\mathbb{E}_{j-2}\,\mathrm{Tr}\exp\left\{\sum_{i\leqslant j-1}\mathbf{V}^*\mathbf{X}_i\mathbf{V}+\mathbf{V}^*\mathbf{A}_j\mathbf{V}\right\}.
\end{aligned}$$

The first step follows from Proposition 2.13.3 and Jensen's inequality, and the second depends on (2.69) and the monotonicity of the trace exponential. Iterate this argument to complete the proof. The main result follows from combining Theorems 2.13.1 and 2.13.2. □

Theorem 2.13.4 (Minimax Laplace Transform). *Consider a finite sequence $\{\mathbf{X}_i\}$ of independent, random, Hermitian matrices with dimension n, and let $k \le n$ be an integer.*

1. *Let $\{\mathbf{A}_i\}$ be a sequence of Hermitian matrices that satisfy the semidefinite relations*

$$\mathbb{E}\left(e^{\theta\mathbf{X}_i}\right) \leqslant e^{g(\theta)\mathbf{A}_i}$$

where $g:(0,\infty)\to[0,\infty)$. Then, for all $t\in\mathbb{R}$,

$$\mathbb{P}\left(\lambda_k\left(\sum_i \mathbf{X}_i\right)\geqslant t\right)\leqslant \inf_{\theta>0}\min_{\mathbf{V}\in\mathbb{V}^n_{n-k+1}}\left[e^{-\theta t}\cdot\mathrm{Tr}\exp\left\{g(\theta)\sum_i\mathbf{V}^*\mathbf{A}_i\mathbf{V}\right\}\right].$$

2. *$\mathbf{A}_i:\mathbb{V}^n_{n-k+1}\to\mathbb{M}^n_H$ be a sequence of functions that satisfy the semidefinite relations*

$$\mathbb{E}\left(e^{\theta\mathbf{V}^*\mathbf{X}_i\mathbf{V}}\right)\leqslant e^{g(\theta)\mathbf{A}_i(\mathbf{V})}$$

for all $\mathbf{V}\in\mathbb{V}^n_{n-k+1}$ where $g:(0,\infty)\to[0,\infty)$. Then, for all $t\in\mathbb{R}$,

$$\mathbb{P}\left(\lambda_k\left(\sum_i \mathbf{X}_i\right) \geqslant t\right) \leqslant \inf_{\theta>0} \min_{\mathbf{V}\in\mathbb{V}_{n-k+1}^n} \left[e^{-\theta t}\cdot \operatorname{Tr}\exp\left\{g\left(\theta\right)\sum_i \mathbf{A}_i\left(\mathbf{V}\right)\right\}\right].$$

The first bound in Theorem 2.13.4 requires less detailed information on how compression affects the summands but correspondingly does not yield as sharp results as the second.

2.14 Chernoff Bounds for Interior Eigenvalues

Classical Chernoff bounds in Sect. 1.1.4 establish that the tails of a sum of independent, nonnegative, random variables decay subexponentially. Tropp [53] develops Chernoff bounds for the maximum and minimum eigenvalues of a *sum of independent, positive-semidefinite matrices*. In particular, sample covariance matrices are positive-semidefinite and the sums of independent, sample covariance matrices are ubiquitous. Following Gittens and Tropp [111], we extend this analysis to study the interior eigenvalues. The analogy with the scalar-valued random variables in Sect. 1.1.4 is aimed at, in this development. At this point, it is insightful if the audience reviews the materials in Sects. 1.1.4 and 1.3.

Intuitively, how concentrated the summands will determine the eigenvalues tail bounds; in other words, if we align the ranges of some operators, the maximum eigenvalue of a sum of these operators varies probably more than that of a sum of operators whose ranges are orthogonal. We are interested in a finite sequence of random summands $\{\mathbf{X}_i\}$. This sequence will concentrate in a given subspace. To measure how much this sequence concentrate, we define a function ψ : $\cup_{1\leqslant k\leqslant n}\mathbb{V}_k^n \to \mathbb{R}$ that has the property

$$\max_i \lambda_{\max}\left(\mathbf{V}^*\mathbf{X}_i\mathbf{V}\right) \leqslant \psi\left(\mathbf{V}\right) \text{ almost surely for each } \mathbf{V} \in \cup_{1\leqslant k\leqslant n}\mathbb{V}_k^n. \quad (2.70)$$

Theorem 2.14.1 (Eigenvalue Chernoff Bounds [111]). *Consider a finite sequence $\{\mathbf{X}_i\}$ of independent, random, positive-semidefinite matrices with dimension n. Given an integer $k \leq n$, define*

$$\mu_k = \lambda_k\left(\sum_i \mathbb{E}\mathbf{X}_i\right),$$

and let $\mathbf{V}_+ \in \mathbb{V}_{n-k+1}^n$ and $\mathbf{V}_- \in \mathbb{V}_k^n$ be isometric embeddings that satisfy

$$\mu_k = \lambda_{\max}\left(\sum_i \mathbf{V}_+^*\mathbb{E}\mathbf{X}_i\mathbf{V}_+\right) = \lambda_{\min}\left(\sum_i \mathbf{V}_-^*\mathbb{E}\mathbf{X}_i\mathbf{V}_-\right).$$

Then

$$\mathbb{P}\left(\lambda_k\left(\sum_i \mathbf{X}_i\right) \geqslant (1+\delta)\,\mu_k\right) \leqslant (n-k+1)\cdot\left[\frac{e^{\delta}}{(1+\delta)^{1+\delta}}\right]^{\mu_k/\psi(\mathbf{V}_+)} \quad \textit{for } \delta>0, \textit{ and}$$

$$\mathbb{P}\left(\lambda_k\left(\sum_i \mathbf{X}_i\right) \leqslant (1-\delta)\,\mu_k\right) \leqslant k\cdot\left[\frac{e^{-\delta}}{(1-\delta)^{1-\delta}}\right]^{\mu_k/\psi(\mathbf{V}_-)} \quad \textit{for } \delta\in[0,1],$$

where ψ is a function that satisfies (2.70).

Practically, if it is difficult to estimate $\psi(\mathbf{V}_+)$ and $\psi(\mathbf{V}_-)$, we can use the weaker estimates

$$\psi(\mathbf{V}_+) \leqslant \max_{\mathbf{V}\in\mathbb{V}^n_{n-k+1}} \max_i \left\|\mathbf{V}^*\mathbf{X}_i\mathbf{V}\right\| = \max_i \left\|\mathbf{X}_i\right\|$$

$$\psi(\mathbf{V}_-) \leqslant \max_{\mathbf{V}\in\mathbb{V}^n_{k}} \max_i \left\|\mathbf{V}^*\mathbf{X}_i\mathbf{V}\right\| = \max_i \left\|\mathbf{X}_i\right\|.$$

The following lemma is due to Ahlswede and Winter [36]; see also [53, Lemma 5.8].

Lemma 2.14.2. *Suppose that* $\mathbf{X}$ *is a random positive-semidefinite matrix that satisfies* $\lambda_{\max}(\mathbf{X}) \leqslant 1$. *Then*

$$\mathbb{E}e^{\theta\mathbf{X}} \leqslant \exp\left(\left(e^{\theta}-1\right)(\mathbb{E}\mathbf{X})\right) \textit{ for } \theta\in\mathbb{R}.$$

Proof of Theorem 2.14.1, upper bound. Without loss of generality, we consider the case $\psi(\mathbf{V}_+)=1$; the general case follows due to homogeneity. Define

$$\mathbf{A}_i(\mathbf{V}_+) = \mathbf{V}_+^*\mathbb{E}\mathbf{X}_i\mathbf{V}_+ \text{ and } g(\theta) = e^{\theta}-1.$$

Using Theorem 2.13.4 and Lemma 2.14.2 gives

$$\mathbb{P}\left(\lambda_k\left(\sum_i \mathbf{X}_i\right) \geqslant (1+\delta)\,\mu_k\right) \leqslant \inf_{\theta>0} e^{-\theta(1+\delta)\mu_k}\cdot \operatorname{Tr}\exp\left\{g(\theta)\sum_i \mathbf{V}_+^*\mathbb{E}\mathbf{X}_i\mathbf{V}_+\right\}.$$

The trace can be bounded by the maximum eigenvalue (since the maximum eigenvalue is nonnegative), by taking into account the reduced dimension of the summands:

$$\begin{aligned}\operatorname{Tr}\exp\left\{g(\theta)\sum_i \mathbf{V}_+^*\mathbb{E}\mathbf{X}_i\mathbf{V}_+\right\} &\leqslant (n-k+1)\cdot\lambda_{\max}\left(\exp\left\{g(\theta)\sum_i \mathbf{V}_+^*\mathbb{E}\mathbf{X}_i\mathbf{V}_+\right\}\right)\\ &= (n-k+1)\cdot\exp\left\{g(\theta)\cdot\lambda_{\max}\left(\sum_i \mathbf{V}_+^*\mathbb{E}\mathbf{X}_i\mathbf{V}_+\right)\right\}.\end{aligned}$$

The equality follows from the spectral mapping theorem (Theorem 1.4.4 at Page 34). We identify the quantity μ_k; then combine the last two inequalities to give

$$\mathbb{P}\left(\lambda_k\left(\sum_i \mathbf{X}_i\right) \geqslant (1+\delta)\,\mu_k\right) \leqslant (n-k+1)\cdot \inf_{\theta>0} e^{[g(\theta)-\theta(1+\delta)]\mu_k}.$$

By choosing $\theta = \log(1+\delta)$, the right-hand side is minimized (by taking care of the infimum), which gives the desired upper tail bound. □

Proof of Theorem 2.14.1, lower bound. The proof of lower bound is very similar to that of upper bound above. As above, consider only $\psi(\mathbf{V}_-) = 1$. It follow from (1.91) (Page 47) that

$$\mathbb{P}\left(\lambda_k\left(\sum_i \mathbf{X}_i\right) \leqslant (1-\delta)\,\mu_k\right) = \mathbb{P}\left(\lambda_{n-k+1}\left(\sum_i -\mathbf{X}_i\right) \geqslant -(1-\delta)\,\mu_k\right). \tag{2.71}$$

Applying Lemma 2.14.2, we find that, for $\theta > 0$,

$$\begin{aligned}\mathbb{E}e^{\theta\left(-\mathbf{V}_{-*}\mathbf{X}_i\mathbf{V}_-\right)} = \mathbb{E}e^{(-\theta)\mathbf{V}_-^*\mathbf{X}_i\mathbf{V}_-} &\leqslant \exp\left(g(\theta)\cdot\left(\mathbb{E}\left[-\mathbf{V}_-^*\mathbf{X}_i\mathbf{V}_-\right]\right)\right)\\ &= \exp\left(g(\theta)\cdot\left(\mathbf{V}_-^*\left(-\mathbb{E}\mathbf{X}_i\right)\mathbf{V}_-\right)\right),\end{aligned}$$

where $g(\theta) = 1-e^{\theta}$. The last equality follows from the linearity of the expectation. Using Theorem 2.13.4, we find the latter probability in (2.71) is bounded by

$$\inf_{\theta>0} e^{\theta(1-\delta)\mu_k}\cdot \operatorname{Tr}\exp\left\{g(\theta)\sum_i \mathbf{V}_-^*\left(-\mathbb{E}\mathbf{X}_i\right)\mathbf{V}_-\right\}.$$

The trace can be bounded by the maximum eigenvalue (since the maximum eigenvalue is nonnegative), by taking into account the reduced dimension of the summands:

$$\begin{aligned}\operatorname{Tr}\exp\left\{g(\theta)\sum_i \mathbf{V}_-^*\left(-\mathbb{E}\mathbf{X}_i\right)\mathbf{V}_-\right\} &\leqslant k\cdot\lambda_{\max}\left(\exp\left\{g(\theta)\sum_i \mathbf{V}_-^*\left(-\mathbb{E}\mathbf{X}_i\right)\mathbf{V}_-\right\}\right)\\ &= k\cdot\exp\left\{-g(\theta)\cdot\lambda_{\min}\left(\sum_i \mathbf{V}_-^*\left(\mathbb{E}\mathbf{X}_i\right)\mathbf{V}_-\right)\right\}\\ &= k\cdot\exp\left\{-g(\theta)\cdot\mu_k\right\}.\end{aligned}$$

The equality follows from the spectral mapping theorem, Theorem 1.4.4 (Page 34), and (1.92) (Page 47). In the second equality, we identify the quantity μ_k. Note that $-g(\theta) \leq 0$. Our argument establishes the bound

$$\mathbb{P}\left(\lambda_k\left(\sum_i \mathbf{X}_i\right) \geqslant (1+\delta)\,\mu_k\right) \leqslant k\cdot \inf_{\theta>0} e^{[\theta(1+\delta)-g(\theta)]\mu_k}.$$

The right-hand side is minimized, (by taking care of the infimum), when $\theta = -\log(1-\delta)$, which gives the desired upper tail bound. □

From the two proofs, we see the property that the maximum eigenvalue is nonnegative is fundamental. Using this property, we convert the trace functional into the maximum eigenvalue functional. Then the Courant-Fischer theorem, Theorem 1.4.22, can be used. The spectral mapping theorem is applied almost everywhere; it is must be recalled behind the mind. The non-commutative property is fundamental in studying random matrices. By using the eigenvalues and their variation property, it is very convenient to think of random matrices as scalar-valued random variables, in which we convert the two dimensional problem into one-dimensional problem—much more convenient to handle.

2.15 Linear Filtering Through Sums of Random Matrices

The linearity of the expectation and the trace is so basic. We must always bear this mind. The trace which is a linear functional converts a random matrix into a scalar-valued random variable; so as the kth interior eigenvalue which is a non-linear functional. Since trace and expectation commute, it follows from (1.87), which says that

$$\mathbb{E}\left(\operatorname{Tr}\mathbf{X}\right) = \operatorname{Tr}\left(\mathbb{E}\mathbf{X}\right). \tag{2.72}$$

As said above, in the left-hand side, $\operatorname{Tr}\mathbf{X}$ is a scalar-valued random variable, so its expectation is treated as our standard textbooks on random variables and processes; remarkably, in the right-hand side, the expectation of a random matrix $\mathbb{E}\mathbf{X}$ is also a matrix whose entries are expected values. After this expectation, a trace functional converts the matrix value into a scalar value. One cannot help replacing $\mathbb{E}\mathbf{X}$ with the empirical average—a sum of random matrices, that is,

$$\mathbb{E}\mathbf{X} \cong \frac{1}{n}\sum_{i=1}^{n}\mathbf{X}_i, \tag{2.73}$$

as we deal with the scalar-valued random variables. This intuition lies at the very basis of modern probability. In this book, one purpose is to prepare us for the intuition of this "approximation" (2.73), for a given n, *large but finite*—the n is taken as it is. We are not interested in the asymptotic limit as $n \to \infty$, rather the non-asymptotic analysis. One natural metric of measure is the kth interior eigenvalues

$$\lambda_k\left(\frac{1}{n}\sum_{i=1}^{n}\mathbf{X}_i - \mathbb{E}\mathbf{X}\right).$$

Note the interior eigenvalues are non-linear functionals. We cannot simply separate the two terms.

We can use the linear trace functional that is the sum of all eigenvalues. As a result, we have

$$\sum_k \lambda_k \left(\frac{1}{n} \sum_{i=1}^{n} \mathbf{X}_i - \mathbb{E}\mathbf{X} \right) = \mathrm{Tr} \left(\frac{1}{n} \sum_{i=1}^{n} \mathbf{X}_i - \mathbb{E}\mathbf{X} \right) = \mathrm{Tr} \left(\frac{1}{n} \sum_{i=1}^{n} \mathbf{X}_i \right) - \mathrm{Tr} \left(\mathbb{E}\mathbf{X} \right)$$
$$= \frac{1}{n} \sum_{i=1}^{n} \mathrm{Tr}\, \mathbf{X}_i - \mathbb{E} \left(\mathrm{Tr}\, \mathbf{X} \right).$$

The linearity of the trace is used in the second and third equality. The property that trace and expectation commute is used in the third equality. Indeed, the linear trace functional is convenient, but a lot of statistical information is contained in the interior eigenvalues. For example, the median of the eigenvalues, rather than the average of the eigenvalues—the trace divided by its dimension can be viewed as the average, is more representative statistically.

We are in particular interested in the signal plus noise model in the matrix setting. We consider instead

$$\mathbb{E} \left(\mathbf{X} + \mathbf{Z} \right) \cong \frac{1}{n} \sum_{i=1}^{n} \left(\mathbf{X}_i + \mathbf{Z}_i \right), \tag{2.74}$$

for

$$\mathbf{X}, \mathbf{Z}, \mathbf{X}_i, \mathbf{Z}_i \geqslant 0 \quad \text{and} \quad \mathbf{X}, \mathbf{Z}, \mathbf{X}_i, \mathbf{Z}_i \in \mathbb{C}^{m \times m},$$

where $\mathbf{X}, \mathbf{X}_i$ represent the signal and $\mathbf{Z}, \mathbf{Z}_i$ the noise. Recall that $\mathbf{A} \geq \mathbf{0}$ means that $\mathbf{A}$ is positive semidefinite (Hermitian and all eigenvalues of $\mathbf{A}$ are nonnegative). Samples covariance matrices of dimensions $m \times m$ are most often used in this context.

Since we have a prior knowledge that $\mathbf{X}, \mathbf{X}_i$ are of low rank, the low-rank matrix recovery naturally fits into this framework. We can choose the matrix dimension m such that enough information of the signal matrices $\mathbf{X}_i$ is recovered, but we don't care if sufficient information of $\mathbf{Z}_i$ can be recovered for this chosen m. For example, only the first dominant k eigenvalues of $\mathbf{X}, \mathbf{X}_i$ are recovered, which will be treated in Sect. 2.10 Low Rank Approximation. We conclude that the sums of random matrices have the fundamental nature of imposing the structures of the data that only exhibit themselves in the matrix setting. The low rank and the positive semi-definite of sample covariance matrices belong to these data structures. When the data is big, we must impose these additional structures for high-dimensional data processing.

The intuition of exploiting (2.74) is as follows: if the estimates of $\mathbf{X}_i$ are so accurate that they are independent and identically distributed $\mathbf{X}_i = \mathbf{X}_0$, then we rewrite (2.74) as

$$\mathbb{E} \left(\mathbf{X} + \mathbf{Z} \right) \cong \frac{1}{n} \sum_{i=1}^{n} \left(\mathbf{X}_i + \mathbf{Z}_i \right) = \frac{1}{n} \sum_{i=1}^{n} \left(\mathbf{X}_0 + \mathbf{Z}_i \right) = \mathbf{X}_0 + \frac{1}{n} \sum_{i=1}^{n} \mathbf{Z}_i. \tag{2.75}$$

Practically, $\mathbf{X}+\mathbf{Z}$ cannot be separated. The exploitation of the additional low-rank structure of the signal matrices allows us to extract the signal matrix $\mathbf{X}_0$. The average of the noise matrices $\frac{1}{n}\sum_{i=1}^{n}\mathbf{Z}_i$ will reduce the total noise power (total variance), while the signal power is kept constant. This process can effectively improve the signal to noise ratio, which is especially critical to detection of extremely weak signals (relative to the noise power).

The basic observation is that the above data processing only involves the **linear operations**. The process is also blind, which means that no prior knowledge of the noise matrix is used. We only take advantage of the rank structure of the underlying signal and noise matrices: the dimensions of the signal space is lower than the dimensions of the noise space. The above process can be extended to more general case: $\mathbf{X}_i$ are more dependent than $\mathbf{Z}_i$, where $\mathbf{X}_i$ are dependent on each other and so are $\mathbf{Z}_i$, but $\mathbf{X}_i$ are independent of $\mathbf{Z}_i$. Thus, we rewrite (2.74) as

$$\mathbb{E}\left(\mathbf{X}+\mathbf{Z}\right) \cong \frac{1}{n}\sum_{i=1}^{n}\left(\mathbf{X}_i+\mathbf{Z}_i\right) = \frac{1}{n}\sum_{i=1}^{n}\mathbf{X}_i + \frac{1}{n}\sum_{i=1}^{n}\mathbf{Z}_i. \tag{2.76}$$

All we care is that, through the sums of random matrices, $\frac{1}{n}\sum_{i=1}^{n}\mathbf{X}_i$ is performing statistically better than $\frac{1}{n}\sum_{i=1}^{n}\mathbf{Z}_i$. For example, we can use the linear trace functional (average operation) and the non-linear median functional. To calculate the median value of $\lambda_k, 1 \le k \le n$,

$$\mathbb{M}\left[\lambda_k\left(\frac{1}{n}\sum_{i=1}^{n}\mathbf{X}_i + \frac{1}{n}\sum_{i=1}^{n}\mathbf{Z}_i\right)\right],$$

where $\mathbb{M}$ is the median value which is a scalar-valued random variable, we need to calculate

$$\lambda_k\left(\frac{1}{n}\sum_{i=1}^{n}\mathbf{X}_i + \frac{1}{n}\sum_{i=1}^{n}\mathbf{Z}_i\right) \quad 1 \le k \le n.$$

The average operation comes down to a trace operation

$$\frac{1}{n}\sum_{i=1}^{n}\operatorname{Tr}\mathbf{X}_i + \frac{1}{n}\sum_{i=1}^{n}\operatorname{Tr}\mathbf{Z}_i,$$

where the linearity of the trace is used. This is simply the standard sum of scalar-valued random variables. It is expected, via the central limit theorem, that their sum approaches to the Gaussian distribution, for a reasonably large n. As pointed out before, this trace operation throws away a lot of statistical information that is available in the random matrices, for example, the matrix structures.

2.16 Dimension-Free Inequalities for Sums of Random Matrices

Sums of random matrices arise in many statistical and probabilistic applications, and hence their concentration behavior is of fundamental significance. Surprisingly, the classical exponential moment method used to derive tail inequalities for scalar random variables carries over to the matrix setting when augmented with certain matrix trace inequalities [68, 113]. Altogether, these results have proven invaluable in constructing and simplifying many probabilistic arguments concerning sums of random matrices.

One deficiency of many of these previous inequalities is their dependence on the explicit matrix dimension, which prevents their application to infinite dimensional spaces that arise in a variety of data analysis tasks, such as kernel based machine learning [114]. In this subsection, we follow [68, 113] to prove analogous results where dimension is replaced with a trace quantity that can be small, even when the explicit matrix dimension is large or infinite. Magen and Zouzias [115] also gives similar results that are complicated and fall short of giving an exponential tail inequality.

We use $\mathbb{E}_i\left[\cdot\right]$ as shorthand for $\mathbb{E}_i\left[\cdot\right] = \mathbb{E}_i\left[\cdot\left|\mathbf{X}_1, \ldots, \mathbf{X}_i\right.\right]$, the conditional expectation. The main idea is to use Theorem 1.4.17: Lieb's theorem.

Lemma 2.16.1 (Tropp [49]). *Let* $\mathbf{I}$ *be the identity matrix for the range of the* $\mathbf{X}_i$*. Then*

$$\mathbb{E}\left[\mathrm{Tr}\left(\exp\left(\sum_{i=1}^{N}\mathbf{X}_i - \sum_{i=1}^{N}\ln\mathbb{E}_i\left[\exp\left(\mathbf{X}_i\right)\right]\right) - \mathbf{I}\right)\right] \leqslant 0. \qquad (2.77)$$

Proof. We follow [113]. The induction method is used for the proof. For $N = 0$, it is easy to check the lemma is correct. For $N \geq 1$, assume as the inductive hypothesis that (2.77) holds with N with replaced with $N-1$. In this case, we have that

$$\begin{aligned}
&\mathbb{E}\left[\mathrm{Tr}\left(\exp\left(\sum_{i=1}^{N}\mathbf{X}_i - \sum_{i=1}^{N}\ln\mathbb{E}_i\left[\exp\left(\mathbf{X}_i\right)\right]\right) - \mathbf{I}\right)\right] \\
&= \mathbb{E}\left[\mathbb{E}_N\left[\mathrm{Tr}\left(\exp\left(\sum_{i=1}^{N-1}\mathbf{X}_i - \sum_{i=1}^{N}\ln\mathbb{E}_i\left[\exp\left(\mathbf{X}_i\right)\right] + \ln\exp\left(\mathbf{X}_N\right)\right) - \mathbf{I}\right)\right]\right] \\
&\leqslant \mathbb{E}\left[\mathbb{E}_N\left[\mathrm{Tr}\left(\exp\left(\sum_{i=1}^{N-1}\mathbf{X}_i - \sum_{i=1}^{N}\ln\mathbb{E}_i\left[\exp\left(\mathbf{X}_i\right)\right] + \ln\mathbb{E}_N\exp\left(\mathbf{X}_N\right)\right) - \mathbf{I}\right)\right]\right] \\
&= \mathbb{E}\left[\mathbb{E}_N\left[\mathrm{Tr}\left(\exp\left(\sum_{i=1}^{N-1}\mathbf{X}_i - \sum_{i=1}^{N-1}\ln\mathbb{E}_i\left[\exp\left(\mathbf{X}_i\right)\right]\right) - \mathbf{I}\right)\right]\right] \\
&\leqslant 0
\end{aligned}$$

where the second line follows from Theorem 1.4.17 and Jensen's inequality. The fifth line follows from the inductive hypothesis. □

While (2.77) gives the trace result, sometimes we need the largest eigenvalue.

Theorem 2.16.2 (Largest eigenvalue—Hsu, Kakade and Zhang [113]). *For any $\alpha \in \mathbb{R}$ and any $t > 0$*

$$\mathbb{P}\left[\lambda_{\max}\left(\alpha\sum_{i=1}^{N}\mathbf{X}_i - \sum_{i=1}^{N}\log\mathbb{E}_i\left[\exp\left(\alpha\mathbf{X}_i\right)\right]\right) > t\right]$$
$$\leqslant \mathrm{Tr}\left(\mathbb{E}\left(-\alpha\sum_{i=1}^{N}\mathbf{X}_i + \sum_{i=1}^{N}\log\mathbb{E}_i\left[\exp\left(\alpha\mathbf{X}_i\right)\right]\right)\right)\cdot\left(e^t - t - 1\right)^{-1}.$$

Proof. Define a new matrix $\mathbf{A} = \alpha\sum\limits_{i=1}^{N}\mathbf{X}_i - \sum\limits_{i=1}^{N}\log\mathbb{E}_i\left[\exp\left(\alpha\mathbf{X}_i\right)\right]$.. Note that $g(x) = e^x - x - 1$ is non-negative for all real x, and increasing for $x \geq 0$. Let $\lambda_i(\mathbf{A})$ be the i-th eigenvalue of the matrix $\mathbf{A}$, we have

$$\begin{aligned}
\mathbb{P}\left[\lambda_{\max}\left(\mathbf{A}\right) > t\right]\left(e^t - t - 1\right) &= \mathbb{E}\left[\mathcal{I}\left(\lambda_{\max}\left(\mathbf{A}\right) > t\right)\left(e^t - t - 1\right)\right] \\
&\leqslant \mathbb{E}\left(e^{\lambda_{\max}(\mathbf{A})} - \lambda_{\max}\left(\mathbf{A}\right) - 1\right) \\
&\leqslant \mathbb{E}\left(\sum_i\left(e^{\lambda_i(\mathbf{A})} - \lambda_i\left(\mathbf{A}\right) - 1\right)\right) \\
&\leqslant \mathbb{E}\left(\mathrm{Tr}\left[\exp\left(\mathbf{A}\right) - \mathbf{A} - \mathbf{I}\right]\right) \\
&\leqslant \mathrm{Tr}\left(\mathbb{E}\left[-\mathbf{A}\right]\right)
\end{aligned}$$

where $\mathcal{I}(x)$ is the indicator function of x. The second line follows from the spectral mapping theorem. The third line follows from the increasing property of the function $g(x)$. The last line follows from Lemma 2.16.1. □

When $\sum\limits_{i=1}^{N}\mathbf{X}_i$ is zero mean, then the first term in Theorem 2.16.2 vanishes, so the trace term

$$\mathrm{Tr}\left(\mathbb{E}\left(\sum_{i=1}^{N}\log\mathbb{E}_i\left[\exp\left(\alpha\mathbf{X}_i\right)\right]\right)\right)$$

can be made small by an appropriate choice of α.

Theorem 2.16.3 (Matrix sub-Gaussian bound—Hsu, Kakade and Zhang [113]). *If there exists $\bar{\sigma} > 0$ and $\bar{\kappa} > 0$ such that for all $i = 1, \ldots, N$,*

$$\mathbb{E}_i\left[\mathbf{X}_i\right] = 0$$

$$\lambda_{\max}\left(\frac{1}{N}\sum_{i=1}^{N}\log \mathbb{E}_i\left[\exp\left(\alpha \mathbf{X}_i\right)\right]\right) \leqslant \frac{\alpha^2\bar{\sigma}^2}{2}$$

$$\mathbb{E}\left(\mathrm{Tr}\left(\frac{1}{N}\sum_{i=1}^{N}\log \mathbb{E}_i\left[\exp\left(\alpha \mathbf{X}_i\right)\right]\right)\right) \leqslant \frac{\alpha^2\bar{\sigma}^2\bar{\kappa}}{2}$$

for all $\alpha > 0$ almost surely, then for any $t > 0$,

$$\mathbb{P}\left[\lambda_{\max}\left(\frac{1}{N}\sum_{i=1}^{N}\mathbf{X}_i\right) > \sqrt{\frac{2\bar{\sigma}^2 t}{N}}\right] \leqslant \bar{\kappa}\cdot t\left(e^t - t - 1\right)^{-1}.$$

The proof is simple. We refer to [113] for a proof.

Theorem 2.16.4 (Matrix Bernstein Bound—Hsu, Kakade and Zhang [113]). *If there exists $\bar{b} > 0$, $\bar{\sigma} > 0$ and $\bar{\kappa} > 0$ such that for all $i = 1, \ldots, N$,*

$$\mathbb{E}_i\left[\mathbf{X}_i\right] = 0$$

$$\lambda_{\max}\left(\mathbf{X}_i\right) \leqslant \bar{b}$$

$$\lambda_{\max}\left(\frac{1}{N}\sum_{i=1}^{N}\mathbb{E}_i\left[\mathbf{X}_i^2\right]\right) \leqslant \bar{\sigma}^2$$

$$\mathbb{E}\left(\mathrm{Tr}\left(\frac{1}{N}\sum_{i=1}^{N}\mathbb{E}_i\left[\mathbf{X}_i^2\right]\right)\right) \leqslant \bar{\sigma}^2\bar{\kappa}$$

almost surely, then for any $t > 0$,

$$\mathbb{P}\left[\lambda_{\max}\left(\frac{1}{N}\sum_{i=1}^{N}\mathbf{X}_i\right) > \sqrt{\frac{2\bar{\sigma}^2 t}{N}} + \frac{\bar{b}t}{3N}\right] \leqslant \bar{\kappa}\cdot t\left(e^t - t - 1\right)^{-1}.$$

The proof is simple. We refer to [113] for a proof.

Explicit dependence on the dimension of the matrix does not allow straightforward use of these results in the infinite-dimensional setting. Minsker [116] deals with this issue by extension of previous results. This new result is of interest to low rank matrix recovery and approximation matrix multiplication. Let $\|\cdot\|$ denote the operator norm $\|\mathbf{A}\| = \max_i \{\lambda_i(\mathbf{A})\}$, where λ_i are eigenvalues of a Hermitian operator $\mathbf{A}$. Expectation $\mathbb{E}\mathbf{X}$ is taken elementwise.

Theorem 2.16.5 (Dimension-free Bernstein inequality—Minsker [116]). *Let $\mathbf{X}_1, \ldots, \mathbf{X}_N$ be a sequence of $n \times n$ independent Hermitian random matrices such* *that* $\mathbb{E}\mathbf{X}_i = 0$ *and* $\|\mathbf{X}_i\| \leq 1$ *almost surely. Denote* $\sigma^2 = \left\|\sum_{i=1}^{N} \mathbb{E}\mathbf{X}_i^2\right\|$. *Then, for any* $t > 0$

$$\mathbb{P}\left(\left\|\sum_{i=1}^{N} \mathbf{X}_i\right\| > t\right) \leqslant 2\frac{\operatorname{Tr}\left(\sum_{i=1}^{N} \mathbb{E}\mathbf{X}_i^2\right)}{\sigma^2} \exp\left(-\Psi_\sigma(t)\right) \cdot r_\sigma(t)$$

where $\Psi_\sigma(t) = \frac{t^2/2}{\sigma^2 + t/3}$ *and* $r_\sigma(t) = 1 + \frac{6}{t^2 \log^2(1+t/\sigma^2)}$.

If $\sum_{i=1}^{N} \mathbb{E}\mathbf{X}_i^2$ is of approximately low rank, i.e., has many small eigenvalues, the number of non-zero eigenvalues are big. The term $\frac{\operatorname{Tr}\left(\sum_{i=1}^{N} \mathbb{E}\mathbf{X}_i^2\right)}{\sigma^2}$, however, is can be much smaller than the dimension n. Minsker [116] has applied Theorem 2.16.5 to the problem of learning the continuous-time kernel.

A concentration inequality for the sums of matrix-valued martingale differences is also obtained by Minsker [116]. Let $\mathbb{E}_{i-1}[\cdot]$ stand for the conditional expectation $\mathbb{E}_{i-1}[\cdot\,|\mathbf{X}_1, \ldots, \mathbf{X}_i]$.

Theorem 2.16.6 (Minsker [116]). *Let* $\mathbf{X}_1, \ldots, \mathbf{X}_N$ *be a sequence of martingale differences with values in the set of* $n \times n$ *independent Hermitian random matrices such that* $\|\mathbf{X}_i\| \leq 1$ *almost surely. Denote* $\mathbf{W}_N = \sum_{i=1}^{N} \mathbb{E}_{i-1}\mathbf{X}_i^2$. *Then, for any* $t > 0$,

$$\mathbb{P}\left(\left\|\sum_{i=1}^{N} \mathbf{X}_i\right\| > t, \lambda_{\max}(\mathbf{W}_N) \leqslant \sigma^2\right) \leqslant 2\operatorname{Tr}\left[p\left(-\frac{t}{\sigma^2}\mathbb{E}\mathbf{W}_N\right)\right] \exp\left(-\Psi_\sigma(t)\right) \cdot \left(1 + \frac{6}{\Psi_\sigma^2(t)}\right),$$

where $p(t) = \min(-t, 1)$.

2.17 Some Khintchine-Type Inequalities

Theorem 2.17.1 (Non-commutative Bernstein-type inequality [53]). *Consider a finite sequence* $\mathbf{X}_i$ *of independent centered Hermitian random* $n \times n$ *matrices. Assume we have for some numbers* K *and* σ *such that*

$$\|\mathbf{X}_i\| \leqslant K \quad \textit{almost surely,} \qquad \left\|\sum_i \mathbb{E}\mathbf{X}_i^2\right\| \leqslant \sigma^2.$$

Then, for every $t \geq 0$, *we have*

$$\mathbb{P}\left(\|\mathbf{X}_i\| \geqslant t\right) \leqslant 2n \cdot \exp\left(\frac{-t^2/2}{\sigma^2 + Kt/3}\right).$$

We say $\xi_1, \ldots, \xi_N$ are independent Bernoulli random variables when each ξ_i takes on the values ± 1 with equal probability. In 1923, in an effort to provide a sharp estimate on the rate of convergence in Borel's strong law of large numbers, Khintchine proved the following inequality that now bears his name.

Theorem 2.17.2 (Khintchine's inequality [117]). *Let $\xi_1, \ldots, \xi_N$ be a sequence of independent Bernoulli random variables, and let $X_1, \ldots, X_N$ be an arbitrary sequence of scalars. Then, for any $N = 1, 2, \ldots$ and $p \in (0, \infty)$, there exists an absolute constant $C_p > 0$ such that*

$$\mathbb{E}\left[\left|\sum_{i=1}^{N} \xi_i X_i\right|^p\right] \leqslant C_p \cdot \left(\sum_{i=1}^{N} |X_i|^2\right)^{p/2}. \tag{2.78}$$

In fact, Khintchine only established the inequality for the case where $p \geq 2$ is an even integer. Since his work, much effort has been spent on determining the optimal value of C_p in (2.78). In particular, it has been shown [117] that for $p \geq 2$, the value

$$C_p^* = \left(\frac{2^p}{\pi}\right)^{1/2} \Gamma\left(\frac{p+1}{2}\right)$$

is the best possible. Here $\Gamma(\cdot)$ is the Gamma function. Using Stirling's formula, one can show [117] that C_p^* is of the order $p^{p/2}$ for all $p \geq 2$.

The Khintchine inequality is extended to the case for arbitrary $m \times n$ matrices. Here $\|\mathbf{A}\|_{S_p}$ denotes the Schatten p-norm of an $m \times n$ matrix $\mathbf{A}$, i.e., $\|\mathbf{A}\|_{S_p} = \|\boldsymbol{\sigma}(\mathbf{A})\|_p$, where $\boldsymbol{\sigma} \in \mathbb{R}^{\min\{m,n\}}$ is the vector of singular values of $\mathbf{A}$, and $\|\cdot\|_p$ is the usual l_p-norm.

Theorem 2.17.3 (Khintchine's inequality for arbitrary $m \times n$ matrices [118]). *Let $\xi_1, \ldots, \xi_N$ be a sequence of independent Bernoulli random variables, and let $\mathbf{X}_1, \ldots, \mathbf{X}_N$ be arbitrary $m \times n$ matrices. Then, for any $N = 1, 2, \ldots$ and $p \geq 2$, we have*

$$\mathbb{E}\left[\left\|\sum_{i=1}^{N} \xi_i \mathbf{X}_i\right\|_{S_p}^p\right] \leqslant p^{p/2} \cdot \left(\sum_{i=1}^{N} \|\mathbf{X}_i\|_{S_p}^2\right)^{p/2}.$$

The normalization $\sum_{i=1}^{N} \|\mathbf{X}_i\|_{S_p}^2$ is not the only one possible in order for a Khintchine-type inequality to hold. In 1986, Lust-Piquard showed another one possibility.

Theorem 2.17.4 (Non-Commutative Khintchine's inequality for arbitrary $m \times n$ matrices [119]). *Let $\xi_1, \ldots, \xi_N$ be a sequence of independent Bernoulli random variables, and let $\mathbf{X}_1, \ldots, \mathbf{X}_N$ be an arbitrary sequence of $m \times n$ matrices. Then, for any $N = 1, 2, \ldots$ and $p \geq 2$, there exists an absolute constant $\gamma_p > 0$ such that*

$$\mathbb{E}\left[\left\|\sum_{i=1}^{N}\xi_i \mathbf{X}_i\right\|_{S_p}^{p}\right] \leqslant \gamma_p \cdot \max\left\{\left\|\left(\sum_{i=1}^{N}\mathbf{X}_i\mathbf{X}_i^T\right)^{1/2}\right\|_{S_p}^{p}, \left\|\left(\sum_{i=1}^{N}\mathbf{X}_i^T\mathbf{X}_i\right)^{1/2}\right\|_{S_p}^{p}\right\}.$$

The proof of Lust-Piquard does not provide an estimate for γ_p. In 1998, Pisier [120] showed that

$$\gamma_p \leqslant \alpha p^{p/2}$$

for some absolute constant $\alpha > 0$. Using the result of Buchholz [121], we have

$$\alpha \leqslant (\pi/e)^{p/2}/2^{p/4} < 1$$

for all $p \geq 2$. We note that Theorem 2.17.4 is valid (with $\gamma_p \leqslant \alpha p^{p/2} < p^{p/2}$) when $\xi_1, \ldots, \xi_N$ are i.i.d. standard Gaussian random variables [121].

Let $\mathbf{C}_i$ be arbitrary $m \times n$ matrices such that

$$\sum_{i=1}^{N}\mathbf{C}_i\mathbf{C}_i^T \leqslant \mathbf{I}_m, \qquad \sum_{i=1}^{N}\mathbf{C}_i^T\mathbf{C}_i \leqslant \mathbf{I}_n. \tag{2.79}$$

So [122] derived another useful theorem.

Theorem 2.17.5 (So [122]). *Let $\xi_1, \ldots, \xi_N$ be independent mean zero random variables, each of which is either (i) supported on [−1,1], or (ii) Gaussian with variance one. Further, let $\mathbf{X}_1, \ldots, \mathbf{X}_N$ be arbitrary $m \times n$ matrices satisfying $\max(m, n) \geqslant 2$ and (2.79). Then, for any $t \geq 1/2$, we have*

$$\text{Prob}\left(\left\|\sum_{i=1}^{N}\xi_i \mathbf{X}_i\right\| \geqslant \sqrt{2e\,(1+t)\ln\max\{m,n\}}\right) \leqslant (\max\{m,n\})^{-t}$$

if $\xi_1, \ldots, \xi_N$ are i.i.d. Bernoulli or standard normal random variables; and

$$\text{Prob}\left(\left\|\sum_{i=1}^{N}\xi_i \mathbf{X}_i\right\| \geqslant \sqrt{8e\,(1+t)\ln\max\{m,n\}}\right) \leqslant (\max\{m,n\})^{-t}$$

if $\xi_1, \ldots, \xi_N$ are independent mean zero random variables supported on [−1,1].

We refer to Sect. 11.2 for a proof.

Here we state a result of [123] that is stronger than Theorem 2.17.4. We deal with bilinear form. Let $\mathbf{X}$ and $\mathbf{Y}$ be two matrices of size $n \times k_1$ and $n \times k_2$ which satisfy

$$\mathbf{X}^T\mathbf{Y} = 0$$

and let $\{\mathbf{x}_i\}$ and $\{\mathbf{y}_i\}$ be row vectors of $\mathbf{X}$ and $\mathbf{Y}$, respectively. Denote ε_i to be a sequence of i.i.d. $\{0/1\}$ Bernoulli random variables with $\mathbb{P}(\varepsilon_i = 1) = \bar{\varepsilon}$. Then, for $p \geq 2$

$$\begin{aligned}\left(\mathbb{E}\left\|\sum_i \varepsilon_i \mathbf{x}_i^T \mathbf{y}_i\right\|_{S_p}^p\right)^{1/p} &\leqslant 2\sqrt{2}\gamma_p^2 \max_i \|\mathbf{x}_i\| \max_i \|\mathbf{y}_i\| + \\ 2\sqrt{\bar{\varepsilon}}\gamma_p \max &\left\{\max_i \|\mathbf{x}_i\| \left\|\sum_i \mathbf{y}_i^T \mathbf{y}_i\right\|_{S_p}^{1/2}, \max_i \|\mathbf{y}_i\| \left\|\sum_i \mathbf{x}_i^T \mathbf{x}_i\right\|_{S_p}^{1/2}\right\},\end{aligned} \tag{2.80}$$

where γ_p is the absolute constant defined in Theorem 2.17.4. This proof of (2.80) uses the following result

$$\left(\mathbb{E}\left\|\sum_i \varepsilon_i \mathbf{x}_i^T \mathbf{x}_i\right\|_{S_p}^p\right)^{1/p} \leqslant 2\gamma_p^2 \max_i \|\mathbf{x}_i\|^2 + \bar{\varepsilon}\left\|\sum_i \mathbf{x}_i^T \mathbf{x}_i\right\|_{S_p}$$

for $p \geq 2$.

Now consider $\mathbf{X}^T\mathbf{X} = \mathbf{I}$, then for $p \geq \log k$, we have [123]

$$\left(\mathbb{E}\left\|\mathbf{I}_{k\times k} - \frac{1}{\bar{\varepsilon}}\sum_{i=1}^n \varepsilon_i \mathbf{x}_i^T \mathbf{x}_i\right\|_{S_p}^p\right)^{1/p} \leqslant C\sqrt{\frac{p}{\bar{\varepsilon}}} \max_i \|\mathbf{x}_i\|,$$

where $C = 2^{3/4}\sqrt{\pi e} \approx 5$. This result guarantees that the invertibility of a submatrix which is formed from sampling a few columns (or rows) of a matrix $\mathbf{X}$.

Theorem 2.17.6 ([124]). *Let $\mathbf{X} \in \mathbb{R}^{n\times n}$, be a random matrix whose entries are independent, zero-mean, random variables. Then, for $p \geq \log n$,*

$$(\mathbb{E}\|\mathbf{X}\|^p)^{1/p} \leqslant c_0 2^{1/p}\sqrt{p}\left(\sqrt{\mathbb{E}\left(\max_i \sum_j X_{ij}^2\right)^p} + \sqrt{\mathbb{E}\left(\max_j \sum_i X_{ij}^2\right)^p}\right)^{1/p},$$

where $c_0 \leqslant 2^{3/4}\sqrt{\pi e} < 5$.

Theorem 2.17.7 ([124]). *Let $\mathbf{A} \in \mathbb{R}^{n\times n}$ be any matrix and $\tilde{\mathbf{A}} \in \mathbb{R}^{n\times n}$ be a random matrix such that*

$$\mathbb{E}\tilde{\mathbf{A}} = \mathbf{A}.$$

Then, for $p \geq \log n$,

$$\left(\mathbb{E}\left\|\mathbf{A}-\tilde{\mathbf{A}}\right\|^p\right)^{1/p} \leqslant c_0 2^{1/p}\sqrt{p}\left(\sqrt{\mathbb{E}\left(\max_i \sum_j \tilde{A}_{ij}^2\right)^p} + \sqrt{\mathbb{E}\left(\max_j \sum_i \tilde{A}_{ij}^2\right)^p}\right)^{1/p},$$

where $c_0 \leqslant 2^{3/4}\sqrt{\pi e} < 5$.

2.18 Sparse Sums of Positive Semi-definite Matrices

Theorem 2.18.1 ([125]). *Let $\mathbf{A}_1, \ldots, \mathbf{A}_N$ be symmetric, positive semidefinite matrices of size $n \times n$ and arbitrary rank. For any $\varepsilon \in (0,1)$, there is a deterministic algorithm to construct a vector $\mathbf{y} \in \mathbb{R}^N$ with $O(n/\varepsilon^2)$ nonzero entries such that $\mathbf{y} \geq \mathbf{0}$ and*

$$\sum_{i=1}^{N} \mathbf{A}_i \leqslant \sum_{i=1}^{N} y_i \mathbf{A}_i \leqslant (1+\varepsilon) \sum_{i=1}^{N} \mathbf{A}_i.$$

The algorithm runs in $O(Nn^3/\varepsilon^2)$ time. Moreover, the result continues to hold if the input matrices $\mathbf{A}_1, \ldots, \mathbf{A}_N$ are Hermitian and positive semi-definite.

Theorem 2.18.2 ([125]). *Let $\mathbf{A}_1, \ldots, \mathbf{A}_N$ be symmetric, positive semidefinite matrices of size $n \times n$ and let $\mathbf{y} = (y_1, \ldots, y_N)^T \in \mathbb{R}^N$ satisfy $\mathbf{y} \geq \mathbf{0}$ and $\sum_{i=1}^{N} y_i = 1$. For any $\varepsilon \in (0,1)$, these exists $\mathbf{x} \geqslant 0$ with $\sum_{i=1}^{N} x_i = 1$ such that $\mathbf{x}$ has $O(n/\varepsilon)$ nonzero entries and*

$$(1-\varepsilon) \sum_{i=1}^{N} y_i \mathbf{A}_i \leqslant \sum_{i=1}^{N} x_i \mathbf{A}_i \leqslant (1+\varepsilon) \sum_{i=1}^{N} y_i \mathbf{A}_i.$$

2.19 Further Comments

This chapter heavily relies on the work of Tropp [53]. Due to its user-friendly nature, we take so much material from it.

Column subsampling of matrices with orthogonal rows is treated in [111]. Exchangeable pairs for sums of dependent random matrices is studied by Mackey, Jordan, Chen, Farrell, Tropp [126]. Learning with integral operators [127, 128] is relevant. Element-wise matrix sparsification by Drineas [129] is interesting. See [130, Page 15] for some matrix concentration inequalities.

Chapter 3
Concentration of Measure

Concentration of measure plays a central role in the content of this book. This chapter gives the first account of this subject. Bernstein-type concentration inequalities are often used to investigate the sums of random variables (scalars, vectors and matrices). In particular, we survey the recent status of sums of random matrices in Chap. 2, which gives us the straightforward impression of the classical view of the subject.

It is safe to say that the modern viewpoint of the subject is the concentration of measure phenomenon through Talagrand's inequality. Lipschitz functions are basic mathematical objects to study. As a result, many complicated quantities can be viewed as Lipschitz functions that can be handled in the framework of Talagrand. This new viewpoint has profound impact on the whole structure of this book. In some sense, the whole book is to prepare the audience to get comfortable with this picture.

3.1 Concentration of Measure Phenomenon

Increase in dimensionality can often help to mathematical analysis. This is called blessing of dimensionality. The regularity of having many "identical" dimensions over which one can "average" is a fundamental tool.

Let $X_1, X_2, \ldots, X_n$ be a sequence of independent random variables taking values ± 1 with equal probability, and set, for example,

$$S_n = X_1 + \cdots + X_n.$$

We think of S_n of the individual variables X_i. The classical law of large number says that S_n is essentially constant (equal to 0). By the central limit theorem, the fluctuations of S_n are of order $\sqrt{n}$ which is hardly zero. But as S_n takes values as large as n, this is the scale at which one should measure S_n, in which case S_n/n indeed essentially zero as expressed by the classical exponential bound [131]

R. Qiu and M. Wicks, *Cognitive Networked Sensing and Big Data*,
DOI 10.1007/978-1-4614-4544-9_3,

$$\mathbb{P}\left(\left\{\frac{|S_n|}{n} \geqslant t\right\}\right) \leqslant 2e^{-nt^2/2}, \quad t \geqslant 0.$$

According to M. Talagrand [131], one probabilistic aspect of measure concentration is that *a random variable that depends (in a smooth way) on the influence of many independent variables is essentially constant*!

Due to concentration of measure, *a Lipschitz function is nearly constant* [132, p. 17]. Even more important, the tails behave at worst like a scalar Gaussian random variable with absolutely controlled mean and variance.

Measure concentration is surprisingly shared by a number of cases that generalized previous examples: (1) by replacing linear functionals (such as sums of independent random variables) by arbitrary Lipschitz functions of the samples; (2) by considering measures that are not of product form. The difference between the concentration phenomenon and the standard probabilistic views on probability inequalities and law of large numbers theorems is made explicit by the extension to Lipschitz (and even Hölder type) functions and more general measures. This insight is simple, yet fundamental. This concept is extended to the matrix setting.

The theory of concentration inequality tries to answer the following question: Given a random vector $\mathbf{x}$ taking value in some measurable space $\mathcal{X}$ (which is usually some high dimensional Euclidean space), and a measurable map $f : \mathcal{X} \to \mathbb{R}$, what is a good explicit bound on $\mathbb{P}(|f(\mathbf{x}) - \mathbb{E}f(\mathbf{x})| \geq t)$? Exact evaluation or accurate approximation is, of course, the **central** purpose of probability theory itself. In situations where exact evaluation or accurate approximation is not possible, concentration inequalities aim to do the **next best job** by providing rapidly decaying tail bounds [133].

3.2 Chi-Square Distributions

The χ^2 distribution is a basic probability distribution. Let us first study the χ^2 distribution to get a feel of concentration in high dimensions.

Lemma 3.2.1. *Let $(X_1, \ldots, Y_n)$ be i.i.d. Gaussian variables, with mean 0 and variance 1. Let $a_1, \ldots, a_n$ be nonnegative. We set*

$$\|\mathbf{a}\|_\infty = \sup_{i=1,\ldots,n} |a_i|, \qquad \|\mathbf{a}\|_2^2 = \sum_{i=1}^{n} a_i^2.$$

Let

$$Z = \sum_{i=1}^{n} a_i \left(X_i^2 - 1\right).$$

Then, the following inequalities hold for any positive t :

$$\begin{aligned}\mathbb{P}\left(Z \geq 2\,\|\mathbf{a}\|_2^2\,\sqrt{t} + 2\|\mathbf{a}\|_\infty t\right) &\leq e^{-t},\\ \mathbb{P}\left(Z \leq -2\,\|\mathbf{a}\|_2\,\sqrt{t}\right) &\leq e^{-t}.\end{aligned} \tag{3.1}$$

As an immediate corollary of Lemma 3.2.1, one obtains an exponential inequality for chi-square distributions. Let Z be a centralized χ^2 statistic with n degrees of freedom [134]. Then for all $t \geq 0$,

$$\begin{aligned}\mathbb{P}\left(Z \geq n + 2\sqrt{nt} + 2t\right) &\leq e^{-t},\\ \mathbb{P}\left(Z \leq n - 2\sqrt{nt}\right) &\leq e^{-t}.\end{aligned} \tag{3.2}$$

The following consequence of this bound is useful [135]: for all $x \geq 1$, we have

$$\mathbb{P}\left(\frac{Z-n}{n} \geq 4x\right) \leq e^{-nx}. \tag{3.3}$$

Starting with the first inequality bound of (3.2), setting $t = nx$ gives

$$\mathbb{P}\left(\frac{Z-n}{n} \geq 2\sqrt{x} + 2x\right) \leq e^{-nx}.$$

Since $4x \geq 2\sqrt{x} + 2x$ for $x \geq 1$, we have $\mathbb{P}\left(\frac{Z-n}{n} \geq 4x\right) \leq e^{-nx}$, for all $x \geq 1$.

Proof. Let X a random variable with $\mathcal{N}(0,1)$ distribution. Let ψ denote the logarithm of the Laplace transform of $X^2 - 1$,

$$\psi(u) = \log\left[\mathbb{E}\left[\exp\left(u\left(X^2-1\right)\right)\right]\right] = -u - \tfrac{1}{2}\log\left(1-2u\right).$$

Then, for $0 < u < \frac{1}{2}$,

$$\psi(u) \leq \frac{u^2}{(1-2u)}.$$

Indeed, considering the power series expansion, we have

$$\psi(u) = 2u^2 \sum_{k\geq 0} \frac{1}{k+2}(2u)^k \text{ and } \frac{u^2}{(1-2u)} = u^2 \sum_{k\geq 0} (2u)^k.$$

Thus,

$$\log\left[\mathbb{E}\left[e^{uZ}\right]\right] = \sum_{i=1}^{n} \log\left[\mathbb{E}\left[\exp\left(\sum_{i=1}^{n} a_i u\left(X_i^2 - 1\right)\right)\right]\right] \leq \sum_{i=1}^{n} \frac{a_i^2 u^2}{1-2a_i u}$$
$$\leq \frac{\|\mathbf{a}\|_2^2}{1-2\|\mathbf{a}\|_\infty u}.$$

We now refer to [136]. It is proved that if

$$\log\left[\mathbb{E}\left[e^{uZ}\right]\right] \leq \frac{vu^2}{2\left(1-cu\right)},$$

then, for any positive t,

$$\mathbb{P}\left(Z \geq ct + 2\sqrt{vt}\right) \leq e^{-t}.$$

The first inequality in (3.1) holds.

In order to prove the second inequality in (3.1), we just note that for $-1/2 < u < 0$, $\psi(u) \leq u^2$. This concludes the proof. □

Given a centralized χ^2-variate X with n degrees of freedom, then for all $t \in (0, 1/2)$, we have

$$\begin{aligned} \mathbb{P}\left(X \geq n\left(1+t\right)\right) &\leq \exp\left(-\frac{3}{16}nt^2\right), \\ \mathbb{P}\left(X \leq n\left(1-t\right)\right) &\leq \exp\left(-\frac{3}{16}nt^2\right), \end{aligned} \tag{3.4}$$

The first bound in (3.4) is taken from [137] and the second one from [134]. Wainwright [138] puts these two bounds together.

For a centralized χ_n^2 variable X with d degrees of freedom, these exists a constant $C > 0$, such that [139]

$$\mathbb{P}\left(X > n\left(1+t\right)\right) \geq \frac{C}{\sqrt{n}} \exp\left(-nt^2/2\right)$$

for all $t \in (0, 1)$.

3.3 Concentration of Random Vectors

Later we need to use the Lipschitz norm (also called Lipschitz constant). For a Lipschitz function $f : \mathbb{R}^n \to \mathbb{R}$, the Lipschitz norm defined as

$$\|f\|_{\mathcal{L}} = \sup_{\mathbf{x},\mathbf{y} \in \mathbb{R}^n} \frac{\left|f\left(\mathbf{x}\right) - f\left(\mathbf{y}\right)\right|}{\|\mathbf{x} - \mathbf{y}\|_2}.$$

We say such a function is $\|f\|_{\mathcal{L}}$-Lipschitz.

For $i = 1, \ldots, n$ let $(X_i, ||\cdot||_i)$, be normed spaces equipped with norm $||\cdot||_i$, let Ω_i be a finite subset of X_i with diameter at most one and let $\mathbb{P}_i$ be a probability measure on Ω_i. Define

$$X = \left(\sum_{i=1}^{n} \oplus X_i \right)_2$$

and

$$\Omega = \Omega_1 \times \Omega_2 \times \cdots \Omega_n \subset X$$

and let

$$\mathbb{P} = \mathbb{P}^{(n)} = \mathbb{P}_1 \times \mathbb{P}_2 \times \cdots \times \mathbb{P}_n$$

be the product probability measure on Ω. For a subset $A \subseteq \Omega$ and $t \in \Omega$ let

$$\phi_A(t) = d(t, \mathrm{conv}\, A)$$

be the distance in X from t to the convex hull of the set A.

Theorem 3.3.1 (Johnson and Schechtman [140]). $\mathbb{E}e^{\frac{1}{4}\phi_A^2(t)} \le \frac{1}{\mathbb{P}(A)}$.

Theorem 3.3.2 (Johnson and Schechtman [140]). *Let* $2 \le p \le \infty$ *and let* f *be a real convex function on the convex hull of the set* Ω*, i.e.,* $\mathrm{conv}\,\Omega$*. Let* σ_p *be the Lipschitz constant. Then, for all t* > 0*,*

$$\mathbb{P}(|f - \mathbb{M}f| > t) \le 4e^{-t^p/4\sigma_p^p} \tag{3.5}$$

where $\mathbb{M}f$ *is the median of* f*. A similar inequality holds with expectation replacing the median*

$$\mathbb{P}(|f - \mathbb{E}f| > t) \le Ke^{-\delta t^p/\sigma_p^p}$$

where one can take $K = 8, \delta = 1/32$.

Applied to sums $S = X_1 + \ldots + X_N$ of real-valued independent random variables $Y_1, \ldots, Y_n$ on some probability space $(\Omega, \mathcal{A}, \mathbb{P})$ such that $u_i \le Y_i \le v_i, i = 1, \ldots, n$, we have a Hoeffding type inequality

$$\mathbb{P}(S \ge \mathbb{E}(S) + t) \le e^{-t^2/2D^2} \tag{3.6}$$

where

$$D^2 \ge \sum_{i=1}^{n} (v_i - u_i)^2.$$

Let $\|\cdot\|_p$ be the L^p-norm. The following functions are norms in $\mathbb{R}^n$

$$\|\mathbf{x}\|_2 = \left(\sum_{i=1}^{n} x_i^2\right)^{1/2}, \|\mathbf{x}\|_\infty = \max_{i=1,\ldots,n} |x_i|, \|\mathbf{x}\|_1 = \sum_{i=1}^{n} |x_i|.$$

Consider a Banach space E with (arbitrary) norm $\|\cdot\|$.

Theorem 3.3.3 (Hoeffding type inequality of [141]). *Let* $\mathbf{x}_1, \ldots, \mathbf{x}_N$ *be independent bounded random vectors in a Banach space* $(E, \|\cdot\|)$, *and let*

$$\mathbf{s} = \mathbf{x}_1 + \ldots + \mathbf{x}_N = \sum_{j=1}^{N} \mathbf{x}_j.$$

For every $t \geq 0$,

$$\mathbb{P}\left(\{|\|\mathbf{s}\| - \mathbb{E}\left(\|\mathbf{s}\|\right)| \geq t\}\right) \leq 2e^{-t^2/2D^2} \tag{3.7}$$

where $D^2 \geq \sum_{j=1}^{N} \|\mathbf{x}_j\|_\infty^2$.

Equation (3.7) is an extension of (3.6).

The Hamming metric is defined as

$$d_{\mathbf{a}}(\mathbf{x}, \mathbf{y}) = \sum_{i=1}^{n} a_i \mathbf{1}_{\{x_i \neq y_i\}}, \quad \mathbf{a} = (a_1, \ldots, a_n) \in \mathbb{R}_+^n,$$

where $\|\mathbf{a}\| = \sum_{i=1}^{n} a_i^2$. Consider a function $f : \Omega_1 \times \Omega_2 \times \cdots \times \Omega_n \to \mathbb{R}$ such that for every $\mathbf{x} \in \Omega_1 \times \Omega_2 \times \cdots \times \Omega_n$ there exists $\mathbf{a} = \mathbf{a}(\mathbf{x}) \in \mathbb{R}_+^n$ with $\|\mathbf{a}\| = 1$ such that for every $\mathbf{y} \in \Omega_1 \times \Omega_2 \times \cdots \times \Omega_n$,

$$f(\mathbf{x}) \leq f(\mathbf{y}) + d_{\mathbf{a}}(\mathbf{x}, \mathbf{y}). \tag{3.8}$$

Theorem 3.3.4 (Corollary 4.7 of [141]). *Let* $\mathbb{P}$ *be a product probability measure on the product space* $\Omega_1 \times \Omega_2 \times \cdots \times \Omega_n$ *and let* $f : \Omega_1 \times \Omega_2 \times \cdots \times \Omega_n \to \mathbb{R}$ *be 1-Lipschitz in the sense of* (3.8). *Then, for every* $t \geq 0$,

$$\mathbb{P}\left(|f - \mathbb{M}f| \geq t\right) \leq 4e^{-t^2/4}$$

where $\mathbb{M}f$ *is a median of* f *for* $\mathbb{P}$.

Replacing f with $-f$, Theorem 3.3.4 applies if (3.8) is replaced by $f(\mathbf{y}) \leq f(\mathbf{x}) + d_{\mathbf{a}}(\mathbf{x}, \mathbf{y})$.

A typical application of Theorem 3.3.4 is involved in supremum of *linear functionals*. Let us study this example. In a probabilistic language, consider *independent* real-valued random variables $Y_1, \ldots, Y_n$ on some probability space $(\Omega, \mathcal{A}, \mathbb{P})$ such that for real numbers $u_i, v_i, i = 1, \ldots, n$,

$$u_i \le Y_i \le v_i, i = 1, \ldots, n.$$

Set

$$Z = \sup_{\mathbf{t} \in \mathcal{T}} \sum_{i=1}^{n} t_i Y_i \tag{3.9}$$

where $\mathcal{T}$ is a finite or countable family of vectors $\mathbf{t} = (t_1, \ldots, t_n) \in \mathbb{R}^n$ such that the "variance" σ is finite

$$\sigma = \sup_{\mathbf{t} \in \mathcal{T}} \left(\sum_{i=1}^{n} t_i^2 \left(v_i^2 - u_i^2\right) \right)^{1/2} < \infty.$$

Now observe that

$$f(\mathbf{x}) = Z = \sup_{\mathbf{t} \in \mathcal{T}} \sum_{i=1}^{n} t_i x_i$$

and apply Theorem 3.3.4 on the product space $\prod_{i=1}^{n} [u_i, v_i]$ under the product probability measure of the laws of the $Y_i, i = 1, \ldots, n$. Let $\mathbf{t}$ achieve the supremum of $f(\mathbf{x})$. Then, for every $\mathbf{x} \in \prod_{i=1}^{n} [u_i, v_i]$,

$$f(\mathbf{x}) = \sum_{i=1}^{n} t_i x_i \le \sum_{i=1}^{n} t_i y_i + \sum_{i=1}^{n} |t_i| \, |x_i - y_i|$$

$$\le f(\mathbf{y}) + \sigma \sum_{i=1}^{n} \frac{|t_i| |u_i - v_i|}{\sigma} \mathbf{1}_{\{x_i \ne y_i\}}.$$

Thus, $\frac{1}{\sigma} f(\mathbf{x})$ satisfies (3.8) with $\mathbf{a} = \mathbf{a}(\mathbf{x}) = \frac{1}{\sigma}(|t_1|, \ldots, |t_n|)$. Combining Theorem 3.3.4 with (3.7), we have the following consequence.

Theorem 3.3.5 (Corollary 4.8 of [141]). *Let Z be defined as* (3.9) *and denote the median of Z by $\mathbb{M}Z$. Then, for every $t \ge 0$,*

$$\mathbb{P}(|Z - \mathbb{M}Z| \ge t) \le 4e^{-t^2/4\sigma^2}.$$

In addition,

$$|\mathbb{E}Z - \mathbb{M}Z| \le 4\sqrt{\pi}\sigma \quad \textit{and} \quad \mathrm{Var}(Z) \le 16\sigma^2.$$

Convex functions are of practical interest. The following result has the central significance to a lot of applications.

Theorem 3.3.6 (Talagrand's concentration inequality [142]). *For every product probability* $\mathbb{P}$ *on* $[-1,1]^n$, *consider a convex and Lipschitz function* $f : \mathbb{R}^n \to \mathbb{R}$ *with Lipschitz constant L. Let* $X_1, \ldots, X_n$ *be independent random variables taking values* $[-1,1]$. *Let* $Y = f(X_1, \ldots, X_n)$ *and let* m *be a median of* Y. *Then for every* $t \geq 0$, *we have*

$$\mathbb{P}(|Y - m| \geq t) \leq 4e^{-t^2/16L^2}.$$

See [142, Theorem 6.6] for a proof. Also see Corollary 4.10 of [141]. Let us see how we can modify Theorem 3.3.6 to have concentration around the mean instead of the median. Following [143], we just notice that by Theorem 3.3.6,

$$\mathbb{E}(Y - m)^2 \leq 64L^2. \tag{3.10}$$

Since $\mathbb{E}(Y - m)^2 \geq \mathrm{Var}(Y)$, this shows that

$$\mathrm{Var}(Y) \leq 64L^2. \tag{3.11}$$

Thus by Chebychev's inequality,

$$\mathbb{P}(|Y - \mathbb{E}[Y]| \geq 16L) \leq \frac{1}{4}.$$

Using the definition of a median, this implies that

$$\mathbb{E}[Y] - 16L \leq m \leq \mathbb{E}[Y] + 16L.$$

Together with Theorem 3.3.6, we have that for any $t \geq 0$,

$$\mathbb{P}(|Y - \mathbb{E}[Y]| \geq 16L + t) \leq 4e^{-t^2/2L^2}.$$

It is essential to point that the eigenvalues of random matrices can be viewed as functions of matrix entries. Note that eigenvalues are not very regular functions of general (non-normal) matrices. For a Hermitian matrix $\mathbf{A} \in \mathbb{R}^{n \times n}$, and the eigenvalues $\lambda_i, i = 1, \ldots, n$ are sorted in decreasing magnitude. The following functions are convex: (1) the largest eigenvalue $\lambda_1(\mathbf{A})$, (2) the sum of first k largest eigenvalues $\sum_{i=1}^{k} \lambda_i(\mathbf{A})$, (3) the sum of the smallest k eigenvalues $\sum_{i=1}^{k} \lambda_{n-i+1}(\mathbf{A})$. However, Theorem 3.3.6 can be applied to these convex functions–not necessarily linear.

According to (3.10) and (3.11), the Lipschitz constant L controls the mean and the variance of the function f. Later on in this book, we have shown how to evaluate this constant. See [23, 144] for related more general result. Here, it suffices to give some examples studied in [145]. The eigenvalues (or singular values) are Lipschitz functions with respect to the matrix elements. In particular, for each $k \in \{1, \ldots, n\}$, the k-th largest singular value $\sigma_k(\mathbf{X})$ (the eigenvalue $\lambda_k(\mathbf{X})$) is Lipschitz with constant 1 if $(X_{ij})_{i,j=1}^n$ is considered as an element of the Euclidean space $\mathbb{R}^{n^2}$ (respectively the submanifold of $\mathbb{R}^{n^2}$ corresponding to the Hermitian matrices). If one insists on thinking of $\mathbf{X}$ as a matrix, this corresponds to considering the underlying Hilbert-Schmidt metric. The Lipschitz constant of $\sigma_k(\mathbf{X}/\sqrt{n})$ is $1/\sqrt{n}$, since the variances of the entries being $1/n$. The trace function $\mathrm{Tr}(\mathbf{X})$ has a Lipschitz constant of $1/n$. Form (3.11), The variance of the trace function is $1/n$ times smaller than that the largest eigenvalue (or the smallest eigenvalue). The same is true to the singular value.

Theorem 3.3.7 (Theorem 4.18 of [141]). *Let $f : \mathbb{R}^n \to \mathbb{R}$ be a real-value function such that $\|f\|_L \le \sigma$ and such that its Lipschitz coefficient with respect to the ℓ_1-metric is less than or equal to κ, that is*

$$|f(\mathbf{x}) - f(\mathbf{y})| \le \kappa \sum_{i=1}^{n} |x_i - y_i|, \quad \mathbf{x}, \mathbf{y} \in \mathbb{R}^n.$$

Then, for every $t > 0$,

$$\mathbb{P}(f \ge M + t) \le C \exp\left(-\frac{1}{K} \min\left(\frac{t}{\kappa}, \frac{t^2}{\sigma^2}\right)\right)$$

for some numerical constant $C > 0$ where M is either a median of f or its mean. Conversely, the concentration result holds.

Let $\mu_1, \ldots, \mu_n$ be arbitrary probability measures on the unit interval $[0, 1]$ and let $\mathbb{P}$ be the product probability measure $\mathbb{P} = \mu_1 \otimes \ldots \otimes \mu_n$ on $[0, 1]^n$. We say a function on $\mathbb{R}^n$ is *separately convex* if it is convex in each coordinate. Recall that a convex function on $\mathbb{R}$ is continuous and almost everywhere differentiable.

Theorem 3.3.8 (Theorem 5.9 of [141]). *Let f be separately convex and 1-Lipschitz on $\mathbb{R}^n$. Then, for every product probability measure $\mathbb{P}$ on $[0, 1]^n$, and every $t \ge 0$,*

$$\mathbb{P}\left(\left\{f \ge \int f d\mathbb{P} + t\right\}\right) \le e^{-t^2/4}.$$

The norm is a convex function. The norm is a supremum of linear functionals. Consider the convex function $f : \mathbb{R}^n \to \mathbb{R}$ defined as

$$f(\mathbf{x}) = \left\|\sum_{i=1}^{n} x_i \mathbf{v}_i\right\|, \quad \mathbf{x} = (x_1, \ldots, x_n) \in \mathbb{R}^n$$

where $\mathbf{v}_i, i = 1, \ldots, n$ are vectors in an arbitrary normed space E with norm $\|\cdot\|$. Then, by duality, for $\mathbf{x}, \mathbf{y} \in \mathbb{R}^n$,

$$\begin{aligned} |f(\mathbf{x}) - f(\mathbf{y})| &\leq \left\| \sum_{i=1}^{n} (x_i - y_i) \mathbf{v}_i \right\| \\ &= \sup_{\|\mathbf{z}\| \leq 1} \sum_{i=1}^{n} (x_i - y_i) \langle \mathbf{z}, \mathbf{v}_i \rangle \leq \sigma \|\mathbf{x} - \mathbf{y}\|_2 \end{aligned}$$

where the last step follows from the Cauchy-Schwarz inequality. So the Lipschitz norm is $\|f\|_{\mathcal{L}} \leq \sigma$. The use of Theorem 3.3.6 or Theorem 3.3.8 gives the following theorem, a main result we are motivated to develop.

Theorem 3.3.9 (Theorem 7.3 of [141]). *Let $\eta_1, \ldots, \eta_n$ be independent (scalar-valued) random variables such that $\eta_i \leq 1$ almost surely, for $i = 1, \ldots, n$, and let $\mathbf{v}_1, \ldots, \mathbf{v}_n$ be vectors in a normed space E with norm $\|\cdot\|$. For every $t \geq 0$,*

$$\mathbb{P}\left(\left\| \sum_{i=1}^{n} \eta_i \mathbf{v}_i \right\| \geq M + t \right) \leq 2e^{-t^2/16\sigma^2}$$

where M is either the mean or a median of $\left\| \sum_{i=1}^{n} \eta_i \mathbf{v}_i \right\|$ and where

$$\sigma^2 = \sup_{\|\mathbf{z}\| \leq 1} \sum_{i=1}^{n} \langle \mathbf{z}, \mathbf{v}_i \rangle^2.$$

Theorem 3.3.9 is an infinite-dimensional extension of the Hoeffding type inequality (3.6).

Let us state a well known result on concentration of measure for a standard Gaussian vector. Let $\gamma = \gamma_n$ be the standard Gaussian measure on $\mathbb{R}^n$ with density $(2\pi)^{-n/2} e^{-|\mathbf{x}|^2/2}$ where $|\mathbf{x}|$ is the usual Euclidean norm for vector $\mathbf{x}$. The expectation of a function is defined as $\mathbb{E} f(\mathbf{x}) = \int_{\mathbb{R}^n} f(\mathbf{x})\, d\gamma_n(\mathbf{x})$.

Theorem 3.3.10 (Equation (1.4) of Ledoux [141]). *Let $f : \mathbb{R}^n \to \mathbb{R}$ be a Lipschitz function and let $\|f\|_{\mathcal{L}}$ be its Lipschitz norm. If $\mathbf{x} \in \mathbb{R}^n$ is a standard Gaussian vector, (a vector whose entries are independent standard Gaussian random variables), then for all $t > 0$*

$$\mathbb{P}\left(f(\mathbf{x}) \geq \mathbb{E} f(\mathbf{x}) + t\sqrt{2} \|f\|_{\mathcal{L}} \right) \leq e^{-t^2}.$$

We are interested in the supremum[1] of a sum of independent random variables $Z_1, Z_2, \ldots, Z_n$ in Banach space

---

[1] The supremum is the least upper bound of a set S defined as a quantity M such that no member of the set exceeds M. It is denoted as $\sup_{x \in S} x$.

$$S = \sup_{g \in \mathcal{G}} \sum_{i=1}^{n} g\left(Z_i\right).$$

Theorem 3.3.11 (Corollary 7.8 of Ledoux [141]). *If $|g| \le \eta$ for every $g \in \mathcal{G}$ and $g\left(Z_1\right), \ldots, g\left(Z_n\right)$ have zero mean for every $g \in \mathcal{G}$. Then, for all $t \ge 0$,*

$$\mathbb{P}\left(|S - \mathbb{E}S| \ge t\right) \le 3 \exp\left(-\frac{t}{C\eta} \log\left(1 + \frac{\eta t}{\sigma^2 + \eta \mathbb{E}\bar{S}}\right)\right), \tag{3.12}$$

where

$$\sigma^2 = \sup_{g \in \mathcal{G}} \sum_{i=1}^{n} \mathbb{E} g^2\left(Z_i\right),$$

$$\bar{S} = \sup_{g \in \mathcal{G}} \sum_{i=1}^{n} \left|g\left(Z_i\right)\right|,$$

and $C > 0$ is a small numerical constant.

Let us see how to apply Theorem 3.3.10.

Example 3.3.12 (Maximum of Correlated Normals—Example 17.8 of [146]). We study the concentration inequality for the maximum of n jointly distributed Gaussian random variables. Consider a Gaussian random vector $\mathbf{x} = \left(X_1, \ldots, X_n\right) \sim \mathcal{N}\left(0, \mathbf{\Sigma}\right)$, where $\mathbf{\Sigma}$ is positive definite, and let $\sigma_i = \mathrm{Var}\left(X_i\right), i = 1, 2, \ldots, n$. Let $\mathbf{\Sigma} = \mathbf{A}\mathbf{A}^T$.

Let us first consider the function $f : \mathbb{R}^n \to \mathbb{R}$ defined by

$$f\left(\mathbf{u}\right) = \max\left\{\left(\mathbf{Au}\right)_1, \ldots, \left(\mathbf{Au}\right)_n\right\}$$

where $\left(\mathbf{Au}\right)_1$ means the first coordinate of the vector $\mathbf{Ay}$ and so on. Let $\sigma_{\max} = \max_i \sigma_i$. Our goal here is to show that f is a Lipschitz function with Lipschitz constant (or norm) by $\sigma_{\max}$. We only consider the case when $\mathbf{A}$ is diagonal; the general case can be treated similarly. For two vectors $\mathbf{x}, \mathbf{y} \in \mathbb{R}^n$, we have

$$\begin{aligned}\left(\mathbf{Au}\right)_1 &= a_{11}u_1 = a_{11}v_1 + \left(a_{11}u_1 - a_{11}v_1\right) \le a_{11}v_1 + \sigma_{\max}\left\|\mathbf{u} - \mathbf{v}\right\| \\ &\le \max\left\{\left(\mathbf{Au}\right)_1, \left(\mathbf{Au}\right)_2, \ldots, \left(\mathbf{Au}\right)_n\right\} + \sigma_{\max}\left\|\mathbf{u} - \mathbf{v}\right\|.\end{aligned}$$

Using the same arguments, for each i, we have

$$\left(\mathbf{Au}\right)_i \le \max\left\{\left(\mathbf{Au}\right)_1, \left(\mathbf{Au}\right)_2, \ldots, \left(\mathbf{Au}\right)_n\right\} + \sigma_{\max}\left\|\mathbf{u} - \mathbf{v}\right\|.$$

Thus,

$$\begin{aligned} f(\mathbf{u}) &= \max\{(\mathbf{Au})_1, \ldots, (\mathbf{Au})_n\} \\ &\leq \max\{(\mathbf{Au})_1, (\mathbf{Au})_2, \ldots, (\mathbf{Au})_n\} + \sigma_{\max}\|\mathbf{u}-\mathbf{v}\| \\ &= f(\mathbf{v}) + \sigma_{\max}\|\mathbf{u}-\mathbf{v}\|. \end{aligned}$$

By switching the roles of $\mathbf{u}$ and $\mathbf{v}$, we have $f(\mathbf{v}) \leq f(\mathbf{u}) + \sigma_{\max}\|\mathbf{u}-\mathbf{v}\|$. Combining both, we obtain

$$|f(\mathbf{u}) - f(\mathbf{v})| \leq \sigma_{\max}\|\mathbf{u}-\mathbf{v}\|,$$

implying that the Lipschitz norm is $\sigma_{\max}$. Next, we observe that

$$\max\{X_1, \ldots, X_n\} = \max\{(\mathbf{Az})_1, (\mathbf{Az})_2, \ldots, (\mathbf{Az})_n\} = f(\mathbf{z})$$

where $\mathbf{z}$ is a random Gaussian vector $\mathbf{z} = (Z_1, \ldots, Z_n) \sim \mathcal{N}(0, \mathbf{\Sigma})$. Since the function $f(\mathbf{z})$ is Lipschitz with constant σ_{max}, we apply Theorem 3.3.10 to obtain

$$\begin{aligned} &\mathbb{P}(|\max\{X_1, \ldots, X_n\} - \mathbb{E}[\max\{X_1, \ldots, X_n\}]| > t) \\ &= \mathbb{P}(|f(\mathbf{z}) - \mathbb{E}[f(\mathbf{z})]| > t) \leq e^{-t^2/2\sigma_{\max}^2}. \end{aligned}$$

It is remarkable that although $X_1, X_2, \ldots, X_n$ are not assumed to be independent, we can still prove an Gaussian concentration inequality by using only the coordinate-wise variances, and the inequality is valid for all n. □

Let us see how to apply Theorem 3.3.11 to study the Frobenius norm bound of the sum of vectors, following [123] closely.

Example 3.3.13 (Frobenius norm bound of the sum of vectors [123]). Let ξ_j be i.i.d. Bernoulli 0/1 random variables with $\mathbb{P}(\xi_j = 1) = d/m$ whose subscript j represents the entry selected from a set $\{1, 2, \ldots, m\}$. In particular, we have

$$S_F = \left\| \sum_{j=1}^{m} \xi_j \mathbf{x}_j^T \mathbf{y}_j \right\|_F$$

where $\mathbf{x}_j$ and $\mathbf{y}_j$ are vectors.

Let $\langle \mathbf{X}, \mathbf{Y} \rangle = \mathrm{Tr}\left(\mathbf{X}^T\mathbf{Y}\right)$ represent the Euclidean inner product between two matrices and $\|\mathbf{X}\|_F = \langle \mathbf{X}, \mathbf{X} \rangle$. It can be easily shown that

$$\|\mathbf{X}\|_F = \sup_{\|\mathbf{G}\|_F=1} \mathrm{Tr}\left(\mathbf{X}^T\mathbf{G}\right) = \sup_{\|\mathbf{G}\|_F=1} \langle \mathbf{X}, \mathbf{G} \rangle.$$

Note that trace and inner product are both linear. For vectors, the only norm we consider is the ℓ_2-norm, so we simply denote the ℓ_2-norm of a vector by $\|\mathbf{x}\|$

which is equal to $\sqrt{\langle \mathbf{x}, \mathbf{x} \rangle}$, where $\langle \mathbf{x}, \mathbf{y} \rangle$ is the Euclidean inner product between two vectors. Like matrices, it is easy to show

$$\|\mathbf{x}\| = \sup_{\|\mathbf{y}\|=1} \langle \mathbf{x}, \mathbf{y} \rangle .$$

See Sect. 1.4.5 for details about norms of matrices and vectors.

Now, let $\mathbf{Z}_j = \xi_i \mathbf{x}_j^T \mathbf{y}$ (rank one matrix), we have

$$S_F = \left\| \sum_{j=1}^{m} \mathbf{Z}_j \right\|_F = \sup_{\|\mathbf{G}\|_F=1} \sum_{j=1}^{m} \langle \mathbf{Z}_j, \mathbf{G} \rangle = \sup_{\|\mathbf{G}\|_F=1} \sum_{j=1}^{m} g\left(\mathbf{Z}_j\right).$$

Since $S_F > 0$, the expected value of S_F is equal to the expected value of $\bar{S}$. That is $\mathbb{E}S_F = \mathbb{E}\bar{S}_F$. We can bound the absolute value of $g\left(\mathbf{Z}_j\right)$ as

$$\left|g\left(\mathbf{Z}_j\right)\right| \le \left|\left\langle \xi_i \mathbf{x}_j^T \mathbf{y}_j, \mathbf{G} \right\rangle\right| \le \left\|\xi_i \mathbf{x}_j^T \mathbf{y}_j\right\|_F \le \left\|\mathbf{x}_j\right\| \left\|\mathbf{y}_j\right\|,$$

where $\|\cdot\|$ is the ℓ_2 norm of a vector. We take $\eta = \max\limits_j \left\|\mathbf{x}_j\right\| \left\|\mathbf{y}_j\right\|$ so that $\left|g\left(\mathbf{Z}_j\right)\right| \le \eta$.

Now we compute the term $\sigma^2 = \sup\limits_{g \in \mathcal{G}} \sum\limits_{i=1}^{n} \mathbb{E}g^2\left(Z_i\right)$, in Theorem 3.3.11. Since

$$\mathbb{E}g^2\left(\mathbf{Z}_j\right) = \mathbb{E}\xi_i \left\langle \mathbf{x}_j^T \mathbf{y}_j, \mathbf{G} \right\rangle^2 = \frac{d}{m} \left\langle \mathbf{x}_j^T \mathbf{y}_j, \mathbf{G} \right\rangle \le \frac{d}{m} \left\|\mathbf{x}_j^T \mathbf{y}_j\right\|_F^2,$$

we have

$$\sum_{j=1}^{m} \mathbb{E}g^2\left(\mathbf{Z}_j\right) \le \frac{d}{m} \sum_{j=1}^{m} \left\|\mathbf{x}_j^T \mathbf{y}_j\right\|_F^2 = \frac{d}{m} \sum_{j=1}^{m} \operatorname{Tr}\left(\mathbf{x}_j^T \mathbf{y}_j \mathbf{y}_j \mathbf{x}_j\right)$$

$$\le \frac{d}{m} \sum_{j=1}^{m} \left\|\mathbf{y}_j\right\|^2 \operatorname{Tr}\left(\mathbf{x}_j^T \mathbf{x}_j\right) \le \frac{d}{m} \max_j \left\|\mathbf{y}_j\right\|^2 \operatorname{Tr}\left(\sum_{j=1}^{m} \mathbf{x}_j^T \mathbf{x}_j\right) = \frac{kd}{m} \max_j \left\|\mathbf{y}_j\right\|^2, \tag{3.13}$$

where $k = \operatorname{Tr}\left(\sum\limits_{j=1}^{m} \mathbf{x}_j^T \mathbf{x}_j\right)$. In the first inequality of the second line, we have used (1.84) that is repeated here for convenience

$$\operatorname{Tr}\left(\mathbf{A} \cdot \mathbf{B}\right) \le \|\mathbf{B}\| \operatorname{Tr}\left(\mathbf{A}\right). \tag{3.14}$$

when $\mathbf{A} \ge 0$ and $\|\mathbf{B}\|$ is the spectrum norm (largest singular value). Prove similarly, we also have

$$\sum_{j=1}^{m} \mathbb{E}g^2(\mathbf{Z}_j) \le \frac{d}{m} \max_j \|\mathbf{x}_j\|^2 \operatorname{Tr}\left(\sum_{j=1}^{m} \mathbf{y}_j^T \mathbf{y}_j\right) = \frac{d\alpha}{m} \max_j \|\mathbf{x}_j\|^2,$$

where $\alpha = \operatorname{Tr}\left(\sum_{j=1}^{m} \mathbf{y}_j^T \mathbf{y}_j\right)$. So we choose

$$\sigma^2 = \sum_{j=1}^{m} \mathbb{E}g^2(\mathbf{Z}_j) \le \frac{d}{m} \max\left\{\alpha \max_j \|\mathbf{x}_j\|^2, k \max_j \|\mathbf{y}_j\|^2\right\}. \tag{3.15}$$

Apply the powerful Talagrand's inequality (3.12) and note that from expectation inequality, $\sigma^2 + \eta\mathbb{E}S_F \le \sigma\mathbb{E}S_F + \eta\mathbb{E}S_F = \mathbb{E}^2 S_F$ we have

$$\begin{aligned}\mathbb{P}(S_F - \mathbb{E}S_F \ge t) &\le 3\exp\left(-\tfrac{t}{C\eta}\log\left(1 + \tfrac{\eta t}{\sigma^2 + \eta\mathbb{E}^2 S_F}\right)\right)\\ &\le 3\exp\left(-\tfrac{1}{C}\tfrac{t^2}{\mathbb{E}^2 S_F}\right).\end{aligned}$$

The last inequality follows from the fact that $\log(1+x) \ge 2x/3$ for $0 \le x \le 1$. Thus, t must be chosen to satisfy $\eta t \le \mathbb{E}^2 S_F$.

Choose $t = C\sqrt{\log\frac{3}{\beta}}\mathbb{E}S_F$ where C is a small numerical constant. By some calculations, we can show that $\eta t \le \mathbb{E}^2 S_F$ as $\eta t \le \mathbb{E}^2 S_F d \ge C^2 \mu m \log\frac{3}{\beta}$. Therefore,

$$\mathbb{P}(S_F - \mathbb{E}S_F \ge t) \le 3\exp\left(-\log\frac{3}{\beta}\right) = \beta.$$

There is a small constant such that $C_1\sqrt{\log\frac{3}{\beta}} = C\sqrt{\log\frac{3}{\beta}} + 1$. Finally, we summarize the result as

$$\mathbb{P}\left(S_F \le C_1\sqrt{\log\frac{3}{\beta}} \cdot \mathbb{E}S_F\right) \ge 1 - \beta.$$

□

Theorem 3.3.14 (Theorem 7.3 of Ledoux [141]). *Let $\xi_1, \ldots, \xi_n$ be a sequence of independent random variable such that $|\xi_i| \le 1$ almost surely with $i = 1, \ldots, n$ and let $\mathbf{x}_1, \ldots, \mathbf{x}_n$ be vectors in Banach space. Then, for every $t \ge 0$,*

$$\mathbb{P}\left(\left\|\sum_{i=1}^{n} \xi_i \mathbf{x}_i\right\| \ge M + t\right) \le 2\exp\left(-\frac{t^2}{16\sigma^2}\right) \tag{3.16}$$

where M is either the mean or median of $\left\| \sum_{i=1}^{n} \xi_i \mathbf{x}_i \right\|$ *and*

$$\sigma^2 = \sup_{\|\mathbf{y}\| \le 1} \sum_{i=1}^{n} \langle \mathbf{y}, \mathbf{x}_i \rangle .$$

The theorem claims that the sum of vectors with random weights is distributed like Gaussian around its mean or median, with standard deviation $2\sqrt{2}\sigma$. This theorem strongly bounds the supremum of a sum of vectors $\mathbf{x}_1, \mathbf{x}_2, \ldots, \mathbf{x}_n$ with random weights in Banach space. For applications such as (3.15), we need to bound $\max_i \|\mathbf{x}_i\|$ and $\max_i \|\mathbf{y}_i\|$. For details, we see [123].

Let $\{a_{ij}\}$ be an $n \times n$ array of real numbers. Let π be chosen uniformly at random from the set of all permutations of $\{1, \ldots, n\}$, and let $X = \sum_{i=1}^{n} a_{i\pi(i)}$. This class of random variables was first studied by Hoeffding [147].

Theorem 3.3.15 ([133]). *Let $\{a_{ij}\}_{i,j=1}^{n}$ be a collection of numbers from $[0, 1]$. Let π be chosen uniformly at random from the set of all permutations of $\{1, \ldots, n\}$, and let $X = \sum_{i=1}^{n} a_{i\pi(i)}$. Let $X = \sum_{i=1}^{n} a_{i\pi(i)}$, where π is drawn from the uniform distribution over the set of all permutations of $\{1, \ldots, n\}$. Then*

$$\mathbb{P}\left(|X - \mathbb{E}X| \ge t \right) \le 2 \exp\left(-\frac{t^2}{4\mathbb{E}X + 2t} \right)$$

for all $t \ge 0$.

3.4 Slepian-Fernique Lemma and Concentration of Gaussian Random Matrices

Following [145], we formulate the eigenvalue problem in terms of the Gaussian process $Z_{\mathbf{u}}$. For $\mathbf{u} \in \mathbb{R}^N$, we define $Z_{\mathbf{u}} = \langle \cdot, \mathbf{u} \rangle$. For a matrix $\mathbf{X}$ and vectors $\mathbf{u}, \mathbf{v} \in \mathbb{R}^n$, we have

$$\langle \mathbf{X}\mathbf{u}, \mathbf{v} \rangle = \mathrm{Tr}\left(\mathbf{X} \left(\mathbf{v} \otimes \mathbf{u} \right) \right) = \langle \mathbf{X}, \mathbf{u} \otimes \mathbf{v} \rangle_{Tr} = Z_{\mathbf{u} \otimes \mathbf{v}}\left(\mathbf{X} \right),$$

where $\mathbf{v} \otimes \mathbf{u}$ stands for the rank one matrix $\left(u_i v_i \right)_{i,j=1}^{n}$, that is, the matrix of the map $\mathbf{x} \to \langle \mathbf{x}, \mathbf{v} \rangle \mathbf{u}$. Here $\langle \mathbf{X}, \mathbf{Y} \rangle_{Tr} = \mathrm{Tr}\left(\mathbf{X}\mathbf{Y}^T \right)$ is the trace duality, often called the Hilbert-Schmidt scalar product, can be also thought of as the usual scalar product on $\mathbb{R}^{n^2}$. The key observation is as follows

$$\|\mathbf{X}\| = \max_{\mathbf{u},\mathbf{v}\in S^{n-1}} \langle \mathbf{X}\mathbf{u}, \mathbf{v}\rangle = \max_{\mathbf{u},\mathbf{v}\in S^{n-1}} Z_{\mathbf{u}\otimes\mathbf{v}}\left(\mathbf{X}\right) \tag{3.17}$$

where $\|\cdot\|$ denotes the operator (or matrix) norm. The Gaussian process $\mathbf{X}_{\mathbf{u},\mathbf{v}} = Z_{\mathbf{u}\otimes\mathbf{v}}\left(\mathbf{X}\right), \mathbf{u}, \mathbf{v} \in S^{n-1}$ is now compared with $\mathbf{Y}_{\mathbf{u},\mathbf{v}} = Z_{(\mathbf{u},\mathbf{v})}$, where $(\mathbf{u},\mathbf{v})$ is regarded as an element of $\mathbb{R}^n \times \mathbb{R}^n = \mathbb{R}^{2n}$. Now we need the Slepian-Fernique lemma.

Lemma 3.4.1 (Slepian-Fernique lemma [145]). *Let $(X_t)_{t\in T}$ and $(Y_t)_{t\in T}$ be two families of jointly Gaussian mean zero random variable such that*

$$(a) \qquad \|X_t - X_{t'}\|_2 \le \|Y_t - Y_{t'}\|_2, \quad fort, t' \in T$$

Then

$$\mathbb{E}\max_{t\in T} X_t \le \mathbb{E}\max_{t\in T} Y_t. \tag{3.18}$$

Similarly, if $T = \cup_{s\in S} T_s$ and

$$(b) \qquad \|X_t - X_{t'}\|_2 \le \|Y_t - Y_{t'}\|_2, \quad if \quad t \in T_s, t' \in T_{s'} \quad with \quad s \ne s'.$$

$$(c) \qquad \|X_t - X_{t'}\|_2 \ge \|Y_t - Y_{t'}\|_2, \quad if \quad t, t' \in T_{s'} \quad for\ some \quad s.$$

then one has

$$\mathbb{E}\max_{t\in S}\min_{t\in T_s} X_{s,t} \le \mathbb{E}\max_{t\in S}\min_{t\in T_s} Y_{s,t}.$$

To see that the Slepian-Fernique lemma applies, we only need to verify that, for $\mathbf{u}, \mathbf{v}, \mathbf{u}', \mathbf{v}' \in S^{n-1}$, where S^{n-1} is a sphere in $\mathbb{R}^n$,

$$|\mathbf{u}\otimes\mathbf{v} - \mathbf{u}'\otimes\mathbf{v}'|^2 \le |(\mathbf{u},\mathbf{v}) - (\mathbf{u}',\mathbf{v}')|^2 = |\mathbf{u}-\mathbf{u}'|^2 + |\mathbf{v}-\mathbf{v}'|^2,$$

where $|\cdot|$ is the usual Euclidean norm. On the other hand, for $(\mathbf{x},\mathbf{y}) \in \mathbb{R}^n \times \mathbb{R}^n$, we have

$$Z_{(\mathbf{u},\mathbf{v})}\left(\mathbf{x},\mathbf{y}\right) = \langle \mathbf{x},\mathbf{u}\rangle + \langle \mathbf{y},\mathbf{v}\rangle$$

so

$$\max_{\mathbf{u},\mathbf{v}\in S^{n-1}} Z_{(\mathbf{u},\mathbf{v})}\left(\mathbf{x},\mathbf{y}\right) = |\mathbf{x}| + |\mathbf{y}|.$$

This is just implying that $\|\cdot\|_{(U\times V)^\circ} = \|\cdot\|_{U^\circ} + \|\cdot\|_{V^\circ}$. (If $K \in \mathbb{R}^n$, one has the gauge, or the Minkowski functional of the polar of K given by $\max_{\mathbf{u}\in K} Z_{\mathbf{u}} = \|\cdot\|_{K^\circ}$.) Therefore, the assertion of Lemma 3.4.1 translates to

$$\sqrt{n}\mathbb{E}\left\|G^{(n)}\right\| \leq 2\int_{\mathbb{R}^n} |\mathbf{x}|\, d\gamma_n\left(\mathbf{x}\right),$$

where $\gamma = \gamma_n$ is the standard Gaussian measure on $\mathbb{R}^n$ with density $(2\pi)^{-n/2}e^{-|\mathbf{x}|^2/2}$ where $|\mathbf{x}|$ is the usual Euclidean norm for vector $\mathbf{x}$. Here $\mathbf{G} = \mathbf{G}^{(n)}$ is the $n \times n$ real Gaussian matrices.

By comparing with the second moment of $|\mathbf{x}|$, the last integral is seen to be $\leq n^{\frac{1}{2}}$. $|\mathbf{x}|^2$ is distributed according to the familiar $\chi^2(n)$ law. The same argument, applied just to symmetric tensors $\mathbf{u} \otimes \mathbf{u}$, allows to analyze $\lambda_1(GOE)$, the largest eigenvalue of the Gaussian orthogonal ensembles (GOE).

Using the Theorem 3.3.10 and remark following it, we have the following theorem. Let $\mathbb{N}$ denote the set of all the natural numbers, and $\mathbb{M}$ the median of the set. As usual, $\Phi(t) = \gamma_1\left((-\infty, t)\right)$ is the cumulative distribution function of the $\mathbb{N}(0,1)$ Gaussian random variable.

Theorem 3.4.2 (Theorem 2.11 of [145]). *Given $n \in \mathbb{N}$, consider the ensembles of $n \times n$ matrices $\mathbf{G}$, and GOE. If the random variable F equals either $\|\mathbf{G}\|$ or $\lambda_1\left(GOE\right)$, then*

$$\mathbb{M}F < \mathbb{E}F < 2,$$

where $\mathbb{M}$ standards for the median operator. As a result, for any $t > 0$,

$$\mathbb{P}\left(F \geq 2 + \kappa t\right) < 1 - \Phi(t) < \exp\left(-nt^2/2\right), \tag{3.19}$$

where $\kappa = 1$ in the case of $\|\mathbf{G}\|$ and $\kappa = \sqrt{2}$ in the case of $\lambda_1\left(GOE\right)$.

For the rectangular matrix of Gaussian matrices with independent entries, we have the following result.

Theorem 3.4.3 (Theorem 2.13 of [145]). *Given $m, n \in \mathbb{N}$, with $m \leq n$, put $\beta = m/n$ and consider the $n \times m$ random matrix $\mathbf{\Gamma}$ whose entries are real, independent Gaussian random variables following $\mathcal{N}(0, 1/n)$ law. Let the singular values be $s_1(\mathbf{\Gamma}), \ldots, s_m(\mathbf{\Gamma})$. Then*

$$1 + \beta < \mathbb{E}s_m(\mathbf{\Gamma}) \leq \mathbb{M}s_1(\mathbf{\Gamma}) \leq \mathbb{E}s_1(\mathbf{\Gamma}) < 1 + \beta$$

and as a result, for any $t > 0$,

$$\max\left\{\mathbb{P}\left(s_1(\mathbf{\Gamma}) \geq 1 + \beta + t\right), \mathbb{P}\left(s_1(\mathbf{\Gamma}) \geq 1 - \beta - t\right)\right\} < 1 - \Phi(t) \leq e^{-nt^2}. \tag{3.20}$$

The beauty of the above result is that the inequality (3.20) is valid for all m, n rather than asymptotically.

The proof of Theorem 3.4.3 is similar to that of Theorem 3.4.2. We use the second part of Lemma 3.4.1.

Complex matrices can be viewed as real matrices with a special structure. Let $\mathbf{G}^{(n)}$ denote the complex non-Hermitian matrix: all the entries are independent and of the form $x + jy$, where x, y are independent real $\mathcal{N}(0, 1/2n)$ Gaussian random variables. We consider

$$\frac{1}{\sqrt{2}}\begin{bmatrix} \mathbf{G} & -\mathbf{G}' \\ \mathbf{G}' & \mathbf{G} \end{bmatrix} = \frac{1}{\sqrt{2}}\left(\begin{bmatrix} 1 & 0 \\ 0 & 1 \end{bmatrix} \otimes \mathbf{G} + \begin{bmatrix} 0 & -1 \\ 1 & 0 \end{bmatrix} \otimes \mathbf{G}'\right)$$

where $\mathbf{G}$ and $\mathbf{G}'$ are independent copies of matrix $\mathbf{G}^{(n)}$.

Another family of interest is product measure. This application requires an additional convexity assumption on the functionals. If μ is a product measure on $\mathbb{R}^n$ with compactly supported factors, a fundamental result of M. Talagrand [148] shows that (3.78) holds for every Lipschitz and convex function. More precisely, assume that $\mu = \mu_1 \otimes \cdots \otimes \mu_n$, where each μ_i is supported on $[a, b]$. Then, for every Lipschitz and convex function $F : \mathbb{R}^n \to \mathbb{R}$,

$$\mu\left(\{|F - \mathbb{M}F| \geq t\}\right) \leq 4e^{-t^2/4(b-a)^2}.$$

By the variational representation of (3.79), the largest eigenvalue of a symmetric (or Hermitian) matrix is clearly a convex function of the entries.[2] The largest eigenvalue is 1-Lipschitz, as pointed above. We get our theorem.

Theorem 3.4.4 (Proposition 3.3 of [149]). *Let* $\mathbf{X}$ *be a real symmetric* $n \times n$ *matrix such that the entries* $X_{ij}, 1 \leq i \leq j \leq n$ *are independent random variables with* $|X_{ij}| \leq 1$. *Then, for any* $t \geq 0$,

$$\mathbb{P}\left(|\lambda_{\max}(\mathbf{X}) - \mathbb{M}\lambda_{\max}(\mathbf{X})| \geq t\right) \leq 4e^{-t^2/32}.$$

Up to some numerical constants, the median $\mathbb{M}$ can be replaced by the mean $\mathbb{E}$. A similar result is expected for all the eigenvalues.

3.5 Dudley's Inequality

We take material from [150, 151] for this presentation. See Sect. 7.6 for some applications. A stochastic process is a collection $X_t, t \in \tilde{T}$, of complex-valued random variables indexed by some set $\tilde{T}$. We are interested in bounding the moments of the supremum of $X_t, t \in \tilde{T}$. To avoid measurability issues, we define, for a subset $T \subset \tilde{T}$, the lattice supremum as

$$\mathbb{E}\sup_{t\in T}|X_t| = \sup\left\{\mathbb{E}\sup_{t\in F}|X_t|, F \subset T, F \text{ finite}\right\}. \tag{3.21}$$

We endow the set $\tilde{T}$ with the pseudometric

$$d(s,t) = \left(\mathbb{E}|X_t - X_s|^2\right)^{1/2}. \tag{3.22}$$

---

[2]They are linear in the entries.

In contrast to a metric, a pseudometric does not need to separate points, i.e., $d(s,t) = 0$ does not necessarily imply $s = t$. We further assume that the increments of the process $X_t, t \in \tilde{T}$ satisfy the concentration property

$$\mathbb{P}\left(|X_t - X_s| \geq ud\left(t, s\right)\right) \leq 2e^{-t^2/2}, \quad u > 0, s, t \in \tilde{T}. \tag{3.23}$$

Now we apply Dudley's inequality for the special case of the Rademacher process of the form

$$X_t = \sum_{i=1}^{M} \varepsilon_i x_i\left(t\right), \quad t \subset \tilde{T}, \tag{3.24}$$

where $\boldsymbol{\varepsilon} = (\varepsilon_1, \ldots, \varepsilon_M)$ is a Rademacher sequence and the $x_i\left(t\right) : \tilde{T} \to \mathbb{C}$ are some deterministic functions. We have

$$\begin{aligned} d(s,t)^2 &= \mathbb{E}|X_t - X_s|^2 = \mathbb{E}\left|\sum_{i=1}^{M} \varepsilon_i\left(x_i\left(t\right) - x_i\left(s\right)\right)\right|^2 \\ &= \sum_{i=1}^{M} \left(x_i\left(t\right) - x_i\left(s\right)\right)^2 = \left\|\mathbf{x}\left(t\right) - \mathbf{x}(s)\right\|_2^2, \end{aligned} \tag{3.25}$$

where $\mathbf{x}\left(t\right) = \left(x_1\left(t\right), \ldots, x_M(t)\right)$ and $\|\cdot\|_2$ is the standard Euclidean norm. So we can rewrite the (pseudo-)metric as

$$d\left(s,t\right) = \left(\mathbb{E}|X_t - X_s|^2\right)^{1/2} = \left\|\mathbf{x}\left(t\right) - \mathbf{x}(s)\right\|_2. \tag{3.26}$$

Hoeffding's inequality shows that the Rademacher process (3.24) satisfies the concentration property (3.23). We deal with the Rademacher process, while the original process was for Gaussian process, see also [27, 81, 141, 151, 152].

For a subset $T \subset \tilde{T}$, the covering number $N\left(T, d, \delta\right)$ is defined as the smallest integer N such that there exists a subset $E \subset \tilde{T}$ with cardinality $|E| = N$ satisfying

$$T \subset \bigcup_{t \in E} \mathbb{B}_d\left(t, \delta\right), \qquad \mathbb{B}_d\left(t, \delta\right) = \left\{s \in \tilde{T}, d(t,s) \leq \delta\right\}. \tag{3.27}$$

In words, T can be covered by N balls of radius δ in the metric d. The diameter of the set T in the metric is defined as

$$D\left(T\right) = \sup_{s,t \in T} d\left(s,t\right).$$

We state the theorem without proof.

Theorem 3.5.1 (Rauhut [30]). *Let $X_t, t \in \tilde{T}$, be a complex-valued process indexed by a pseudometric space $(\tilde{T}, d)$ with pseudometric defined by* (3.22) *which satisfies* (3.23) . *Then, for a subset $T \subset \tilde{T}$ and any point $t_0 \in T$ it holds*

$$\mathbb{E} \sup_{t \in T} |X_t - X_{t_0}| \le 16.51 \cdot \int_0^{D(T)} \sqrt{\ln (N (T, d, u))} du + 4.424 \cdot D (T) . \quad (3.28)$$

Further, for $p \ge 2$,

$$\left(\mathbb{E} \sup_{t \in T} |X_t - X_{t_0}|^p \right)^{1/p} \le 6.028^{1/p} \left(14.372 \int_0^{D(T)} \sqrt{\ln (N (T, d, u))} du + 5.818 \cdot D (T) \right) . \quad (3.29)$$

The main proof ingredients are the covering number arguments and the concentration of measure. The estimate (3.29) is also valid for $1 \le p \le 2$ with possibly slightly different constants: this can be seen, for instance, from interpolation between $p = 1$ and $p = 2$. The theorem and its proof easily extend to Banach space valued processes satisfying

$$\mathbb{P} (|X_t - X_s| \ge ud (t, s)) \le 2e^{-t^2/2}, \quad u > 0, s, t \in \tilde{T}.$$

Inequality (3.29) for the increments of the process can be used in the following way to bound the supremum

$$\begin{aligned} \left(\mathbb{E} \sup_{t \in T} |X_t|^p \right)^{1/p} &\le \inf_{t_0 \in T} \left[\left(\mathbb{E} \sup_{t \in T} |X_t - X_{t_0}|^p \right)^{1/p} + (\mathbb{E}|X_{t_0}|^p)^{1/p} \right] \\ &\le 6.028^{1/p} \sqrt{p} \left(14.372 \int_0^{D(T)} \sqrt{\ln (N (T, d, u))} du + 5.818 \cdot D (T) \right) \\ &\quad + \inf_{t_0 \in T} (\mathbb{E}|X_{t_0}|^p)^{1/p}. \end{aligned} \quad (3.30)$$

The second term is often easy to estimate. Also, for a centered real-valued process, that is $\mathbb{E}X_t = 0$, for all $t \in \tilde{T}$, we have

$$\mathbb{E} \sup_{t \in T} X_t = \mathbb{E} \sup_{t \in T} (X_t - X_{t_0}) \le \mathbb{E} \sup_{t \in T} |X_t - X_{t_0}| . \quad (3.31)$$

For completeness we also state the usual version of Dudley's inequality.

Corollary 3.5.2. *Let $X_t, t \in T$, be a real-valued centered process indexed by a pseudometric space (T, d) such that* (3.23) *holds. Then*

$$\mathbb{E} \sup_{t \in T} X_t \le 30 \int_0^{D(T)} \sqrt{\ln (N (T, d, u))} du. \quad (3.32)$$

Proof. Without loss of generality, we assume that $D(T) = 1$. Then, it follows that $N(T, d, \delta) \geq 2$, for all $u < 1/2$. Indeed, if $N(T, d, \delta) = 1$, for some $u < 1/2$ then, for any $\delta > 0$, there would be two points of distance at least $1 - \delta$ that are covered by one ball of radius u. This is a contradiction to the triangle inequality. So,

$$\int_0^{D(T)} \sqrt{\ln(N(T, d, u))} du \geq \int_0^{1/2} \sqrt{\ln(2)} du = \frac{\sqrt{\ln 2}}{2} D(T).$$

Therefore, (3.32) follows from (3.31) and the estimate

$$16.51 + \frac{2 \times 4.424}{\sqrt{\ln 2}} < 30.$$

Generalizations of Dudley's inequality are contained in [27, 81]. □

3.6 Concentration of Induced Operator Norms

For a vector $\mathbf{x} \in \mathbb{R}^n$, we use $\|\mathbf{x}\|_p = \left(\sum_{i=1}^{n} |x_i|^p \right)^{1/p}$ to denote its ℓ_p-norm. For a matrix, we use $\|\mathbf{X}\|_{p \to q}$ to denote the matrix operator norm induced by vectors norms ℓ_p and ℓ_q. More precisely,

$$\|\mathbf{A}\|_{p \to q} = \max_{\|\mathbf{x}\|_q = 1} \|\mathbf{A}\mathbf{x}\|_p.$$

The spectral norm for a matrix $\mathbf{A} \in \mathbb{R}^{m \times n}$ is given by

$$\|\mathbf{A}\|_{2 \to 2} = \max_{\|\mathbf{x}\|_2 = 1} \|\mathbf{A}\mathbf{x}\|_2 = \max_{i=1,\ldots,m} \{\sigma_i(\mathbf{A})\},$$

where $\sigma_i(\mathbf{A})$ is the singular value of matrix $\mathbf{A}$. The ℓ_∞-operator norm is given by

$$\|\mathbf{A}\|_{\infty \to \infty} = \max_{\|\mathbf{x}\|_\infty = 1} \|\mathbf{A}\mathbf{x}\|_\infty = \max_{i=1,\ldots,m} \sum_{j=1}^{n} |A_{ij}|,$$

where A_{ij} are the entries of matrix $\mathbf{A}$. Also we have the norm $\|\mathbf{X}\|_{1 \to 2}$

$$\begin{aligned}
\|\mathbf{A}\|_{1 \to 2} &= \sup_{\|\mathbf{u}\|_1 = 1} \|\mathbf{A}\mathbf{u}\|_2 \\
&= \sup_{\|\mathbf{v}\|_2 = 1} \sup_{\|\mathbf{u}\|_1 = 1} \mathbf{v}^T \mathbf{A}\mathbf{u} \\
&= \max_{i=1,\ldots,d} \|\mathbf{A}\|_2.
\end{aligned}$$

The matrix inner product for two matrices is defined as

$$\langle \mathbf{A}, \mathbf{B} \rangle = \mathrm{Tr}\left(\mathbf{A}\mathbf{B}^T\right) = \mathrm{Tr}\left(\mathbf{A}^T\mathbf{B}\right) = \sum_{i,j} X_{ij} Y_{ij}.$$

The inner product induces the Hilbert-Schmidt norm (or Frobenius) norm

$$\|\mathbf{A}\|_F = \|\mathbf{A}\|_{HS} = \sqrt{\langle \mathbf{A}, \mathbf{A} \rangle}.$$

A widely studied instance is the standard Gaussian ensemble. Consider a random Gaussian matrix $\mathbf{X} \in \mathbb{R}^{n \times d}$, formed by drawing each row $\mathbf{x}_i \in \mathbb{R}^d$ i.i.d. from an $\mathcal{N}(0, \mathbf{\Sigma})$. Or

$$\mathbf{X} = \begin{bmatrix} \mathbf{x}_1 \\ \vdots \\ \mathbf{x}_n \end{bmatrix}.$$

Our goal here is to derive the concentration inequalities of $\frac{1}{\sqrt{n}}\|\mathbf{X}\mathbf{v}\|_2$. The emphasis is on the standard approach of using Slepian-lemma [27,141] as well as an extension due to Gordon [153]. We follow [135] closely for our exposition of this approach. See also Sect. 3.4 and the approach used for the proof of Theorem 3.8.4.

Given some index set $U \times V$, let $\{Y_{\mathbf{u},\mathbf{v}}, (\mathbf{u}, \mathbf{v}) \in U \times V\}$ and $\{Z_{\mathbf{u},\mathbf{v}}, (\mathbf{u}, \mathbf{v}) \in U \times V\}$ be a pair of zero-mean Gaussian processes. Given the semi-norm on the processes defined via $\sigma(X) = \left(\mathbb{E}\left[X^2\right]\right)^{1/2}$, Slepian's lemma states that if

$$\sigma\left(Y_{\mathbf{u},\mathbf{v}} - Y_{\mathbf{u}',\mathbf{v}'}\right) \le \sigma\left(Z_{\mathbf{u},\mathbf{v}} - Z_{\mathbf{u}',\mathbf{v}'}\right) \text{ for all } (\mathbf{u}, \mathbf{v}) \text{ and } (\mathbf{u}', \mathbf{v}') \text{ in } U \times V, \tag{3.33}$$

then

$$\mathbb{E} \sup_{(\mathbf{u},\mathbf{v}) \in U \times V} Y_{\mathbf{u},\mathbf{v}} \le \mathbb{E} \sup_{(\mathbf{u},\mathbf{v}) \in U \times V} Z_{\mathbf{u},\mathbf{v}}. \tag{3.34}$$

One version of Gordon's extension [153] asserts that if the inequality (3.33) holds for for all $(\mathbf{u}, \mathbf{v})$ and $(\mathbf{u}', \mathbf{v}')$ in $U \times V$, and holds with *equality* when $\mathbf{v} = \mathbf{v}'$, then

$$\mathbb{E}\left[\sup_{\mathbf{u} \in U} \inf_{\mathbf{v} \in V} Y_{\mathbf{u},\mathbf{v}}\right] \le \mathbb{E}\left[\sup_{\mathbf{u} \in U} \inf_{\mathbf{v} \in V} Z_{\mathbf{u},\mathbf{v}}\right]. \tag{3.35}$$

Now let us turn to the problem at hand. Any random matrix $\mathbf{X}$ from the given ensemble can be written as $\mathbf{W}\mathbf{\Sigma}^{1/2}$, where $\mathbf{W} \in \mathbb{R}^{n \times d}$ is a matrix with i.i.d. $\mathcal{N}(0, 1)$ entries, and $\mathbf{\Sigma}^{1/2}$ is the symmetric matrix square root. We choose the set U as the unit ball

$$S^{n-1} = \{\mathbf{u} \in \mathbb{R}^n : \|\mathbf{u}\|_2 = 1\},$$

and for some radius r, we choose V as the set

$$V(r) = \left\{ \mathbf{v} \in \mathbb{R}^n : \left\| \mathbf{\Sigma}^{1/2} \mathbf{v} \right\|_2 = 1, \|\mathbf{u}\|_q^q \leq r \right\}.$$

For any $\mathbf{v} \in V(r)$, we use the shorthand $\tilde{\mathbf{v}} = \mathbf{\Sigma}^{1/2}\mathbf{v}$.

Consider the centered Gaussian processes $Y_{\mathbf{u},\mathbf{v}} = \mathbf{u}^T \mathbf{W} \mathbf{v}$ indexed by the set $S^{n-1} \times V(r)$. Given two pairs $(\mathbf{u}, \mathbf{v})$ and $(\mathbf{u}', \mathbf{v}')$ in the set $S^{n-1} \times V(r)$, we have

$$\begin{aligned}
\sigma^2 \left(Y_{\mathbf{u},\mathbf{v}} - Y_{\mathbf{u}',\mathbf{v}'} \right) &= \left\| \mathbf{u}\tilde{\mathbf{v}}^T - \mathbf{u}'(\tilde{\mathbf{v}}')^T \right\|_F^2 \\
&= \left\| \mathbf{u}\tilde{\mathbf{v}}^T - \mathbf{u}'\tilde{\mathbf{v}}^T + \mathbf{u}'\tilde{\mathbf{v}}^T - \mathbf{u}'(\tilde{\mathbf{v}}')^T \right\|_F^2 \\
&= \|\tilde{\mathbf{v}}\|_2^2 \|\mathbf{u} - \mathbf{u}'\|_2^2 + \|\mathbf{u}'\|_2^2 \|\tilde{\mathbf{v}} - \tilde{\mathbf{v}}'\|_2^2 + 2 \left(\mathbf{u}^T \mathbf{u}' - \|\mathbf{u}'\|_2^2 \right) \left(\|\tilde{\mathbf{v}}\|_2^2 - \tilde{\mathbf{v}}^T \tilde{\mathbf{v}}' \right)
\end{aligned} \tag{3.36}$$

Now we use the Cauchy-Schwarz inequality and the equalities: $\|\mathbf{u}\|_2 = \|\mathbf{u}'\|_2 = 1$ and $\|\tilde{\mathbf{v}}\|_2 = \|\tilde{\mathbf{v}}'\|_2$, we have $\mathbf{u}^T\mathbf{u}' - \|\mathbf{u}\|_2^2 \leq 0$, and $\|\tilde{\mathbf{v}}\|_2^2 - \tilde{\mathbf{v}}^T\tilde{\mathbf{v}}' \geq 0$. As a result, we may conclude that

$$\sigma^2 \left(Y_{\mathbf{u},\mathbf{v}} - Y_{\mathbf{u}',\mathbf{v}'} \right) \leq \|\mathbf{u} - \mathbf{u}'\|_2^2 + \|\tilde{\mathbf{v}} - \tilde{\mathbf{v}}'\|_2^2. \tag{3.37}$$

We claim that the Gaussian process $Y_{\mathbf{u},\mathbf{v}}$ satisfies the conditions of Gordon's lemma in terms of the zero-mean Gaussian process $Z_{\mathbf{u},\mathbf{v}}$ given by

$$Z_{\mathbf{u},\mathbf{v}} = \mathbf{g}^T \mathbf{u} + \mathbf{h}^T \left(\mathbf{\Sigma}^{1/2} \mathbf{v} \right), \tag{3.38}$$

where $\mathbf{g} \in \mathbb{R}^n$ and $\mathbf{h} \in \mathbb{R}^d$ are both standard Gaussian vectors (i.e., with i.i.d. $\mathcal{N}(0, 1)$ entries). To prove the claim, we compute

$$\begin{aligned}
\sigma^2 \left(Z_{\mathbf{u},\mathbf{v}} - Z_{\mathbf{u}',\mathbf{v}'} \right) &= \|\mathbf{u} - \mathbf{u}'\|_2^2 + \left\| \mathbf{\Sigma}^{1/2} (\mathbf{v} - \mathbf{v}') \right\|_2^2 \\
&= \|\mathbf{u} - \mathbf{u}'\|_2^2 + \|\tilde{\mathbf{v}} - \tilde{\mathbf{v}}'\|_2^2.
\end{aligned}$$

From (3.37), we see that

$$\sigma^2 \left(Y_{\mathbf{u},\mathbf{v}} - Y_{\mathbf{u}',\mathbf{v}'} \right) \leq \sigma^2 \left(Z_{\mathbf{u},\mathbf{v}} - Z_{\mathbf{u}',\mathbf{v}'} \right),$$

says that the Slepian's condition (3.33) holds. On the other hand, when $\mathbf{v} = \mathbf{v}'$, we see Eq. (3.36) that

$$\sigma^2 \left(Y_{\mathbf{u},\mathbf{v}} - Y_{\mathbf{u}',\mathbf{v}'} \right) = \|\mathbf{u} - \mathbf{u}'\|_2^2 = \sigma^2 \left(Z_{\mathbf{u},\mathbf{v}} - Z_{\mathbf{u}',\mathbf{v}'} \right),$$

so that the equality required for Gordon's inequality (3.34) also holds.

Upper Bound: Since all the conditions required for Gordon's inequality (3.34) are satisfied, we have

$$\begin{aligned}\mathbb{E}\left[\sup_{\mathbf{v}\in V(r)}\|\mathbf{X}\mathbf{v}\|_2^2\right] &= \mathbb{E}\left[\sup_{(\mathbf{u},\mathbf{v})\in S^{n-1}\times V(r)}\mathbf{u}^T\mathbf{X}\mathbf{v}\right]\\ &\leq \mathbb{E}\left[\sup_{(\mathbf{u},\mathbf{v})\in S^{n-1}\times V(r)}Z_{\mathbf{u},\mathbf{v}}\right]\\ &= \mathbb{E}\left[\sup_{\|\mathbf{u}\|_2=1}\mathbf{g}^T\mathbf{u}\right] + \mathbb{E}\left[\sup_{\mathbf{v}\in V(r)}\mathbf{h}^T\left(\mathbf{\Sigma}^{1/2}\mathbf{v}\right)\right]\\ &\leq \mathbb{E}\left[\|\mathbf{g}\|_2\right] + \mathbb{E}\left[\sup_{\mathbf{v}\in V(r)}\mathbf{h}^T\left(\mathbf{\Sigma}^{1/2}\mathbf{v}\right)\right].\end{aligned}$$

By convexity, we have

$$\mathbb{E}\left[\|\mathbf{g}\|_2\right] \leq \sqrt{\mathbb{E}\left[\|\mathbf{g}\|_2^2\right]} = \sqrt{\mathbb{E}\,\mathrm{Tr}\left(\mathbf{g}\mathbf{g}^T\right)} = \sqrt{\mathrm{Tr}\,\mathbb{E}\left(\mathbf{g}^T\mathbf{g}\right)} = \sqrt{n},$$

since $\mathbb{E}\left(\mathbf{g}^T\mathbf{g}\right) = \mathbf{I}_{n\times n}$. From this, we obtain that

$$\mathbb{E}\left[\sup_{\mathbf{v}\in V(r)}\|\mathbf{X}\mathbf{v}\|_2^2\right] \leq \sqrt{n} + \mathbb{E}\left[\sup_{\mathbf{v}\in V(r)}\mathbf{h}^T\left(\mathbf{\Sigma}^{1/2}\mathbf{v}\right)\right]. \tag{3.39}$$

Turning to the remaining term, we have

$$\sup_{\mathbf{v}\in V(r)}\left|\mathbf{h}^T\left(\mathbf{\Sigma}^{1/2}\mathbf{v}\right)\right| \leq \sup_{\mathbf{v}\in V(r)}\|\mathbf{v}\|_1\left\|\mathbf{\Sigma}^{1/2}\mathbf{v}\right\|_\infty \leq r\left\|\mathbf{\Sigma}^{1/2}\mathbf{v}\right\|_\infty.$$

Since each element $\left(\mathbf{\Sigma}^{1/2}\mathbf{v}\right)_i$ is zero-mean Gaussian with variance at most $\rho(\mathbf{\Sigma}) = \max_i \Sigma_{ii}$, standard results on Gaussian maxima (e.g., [27]) imply that

$$\mathbb{E}\left[\left\|\mathbf{\Sigma}^{1/2}\mathbf{v}\right\|_\infty\right] \leq \sqrt{3\rho(\mathbf{\Sigma})\log d}.$$

Putting all the pieces together, we conclude that for $q = 1$

$$\mathbb{E}\left[\sup_{\mathbf{v}\in V(r)} \|\mathbf{X}\mathbf{v}\|_2/\sqrt{n}\right] \leq 1 + [3\rho(\mathbf{\Sigma})\log d/n]^{1/2} r. \tag{3.40}$$

Having controlled the expectation, in the standard two-step approach to establish concentration inequality, it remains to establish sharp concentration around its expectation. Let $f : \mathbb{R}^D \to \mathbb{R}$ be Lipschitz function with constant L with respect to the ℓ_1-norm. Thus if $\mathbf{w} \sim \mathcal{N}(0, \mathbf{I}_{D\times D})$ is standard normal, we are guaranteed [141] that for all $t > 0$,

$$\mathbb{P}\left(|f(\mathbf{w}) - \mathbb{E}[f(\mathbf{w})]| \geq t\right) \leq 2\exp\left(-\frac{t^2}{2L^2}\right). \tag{3.41}$$

Note the *dimension-independent* nature of this inequality. Now we use it to the random matrix $\mathbf{W} \in \mathbb{R}^{n\times d}$, which is viewed as a standard normal random vector in $D = nd$ dimensions. Let us consider the function

$$f(\mathbf{W}) = \sup_{\mathbf{v}\in V(r)} \left\|\mathbf{W}\mathbf{\Sigma}^{1/2}\mathbf{v}\right\|_2/\sqrt{n},$$

we obtain that

$$\begin{aligned}
\sqrt{n}\left[f(\mathbf{W}) - f(\mathbf{W}')\right] &= \sup_{\mathbf{v}\in V(r)} \left\|\mathbf{W}\mathbf{\Sigma}^{1/2}\mathbf{v}\right\|_2 - \sup_{\mathbf{v}\in V(r)} \left\|\mathbf{W}'\mathbf{\Sigma}^{1/2}\mathbf{v}\right\|_2 \\
&\leq \sup_{\mathbf{v}\in V(r)} \left\|\mathbf{\Sigma}^{1/2}\mathbf{v}\right\|_2 \|\mathbf{W} - \mathbf{W}'\|_F \\
&= \|\mathbf{W} - \mathbf{W}'\|_F
\end{aligned}$$

since $\left\|\mathbf{\Sigma}^{1/2}\mathbf{v}\right\|_2 = 1$ for all $\mathbf{v} \in V(r)$. We have thus shown that the Lipschitz constant $L \leq 1/\sqrt{n}$. Following the rest of the derivations in [135], we conclude that

$$\frac{1}{\sqrt{n}}\|\mathbf{X}\mathbf{v}\|_2 \leq 3\left\|\mathbf{\Sigma}^{1/2}\mathbf{v}\right\|_2 + 6\left[\frac{1}{n}\rho(\mathbf{\Sigma})\log d\right]^{1/2} \|\mathbf{v}\|_1 \quad \text{for all} \quad \mathbf{v} \in \mathbb{R}^d. \tag{3.42}$$

Lower Bound: We use Gordon's inequality to show the lower bound. We have

$$-\inf_{\mathbf{v}\in V(r)} \|\mathbf{X}\mathbf{v}\|_2 = \sup_{\mathbf{v}\in V(r)} -\|\mathbf{X}\mathbf{v}\|_2 = \sup_{\mathbf{v}\in V(r)} \inf_{\mathbf{u}\in U} \mathbf{u}^T\mathbf{X}\mathbf{v}.$$

Applying Gordon's inequality, we obtain

$$\mathbb{E}\left[\sup_{\mathbf{v}\in V(r)} -\|\mathbf{X}\mathbf{v}\|_2\right] \leq \mathbb{E}\left[\sup_{\mathbf{v}\in V(r)} \inf_{\mathbf{u}\in S^{n-1}} Z_{\mathbf{u},\mathbf{v}}\right]$$

$$= \mathbb{E}\left[\inf_{\mathbf{u}\in S^{n-1}} \mathbf{g}^T\mathbf{u}\right] + \mathbb{E}\left[\sup_{\mathbf{v}\in V(r)} \mathbf{h}^T\mathbf{\Sigma}^{1/2}\mathbf{v}\right]$$

$$\leq -\mathbb{E}\left[\|\mathbf{g}\|_2\right] + \left[3\rho\left(\mathbf{\Sigma}\right)\log d\right]^{1/2} r.$$

where we have used previous derivation to upper bound $\mathbb{E}\left[\sup_{\mathbf{v}\in V(r)} \mathbf{h}^T\mathbf{\Sigma}^{1/2}\mathbf{v}\right]$. Since $|\mathbb{E}\|\mathbf{g}\|_2 - \sqrt{n}| = o(\sqrt{n})$, $\mathbb{E}\|\mathbf{g}\|_2 \geq \sqrt{n}/2$ for all $n \geq 1$. We divide by $\sqrt{n}$ and add 1 to both sides so that

$$\mathbb{E}\left[\sup_{\mathbf{v}\in V(r)} \left(1 - \|\mathbf{X}\mathbf{v}\|_2\right)/\sqrt{n}\right] \leq 1/2 + \left[3\rho\left(\mathbf{\Sigma}\right)\log d\right]^{1/2} r. \tag{3.43}$$

Defining

$$f\left(\mathbf{W}\right) = \sup_{\mathbf{v}\in V(r)} \left(1 - \|\mathbf{X}\mathbf{v}\|_2\right)/\sqrt{n},$$

we can use the same arguments to show that its Lipschitz constant is at most $1/\sqrt{n}$. Following the rest of arguments in [135], we conclude that

$$\frac{1}{\sqrt{n}}\|\mathbf{X}\mathbf{v}\|_2 \geq \frac{1}{2}\left\|\mathbf{\Sigma}^{1/2}\mathbf{v}\right\|_2 - 6\left[\frac{1}{n}\rho\left(\mathbf{\Sigma}\right)\log d\right]^{1/2} \|\mathbf{v}\|_1 \quad \text{for all} \quad \mathbf{v}\in\mathbb{R}^d.$$

For convenience, we summarize this result in a theorem.

Theorem 3.6.1 (Proposition 1 of Raskutti, Wainwright and Yu [135]). *Consider a random matrix* $\mathbf{X} \in \mathbb{R}^{n\times d}$ *formed by drawing each row from* $\mathbf{x}_i \in \mathbb{R}^d, i = 1, 2, \ldots, n$ *i.i.d. from an* $\mathcal{N}\left(0, \mathbf{\Sigma}\right)$ *distribution. Then for some numerical constants* $c_k \in (0, \infty), k = 1, 2,$ *we have*

$$\frac{1}{\sqrt{n}}\|\mathbf{X}\mathbf{v}\|_2 \leq 3\left\|\mathbf{\Sigma}^{1/2}\mathbf{v}\right\|_2 + 6\left[\frac{1}{n}\rho\left(\mathbf{\Sigma}\right)\log d\right]^{1/2} \|\mathbf{v}\|_1 \quad \textit{for all} \quad \mathbf{v}\in\mathbb{R}^d,$$

and

$$\frac{1}{\sqrt{n}}\|\mathbf{X}\mathbf{v}\|_2 \geq \frac{1}{2}\left\|\mathbf{\Sigma}^{1/2}\mathbf{v}\right\|_2 - 6\left[\frac{1}{n}\rho\left(\mathbf{\Sigma}\right)\log d\right]^{1/2} \|\mathbf{v}\|_1 \quad \textit{for all} \quad \mathbf{v}\in\mathbb{R}^d,$$

with probability $1 - c_1\exp\left(-c_2 n\right)$.

The notions of sparsity can be defined precisely in terms of the ℓ_p-balls[3] for $p \in (0, 1]$, defined as [135]

$$\mathbb{B}_p\left(R_p\right) = \left\{ \mathbf{z} \in \mathbb{R}^n : \quad \|\mathbf{z}\|_p^p = \sum_{i=1}^{n} |z_i|^p \leq R_p \right\}, \tag{3.44}$$

where $\mathbf{z} = (z_1, z_2, \ldots, z_n)^T$. In the limiting case of $p = 0$, we have the ℓ_0-ball

$$\mathbb{B}_0\left(k\right) = \left\{ \mathbf{z} \in \mathbb{R}^n : \quad \sum_{i=1}^{n} I\left[Z_i \neq 0\right] \leq k \right\}, \tag{3.45}$$

where I is the indicator function and $\mathbf{z}$ has exactly k non-zero entries, where $k \ll n$. We see Sect. 8.7 for its application in linear regression.

To illustrate the discretization arguments of the set, we consider another result, taken also from Raskutti, Wainwright and Yu [135].

Theorem 3.6.2 (Lemma 6 of Raskutti, Wainwright and Yu [135]). *Consider a random matrix* $\mathbf{X} \in \mathbb{R}^{N \times n}$ *with the* ℓ_2*-norm upper-bounded by* $\frac{\|\mathbf{Xz}\|_2}{\sqrt{N}\|\mathbf{z}\|_2} \leq \kappa$ *for all sparse vectors with exactly* $2s$ *non-zero entries* $\mathbf{z} \in \mathbb{B}_0\left(2s\right)$*, i.e.*

$$\mathbb{B}_0\left(2s\right) = \left\{ \mathbf{z} \in \mathbb{R}^n : \quad \sum_{i=1}^{n} I\left[Z_i \neq 0\right] \leq 2s \right\}, \tag{3.46}$$

and a zero-mean white Gaussian random vector $\mathbf{w} \in \mathbb{R}^n$ *with variance* σ^2*, i.e.,* $\mathbf{w} \sim \mathcal{N}\left(0, \sigma^2 \mathbf{I}_{n \times n}\right)$*. Then, for any radius* $R > 0$*, we have*

$$\mathbb{P}\left(\sup_{\|\mathbf{z}\|_0 \leq 2s, \quad \|\mathbf{z}\|_2 \leq R} \frac{1}{N} \left|\mathbf{w}^T \mathbf{Xz}\right| \geq 6\sigma R \kappa \sqrt{\frac{s \log\left(n/s\right)}{N}} \right)$$
$$\leq c_1 \exp\left(-c_2 \min\left\{N, s \log\left(n{-}s\right)\right\}\right). \tag{3.47}$$

In other words, we have

$$\sup_{\|\mathbf{z}\|_0 \leq 2s, \quad \|\mathbf{z}\|_2 \leq R} \frac{1}{n} \left|\mathbf{w}^T \mathbf{Xz}\right| \leq 6\sigma R \kappa \sqrt{\frac{s \log\left(n/s\right)}{N}} \tag{3.48}$$

with probability greater than $1 - c_1 \exp\left(-c_2 \min\left\{N, s \log\left(n - s\right)\right\}\right)$.

Proof. For a given radius $R > 0$, define the set

$$\mathbb{S}\left(s, R\right) = \left\{ \mathbf{z} \in \mathbb{R}^n : \quad \|\mathbf{z}\|_0 \leq 2s, \quad \|\mathbf{z}\|_2 \leq R \right\},$$

and the random variable $Z_N = Z_N\left(s, R\right)$ given by

[3] Strictly speaking, these sets are not "balls" when $p < 1$, since they fail to be convex.

$$Z_N = \sup_{\mathbf{z} \in \mathbb{S}(s,R)} \frac{1}{N} \left| \mathbf{w}^T \mathbf{X} \mathbf{z} \right| .$$

For a given $\varepsilon \in (0,1)$ to be chosen later, let us upper bound the minimal cardinality of a set that covers the set $\mathbb{S}(s,R)$ up to $(R\varepsilon)$-accuracy in ℓ_2-norm. Now we claim that we may find a covering set $\left\{ \mathbf{z}^1, \ldots, \mathbf{z}^N \right\} \subset \mathbb{S}(s,R)$ with cardinality $K = K(s,R,\varepsilon)$

$$\log K(s,R,\varepsilon) \le \log \binom{n}{2s} + 2s \log(1/\varepsilon) .$$

To establish the claim, we note that there are $\binom{n}{2s}$ subsets of size $2s$ within the set $\{1,2,\ldots,n\}$. Also, for any $2s$-sized subset, there is an $(R\epsilon)$-covering in ℓ_2-norm of the ball $\mathbb{B}(R)$ (radius R) with at most $2^{2s\log(1/\varepsilon)}$ elements [154].

As a result, for each $\mathbf{z} \subset \mathbb{S}(s,R)$, we may find some $\|\mathbf{z}^l\|_2$ such that $\left\| \mathbf{z} - \mathbf{z}^l \right\|_2 \le R\varepsilon$. By triangle inequality, we have

$$\begin{aligned} \tfrac{1}{N} \left| \mathbf{w}^T \mathbf{X} \mathbf{z} \right| &\le \tfrac{1}{N} \left| \mathbf{w}^T \mathbf{X} \mathbf{z}^l \right| + \tfrac{1}{N} \left| \mathbf{w}^T \mathbf{X} \left(\mathbf{z} - \mathbf{z}^i \right) \right| \\ &\le \tfrac{1}{N} \left| \mathbf{w}^T \mathbf{X} \mathbf{z}^l \right| + \tfrac{\|\mathbf{w}\|_2}{\sqrt{N}} \tfrac{\left\| \mathbf{w}^T \mathbf{X} \left(\mathbf{z} - \mathbf{z}^i \right) \right\|_2}{\sqrt{N}} . \end{aligned}$$

Using the assumption on $\mathbf{X}$, we have

$$\left\| \mathbf{w}^T \mathbf{X} \left(\mathbf{z} - \mathbf{z}^i \right) \right\|_2 / \sqrt{N} \le \kappa \left\| \left(\mathbf{z} - \mathbf{z}^i \right) \right\|_2 \le \kappa R \varepsilon.$$

Also, since the variate $\frac{\|\mathbf{w}^2\|_2}{\sqrt{N}}$ is χ^2 distribution with N degrees of freedom, we have $\|\mathbf{w}\|_2 / \sqrt{N} \le 2\sigma$ with probability at least $1 - c_1 \exp(-c_2 N)$, using standard tail bounds (See Sect. 3.2). Putting together the pieces, we have that

$$\frac{1}{N} \left| \mathbf{w}^T \mathbf{X} \mathbf{z} \right| \le \frac{1}{N} \left| \mathbf{w}^T \mathbf{X} \mathbf{z}^l \right| + 2\kappa R \varepsilon \sigma$$

with high probability. Taking the supremum over $\mathbf{z}$ on both sides gives

$$Z_N \le \max_{l=1,\ldots,N} \frac{1}{N} \left| \mathbf{w}^T \mathbf{X} \mathbf{z}^l \right| + 2\kappa R \varepsilon \sigma.$$

We need to bound the finite maximum over the covering set. See Sect. 1.10. First we observe that each variate $\mathbf{w}^T \mathbf{X} \mathbf{z}^l / N$ is zero mean Gaussian with variance $\sigma^2 \left\| \mathbf{X} \mathbf{z}^l \right\|_2^2 / N^2$. Under the assumed conditions on $\mathbf{z}^l$ and $\mathbf{X}$, this variance is at most $\sigma^2 \kappa^2 R^2 / N$, so that by standard Gaussian tail bounds, we conclude that

$$\begin{aligned} Z_N &\le \sigma\kappa R\sqrt{\tfrac{K(s,R,\varepsilon)}{N}} + 2\kappa R\varepsilon\sigma. \\ &= \sigma\kappa R\left(\sqrt{\tfrac{K(s,R,\varepsilon)}{N}} + 2\varepsilon\right). \end{aligned} \tag{3.49}$$

with probability at least $1 - c_1 \exp\left(-c_2 \log K\left(s, R, \varepsilon\right)\right)$.

Finally, suppose that $\varepsilon = \sqrt{\frac{s\log(n/2s)}{N}}$. With this choice and recalling that $N \le n$ by assumption, we have

$$\begin{aligned} \frac{\log K(s,R,\varepsilon)}{N} &\le \frac{\log\begin{pmatrix} n \\ 2s \end{pmatrix}}{N} + \frac{s\log\frac{N}{s\log(n/2s)}}{N} \\ &\le \frac{\log\begin{pmatrix} n \\ 2s \end{pmatrix}}{N} + \frac{s\log(n/s)}{N} \\ &\le \frac{2s+s\log(n/s)}{N} + \frac{s\log(n/s)}{N}, \end{aligned}$$

where the last line uses standard bounds on binomial coefficients. Since $n/s \ge 2$ by assumption, we conclude that our choice of ε guarantees that $\frac{\log K(s,R,\varepsilon)}{N} \le 5s\log\left(n/s\right)$. Substituting these relations into the inequality (3.49), we conclude that

$$Z_N \le \sigma R\kappa\left\{4\sqrt{\frac{s\log\left(n/2s\right)}{N}} + 2\sqrt{\frac{s\log\left(n/2s\right)}{N}}\right\},$$

as claimed. Since $\log K\left(s, R, \varepsilon\right) \ge s\log\left(n - 2s\right)$, this event occurs with probability at least

$$1 - c_1 \exp\left(-c_2 \min\left\{N, s\log\left(n-s\right)\right\}\right),$$

as claimed. □

3.7 Concentration of Gaussian and Wishart Random Matrices

Theorem 3.7.1 (Davidson and Szarek [145]). *For $k \le n$, let $\mathbf{X} \in \mathbb{R}^{n\times k}$ be a random matrix from a standard Gaussian ensemble, i.e., ($X_{ij} \sim \mathcal{N}(0,1)$, i.i.d.). Then, for all $t > 0$,*

$$\mathbb{P}\left(\left\|\frac{1}{n}\mathbf{X}^T\mathbf{X}-\mathbf{I}_{k\times k}\right\|_2>2\left(\sqrt{\frac{k}{n}}+t\right)+\left(\sqrt{\frac{k}{n}}+t\right)^2\right)+\leq 2\exp\left(-nt^2/2\right). \tag{3.50}$$

We can extend to more general Gaussian ensembles. In particular, for a positive definite matrix $\mathbf{\Sigma}\in\mathbb{R}^{k\times k}$, setting $\mathbf{Y}=\mathbf{X}\sqrt{\mathbf{\Sigma}}$ gives $n\times k$ matrix with i.i.d. rows, $\mathbf{x}_i\sim\mathcal{N}(0,\mathbf{\Sigma})$. Then

$$\left\|\frac{1}{n}\mathbf{Y}^T\mathbf{Y}-\mathbf{\Sigma}\right\|_2=\left\|\sqrt{\mathbf{\Sigma}}\left(\frac{1}{n}\mathbf{X}^T\mathbf{X}-\mathbf{I}_{k\times k}\right)\sqrt{\mathbf{\Sigma}}\right\|_2$$

is upper-bounded by $\lambda_{\max}(\mathbf{\Sigma})\left\|\frac{1}{n}\mathbf{Y}^T\mathbf{Y}-\mathbf{I}_{k\times k}\right\|_2$. So the claim (3.51) follows from the basic bound (3.50). Similarly, we have

$$\begin{aligned}\left\|\left(\frac{1}{n}\mathbf{Y}^T\mathbf{Y}\right)^{-1}-\mathbf{\Sigma}^{-1}\right\|_2&=\left\|\mathbf{\Sigma}^{-1/2}\left(\left(\frac{1}{n}\mathbf{X}^T\mathbf{X}\right)^{-1}-\mathbf{I}_{k\times k}\right)\mathbf{\Sigma}^{-1/2}\right\|_2\\&\leq\left\|\left(\frac{1}{n}\mathbf{X}^T\mathbf{X}\right)^{-1}-\mathbf{I}_{k\times k}\right\|_2\frac{1}{\lambda_{\min}(\mathbf{\Sigma})}\end{aligned}$$

so that the claim (3.52) follows from the basic bound (3.50).

Theorem 3.7.2 (Lemma 9 of Negahban and Wainwright [155]). *For $k\leq n$, let $\mathbf{Y}\in\mathbb{R}^{n\times k}$ be a random matrix having i.i.d. rows, $\mathbf{x}_i\sim\mathcal{N}(0,\mathbf{\Sigma})$.*

1. *If the covariance matrix $\mathbf{\Sigma}$ has maximum eigenvalue $\lambda_{\max}(\mathbf{\Sigma})<+\infty$, then for all $t>0$,*

$$\begin{aligned}\mathbb{P}\left(\left\|\frac{1}{n}\mathbf{Y}^T\mathbf{Y}-\mathbf{\Sigma}\right\|_2>\lambda_{\max}(\mathbf{\Sigma})\left(2\left(\sqrt{\frac{n}{N}}+t\right)+\left(\sqrt{\frac{n}{N}}+t\right)^2\right)\right)\\\leq 2\exp\left(-nt^2/2\right).\end{aligned} \tag{3.51}$$

2. *If the covariance matrix $\mathbf{\Sigma}$ has minimum eigenvalue $\lambda_{\min}(\mathbf{\Sigma})>0$, then for all $t>0$,*

$$\begin{aligned}\mathbb{P}\left(\left\|\left(\frac{1}{n}\mathbf{Y}^T\mathbf{Y}\right)^{-1}-\mathbf{\Sigma}^{-1}\right\|_2>\frac{1}{\lambda_{\min}(\mathbf{\Sigma})}\left(2\left(\sqrt{\frac{n}{N}}+t\right)+\left(\sqrt{\frac{n}{N}}+t\right)^2\right)\right)\\\leq 2\exp\left(-nt^2/2\right).\end{aligned} \tag{3.52}$$

For $t=\sqrt{\frac{k}{n}}$, then since $k/n\leq 1$, we have

$$\gamma = \left(2\left(\sqrt{\frac{n}{N}} + t\right) + \left(\sqrt{\frac{n}{N}} + t\right)^2\right) = 4\left\{\sqrt{\frac{k}{n}} + \frac{k}{n}\right\} \leq 8\sqrt{\frac{k}{n}}.$$

Let us consider applying Theorem 3.7.2, following [138]. As a result, we have a specialized version of (3.51)

$$\mathbb{P}\left(\left\|\frac{1}{n}\mathbf{Y}^T\mathbf{Y} - \Sigma\right\|_2 > 8\lambda_{\max}(\Sigma)\sqrt{\frac{k}{n}}\right) \leq 2\exp(-k/2).$$

$$\mathbb{P}\left(\left\|\left(\frac{1}{n}\mathbf{Y}^T\mathbf{Y}\right)^{-1} - \Sigma^{-1}\right\|_2 > \frac{8}{\lambda_{\min}(\Sigma)}\sqrt{\frac{k}{n}}\right) \leq 2\exp(-k/2). \quad (3.53)$$

Example 3.7.3 (Concentration inequality for random matrix (sample covariance matrix)[138]). We often need to deal with the random matrix (sample covariance matrix) $\left(\mathbf{Y}^T\mathbf{Y}/n\right)^{-1}$, where $\mathbf{Y} \in \mathbb{R}^{n\times k}$ is a random matrix whose entries are i.i.d. elements $Y_{ij} \sim \mathcal{N}(0,1)$. Consider the eigen decomposition

$$\left(\mathbf{Y}^T\mathbf{Y}/n\right)^{-1} - \mathbf{I}_{k\times k} = \mathbf{U}^T\mathbf{D}\mathbf{U},$$

where $\mathbf{D}$ is diagonal and $\mathbf{U}$ is unitary. Since the distribution of $\mathbf{Y}$ is invariant to rotations, the matrices $\mathbf{D}$ and $\mathbf{U}$ are *independent.* Since $\|\mathbf{D}\|_2 = \left\|\left(\mathbf{Y}^T\mathbf{Y}/n\right)^{-1} - \mathbf{I}_{k\times k}\right\|_2$, the random matrix bound (3.53) implies that

$$\mathbb{P}\left(\|\mathbf{D}\|_2 > 8\sqrt{k/n}\right) \leq 2\exp(-k/2).$$

Below, we condition on the event $\|\mathbf{D}\|_2 < 8\sqrt{k/n}$.

Let $\mathbf{e}_j$ denote the unit vector with 1 in position j, and $\mathbf{z} = (z_1, \ldots, z_k) \in \mathbb{R}^k$ be a fixed vector $\mathbf{z} \in \mathbb{R}^k$. Define, for each $i = 1, \ldots, k$, the random variable of interest

$$V_i = \mathbf{e}_i^T\mathbf{U}^T\mathbf{D}\mathbf{U}\mathbf{z} = z_i\mathbf{u}_i^T\mathbf{D}\mathbf{u}_i + \mathbf{u}_i^T\mathbf{D}\left[\sum_{l\neq i} z_l\mathbf{u}_l\right]$$

where $\mathbf{u}_j$ is the j-th column of the unitary matrix $\mathbf{U}$. As an example, consider the variable $\max_i |V_i|$. Since V_i is identically distributed, it is sufficient to obtain an exponential tail bound on $\{V_1 > t\}$.

Under the conditioned event on $\|\mathbf{D}\|_2 < 8\sqrt{k/n}$, we have

$$|V_1| \leq 8\sqrt{k/n}\,|z_1| + \mathbf{u}_1^T\mathbf{D}\left[\sum_{l=2}^{k} z_l\mathbf{u}_l\right]. \quad (3.54)$$

As a result, it is sufficient to obtain a sharp tail bound on the second term. Conditioned on $\mathbf{D}$ and the vector $\mathbf{w} = \left[\sum_{l=2}^{k} z_l \mathbf{u}_l\right]$, the random vector $\mathbf{u}_1 \in \mathbb{R}^k$ is uniformly distributed over a sphere in $k-1$ dimensions: one dimension is lost since $\mathbf{u}_1$ must be orthogonal to $\mathbf{w} \in \mathbb{R}^k$. Now consider the function

$$F(\mathbf{u}_1) = \mathbf{u}_1^T \mathbf{D}\mathbf{w};$$

we can show that this function is Lipschitz (with respect to the Euclidean norm) with constant at most $\|F\|_{\mathcal{L}} \le 8\sqrt{k/n}\sqrt{k-1}\|z\|_\infty$. For any pair of vectors $\mathbf{u}_1$ and $\mathbf{u}_1^{'}$, we have

$$\begin{aligned}
\left|F(\mathbf{u}_1) - F\left(\mathbf{u}_1^{'}\right)\right| &= \left|\left(\mathbf{u}_1 - \mathbf{u}_1^{'}\right)^T \mathbf{D}\mathbf{w}\right| \\
&\le \left\|\mathbf{u}_1 - \mathbf{u}_1^{'}\right\|_2 \|\mathbf{D}\|_2 \|\mathbf{w}\|_2 \\
&\le 8\sqrt{k/n}\sqrt{\sum_{l=2}^{k} z_l^2}\left\|\mathbf{u}_1 - \mathbf{u}_1^{'}\right\|_2 \\
&= 8\sqrt{k/n}\sqrt{k-1}\|\mathbf{z}\|_\infty \left\|\mathbf{u}_1 - \mathbf{u}_1^{'}\right\|_2
\end{aligned}$$

where we have used the fact that $\|\mathbf{w}\|_2 = \sqrt{\sum_{l=2}^{k} z_l^2}$, by the orthonormality of the $\{\mathbf{u}_l\}$ vectors. Since $\mathbb{E}\left[F(\mathbf{u}_1)\right] = 0$, by concentration of measure for Lipschitz functions on the sphere [141], for all $t > 0$, we have

$$\begin{aligned}
\mathbb{P}\left(|F(\mathbf{u}_1)| > t\|\mathbf{z}\|_\infty\right) &\le 2\exp\left(-c_1 (k-1)\frac{t^2}{128\frac{k}{n}(k-1)}\right) \\
&\le 2\exp\left(-c_1 \frac{nt^2}{128k}\right).
\end{aligned}$$

Taking union bound, we have

$$\mathbb{P}\left(\max_{i=1,2,\ldots,k} |F(\mathbf{u}_i)| > t\|\mathbf{z}\|_\infty\right) \le 2k\exp\left(-c_1\frac{nt^2}{128k}\right) = 2\exp\left(-c_1\frac{nt^2}{128k} + \log k\right).$$

Consider $\log(p-k) > \log k$, if we set $t = \frac{256k\log(p-k)}{c_1 n}$, then this probability vanishes at rate $2\exp\left(-c_2\log(p-k)\right)$. If we assume $n = \Omega\left(k\log(p-k)\right)$, the quantity t is order one. □

We can summarize the above example in a formal theorem.

Theorem 3.7.4 (Lemma 5 of [138]). *Consider a fixed nonzero vector* $\mathbf{z} \in \mathbb{R}^k$ *and a random matrix* $\mathbf{Y} \in \mathbb{R}^{n\times k}$ *with i.i.d. elements* $Y_{ij} \sim \mathcal{N}(0,1)$*. Under the scaling*

$n = \Omega\left(k \log\left(p-k\right)\right)$, *there are positive constants* c_1 *and* c_2 *such that for all* $t > 0$

$$\mathbb{P}\left(\left\|\left[\left(\mathbf{Y}^T\mathbf{Y}/n\right)^{-1} - \mathbf{I}_{k\times k}\right]\mathbf{z}\right\|_\infty \geq c_1\|\mathbf{z}\|_\infty\right) \leq 4\exp\left(-c_1 \min\left\{k, \log\left(p-k\right)\right\}\right).$$

Example 3.7.5 (Feature-based detection). In our previous work in the data domain [156, 157] and the kernel domain [158–160], features of a signal are used for detection. For hypothesis $\mathcal{H}_0$, there is only white Gaussian noise, while for $\mathcal{H}_1$, there is a signal (with some detectable features) in presence of the white Gaussian noise. For example, we can use the leading eigenvector of the covariance matrix $\mathbf{R}$ as the feature which is a fixed vector $\mathbf{z}$. The inverse of the sample covariance matrix is considered $\hat{\mathbf{R}}^{-1} = \left(\mathbf{Y}^T\mathbf{Y}/n\right)^{-1} - \mathbf{I}_{k\times k}$, where $\mathbf{Y}$ is as assumed in Theorem 3.7.4. Consequently, the problem boils down to

$$\hat{\mathbf{R}}^{-1}\mathbf{z} = \left[\left(\mathbf{Y}^T\mathbf{Y}/n\right)^{-1} - \mathbf{I}_{k\times k}\right]\mathbf{z}.$$

We can bound the above expression. □

The following theorem shows sharp concentration of a Lipschitz function of Gaussian random variables around its mean. Let $||\mathbf{y}||_2$ be the ℓ_2-norm of an arbitrary Gaussian vector $\mathbf{y}$.

Theorem 3.7.6 ([141, 161]). *Let a random vector* $\mathbf{x} \in \mathbb{R}^n$ *have i.i.d.* $\mathcal{N}(0,1)$ *entries, and let* $f : \mathbb{R}^n \to \mathbb{R}$ *be Lipschitz with constant L, (i.e.,* $|f(\mathbf{x}) - f(\mathbf{y})| \leq L\|\mathbf{x}-\mathbf{y}\|_2^2, \forall \mathbf{x},\mathbf{y} \in \mathbb{R}^n$*). Then, for all* $t > 0$*, we have*

$$\mathbb{P}\left(\left|f(\mathbf{x}) - f(\mathbf{y})\right| > t\right) \leq 2\exp\left(-\frac{t^2}{2L^2}\right).$$

Let σ_1 be the largest singular value of a rectangular complex matrix $\mathbf{A}$. Let $\|\mathbf{A}\|_{op} = \sigma_1$ is the operator norm of matrix $\mathbf{A}$. Let $\sqrt{\mathbf{R}}$ be the symmetric matrix square root, and consider the function

$$f(\mathbf{x}) = \|\mathbf{R}\mathbf{x}\|_2/\sqrt{n}.$$

Since it is Lipschitz with constant $\|\mathbf{R}\|_{op}/\sqrt{n}$, Theorem 3.7.6 implies that

$$\mathbb{P}\left(\left|\|\mathbf{R}\mathbf{x}\|_2 - \mathbb{E}\|\mathbf{R}\mathbf{x}\|_2\right| > \sqrt{n}t\right) \leq 2\exp\left(-\frac{nt^2}{2\|\mathbf{R}\|_{op}}\right), \tag{3.55}$$

for all $t > 0$. By integrating this tail bound, we find the variable $Z = \|\mathbf{R}\mathbf{x}\|_2/\sqrt{n}$ satisfies the bound

$$\mathrm{var}(Z) \leq 4\|\mathbf{R}\|_{op}/\sqrt{n}.$$

So that

$$\left|\mathbb{E}Z^2 - |\mathbb{E}Z|\right| = \left|\sqrt{\mathrm{Tr}\,(\mathbf{R})\,/n} - \mathbb{E}\left(\|\mathbf{R}\mathbf{x}\|_2\,/\sqrt{n}\right)\right| \le 2\sqrt{\|\mathbf{R}\|_{op}}/\sqrt{n}. \quad (3.56)$$

Combining (3.56) with (3.55), we obtain

$$\mathbb{P}\left(\left|\frac{1}{n}\left\|\sqrt{\mathbf{R}}\mathbf{x}\right\|_2 - \sqrt{\mathrm{Tr}\,(\mathbf{R})}\right| \ge t + 2\sqrt{\frac{\|\mathbf{R}\|_{op}}{n}}\right) \le 2\exp\left(-\frac{nt^2}{2\|\mathbf{R}\|_{op}}\right) \quad (3.57)$$

for all $t > 0$. Setting $\tau = (t - 2/\sqrt{n})\sqrt{\|\mathbf{R}\|_{op}}$ in the bound (3.57) gives that

$$\mathbb{P}\left(\left|\frac{1}{n}\left\|\sqrt{\mathbf{R}}\mathbf{x}\right\|_2 - \sqrt{\mathrm{Tr}\,(\mathbf{R})}\right| \ge \tau\sqrt{\|\mathbf{R}\|_{op}}\right) \le 2\exp\left(-\frac{1}{2}n\left(t - \frac{2}{\sqrt{n}}\right)^2\right). \quad (3.58)$$

Similarly, considering $t = \sqrt{\|\mathbf{R}\|_{op}}$ in the bound (3.57) gives that with probability greater than $1 - 2\exp(-n/2)$, we have

$$\left|\frac{\|\mathbf{x}\|_2}{\sqrt{n}} + \sqrt{\frac{\mathrm{Tr}\,(\mathbf{R})}{n}}\right| \le \sqrt{\frac{\mathrm{Tr}\,(\mathbf{R})}{n}} + 3\sqrt{\|\mathbf{R}\|_{op}} \le 4\sqrt{\|\mathbf{R}\|_{op}}. \quad (3.59)$$

Using the two bounds, we have

$$\left|\frac{\|\mathbf{x}\|_2^2}{\sqrt{n}} - \sqrt{\frac{\mathrm{Tr}\,(\mathbf{R})}{n}}\right| = \left|\frac{\|\mathbf{x}\|_2}{\sqrt{n}} - \sqrt{\frac{\mathrm{Tr}\,(\mathbf{R})}{n}}\right|\left|\frac{\|\mathbf{x}\|_2}{\sqrt{n}} + \sqrt{\frac{\mathrm{Tr}\,(\mathbf{R})}{n}}\right| \le 4\tau\sqrt{\|\mathbf{R}\|_{op}}.$$

We summarize the above result here.

Theorem 3.7.7 (Lemma I.2 of Negahban and Wainwright [155]). *Given a Gaussian random vector* $\mathbf{x} \sim \mathcal{N}(0, \mathbf{R})$, *for all* $t > 2/\sqrt{n}$, *we have*

$$\mathbb{P}\left[\frac{1}{n}\left|\|\mathbf{x}\|_2^2 - \mathrm{Tr}\,(\mathbf{R})\right| > 4t\|\mathbf{R}\|_{op}\right] \le 2\exp\left(-\tfrac{1}{2}n\left(t - \frac{2}{\sqrt{n}}\right)\right) + 2\exp(-n/2).$$

We take material from [139]. Defining the standard Gaussian random matrix $\mathbf{G} = (G_{ij})_{1 \le i \le n, 1 \le j \le p} \in \mathbb{R}^{n \times p}$, we have the $p \times p$ Wishart random matrix

$$\mathbf{W} = \frac{1}{n}\mathbf{G}^T\mathbf{G} - \mathbf{I}_p, \quad (3.60)$$

where $\mathbf{I}_p$ is the $p \times p$ identity matrix. We essentially deal with the "sums of Gaussian product" random variates. Let Z_1 and Z_2 be independent Gaussian random variables, we consider the sum $\sum_{i=1}^{n} X_i$ where $X_i \overset{i.i.d.}{\sim} Z_1 Z_2, 1 \le i \le n$. The following tails bound is also known [162, 163]

$$\mathbb{P}\left(\left|\frac{1}{n}\sum_{i=1}^{n}X_i\right|>t\right)\leq C\exp\left(-3nt^2/2\right)\ \text{as}\ t\to 0; \tag{3.61}$$

Let $\mathbf{w}_i^T$ be the i-th row of Wishart matrix $\mathbf{W}\in\mathbb{R}^{p\times p}$, and $\mathbf{g}_i^T$ be the i-th row of data matrix $\mathbf{G}\in\mathbb{R}^{n\times p}$. The linear combination of off-diagonal entries of the first row $\mathbf{w}_1$

$$\langle\mathbf{a},\mathbf{w}_1\rangle=\sum_{j=2}^{n}a_jW_{1j}=\frac{1}{n}\sum_{i=1}^{n}G_{1i}\sum_{j=2}^{p}G_{ij}a_j,$$

for a vector $\mathbf{a}=(a_2,\ldots,a_p)\in\mathbb{R}^{p-1}$. Let

$$\xi_i=\frac{1}{\|\mathbf{a}\|_2}\langle\mathbf{a},\mathbf{g}_i\rangle=\frac{1}{\|\mathbf{a}\|_2}\sum_{j=2}^{p}G_{ij}a_j.$$

Note that $\{\xi_i\}_{i=1}^n$ is a collection of independent standard Gaussian random variables. Also, $\{\xi_i\}_{i=1}^n$ are independent of $\{G_{i1}\}_{i=1}^n$. Now we have

$$\langle\mathbf{a},\mathbf{w}_1\rangle=\frac{1}{n}\|\mathbf{a}\|_2\sum_{i=1}^{n}G_{1i}\xi_i,$$

which is a (scaled) sum of Gaussian products. Using (3.61), we obtain

$$\mathbb{P}\left(|\langle\mathbf{a},\mathbf{w}_1\rangle|>t\right)\leq C\exp\left(-3nt^2/2\,\|\mathbf{a}\|_2^2\right). \tag{3.62}$$

Combining (3.4) and (3.62), we can bound a linear combination of first-row entries. Noting that $W_{11}=\frac{1}{n}\sum_{i=1}^{n}(G_{1i})^2-1$ is a centered χ_n^2, we have

$$\begin{aligned}\mathbb{P}\left(|\mathbf{w}_i\mathbf{x}|>t\right)=\mathbb{P}\left(\left|\sum_{j=1}^{p}W_{ij}x_j\right|>t\right)&\leq\mathbb{P}\left(|x_1W_{11}|+\left|\sum_{j=2}^{p}W_{ij}x_j\right|>t\right)\\&\leq\mathbb{P}\left(|x_1W_{11}|>t/2\right)+\mathbb{P}\left(\left|\sum_{j=2}^{p}W_{ij}x_j\right|>t/2\right)\\&\leq 2\exp\left(-\frac{3nt^2}{16\cdot 4x_1^2}\right)+C\exp\left(-\frac{3nt^2}{2\cdot 4\sum_{i=2}^{p}x_j^2}\right)\\&\leq 2\max(2,C)\exp\left(-\frac{3nt^2}{16\cdot 4\sum_{i=1}^{p}x_j^2}\right).\end{aligned}$$

There is nothing special about the "first" row, we conclude the following. Note that the inner product is $\langle \mathbf{w}_i, \mathbf{x} \rangle = \mathbf{w}_i^T \mathbf{x} = \sum_{j=1}^{p} W_{ij} x_j, i = 1, \ldots, p$ for a vector $\mathbf{x} \in \mathbb{R}^p$.

Theorem 3.7.8 (Lemma 15 of [139]). *Let $\mathbf{w}_i^T$ be the j-th row of Wishart matrix $\mathbf{W} \in \mathbb{R}^{p \times p}$, defined as (3.60). For $t > 0$ small enough, there are (numerical constants) $c > 0$ and $C > 0$ such that for all $\mathbf{x} \in \mathbb{R}^n \setminus \{0\}$,*

$$\mathbb{P}\left(\left| \mathbf{w}_i^T \mathbf{x} \right| > t \right) \le C \exp\left(-cnt^2 / \|\mathbf{x}\|^2 \right), \quad i = 1, \ldots, p.$$

3.8 Concentration of Operator Norms

For a vector $\mathbf{a}$, $\|\mathbf{a}\|_p$ is the ℓ_p norm. For a matrix $\mathbf{A} \in \mathbb{R}^{m \times n}$, the singular values are ordered decreasingly as $\sigma_1(\mathbf{A}) \ge \sigma_2(\mathbf{A}) \ge \cdots \ge \sigma_{\min(m,n)}(\mathbf{A})$. Then we have $\sigma_{\max}(\mathbf{A}) = \sigma_1(\mathbf{A})$, and $\sigma_{\min}(\mathbf{A}) = \sigma_{\min(m,n)}(\mathbf{A})$. Let the operator norm of the matrix $\mathbf{A}$ be defined as $\|\mathbf{A}\|_{op} = \sigma_1(\mathbf{A})$. The nuclear norm is defined as

$$\|\mathbf{A}\|_* = \sum_{i=1}^{\min(m,n)} \sigma_i(\mathbf{A}),$$

while the Frobenius norm is defined as

$$\|\mathbf{A}\|_F = \sqrt{\mathrm{Tr}\left(\mathbf{A}^T \mathbf{A}\right)} = \sqrt{\sum_{i=1}^{\min(m,n)} \sigma_i^2(\mathbf{A})}.$$

For a matrix $\mathbf{A} \in \mathbb{R}^{m_1 \times m_2}$, we use vector $\mathrm{vec}(\mathbf{A}) \in \mathbb{R}^M, M = m_1 m_2$. Given a symmetric positive definite matrix $\boldsymbol{\Sigma} \in \mathbb{R}^{M \times M}$, we say that the random matrix $\mathbf{X}_i$ is sampled from the $\boldsymbol{\Sigma}$-ensemble if

$$\mathrm{vec}(\mathbf{X}_i) \sim \mathcal{N}(0, \boldsymbol{\Sigma}).$$

We define the quantity

$$\rho^2(\boldsymbol{\Sigma}) = \sup_{\|\mathbf{u}\|_1 = 1, \|\mathbf{v}\|_1 = 1} \mathrm{var}\left(\mathbf{u}^T \mathbf{X} \mathbf{v}\right),$$

where the random matrix $\mathbf{X} \in \mathbb{R}^{m_1 \times m_2}$ is sampled from the $\boldsymbol{\Sigma}$-ensemble. For the special case (white Gaussian random vector) $\boldsymbol{\Sigma} = \mathbf{I}$, we have $\rho^2(\boldsymbol{\Sigma}) = 1$.

Now we are ready to study the concentration of measure for the operator norm.

Theorem 3.8.1 (Negahban and Wainwright [155]). *Let* $\mathbf{X} \in \mathbb{R}^{m_1 \times m_2}$ *be a random sample from the* $\mathbf{\Sigma}$*-ensemble. Then we have*

$$\mathbb{E}\left[\|\mathbf{X}\|_{op}\right] \le 12\rho\left(\mathbf{\Sigma}\right)\left[\sqrt{m_1} + \sqrt{m_2}\right] \tag{3.63}$$

and moreover

$$\mathbb{P}\left(\|\mathbf{X}\|_{op} \ge \mathbb{E}\left[\|\mathbf{X}\|_{op}\right] + t\right) \le \exp\left(-\frac{t^2}{2\rho^2\left(\mathbf{\Sigma}\right)}\right). \tag{3.64}$$

Proof. The variational representation

$$\|\mathbf{X}\|_{op} = \sup_{\|\mathbf{u}\|_1=1, \|\mathbf{v}\|_1=1} \mathbf{u}^T\mathbf{X}\mathbf{v}$$

is the starting point. Since each (bi-linear) variable $\mathbf{u}^T\mathbf{X}\mathbf{v}$ is zero-mean Gaussian, thus we find that the operator norm $\|\mathbf{X}\|_{op}$ is the supremum of a Gaussian process. The bound (3.64) follows from Ledoux [141, Theorem 7.1].

We now use a simple covering argument to establish the upper bound (3.63). For more details, we refer to [155]. □

Theorem 3.8.2 (Lemma C.1 of Negahban and Wainwright [155]). *The random matrix* $\{\mathbf{X}_i\}_{i=1}^N$ *are drawn i.i.d. from the* $\mathbf{\Sigma}$*-Gaussian ensemble, i.e.,* $\mathrm{vec}(\mathbf{X}_i) \sim \mathcal{N}\left(0, \mathbf{\Sigma}\right)$. *For a random vector* $\boldsymbol{\epsilon} = (\epsilon_1, \ldots, \epsilon_N)$, *if* $\|\boldsymbol{\epsilon}\|_2 \le 2\nu\sqrt{N}$, *then there are universal constants* $c_0, c_1, c_2 > 0$ *such that*

$$\mathbb{P}\left(\left\|\frac{1}{N}\sum_{i=1}^N \epsilon_i \mathbf{X}_i\right\|_{op} \ge c_0\nu\rho\left(\mathbf{\Sigma}\right)\left(\sqrt{\frac{m_1}{N}} + \sqrt{\frac{m_2}{N}}\right)\right) \le c_1 \exp\left(-c_2\left(m_1 + m_2\right)\right).$$

Proof. Define the random matrix $\mathbf{Z}$ as

$$\mathbf{Z} = \frac{1}{N}\sum_{i=1}^N \varepsilon_i \mathbf{X}_i.$$

Since the random matrices $\{\mathbf{X}_i\}_{i=1}^N$ are i.i.d. Gaussian, if the sequence $\{\epsilon_i\}_{i=1}^N$ are fixed (by conditioning as needed), then the random matrix $\mathbf{Z}$ is a sample from the $\mathbf{\Gamma}$-Gaussian ensemble with the covariance matrix $\mathbf{\Gamma} = \frac{\|\boldsymbol{\epsilon}\|_2}{N^2}\mathbf{\Sigma}$. So, if $\tilde{\mathbf{Z}} \in \mathbb{R}^{m_1 \times m_2}$ is a random matrix drawn from the $\frac{\|\boldsymbol{\epsilon}\|_2}{N^2}\mathbf{\Sigma}$-ensemble, we have

$$\mathbb{P}\left(\|\mathbf{Z}\|_{op} \ge t\right) \le \mathbb{P}\left(\left\|\tilde{\mathbf{Z}}\right\|_{op} \ge t\right).$$

Using Theorem 3.8.1, we have

$$\mathbb{E}\left[\left\|\tilde{\mathbf{Z}}\right\|_{op}\right] \leq \frac{12\sqrt{2}\nu\rho\left(\mathbf{\Sigma}\right)}{\sqrt{N}}\left(\sqrt{m_1}+\sqrt{m_2}\right)$$

and

$$\mathbb{P}\left(\left\|\tilde{\mathbf{Z}}\right\|_{op} \geq \mathbb{E}\left[\left\|\tilde{\mathbf{Z}}\right\|_{op}\right]+t\right) \leq \exp\left(-c_1\frac{Nt^2}{\nu^2\rho^2\left(\mathbf{\Sigma}\right)}\right)$$

for a universal constant c_1. Setting $t^2 = \Omega\left(N^{-1}\nu^2\rho^2\left(\mathbf{\Sigma}\right)\left(\sqrt{m_1}+\sqrt{m_2}\right)^2\right)$ gives the claim. □

This following result follows by adapting known concentration results for random matrices (see [138] for details):

Theorem 3.8.3 (Lemma 2 of Negahban and Wainwright [155]). *Let $\mathbf{X} \in \mathbb{R}^{m\times n}$ be a random matrix with i.i.d. rows sampled from a n-variate $\mathcal{N}\left(0,\mathbf{\Sigma}\right)$ distribution. Then for $m \geq 2n$, we have*

$$\mathbb{P}\left(\sigma_{\min}\left(\frac{1}{n}\mathbf{X}^T\mathbf{X}\right) \geq \frac{1}{9}\sigma_{\min}\left(\mathbf{\Sigma}\right), \sigma_{\max}\left(\frac{1}{n}\mathbf{X}^T\mathbf{X}\right) \geq 9\sigma_{\max}\left(\mathbf{\Sigma}\right)\right) \geq 1-4\exp\left(-n/2\right).$$

Consider zero-mean Gaussian random vectors $\mathbf{w}_i$ defined as $\mathbf{w}_i \sim \mathcal{N}\left(0,\nu^2\mathbf{I}_{m_1\times m_1}\right)$ and random vectors $\mathbf{x}_i$ defined as $\mathbf{x}_i \sim \mathcal{N}\left(0,\mathbf{\Sigma}\right)$. We define random matrices $\mathbf{X},\mathbf{W}$ as

$$\mathbf{X} = \begin{bmatrix}\mathbf{x}_1^T\\ \mathbf{x}_2^T\\ \vdots\\ \mathbf{x}_n^T\end{bmatrix} \in \mathbb{R}^{n\times m_2} \quad\text{and}\quad \mathbf{W} = \begin{bmatrix}\mathbf{w}_1^T\\ \mathbf{w}_2^T\\ \vdots\\ \mathbf{w}_n^T\end{bmatrix} \in \mathbb{R}^{n\times m_1}. \tag{3.65}$$

Theorem 3.8.4 (Lemma 3 of Negahban and Wainwright [155]). *For random matrices $\mathbf{X},\mathbf{W}$ defined in (3.65), there are constants $c_1,c_2 > 0$ such that*

$$\mathbb{P}\left(\frac{1}{n}\left\|\mathbf{X}^T\mathbf{W}\right\|_{op} \geq 5\nu\sqrt{\sigma_{\max}\left(\mathbf{\Sigma}\right)}\sqrt{\frac{m_1+m_2}{n}}\right) \leq c_1\exp\left(-c_2\left(m_1+m_2\right)\right).$$

The following proof, taken from [155], will illustrate the standard approach: arguments based on Gordon-Slepian lemma (see also Sects. 3.4 and 3.6) and Gaussian concentration of measure [27, 141].

Proof. Let $\mathcal{S}^{m-1} = \{\mathbf{u} \in \mathbb{R}^m : \left\|\mathbf{u}\right\|_2 = 1\}$ denote the Euclidean sphere in m-dimensional space. The operator norm of interest has the variation representation

$$\frac{1}{n}\left\|\mathbf{X}^T\mathbf{W}\right\|_{op} = \frac{1}{n}\sup_{\mathbf{u}\in\mathcal{S}^{m_1-1}}\sup_{\mathbf{v}\in\mathcal{S}^{m_2-1}}\mathbf{v}^T\mathbf{X}^T\mathbf{W}\mathbf{u}. \tag{3.66}$$

For positive scalars a and b, define the random quantity

$$\Psi\left(a,b\right) = \frac{1}{n}\sup_{\mathbf{u}\in a\mathcal{S}^{m_1-1}}\sup_{\mathbf{v}\in b\mathcal{S}^{m_2-1}}\mathbf{v}^T\mathbf{X}^T\mathbf{W}\mathbf{u}.$$

Our goal is to upper bound $\Psi\left(1,1\right)$. Note that $\Psi\left(a,b\right) = ab\Psi\left(1,1\right)$ due to the bi-linear property of the right-hand side of (3.66).

Let

$$\mathcal{A} = \left\{\mathbf{u}^1,\ldots,\mathbf{u}^A\right\}, \mathcal{B} = \left\{\mathbf{u}^1,\ldots,\mathbf{u}^B\right\}$$

denote the 1/4 coverings of $\mathcal{S}^{m_1-1}$ and $\mathcal{S}^{m_2-1}$, respectively. We now claim that the upper bound

$$\Psi\left(1,1\right) \le 4\max_{\mathbf{u}^a\in\mathcal{A},\mathbf{v}^b\in\mathcal{B}}\left\langle \mathbf{X}\mathbf{v}^b,\mathbf{W}\mathbf{u}^a\right\rangle. \tag{3.67}$$

is valid. To establish the claim, since we note that the sets $\mathcal{A}$ and $\mathcal{B}$ are 1/4-covers, for any pair $(\mathbf{u},\mathbf{v}) \in \mathcal{S}^{m_1-1}\times\mathcal{S}^{m_2-1}$, there exists a pair $\left(\mathbf{u}^a,\mathbf{v}^b\right) \in \mathcal{A}\times\mathcal{B}$, such that $\mathbf{u} = \mathbf{u}^a + \Delta\mathbf{u}$ and $\mathbf{v} = \mathbf{v}^b + \Delta\mathbf{v}$, with

$$\max\left\{\left\|\Delta\mathbf{u}\right\|_2,\left\|\Delta\mathbf{v}\right\|_2\right\} \le 1/4.$$

Consequently, due to the linearity of the inner product, we have

$$\left\langle \mathbf{X}\mathbf{v},\mathbf{W}\mathbf{u}\right\rangle = \left\langle \mathbf{X}\mathbf{v}^b,\mathbf{W}\mathbf{u}^a\right\rangle + \left\langle \mathbf{X}\mathbf{v}^b,\mathbf{W}\Delta\mathbf{u}\right\rangle + \left\langle \mathbf{X}\Delta\mathbf{v},\mathbf{W}\mathbf{u}^a\right\rangle + \left\langle \mathbf{X}\Delta\mathbf{u},\mathbf{W}\Delta\mathbf{v}\right\rangle. \tag{3.68}$$

By construction, we have the bound

$$\left|\left\langle \mathbf{X}\mathbf{v}^b,\mathbf{W}\Delta\mathbf{u}\right\rangle\right| \le \Psi\left(1,1/4\right) = \frac{1}{4}\Psi\left(1,1\right),$$

and similarly

$$\left|\left\langle \mathbf{X}\Delta\mathbf{v},\mathbf{W}\mathbf{u}^a\right\rangle\right| \le \frac{1}{4}\Psi\left(1,1\right),$$

as well as

$$\left|\left\langle \mathbf{X}\Delta\mathbf{u},\mathbf{W}\Delta\mathbf{v}\right\rangle\right| \le \frac{1}{16}\Psi\left(1,1\right).$$

Substituting these bounds into (3.68) and taking suprema over the left and right-hand sides, we conclude that

$$\Psi(1,1) \leq \max_{\mathbf{u}^a \in \mathcal{A}, \mathbf{v}^b \in \mathcal{B}} \left\langle \mathbf{X}\mathbf{v}^b, \mathbf{W}\mathbf{u}^a \right\rangle + \frac{9}{16}\Psi(1,1)$$

from which (3.67) follows.

Now we need to control the discrete maximum. See Sect. 1.10 related theorems. According to [27, 154], there exists a 1/4 covering of spheres $\mathcal{S}^{m_1-1}$ and $\mathcal{S}^{m_2-1}$ with at most $A \leq 8^{m_1}$ and $B \leq 8^{m_2}$ elements, respectively. As a result, we have

$$\mathbb{P}\left(|\Psi(1,1)| \geq 4\delta n\right) \leq 8^{m_1+m_2} \max_{\mathbf{u}^a \in \mathcal{A}, \mathbf{v}^b \in \mathcal{B}} \mathbb{P}\left(\frac{1}{n}\left\langle \mathbf{X}\mathbf{v}^b, \mathbf{W}\mathbf{u}^a \right\rangle \geq \delta\right). \quad (3.69)$$

The rest is to do obtain a good bound on the quantity

$$\frac{1}{n}\langle \mathbf{X}\mathbf{v}, \mathbf{W}\mathbf{u}\rangle = \frac{1}{n}\sum_{i=1}^{n} \langle \mathbf{v}, \mathbf{x}_i\rangle \langle \mathbf{u}, \mathbf{w}_i\rangle$$

where $(\mathbf{u}, \mathbf{v}) \in \mathcal{S}^{m_1-1} \times \mathcal{S}^{m_2-1}$ are arbitrary but fixed. Here, $\mathbf{x}_i$ and $\mathbf{w}_i$ are, respectively, the i-th row of matrices $\mathbf{X}$ and $\mathbf{W}$. Since $\mathbf{x}_i \in \mathbb{R}^{m_1}$ has i.i.d. $\mathcal{N}\left(0, \nu^2\right)$ elements and $\mathbf{u}$ is fixed, we have

$$Z_i = \langle \mathbf{u}, \mathbf{w}_i\rangle \sim \mathcal{N}\left(0, \nu^2\right), \quad i = 1, \ldots, n.$$

These variables $\{Z_i\}_{i=1}^n$ are independent from each other, and of the random matrix $\mathbf{X}$. So, conditioned on $\mathbf{X}$, the sum

$$Z = \frac{1}{n}\sum_{i=1}^{n} \langle \mathbf{v}, \mathbf{x}_i\rangle \langle \mathbf{u}, \mathbf{w}_i\rangle$$

is zero-mean Gaussian with variance

$$\alpha^2 = \frac{\nu^2}{n}\left(\frac{1}{n}\|\mathbf{X}\mathbf{v}\|_2^2\right) \leq \frac{\nu^2}{n}\left\|\mathbf{X}^T\mathbf{X}/n\right\|_{op}.$$

Define the event

$$\mathcal{T} = \left\{\alpha^2 \leq \frac{9\nu^2}{n}\|\boldsymbol{\Sigma}\|_{op}\right\}.$$

Using Theorem 3.8.3, we have

$$\left\|\mathbf{X}^T\mathbf{X}/n\right\|_{op} \leq 9\sigma_{\max}(\boldsymbol{\Sigma})$$

with probability at least $1 - 2\exp(-n/2)$, which implies that $\mathbb{P}(\mathcal{T}^c) \leq 2\exp(-n/2)$. Therefore, conditioned on the event $\mathcal{T}$ and its complement $\mathcal{T}^c$, we have

$$\mathbb{P}\left(|Z| \geq t\right) \leq \mathbb{P}\left(|Z| \geq t \,|\mathcal{T}\right) + \mathbb{P}\left(\mathcal{T}^c\right)$$
$$\leq \exp\left(-n\frac{t^2}{2\nu^2\left(4+\|\mathbf{\Sigma}\|_{op}\right)}\right) + 2\exp\left(-n/2\right).$$

Combining this tail bound with the upper bound (3.69), we obtain

$$\mathbb{P}\left(|\Psi\left(1,1\right)| \geq 4\delta n\right) \leq 8^{m_1+m_2}\exp\left(-n\frac{t^2}{2\nu^2\left(4+\|\mathbf{\Sigma}\|_{op}\right)}\right) + 2\exp\left(-n/2\right). \tag{3.70}$$

Setting $t^2 = 20\nu^2\|\mathbf{\Sigma}\|_{op}\frac{m_1+m_2}{n}$, this probability vanishes as long as $n > 16\left(m_1+m_2\right)$. □

Consider the vector random operator $\varphi\left(\mathbf{A}\right) : \mathbb{R}^{m_1\times m_2} \to \mathbb{R}^N$ with $\varphi\left(\mathbf{A}\right) = \left(\varphi_1\left(\mathbf{A}\right), \ldots, \varphi_N\left(\mathbf{A}\right)\right) \in \mathbb{R}^N$. The scalar random operator $\varphi_i\left(\mathbf{A}\right)$ is defined by

$$\varphi_i\left(\mathbf{A}\right) = \left\langle \mathbf{X}_i, \mathbf{A}\right\rangle, \quad i = 1, \ldots, N, \tag{3.71}$$

where the matrices $\{\mathbf{X}_i\}_{i=1}^N$ are formed from the $\mathbf{\Sigma}$-ensemble, i.e., vec$(\mathbf{X}_i) \sim \mathcal{N}\left(0, \mathbf{\Sigma}\right)$.

Theorem 3.8.5 (Proposition 1 of Negahban and Wainwright [155]). *Consider the random operator $\varphi(\mathbf{A})$ defined in* (3.71). *Then, for all $\mathbf{A} \in \mathbb{R}^{m_1\times m_2}$, the random operator $\varphi(\mathbf{A})$ satisfies*

$$\frac{1}{\sqrt{N}}\|\varphi\left(\mathbf{A}\right)\|_2 \geq \frac{1}{4}\left\|\sqrt{\mathbf{\Sigma}}\,\mathrm{vec}\left(\mathbf{A}\right)\right\|_2 - 12\rho\left(\mathbf{\Sigma}\right)\left(\sqrt{\frac{m_1}{N}} + \sqrt{\frac{m_2}{N}}\right)\|\mathbf{A}\|_1 \tag{3.72}$$

with probability at least $1 - 2\exp\left(-N/32\right)$. *In other words, we have*

$$\mathbb{P}\left(\frac{1}{\sqrt{N}}\|\varphi\left(\mathbf{A}\right)\|_2 \leq \frac{1}{4}\left\|\sqrt{\mathbf{\Sigma}}\,\mathrm{vec}\left(\mathbf{A}\right)\right\|_2 - 12\rho\left(\mathbf{\Sigma}\right)\left(\sqrt{\frac{m_1}{N}} + \sqrt{\frac{m_2}{N}}\right)\|\mathbf{A}\|_1\right) \leq 2\exp\left(-N/32\right). \tag{3.73}$$

The proof of Theorem 3.8.5 follows from the use of Gaussian comparison inequalities [27] and concentration of measure [141]. Its proof is similar to the proof of Theorem 3.8.4 above. We see [155] for details.

3.9 Concentration of Sub-Gaussian Random Matrices

We refer to Sect. 1.7 on sub-Gaussian random variables and Sect. 1.9 for the background on exponential random variables. Given a zero-mean random variable Y, we refer to

$$\|Y\|_{\psi_1} = \sup_{l \geq 1} \frac{1}{l} \left(\mathbb{E}|Y|^l\right)^{1/l}$$

as its sub-exponential parameter. The finiteness of this quantity guarantees existence of all moments, and hence large-deviation bounds of the Bernstein type.

We say that a random matrix $\mathbf{X} \in \mathbb{R}^{n \times p}$ is *sub-Gaussian* with parameters $(\mathbf{\Sigma}, \sigma^2)$ if

1. Each row $\mathbf{x}_i^T \in \mathbb{R}^p, i = 1, \ldots, n$ is sampled *independently* from a zero-mean distribution with covariance $\mathbf{\Sigma}$, and
2. For any unit vector $\mathbf{u} \in \mathbb{R}^p$, the random variable $\mathbf{u}^T \mathbf{x}_i$ is sub-Gaussian with parameter at most σ.

If we a random matrix by drawing each row independently from the distribution $\mathcal{N}(0, \mathbf{\Sigma})$, then the resulting matrix $\mathbf{X} \in \mathbb{R}^{n \times p}$ is a sub-Gaussian matrix with parameters $\left(\mathbf{\Sigma}, \|\mathbf{\Sigma}\|_{op}\right)$, where $\|\mathbf{A}\|_{op}$ is the operator norm of matrix $\mathbf{A}$.

By Lemma 1.9.1, if a (scalar valued) random variable X is a zero-mean sub-Gaussian with parameter σ, then the random variable $Y = X^2 - \mathbb{E}\left(X^2\right)$ is sub-exponential with $\|Y\|_{\psi_1} \leq 2\sigma^2$. It then follows that if $X_1, \ldots, X_n$ are zero-mean i.i.d. sub-Gaussian random variables, we have the deviation inequality

$$\mathbb{P}\left(\frac{1}{N}\left|\sum_{i=1}^{N} X_i^2 - \mathbb{E}\left(X_i^2\right)\right| \geq t\right) \leq 2\exp\left[-c\min\left(\frac{nt^2}{4\sigma^2}, \frac{nt}{2\sigma^2}\right)\right]$$

for all $t \geq 0$ where $c > 0$ is a universal constant (see Corollary 1.9.3). This deviation bound may be used to obtain the following result.

Theorem 3.9.1 (Lemma 14 of [164]). *If* $\mathbf{X} \in \mathbb{R}^{n \times p_1}$ *is a zero-mean sub-Gaussian matrix with parameters* $(\mathbf{\Sigma}_x, \sigma_x^2)$, *then for any fixed (unit) vector* $\mathbf{v} \in \mathbb{R}^p$, *we have*

$$\mathbb{P}\left(\left|\|\mathbf{Xv}\|_2^2 - \mathbb{E}\|\mathbf{Xv}\|_2^2\right| \geq nt\right) \leq 2\exp\left(-cn\min\left(\frac{t^2}{\sigma_x^4}, \frac{t}{\sigma_x^2}\right)\right). \tag{3.74}$$

Moreover, if $\mathbf{Y} \in \mathbb{R}^{n \times p_2}$ *is a zero-mean sub-Gaussian matrix with parameters* $(\mathbf{\Sigma}_y, \sigma_y^2)$, *then*

$$\mathbb{P}\left(\left\|\frac{1}{n}\mathbf{Y}^T\mathbf{X} - \operatorname{cov}(\mathbf{y}_i, \mathbf{x}_i)\right\|_{\max} \geq t\right) \leq 6p_1p_2\exp\left(-cn\min\left(\frac{t^2}{(\sigma_x\sigma_y)^2}, \frac{t}{\sigma_x\sigma_y}\right)\right), \tag{3.75}$$

where $\mathbf{X}_i$ *and* $\mathbf{Y}_i$ *are the i-th rows of* $\mathbf{X}$ *and* $\mathbf{Y}$, *respectively. In particular, if* $n \gtrsim \log p$, *then*

$$\mathbb{P}\left(\left\|\frac{1}{n}\mathbf{Y}^T\mathbf{X} - \text{cov}\left(\mathbf{y}_i, \mathbf{x}_i\right)\right\|_{\max} \geq t\right) \leq c_0\sigma_x\sigma_y\sqrt{\frac{\log p}{n}} \leq c_1 \exp\left(-c_2 \log p\right). \tag{3.76}$$

The ℓ_1 balls are defined as

$$\mathbb{B}_l\left(s\right) = \left\{\mathbf{v} \in \mathbb{R}^p : \left\|\mathbf{v}\right\|_l \leq s, l = 0, 1, 2\right\}.$$

For a parameter $s \geq 1$, use the notation

$$\mathbb{K}\left(s\right) = \left\{\mathbf{v} \in \mathbb{R}^p : \left\|\mathbf{v}\right\|_2 \leq 1, \left\|\mathbf{v}\right\|_0 \leq s\right\},$$

the ℓ_0 norm $\left\|\mathbf{x}\right\|_0$ stands for the non-zeros of vector $\mathbf{x}$. The sparse set is $\mathbb{K}\left(s\right) = \mathbb{B}_0\left(s\right) \cap \mathbb{B}_2\left(1\right)$ and the cone set is

$$\mathbb{C}\left(s\right) = \left\{\mathbf{v} \in \mathbb{R}^p : \left\|\mathbf{v}\right\|_1 \leq \sqrt{s}\left\|\mathbf{v}\right\|_2\right\}.$$

We use the following result to control deviations uniformly over vectors in $\mathbb{R}^p$.

Theorem 3.9.2 (Lemma 12 of [164]). *For a fixed matrix $\mathbf{\Gamma} \in \mathbb{R}^{p\times p}$, parameter $s \geq 1$, and tolerance $\varepsilon > 0$, suppose we have the deviation condition*

$$\left|\mathbf{v}^T\mathbf{\Gamma}\mathbf{v}\right| \leq \varepsilon \qquad \forall \mathbf{v} \in \mathbb{K}\left(s\right).$$

Then

$$\left|\mathbf{v}^T\mathbf{\Gamma}\mathbf{v}\right| \leq 27\left(\left\|\mathbf{v}\right\|_2^2 + \frac{1}{s}\left\|\mathbf{v}\right\|_1^2\right) \qquad \forall \mathbf{v} \in \mathbb{R}^p.$$

Theorem 3.9.3 (Lemma 13 of [164]). *Suppose $s \geq 1$ and $\hat{\mathbf{\Gamma}}$ is an estimator of $\mathbf{\Sigma}_x$ satisfying the deviation condition*

$$\left|\mathbf{v}^T\left(\hat{\mathbf{\Gamma}} - \mathbf{\Sigma}_x\right)\mathbf{v}\right| \leq \frac{\lambda_{\min}\left(\mathbf{\Sigma}_x\right)}{54} \qquad \forall \mathbf{v} \in \mathbb{K}\left(2s\right).$$

Then we have the lower-restricted eigenvalue condition

$$\left|\mathbf{v}^T\left(\hat{\mathbf{\Gamma}} - \mathbf{\Sigma}_x\right)\mathbf{v}\right| \geq \frac{1}{2}\lambda_{\min}\left(\mathbf{\Sigma}_x\right)\left\|\mathbf{v}\right\|_2^2 - \frac{1}{2s}\lambda_{\min}\left(\mathbf{\Sigma}_x\right)\left\|\mathbf{v}\right\|_1^2$$

and the upper-restricted eigenvalue condition

$$\left|\mathbf{v}^T\left(\hat{\mathbf{\Gamma}} - \mathbf{\Sigma}_x\right)\mathbf{v}\right| \leq \frac{3}{2}\lambda_{\max}\left(\mathbf{\Sigma}_x\right)\left\|\mathbf{v}\right\|_2^2 + \frac{1}{2s}\lambda_{\max}\left(\mathbf{\Sigma}_x\right)\left\|\mathbf{v}\right\|_1^2.$$

We combine Theorem 3.9.1 with a discretization argument and union bound to obtain the next result.

Theorem 3.9.4 (Lemma 15 of [164]). *If $\mathbf{X} \in \mathbb{R}^{n \times p}$ is a zero-mean sub-Gaussian matrix with parameters $(\boldsymbol{\Sigma}, \sigma^2)$, then there is a universal constant $c > 0$ such that*

$$\mathbb{P}\left(\sup_{\mathbf{v} \in \mathbb{K}(2s)} \left|\frac{1}{n}\|\mathbf{X}\mathbf{v}\|_2^2 - \mathbb{E}\left[\frac{1}{n}\|\mathbf{X}\mathbf{v}\|_2^2\right]\right| \geq t\right) \leq 2\exp\left(-cn\min\left(\frac{t^2}{\sigma^4}, \frac{t}{\sigma^2}\right) + 2s\log p\right).$$

We consider the dependent data. The rows of $\mathbf{X}$ are drawn from a stationary vector autoregressive (AR) process [155] according to

$$\mathbf{x}_{i+1} = \mathbf{A}\mathbf{x}_i + \mathbf{v}_i, \qquad i = 1, 2, \ldots, n-1, \tag{3.77}$$

where $\mathbf{v}_i \in \mathbb{R}^p$ is a zero-mean noise vector with covariance matrix $\boldsymbol{\Sigma}_v$, and $\mathbf{A} \in \mathbb{R}^{p \times p}$ is a driving matrix with spectral norm $\|\mathbf{A}\|_2 < 1$. We assume the rows of $\mathbf{X}$ are drawn from a Gaussian distribution with $\boldsymbol{\Sigma}_x$, such that

$$\boldsymbol{\Sigma}_x = \mathbf{A}\boldsymbol{\Sigma}_x\mathbf{A}^T + \boldsymbol{\Sigma}_v.$$

Theorem 3.9.5 (Lemma 16 of [164]). *Suppose $\mathbf{y} = [Y_1, Y_2, \ldots, Y_n] \in \mathbb{R}^n$ is a mixture of multivariate Gaussians $\mathbf{y}_i \sim \mathcal{N}(0, \mathbf{Q}_i)$, and let $\sigma^2 = \sup_j \|\mathbf{Q}_j\|_{op}$. Then for all $t > \frac{2}{\sqrt{n}}$, we have*

$$\mathbb{P}\left(\left|\frac{1}{n}\|\mathbf{y}\|_2^2 - \mathbb{E}\left[\frac{1}{n}\|\mathbf{y}\|_2^2\right]\right| \geq 4t\sigma^2\right) \leq 2\exp\left(-\frac{1}{2}n\left(t - \frac{2}{\sqrt{n}}\right)^2\right) + +2\exp\left(-n/2\right).$$

This result is a generalization of Theorem 3.7.7. It follows from the concentration of Lipschitz functions of Gaussian random vectors [141]. By definition, the random vector $\mathbf{y}$ is a mixture of random vectors of the form $\sqrt{\mathbf{Q}_i}\mathbf{x}_i$, where $\mathbf{x}_i \sim \mathcal{N}(0, \mathbf{I}_n)$. The key idea is to study the function

$$f_j(\mathbf{x}) = \|\mathbf{Q}_j\mathbf{x}\|_2 / \sqrt{n}$$

and obtain the Lipschitz constant as $\|\mathbf{Q}_j\|_{op}/\sqrt{n}$. Also note that $f_j(\mathbf{x})$ is a sub-Gaussian random variable with parameter $\sigma_j^2 = \|\mathbf{Q}_j\|_{op}/\sqrt{n}$. So the mixture $\frac{1}{n}\|\mathbf{y}\|_2$ is sub-Gaussian with parameter $\sigma^2 = \frac{1}{n}\sup_j \|\mathbf{Q}_j\|_{op}$. The rest follows from [155].

Example 3.9.6 (Additive noise [164]). Suppose we observe

$$\mathbf{Z} = \mathbf{X} + \mathbf{W},$$

where $\mathbf{W}$ is a random matrix independent of $\mathbf{X}$, with the rows $\mathbf{w}_i$ drawn from a zero-mean distribution with known covariance matrix $\boldsymbol{\Sigma}_w$. We define

$$\hat{\mathbf{\Gamma}}_{add} = \frac{1}{n}\mathbf{Z}^T\mathbf{Z} - \mathbf{\Sigma}_w.$$

Note that $\hat{\mathbf{\Gamma}}_{add}$ is not positive semidefinite. □

Example 3.9.7 (Missing data [164]). The entries of matrix $\mathbf{X}$ are missing at random. We observe the matrix $\mathbf{Z} \in \mathbb{R}^{n\times p}$ with entries

$$Z_{ij} = \begin{cases} X_{ij} \text{ with probability } 1-\rho, \\ 0 \qquad \text{otherwise.} \end{cases}$$

Given the observed matrix $\mathbf{Z} \in \mathbb{R}^{n\times p}$, we use

$$\hat{\mathbf{\Gamma}}_{miss} = \frac{1}{n}\tilde{\mathbf{Z}}^T\tilde{\mathbf{Z}} - \rho\,\text{diag}\left(\frac{1}{n}\tilde{\mathbf{Z}}^T\tilde{\mathbf{Z}}\right)$$

where $\tilde{Z}_{ij} = Z_{ij}/\left(1-\rho\right)$. □

Theorem 3.9.8 (Lemma 17 of [164]). *Let* $\mathbf{X} \in \mathbb{R}^{n\times p}$ *is a Gaussian random matrix, with rows* $\mathbf{x}_i$ *generated according to a vector autoregression* (3.77) *with driving matrix* $\mathbf{A}$. *Let* $\mathbf{v} \in \mathbb{R}^p$ *be a fixed vector with unit norm. Then for all* $t > \frac{2}{\sqrt{n}}$,

$$\mathbb{P}\left(\left|\mathbf{v}^T\left(\hat{\mathbf{\Gamma}} - \mathbf{\Sigma}_x\right)\mathbf{v}\right| \geq 4t\varsigma^2\right) \leq 2\exp\left(-\frac{1}{2}n\left(t-\frac{2}{\sqrt{n}}\right)^2\right) + 2\exp\left(-n/2\right),$$

where

$$\varsigma^2 = \begin{cases} \left\|\mathbf{\Sigma}_w\right\|_{op} + \frac{2\left\|\mathbf{\Sigma}_x\right\|_{op}}{1-\left\|\mathbf{A}\right\|_{op}} \ (\textit{additive noise case}). \\ \frac{1}{(1-\rho_{\max})^2}\frac{2\left\|\mathbf{\Sigma}_x\right\|_{op}}{1-\left\|\mathbf{A}\right\|_{op}} \quad (\textit{missing data case}). \end{cases}$$

Theorem 3.9.9 (Lemma 18 of [164]). *Let* $\mathbf{X} \in \mathbb{R}^{n\times p}$ *is a Gaussian random matrix, with rows* $\mathbf{x}_i$ *generated according to a vector autoregression* (3.77) *with driving matrix* $\mathbf{A}$. *Let* $\mathbf{v} \in \mathbb{R}^p$ *be a fixed vector with unit norm. Then for all* $t > \frac{2}{\sqrt{n}}$,

$$\mathbb{P}\left(\sup_{\mathbf{v}\in\mathbb{K}(2s)}\left|\mathbf{v}^T\left(\hat{\mathbf{\Gamma}} - \mathbf{\Sigma}_x\right)\mathbf{v}\right| \geq 4t\varsigma^2\right) \leq 4\exp\left(-cn\left(t-\frac{2}{\sqrt{n}}\right)^2 + 2s\log p\right),$$

where ς *is defined as in Lemma 3.9.8.*

3.10 Concentration for Largest Eigenvalues

We draw material from [149] in this section. Let μ be the standard Gaussian measure on $\mathbb{R}^n$ with density $(2\pi)^{-n/2}e^{-|\mathbf{x}|^2/2}$ with respect to Lebesgue measure. Here $|\mathbf{x}|$ is the usual Euclidean norm for vector $\mathbf{x}$. One basic concentration property [141] indicates that for every Lipschitz function $F : \mathbb{R}^n \to \mathbb{R}$ with the Lipschitz constant $||F||_{\mathcal{L}} \leq 1$, and every $t \geq 0$, we have

$$\mu\left(\left\{\left|F - \int F d\mu\right| \geq t\right\}\right) \leq 2e^{-t^2/2}. \tag{3.78}$$

The same holds for a median of F instead of the mean. One fundamental property of (3.78) is its independence of dimension n of the underlying state space. Later on, we find that (3.78) holds for non-Gaussian classes of random variables. Eigenvalues are the matrix functions of interest.

Let us illustrate the approach by studying concentration for largest eigenvalues. For example, consider the Gaussian unitary ensemble (GUE) $\mathbf{X}$: For each integer $n \geq 1$, $\mathbf{X} = (X_{ij})_{1 \leq i,j \leq n}$ is an $n \times n$ Hermitian centered Gaussian random matrix with variance σ^2. Equivalently, the random matrix $\mathbf{X}$ is distributed according to the probability distribution

$$\mathbb{P}(d\mathbf{X}) = \frac{1}{Z}\exp\left(-\operatorname{Tr}\left(\mathbf{X}^2\right)/2\sigma^2\right) d\mathbf{X}$$

on the space $\mathcal{H}_n \cong \mathbb{R}^{n^2}$ of $n \times n$ Hermitian matrices where

$$d\mathbf{X} = \prod_{1 \leq i \leq n} dX_{ii} \prod_{1 \leq i,j \leq n} d\operatorname{Re}(X_{ij})\, d\operatorname{Im}(X_{ij})$$

is Lebesgue measure on $\mathcal{H}_n$ and Z is the normalization constant. This probability measure is invariant under the action of the unitary group on $\mathcal{H}_n$ in the sense that $\mathbf{U}\mathbf{X}\mathbf{U}^H$ has the same law as $\mathbf{X}$ for each unitary element $\mathbf{U}$ of $\mathcal{H}_n$. The random matrix $\mathbf{X}$ is then said to be an element of the Gaussian unitary ensemble (GUE) ("ensemble" for probability distribution).

The variational characterization is critical to the largest eigenvalue

$$\lambda_{\max}(\mathbf{X}) = \sup_{|\mathbf{u}|=1} \mathbf{u}\mathbf{X}\mathbf{u}^H \tag{3.79}$$

where the function $\lambda_{\max}$ is linear in $\mathbf{X}$. The expression is the quadratic form. Later on in Chap. 4, we study the concentration of the quadratic forms. $\lambda_{\max}(\mathbf{X})$ is easily seen (see Chap. 4) to be a 1-Lipschitz map of the n^2 independent real and imaginary entries

$$X_{ii}, 1 \leq i \leq n, \mathrm{Re}\,(X_{ij})\,/\sqrt{2}, 1 \leq i \leq n, \mathrm{Im}\,(X_{ij})\,/\sqrt{2}, 1 \leq i \leq n,$$

of matrix $\mathbf{X}$. Using Theorem (3.3.10) together with the scaling of the variance $\sigma^2 = \frac{1}{4n}$, we get the following concentration inequality on $\lambda_{\max}(\mathbf{X})$.

Theorem 3.10.1. *For all $n \geq 1$ and $t \geq 0$,*

$$\mathbb{P}\left(\{|\lambda_{\max}(\mathbf{X}) - \mathbb{E}\lambda_{\max}(\mathbf{X})| \geq t\}\right) \leq 2e^{-2nt^2}.$$

As a consequence, note that

$$\mathrm{var}\,(\lambda_{\max}(\mathbf{X})) \leq C/n.$$

Using Random Matrix Theory [145], this variance is

$$\mathrm{var}\,(\lambda_{\max}(\mathbf{X})) \leq C/n^{4/3}.$$

Viewing the largest eigenvalue as one particular example of Lipschitz function of the entries of the matrix does not reflect enough the structure of the model. This comment more or less applies to all the results presented in this book deduced from the concentration principle.

3.10.1 Talagrand's Inequality Approach

Let us estimate $\mathbb{E}\lambda_{\max}(\mathbf{X})$. We emphasize the approach used here. Consider the real-valued Gaussian process

$$G_{\mathbf{u}} = \mathbf{u}\mathbf{X}\mathbf{u}^H = \sum_{i,j=1}^{n} X_{ij}u_i\bar{u}_j, \quad |\mathbf{u}| = 1,$$

where $\mathbf{u} = (u_1, \ldots, u_n) \in \mathbb{C}^n$. We have that for $\mathbf{u}, \mathbf{v} \in \mathbb{C}^n$,

$$\mathbb{E}\left(|G_{\mathbf{u}} - G_{\mathbf{v}}|^2\right) = \sigma^2 \sum_{i,j=1}^{n} |u_i\bar{u}_j - v_i\bar{v}_j|.$$

We define the Gaussian random processes indexed by the vector $\mathbf{u} \in \mathbb{C}^n, |\mathbf{u}| = 1$. We have that for $\mathbf{u} \in \mathbb{C}^n, |\mathbf{u}| = 1$

$$H_{\mathbf{u}} = \sum_{i=1}^{n} g_i\,\mathrm{Re}\,(u_i) + \sum_{i=1}^{n} h_i\,\mathrm{Im}\,(u_i)$$

where $g_1, \ldots, g_n, h_1, \ldots, h_n$ are *independent* standard Gaussian variables. It follows that for every $\mathbf{u}, \mathbf{v} \in \mathbb{C}^n$, such that $|\mathbf{u}| = |\mathbf{v}| = 1$,

$$\mathbb{E}\left(|G_{\mathbf{u}} - G_{\mathbf{v}}|^2\right) \leq 2\sigma^2 \mathbb{E}\left(|H_{\mathbf{u}} - H_{\mathbf{v}}|^2\right).$$

By the Slepian-Fernique lemma [27] we have that

$$\mathbb{E}\left(\sup_{|\mathbf{u}|=1} G_{\mathbf{u}}\right) \leq \sqrt{2}\sigma \mathbb{E}\left(\sup_{|\mathbf{u}|=1} G_{\mathbf{u}}\right) \leq 2\sqrt{2}\sigma \mathbb{E}\left(\left[\sum_{i=1}^{n} g_i^2\right]^{1/2}\right).$$

When $\sigma^2 = \frac{1}{4n}$, we thus have that

$$\mathbb{E}\lambda_{\max}(\mathbf{X}) \leq \sqrt{2}. \tag{3.80}$$

Equation (3.80) extends to the class of *sub-Gaussian* distributions including random matrices with symmetric Bernoulli entries.

Combining (3.80) with Theorem 3.10.1, we have that for every $t \geq 0$,

$$\mathbb{P}\left(\left\{\lambda_{\max}(\mathbf{X}) \geq \sqrt{2} + t\right\}\right) \leq 2e^{-2nt^2}.$$

3.10.2 Chaining Approach

Based on the supremum representation (3.79) of the largest eigenvalue, we can use another chaining approach [27, 81]. The supremum of Gaussian or more general process $(Z_t)_{t \in T}$ is considered. We study the random variable $\sup_{t \in T} Z_t$ as a function of the set T and its expectation $\mathbb{E}\left(\sup_{t \in T} Z_t\right)$. Then we can study the probability $\mathbb{P}\left(\sup_{t \in T} Z_t \geq r\right), r \geq 0$.

For real symmetric matrices $\mathbf{X}$, we have

$$Z_{\mathbf{u}} = \mathbf{u}\mathbf{X}\mathbf{u}^T = \sum_{i,j=1}^{n} X_{ij} u_i u_j, \quad |\mathbf{u}| = 1,$$

where $\mathbf{u} = (u_1, \ldots, u_n) \in \mathbb{R}^n$ and $X_{ij}, 1 \leq i \leq j \leq n$ are independent centered either Gaussian or Bernoulli random variables. Note $Z_{\mathbf{u}}$ is *linear* in the entries of X_{ij}, u_i, u_j. Basically, $Z_{\mathbf{u}}$ is the sum of independent random variables. To study the size of the unit sphere $|\mathbf{u}| = 1$ under the L^2-metric, we have that

$$\mathbb{E}|Z_{\mathbf{u}} - Z_{\mathbf{v}}|^2 = \sum_{i,j=1}^{n} |u_i u_j - v_i v_j|, \quad |\mathbf{u}| = |\mathbf{v}| = 1.$$

3.10.3 General Random Matrices

The main interest of the theory is to apply concentration of measure to general families of random matrices. For measures μ on $\mathbb{R}^n$, the dimension-free inequality is defined as

$$\mu\left(\left\{\left|F - \int F d\mu\right| \geq t\right\}\right) \leq Ce^{-t^2/C}, \quad t \geq 0 \tag{3.81}$$

for some constant $C > 0$ independent of dimension n and every 1-Lipschitz function $F : \mathbb{R}^n \to \mathbb{R}$. The mean $\int F d\mu$ can be replaced by a median of F. The primary message is that (3.81) is valid for non-Gaussian random variables. Consider the example of independent uniform entries. If $\mathbf{X} = (X_{ij})_{1 \leq i,j \leq n}$ is a real symmetric $n \times n$ matrix, then its eigenvalues are 1-Lipschitz functions of the entries.

Theorem 3.10.2 (Proposition 3.2 of [149]). *Let* $\mathbf{X} = (X_{ij})_{1 \leq i,j \leq n}$ *be a real symmetric* $n \times n$ *random matrix and* $\mathbf{Y} = (Y_{ij})_{1 \leq i,j \leq n}$ *be a real* $n \times n$ *random matrix. Assume that the distributions of the random vector* $X_{ij}, 1 \leq i \leq j \leq n$ *and* $Y_{ij}, 1 \leq i \leq j \leq n$ *in* $\mathbb{R}^{n(n+1)/2}$ *and, respectively,* $\mathbb{R}^{n^2}$ *satisfy the dimension-free concentration property* (3.81). *Then, if* τ *is any eigenvalue of* $\mathbf{X}$*, and singular value of* $\mathbf{Y}$*, respectively, for every* $t \geq 0$*,*

$$\mathbb{P}\left(\left|\tau\left(\mathbf{X}\right) - \mathbb{E}\tau\left(\mathbf{X}\right)\right| \geq t\right) \leq Ce^{-t^2/2C}, \textit{respectively,} Ce^{-t^2/C}.$$

Below we give two examples of distributions satisfying concentration inequalities of the type (3.81). The first class is measures satisfying a logarithmic Sobolev inequality that is a natural extension of the Gaussian example. A probability measure μ on $\mathbb{R}$, or $\mathbb{R}^n$ is said to satisfy a logarithmic Sobolev inequality if for some constant $C > 0$,

$$\int_{\mathbb{R}^n} f^2 \log f^2 d\mu \leq 2C \int_{\mathbb{R}^n} \left|\nabla f\right|^2 d\mu \tag{3.82}$$

for every smooth enough function $f : \mathbb{R}^n \to \mathbb{R}$ such that $\int f^2 d\mu = 1$.

The prototype example is the standard Gaussian measure on $\mathbb{R}^n$ which satisfies (3.82) with $C = 1$. Another example consists of probability measures on $\mathbb{R}^n$ of the type

$$d\mu\left(\mathbf{x}\right) = e^{-V(\mathbf{x})} d\mathbf{x}$$

where $V - c\left(\left|\mathbf{x}\right|^2/2\right)$ is a convex function for some constant $c > 0$. The measures satisfy (3.82) for $C = 1/c$.

Regarding the logarithmic Sobolev inequality, an important point to remember is its stability by product that gives dimension-free constants. If $\mu_1, \ldots, \mu_n$ are probability measures on $\mathbb{R}^n$ satisfying the logarithmic Sobolev inequality (3.82) with the same constant C, then the product measure $\mu_1 \otimes \cdots \otimes \mu_n$ also satisfies it (on $\mathbb{R}^n$) with the same constant.

By the so-called Herbst argument, we can apply the logarithmic Sobolev inequality (3.82) to study concentration of measure. If μ satisfies (3.82), then for any 1-Lipschitz function $F : \mathbb{R}^n \to \mathbb{R}$ and any $t \in \mathbb{R}$,

$$\int e^{tF} d\mu \le e^{t \int F d\mu + Ct^2/2}.$$

In particular, by a simple use of Markov's exponential inequality (for both F and $-F$), for any $t > 0$,

$$\mu\left(\left\{\left|F - \int F d\mu\right| \ge t\right\}\right) \le 2e^{-t^2/2C},$$

so that the dimension-free concentration property (3.81) holds. We refer to [141] for more details. Related Poincare inequalities may be also considered similarly in this context.

3.11 Concentration for Projection of Random Vectors

The goal here is to apply the concentration of measure. For a random vector $\mathbf{x} \in \mathbb{R}^d$, we study its projections to subspaces. The central problem here is to show that for most subspaces, the resulting distributions are about the same, approximately Gaussian, and to determine how large the dimension k of the subspace may be, relative to d, for this phenomenon to persist.

The Euclidean length of a vector $\mathbf{x} \in \mathbb{R}^d$ is defined by $\|\mathbf{x}\| = \sqrt{x_1^2 + \ldots, x_d^2}$. The Stiefel manifold[4] $\mathcal{Z}_{d,k} \in \mathbb{R}^{k \times d}$ is defined by

$$\mathcal{Z}_{d,k} = \left\{ \mathbf{Z} = (\mathbf{z}_1, \ldots, \mathbf{z}_k) : \mathbf{z}_i \in \mathbb{R}^d, \langle \mathbf{z}_i, \mathbf{z}_j \rangle = \delta_{ij} \quad \forall 1 \le i, j \le k \right\},$$

with metric $\rho(\mathbf{Z}, \mathbf{Z}')$ between a pair of two matrices $\mathbf{Z}$ and $\mathbf{Z}'$—two points in the manifold $\mathcal{Z}_{d,k}$—defined by

$$\rho(\mathbf{Z}, \mathbf{Z}') = \left(\sum_{i=1}^{k} \|\mathbf{z} - \mathbf{z}_i'\|^2 \right)^{1/2}.$$

The manifold $\mathcal{Z}_{d,k}$ preserves a rotation-invariant (Haar) probability measure.

[4] A manifold of dimension n is a topological space that near each point resembles n-dimensional Euclidean space. More precisely, each point of an n-dimensional manifold has a neighborhood that is homeomorphic to the Euclidean space of dimension n.

One version of the concentration of measure [165] is included here. We need some notation first. A modulus of continuity [166] is a function $\omega : [0, \infty] \to [0, \infty]$ used to measure quantitatively the uniform continuity of functions. So, a function $f : I \to \mathbb{R}$ admits ω as a modulus of continuity if and only if

$$|f(\mathbf{x}) - f(\mathbf{y})| \leq \omega(|x - y|)$$

for all x and y in the domain of f. Since moduli of continuity are required to be infinitesimal at 0, a function turns out to be uniformly continuous if and only if it admits a modulus of continuity. Moreover, relevance to the notion is given by the fact that sets of functions sharing the same modulus of continuity are exactly equicontinuous families. For instance, the modulus $\omega(t) = Lt$ describes the L-Lipschitz functions, the moduli $\omega(t) = Lt^\alpha$ describe the Hölder continuity, the modulus $\omega(t) = Lt(|log(t)| + 1)$ describes the almost Lipschitz class, and so on.

Theorem 3.11.1 (Milman and Schechtman [167]). *For any $F : \mathcal{Z}_{n,k} \to \mathbb{R}$ with the median $\mathbb{M}F$ and modulus of continuity $\omega_F(t), t > 0$*

$$\mathbb{P}(|F(\mathbf{z}_1, \ldots, \mathbf{z}_k) - \mathbb{M}F(\mathbf{z}_1, \ldots, \mathbf{z}_k)| > \omega_F(t)) < \sqrt{\frac{\pi}{2}} e^{-nt^2/8}, \tag{3.83}$$

where $\mathbb{P}$ is the rotation-invariant probability measure on the span of $\mathbf{z}_1, \ldots, \mathbf{z}_k$.

Let $\mathbf{x}$ be a random vector in $\mathbb{R}^d$ and let $\mathbf{Z} \in \mathcal{Z}_{d,k}$. Let

$$\mathbf{x}_z = (\langle \mathbf{x}, \mathbf{z}_1 \rangle, \ldots, \langle \mathbf{x}, \mathbf{z}_k \rangle) \in \mathbb{R}^k;$$

that is, $\mathbf{x}_z$ is the projection of the vector $\mathbf{x}$ onto the span of $\mathbf{Z}$. $\mathbf{x}_z$ is a projection from dimension d to dimension k.

The bounded-Lipschitz distance between two random vectors $\mathbf{x}$ and $\mathbf{y}$ is defined by

$$d_{BL}(\mathbf{x}, \mathbf{y}) = \sup_{\|f\|_1 \leq 1} \|\mathbb{E}f(\mathbf{x}) - \mathbb{E}f(\mathbf{y})\|,$$

where

$$\|f\|_1 = \max\{\|f\|_\infty, \|f\|_L\}$$

with the Lipschitz constant of f defined by $\|f\|_L = \sup\limits_{\mathbf{x} \neq \mathbf{y}} \frac{\|f(\mathbf{x}) - f(\mathbf{y})\|}{\|\mathbf{x} - \mathbf{y}\|}$.

Theorem 3.11.2 (Meckes [165]). *Let $\mathbf{x}$ be a random vector in $\mathbb{R}^d$, with $\mathbb{E}\mathbf{x} = 0$, $\mathbb{E}\left[\|\mathbf{x}\|^2\right] = \sigma^2 d$, and let $\alpha = \mathbb{E}\left[\left|\|\mathbf{x}\|^2/\sigma^2 - d\right|\right]$. If $\mathbf{Z}$ is a random point of the manifold $\mathcal{Z}_{d,k}$, $\mathbf{x}_z$ is defined as above, and $\mathbf{w}$ is a standard Gaussian random vector, then*

$$d_{BL}(\mathbf{x}_z, \sigma\mathbf{w}) \leq \frac{\sigma\sqrt{k}(\alpha + 1) + \sigma k}{d + 1}.$$

Theorem 3.11.3 (Meckes [165]). *Suppose that β is defined by $\beta = \sup_{\mathbf{y} \in \mathbb{S}^{d-1}} \mathbb{E}\langle \mathbf{x}, \mathbf{y} \rangle^2$. For $\mathbf{z}_i \in \mathbb{R}^d, i = 1, \ldots, k$ and $\mathbf{Z} = (\mathbf{z}_1, \ldots, \mathbf{z}_k) \in \mathcal{Z}_{d,k}$, let*

$$d_{BL}(\mathbf{x}_z, \sigma\mathbf{w}) = \sup_{\|f\|_1 \le 1} \|\mathbb{E}f(\langle \mathbf{x}, \mathbf{z}_1 \rangle, \ldots, \langle \mathbf{x}, \mathbf{z}_k \rangle) - \mathbb{E}f(\sigma\mathbf{w}_1, \ldots, \sigma\mathbf{w}_k)\|;$$

That is, $d_{BL}(\mathbf{x}_z, \sigma\mathbf{w})$ is the conditional bounded-Lipschitz distance from the random point $\mathbf{x}_z$ (a random vector in $\mathbb{R}^k$) to the standard Gaussian random vector $\sigma\mathbf{w}$, conditioned on the matrix $\mathbf{Z}$. Then, for $t > 2\pi\sqrt{\frac{\beta}{d}}$ and $\mathbf{Z}$ a random point of the manifold (a random matrix in $\mathbb{R}^{k\times d}$) $\mathcal{Z}_{d,k}$,

$$\mathbb{P}(|d_{BL}(\mathbf{x}_z, \sigma\mathbf{w}) - \mathbb{E}d_{BL}(\mathbf{x}_z, \sigma\mathbf{w})| > t) \le \sqrt{\frac{\pi}{2}} e^{-dt^2/32\beta}.$$

Theorem 3.11.4 (Meckes [168]). *With the notation as in the previous theorems, we have*

$$\mathbb{E}d_{BL}(\mathbf{x}_z, \sigma\mathbf{w}) \le C\left[\frac{(k\beta + \beta\log d)\,\beta^{2/(9k+12)}}{k^{2/3}\beta^{2/3}d^{2/(3k+4)}} + \frac{\sigma\left(\sqrt{k}(\alpha+1)+k\right)}{d-1}\right].$$

In particular, under the additional assumptions that $\alpha \le C_0\sqrt{d}$ and $\beta = 1$, then

$$\mathbb{E}d_{BL}(\mathbf{x}_z, \sigma\mathbf{w}) \le C\frac{k + \log(d)}{k^{2/3}d^{2/(3k+4)}}.$$

The assumption that $\beta = 1$ is automatically satisfies, if the covariance matrix of the random vector $\mathbf{x} \in \mathbb{R}^d$ is the identity, i.e., $\mathbb{E}\left[\mathbf{x}\mathbf{x}^T\right] = \mathbf{I}_{d\times d}$; in the language of geometry, this is simply the case that the random vector $\mathbf{x}$ is isotropic. The assumption that $\alpha = O(\sqrt{d})$ is a geometrically natural one that arise, for example, if $\mathbf{x}$ is distributed uniformly on the isotropic dilate of the ℓ_1 ball in $\mathbb{R}^d$. The key observation in the proof is to view the distance as the supremum of a stochastic process.

Proof of Theorem 3.11.2. This proof follows [165], changing to our notation. Define the function $F : \mathcal{Z}_{d,k} \to \mathbb{R}$ by

$$F(\mathbf{Z}) = \sup_{\|f\|_1 \le 1} \|\mathbb{E}_{\mathbf{x}}f(\mathbf{x}_z) - \mathbb{E}f(\sigma\mathbf{w})\|,$$

where $\mathbb{E}_{\mathbf{x}}$ denotes the expectation with respect to the distribution of the random vector $\mathbf{x}$ only; that is,

$$\mathbb{E}_{\mathbf{x}}f(\mathbf{x}_z) = \mathbb{E}\left[f(\mathbf{x}_z)\,|\mathbf{Z}\right].$$

The goal here is to apply the concentration of measure. We use the standard method here. We need to find the Lipschitz constant first. For a pair of random vectors $\mathbf{x}$ and $\mathbf{x}'$, which will be projected to the same span of $\mathbf{Z}$, we observe that for f with $\|f\|_1 \le 1$ given,

$$\begin{aligned}
&\|\mathbb{E}_{\mathbf{x}} f(\mathbf{x}_z) - \mathbb{E} f(\sigma \mathbf{w})\| - \|\mathbb{E}_{\mathbf{x}} f(\mathbf{x}'_z) - \mathbb{E} f(\sigma \mathbf{w})\| \\
&\quad \le \|\mathbb{E}_{\mathbf{x}} f(\mathbf{x}'_z) - \mathbb{E}_{\mathbf{x}} f(\mathbf{x}_z)\| \\
&\quad = \mathbb{E}\left[\|f(\langle \mathbf{x}, \mathbf{z}'_1\rangle, \ldots, \langle \mathbf{x}, \mathbf{z}'_k\rangle) - f(\langle \mathbf{x}, \mathbf{z}_1\rangle, \ldots, \langle \mathbf{x}, \mathbf{z}_k\rangle)\| \,|\mathbf{Z}, \mathbf{Z}'\right] \\
&\quad \le \mathbb{E}\left[\|f(\langle \mathbf{x}, \mathbf{z}'_1 - \mathbf{z}_1\rangle, \ldots, \langle \mathbf{x}, \mathbf{z}'_k - \mathbf{z}_k\rangle)\| \,|\mathbf{Z}, \mathbf{Z}'\right] \\
&\quad \le \sqrt{\sum_{i=1}^{k} \|\mathbf{z}'_i - \mathbf{z}_i\|^2 \mathbb{E}\left\langle \mathbf{x}, \frac{\mathbf{z}'_i - \mathbf{z}_i}{\|\mathbf{z}'_i - \mathbf{z}_i\|}\right\rangle^2} \\
&\quad \le \rho(\mathbf{Z}, \mathbf{Z}')\sqrt{\beta}.
\end{aligned}$$

It follows that

$$\|d_{BL}(\mathbf{x}_z, \sigma \mathbf{w}) - d_{BL}(\mathbf{x}_{z'}, \sigma \mathbf{w})\| \tag{3.84}$$

$$= \left| \sup_{\|f\|_1 \le 1} \|\mathbb{E}_{\mathbf{x}} f(\mathbf{x}_z) - \mathbb{E} f(\sigma \mathbf{w})\| - \sup_{\|f\|_1 \le 1} \|\mathbb{E}_{\mathbf{x}} f(\mathbf{x}_{z'}) - \mathbb{E} f(\sigma \mathbf{w})\| \right| \tag{3.85}$$

$$\le \sup_{\|f\|_1 \le 1} \left|\|\mathbb{E}_{\mathbf{x}} f(\mathbf{x}_z) - \mathbb{E} f(\sigma \mathbf{w})\| - \|\mathbb{E}_{\mathbf{x}} f(\mathbf{x}_{z'}) - \mathbb{E} f(\sigma \mathbf{w})\|\right| \tag{3.86}$$

$$\le \rho(\mathbf{Z}, \mathbf{Z}')\sqrt{\beta}. \tag{3.87}$$

$$\tag{3.88}$$

Thus, $d_{BL}(\mathbf{x}_z, \sigma \mathbf{w})$ is a Lipschitz function with Lipschitz constant $\sqrt{\beta}$.

Applying the concentration of measure inequality (3.83), then we have that

$$\mathbb{P}(|F(\mathbf{z}_1, \ldots, \mathbf{z}_k) - \mathbb{M}F(\mathbf{z}_1, \ldots, \mathbf{z}_k)| > t) \le \sqrt{\frac{\pi}{2}} e^{-t^2 d/8\beta}.$$

Now, if $\mathbf{Z} = (\mathbf{z}_1, \ldots, \mathbf{z}_k)$ is a Haar-distributed random point of the manifold $\mathcal{Z}_{d,k}$, then

$$|\mathbb{E}F(\mathbf{Z}) - \mathbb{M}F(\mathbf{Z})| \le \mathbb{E}|F(\mathbf{Z}) - \mathbb{M}F(\mathbf{Z})| = \int_0^\infty \mathbb{P}(|\mathbb{E}F(\mathbf{Z}) - \mathbb{M}F(\mathbf{Z})| > t)dt$$

$$\le \int_0^\infty \sqrt{\frac{\pi}{2}} e^{-t^2 d/8\beta} dt = \pi \sqrt{\frac{\beta}{d}}.$$

So as long as $t > 2\pi\sqrt{\frac{\beta}{d}}$, replacing the median of F with its mean only changes the constants:

$$\begin{aligned}\mathbb{P}\left(\left|F\left(\mathbf{Z}\right)-\mathbb{E}F\left(\mathbf{Z}\right)\right|>t\right) &\le \mathbb{P}\left(\left|F\left(\mathbf{Z}\right)-\mathbb{M}F\left(\mathbf{Z}\right)\right|>t-\left|\mathbb{M}F\left(\mathbf{Z}\right)-\mathbb{E}F\left(\mathbf{Z}\right)\right|\right)\\ &\le \mathbb{P}\left(\left|F\left(\mathbf{Z}\right)-\mathbb{M}F\left(\mathbf{Z}\right)\right|>t/2\right) \le \sqrt{\tfrac{\pi}{2}}e^{-dt^2/32\beta}.\end{aligned}$$

3.12 Further Comments

The standard reference for concentration of measure is [141] and [27]. We only provide necessary results that are needed for the later chapters. Although some results are recent, no attempt has been made to survey the latest results in the literature.

We follow closely [169, 170] and [62, 171], which versions are highly accessible. The entropy method is introduced by Ledoux [172] and further refined by Massart [173] and Rio [174]. Many applications are considered [62, 171, 175–178].

Chapter 4
Concentration of Eigenvalues and Their Functionals

Chapters 4 and 5 are the core of this book. Talagrand's concentration inequality is a very powerful tool in probability theory. Lipschitz functions are the mathematics objects. Eigenvalues and their functionals may be shown to be Lipschitz functions so the Talagrand's framework is sufficient. Concentration inequalities for many complicated random variables are also surveyed here from the latest publications. As a whole, we bring together concentration results that are motivated for future engineering applications.

4.1 Supremum Representation of Eigenvalues and Norms

Eigenvalues and norms are the butter and bread when we deal with random matrices. The supremum of a stochastic process [27, 82] has become a basic tool. The aim of this section to make connections between the two topics: we can represent eigenvalues and norms in terms of the supremum of a stochastic process.

The standard reference for our matrices analysis is Bhatia [23]. The **inner product** of two finite-dimensional vectors in a Hilbert space $\mathbb{H}$ is denoted by $\langle \mathbf{u}, \mathbf{v} \rangle$. The form of a vector is denoted by $\|\mathbf{u}\| = \langle \mathbf{u}, \mathbf{u} \rangle^{1/2}$. A matrix is **self-adjoint** or **Hermitian** if $\mathbf{A}^* = \mathbf{A}$, **skew-Hermitian** if $\mathbf{A}^* = -\mathbf{A}$, **unitary** if $\mathbf{A}^*\mathbf{A} = I = \mathbf{A}\mathbf{A}^*$, and **normal** if $\mathbf{A}^*\mathbf{A} = \mathbf{A}\mathbf{A}^*$.

Every complex matrix can be decomposed into

$$\mathbf{A} = \mathrm{Re}\,\mathbf{A} + i\,\mathrm{Im}\,\mathbf{A}$$

where $\mathrm{Re}\,\mathbf{A} = \frac{\mathbf{A}+\mathbf{A}^*}{2}$ and $\mathrm{Im}\,\mathbf{A} = \frac{\mathbf{A}-\mathbf{A}^*}{2}$. This is called the **Cartesian Decomposition** of $\mathbf{A}$ into its "real" and "imaginary" parts. The matrices $\mathrm{Re}\,\mathbf{A}$ and $\mathrm{Im}\,\mathbf{A}$ are both Hermitian.

The norm of a matrix $\mathbf{A}$ is defined as

$$\|\mathbf{A}\| = \sup_{\|\mathbf{x}\|=1} \|\mathbf{A}\mathbf{x}\|.$$

R. Qiu and M. Wicks, *Cognitive Networked Sensing and Big Data*,
DOI 10.1007/978-1-4614-4544-9_4,

We also have the inner product version

$$\|\mathbf{A}\| = \sup_{\|\mathbf{x}\|=\|\mathbf{y}\|=1} |\langle \mathbf{y}, \mathbf{A}\mathbf{x}\rangle| \,.$$

When $\mathbf{A}$ is Hermitian, we have

$$\|\mathbf{A}\| = \sup_{\|\mathbf{x}\|=1} |\langle \mathbf{x}, \mathbf{A}\mathbf{x}\rangle| \,.$$

For every matrix $\mathbf{A}$, we have

$$\|\mathbf{A}\| = \sigma_1 (\mathbf{A}) = \|\mathbf{A}^*\mathbf{A}\|^{1/2}.$$

When $\mathbf{A}$ is normal, we have

$$\|\mathbf{A}\| = \max \{|\lambda_i (\mathbf{A})|\} \,.$$

Another useful norm is the norm

$$\|\mathbf{A}\|_F = \left(\sum_{i=1}^{n} \sigma_i^2 (\mathbf{A}) \right)^{1/2} = (\mathrm{Tr}\mathbf{A}^*\mathbf{A})^{1/2}.$$

If a_{ij} are entries of a matrix $\mathbf{A}$, then

$$\|\mathbf{A}\|_F = \left(\sum_{i,j=1}^{n} |a_{ij}|^2 \right)^{1/2} .$$

This makes this norm useful in calculations with matrices. This is called **Frobenius norm** or **Schatten 2-norm** or the **Hilbert-Schmidt norm**, or **Euclidean norm**.

Both $\|\mathbf{A}\|$ and $\|\mathbf{A}\|_F$ have an important invariance property called **unitary invariant**: we have $\|\mathbf{UAV}\| = \|\mathbf{A}\|$ and $\|\mathbf{UAV}\|_F = \|\mathbf{A}\|_F$ for all unitary $\mathbf{U}, \mathbf{V}$.

Any two norms on a finite-dimensional space are equivalent. For the norms $\|\mathbf{A}\|$ and $\|\mathbf{A}\|_F$, it follows from the properties above that

$$\|\mathbf{A}\| \leqslant \|\mathbf{A}\|_F \leqslant \sqrt{n} \, \|\mathbf{A}\| \tag{4.1}$$

for every $\mathbf{A}$. Equation (4.1) is the central result we want to revisit here.

Exercise 4.1.1 (Neumann series). If $\|\mathbf{A}\| < 1$, then $\mathbf{I} - \mathbf{A}$ is invertible and

$$(\mathbf{I} - \mathbf{A})^{-1} = \mathbf{I} + \mathbf{A} + \mathbf{A}^2 + \cdots + \mathbf{A}^k + \cdots$$

is a convergent power series. This is called the **Neumann series**.

Exercise 4.1.2 (Matrix Exponential). For any matrix $\mathbf{A}$ the series

$$\exp \mathbf{A} = e^{\mathbf{A}} = \mathbf{I} + \mathbf{A} + \frac{1}{2!}\mathbf{A}^2 + \cdots + \frac{1}{k!}\mathbf{A}^k + \cdots$$

converges. The matrix $\exp \mathbf{A}$ is always convertible

$$\left(e^{\mathbf{A}}\right)^{-1} = e^{-\mathbf{A}}.$$

Conversely, every invertible matrix can be expressed as the exponential of some matrix. Every unitary matrix can be expressed as the exponential of a skew-Hermitian matrix.

The number $w(\mathbf{A})$ defined as

$$w\left(\mathbf{A}\right) = \sup_{\|\mathbf{x}\|=1} \left|\langle \mathbf{x}, \mathbf{Ax}\rangle\right|$$

is called the **numerical radius** of $\mathbf{A}$. The spectral radius of a matrix $\mathbf{A}$ is defined as

$$\operatorname{spr}\left(\mathbf{A}\right) = \max\left\{\left|\lambda\left(\mathbf{A}\right)\right|\right\}.$$

We note that $\operatorname{spr}\left(\mathbf{A}\right) \leqslant w\left(\mathbf{A}\right) \leqslant \|\mathbf{A}\|$. They three are equal if (but not only if) the matrix is normal.

Let $\mathbf{A}$ be Hermitian with eigenvalues $\lambda_1 \geqslant \lambda_2 \geqslant \cdots \geqslant \lambda_n$. We have

$$\begin{aligned}\lambda_1 &= \max\left\{\langle \mathbf{x}, \mathbf{Ax}\rangle : \|\mathbf{x}\| = 1\right\},\\ \lambda_n &= \min\left\{\langle \mathbf{x}, \mathbf{Ax}\rangle : \|\mathbf{x}\| = 1\right\}.\end{aligned} \tag{4.2}$$

The inner product of two finite-dimensional vectors in a Hilbert space $\mathbb{H}$ is denoted by $\langle \mathbf{u}, \mathbf{v}\rangle$. The form of a vector is denoted by $\|\mathbf{u}\| = \langle \mathbf{u}, \mathbf{u}\rangle^{1/2}$.

For every $k = 1, 2, \ldots, n$

$$\sum_{i=1}^{k} \lambda_i\left(\mathbf{A}\right) = \max \sum_{i=1}^{k} \langle \mathbf{x}_i, \mathbf{Ax}_i\rangle,$$

$$\sum_{i=n-k+1}^{k} \lambda_i\left(\mathbf{A}\right) = \min \sum_{i=1}^{k} \langle \mathbf{x}_i, \mathbf{Ax}_i\rangle,$$

where the maximum and the minimum are taken over all choices of orthogonal k-tuples $(\mathbf{x}_1, \ldots, \mathbf{x}_k)$ in $\mathbb{H}$. The first statement is referred to as **Ky Fan Maximum Principle**.

If $\mathbf{A}$ is positive, then for every $k = 1, 2, \ldots, n$,

$$\prod_{i=n-k+1}^{n} \lambda_i (\mathbf{A}) = \min \prod_{i=n-k+1}^{n} \langle \mathbf{x}_i, \mathbf{A}\mathbf{x}_i \rangle ,$$

where the minimum is taken over all choices of orthogonal k-tuples $(\mathbf{x}_1, \ldots, \mathbf{x}_k)$ in $\mathbb{H}$.

4.2 Lipschitz Mapping of Eigenvalues

The following lemma is from [152] but we follow the exposition of [69]. Let G denote the Gaussian distribution on $\mathbb{R}^n$ with density

$$\frac{dG(\mathbf{x})}{d\mathbf{x}} = \frac{1}{(2\pi\sigma^2)^n} \exp\left(-\frac{\|\mathbf{x}\|^2}{2\sigma^2} \right),$$

and $\|\mathbf{x}\|^2 = x_1^2 + \cdots + x_n^2$ is the Euclidean norm of $\mathbf{x}$. Furthermore, for a K-Lipschitz function $F : \mathbb{R}^n \to \mathbb{R}$, we have

$$|F(\mathbf{x}) - F(\mathbf{y})| \leqslant K \|\mathbf{x} - \mathbf{y}\|, \quad \mathbf{x}, \mathbf{y} \in \mathbb{R}^n,$$

for some positive Lipschitz constant K. Then for any positive number t, we have that

$$G(\{\mathbf{x} \in \mathbb{R}^n : |F(\mathbf{x}) - F(\mathbf{y})| > t\}) \leqslant 2\exp\left(-\frac{ct^2}{K^2\sigma^2} \right)$$

where $\mathbb{E}(F(\mathbf{x})) = \int_{\mathbb{R}^n} F(\mathbf{x})\, dG(\mathbf{x})$, and $c = \frac{2}{\pi^2}$.

The case of $\sigma = 1$ is proven in [152]. The general case follows by using the following mapping. Under the mapping $\mathbf{x} \mapsto \sigma\mathbf{x} : \mathbb{R}^n \mapsto \mathbb{R}^n$, the composed function $\mathbf{x} \mapsto F(\sigma\mathbf{x})$ satisfies a Lipschitz condition with constant $K\sigma$.

Now let us consider the Hilbert-Schmidt norm (also called Frobenius form and Euclidean norm) $\|\cdot\|_F$ under the Lipschitz functional mapping. Let $f : \mathbb{R} \mapsto \mathbb{R}$ be a function that satisfies the Lipschitz condition

$$|f(s) - f(t)| \leqslant K |s - t|, \quad s, t \in \mathbb{R}.$$

Then for any n in $\mathbb{N}$, and all complex Hermitian matrices $\mathbf{A}, \mathbf{B} \in \mathbb{C}^{n\times n}$. We have that

$$\|f(\mathbf{A}) - f(\mathbf{B})\|_F \leq K\|\mathbf{A} - \mathbf{B}\|_F,$$

where

$$\|\mathbf{C}\|_F = (\mathrm{Tr}\,(\mathbf{C}^*\mathbf{C}))^{1/2} = \left(\mathrm{Tr}\,(\mathbf{C}^2)\right)^{1/2},$$

for all complex Hermitian matrix $\mathbf{C}$.

A short proof follows from [179] but we follow closely [69] for the exposition here. We start with the spectral decomposition

$$\mathbf{A} = \sum_{i=1}^{n} \lambda_i \mathbf{E}_i, \qquad \mathbf{B} = \sum_{i=1}^{n} \mu_i \mathbf{F}_i,$$

where λ_i and μ_i are eigenvalues of $\mathbf{A}$, and $\mathbf{B}$, respectively, and where $\mathbf{E}_i$ and $\mathbf{F}_i$ are two families of mutually orthogonal one-dimensional projections (adding up to $\mathbf{I}_n$). Using $\mathrm{Tr}\,(\mathbf{E}_i\mathbf{F}_j) \geqslant 0$ for all i, j, we obtain that

$$\begin{aligned}
\|f(\mathbf{A}) - f(\mathbf{B})\|_F^2 &= \mathrm{Tr}\left(f(\mathbf{A})^2\right) + \mathrm{Tr}\left(f(\mathbf{B})^2\right) - 2\mathrm{Tr}\,(f(\mathbf{A})\, f(\mathbf{B})) \\
&= \sum_{i,j=1}^{n} (f(\lambda_i) - f(\mu_i))^2 \cdot \mathrm{Tr}\,(\mathbf{E}_i\mathbf{F}_j) \\
&\leqslant K^2 \cdot \sum_{i,j=1}^{n} (\lambda_i - \mu_i)^2 \cdot \mathrm{Tr}\,(\mathbf{E}_i\mathbf{F}_j) \\
&= K^2 \,\|\mathbf{A} - \mathbf{B}\|_F^2 .
\end{aligned}$$

4.3 Smoothness and Convexity of the Eigenvalues of a Matrix and Traces of Matrices

The following lemma (Lemma 4.3.1) is at the heart of the results. First we recall that

$$\mathrm{Tr}\,(f(\mathbf{A})) = \sum_{i=1}^{n} f(\lambda_i(\mathbf{A})), \tag{4.3}$$

where $\lambda_i(\mathbf{A})$ are the eigenvalues of $\mathbf{A}$. Consider a Hermitian $n \times n$ matrix $\mathbf{A}$. Let f be a real valued function on $\mathbb{R}$. We can study the function of the matrix, $f(\mathbf{A})$. If $\mathbf{A} = \mathbf{UDU}^*$, for a diagonal real matrix $\mathbf{D} = \mathrm{diag}\,(\lambda_1, \ldots, \lambda_n)$ and a unitary matrix $\mathbf{U}$, then

$$f(\mathbf{A}) = \mathbf{U} f(\mathbf{D})\, \mathbf{U}^*$$

where $f(\mathbf{D})$ is the diagonal matrix with entries $f(\lambda_1), \ldots, f(\lambda_n)$ and $\mathbf{U}^*$ denotes the conjugate, transpose of $\mathbf{U}$.

Lemma 4.3.1 (Guionnet and Zeitoumi [180]).

1. *If* f *is a real-valued convex function on* $\mathbb{R}$, *it holds that* $\mathrm{Tr}\left(f\left(\mathbf{A}\right)\right) = \sum_{i=1}^{n} f\left(\lambda_i\left(\mathbf{A}\right)\right)$ *is convex.*
2. *If* f *is a Lipschitz function on* $\mathbb{R}$, $\mathbf{A} \to \mathrm{Tr}\left(f\left(\mathbf{A}\right)\right)$ *is a Lipschitz function on* $\mathbb{R}^{n^2}$ *with Lipschitz constant bounded by* $\sqrt{n}|f|_{\mathcal{L}}$.

Theorem 4.3.2 (Lidskii [18]). *Let* $\mathbf{A}, \mathbf{B}$ *be Hermitian matrices. Then, there is a doubly stochastic matrix* $\mathbf{E}$ *such that*

$$\lambda_i\left(\mathbf{A}+\mathbf{B}\right) - \lambda_i\left(\mathbf{A}\right) \leqslant \sum_{m=1}^{n} E_{i,m}\lambda_i\left(\mathbf{B}\right)$$

In particular,

$$\sum_{i=1}^{n} \left|\lambda_i\left(\mathbf{A}\right) - \lambda_i\left(\mathbf{B}\right)\right|^2 \leqslant \left\|\mathbf{A}-\mathbf{B}\right\|_F^2 \leqslant \sum_{i=1}^{n} \left|\lambda_i\left(\mathbf{A}\right) - \lambda_{n-i+1}\left(\mathbf{B}\right)\right|^2. \quad (4.4)$$

For all integer k, the functional

$$\left(A_{ij}\right)_{1\leqslant i,j\leqslant n} \in \mathbb{R}^{n^2} \to \lambda_k\left(\mathbf{A}\right)$$

is Lipschitz with constant one, following (4.4). With the aid of Lidskii's theorem [18, p. 657] (Theorem 4.3.2 here), we have

$$\begin{aligned}\left|\sum_{i=1}^{n} f\left(\lambda_i\left(\mathbf{A}\right)\right) - \sum_{i=1}^{n} f\left(\lambda_i\left(\mathbf{B}\right)\right)\right| &\leqslant |f|_{\mathcal{L}} \sum_{i=1}^{n} \left|\lambda_i\left(\mathbf{A}\right) - \lambda_i\left(\mathbf{B}\right)\right| \\ &\leqslant \sqrt{n}|f|_{\mathcal{L}} \left(\sum_{i=1}^{n} \left|\lambda_i\left(\mathbf{A}\right) - \lambda_i\left(\mathbf{B}\right)\right|^2\right)^{1/2} \\ &\leqslant \sqrt{n}|f|_{\mathcal{L}} \left\|\mathbf{A}-\mathbf{B}\right\|_F. \end{aligned} \quad (4.5)$$

We use the definition of a Lipschitz function in the first inequality. The second step follows from Cauchy-Schwartz's inequality [181, p. 31]: for arbitrary real numbers $a_i, b_i \in \mathbb{R}$

$$\left|a_1 b_1 + a_2 b_2 + \cdots + a_n b_n\right| \leqslant \sqrt{a_1^2 + a_2^2 + \cdots + a_n^2}\sqrt{b_1^2 + b_2^2 + \cdots + b_n^2}. \quad (4.6)$$

In particular, we have used $b_i = 1$. The second inequality is a direct consequence of Lidskii's theorem, Theorem 4.3.2. In other words, we have shown that the function

$$\left(A_{ij}\right)_{1\leqslant i,j\leqslant n} \in \mathbb{R}^{n^2} \to \sum_{k=1}^{n} f\left(\lambda_k\left(\mathbf{A}\right)\right)$$

is Lipschitz with a constant bounded above by $\sqrt{n}|f|_{\mathcal{L}}$. Observe that using $\sum_{k=1}^{n} f(\lambda_k(\mathbf{A}))$ rather than $\lambda_k(\mathbf{A})$ increases the Lipschitz constant from 1 to $\sqrt{n}|f|_{\mathcal{L}}$. This observation is useful later in Sect. 4.5 when the tail bound is considered.

Lemma 4.3.3 ([182]). *For a given $n_1 \times n_2$ matrix $\mathbf{X}$, let $\sigma_i(\mathbf{X})$ the i-th largest singular value. Let $f(\mathbf{X})$ be a function on matrices in the following form: $f(\mathbf{X}) = \sum_{i=1}^{m} a_i \sigma_i(\mathbf{X})$ for some real constants $\{a_i\}_{i=1}^{m}$. Then $f(\mathbf{X})$ is a Lipschitz function with a constant of $\sqrt{\sum_{i=1}^{m} a_i^2}$.*

Let us consider the special case $f(t) = t^k$. The power series $(\mathbf{A} + \varepsilon\mathbf{B})^k$ is expanded as

$$\mathrm{Tr}\left((\mathbf{A} + \varepsilon\mathbf{B})^k\right) = \mathrm{Tr}\left(\mathbf{A}^k\right) + \varepsilon k\,\mathrm{Tr}\left(\mathbf{A}^{k-1}\mathbf{B}\right) + O\left(\varepsilon^2\right).$$

Or for small ε we have

$$\mathrm{Tr}\left((\mathbf{A} + \varepsilon\mathbf{B})^k\right) - \mathrm{Tr}\left(\mathbf{A}^k\right) = \varepsilon k\,\mathrm{Tr}\left(\mathbf{A}^{k-1}\mathbf{B}\right) + O\left(\varepsilon^2\right) \to 0, \text{ as } \varepsilon \to 0. \quad (4.7)$$

Recall that the trace function is linear.

$$\lim_{\varepsilon\to 0} \frac{1}{\varepsilon}\left[\mathrm{Tr}\left((\mathbf{A} + \varepsilon\mathbf{B})^k\right) - \mathrm{Tr}\left(\mathbf{A}^k\right)\right] = k\,\mathrm{Tr}\left(\mathbf{A}^{k-1}\mathbf{B}\right) = \mathrm{Tr}\left(\left(\mathbf{A}^k\right)'\mathbf{B}\right)$$

where $(\cdot)'$ denotes the derivative. More generally, if f is continuously differentiable, we have

$$\lim_{\varepsilon\to 0} \frac{1}{\varepsilon}\left[\sum_{i=1}^{n} f\left(\lambda_i\left(\mathbf{A} + \varepsilon\mathbf{B}\right)\right) - \sum_{i=1}^{n} f\left(\lambda_i\left(\mathbf{A}\right)\right)\right] = \mathrm{Tr}\left(f'\left(\mathbf{A}\right)\mathbf{B}\right).$$

Recall that $\sum_{i=1}^{n} f(\lambda_i(\mathbf{A}))$ is equal to $\mathrm{Tr}(f(\mathbf{A}))$. $\lambda_1(\mathbf{X})$ is convex and $\lambda_n(\mathbf{X})$ is concave.

Let us consider the function of the sum of the first largest eigenvalues: for $k = 1, 2, \ldots, n$

$$F_k(\mathbf{A}) = \sum_{i=1}^{k} \lambda_i(\mathbf{A})$$

$$G_k(\mathbf{A}) = \sum_{i=1}^{k} \lambda_{n-i+1}(\mathbf{A}) = \mathrm{Tr}(\mathbf{A}) - F_k(\mathbf{A}). \quad (4.8)$$

The trace function of (4.3) is the special case of these two function with $k = n$.

Now we follow [183] to derive the Lipschitz constant of functions defined in (4.8), which turns out to be $C|f|_{\mathcal{L}}\sqrt{k}$, where $C < 1$. Recall the Euclidean norm or the Frobenius norm is defined as

$$\|\mathbf{X}\|_F = \left(\sum_{i,j=1}^{n} |X_{ij}|^2 \right)^{1/2} = \sqrt{2} \left(\sum_{i=1}^{n} \left| \frac{X_{ii}}{\sqrt{2}} \right|^2 + \sum_{1 \leqslant i \leqslant j \leqslant n,} |X_{ij}|^2 \right)^{1/2}.$$

Recall also that

$$\|\mathbf{X}\|_F = \left(\sum_{i=1}^{n} \lambda_i^2 (\mathbf{X}) \right)^{1/2},$$

which implies from (4.1) that $\lambda_i(\mathbf{X})$ is a 1-Lipschitz function of $\mathbf{X}$ with respect to $\|\mathbf{X}\|_F$.

F_k is positively homogeneous (of degree 1) and $F_k(-\mathbf{A}) = -G_k(\mathbf{A})$. From this we have that

$$\begin{aligned} |F_k(\mathbf{A}) - F_k(\mathbf{B})| &\leqslant \max\{F_k(\mathbf{A}-\mathbf{B}), -G_k(\mathbf{A}-\mathbf{B})\} \leqslant \sqrt{k}\|\mathbf{A}-\mathbf{B}\|_F, \\ |G_k(\mathbf{A}) - G_k(\mathbf{B})| &\leqslant \max\{G_k(\mathbf{A}-\mathbf{B}), -F_k(\mathbf{A}-\mathbf{B})\} \leqslant \sqrt{k}\|\mathbf{A}-\mathbf{B}\|_F. \end{aligned} \tag{4.9}$$

In other words, the functions $F_k(\mathbf{A}), G_k(\mathbf{A}) : \mathbb{R}^n \mapsto \mathbb{R}$ are Lipschitz continuous with the Lipschitz constant $\sqrt{k}$. For a trace function, we have $k = n$. Moreover, $F_k(\mathbf{A})$ is convex and $G_k(\mathbf{A})$ is concave. This follows from Ky Fan's maximum principle in (4.2) or Davis' characterization [184] of all convex unitarily invariant functions of a self-adjoint matrix.

Let us give our version of the proof of (4.9). There are no details about this proof in [183]. When $\mathbf{A} \geq \mathbf{B}$, implying that $\lambda_i(\mathbf{A}) \geqslant \lambda_i(\mathbf{B})$, we have

$$\begin{aligned} |F_k(\mathbf{A}) - F_k(\mathbf{B})| &= \left| \sum_{i=1}^{k} \lambda_i(\mathbf{A}) - \sum_{i=1}^{k} \lambda_i(\mathbf{B}) \right| = \left| \sum_{i=1}^{k} (\lambda_i(\mathbf{A}) - \lambda_i(\mathbf{B})) \right| \\ &\leqslant \sum_{i=1}^{k} |\lambda_i(\mathbf{A}) - \lambda_i(\mathbf{B})| = \sum_{i=1}^{k} (\lambda_i(\mathbf{A}) - \lambda_i(\mathbf{B})) \leqslant \sum_{i=1}^{k} \lambda_i(\mathbf{A}-\mathbf{B}) = F_k(\mathbf{A}-\mathbf{B}). \end{aligned}$$

In the second line, we have used the Ky Fan inequality (1.28)

$$\lambda_1(\mathbf{C}+\mathbf{D}) + \cdots + \lambda_k(\mathbf{C}+\mathbf{D}) \leqslant \lambda_1(\mathbf{C}) + \cdots + \lambda_k(\mathbf{C}) + \lambda_1(\mathbf{D}) + \cdots + \lambda_k(\mathbf{D}) \tag{4.10}$$

where $\mathbf{C}$ and $\mathbf{D}$ are Hermitian, by identifying $\mathbf{C} = \mathbf{B}, \mathbf{D} = \mathbf{A} - \mathbf{B}, \mathbf{C} + \mathbf{D} = \mathbf{A}$..

If $\mathbf{A} < \mathbf{B}$, implying that $\lambda_i(\mathbf{B}) \geqslant \lambda_i(\mathbf{A})$, we have

$$|F_k(\mathbf{A}) - F_k(\mathbf{B})| = \left|\sum_{i=1}^{k} \lambda_i(\mathbf{A}) - \sum_{i=1}^{k} \lambda_i(\mathbf{B})\right| = \left|\sum_{i=1}^{k} (\lambda_i(\mathbf{A}) - \lambda_i(\mathbf{B}))\right|$$

$$\leqslant \sum_{i=1}^{k} |\lambda_i(\mathbf{A}) - \lambda_i(\mathbf{B})| = \sum_{i=1}^{k} (\lambda_i(\mathbf{B}) - \lambda_i(\mathbf{A})) \leqslant \sum_{i=1}^{k} \lambda_i(\mathbf{B}-\mathbf{A}) = F_k(\mathbf{B}-\mathbf{A}) = -G_k(\mathbf{A}-\mathbf{B}).$$

Let us define the following function: for $k = 1, 2, \ldots, n$

$$\varphi_k(\mathbf{A}) = \sum_{i=1}^{k} f(\lambda_i(\mathbf{A})) \tag{4.11}$$

where $f : \mathbb{R} \mapsto \mathbb{R}$ is the Lipschitz function with constant $|f|_{\mathcal{L}}$. We can compare (4.8) and (4.11). We can show [185] that $\varphi_k(\mathbf{A}) = \sum_{i=1}^{k} f(\lambda_i(\mathbf{A}))$ is a Lipschitz function with a constant bounded by $\sqrt{k}|f|_{\mathcal{L}}$. It follows from [185] that

$$\begin{aligned}
&\left|\sum_{i=1}^{k} f(\lambda_i(\mathbf{A})) - \sum_{i=1}^{k} f(\lambda_i(\mathbf{B}))\right| \\
&= \left|\sum_{i=1}^{k} [f(\lambda_i(\mathbf{A})) - f(\lambda_i(\mathbf{B}))]\right| \\
&\leqslant \sum_{i=1}^{k} |f(\lambda_i(\mathbf{A})) - f(\lambda_i(\mathbf{B}))| \\
&\leqslant |f|_{\mathcal{L}} \sum_{i=1}^{k} |\lambda_i(\mathbf{A}) - \lambda_i(\mathbf{B})| \\
&\leqslant |f|_{\mathcal{L}} \sqrt{k} \left(\sum_{i=1}^{k} |\lambda_i(\mathbf{A}) - \lambda_i(\mathbf{B})|^2\right)^{1/2} \\
&\leqslant C|f|_{\mathcal{L}} \sqrt{k} \left(\sum_{i=1}^{n} |\lambda_i(\mathbf{A}) - \lambda_i(\mathbf{B})|^2\right)^{1/2} \\
&\leqslant C|f|_{\mathcal{L}} \sqrt{k} \|\mathbf{A} - \mathbf{B}\|_F,
\end{aligned}$$

where

$$C = \frac{\left(\sum_{i=1}^{k} \left|\lambda_i\left(\mathbf{A}\right) - \lambda_i\left(\mathbf{B}\right)\right|^2\right)^{1/2}}{\left(\sum_{i=1}^{n} \left|\lambda_i\left(\mathbf{A}\right) - \lambda_i\left(\mathbf{B}\right)\right|^2\right)^{1/2}} \leqslant 1.$$

The third line follows from the triangle inequality for complex numbers [181, 30]: for n complex numbers

$$\left|\sum_{i=1}^{n} z_i\right| = |z_1 + z_2 + \cdots + z_n| \leqslant |z_1| + |z_2| + \cdots + |z_n| = \sum_{i=1}^{n} |z_i|. \quad (4.12)$$

In particular, we set $z_i = \lambda_i\left(\mathbf{A}\right) - \lambda_i\left(\mathbf{B}\right)$. The fourth line follows from the definition for a Lipschitz function: for $f : \mathbb{R} \to \mathbb{R}$, $|f(s) - f(t)| \leqslant |f|_{\mathcal{L}} |s - t|$. The fifth line follows from Cauchy-Schwartz's inequality (4.6) by identifying $a_i = |\lambda_i\left(\mathbf{A}\right) - \lambda_i\left(\mathbf{B}\right)|, \quad b_i = 1$. The final line follows from Liskii's theorem (4.4).

Example 4.3.4 (Standard hypothesis testing problem revisited: Moments as a function of SNR). Our standard hypothesis testing problem is expressed as

$$\begin{aligned} \mathcal{H}_0 &: \mathbf{y} = \mathbf{n} \\ \mathcal{H}_1 &: \mathbf{y} = \sqrt{SNR}\mathbf{x} + \mathbf{n} \end{aligned} \quad (4.13)$$

where $\mathbf{x}$ is the signal vector in $\mathbb{C}^n$ and $\mathbf{n}$ the noise vector in $\mathbb{C}^n$. We assume that $\mathbf{x}$ is independent of $\mathbf{n}$. SNR is the dimensionless real number representing the signal to noise ratio. It is assumed that N independent realizations of these vector valued random variables are observed.

Representing

$$\mathbf{Y} = \begin{bmatrix} \mathbf{y}_1^T \\ \vdots \\ \mathbf{y}_N^T \end{bmatrix}, \quad \mathbf{X} = \begin{bmatrix} \mathbf{x}_1^T \\ \vdots \\ \mathbf{x}_N^T \end{bmatrix}, \quad \mathbf{N} = \begin{bmatrix} \mathbf{n}_1^T \\ \vdots \\ \mathbf{n}_N^T \end{bmatrix},$$

we can rewrite (4.13) as

$$\begin{aligned} \mathcal{H}_0 &: \mathbf{Y} = \mathbf{N} \\ \mathcal{H}_1 &: \mathbf{Y} = \sqrt{SNR} \cdot \mathbf{X} + \mathbf{N} \end{aligned} \quad (4.14)$$

We form the sample covariance matrices as follows:

$$\mathbf{S}_y = \frac{1}{N}\sum_{i=1}^{N} \mathbf{y}_i\mathbf{y}_i^*, \quad \mathbf{S}_x = \frac{1}{N}\sum_{i=1}^{N} \mathbf{x}_i\mathbf{x}_i^*, \quad \mathbf{S}_n = \frac{1}{N}\sum_{i=1}^{N} \mathbf{n}_i\mathbf{n}_i^*. \quad (4.15)$$

we rewrite (4.15) as

$$\begin{aligned}\mathcal{H}_0 &: N\mathbf{S}_n = \mathbf{N}\mathbf{N}^* \\ \mathcal{H}_1 &: N\mathbf{S}_y = \mathbf{Y}\mathbf{Y}^* = \left(\sqrt{SNR}\cdot\mathbf{X}+\mathbf{N}\right)\left(\sqrt{SNR}\cdot\mathbf{X}+\mathbf{N}\right)^* \\ &= SNR\cdot\mathbf{X}\mathbf{X}^* + \mathbf{N}\mathbf{N}^* + \sqrt{SNR}\cdot\left(\mathbf{X}\mathbf{N}^*+\mathbf{N}\mathbf{X}^*\right)\end{aligned} \tag{4.16}$$

$\mathcal{H}_0$ can be viewed as a function of $\mathcal{H}_1$ as we take the limit $SNR \to 0$. This function is apparently continuous as a function of SNR. This way we can consider one function of $\mathcal{H}_1$ only and take the limit for another hypothesis. In the ideal case of $N \to \infty$, the true covariance version of the problem is as follows

$$\begin{aligned}\mathcal{H}_0 &: \mathbf{R}_y = \mathbf{R}_n \\ \mathcal{H}_1 &: \mathbf{R}_y = SNR\cdot\mathbf{R}_x + \mathbf{R}_n.\end{aligned}$$

Let $\mathbf{N}\mathbf{N}^*$ and $\mathbf{N}'\mathbf{N}'^*$ are two independent copies of underlying random matrices. When the sample number N goes large, we expect that the sample covariance matrices approach the true covariance matrix, as close as we desire. In other words, $\frac{1}{N}\mathbf{N}\mathbf{N}^*$ and $\frac{1}{N}\mathbf{N}'\mathbf{N}'^*$ are very close to each other. With this intuition, we may consider the following matrix function

$$\mathbf{S}(SNR,\mathbf{X}) = SNR\cdot\frac{1}{N}\mathbf{X}\mathbf{X}^* + \frac{1}{N}\sqrt{SNR}\cdot(\mathbf{X}\mathbf{N}^*+\mathbf{N}\mathbf{X}^*) + \frac{1}{N}\left(\mathbf{N}\mathbf{N}^* - \mathbf{N}'\mathbf{N}'^*\right)$$

Taking the trace of both sides, we reach a more convenient form

$$f\left(SNR,\mathbf{X}\right) = \mathrm{Tr}\left(\mathbf{S}\mathbf{S}^*\right).$$

$f\left(SNR,\mathbf{X}\right)$ is apparently a continuous function of SNR. $f\left(SNR,\mathbf{X}\right)$ represents hypothesis $\mathcal{H}_0$ as we take the limit $SNR \to 0$. Note that the trace $rmTr$ is a linear function. More generally we may consider the k-th moment

$$g\left(SNR,\mathbf{X}\right) = \mathrm{Tr}\left[\left(\mathbf{S}\mathbf{S}^*\right)^k\right].$$

$\mathbf{S}\mathbf{S}^*$ can be written as a form

$$\mathbf{S}\mathbf{S}^* = \mathbf{A} + \varepsilon\left(SNR\right)\mathbf{B},$$

where $\varepsilon\left(SNR\right) \to 0, SNR \to 0$. It follows from (4.7) that

$$\mathrm{Tr}\left(\left(\mathbf{A}+\varepsilon\mathbf{B}\right)^k\right) - \mathrm{Tr}\left(\mathbf{A}^k\right) = \varepsilon k\,\mathrm{Tr}\left(\mathbf{A}^{k-1}\mathbf{B}\right) + O\left(\varepsilon^2\right) \to 0, \text{ as } SNR \to 0.$$

□

Exercise 4.3.5. Show that $\mathbf{SS}^*$ has the following form $\mathbf{SS}^* = \mathbf{A} + \varepsilon\,(SNR)\,\mathbf{B}$.

Example 4.3.6 (Moments of Random Matrices [23]). The special case of $f(t) = t^k$ for integer k is Lipschitz continuous. Using the fact that [48, p. 72]

$$f(\mathbf{x}) = \log\left(e^{x_1} + e^{x_2} + \cdots + e^{x_n}\right)$$

is a convex function of $\mathbf{x}$, and setting $\lambda_i^k = e^{\log \lambda_i^k} = x_i$, we have

$$f(\mathbf{x}) = \log\left(\lambda_1^k + \cdots + \lambda_n^k\right) = \log\left(\sum_{i=1}^{n} \lambda_i^k\right)$$

is also a convex function.

Jensen's inequality says that for a convex function ϕ

$$\phi\left(\sum_i a_i x_i\right) \leqslant \sum_i a_i \phi(x_i), \quad \sum_i a_i = 1, \quad a_i \geqslant 0.$$

It follows from Jensen's inequality that

$$f(\mathbf{x}) = \log\left(\sum_{i=1}^{n} \lambda_i^k\right) \leqslant \sum_{i=1}^{n} \log \lambda_i^k = k \sum_{i=1}^{n} \log \lambda_i \leqslant k \sum_{i=1}^{n} (\lambda_i - 1).$$

In the last step, we use the inequality that $\log x \leqslant x - 1, x > 0$.

Since $\lambda_i(\mathbf{A})$ is 1-Lipschitz, $f(\mathbf{x})$ is also Lipschitz.

Another direct approach is to use Lemma 4.3.1: If $f : \mathbb{R} \to \mathbb{R}$ is a Lipschitz function, then

$$F = \frac{1}{\sqrt{n}} \sum_{i=1}^{n} f(\lambda_i)$$

is a Lipschitz function of (real and imaginary) entries of $\mathbf{A}$. If f is convex on the real line, then F is convex on the space of matrices (Klein's lemma). Clearly $f(t) = t^a, a \geq 1$ is a convex function from $\mathbb{R}$ to $\mathbb{R}$. $f(t) = t^a$ is also Lipschitz.

Since $\frac{1}{a}\log f(t) = \log(t) \leqslant t - 1, t > 0$ we have

$$\frac{1}{a} \log |f(x) - f(y)| \leqslant |x - y|$$

where $f(t) = t^a$. □

4.4 Approximation of Matrix Functions Using Matrix Taylor Series

When $f(t)$ is a polynomial or rational function with scalar coefficients a_i and a scalar argument t, it is natural to define [20] $f(\mathbf{A})$ by substituting $\mathbf{A}$ for t, replacing division by matrix inverse and replacing 1 by the identity matrix. Then, for example

$$f(t) = \frac{1+t^2}{1-t} \Rightarrow f(\mathbf{A}) = (\mathbf{I} - \mathbf{A})^{-1}\left(\mathbf{I} + \mathbf{A}^2\right) \quad \text{if } 1 \notin \Omega(\mathbf{A}).$$

Here $\Omega(\mathbf{A})$ is the set of eigenvalues of $\mathbf{A}$ (also called the spectrum of $\mathbf{A}$). Note that rational functions of a matrix commute, so it does not matter whether we write $(\mathbf{I} - \mathbf{A})^{-1}\left(\mathbf{I} + \mathbf{A}^2\right)$ or $\left(\mathbf{I} + \mathbf{A}^2\right)(\mathbf{I} - \mathbf{A})^{-1}$. If f has a convergent power series representation, such as

$$\log(1+t) = t - \frac{t^2}{2} + \frac{t^3}{3} - \frac{t^4}{4} + \cdots, |t| < 1,$$

we can again simply substitute $\mathbf{A}$ for t to define

$$\log(1+\mathbf{A}) = \mathbf{A} - \frac{\mathbf{A}^2}{2} + \frac{^3\mathbf{A}}{3} - \frac{\mathbf{A}^4}{4} + \cdots, \rho(\mathbf{A}) < 1.$$

Here, ρ is the spectral radius of the condition $\rho(\mathbf{A}) < 1$ ensures convergence of the matrix series. We can consider the polynomial

$$f(t) = P_n(t) = a_0 + a_1 t + \cdots + a_n t^n.$$

For we have

$$f(\mathbf{A}) = P_n(\mathbf{A}) = a_0 + a_1\mathbf{A} + \cdots + a_n\mathbf{A}^n.$$

A basic tool for approximating matrix functions is the Taylor series. We state a theorem from [20, Theorem 4.7] that guarantees the validity of a matrix Taylor series if the eigenvalues of the "increment" lie within the radius of convergence of the associated scalar Taylor series.

Theorem 4.4.1 (convergence of matrix Taylor series). *Suppose f has a Taylor series expansion*

$$f(z) = \sum_{k=1}^{\infty} a_k (z-\alpha)^k \qquad \left(a_k = \frac{f^{(k)}(\alpha)}{k!}\right) \tag{4.17}$$

with radius of convergence r. If $\mathbf{A} \in \mathbb{C}^{n \times n}$, then $f(\mathbf{A})$ is defined and is given by

$$f(\mathbf{A}) = \sum_{k=1}^{\infty} a_k(\mathbf{A} - \alpha\mathbf{I})^k$$

if and only if each of the distinct eigenvalues $\lambda_1, \ldots, \lambda_s$ *of* $\mathbf{A}$ *satisfies one of the conditions*

1. $|\lambda_i - \alpha| \leqslant r$,
2. $|\lambda_i - \alpha| = r$, *and the series for* $f^{(n_i-1)}(\lambda)$ *(where* n_i *is the index of* λ_i*) is convergent at the points* $\lambda = \lambda_i, i = 1, 2, \ldots, s$.

The four most important matrix Taylor series are

$$\exp(\mathbf{A}) = \mathbf{I} + \mathbf{A} + \frac{\mathbf{A}^2}{2!} + \frac{\mathbf{A}^3}{3!} + \cdots,$$

$$\cos(\mathbf{A}) = \mathbf{I} - \frac{\mathbf{A}^2}{2!} + \frac{\mathbf{A}^4}{4!} - \frac{\mathbf{A}^6}{6!} + \cdots,$$

$$\sin(\mathbf{A}) = \mathbf{I} - \frac{\mathbf{A}^3}{3!} + \frac{\mathbf{A}^5}{5!} - \frac{\mathbf{A}^7}{7!} + \cdots,$$

$$\log(\mathbf{I} + \mathbf{A}) = \mathbf{A} - \frac{\mathbf{A}^2}{2!} + \frac{\mathbf{A}^3}{3!} - \frac{\mathbf{A}^4}{4!} + \cdots, \quad \rho(\mathbf{A}) < 1,$$

the first three series having infinite radius of convergence. These series can be used to approximate the respective functions, by summing a suitable finite number of terms. Two types of errors arise: truncated errors, and rounding errors in the floating point evaluation. Truncated errors are bounded in the following result from [20, Theorem 4.8].

Theorem 4.4.2 (Taylor series truncation error bound). *Suppose* f *has the Taylor series expansion* (4.17) *with radius of convergence* r. *If* $\mathbf{A} \in \mathbb{C}^{n \times n}$ *with* $\rho(\mathbf{A} - \alpha\mathbf{I}) < r$, *then for any matrix norm*

$$\left\| f(\mathbf{A}) - \sum_{k=1}^{K} a_k(\mathbf{A} - \alpha\mathbf{I})^k \right\| \leqslant \frac{1}{K!} \max_{0 \leqslant t \leqslant 1} \left\| (\mathbf{A} - \alpha\mathbf{I})^K f^{(K)}(\alpha\mathbf{I} + t(\mathbf{A} - \alpha\mathbf{I})) \right\|. \tag{4.18}$$

In order to apply this theorem, we need to bound the term $\max_{0 \leqslant t \leqslant 1} \left\| (\mathbf{A} - \alpha\mathbf{I})^K f^{(K)} \right.$ $\left. (\alpha\mathbf{I} + t(\mathbf{A} - \alpha\mathbf{I})) \right\|$. For certain function f this is easy. We illustrate using the cosine function, with $\alpha = 0$, and $K = 2k + 2$, and

$$T_{2k}(\mathbf{A}) = \sum_{i=0}^{2k} \frac{(-1)^i}{(2i)!} \mathbf{A}^{2i},$$

the bound of Theorem 4.4.2 is, for the ∞-norm,

$$\begin{aligned}\|\cos(\mathbf{A}) - T_{2k}(\mathbf{A})\|_\infty &= \tfrac{1}{(2k+2)!} \max_{0\le t\le 1} \left\|\mathbf{A}^{2k+2}\cos^{(2k+2}(t\mathbf{A})\right\|_\infty \\ &= \tfrac{1}{(2k+2)!}\left\|\mathbf{A}^{2k+2}\right\|_\infty \max_{0\le t\le 1}\left\|\cos^{(2k+2}(t\mathbf{A})\right\|_\infty .\end{aligned}$$

Now

$$\begin{aligned}&\max_{0\le t\le 1}\left\|\cos^{(2k+2}(t\mathbf{A})\right\|_\infty = \max_{0\le t\le 1}\|\cos(t\mathbf{A})\|_\infty \\ &\le 1 + \tfrac{\|\mathbf{A}\|_\infty}{2!} + \tfrac{\|\mathbf{A}\|_\infty^4}{4!} + \cdots = \cosh\left(\|\mathbf{A}\|_\infty\right),\end{aligned}$$

and thus the error is the truncated Taylor series approximation to the matrix cosine has the bound

$$\|\cos(\mathbf{A}) - T_{2k}(\mathbf{A})\|_\infty \le \frac{\left\|\mathbf{A}^{2k+2}\right\|_\infty}{(2k+2)!}\cosh\left(\|\mathbf{A}\|_\infty\right).$$

Consider the characteristic function

$$\det(\mathbf{A} - t\mathbf{I}) = t^n + c_1 t^{n-1} + \cdots + c_n.$$

By the Cayley-Hamilton theorem,

$$\mathbf{A}^{-1} = -\frac{1}{c_n}\left(\mathbf{A}^{n-1} + \sum_{i=1}^{n-1} c_i \mathbf{A}^{n-i-1}\right).$$

The c_i can be obtained by computing the moments

$$m_k = \mathrm{Tr}\left(\mathbf{A}^k\right), k = 1, \ldots, n,$$

and then solving the Newton identities [20, p. 90]

$$\begin{bmatrix} 1 & & & & & \\ m_1 & 2 & & & & \\ m_2 & m_1 & 3 & & & \\ m_3 & m_2 & m_1 & 4 & & \\ \vdots & \vdots & \vdots & \vdots & \ddots & \\ m_{n-1} & \cdots & m_3 & m_2 & m_1 & n \end{bmatrix} \begin{bmatrix} c_1 \\ c_2 \\ \vdots \\ \vdots \\ \vdots \\ c_n \end{bmatrix} = \begin{bmatrix} m_1 \\ m_2 \\ \vdots \\ \vdots \\ \vdots \\ m_n \end{bmatrix}.$$

Let us consider a power series of random matrices $\mathbf{X}^k, k = 1, \ldots, K$. We define a matrix function $F(\mathbf{X})$ as

$$F(\mathbf{X}) = a_0\mathbf{I} + a_1\mathbf{X} + \cdots + a_K\mathbf{X}^K = \sum_{k=0}^{K} a_k\mathbf{X}^k,$$

where a_k are scalar-valued coefficients. Often we are interested in the trace of this matrix function

$$\text{Tr}(F(\mathbf{X})) = a_0\text{Tr}(\mathbf{I}) + a_1\text{Tr}(\mathbf{X}) + \cdots + a_K\text{Tr}\left(\mathbf{X}^K\right) = \sum_{k=0}^{K} a_k\text{Tr}\left(\mathbf{X}^k\right).$$

Taking the expectation from both sides gives

$$\text{Tr}(\mathbb{E}(F(\mathbf{X}))) = \mathbb{E}(\text{Tr}(F(\mathbf{X}))) = \mathbb{E}\sum_{k=0}^{K} a_k\text{Tr}\left(\mathbf{X}^k\right) = \sum_{k=0}^{K} a_k\mathbb{E}\left(\text{Tr}\left(\mathbf{X}^K\right)\right).$$

Let us consider the fluctuation of this trace function around its expectation

$$\text{Tr}(F(\mathbf{X})) - \text{Tr}(\mathbb{E}F(\mathbf{X})) = \sum_{k=0}^{K} a_k\text{Tr}\left(\mathbf{X}^k\right) - \sum_{k=0}^{K} a_k\mathbb{E}\left(\text{Tr}\left(\mathbf{X}^K\right)\right)$$

$$= \sum_{k=0}^{K} a_k\left[\text{Tr}\left(\mathbf{X}^K\right) - \mathbb{E}\left(\text{Tr}\left(\mathbf{X}^K\right)\right)\right].$$

If we are interested in the absolute of this fluctuation, or called the distance $\text{Tr}(F(\mathbf{X}))$ from $\text{Tr}(\mathbb{E}F(\mathbf{X}))$, it follows that

$$\begin{aligned}|\text{Tr}(F(\mathbf{X})) - \text{Tr}(\mathbb{E}F(\mathbf{X}))| &\leqslant \sum_{k=0}^{K} a_k\left|\text{Tr}\left(\mathbf{X}^K\right) - \mathbb{E}\left(\text{Tr}\left(\mathbf{X}^K\right)\right)\right|, \\ &\leqslant \sum_{k=0}^{K} a_k\left\{\left|\text{Tr}\left(\mathbf{X}^K\right)\right| + \left|\mathbb{E}\left(\text{Tr}\left(\mathbf{X}^K\right)\right)\right|\right\}, \end{aligned} \tag{4.19}$$

since, for two complex scalars a, b, $a - b \leqslant |a - b| \leqslant |a| + |b|$. $\mathbb{E}\left(\text{Tr}\left(\mathbf{X}^K\right)\right)$ can be calculated. In fact, $|\text{Tr}(\mathbf{A})|$ is a seminorm (but not a norm [23, p. 101]) of $\mathbf{A}$. Another relevant norm is weakly unitarily invariant norm

$$\tau(\mathbf{A}) = \tau(\mathbf{U}^*\mathbf{A}\mathbf{U}), \text{ for all } \mathbf{A}, \mathbf{U} \in \mathbb{C}^{n \times n}.$$

The closed form of expected moments $\mathbb{E}\left(\text{Tr}\left(\mathbf{X}^K\right)\right)$ is extensively studied and obtained [69]. $\text{Tr}\left(\mathbf{X}^K\right)$ can be obtained numerically.

More generally, we can study $F(\mathbf{X}) - \mathbb{E}F(\mathbf{X})$ rather than its trace function; we have the matrix-valued fluctuation

$$F(\mathbf{X}) - \mathbb{E}(F(\mathbf{X})) = \sum_{k=0}^{K} a_k \mathbf{X}^k - \mathbb{E}\left(\sum_{k=0}^{K} a_k \mathbf{X}^k\right) = \sum_{k=0}^{K} a_k \mathbf{X}^k - \sum_{k=0}^{K} a_k \mathbb{E}\left(\mathbf{X}^k\right)$$

$$= \sum_{k=0}^{K} a_k \left(\mathbf{X}^k - \mathbb{E}\left(\mathbf{X}^k\right)\right). \tag{4.20}$$

According to (4.20), the problem boils down to a sum of random matrices, which are already treated in other chapters of this book. The random matrices $\mathbf{X}^k - \mathbb{E}\left(\mathbf{X}^k\right)$ play a fundamental role in this problem.

Now we can define the distance as the unitarily invariant norm [23, p. 91] $|||\cdot|||$ of the matrix fluctuation. We have

$$|||\mathbf{UAV}||| = |||\mathbf{A}|||$$

for all $\mathbf{A}$ of $n \times n$ and for unitary $\mathbf{U}, \mathbf{V}$.

For any complex matrix $\mathbf{A}$, we use $\|\mathbf{A}\|_+ = (\mathbf{A}^*\mathbf{A})^{1/2}$ to denote the positive semidefinite matrix [16, p. 235]. The main result, due to R. Thompson, is that for any square complex matrices $\mathbf{A}$ and $\mathbf{B}$ of the same size, there exist two unitary matrices $\mathbf{U}$ and $\mathbf{V}$ such that

$$\|\mathbf{A}+\mathbf{B}\|_+ \leqslant \mathbf{U}^*\|\mathbf{A}\|_+\mathbf{U} + \mathbf{V}^*\|\mathbf{B}\|_+\mathbf{V}. \tag{4.21}$$

Note that it is false to write $\|\mathbf{A}+\mathbf{B}\|_+ \leqslant \|\mathbf{A}\|_+ + \|\mathbf{B}\|_+$. However, we can take the trace of both sides and use the linearity of the trace to get

$$\mathrm{Tr}\|\mathbf{A}+\mathbf{B}\|_+ \leqslant \mathrm{Tr}\|\mathbf{A}\|_+ + \mathrm{Tr}\|\mathbf{B}\|_+, \tag{4.22}$$

since $\mathrm{Tr}\mathbf{U}^*\mathbf{U} = \mathbf{I}$ and $\mathrm{Tr}\mathbf{V}^*\mathbf{V} = \mathbf{I}$. Thus we have

$$\mathrm{Tr}\left\|\mathbf{X}^k + \mathbb{E}\left(-\mathbf{X}^k\right)\right\|_+ \leqslant \mathrm{Tr}\left\|\mathbf{X}^k\right\|_+ + \mathrm{Tr}\left\|\mathbb{E}\left(-\mathbf{X}^k\right)\right\|_+. \tag{4.23}$$

Inserting (4.23) into (4.20) yields

$$\mathrm{Tr}\|F(\mathbf{X}) - \mathbb{E}(F(\mathbf{X}))\|_+ \leqslant \sum_{k=0}^{K} a_k \left[\mathrm{Tr}\left\|\mathbf{X}^k\right\|_+ + \mathrm{Tr}\left\|\mathbb{E}\left(-\mathbf{X}^k\right)\right\|_+\right]. \tag{4.24}$$

We can choose the coefficients a_k to minimize the right-hand-side of (4.19) or (4.24). It is interesting to compare (4.19) with (4.24). If $\mathbf{X}^k$ are positive semi-definite, $\mathbf{X}^k \geqslant 0$, both are almost identical. If $\mathbf{X} \geq 0$, then $\mathbf{X}^k \geq 0$. Also note that

$$\mathrm{Tr}\mathbf{A}^k = \sum_{i=1}^{n} \lambda_i^k(\mathbf{A}),$$

where λ_i are the eigenvalues of $\mathbf{A}$.

4.5 Talagrand Concentration Inequality

Let $\mathbf{x} = (X_1, \ldots, X_n)$ be a random vector consisting of n random variables. We say a function $f : \mathbb{R}^n \to \mathbb{R}$ is Lipschitz with constant K or K-Lipschitz, if

$$|f(\mathbf{x}) - f(\mathbf{y})| \leqslant K \|\mathbf{x} - \mathbf{y}\|$$

for all $\mathbf{x}, \mathbf{y} \in \mathbb{R}^n$. Here we use the Euclidean norm $\|\cdot\|$ on $\mathbb{R}^n$.

Theorem 4.5.1 (Talagrand concentration inequality [148]). *Let $\kappa > 0$, and let $X_1, \ldots, X_n$ be independent complex variables with $|X_i| \leqslant \kappa$ for all $1 \leq i \leq n$. Let $f : \mathbb{C}^n \to \mathbb{R}$ be a 1-Lipschitz and convex function (where we identify $\mathbb{C}^n$ with $\mathbb{R}^{2n}$ for the purposes of defining "Lipschitz" and "convex"). Then, for any t, one has*

$$\mathbb{P}(|f(\mathbf{x}) - \mathbb{M}f(\mathbf{x})| \geqslant \kappa t) \leqslant Ce^{-ct^2} \tag{4.25}$$

and

$$\mathbb{P}(|f(\mathbf{x}) - \mathbb{E}f(\mathbf{x})| \geqslant \kappa t) \leqslant Ce^{-ct^2} \tag{4.26}$$

for some absolute constants $C, c > 0$, where $\mathbb{M}f(\mathbf{x})$ is the median of $f(\mathbf{x})$.

See [63] for a proof. Let us illustrate how to use this theorem, by considering the operator norm of a random matrix.

The operator (or matrix) norm $\|\mathbf{A}\|_{op}$ is the most important statistic of a random matrix $\mathbf{A}$. It is a basic statistic at our disposal. We define

$$\|\mathbf{A}\|_{op} = \sup_{\mathbf{x} \in \mathbb{C}^n : \|\mathbf{x}\| = 1} \|\mathbf{A}\mathbf{x}\|$$

where $\|\mathbf{x}\|$ is the Euclidean norm of vector $\mathbf{x}$. The operator norm is the basic upper bound for many other quantities.

The operator norm $\|\mathbf{A}\|_{op}$ is also the largest singular value $\sigma_{\max}(\mathbf{A})$ or $\sigma_1(\mathbf{A})$ assuming that all singular values are sorted in an non-increasing order. $\|\mathbf{A}\|_{op}$ dominates the other singular values; similarly, all eigenvalues $\lambda_i(\mathbf{A})$ of $\mathbf{A}$ have magnitude at most $\|\mathbf{A}\|_{op}$.

Suppose that the coefficients ξ_{ij} of $\mathbf{A}$ are independent, have mean zero, and uniformly bounded in magnitude by 1 ($\kappa = 1$). We consider $\sigma_1(\mathbf{A})$ as a function $f\left((\xi_{ij})_{1 \leqslant i,j \leqslant n}\right)$ of the independent complex ξ_{ij}, thus f is a function from $\mathbb{C}^{n^2}$ to $\mathbb{R}$. The convexity of the operator norm tells us that f is convex. The elementary bound is

$$\|\mathbf{A}\| \leqslant \|\mathbf{A}\|_F \text{ or } \sigma_1(\mathbf{A}) \leqslant \|\mathbf{A}\|_F \text{ or } \sigma_{\max}(\mathbf{A}) \leqslant \|\mathbf{A}\|_F \tag{4.27}$$

where

$$\|\mathbf{A}\|_F = \left(\sum_{i=1}^{n} \sum_{j=1}^{n} |\xi_{ij}|^2 \right)^{1/2}$$

is the *Frobenius norm*, also known as the *Hilbert-Schmidt norm* or *2-Schatten norm*. Combining the triangle inequality with (4.27) tells us that f is Lipschitz with constant 1 ($K = 1$). Now, we are ready to apply Talagrand's inequality, Theorem 4.5.1. We thus obtain the following: For any $t > 0$, one has

$$\mathbb{P}\left(\sigma_1(\mathbf{A}) - \mathbb{M}\sigma_1(\mathbf{A}) \geqslant t\right) \leqslant Ce^{-ct^2}$$

and

$$\mathbb{P}\left(\sigma_1(\mathbf{A}) - \mathbb{E}\sigma_1(\mathbf{A}) \geqslant t\right) \leqslant Ce^{-ct^2} \tag{4.28}$$

for some constants $C, c > 0$.

If $f : \mathbb{R} \to \mathbb{R}$ is a Lipschitz function, then

$$F = \frac{1}{\sqrt{n}} \sum_{i=1}^{n} f(\lambda_i)$$

is also a Lipschitz function of (real and imaginary) entries of $\mathbf{A}$. If f is convex on the real line, then F is convex on the space of matrices (Klein's lemma). As a result, we can use the general concentration principle to functions of the eigenvalues $\lambda_i(\mathbf{A})$. By applying Talagrand's inequality, Theorem 4.5.1, it follows that

$$\mathbb{P}\left(\frac{1}{n} \sum_{i=1}^{n} f(\lambda_i(\mathbf{A})) - \mathbb{E}\frac{1}{n} \sum_{i=1}^{n} f(\lambda_i(\mathbf{A})) \geqslant t \right) \leqslant Ce^{-ct^2} \tag{4.29}$$

Talagrand's inequality, as formulated in Theorem 4.5.1, heavily relies on convexity. As a result, we cannot apply it directly to non-convex matrix statistics, such as singular values $\sigma_i(\mathbf{A})$ other than the largest singular value $\sigma_1(\mathbf{A})$. The partial sum $\sum_{i=1}^{k} \sigma_i(\mathbf{A})$, the sum of the first k singular values, is convex. See [48] for more convex functions based on singular values.

The eigenvalue stability inequality is

$$|\lambda_i(\mathbf{A} + \mathbf{B}) - \lambda_i(\mathbf{A})| \leqslant \|\mathbf{B}\|_{op}$$

The spectrum of $\mathbf{A} + \mathbf{B}$ is close to that of $\mathbf{A}$ if $\mathbf{B}$ is small in operator norm. In particular, we see that the map $\mathbf{A} \to \lambda_i(\mathbf{A})$ is Lipschitz continuous on the space of Hermitian matrices, for fixed $1 \leq i \leq n$.

It is easy to observe that the operator norm of a matrix $\mathbf{A} = (\xi_{ij})$ bounds the magnitude of any of its coefficients, thus

$$\sup_{1 \leqslant i,j \leqslant n} |\xi_{ij}| \leqslant \sigma_1(\mathbf{A})$$

or, equivalently

$$\mathbb{P}(\sigma_1(\mathbf{A}) \leqslant t) \leqslant \mathbb{P}\left(\bigcup_{1 \leqslant i,j \leqslant n} |\xi_{ij}| \leqslant t\right).$$

We can view the upper tail event $\sigma_1(\mathbf{A}) \leqslant t$ as a union of many simpler events $|\xi_{ij}| \leqslant t$. In the i.i.d. case $\xi_{ij} \equiv \xi$, and setting $t = \alpha\sqrt{n}$ for some fixed α independent of n, we have

$$\mathbb{P}(\sigma_1(\mathbf{A}) \leqslant t) \leqslant \left\{\mathbb{P}\left(|\xi| \leqslant \alpha\sqrt{n}\right)\right\}^{n^2}.$$

4.6 Concentration of the Spectral Measure for Wigner Random Matrices

We follow [180, 186]. A Hermitian Wigner matrix is an $n \times n$ matrix $\mathbf{H} = (h_{ij})_{1 \leqslant i \leqslant j \leqslant n}$ such that

$$h_{ij} = \frac{1}{\sqrt{n}}(x_{ij} + jy_{ij}) \quad \text{for all} \quad 1 \leqslant i \leqslant j \leqslant n$$
$$h_{ii} = \frac{1}{\sqrt{n}}x_{ii} \qquad \text{for all} \quad 1 \leqslant i \leqslant n$$

where $\{x_{ij}, y_{ij}, x_{ii}\}$ are a collection of real independent, identically distributed random variables with $\mathbb{E}x_{ij} = 0$ and $\mathbb{E}x_{ij}^2 = 1/2$.

The diagonal elements are often assumed to have a different distribution, with $\mathbb{E}x_{ij} = 0$ and $\mathbb{E}x_{ij}^2 = 1$. The entries scale with the dimension n. The scaling is chosen such that, in the limit $n \to \infty$, all eigenvalues of $\mathbf{H}$ remain bounded. To see this, we use

$$\mathbb{E}\sum_{k=1}^{n} \lambda_k^2 = \mathbb{E}\,\mathrm{Tr}\,\mathbf{H}^2 = \mathbb{E}\sum_{i=1}^{n}\sum_{j=1}^{n} |h_{ij}|^2 = n^2\mathbb{E}|h_{ij}|^2$$

where $\lambda_k, k = 1, \ldots, n$ are the eigenvalues of $\mathbf{H}$. If all λ_k stay bounded and of order one in the limit of large dimension n, we must have $\mathbb{E}\,\mathrm{Tr}\,\mathbf{H}^2 \simeq n$ and therefore $\mathbb{E}|h_{ij}|^2 \simeq \frac{1}{n}$. Note that a trace function is linear.

In Sect. 4.3, we observe that using $\sum_{k=1}^{n} f(\lambda_k(\mathbf{A}))$ rather than $\lambda_k(\mathbf{A})$ increases the Lipschitz constant from 1 to $\sqrt{n}|f|_{\mathcal{L}}$.

Theorem 4.6.1 (Guionnet and Zeitouni [180]). *Suppose that the laws of the entries $\{x_{ij}, y_{ij}, x_{ii}\}$ satisfies the logarithmic Sobolev inequality with constant $c > 0$. Then, for any Lipschitz function $f : \mathbb{R} \to \mathbb{C}$, with Lipschitz constant $|f|_{\mathcal{L}}$ and $t > 0$, we have that*

$$\mathbb{P}\left(\left|\frac{1}{n}\operatorname{Tr} f(\mathbf{H}) - \mathbb{E}\frac{1}{n}\operatorname{Tr} f(\mathbf{H})\right| \geqslant t\right) \leqslant 2e^{-\frac{n^2t^2}{4c|f|_{\mathcal{L}}^2}}. \tag{4.30}$$

Moreover, for any $k = 1, \ldots, n$, we have

$$\mathbb{P}\left(\left|f(\lambda_k) - \mathbb{E}f(\lambda_k)\right| \geqslant t\right) \leqslant 2e^{-\frac{nt^2}{4c|f|_{\mathcal{L}}^2}}. \tag{4.31}$$

In order to prove this theorem, we want to use of the observation of Herbst that Lipschitz functions of random matrices satisfying the log-Sobolev inequality exhibit Gaussian concentration.

Theorem 4.6.2 (Herbst). *Suppose that $\mathbb{P}$ satisfies the log-Sobolev inequalities on $\mathbb{R}^n$ with constant c. Let $G : \mathbb{R}^m \to \mathbb{R}$ be a Lipschitz function with constant $|G|_{\mathcal{L}}$. Then, for every $t > 0$,*

$$\mathbb{P}\left(\left|g(\mathbf{x}) - \mathbb{E}g(\mathbf{x})\right| \geqslant t\right) \leqslant 2e^{-\frac{t^2}{2c|G|_{\mathcal{L}}}}.$$

See [186] for a proof.

To prove the theorem, we need another lemma.

Lemma 4.6.3 (Hoffman-Wielandt). *Let $\mathbf{A}, \mathbf{B}$ be $n \times n$ Hermitian matrices, with eigenvalues*

$$\sum_{i=1}^{n} \left|\lambda_i(\mathbf{A}) - \lambda_i(\mathbf{B})\right|^2 \leqslant \operatorname{Tr}(\mathbf{A} - \mathbf{B})^2.$$

We see [187] for a proof.

Corollary 4.6.4. *Let $\mathbf{X} = (\{x_{ij}, y_{ij}, x_{ii}\}) \in \mathbb{R}^{n^2}$ and let $\lambda_k(\mathbf{X}), k = 1, \ldots, n$ be the eigenvalues of the Wigner matrix $\mathbf{X} = \mathbf{H}(\mathbf{X})$. Let $g : \mathbb{R}^n \to \mathbb{R}$ be a Lipschitz function with constant $|g|_{\mathcal{L}}$. Then the map $\mathbf{X} \in \mathbb{R}^{n^2} \to g(\lambda_1(\mathbf{X}), \ldots, \lambda_n(\mathbf{X})) \in \mathbb{R}$ is a Lipschitz function with coefficient $\sqrt{2/n}|g|_{\mathcal{L}}$. In particular, if $f : \mathbb{R} \to \mathbb{R}$ is a Lipschitz function with constant $|f|_{\mathcal{L}}$, the map $\mathbf{X} \in \mathbb{R}^{n^2} \to \sum_{i=1}^{n} f(\lambda_k) \in \mathbb{R}$ is a Lipschitz function with constant $\sqrt{2}|f|_{\mathcal{L}}$.*

Proof. Let $\Lambda = (\lambda_1, \ldots, \lambda_n)$. Observe that

$$
\begin{aligned}
|g(\Lambda(\mathbf{X})) - g(\Lambda(\mathbf{X}'))| &\leqslant |g|_{\mathcal{L}} \|\Lambda(\mathbf{X}) - \Lambda(\mathbf{X}')\|_2 \\
&= |g|_{\mathcal{L}} \sqrt{\sum_{i=1}^{n} |\lambda_i(\mathbf{X}) - \lambda_i(\mathbf{X}')|} \leqslant |g|_{\mathcal{L}} \sqrt{\operatorname{Tr}(\mathbf{H}(\mathbf{X}) - \mathbf{H}(\mathbf{X}'))^2} \\
&= |g|_{\mathcal{L}} \sqrt{\sum_{i=1}^{n}\sum_{j=1}^{n} |h_{ij}(\mathbf{X}) - h_{ij}(\mathbf{X}')|^2} \leqslant \sqrt{2/n}|g|_{\mathcal{L}} \|\mathbf{X} - \mathbf{X}'\|_{\mathbb{R}^{n^2}}.
\end{aligned}
\tag{4.32}
$$

The first inequality of the second line follows from the lemma of Hoffman-Wielandt above. $\|\mathbf{X} - \mathbf{X}'\|_{\mathbb{R}^{n^2}}$ is also the Frobenius norm.

Since $g(\Lambda) = \operatorname{Tr} f(\mathbf{H}) = \sum_{i=1}^{n} f(\lambda_k)$ is such that

$$
|g(\Lambda) - g(\Lambda')| \leqslant |f|_{\mathcal{L}} \sum_{i=1}^{n} \left|\lambda_i - \lambda_i^{'}\right| \leqslant \sqrt{n}|f|_{\mathcal{L}} \|\mathbf{X} - \mathbf{X}'\|_{\mathbb{R}^n}
$$

it follows that g is a Lipschitz function on $\mathbb{R}^n$ with constant $\sqrt{n}|f|_{\mathcal{L}}$. Combined with (4.32), we complete the proof of the corollary. □

Now we have all the ingredients to prove Theorem 4.6.1.

Proof of Theorem 4.6.1. Let $\mathbf{X} = (\{x_{ij}, y_{ij}, x_{ii}\}) \in \mathbb{R}^{n^2}$. Let $G(\mathbf{X}) = \operatorname{Tr} f(\mathbf{H}(\mathbf{X}))$. Then the matrix function G is Lipschitz with constant $\sqrt{2}|f|_{\mathcal{L}}$. By Theorem 4.6.2, it follows that

$$
\mathbb{P}\left(\left|\frac{1}{n}\operatorname{Tr} f(\mathbf{H}) - \mathbb{E}\frac{1}{n}\operatorname{Tr} f(\mathbf{H})\right| \geqslant t\right) \leqslant 2e^{-\frac{n^2t^2}{4c|f|_{\mathcal{L}}^2}}.
$$

To show (4.31), we see that, using Corollary 4.6.4, the matrix function $G(\mathbf{X}) = f(\lambda_k)$ is Lipschitz with constant $\sqrt{2/n}|f|_{\mathcal{L}}$. By Theorem 4.6.2, we find that

$$
\mathbb{P}(|f(\lambda_k) - \mathbb{E}f(\lambda_k)| \geqslant t) \leqslant 2e^{-\frac{nt^2}{4c|f|_{\mathcal{L}}^2}},
$$

which is (4.31). □

Example 4.6.5 (Applications of Theorem 4.6.1). We consider a special case of $f(s) = s$. Thus $|f|_{\mathcal{L}}^2 = 1$. From (4.31) for the k-th eigenvalue λ_k of random matrix $\mathbf{H}$, we see at once that, for any $k = 1, \ldots, n$,

$$
\mathbb{P}(|\lambda_k - \mathbb{E}\lambda_k| \geqslant t) \leqslant 2e^{-\frac{nt^2}{4c}}.
$$

For the trace function $\frac{1}{n}\operatorname{Tr}(\mathbf{H})$, on the other hand, from (4.30), we have

$$
\mathbb{P}\left(\left|\frac{1}{n}\operatorname{Tr}(\mathbf{H}) - \mathbb{E}\frac{1}{n}\operatorname{Tr}(\mathbf{H})\right| \geqslant t\right) \leqslant 2e^{-\frac{n^2t^2}{4c}}.
$$

The right-hand-side of the above two inequalities has the same Gaussian tail but with different "variances". The k-th eigenvalue has a variance of $\sigma^2 = 2c/n$, while the normalized trace function has that of $\sigma^2 = 2c/n^2$. The variance of the normalized trace function is $1/n$ times that of the k-th eigenvalue. For example, when $n = 100$, this factor is 0.01. In other words, for the normalized trace function—viewed as a statistical average of n eigenvalues (random variables)—reduces the variance by 20 dB, compared with the each individual eigenvalue. In [14], this phenomenon has been used for signal detection of extremely low signal to noise ratio, such as $SNR = -34\,\text{dB}$.

The Wishart matrix is involved in [14] while here the Wigner matrix is used. In Sect. 4.8, it is shown that for the normalized trace function of a Wishart matrix, we have the tail bound of $2e^{-\frac{n^2t^2}{4c}}$, similar to a Wigner matrix. On the other hand, for the largest singular value (operator norm)—see Theorem 4.8.11, the tail bound is Ce^{-cnt^2}.

For hypothesis testing such as studied in [14], reducing the variance of the statistic metrics (such as the k-th eigenvalue and the normalized trace function) is critical to algorithms. From this perspective, we can easily understand why the use of the normalized trace as statistic metric in hypothesis testing leads to much better results than algorithms that use the k-th eigenvalue [188] (in particular, $k = 1$ and $k = n$ for the largest and smallest eigenvalues, respectively). Another big advantage is that the trace function is linear. It is insightful to view the trace function as a statistic average $\frac{1}{n}\operatorname{Tr}(\mathbf{H}) = \frac{1}{n}\sum\limits_{k=1}^{n}\lambda_k$, where λ_k is a random variable. The statistical average of n random variables, of course, reduces the variance, a classical result in probability and statistics. Often one deals with sums of independent random variables [189]. But here the n eigenvalues are not necessarily independent. Techniques like Stein's method [87] can be used to approximate this sum using Gaussian distribution. Besides, the reduction of the variance by a factor of n by using the trace function rather than each individual eigenvalue is not obvious to observe, if we use the classical techniques since it is very difficult to deal with sums of n dependent random variables.

For the general case we have the variance of $\sigma^2 = 2c\left|f\right|_{\mathcal{L}}^2/n$ and $\sigma^2 = 2c\left|f\right|_{\mathcal{L}}^2/n^2$ for the k-th eigenvalue and normalized trace, respectively. The general Lipschitz function f with constant $\left|f\right|_{\mathcal{L}}$ increases the variance by a factor of $\left|f\right|_{\mathcal{L}}^2$.

If we seek to find a statistic metric (viewed as a matrix function) that has a minimum variance, we find here that the normalized trace function $\frac{1}{n}\operatorname{Tr}(\mathbf{H})$ is optimum in the sense of minimizing the variance. This finding is in agreement with empirical results in [14]. To a large degree, a large part of this book was motivated to understand why the normalized trace function $\frac{1}{n}\operatorname{Tr}(\mathbf{H})$ always gave us the best results in Monte Carlo simulations. It seemed that we could not find a better matrix function for the statistic metrics. The justification for recommending the normalized trace function $\frac{1}{n}\operatorname{Tr}(\mathbf{H})$ is satisfactory to the authors. □

4.7 Concentration of Noncommutative Polynomials in Random Matrices

Following [190], we also refer to [191, 192]. The Schatten p-norm of a matrix $\mathbf{A}$ is defined as $\|\mathbf{A}\|_p = \left(\operatorname{Tr}\left(\mathbf{A}^H\mathbf{A}\right)^{p/2}\right)^{1/p}$. The limiting case $p = \infty$ corresponds to the operator (or spectral) norm, while $p = 2$ leads to the Hilbert-Schmidt (or Frobenius) norm. We also denote by $\|\cdot\|_p$ the L_p-norm of a real or complex random variable, or ℓ_p-norm of a vector in $\mathbb{R}^n$ or $\mathbb{C}^n$.

A random vector $\mathbf{x}$ in a normed space V satisfies the (sub-Gaussian) convex concentration property (CCP), or in the class CCP, if

$$\mathbb{P}\left[\left|f\left(\mathbf{x}\right) - \mathbb{M}f\left(\mathbf{x}\right)\right| \geqslant t\right] \leqslant Ce^{-ct^2} \tag{4.33}$$

for every $t > 0$ and every convex 1-Lipschitz function $f : V \to \mathbb{R}$, where $C, c > 0$ are constants (parameters) independent of f and t, and $\mathbb{M}$ denotes a median of a random variable.

Theorem 4.7.1 (Meckes [190]). *Let $\mathbf{X}_1, \ldots, \mathbf{X}_m \in \mathbb{C}^{n\times n}$ be independent, centered random matrices that satisfy the convex concentration property* (4.33) *(with respect to the Frobenius norm on $\mathbb{C}^{n\times n}$) and let $k \geq 1$ be an integer. Let P be a noncommutative $*$-polynomial in m variables of degree at most k, normalized so that its coefficients have modulus at most 1. Define the complex random variable*

$$Z_P = \operatorname{Tr} P\left(\frac{1}{\sqrt{n}}\mathbf{X}_1, \ldots, \frac{1}{\sqrt{n}}\mathbf{X}_m\right).$$

Then, for $t > 0$,

$$\mathbb{P}\left[\left|Z_P - \mathbb{E}Z_P\right| \geqslant t\right] \leqslant C_{m,k}\exp\left[-c_{m,k}\min\left\{t^2, nt^{2/k}\right\}\right].$$

The conclusion holds also for non-centered random matrices if—when $k \geq 2$—we assume that $\|\mathbb{E}\mathbf{X}_i\|_{2(k-1)} \leqslant Cn^{k/2(k-1)}$ for all i.

We also have that for $q \geq 1$

$$\left\|Z_P - \mathbb{E}Z_P\right\|_q \leqslant C_{m,k}^{'}\max\left\{\sqrt{q}, \left(\frac{q}{m}\right)^{k/2}\right\}.$$

Consider a special case. Let $\mathbf{X} \in \mathbb{C}^{n\times n}$ be a random Hermitian matrix which satisfies the convex concentration property (4.33) (with respect to the Frobenius norm on the Hermitian matrix space), let $k \geq 1$ be an integer, and suppose—when $k \geq 2$—that $\operatorname{Tr}\left(\frac{1}{\sqrt{n}}\mathbf{X}\right)^{2(k-1)} \leqslant Cn$. Then, for $t > 0$,

$$\mathbb{P}\left[\left|\operatorname{Tr}\left(\frac{1}{\sqrt{n}}\mathbf{X}\right)^k - \mathbb{M}\operatorname{Tr}\left(\frac{1}{\sqrt{n}}\mathbf{X}\right)^k\right| \geqslant t\right] \leqslant C\exp\left[-\min\left\{c^k t^2, cnt^{2/k}\right\}\right].$$

Consider non-Hermtian matrices.

Theorem 4.7.2 (Meckes [190]). *Let* $\mathbf{X}\in\mathbb{C}^{n\times n}$ *be a random matrix which satisfies the convex concentration property* (4.33) *(with respect to the Frobenius norm on* $\mathbb{C}^{n\times n}$*), let* $k \geq 1$ *be an integer, and suppose—when* $k \geq 2$*—that* $\|\mathbb{E}\mathbf{X}\|_{2(k-1)} \leqslant cn^{k/2(k-1)}$. *Then, for* $t>0$,

$$\mathbb{P}\left[\left|\operatorname{Tr}\left(\frac{1}{\sqrt{n}}\mathbf{X}\right)^k - \mathbb{E}\operatorname{Tr}\left(\frac{1}{\sqrt{n}}\mathbf{X}\right)^k\right| \geqslant t\right] \leqslant C\,(k+1)\exp\left[-\min\left\{c^k t^2, cnt^{2/k}\right\}\right].$$

4.8 Concentration of the Spectral Measure for Wishart Random Matrices

Two forms of approximations have central importance in statistical applications [193]. In one form, one random variable is approximated by another random variable. In the other, a given distribution is approximated by another.

We consider the Euclidean operator (or matrix) norm

$$\|(a_{ij})\| = \|(a_{ij})\|_{l^2\to l^2} = \sup\left\{\sum_{i,j} a_{ij}x_i y_j : \sum_i x_i^2 \leqslant 1, \sum_j y_j^2 \leqslant 1\right\}$$

of random matrices whose entries are independent random variables.

Seginer [107] showed that for any $m \times n$ random matrix $\mathbf{X} = (X_{ij})_{i\leqslant m, j\leqslant n}$ with iid mean zero entries

$$\mathbb{E}\,\|(X_{ij})\| \leqslant C\left(\mathbb{E}\max_{i\leqslant m}\sqrt{\sum_{j\leqslant n} X_{ij}^2} + \mathbb{E}\max_{j\leqslant n}\sqrt{\sum_{i\leqslant m} X_{ij}^2}\right),$$

where C is a universal constant.

For any random matrix with independent mean zero entries, Latala [194] showed that

$$\mathbb{E}\,\|(X_{ij})\| \leqslant C\left(\max_i\sqrt{\sum_j \mathbb{E}X_{ij}^2} + \max_j\sqrt{\sum_i \mathbb{E}X_{ij}^2} + \left(\sum_{i,j}\mathbb{E}X_{ij}^4\right)^{1/4}\right),$$

where C is some universal constant.

Reference [183] is relevant in the context.

For a symmetric $n \times n$ matrix $\mathbf{M}$, $F_{\boldsymbol{\lambda}}(\lambda)$ is the cumulative distribution function (CDF) of the spectral distribution of matrix $\mathbf{M}$

$$F_{\mathbf{M}}(\lambda) = \frac{1}{n}\sum_{i=1}^{n}\{\lambda_i(\mathbf{M}) \leqslant \lambda\}, \quad \lambda \in \mathbb{R}.$$

The integral of a function $f(\cdot)$ with respect to the measure induced by $F_{\mathbf{M}}$ is denoted by the function

$$F_{\mathbf{M}}(f) = \frac{1}{n}\sum_{i=1}^{n} f(\lambda_i(\mathbf{M})) = \frac{1}{n}\operatorname{Tr} f(\mathbf{M}).$$

For certain classes of random matrices $\mathbf{M}$ and certain classes of functions f, it can be shown that $F_{\mathbf{M}}(f)$ is concentrated around its expectation $\mathbb{E}F_{\mathbf{M}}(f)$ or around any median $\mathbb{M}F_{\mathbf{M}}(f)$.

For a Lipschitz function g, we write $||g||_{\mathcal{L}}$ for its Lipschitz constant. To state the result, we also need to define bounded functions: $f : (a,b) \mapsto \mathbb{R}$ are of bounded variation on (a,b) (where $-\infty \le a \le b \le \infty$), in the sense that

$$V_f(a,b) = \sup_{n \geqslant 1} \sup_{a < x_0 \leqslant x_1 \leqslant \cdots \leqslant x_n} \sum_{k=1}^{n} |f(x_k) - f(x_{k-1})|$$

is finite [195, Sect. X.1]. A function is bounded variation if and only if it can be written as the difference of two bounded monotone functions on (a,b). The indicator function $g : x \mapsto \{x \leqslant \lambda\}$ is of bounded variation on $\mathbb{R}$ with $V_g(\mathbb{R}) = 1$ for each $\lambda \in \mathbb{R}$.

Theorem 4.8.1 (Guntuboyina and Lebb [196]). *Let $\mathbf{X}$ be an $m \times n$ matrix whose row-vectors are independent, set a Wishart matrix $\mathbf{S} = \mathbf{X}^T\mathbf{X}/m$, and fix $f : \mathbb{R} \to \mathbb{R}$.*

1. *Suppose that f is such that the mapping $x \mapsto f(x^2)$ is convex and Lipschitz, and suppose that $|X_{i,j}| \le 1$ for each i and j. For all $t > 0$, we then have*

$$\mathbb{P}\left(\left|\frac{1}{n}\operatorname{Tr} f(\mathbf{S}) - \mathbb{M}\frac{1}{n}\operatorname{Tr} f(\mathbf{S})\right| \geqslant t\right) \leqslant 4\exp\left[-\frac{nm}{n+m}\frac{t^2}{8\left\|f(\cdot^2)\right\|_{\mathcal{L}}^2}\right]. \tag{4.34}$$

   where $\mathbb{M}$ stands for the median. [From the upper bound (4.34), one can also obtain a similar bound for $\mathbb{P}\left(\left|\frac{1}{n}\operatorname{Tr} f(\mathbf{S}) - \mathbb{E}\frac{1}{n}\operatorname{Tr} f(\mathbf{S})\right| \geqslant t\right)$ using standard methods (e.g.[197])]
2. *Suppose that f is of bounded variation on $\mathbb{R}$. For all $t > 0$, we then have*

$$\mathbb{P}\left(\left|\frac{1}{n}\operatorname{Tr} f\left(\mathbf{S}\right)-\mathbb{E}\frac{1}{n}\operatorname{Tr} f\left(\mathbf{S}\right)\right|\geqslant t\right)\leqslant 2\exp\left[-\frac{n^2}{m}\frac{2t^2}{V_f^2\left(\mathbb{R}\right)}\right].$$

In particular, for each $\lambda \in \mathbb{R}$ *and all* $t > 0$, *we have*

$$\mathbb{P}\left(\left|\frac{1}{n}\operatorname{Tr} f\left(\mathbf{S}\right)-\mathbb{E}\frac{1}{n}\operatorname{Tr} f\left(\mathbf{S}\right)\right|\geqslant t\right)\leqslant 2\exp\left[-\frac{2n^2}{m}t^2\right].$$

These bounds cannot be improved qualitatively without imposing additional assumptions. This theorem only requires that the rows of $\mathbf{X}$ are independent while allowing for dependence *within* each row of $\mathbf{X}$.

Let us consider the case that is more general than Theorem 4.8.1. Let $\mathbf{X}$ be a random symmetric $n \times n$ matrix. Let $\mathbf{X}$ be a function of m *independent* random quantities $\mathbf{y}_1, \ldots, \mathbf{y}_m$, i.e., $\mathbf{X} = \mathbf{X}\left(\mathbf{y}_1, \ldots, \mathbf{y}_m\right)$. Write

$$\mathbf{X}_{(i)} = \mathbf{X}\left(\mathbf{y}_1, \ldots, \mathbf{y}_{i-1}, \tilde{\boldsymbol{y}}_i, \mathbf{y}_{i+1}, \ldots, \mathbf{y}_m\right) \tag{4.35}$$

where $\tilde{\mathbf{y}}_i$ is distributed the same as $\mathbf{y}_i$ and represents an independent copy of $\mathbf{y}_i$. $\tilde{\mathbf{y}}_i, i = 1, \ldots, m$ are independent of $\mathbf{y}_1, \ldots, \mathbf{y}_m$. Assume that

$$\left\|\frac{1}{n}\operatorname{Tr} f\left(\mathbf{X}/\sqrt{m}\right)-\frac{1}{n}\operatorname{Tr} f\left(\mathbf{X}_{(i)}/\sqrt{m}\right)\right\|\leqslant \frac{r}{n} \tag{4.36}$$

holds almost surely for each $i = 1, \ldots, m$ and for some (fixed) integer r.

Theorem 4.8.2 (Guntuboyina and Lebb [196]). *Assume* (4.36) *is satisfied for each* $i = 1, \ldots, m$. *Assume* $f : \mathbb{R} \to \mathbb{R}$ *is of bounded variation on* $\mathbb{R}$. *For any* $t > 0$, *we have that*

$$\mathbb{P}\left(\left|\frac{1}{n}\operatorname{Tr} f\left(\mathbf{X}/\sqrt{m}\right)-\mathbb{E}\frac{1}{n}\operatorname{Tr} f\left(\mathbf{X}/\sqrt{m}\right)\right|\geqslant t\right)\leqslant 2\exp\left[-\frac{n^2}{m}\frac{2t^2}{r^2V_f^2\left(\mathbb{R}\right)}\right]. \tag{4.37}$$

To estimate the integer r in (4.35), we need the following lemma.

Lemma 4.8.3 (Lemma 2.2 and 2.6 of [198]). *Let* $\mathbf{A}$ *and* $\mathbf{B}$ *be symmetric* $n \times n$ *matrices and let* $\mathbf{C}$ *and* $\mathbf{D}$ *be* $m \times n$ *matrices. We have*

$$\left\|\frac{1}{n}\operatorname{Tr} f\left(\mathbf{A}\right)-\frac{1}{n}\operatorname{Tr} f\left(\mathbf{B}\right)\right\|_\infty \leqslant \frac{\operatorname{rank}\left(\mathbf{A}-\mathbf{B}\right)}{n},$$

and

$$\left\|\frac{1}{n}\operatorname{Tr} f\left(\mathbf{C}^T\mathbf{C}\right)-\frac{1}{n}\operatorname{Tr} f\left(\mathbf{D}^T\mathbf{D}\right)\right\|_\infty \leqslant \frac{\operatorname{rank}\left(\mathbf{C}-\mathbf{D}\right)}{n}.$$

Now we are ready to consider an example to illustrate the application.

Example 4.8.4 (Sample covariance matrix of vector moving average (MA) processes [196]). Consider an $m \times n$ matrix $\mathbf{X}$ whose row-vectors follow a vector MA process of order 2, i.e.,

$$(\mathbf{X}_{i,\cdot})^T = \mathbf{y}_{i+1} + \mathbf{B}\mathbf{y}_i, \quad i = 1, \ldots, m$$

where $\mathbf{y}_1, \ldots, \mathbf{y}_m \in \mathbb{R}^n$ are $m+1$ *independent* n-vectors and $\mathbf{B}$ is some fixed $n \times n$ matrix.

Now for the innovations $\mathbf{y}_i$, we assume that $\mathbf{y}_i = \mathbf{H}\mathbf{z}_i, i = 1, \ldots, m+1$, where $\mathbf{H}$ is a fixed $n \times n$ matrix. Write $\mathbf{Z}$ as

$$\mathbf{Z} = (\mathbf{z}_1, \ldots, \mathbf{z}_{m+1})^T = (Z_{ij}), \quad i = 1, \ldots, m+1, j = 1, \ldots, n$$

where the entries Z_{ij} of matrix $\mathbf{Z}$ are independent and satisfy $|Z_{ij}| \leq 1$. The (random) sample covariance matrix is $\mathbf{S} = \mathbf{X}^T\mathbf{X}/m$. For a function f such that the mapping $x \mapsto f(x^2)$ is convex and Lipschitz, we then have that, for all $t > 0$

$$\mathbb{P}\left(\left|\frac{1}{n}\operatorname{Tr} f(\mathbf{S}) - \mathrm{M}\frac{1}{n}\operatorname{Tr} f(\mathbf{S})\right| \geqslant t\right) \leqslant 4\exp\left[-\frac{n^2 m}{n+m}\frac{t^2}{8C^2\left\|f(\cdot^2)\right\|_{\mathcal{L}}^2}\right]. \tag{4.38}$$

where $C = (1 + \|\mathbf{B}\|)\|\mathbf{H}\|$ with $\|\cdot\|$ denoting the operator (or matrix) norm.

□

We focus on Wishart random matrices (sample covariance matrix), that is $\mathbf{S} = \frac{1}{n}\mathbf{Y}\mathbf{Y}^H$, where $\mathbf{Y}$ is a rectangular $p \times n$ matrix. Our objective is to use new exponential inequalities of the form

$$Z = g(\lambda_1, \cdots, \lambda_p)$$

where $(\lambda_i)_{1 \leqslant i \leqslant p}$ is the set of eigenvalues of $\mathbf{S}$. These inequalities will be upper bounds on $\mathbb{E}\left[e^{Z - \mathbb{E}Z}\right]$ and lead to natural bounds $\mathbb{P}(|Z - \mathbb{E}Z| \geqslant t)$ for large values of t.

What is new is following [199]: (i) g is a once or twice differentiable function (i.e., not necessarily of the form (4.39)); (ii) the entries of $\mathbf{Y}$ are possibly dependent; (iii) they are not necessarily identically distributed; (iv) the bound is instrumental when both p and n are large. In particular, we consider

$$g(\lambda_1, \cdots, \lambda_p) = \sum_{k=1}^{p} \varphi(\lambda_k). \tag{4.39}$$

A direct application is to spectrum sensing. $\mathbb{E}Z$ is used to set the threshold for hypothesis detection [14, 200]. Only simple trace functional of the form (4.39) is used in [14]. We propose to investigate this topic using new results from statistical literature [180, 199, 201].

For a vector $\mathbf{x}$, $\|\mathbf{x}\|$ stands for the Euclidean norm of the vector $\mathbf{x}$. $\|X\|_p$ is the usual L_p-norm of the real random variable X.

Theorem 4.8.5 (Deylon [199]). *Let* $\mathbf{Q}$ *be a* $N \times N$ *deterministic matrix,* $\mathbf{X} = (X_{ij}), 1 \leqslant i \leqslant N, 1 \leqslant j \leqslant n$ *be a matrix of random independent entries, and set*

$$\mathbf{Y} = \frac{1}{n}\mathbf{Q}\mathbf{X}\mathbf{X}^T\mathbf{Q}.$$

Let $\boldsymbol{\lambda} \to g(\boldsymbol{\lambda})$ *be a twice differentiable symmetric function on* $\mathbb{R}^N$ *and define the random variable* $Z = g(\lambda_1, \ldots, \lambda_N)$ *where* $(\lambda_1, \ldots, \lambda_N)$ *is the vector of the eigenvalues of* $\mathbf{Y}$. *Then*

$$\mathbb{E}\left[e^{Z-\mathbb{E}Z}\right] \leqslant \exp\left(\frac{64N}{n}\xi^4\left(\gamma_1 + \frac{N}{n}\xi^2\gamma_2\right)^2\right),$$

where

$$\xi = \|\mathbf{Q}\| \sup_{i,j} \|X_{ij}\|_\infty$$

$$\gamma_1 = \sup_{k,\boldsymbol{\lambda}} \left|\frac{\partial g}{\partial \lambda_k}(\boldsymbol{\lambda})\right|$$

$$\gamma_2 = \sup_{\boldsymbol{\lambda}} \left\|\nabla^2 g(\lambda)\right\| \quad (\textit{matrix norm}).$$

In particular, for any $t > 0$

$$\mathbb{P}(|Z - \mathbb{E}Z| > t) \leqslant 2\exp\left(-\frac{n}{256N}t^2\xi^{-4}\left(\gamma_1 + \frac{N}{n}\xi^2\gamma_2\right)^{-2}\right).$$

When dealing with the matrices with independent columns, we have $\mathbf{Q} = \mathbf{I}$, where $\mathbf{I}$ is the identity matrix.

Theorem 4.8.6 (Deylon [199]). *Let* $\mathbf{X} = (X_{ij}), 1 \leqslant i \leqslant N, 1 \leqslant j \leqslant n$ *be a matrix of random independent entries, and set*

$$\mathbf{Y} = \frac{1}{n}\mathbf{X}\mathbf{X}^T.$$

Let $\boldsymbol{\lambda} \to g(\boldsymbol{\lambda})$ *be a twice differentiable symmetric function on* $\mathbb{R}^N$ *and define the random variable*

$$Z = g(\lambda_1, \ldots, \lambda_N)$$

where $(\lambda_1, \ldots, \lambda_N)$ *is the vector of the eigenvalues of* $\mathbf{Y}$. *Then*

$$\mathbb{E}\left[e^{Z-\mathbb{E}Z}\right] \leqslant \exp\left(\frac{64a^2\gamma_1^2}{n}\right),$$

where

$$\gamma_1 = \sup_{\lambda}\left|\frac{\partial g}{\partial \lambda_1}(\lambda)\right|$$

$$a = \sup_{j}\left\|\sum_i X_{ij}^2\right\|_{\infty}^{1/2}.$$

In particular, for any $t > 0$

$$\mathbb{E}\left[e^{Z-\mathbb{E}Z}\right] \leqslant 2\exp\left(-\frac{nt^2}{64a^4\gamma_1^2}\right).$$

Random matrices can be expressed as functions of independent random variables. Then we think of the linear statistics of eigenvalues as functions of these independent random variables. The central idea is pictorially represented as

$$\text{large vector } \mathbf{x} \Rightarrow \text{matrix } \mathbf{A}(\mathbf{x}) \Rightarrow \text{linear statistic } \sum_i f\left(\lambda_i\left(\mathbf{A}\right)\right) = g\left(\mathbf{x}\right).$$

For each $c_1, c_2 > 0$, let $\mathcal{L}(c_1, c_2)$ be the class of probability measures on $\mathbb{R}$ that arise as laws for random variables like $u(Z)$, where Z is a standard Gaussian random variable and u is a twice continuously differentiable function such that for all $x \in \mathbb{R}$

$$|u'(x) \leqslant c_1| \text{ and } |u''(x) \leqslant c_2|.$$

For example, the standard Gaussian law is in $\mathcal{L}(1,0)$. Again, taking u = the Gaussian cumulative distribution, we see that the uniform distribution on the unit interval is in $\mathcal{L}\left((2\pi)^{-1/2}, (2\pi e)^{-1/2}\right)$. A random variable is said to be "in $\mathcal{L}(c_1, c_2)$" instead of the more elaborate statement that "the distribution of X belongs to $\mathcal{L}(c_1, c_2)$." For two random variables X and Y, the supremum of $|\mathbb{P}(X \in B) - \mathbb{P}(Y \in B)|$ as B ranges from all Borel sets is called the *total variation distance* between the laws of X and Y, often denoted simply by $d_{TV}(X,Y)$.

The following theorem gives normal approximation bounds for general smooth functions of independent random variables whose law are in $\mathcal{L}(c_1, c_2)$ for some finite c_1, c_2.

Theorem 4.8.7 (Theorem 2.2 of Chatterjee [201]). *Let* $\mathbf{x} = (X_1, \ldots, X_n)$ *be a vector of independent random variable in* $\mathcal{L}(c_1, c_2)$ *for some finite* c_1, c_2. *Take any* $g \in C^2(\mathbb{R}^n)$ *and let* ∇g *and* $\nabla^2 g$ *denote the gradient and Hessian of* g. *Let*

$$\kappa_0 = \left(\mathbb{E} \sum_{i=1}^{n} \left| \frac{\partial g}{\partial x_i} (X) \right|^4 \right)^{1/2}, \quad \kappa_1 = \left(\mathbb{E} \| \nabla g (X) \|^4 \right)^{1/4}, \quad \kappa_2 = \left(\mathbb{E} \| \nabla^2 g (X) \|^4 \right)^{1/4}.$$

Suppose $W = g(\mathbf{x})$ *has a finite fourth moment and let* $\sigma^2 = \mathrm{Var}(W)$. *Let* Z *be a normal random variable with the same mean and variance as* W. *Then*

$$d_{TV}(W, Z) \leqslant \frac{2\sqrt{5} \left(c_1 c_2 \kappa_0 + c_1^3 \kappa_1 \kappa_2 \right)}{\sigma^2}.$$

If we slightly change the setup by assuming that $\mathbf{x}$ *is a Gaussian random vector with mean 0 and covariance matrix* $\mathbf{\Sigma}$, *keeping all other notations the same, then the corresponding bound is*

$$d_{TV}(W, Z) \leqslant \frac{2\sqrt{5} \|\Sigma\|^{3/2} \kappa_1 \kappa_2}{\sigma^2}.$$

The cornerstone of Chatterjee[201] is Stein's method [202]. Let us consider a particular function f. Let n be a fixed positive integer and $\mathcal{J}$ be a finite indexing set. Suppose that for each $1 \leq i, j, \leq n$, we have a C^2 map $a_{ij} : \mathbb{R}^2 \to \mathbb{C}$. For each $\mathbf{x} \in \mathbb{R}^{\mathcal{J}}$, let $\mathbf{A}(\mathbf{x})$ be the complex $n \times n$ matrix whose (i, j)-th element is $a_{ij}(\mathbf{x})$. Let

$$f(z) = \sum_{k=0}^{\infty} b_k z^k$$

be an analytic function on the complex plane. Let $\mathbf{X} = (X_u)_{u \in \mathcal{J}}$ be a collection of independent random variables in $\mathcal{L}(c_1, c_2)$ for some finite c_1, c_2. Under this very general setup, we give an explicit bound on the total variation distance between the laws of the random variable $\mathrm{Re}\,\mathrm{Tr}\, f(\mathbf{A}(\mathbf{x}))$ and a Gaussian random variable with matching mean and variance. (Here $\mathrm{Re}\, z$ denotes the real part of a complex number z.)

Theorem 4.8.8 (Theorem 3.1 of Chatterjee [201]). *Let all notations be as above. Suppose* $W = \mathrm{Re}\,\mathrm{Tr}\, f(\mathbf{A}(\mathbf{x}))$ *has finite fourth moment and let* $\sigma^2 = \mathrm{Var}(W)$. *Let* Z *be a normal random variable with the same mean and variance as* W. *Then*

$$d_{TV}(W, Z) \leqslant \frac{2\sqrt{5} \left(c_1 c_2 \kappa_0 + c_1^3 \kappa_1 \kappa_2 \right)}{\sigma^2}.$$

If we slightly change the setup by assuming that $\mathbf{x}$ *is a Gaussian random vector with mean 0 and covariance matrix* $\mathbf{\Sigma}$, *keeping all other notations the same, then the corresponding bound is*

$$d_{TV}(W, Z) \leqslant \frac{2\sqrt{5} \|\Sigma\|^{3/2} \kappa_1 \kappa_2}{\sigma^2}.$$

Let us consider the Wishart matrix or sample covariance matrix. Let $N \leq n$ be two positive integers, and let $\mathbf{X} = (X_{ij})_{1 \leqslant i \leqslant N, 1 \leqslant j \leqslant n}$ be a collection of independent random variables in $\mathcal{L}(c_1, c_2)$ for some finite c_1, c_2. Let

$$\mathbf{A} = \frac{1}{n}\mathbf{X}\mathbf{X}^T.$$

Theorem 4.8.9 (Proposition 4.6 of Chatterjee [201]). *Let λ be the largest eigenvalue of $\mathbf{A}$. Take any entire function f and define f_1 and f_2 as in Theorem 4.8.8. Let $a = \left(\mathbb{E}\left(f_1(\lambda)^4\lambda^2\right)\right)^{1/4}$ and $b = \left(\mathbb{E}(f_1(\lambda) + 2N^{-1/2}\lambda f_2(\lambda))^4\right)^{1/4}$. Suppose $W = \mathrm{Re}\,\mathrm{Tr}\, f(\mathbf{A}(\mathbf{x}))$ has finite fourth moment and let $\sigma^2 = \mathrm{Var}(W)$. Let Z be a normal random variable with the same mean and variance as W. Then*

$$d_{TV}(W, Z) \leqslant \frac{8\sqrt{5}}{\sigma^2}\left(\frac{c_1c_2a^2\sqrt{N}}{n} + \frac{c_1^3abN}{n^{3/2}}\right).$$

If we slightly change the setup by assuming that the entries of $\mathbf{x}$ is jointly Gaussian with mean 0 and $nN \times nN$ covariance matrix $\boldsymbol{\Sigma}$, keeping all other notation the same, then the corresponding bound is

$$d_{TV}(W, Z) \leqslant \frac{8\sqrt{5}\|\Sigma\|^{3/2}abN}{\sigma^2 n^{3/2}}.$$

Double Wishart matrices are important in statistical theory of canonical correlations [203, Sect. 2.2]. Let $N \leqslant n \leqslant M$ be three positive integers. Let $\mathbf{X} = (X_{ij})_{1 \leqslant i \leqslant N, 1 \leqslant j \leqslant n}$ and $\mathbf{Y} = (X_{ij})_{1 \leqslant i \leqslant N, 1 \leqslant j \leqslant M}$ be a collection of independent random variables in $\mathcal{L}(c_1, c_2)$ for some finite c_1, c_2. Define the double Wishart matrix as

$$\mathbf{A} = \mathbf{X}\mathbf{X}^T\left(\mathbf{Y}\mathbf{Y}^T\right)^{-1}.$$

A theorem similar to Theorem 4.8.9 is obtained in [201]. In [204], a central limit theorem is proven for the Jacob matrix ensemble. A Jacob matrix is defined as $\mathbf{A} = \mathbf{X}\mathbf{X}^T\left(\mathbf{X}\mathbf{X}^T + \mathbf{Y}\mathbf{Y}^T\right)^{-1}$, when the matrices $\mathbf{X}$ and $\mathbf{Y}$ have independent standard Gaussian entries.

Let μ and ν be two probability measures on $\mathbb{C}$ (or $\mathbb{R}^2$). We define [59]

$$\rho(\mu, \nu) = \sup_{\|f\|_{\mathcal{L}} \leqslant 1}\left|\int_{\mathbb{C}} f(x)\mu(dx) - \int_{\mathbb{C}} f(x)\nu(dx)\right|, \tag{4.40}$$

where f above is a bounded Lipschitz function defined on $\mathbb{C}$ with $\|f\| = \sup_{x\in\mathbb{C}} |f(x)|$ and $\|f\|_{\mathcal{L}} = \|f\| + \sup_{x\neq y} \frac{|f(x)-f(y)|}{|x-y|}$.

Theorem 4.8.10 (Jiang [204]). *Let* $\mathbf{X}$ *be an* $n\times n$ *matrix with complex entries, and* $\lambda_1, \ldots, \lambda_n$ *are its eigenvalues. Let* $\mu_{\mathbf{X}}$ *be the empirical law of* $\lambda_i, 1 \leq i \leq n$. *Let* μ *be a probability measure. Then* $\rho(\mu_{\mathbf{X}}, \mu)$ *is a continuous function in the entries of matrix* $\mathbf{X}$, *where* ρ *is defined as in* (4.40).

Proof. The proof of [204] illustrates a lot of insights into the function $\rho(\mu_{\mathbf{X}}, \mu)$, so we include their proof here for convenience. The eigenvalues of $\mathbf{X}$ is denoted by $\lambda_i, 1 \leq i \leq n$. First, we observe that

$$\int\limits_{\mathbb{C}} f(x)\mu_{\mathbf{X}}(dx) = \frac{1}{n}\sum_{i=1}^{n} f(\lambda_i(\mathbf{X})).$$

Then by triangle inequality, for any permutation π of $1, 2, \ldots, n$,

$$\begin{aligned}
|\rho(\mu_{\mathbf{X}}, \mu) - \rho(\mu_{\mathbf{Y}}, \mu)| &\leqslant \tfrac{1}{n} \sup_{\|f\|_{\mathcal{L}}\leqslant 1} \left| \sum_{i=1}^{n} f(\lambda_i(\mathbf{X})) - \sum_{i=1}^{n} f(\lambda_i(\mathbf{Y}))\right| \\
&\leqslant \max_{1\leqslant i\leqslant n} \sup_{\|f\|_{\mathcal{L}}\leqslant 1} \left| \sum_{i=1}^{n} f(\lambda_{\pi(i)}(\mathbf{X})) - \sum_{i=1}^{n} f(\lambda_i(\mathbf{Y}))\right| \\
&\leqslant \max_{1\leqslant i\leqslant n} \left|\lambda_{\pi(i)}(\mathbf{X}) - \lambda_i(\mathbf{Y})\right|,
\end{aligned}$$

where in the last step we use the Lipschitz property of $f : |f(x) - f(y)| \leqslant |x - y|$ for any x and y. Since the above inequality is true for any permutation π, we have that

$$|\rho(\mu_{\mathbf{X}}, \mu) - \rho(\mu_{\mathbf{Y}}, \mu)| \leqslant \min_{\pi} \max_{1\leqslant i\leqslant n} \left|\lambda_{\pi(i)}(\mathbf{X}) - \lambda_i(\mathbf{Y})\right|.$$

Using Theorem 2 of [205], we have that

$$\min_{\pi} \max_{1\leqslant i\leqslant n} \left|\lambda_{\pi(i)}(\mathbf{X}) - \lambda_i(\mathbf{Y})\right| \leqslant 2^{2-1/n} \|\mathbf{X} - \mathbf{Y}\|_2^{1/n} \left(\|\mathbf{X}\|_2 + \|\mathbf{Y}\|_2\right)^{1-1/n}$$

where $\|\mathbf{X}\|_2$ is the operator norm of $\mathbf{X}$. Let $\mathbf{X} = (x_{ij})$ and $\mathbf{Y} = (y_{ij})$. We know that $\|\mathbf{X}\|_2 \leqslant \sum_{1\leqslant i,j\leqslant n} |x_{ij}|^2$. Therefore

$$|\rho(\mu_{\mathbf{X}}, \mu) - \rho(\mu_{\mathbf{Y}}, \mu)| \leqslant 2^{2-1/n} \|\mathbf{X} - \mathbf{Y}\|_2^{1/n} \cdot \left(\sum_{1\leqslant i,j\leqslant n} |x_{ij} - y_{ij}|^2\right)^{1/(2n)}.$$

□

For a subset $\mathcal{A} \subseteq \mathbb{C}$, the space of Lipschitz functions $f : \mathcal{A} \to \mathbb{R}$ is denoted by Lip $(\mathcal{A})$. Denote by $\mathcal{P}(\mathbf{A})$ the space of all probability measures supported in $\mathbf{A}$, and by $\mathcal{P}_p(\mathbf{A})$ be the space of all probability measures with finite p-th moment, equipped with the L_p Wasserstein distance d_p defined by

$$d_p(\mu, \nu) = \inf_{\pi} \left(\int \|\mathbf{x} - \mathbf{y}\|^p d\pi\left(\mathbf{x}, \mathbf{y}\right) \right)^{1/p}. \tag{4.41}$$

$\|\cdot\|$ is the Euclidean length (norm) of a vector. The infimum above is over probability measures π on $\mathcal{A} \times \mathcal{A}$ with marginals μ and ν. Note that $d_p \le d_q$ when $p \le q$. The L_1 Wasserstein distance can be equivalently defined [59] by

$$d_1\left(\mu, \nu\right) = \sup_{f} \int \left[f(\mathbf{x}) - f(\mathbf{y})\right] d\mu\left(\mathbf{x}\right) d\nu(\mathbf{y}), \tag{4.42}$$

where the supremum is over f in the unit ball $\mathcal{B}\left(\text{Lip}\left(\mathcal{A}\right)\right)$ of Lip $(\mathcal{A})$.

Denote the space of $n \times n$ Hermitian matrices by $\mathbb{M}_{n\times n}^{sa}$. Let $\mathbf{A}$ be a random $n \times n$ Hermitian matrix. An essential condition on some of random matrices used in the construction below is the following. Suppose that for some $C, c > 0$,

$$\mathbb{P}\left(\left|F(\mathbf{A}) - \mathbb{E}F(\mathbf{A})\right| \geqslant t\right) \leqslant C \exp\left[-ct^2\right] \tag{4.43}$$

for every $t > 0$ and $F : \mathbb{M}_{n\times n}^{sa} \to \mathbb{R}$ which is 1-Lipschitz with respect to the Hilbert-Schmidt norm (or Frobenius norm). Examples in which (4.43) is satisfied include:

1. The diagonal and upper-diagonal entries of matrix $\mathbf{M}$ are independent and each satisfy a quadratic transportation cost inequality with constant $c/\sqrt{n}$. This is slightly more general than the assumption of a log-Sobolev inequality [141, Sect. 6.2], and is essentially the most general condition with independent entries [206]. It holds, e.g., for Gaussian entries and, more generally, for entries with densities of the form $e^{-nu_{ij}(x)}$ where $u_{ij}''(x) \ge c > 0$.
2. The distribution of $\mathbf{M}$ itself has a density proportional to $e^{-n\,\mathrm{Tr}\,u(\mathbf{M})}$ with $u : \mathbb{R} \to \mathbb{R}$ such that $u''(x) \ge c > 0$. This is a subclass of the unitarily invariant ensembles [207]. The hypothesis on u guarantees that $\mathbf{M}$ satisfies a log-Sobolev inequality [208, Proposition 4.4.26].

The first model is as follows. Let $\mathbb{U}(n)$ be the group of $n \times n$ unitary matrices. Let $\mathbf{U} \in \mathbb{U}(n)$ distributed according to Haar measure, independent of $\mathbf{A}$, and let $\mathbf{P}_k$ denote the projection of $\mathbb{R}^n$ onto the span of the first k basis elements. Define a random matrix $\mathbf{M}$ by

$$\mathbf{M} = \mathbf{P}_k\mathbf{U}\mathbf{A}\mathbf{U}^H\mathbf{P}_k^H \tag{4.44}$$

Then $\mathbf{M}$ is a compression of $\mathbf{A}$ (as an operator on $\mathbb{R}^n$) to a random k-dimensional subspace chosen independently of $\mathbf{A}$. This compression is relevant to the spectrum sensing based on principal component analysis [156, 157] and kernel principal component analysis [159]. The central idea is based on an observation that under extremely signal to noise ratios the first several eigenvectors (or features)—corresponding to the first k basis elements above—are found to be robust.

For an $N \times N$ Hermitian matrix $\mathbf{X}$, we define the empirical spectral measure as

$$\mu_{\mathbf{X}} = \frac{1}{N}\sum_{i=1}^{N} \delta_{\lambda_i(\mathbf{X})},$$

where $\lambda_i(\mathbf{X})$ are eigenvalues of matrix $\mathbf{X}$, and δ is the Dirac function. The empirical spectral measure of random matrix $\mathbf{M}$ is denoted by $\mu_{\mathbf{M}}$, while its expectation is denoted by $\mu = \mathbb{E}\mu_{\mathbf{M}}$.

Theorem 4.8.11 (Meckes, E.S. and Meckes, M.W. [192]). *Suppose that matrix* $\mathbf{A}$ *satisfies* (4.43) *for every 1-Lipschitz function* $F : \mathbb{M}_{n\times n}^{sa} \to \mathbb{R}$.

1. *If* $F : \mathbb{M}_{n\times n}^{sa} \to \mathbb{R}$ *is 1-Lipschitz, then for* $\mathbf{M} = \mathbf{P}_k\mathbf{U}\mathbf{A}\mathbf{U}^H\mathbf{P}_k^H$,

$$\mathbb{P}\left(|F(\mathbf{M}) - \mathbb{E}F(\mathbf{M})| \geqslant t\right) \leqslant C\exp\left[-cnt^2\right]$$

   for every $t > 0$.
2. *In particular,*

$$\mathbb{P}\left(\left|\|\mathbf{M}\|_{op} - \mathbb{E}\|\mathbf{M}\|_{op}\right| \geqslant t\right) \leqslant C\exp\left[-cnt^2\right]$$

   for every $t > 0$.
3. *For any fixed probability measure* $\mu \in \mathcal{P}_2(\mathbb{C})$, *and 1-Lipschitz* $f : \mathbb{R} \to \mathbb{R}$, *if*

$$Z_f = \int f d\mu_{\mathbf{M}} - \int f d\mu,$$

   then

$$\mathbb{P}\left(|Z_f - \mathbb{E}Z_f| \geqslant t\right) \leqslant C\exp\left[-cknt^2\right]$$

   for every $t > 0$.
4. *For any fixed probability measure* $\mu \in \mathcal{P}_2(\mathbb{C})$, *and* $1 \leq p \leq 2$,

$$\mathbb{P}\left(|d_p(\mu,\nu) - \mathbb{E}d_p(\mu,\nu)| \geqslant t\right) \leqslant C\exp\left[-cknt^2\right]$$

   for every $t > 0$.

Let us consider the expectation $\mathbb{E}d_1\left(\mu_{\mathbf{M}}, \mathbb{E}\mu_{\mathbf{M}}\right)$.

Theorem 4.8.12 (Meckes, E.S. and Meckes, M.W. [192]). *Suppose that matrix* $\mathbf{A}$ *satisfies* (4.43) *for every 1-Lipschitz function* $F : \mathbb{M}_{n\times n}^{sa} \to \mathbb{R}$. *Let* $\mathbf{M} = \mathbf{P}_k\mathbf{U}\mathbf{A}\mathbf{U}^H\mathbf{P}_k^H$, *and let* $\mu_\mathbf{M}$ *denote the empirical spectral distribution of random matrix* $\mathbf{M}$ *with* $\mu = \mathbb{E}\mu_\mathbf{M}$. *Then,*

$$\mathbb{E}d_1\left(\mu_\mathbf{M}, \mathbb{E}\mu_\mathbf{M}\right) \leqslant \frac{C''\left(\mathbb{E}\|\mathbf{M}\|_{op}\right)^{1/3}}{(kn)^{1/3}} \leqslant \frac{C'''}{(kn)^{1/3}},$$

and so

$$\mathbb{P}\left(d_1\left(\mu_\mathbf{M}, \mathbb{E}\mu_\mathbf{M}\right) \geqslant \frac{C'''}{(kn)^{1/3}} + t\right) \leqslant C\exp\left[-cknt^2\right]$$

for every $t > 0$.

Let us consider the second model that has been considered in free probability [209]. Let $\mathbf{A}, \mathbf{B}$ be $n \times n$ Hermitian matrices and satisfy (4.43). Let $\mathbf{U}$ be Haar distributed, with $\mathbf{A}, \mathbf{B}, \mathbf{U}$ independent. Define [192]

$$\mathbf{M} = \mathbf{U}\mathbf{A}\mathbf{U}^H + \mathbf{B}, \tag{4.45}$$

the "randomized sum" of $\mathbf{A}$ and $\mathbf{B}$.

Theorem 4.8.13 (Meckes, E.S. and Meckes, M.W. [192]). *Let* $\mathbf{A}, \mathbf{B}$ *satisfies* (4.43) *and let* $\mathbf{U} \in \mathbb{U}(n)$ *be Haar-distributed with* $\mathbf{A}, \mathbf{B}, \mathbf{U}$ *independent of each other. Define* $\mathbf{M} = \mathbf{U}\mathbf{A}\mathbf{U}^H + \mathbf{B}$.

1. *There exist* C, c *depending only on the constants in* (4.43) *for* $\mathbf{A}$ *and* $\mathbf{B}$, *such that if* $F : \mathbb{M}_{k\times k}^{sa} \to \mathbb{R}$ *is 1-Lipschitz, then*

$$\mathbb{P}\left(\left|F(\mathbf{M}) - \mathbb{E}F(\mathbf{M})\right| \geqslant t\right) \leqslant C\exp\left[-cnt^2\right]$$

   for every $t > 0$.
2. *In particular,*

$$\mathbb{P}\left(\left|\|\mathbf{M}\|_{op} - \mathbb{E}\|\mathbf{M}\|_{op}\right| \geqslant t\right) \leqslant C\exp\left[-cnt^2\right]$$

   for every $t > 0$.
3. *For any fixed probability measure* $\rho \in \mathcal{P}_2(\mathbb{R})$, *and 1-Lipschitz* $f : \mathbb{R} \to \mathbb{R}$, *if*

$$Z_f = \int f d\mu_\mathbf{M} - \int f d\rho,$$

then

$$\mathbb{P}\left(|Z_f - \mathbb{E}Z_f| \geqslant t\right) \leqslant C \exp\left[-cn^2t^2\right]$$

for every $t > 0$.

4. *For any fixed probability measure* $\rho \in \mathcal{P}_2(\mathbb{R})$, *and* $1 \leq p \leq 2$,

$$\mathbb{P}\left(|d_p(\mu,\nu) - \mathbb{E}d_p(\mu,\nu)| \geqslant t\right) \leqslant C \exp\left[-cn^2t^2\right]$$

for every $t > 0$.

Theorem 4.8.14 (Meckes, E.S. and Meckes, M.W. [192]). *In the setting of Theorem 4.8.13, there are constants* c, C, C', C'' *depending only on the concentration hypothesis for* $\mathbf{A}$ *and* $\mathbf{B}$, *such that*

$$\mathbb{E}d_1\left(\mu_{\mathbf{M}}, \mathbb{E}\mu_{\mathbf{M}}\right) \leqslant \frac{C\left(\mathbb{E}\|\mathbf{M}\|_{op}\right)^{1/3}}{(n)^{2/3}} \leqslant \frac{C'}{(n)^{2/3}},$$

and so

$$\mathbb{P}\left(d_1\left(\mu_{\mathbf{M}}, \mathbb{E}\mu_{\mathbf{M}}\right) \geqslant \frac{C'}{(n)^{2/3}} + t\right) \leqslant C'' \exp\left[-cn^2t^2\right]$$

for every $t > 0$.

Theorem 4.8.15 (Meckes, E.S. and Meckes, M.W. (2011) [192]). *For each* n, *let* $\mathbf{A}_n, \mathbf{B}_n \in \mathbb{M}_{n\times n}^{sa}$ *be **fixed** matrices with spectra bounded independently of* n. *Let* $\mathbf{U} \in \mathbb{U}(n)$ *be Haar-distributed. Let* $\mathbf{M}_n = \mathbf{U}\mathbf{A}_n\mathbf{U}^H + \mathbf{B}_n$ *and let* $\mu_n = \mathbb{E}\mu_{\mathbf{M}_n}$. *Then with probability 1,*

$$d_1\left(\mu_{\mathbf{M}_n}, \mathbb{E}\mu_{\mathbf{M}_n}\right) \leqslant \frac{C}{n^{2/3}}$$

for all sufficiently large n, *where* C *depends only on the bounds on the sizes of the spectra of* $\mathbf{A}_n$ *and* $\mathbf{B}_n$.

4.9 Concentration for Sums of Two Random Matrices

We follow [210]. If $\mathbf{A}$ and $\mathbf{B}$ are two Hermitian matrices with a known spectrum (the set of eigenvalues), it is a classical problem to determine all possibilities for the spectrum of $\mathbf{A} + \mathbf{B}$. The problem goes back at least to H. Weyl [211]. Later, Horn [212] suggested a list of inequalities, which must be satisfied by eigenvalues of $\mathbf{A}+\mathbf{B}$. For larger matrices, it is natural to consider the probabilistic analogue of this problem, when matrices $\mathbf{A}$ and $\mathbf{B}$ are "in general position." Let

$$\mathbf{H} = \mathbf{A} + \mathbf{U}\mathbf{B}\mathbf{U}^H,$$

where $\mathbf{A}$ and $\mathbf{B}$ are two fixed $n \times n$ Hermitian matrices and $\mathbf{U}$ is a random unitary matrix with the Haar distribution on the unitary group $\mathcal{U}(n)$. Then, the eigenvalues of $\mathbf{H}$ are random and we are interested in their joint distribution. Let $\lambda_1(\mathbf{A}) \geqslant \cdots \geqslant \lambda_n(\mathbf{A})$ (repeated by multiplicities) denote the eigenvalues of $\mathbf{A}$, and define the *spectral measure* of $\mathbf{A}$ as

$$\mu_{\mathbf{A}} = \frac{1}{n}\sum_{i=1}^{n} \delta_{\lambda_i(\mathbf{A})}. \tag{4.46}$$

Define similarly for $\mu_{\mathbf{A}}$ and $\mu_{\mathbf{H}}$. The empirical spectral cumulative distribution function of $\mathbf{A}$ is defined as

$$F_{\mathbf{A}}(x) = \frac{\#\left\{i : \lambda_i \leqslant x\right\}}{n}.$$

By an ingenious application of the Stein's method [87], Chatterjee [213] proved that for every $x \in \mathbb{R}$

$$\mathbb{P}\left(\left|F_{\mathbf{H}}(x) - \mathbb{E}F_{\mathbf{H}}(x)\right| \geqslant t\right) \leqslant 2\exp\left(-ct^2\frac{n}{\log n}\right),$$

where c is a numerical constant. The decay rate of tail bound is sub-linear in n. Kargin [210] greatly improved the decay rate of tail bound to n^2. To state his result, we need to define the *free deconvolution* [209].

When n is large, it is natural to define

$$\mu_{\mathbf{A}} \boxplus \mu_{\mathbf{B}},$$

where $\boxplus$ denotes free convolution, a non-linear operation on probability measures introduced by Voiculescu [66].

Assumption 4.9.1. The measure $\mu_{\mathbf{A}} \boxplus \mu_{\mathbf{B}}$ is absolutely continuous everywhere on $\mathbb{R}$, and its density is bounded by a constant C.

Theorem 4.9.2 (Kargin [210]). *Suppose that Assumption 4.9.1 holds. Let* $F_{\mathbf{H}}(x)$ *and* $F_{\boxplus}(x)$ *be cumulative distribution functions for the eigenvalues of* $\mathbf{H} = \mathbf{A} + \mathbf{U}\mathbf{B}\mathbf{U}^H$ *and for* $\mu_{\mathbf{A}} \boxplus \mu_{\mathbf{B}}$, *respectively. For all* $n \geqslant \exp\left((c_1/t)^{4/\varepsilon}\right)$,

$$\mathbb{P}\left(\sup_x \left|F_{\mathbf{H}}(x) - F_{\boxplus}(x)\right| \geqslant t\right) \leqslant \exp\left(-c_2 t^2 \frac{n^2}{(\log n)^{\varepsilon}}\right),$$

where c_1 *and* c_2 *are positive and depend only on* C, $\varepsilon \in (0,2]$, *and* $K = \max\{\|\mathbf{A}\|, \|\mathbf{B}\|\}$.

The main tools in the proof of this theorem are the Stieltjes transform method and standard concentration inequalities.

4.10 Concentration for Submatrices

Let $\mathbf{M}$ be a square matrix of order n. For any two sets of integers $i_1, \ldots, i_k$ and $j_1, \ldots, j_l$ between 1 and n, $\mathbf{M}(i_1, \ldots, i_k; j_1, \ldots, j_l)$ denotes the submatrix of $\mathbf{M}$ formed by deleting all rows except rows $i_1, \ldots, i_k$ and all columns except columns $j_1, \ldots, j_l$. A submatrix like $\mathbf{M}(i_1, \ldots, i_k; j_1, \ldots, j_l)$ is called a principal submatrix.

We define $F_{\mathbf{A}}(x)$ the empirical spectral cumulative distribution function of matrix $\mathbf{A}$ as (4.46). The following result shows that given $1 \leq k \leq n$ and any Hermitian matrix $\mathbf{M}$ of order n, the empirical spectral distribution is *almost the same* for *almost every* principal submatrix $\mathbf{M}$ of order k.

Theorem 4.10.1 (Chatterjee and Ledoux [214]). *Take any* $1 \leq k \leq n$ *and a Hermitian matrix* $\mathbf{M}$ *of order* n. *Let* $\mathbf{A}$ *be a principal submatrix of* $\mathbf{M}$ *chosen uniformly at random from the set of all* $k \times k$ *principal submatrices of* $\mathbf{M}$. *Let* F *be the expected spectral cumulative distribution function, that is,* $F(x) = \mathbb{E}F_{\mathbf{A}}(x)$. *Then, for each* $t \geq 0$,

$$\mathbb{P}\left(\|F_{\mathbf{A}} - F\|_\infty \geqslant \frac{1}{\sqrt{k}} + t\right) \leqslant 12\sqrt{k}e^{-t\sqrt{k/8}}.$$

Consequently, we have

$$\mathbb{E}\|F_{\mathbf{A}} - F\|_\infty \leqslant \frac{13 + \sqrt{8}\log k}{\sqrt{k}}.$$

Exactly the same results hold if $\mathbf{A}$ *is a* $k \times n$ *submatrix of* $\mathbf{M}$ *chosen uniformly at random, and* $F_{\mathbf{A}}$ *is the empirical cumulative distribution function of the singular values of* $\mathbf{A}$. *Moreover, in this case* $\mathbf{M}$ *need not be Hermitian.*

Theorem 4.10.1 can be used to sample the matrix from a larger database. In the example [214] of an $n \times n$ covariance matrix with $n = 100$, they chose $k = 20$ and picked two $k \times k$ principal submatrices $\mathbf{A}$ and $\mathbf{B}$ of $\mathbf{M}$, uniformly and independently at random. The two distributions $F_{\mathbf{A}}$ and $F_{\mathbf{B}}$ are statistically distinguishable.

4.11 The Moment Method

We follow [63] and [3] for our exposition of Wigner's trace method. The idea of using Wigner's trace method to obtain an upper bound for the eigenvalue of $\mathbf{A}$, $\lambda(\mathbf{A})$, was initiated in [215]. The standard linear algebra identity has

$$\mathrm{Tr}(\mathbf{A}) = \sum_{i=1}^{N} \lambda_i(\mathbf{A}).$$

The trace of a matrix is the sum of its eigenvalues. The trace is a linear functional. More generally, we have

$$\mathrm{Tr}\left(\mathbf{A}^k\right) = \sum_{i=1}^{N} \lambda_i^k(\mathbf{A}).$$

The linearity of expectation implies

$$\sum_{i=1}^{n} \mathbb{E}\left(\lambda_i(\mathbf{A})^k\right) = \mathbb{E}\left(\mathrm{Tr}\,\mathbf{A}^k\right).$$

For an even integer k, the geometric average of the high-order moment $\left(\mathrm{Tr}\left(\mathbf{A}^k\right)\right)^{1/k}$ is the l^k norm of these eigenvalues, and we have

$$\sigma_1(\mathbf{A})^k \leqslant \mathrm{Tr}\left(\mathbf{A}^k\right) \leqslant n\sigma_1(\mathbf{A})^k \tag{4.47}$$

The knowledge of the k-th moment $\mathrm{Tr}\left(\mathbf{A}^k\right)$ controls the operator norm (the also the largest singular value) up to a multiplicative factor of $n^{1/k}$. Taking larger and larger k, we should obtain more accurate control on the operator norm.

Let us see how the moment method works in practice. The simplest case is that of the second moment $\mathrm{Tr}\left(\mathbf{A}^2\right)$, which in the Hermitian case works out to

$$\mathrm{Tr}\left(\mathbf{A}^2\right) = \sum_{i=1}^{n}\sum_{j=1}^{n} \left|\xi_{ij}\right|^2 = \|\mathbf{A}\|_F^2.$$

The expression $\sum_{i=1}^{n}\sum_{j=1}^{n} \left|\xi_{ij}\right|^2$ is easy to compute in practice. For instance, for the symmetric matrix $\mathbf{A}$ consisting of Bernoulli random variables taking random values of ± 1, this expression is exactly equal to n^2.

From the weak law of large numbers, we have

$$\sum_{i=1}^{n}\sum_{j=1}^{n}\left|\xi_{ij}\right|^2=(1+o(1))\,n^2 \tag{4.48}$$

asymptotically almost surely. In fact, if the ξ_{ij} have uniformly sub-exponential tail, we have (4.48) with overwhelming probability.

Applying (4.48) , we have the bounds

$$(1+o(1))\sqrt{n}\leqslant\sigma_1\left(\mathbf{A}\right)\leqslant(1+o(1))\,n \tag{4.49}$$

asymptotically almost surely. The median of $\sigma_1\left(\mathbf{A}\right)$ is at least $(1+o(1))\sqrt{n}$. But the upper bound here is terrible. We need to move to higher moments to improve it.

Let us move to the fourth moment. For simplicity, all entries ξ_{ij} have zero mean and unit variance. To control moments beyond the second moments, we also assume that all entries are bounded in magnitude by some K. We expand

$$\mathrm{Tr}\left(\mathbf{A}^4\right)=\sum_{1\leqslant i_1,i_2,i_3,i_4\leqslant n}^{n}\xi_{i_1i_2}\xi_{i_2i_3}\xi_{i_3i_4}\xi_{i_4i_1}.$$

To understand this expression, we take expectations

$$\mathbb{E}\,\mathrm{Tr}\left(\mathbf{A}^4\right)=\sum_{1\leqslant i_1,i_2,i_3,i_4\leqslant n}^{n}\mathbb{E}\xi_{i_1i_2}\xi_{i_2i_3}\xi_{i_3i_4}\xi_{i_4i_1}.$$

One can view this sum graphically, as a sum over length four cycles in the vertex set $\{1,\ldots,n\}$.

Using the combinatorial arguments [63], we have

$$\mathbb{E}\,\mathrm{Tr}\left(\mathbf{A}^4\right)\leqslant O\left(n^3\right)+O\left(n^2K^2\right).$$

In particular, if we have the assumption $K=O(\sqrt{n})$, then we have

$$\mathbb{E}\,\mathrm{Tr}\left(\mathbf{A}^4\right)\leqslant O\left(n^3\right).$$

Consider the general k-th moment

$$\mathbb{E}\,\mathrm{Tr}\left(\mathbf{A}^k\right)=\sum_{1\leqslant i_1,\ldots,i_k\leqslant n}^{n}\mathbb{E}\xi_{i_1i_2}\cdots\xi_{i_ki_1}.$$

We have

$$\mathbb{E}\,\mathrm{Tr}\left(\mathbf{A}^k\right)\leqslant(k/2)^k n^{k/2+1}\max\left(1,K/\sqrt{n}\right)^{k-2}.$$

With the aid of (4.47), one has

$$\mathbb{E}\sigma_1(\mathbf{A})^k \leqslant (k/2)^k n^{k/2+1} \max\left(1, K/\sqrt{n}\right)^{k-2},$$

and so by Markov's inequality, we have

$$\mathbb{P}\left(\sigma_1\left(\mathbf{A}\right) \geqslant t\right) \leqslant \frac{1}{t^k}(k/2)^k n^{k/2+1} \max\left(1, K/\sqrt{n}\right)^{k-2}$$

for all $t > 0$. This gives the median of $\sigma_1\left(\mathbf{A}\right)$ at

$$O\left(n^{1/k} k\sqrt{n} \max\left(1, K/\sqrt{n}\right)\right).$$

We can optimize this in k by choosing k to be comparable to $\log n$, and then we obtain an upper bound $O\left(\sqrt{n}\log n \max\left(1, K/\sqrt{n}\right)\right)$ for the median. After a slight tweaking of the constants, we have

$$\sigma_1\left(\mathbf{A}\right) = O\left(\sqrt{n}\log n \max\left(1, K/\sqrt{n}\right)\right)$$

with high probability.

We can summarize the above results into the following:

Proposition 4.11.1 (Weak upper bound). *Let* $\mathbf{A}$ *be a random Hermitian matrix, with the upper triangular entries* $\xi_{ij}, i \leq j$ *being independent with mean zero and variance at most 1, and bounded in magnitude by* K. *Then,*

$$\sigma_1\left(\mathbf{A}\right) = O\left(\sqrt{n}\log n \max\left(1, K/\sqrt{n}\right)\right)$$

with high probability.

When $K \leq \sqrt{n}$, this gives an upper bound of $O\left(\sqrt{n}\log n\right)$, which is still off by a logarithmic factor from the expected bound of $O\left(\sqrt{n}\right)$. We will remove this factor later.

Now let us consider the case when $K = o\left(\sqrt{n}\right)$, and each entry has variance exactly[1] 1. We have the upper bound

$$\mathbb{E}\,\mathrm{Tr}\left(\mathbf{A}^k\right) \leqslant (k/2)^k n^{k/2+1}.$$

We later need the classical formula for the Catalan number

$$C_{n+1} = \sum_{i=0}^{n} C_i C_{n-i}$$

[1]Later we will relax this to "at most 1".

for all $n \geq 1$ with $C_0 = 1$, and use this to deduce that

$$C_{k/2} = \frac{k!}{(k/2+1)!\,(k/2)!} \tag{4.50}$$

for all $k = 2, 4, 6, \ldots$.

Note that $n\,(n-1)\cdots(n-k/2) = (1+o_k\,(1))\,n^{k/2+1}$. After putting all the above computations we conclude

Theorem 4.11.2 (Moment computation). *Let* $\mathbf{A}$ *be a real symmetric random matrix, with the upper triangular elements* $\xi_{ij}, i \leq j$ *jointly independent with mean zero and variance one, and bounded in magnitude by* $o(\sqrt{n})$. *Let* k *be a positive even integer. Then we have*

$$\mathbb{E}\,\mathrm{Tr}\left(\mathbf{A}^k\right) = \left(C_{k/2} + o_k\,(1)\right) n^{k/2+1}$$

where $C_{k/2}$ *is given by* (4.50).

If we allow the ξ_{ij} to have variance at most one, rather than equal to one, we obtain the upper bound

$$\mathbb{E}\,\mathrm{Tr}\left(\mathbf{A}^k\right) \leqslant \left(C_{k/2} + o_k\,(1)\right) n^{k/2+1}.$$

Theorem 4.11.2 is also valid for Hermitian random matrices.

Theorem 4.11.2 can be compared with the formula

$$\mathbb{E}S^k = \left(C'_{k/2} + o_k\,(1)\right) n^{k/2}$$

derived in [63, Sect. 2.1], where

$$S = X_1 + \cdots + X_n$$

is the sum of n i.i.d. random variables of mean zero and variance one, and

$$C'_{k/2} = \frac{k!}{2^{k/2}\,(k/2)!}.$$

Combining Theorem 4.11.2 with (4.47) we obtain a lower bound

$$\mathbb{E}\sigma_1(\mathbf{A})^k \geqslant \left(C_{k/2} + o_k\,(1)\right) n^{k/2}.$$

Proposition 4.11.3 (Lower Bai-Yin theorem). *Let* $\mathbf{A}$ *be a real symmetric random matrix, with the upper triangular elements* $\xi_{ij}, i \leq j$ *jointly independent with mean zero and variance one, and bounded in magnitude by* $O(1)$. *Then the median (or mean) of* $\sigma_1\,(\mathbf{A})$ *at least* $(2-o(1))\,\sqrt{n}$.

Let us remove the logarithmic factor in the following theorem.

Theorem 4.11.4 (Improved moment bound). *Let* $\mathbf{A}$ *be a real symmetric random matrix, with the upper triangular elements* $\xi_{ij}, i \le j$ *jointly independent with mean zero and variance one, and bounded in magnitude by* $O(n^{0.49})$ *(say). Let* k *be a positive even integer of size* $k = O\left(\log^2 n\right)$ *(say). Then we have*

$$\mathbb{E}\,\mathrm{Tr}\left(\mathbf{A}^k\right) \leqslant C_{k/2} n^{k/2+1} + O\left(k^{O(1)} 2^k n^{k/2+0.98}\right)$$

where $C_{k/2}$ *is given by* (4.50). *In particular, from the trivial bound* $C_{k/2} \leqslant 2^k$ *one has*

$$\mathbb{E}\,\mathrm{Tr}\left(\mathbf{A}^k\right) \leqslant (2 + o(1))^k n^{k/2+1}.$$

We refer to [63] for the proof.

Theorem 4.11.5 (Weak Bai-Yin theorem, upper bound). *Let* $\mathbf{A} = (\xi_{ij})_{1 \leqslant i,j \leqslant n}$ *be a real symmetric random matrix, whose entries all have the same distribution* ξ, *with mean zero, variance one, and fourth moment* $O(1)$. *Then for every* $\varepsilon > 0$ *independent of* n, *one has* $\sigma_1(\mathbf{A}) \leqslant (2+\varepsilon)\sqrt{n}$ *asymptotically almost surely. In particular,* $\sigma_1(\mathbf{A}) \leqslant (2 + o(1))\sqrt{n}$ *asymptotically almost surely; as a consequence, the median of* $\sigma_1(\mathbf{A})$ *is at most* $(2 + o(1))\sqrt{n}$. *(If* ξ *is bounded, we see, in particular, from Proposition 4.11.3 that the median is in fact equal to* $(2 + o(1))\sqrt{n}$.*)*

The fourth moment hypothesis is best possible.

Theorem 4.11.6 (Strong Bai-Yin theorem, upper bound). *Let* ξ *be a real random variable with mean zero, variance one and finite fourth moment, and for all* $1 \le i \le j$, *let* ξ_{ij} *be an i.i.d. sequence with distribution* ξ, *and set* $\xi_{ij} = \xi_{ji}$. *Let* $\mathbf{A} = (\xi_{ij})_{1 \leqslant i,j \leqslant n}$ *be the random matrix formed by the top left* $n \times n$ *block. Then almost surely, one has*

$$\lim_{n\to\infty} \sup \frac{\sigma_1(\mathbf{A})}{\sqrt{n}} \leqslant 2.$$

It is a routine matter to generalize the Bai-Yin result from real symmetric matrices to Hermitian matrices. We use a substitute of (4.47), names the bounds

$$\sigma_1(\mathbf{A})^k \leqslant \mathrm{Tr}\left((\mathbf{A}\mathbf{A}^*)^{k/2}\right) \leqslant n\sigma_1(\mathbf{A})^k,$$

valid for any $n \times n$ matrix $\mathbf{A}$ with complex entries and every positive integer k.

It is possible to adapt all of the above moment calculations for $\mathrm{Tr}\left(\mathbf{A}^k\right)$ in the symmetric or Hermitian cases to give analogous cases for $\mathrm{Tr}\left((\mathbf{A}\mathbf{A}^*)^{k/2}\right)$ in the non-symmetric cases. Another approach is to use the augmented matrix defined as

$$\tilde{\mathbf{A}} = \begin{bmatrix} 0 & \mathbf{A} \\ \mathbf{A}^* & 0 \end{bmatrix},$$

which is $2n \times 2n$ Hermitian matrix. If $\mathbf{A}$ has singular values $\sigma_1(\mathbf{A}), \ldots, \sigma_n(\mathbf{A})$, then $\tilde{\mathbf{A}}$ has eigenvalues $\pm\sigma_1(\mathbf{A}), \ldots, \pm\sigma_n(\mathbf{A})$.

$\sigma_1(\mathbf{A})$ is concentrated in the range of $\left[2\sqrt{n} - O\left(n^{-1/6}\right), 2\sqrt{n} + O\left(n^{-1/6}\right)\right]$, and even to get a universal distribution for the normalized expression $\left(\sigma_1(\mathbf{A}) - 2\sqrt{n}\right) n^{1/6}$, known as the *Tracy-Widom law* [216]. See [217–221].

4.12 Concentration of Trace Functionals

Consider a random $n \times n$ Hermitian matrix $\mathbf{X}$. The sample covariance matrix has the form $\mathbf{XDX}^*$ where $\mathbf{D}$ is a diagonal matrix. Our goal here is to study the non-asymptotic deviations, following [180]. We give concentration inequalities for functions of the empirical measure of eigenvalues for large, random, Hermitian matrices $\mathbf{X}$, with not necessarily Gaussian entries. The results apply in particular to non-Gaussian Wigner and Wishart matrices.

Consider

$$\mathbf{X} = \left((\mathbf{X})_{ij}\right)_{1 \leqslant i,j \leqslant n}, \mathbf{X} = \mathbf{X}^*, \mathbf{X}_{ij} = \frac{1}{\sqrt{n}}\mathbf{A}_{ij}\omega_{ij}$$

with

$$\omega := \left(\omega^R + j\omega^I\right) = \left(\omega_{ij}\right)_{1 \leqslant i,j \leqslant n}, \omega_{ij} = \bar{\omega}_{ij},$$

$$\mathbf{A} = \left((\mathbf{A})_{ij}\right)_{1 \leqslant i,j \leqslant n}, \mathbf{A} = \mathbf{A}^*.$$

Here, $\{\omega_{ij}\}_{1 \leqslant i,j \leqslant n}$ are independent complex random variables with laws $\{P_{ij}\}_{1 \leqslant i,j \leqslant n}$, P_{ij} being a probability measure on $\mathbb{C}$ with

$$P_{ij}\left(\omega_{ij} \in \Theta\right) = \int 1_{u+jv \in \Theta} P_{ij}^R(du)\, P_{ij}^I(dv),$$

and $\mathbf{A}$ is a non-random complex matrix with entires $\left\{(\mathbf{A})_{ij}\right\}_{1 \leqslant i,j \leqslant n}$ uniformly bounded by, say, a.

We consider a real-valued function on $\mathbb{R}$. For a compact set $\mathcal{K}$, denoted by $|\mathcal{K}|$ its diameter, that is the maximal distance between two points of $\mathcal{K}$. For a Lipschitz function $f : \mathbb{R}^k \mapsto \mathbb{R}$, we define the Lipschitz constant $|f|_{\mathcal{L}}$ by

$$|f|_{\mathcal{L}} = \sup_{x,y} \frac{|f(x) - f(y)|}{\|x - y\|},$$

where $\|\cdot\|$ denotes the Euclidean norm on $\mathbb{R}^k$.

We say that a measure μ on $\mathbb{R}$ satisfies the logarithmic Sobolev inequality with (not necessarily optimal) constant c if, for any differentiable function f,

$$\int f^2 \log \frac{f^2}{\int f^2 d\mu} \leqslant 2c \int \left| f' \right|^2 d\mu.$$

A measure μ satisfying the logarithmic Sobolev inequality has sub-Gaussian tails [197].

Theorem 4.12.1 (Guionnet and Zeitouni [180]).

(a) Assume that the $(P_{ij}, i \leqslant j \in \mathbb{N})$ are uniformly compactly supported, that is that there exists a compact set $\mathcal{K} \in \mathbb{C}$ so that for any $1 \leqslant i \leqslant j \leqslant n$, $P_{ij}(\mathcal{K}^c) = 0$. Assume f is convex and Lipschitz. Then, for any $t > t_0(n) = 8|\mathcal{K}|\sqrt{\pi} a |f|_{\mathcal{L}}/n > 0$,

$$\mathbb{P}\left(\left|\mathrm{Tr}\left(f\left(\mathbf{X}\right)\right) - \mathbb{E}\,\mathrm{Tr}\left(f\left(\mathbf{X}\right)\right)\right| \geqslant t\right) \leqslant 4n \exp\left\{ -\frac{n^2 (t - t_0(n))^2}{16 |\mathcal{K}|^2 a^2 |f|_{\mathcal{L}}^2} \right\}.$$

(b) If $\left(P_{ij}^R, P_{ij}^I, i \leqslant j \in \mathbb{N}\right)$ satisfy the logarithmic Sobolev inequality with uniform constant c, then for any Lipschitz function f, for any $t > 0$,

$$\mathbb{P}\left(\left|\mathrm{Tr}\left(f\left(\mathbf{X}\right)\right) - \mathbb{E}\,\mathrm{Tr}\left(f\left(\mathbf{X}\right)\right)\right| \geqslant t\right) \leqslant 2n \exp\left\{ -\frac{n^2 t^2}{8ca^2 |f|_{\mathcal{L}}^2} \right\}.$$

We regard $\mathrm{Tr}\left(f\left(\mathbf{X}\right)\right)$ as a function of the entries $\left(\omega_{ij}^R, \omega_{ij}^I\right)_{1 \leqslant i,j \leqslant n}$.

4.13 Concentration of the Eigenvalues

Concentration of eigenvalues are treated in [1–3]. Trace functionals are easier to handle since trace is a linear operator. The k-th largest eigenvalue is a nonlinear functional. We consider the quite general model of random symmetric matrices. Let $x_{ij}, 1 \leq i \leq j \leq n$, be independent, real random variables with absolute value at most 1. Define $x_{ij} = x_{ji}$ and consider the symmetric random matrix $\mathbf{X} = \left(\mathrm{x}_{ij}\right)_{i,j=1}^n$. We consider λ_k the k-th largest eigenvalue of $\mathbf{X}$.

Theorem 4.13.1 (Alon, Krivelevich, and Vu [1]). *For every $1 \leq k \leq n$, the probability that $\lambda_k(\mathbf{X})$ deviates from its median by more than t is at most $4e^{-t^2/32k^2}$. The same estimate holds for the probability that $\lambda_{n-k+1}(\mathbf{X})$ deviates from its median by more than t.*

In practice, we can replace the median $\mathbb{M}$ with the expectation $\mathbb{E}$.

Let $x_{ij}, 1 \leq i \leq j \leq n$, be independent—but not necessarily identical—random variables:

- $|x_{ij}| \leqslant K$ for all $1 \leqslant i \leqslant j \leqslant n$;
- $\mathbb{E}(x_{ij}) = 0,$ for all $1 \leqslant i \leqslant j \leqslant n$;
- $\mathrm{Var}(x_{ij}) = \sigma^2$, for all $1 \leqslant i \leqslant j \leqslant n$.

Define $x_{ij} = x_{ji}$ and consider the symmetric random matrix $\mathbf{X} = (\mathrm{x}_{ij})_{i,j=1}^{n}$. We consider the largest eigenvalue of $\mathbf{X}$, defined as

$$\lambda_{\max}(\mathbf{X}) = \sup_{\mathbf{v} \in \mathbb{R}^n, \|\mathbf{v}\|=1} \left|\mathbf{v}^T \mathbf{X} \mathbf{v}\right|.$$

The most well known estimate on $\lambda_{\max}(\mathbf{X})$ is perhaps the following theorem [222].

Theorem 4.13.2 (Füredi and Komlós [222]). *For a random matrix* $\mathbf{X}$ *as above, there is a positive constant* $c = c(\sigma, K)$ *such that*

$$2\sigma\sqrt{n} - cn^{1/3}\ln n \leqslant \lambda_{\max}(\mathbf{X}) \leqslant 2\sigma\sqrt{n} + cn^{1/3}\ln n,$$

holds almost surely.

Theorem 4.13.3 (Krivelevich and Vu [223]). *For a random matrix* $\mathbf{X}$ *as above, there is a positive constant* $c = c(K)$ *such that for any* $t > 0$

$$\mathbb{P}(|\lambda_{\max}(\mathbf{X}) - \mathbb{E}\lambda_{\max}(\mathbf{X})| \geqslant ct) \leqslant 4e^{-t^2/32}.$$

In this theorem, c does not depend on σ; we do not have to assume anything about the variances.

Theorem 4.13.4 (Vu [3]). *For a random matrix* $\mathbf{X}$ *as above, there is a positive constant* $c = c(\sigma, K)$ *such that*

$$\lambda_{\max}(\mathbf{X}) \leqslant 2\sigma\sqrt{n} + cn^{1/4}\ln n,$$

holds almost surely.

Theorem 4.13.5 (Vu [3]). *There are constants* C *and* C' *such that the following holds. Let* $x_{ij}, 1 \leq i \leq j \leq n$, *be independent random variables, each of which has mean 0 and variance* σ^2 *and is bounded in absolute value* K, *where* $\sigma \geqslant C' n^{-1/2} K \ln^2 n$. *Then, almost surely,*

$$\lambda_{\max}(\mathbf{X}) \leqslant 2\sigma\sqrt{n} + C(K\sigma)^{1/2} n^{1/4}\ln n.$$

When the entries of $\mathbf{X}$ is i.i.d. symmetric random variables, there are sharper bounds. The best current bound we know of is due to Soshnikov [216], which shows that the error term in Theorem 4.13.4 can be reduced to $n^{1/6+o(1)}$.

4.14 Concentration for Functions of Large Random Matrices: Linear Spectral Statistics

The general concentration principle do not yield the correct small deviation rate for the largest eigenvalues. They, however, apply to large classes of Lipschitz functions.

For $M \in \mathbb{N}$, we denote by $\langle \cdot, \cdot \rangle$ the Euclidean scalar product on $\mathbb{R}^M$ (or $\mathbb{C}^M$). For two vectors $\mathbf{x} = (x_1, \ldots, x_M)$ and $\mathbf{y} = (y_1, \ldots, y_M)$, we have $\langle \mathbf{x}, \mathbf{y} \rangle = \sum_{i=1}^{M} x_i y_i$ (or $\langle \mathbf{x}, \mathbf{y} \rangle = \sum_{i=1}^{M} x_i y_i^*$). The Euclidean norm is defined as $\|\mathbf{x}\| = \sqrt{\langle \mathbf{x}, \mathbf{x} \rangle}$.

For a Lipschitz function $f : \mathbb{R}^M \to \mathbb{R}$, we define the Lipschitz constant $|f|_{\mathcal{L}}$ by

$$|f|_{\mathcal{L}} = \sup_{\mathbf{x} \neq \mathbf{y} \in \mathbb{R}^M} \frac{|f(\mathbf{x}) - f(\mathbf{y})|}{\|\mathbf{x} - \mathbf{y}\|},$$

where $\|\cdot\|$ denotes the Euclidean norm on $\mathbb{R}^M$. In other words, we have

$$|f(\mathbf{x}) - f(\mathbf{y})| \leqslant |f|_{\mathcal{L}} \sum_{i=1}^{M} |x_i - y_i|,$$

where x_i, y_i are elements of the vectors $\mathbf{x}, \mathbf{y}$ on $\mathbb{R}^M$ (or $\mathbb{C}^M$).

Consider $\mathbf{X}$ being a $n \times n$ Hermitian matrix whose entries are centered, independent Gaussian random variables with variance σ^2. In other words, $\mathbf{X}$ is a Hermitian matrix such that the entries above the diagonal are independent, complex (real on the diagonal) Gaussian random variables with zero mean and variance σ^2. This is so called the Gaussian Unitary Ensembles (GUE).

Lemma 4.3.1 says that: If $f : \mathbb{R} \to \mathbb{R}$ is a Lipschitz function, then

$$F = \frac{1}{\sqrt{n}} \sum_{i=1}^{n} f(\lambda_i)$$

is a Lipschitz function of (real and imaginary) entries of $\mathbf{X}$. If f is convex on the real line, then F is convex on the space of matrices (Klein's lemma). As a result, we can use the general concentration principle to functions of the eigenvalues. For example, if $\mathbf{X}$ is a GUE random matrix with variance $\sigma^2 = \frac{1}{4n}$, and if $f : \mathbb{R} \to \mathbb{R}$ is 1-Lipschitz, for any $t \geq 0$,

$$\mathbb{P}\left(\left\{ \frac{1}{n} \sum_{i=1}^{n} f(\lambda_i) - \int f d\mu \right\} \geqslant t \right) \leqslant 2e^{-n^2 t^2}, \tag{4.51}$$

where $\mu = \mathbb{E}\left(\frac{1}{n}\sum_{i=1}^{n}\delta_{\lambda_i}\right)$ is the mean spectral measure. Inequality (4.51) has the n^2 speed of the large deviation principles for spectral measures. With the additional assumption of convexity on f, similar inequalities hold for real or complex matrices with the entries that are independent with bounded support.[2]

Lemma 4.14.1 ([180, 224]). *Let $f : \mathbb{R} \to \mathbb{R}$ be Lipschitz with Lipschitz constant $|f|_{\mathcal{L}}$. $\mathbf{X}$ denotes the Hermitian (or symmetric) matrix with entries $(X_{ij})_{1\leqslant i,j\leqslant n}$, the map*

$$(X_{ij})_{1\leqslant i,j\leqslant n} \to \mathrm{Tr}\left(f\left(\mathbf{X}\right)\right)$$

is a Lipschitz function with constant $\sqrt{n}|f|_{\mathcal{L}}$. If the joint law of $(X_{ij})_{1\leqslant i,j\leqslant n}$ is "good", there is $\alpha > 0$, constant $c > 0$ and $C < \infty$ so that for all $n \in \mathbb{N}$

$$\mathbb{P}\left(\left|\mathrm{Tr}\left(f\left(\mathbf{X}\right)\right) - \mathbb{E}\,\mathrm{Tr}\left(f\left(\mathbf{X}\right)\right)\right| \geqslant t|f|_{\mathcal{L}}\right) \leqslant Ce^{-c|t|^{\alpha}}.$$

"Good" here has, for instance, the meaning that the law satisfies a log-Sobolev inequality; an example is when $(X_{ij})_{1\leqslant i,j\leqslant N}$ are independent, Gaussian random variables with uniformly bounded covariance.

The significance of results such as Lemma 4.14.1 is that they provide bounds on deviations that do not depend on the *dimension* n of the random matrix. They can be used to prove *law of large numbers*—reducing the proof of the almost sure convergence to the proof of the *convergence in expectation* $\mathbb{E}$. They can also be used to relax the proof of a central limit theorem: when $\alpha = 2$, Lemma 4.14.1 says that the random variable $\mathrm{Tr}\left(f\left(\mathbf{X}\right)\right) - \mathbb{E}\,\mathrm{Tr}\left(f\left(\mathbf{X}\right)\right)$ has a sub-Gaussian tail, providing tightness augments for free.

Theorem 4.14.2 ([225]). *Let $\|f\|_{TV}$ be the total variation norm,*

$$\|f\|_{TV} = \sup_{x_1<\cdots<x_m} \sum_{i=2}^{m} \left|f\left(x_i\right) - f\left(x_{i-1}\right)\right|.$$

$\mathbf{X}$ is either Wigner or the Wishart matrices. Then, for any $t > 0$ and any function f with finite total variation norm so that $\mathbb{E}\left|\frac{1}{n}\sum_{i=1}^{n} f\left(\lambda_i\left(\mathbf{X}\right)\right)\right| < \infty$,

$$\mathbb{P}\left(\left|\frac{1}{n}\sum_{i=1}^{n} f\left(\lambda_i\left(\mathbf{X}\right)\right) - \mathbb{E}\left[\frac{1}{n}\sum_{i=1}^{n} f\left(\lambda_i\left(\mathbf{X}\right)\right)\right]\right| \geqslant t\|f\|_{TV}\right) \leqslant 2e^{-\frac{n}{8c_X}t^2},$$

[2]The support of a function is the set of points where the function is not zero-valued, or the closure of that set. In probability theory, the support of a probability distribution can be loosely thought of as the closure of the set of possible values of a random variable having that distribution.

where $c_X = 1$ for Wigner's matrices and M/n for Wishart matrices.

Recall that $\sum_{i=1}^{n} f(\lambda_i(\mathbf{X})) = \mathrm{Tr}(f(\mathbf{X}))$. The above speed is not optimal for laws which have sufficiently fast decaying rates: $\sum_{i=1}^{n} f(\lambda_i(\mathbf{X})) - \mathbb{E}\left[\sum_{i=1}^{n} f(\lambda_i(\mathbf{X}))\right]$ is of order one. However, it is optimal rate for instance for matrices whose entries have heavy tails where the central limit theorem holds for

$$\frac{1}{\sqrt{n}}\left(\sum_{i=1}^{n} f(\lambda_i(\mathbf{X})) - \mathbb{E}\left[\frac{1}{n}\sum_{i=1}^{n} f(\lambda_i(\mathbf{X}))\right]\right).$$

In Theorem 4.14.2, we only require independence of the *vectors*, rather than the entries.

A probability measure $\mathbb{P}$ on $\mathbb{E}^n$ is said to satisfy the logarithmic Sobolev inequality with constant c, if for any differentiable function $f:\mathbb{R}^n \to \mathbb{R}$, we have

$$\int f^2\left(\log f^2 - \log\int f^2 d\mathbb{P}\right) \leqslant 2c\int\sum_{i=1}^{n}\left(\frac{\partial f}{\partial x_i}\right)^2 d\mathbb{P}. \qquad (4.52)$$

(4.52) implies sub-Gaussian tails, which we use commonly. See [141, 226] for a general study. It is sufficient to know that Gaussian law satisfies the logarithmic Sobolev inequality (4.52).

Lemma 4.14.3 (Herbst). *Assume that $\mathbb{P}$ satisfies the logarithmic Sobolev inequality (4.52) on $\mathbb{R}^n$ with constant c. Let g be a Lipschitz function on $\mathbb{R}^n$ with Lipschitz constant $|g|_{\mathcal{L}}$. Then, for all $t \in \mathbb{R}$, we have*

$$\int e^{t(g-\mathbb{E}_P(g))} d\mathbb{P} \leqslant e^{ct^2|f|_{\mathcal{L}}^2/2},$$

and so for all $t > 0$

$$\mathbb{P}\left(|g - \mathbb{E}_P(g)| \geqslant t\right) \leqslant 2e^{-ct^2|f|_{\mathcal{L}}^2/2}.$$

Lemma 4.14.3 implies that $\mathbb{E}_P(g)$ is finite.

Klein's lemma is given in [227].

Lemma 4.14.4 (Klein's lemma [227]). *Let $f:\mathbb{R}\to\mathbb{R}$ be a convex function. Then, if $\mathbf{A}$ is the $n \times n$ Hermitian matrix with entries $(A_{ij})_{1\leqslant i,j\leqslant n}$ on the above the diagonal, the map (function) ψ_f*

$$\psi_f : (A_{ij})_{1\leqslant i,j\leqslant n} \in \mathbb{C}^{n^2} \to \sum_{i=1}^{n} f(\lambda_i(\mathbf{A}))$$

is convex. If f is twice continuously differentiable with $f''(x) \geqslant c$ for all x, ψ_f is also twice continuously differentiable with Hessian bounded below by $c\mathbf{I}$.

See [224] for a proof.

4.15 Concentration of Quadratic Forms

We follow [228] for this development. Let $\mathbf{x} \in \mathbb{C}^n$ be a vector of independent random variables $\xi_1, \ldots, \xi_n$ of mean zero and variance σ^2, obeying the uniform subexponential delay bound $\mathbb{E}\left(|\xi_i| \geqslant t^{c_0}\sigma\right) \leqslant e^{-t}$ for all $t \geq c_1$ and $1 \leq i \leq n$, and some $c_0, c_1 > 0$ independent of n. Let $\mathbf{A}$ be an $n \times n$ matrix. Then for any $t > 0$, one has

$$\mathbb{P}\left(\left|\mathbf{x}^*\mathbf{A}\mathbf{x} - \sigma^2 \operatorname{Tr}\mathbf{A}\right| \geqslant t\sigma^2(\operatorname{Tr}\mathbf{A}^*\mathbf{A})^{1/2}\right) \leqslant Ce^{-ct^c}.$$

Thus

$$\mathbf{x}^*\mathbf{A}\mathbf{x} = \sigma^2\left[\operatorname{Tr}\mathbf{A} + O\left(t(\operatorname{Tr}\mathbf{A}^*\mathbf{A})^{1/2}\right)\right]$$

outside of an event of probability $O\left(e^{-ct^c}\right)$.

We consider an example (Likelihood Ratio Testing for Detection) : $\mathcal{H}_1$ is claimed if

$$\mathbf{y}^*\mathbf{R}^{-1}\mathbf{y} > \gamma_{LRT}.$$

If we have a number of random vectors $\mathbf{x}_i$, we can consider the following quadratic form

$$\sum_i \mathbf{x}_i^*\mathbf{A}\mathbf{x}_i = \operatorname{Tr}\left(\sum_i \mathbf{x}_i^*\mathbf{A}\mathbf{x}_i\right) = \sum_i \operatorname{Tr}\left(\mathbf{x}_i^*\mathbf{A}\mathbf{x}_i\right) = \sum_i \operatorname{Tr}\left(\mathbf{A}\mathbf{x}_i\mathbf{x}_i^*\right)$$

$$= \operatorname{Tr}\sum_i \left(\mathbf{A}\mathbf{x}_i\mathbf{x}_i^*\right) = \operatorname{Tr}\left[\mathbf{A}\sum_i \left(\mathbf{x}_i\mathbf{x}_i^*\right)\right] = \operatorname{Tr}\left[\mathbf{A}\mathbf{X}\right],$$

where $\mathbf{X} = \sum_i \left(\mathbf{x}_i\mathbf{x}_i^*\right)$.

Theorem 4.15.1. *Let $\mathbf{x}$ and σ as the previous page, and let $\mathbf{V}$ a d-dimensional complex subspace of $\mathbb{C}^n$. Let $\mathbf{P}_V$ be the orthogonal projection to $\mathbf{V}$. Then one has*

$$0.9d\sigma^2 \leqslant \left\|\mathbf{P}_V\left(\mathbf{x}\right)\right\|^2 \leqslant 1.1d\sigma^2$$

outside of an event of probability $O\left(e^{-cd^c}\right)$.

Proof. Apply the preceding proposition with $\mathbf{A} = \mathbf{P}_V$;(so $\mathrm{Tr}\,\mathbf{A} = \mathrm{Tr}\,\mathbf{A}^*\mathbf{A} = d$) and $t = d^{1/2}/10$. □

Example 4.15.2 ($f(\mathbf{x}) = \|\mathbf{Xv}\|_2$). Consider the $n \times n$ sample covariance matrix $\mathbf{R} = \frac{1}{n}\mathbf{X}^T\mathbf{X}$, where $\mathbf{X}$ is the data matrix of $n \times p$. We closely follow [229]. We note the quadratic form

$$\mathbf{v}^T\mathbf{R}\mathbf{v} = \left\|\frac{1}{\sqrt{n}}\mathbf{X}\mathbf{v}\right\|_2^2,$$

where $\mathbf{v}$ is a column vector. We want to show that this function $f(\mathbf{x}) = \|\mathbf{Xv}\|_2$ is convex and 1-Lipschitz with respect to the Euclidean norm. The function $f(\mathbf{x})$ maps a vector $\mathbb{R}^{np}$ to $\mathbb{R}$ and is defined by turning the vector $\mathbf{x}$ into the matrix $\mathbf{X}$, by first filling the rows of $\mathbf{X}$, and then computing the Euclidean norm of $\mathbf{Xv}$. In fact, for $\theta \in [0,1]$ and $\mathbf{x}, \mathbf{z} \in \mathbb{R}^{np}$,

$$\begin{aligned} f(\theta\mathbf{x} + (1-\theta)\,\mathbf{z}) &= \|(\theta\mathbf{X} + (1-\theta)\,\mathbf{Z})\,\mathbf{v}\|_2 \leqslant \|\theta\mathbf{X}\mathbf{v}\|_2 + \|(1-\theta)\,\mathbf{Z}\mathbf{v}\|_2 \\ &= \theta f(\mathbf{x}) + (1-\theta)\,f(\mathbf{z}). \end{aligned}$$

where we have used the triangular inequality of the Euclidean norm. This says that the function $f(\mathbf{x})$ is convex. Similarly,

$$\begin{aligned} |f(\mathbf{x}) - f(\mathbf{z})| &= |\|\mathbf{Xv}\|_2 - \|\mathbf{Zv}\|_2| \leqslant \|(\mathbf{X}-\mathbf{Z})\,\mathbf{v}\|_2 \leqslant \|\mathbf{X}-\mathbf{Z}\|_F\|\mathbf{v}\|_2 \\ &= \|\mathbf{x}-\mathbf{z}\|_2, \end{aligned} \tag{4.53}$$

using the Cauchy-Schwartz inequality and the fact that $\|\mathbf{v}\|_2^2 = 1$. Equation (4.53) implies that the function $f(\mathbf{x})$ is 1-Lipschitz. □

If $\mathbf{Y}$ is an $n \times p$ matrix, we naturally denote by $Y_{i,j}$ its (i,j) entry and call $\bar{\mathbf{Y}}$ the matrix whose jth column is constant and equal to $\bar{Y}_{\cdot,j}$. The sample covariance matrix of the data stored in matrix $\mathbf{Y}$ is

$$\mathbf{S}_p = \frac{1}{n-1}\left(\mathbf{Y} - \bar{\mathbf{Y}}\right)^T\left(\mathbf{Y} - \bar{\mathbf{Y}}\right).$$

We have

$$\mathbf{Y} - \bar{\mathbf{Y}} = \left(\mathbf{I}_n - \frac{1}{n}\mathbf{1}\mathbf{1}^T\right)\mathbf{Y} = \left(\mathbf{I}_n - \frac{1}{n}\mathbf{1}\mathbf{1}^T\right)\mathbf{XG}$$

where $\left(\mathbf{I}_n - \frac{1}{n}\mathbf{1}\mathbf{1}^T\right)$ is the centering matrix. Here $\mathbf{1}$ is a column vector of ones and $\mathbf{I}_n$ is the identity matrix of $n \times n$. We often encounter the quadratic form

$$\frac{1}{n-1}\mathbf{v}^T\mathbf{X}^T\left(\mathbf{I}_n - \frac{1}{n}\mathbf{1}\mathbf{1}^T\right)\mathbf{X}\mathbf{v}.$$

So the same strategy as above can be used, with the f defined as

$$f(\mathbf{x}) = f(\mathbf{X}) = \left\|\left(\mathbf{I}_n - \frac{1}{n}\mathbf{1}\mathbf{1}^T\right)\mathbf{X}\mathbf{v}\right\|_2.$$

Now we use this function as another example to illustrate how to show a function is convex and Lipschitz, which is required to apply the Talagrand's inequality.

Example 4.15.3 ($f(\mathbf{x}) = f(\mathbf{X}) = \left\|\left(\mathbf{I}_n - \frac{1}{n}\mathbf{1}\mathbf{1}^T\right)\mathbf{X}\mathbf{v}\right\|_2$). Convexity is a simple consequence of the fact that norms are convex.

The Lipschitz coefficient is $\|\mathbf{v}\|_2\left\|\mathbf{I}_n - \frac{1}{n}\mathbf{1}\mathbf{1}^T\right\|_2$. The eigenvalues of the matrix $\left(\mathbf{I}_n - \frac{1}{n}\mathbf{1}\mathbf{1}^T\right)$ are $n-1$ ones and one zero, i.e.,

$$\lambda_i\left(\mathbf{I}_n - \frac{1}{n}\mathbf{1}\mathbf{1}^T\right) = 1, i = 1, \ldots, n-1, \text{ and } \lambda_n\left(\mathbf{I}_n - \frac{1}{n}\mathbf{1}\mathbf{1}^T\right) = 0.$$

As a consequence, its operator norm, the largest singular value, is therefore 1, i.e.,

$$\sigma_{\max}\left(\mathbf{I}_n - \frac{1}{n}\mathbf{1}\mathbf{1}^T\right) = \left\|\mathbf{I}_n - \frac{1}{n}\mathbf{1}\mathbf{1}^T\right\|_2 = 1.$$

□

We now apply the above results to justify the centering process in statistics. In (statistical) practice, we almost always use the centered sample covariance matrix

$$\mathbf{S}_p = \frac{1}{n-1}\left(\mathbf{Y} - \bar{\mathbf{Y}}\right)^T\left(\mathbf{Y} - \bar{\mathbf{Y}}\right),$$

in contrast to the (uncenterted) correlation matrix

$$\tilde{\mathbf{S}}_p = \frac{1}{n}\mathbf{Y}^T\mathbf{Y}.$$

Sometimes there are debates as to whether one should use the correlation matrix the data or their covariance matrix. It is therefore important for practitioners to understand the behavior of correlation matrices in high dimensions. The matrix norm (or operator norm) is the largest singular value. In general, the operator norm $\left\|\mathbf{S}_p - \tilde{\mathbf{S}}_p\right\|_{op}$ does not go to zero. So a course bound of the type

$$\left|\lambda_1(\mathbf{S}_p) - \lambda_1\left(\tilde{\mathbf{S}}_p\right)\right| \leqslant \left\|\mathbf{S}_p - \tilde{\mathbf{S}}_p\right\|_{op}$$

is nor enough to determine the behavior of $\lambda_1(\mathbf{S}_p)$ from that of $\lambda_1\left(\tilde{\mathbf{S}}_p\right)$.

Letting the centering matrix $\mathbf{H} = \mathbf{I}_n - \frac{1}{n}\mathbf{1}\mathbf{1}^T$, we see that

$$\mathbf{S}_p - \tilde{\mathbf{S}}_p = \mathbf{H}_n\mathbf{Y}$$

which is a linear matrix problem: a product of a deterministic matrix $\mathbf{H}_n$ and a random matrix $\mathbf{Y}$. Therefore, since the largest singular value $\sigma_{\max}$ is a matrix norm, we have

$$\sigma_{\max}\left(\mathbf{Y} - \bar{\mathbf{Y}}\right)/\sqrt{n} \leqslant \sigma_{\max}\left(\mathbf{H}_n\right)\sigma_{\max}\left(\mathbf{Y}/\sqrt{n}\right) = \sigma_{\max}\left(\mathbf{Y}/\sqrt{n}\right)$$

since $\mathbf{H}_n$ is a symmetric (thus Hermitian) matrix with $(n-1)$ eigenvalues equal to 1 and one eigenvalue equal to 0.

The correlation matrix $\frac{1}{n}\mathbf{Y}^T\mathbf{Y}$ and covariance matrix $\frac{1}{n-1}\left(\mathbf{Y} - \bar{\mathbf{Y}}\right)^T\left(\mathbf{Y} - \bar{\mathbf{Y}}\right)$ have, asymptotically, the same spectral distribution, see [229]. Letting l_1 denote the right endpoint of the support of this limiting distribution (if it exists), we have that

$$\liminf \sigma_{\max}\left(\mathbf{Y} - \bar{\mathbf{Y}}\right)/\sqrt{n-1} \geqslant l_1.$$

So, when $\left\|\frac{1}{n}\mathbf{Y}^T\mathbf{Y}\right\|_{op} \to l_1$, we have

$$\left\|\frac{1}{n-1}\left(\mathbf{Y} - \bar{\mathbf{Y}}\right)^T\left(\mathbf{Y} - \bar{\mathbf{Y}}\right)\right\|_{op} \to l_1.$$

This justifies the assertion that when the norm of a sample covariance matrix (which is not re-centered) whose entries have mean 0 converges to the right endpoint of the support of its limiting spectral distribution, so does the norm of the centered sample covariance matrix.

When dealing with $\mathbf{S}_p$, the mean of the entries of $\mathbf{Y}$ does not matter, so we can assume, without loss of generality, that the mean is zero.

Let us deal with another type of concentration in the quadratic form. Suppose that the random vector $\mathbf{x} \in \mathbb{R}^n$ has the property: For any convex and 1-Lipschitz function (with respect to the Euclidean norm) $F : \mathbb{R}^n \mapsto \mathbb{R}$, we have,

$$\mathbb{P}\left(|F(\mathbf{x}) - m_F| > t\right) \leqslant C\exp\left(-c(n)t^2\right),$$

where m_F is the median of $F(\mathbf{x})$, and C and $c(n)$ are independent of the dimension n. We allow $c(n)$ to be a constant or to go to zero with n like $n^{-\alpha}, 0 \leqslant \alpha \leqslant 1$. Suppose, further, that the random vector has zero mean and variance $\mathbf{\Sigma}$, $\mathbb{E}(\mathbf{x}) = 0, \mathbb{E}(\mathbf{x}\mathbf{x}^*) = \mathbf{\Sigma}$, with the bound of the operator norm (the largest singular value) given by $\|\mathbf{\Sigma}\|_{\text{op}} \leqslant \log(n)$.

Consider a complex deterministic matrix $\mathbf{M}$ such that $\|\mathbf{M}\|_{\text{op}} \leqslant \kappa$, where κ is independent of dimension n, then the quadratic form $\frac{1}{n}\mathbf{x}^*\mathbf{M}\mathbf{x}$ is strongly concentrated around its mean $\frac{1}{n}\text{Tr}\left(\mathbf{M}\mathbf{\Sigma}\right)$.

In particular, if for $\varepsilon > 0$,

$$\log\left\{\mathbb{P}\left(\left|\frac{1}{n}\mathbf{x}^*\mathbf{M}\mathbf{x} - \frac{1}{n}\mathrm{Tr}\left(\mathbf{M}\boldsymbol{\Sigma}\right)\right| > t_n\left(\varepsilon\right)\right)\right\} \asymp -\log\left(n\right)^{1+2\varepsilon}. \tag{4.54}$$

where $\asymp$ denotes the asymptotics for large n. If there is a non-zero expectation $\mathbb{E}\left(\mathbf{x}\right) \neq 0$, then the same results are true when one replaces $\mathbf{x}$ with $\mathbf{x} - \mathbb{E}\left(\mathbf{x}\right)$ everywhere and $\boldsymbol{\Sigma}$ is the covariance of $\mathbf{x}$.

If the bounding parameter κ is allowed to vary with the dimension n, then the same results hold when one replace $t_n\left(\varepsilon\right)$ by $t_n\left(\varepsilon\right)\kappa$, or equivalently, divides $\mathbf{M}$ by ε.

Proof. We closely follow [229], changing to our notation. A complex matrix $\mathbf{M}$ can be decomposed into the real part and imaginary part $\mathbf{M} = \mathbf{M}_r + j\mathbf{M}_i$, where $\mathbf{M}_r$ and $\mathbf{M}_i$ are real matrices. On the other hand, the spectral norm of those matrices is upper bounded by κ. This is of course true for the real part since $\mathbf{M}_r = \left(\mathbf{M} + \mathbf{M}^*\right)/2$.

For a random vector $\mathbf{x}$, the quadratic form $\mathbf{x}^*\mathbf{A}\mathbf{x} = \mathrm{Tr}\left(\mathbf{A}\mathbf{x}\mathbf{x}^*\right)$ is a *linear* functional of the complex matrix $\mathbf{A}$ and the rank-1 random matrix $\mathbf{x}\mathbf{x}^*$. Note that the trace function Tr is a linear functional. Let $\mathbf{A} = \mathbf{M}_r + j\mathbf{M}_i$. It follows that

$$\begin{aligned}\mathbf{x}^*\left(\mathbf{M}_r + j\mathbf{M}_i\right)\mathbf{x} &= \mathrm{Tr}\left(\left(\mathbf{M}_r + j\mathbf{M}_i\right)\mathbf{x}\mathbf{x}^*\right)\\ &= \mathrm{Tr}\left(\mathbf{M}_r\mathbf{x}\mathbf{x}^*\right) + j\mathrm{Tr}\left(\mathbf{M}_r\mathbf{x}\mathbf{x}^*\right)\\ &= \mathbf{x}^*\mathbf{M}_r\mathbf{x} + j\mathbf{x}^*\mathbf{M}_i\mathbf{x}.\end{aligned}$$

In other words, strong concentration for $\mathbf{x}^*\mathbf{M}_r\mathbf{x}$ and $\mathbf{x}^*\mathbf{M}_i\mathbf{x}$ will imply strong concentration for the sum of those two terms (quadratic forms). $\mathbf{x}^*\mathbf{A}\mathbf{x} = \sum\limits_i x_i^* A_{ii} x_i$, which is real for real numbers A_{ii}. So $\mathbf{x}^*\mathbf{M}_r\mathbf{x}$ is real, $\left(\mathbf{x}^*\mathbf{M}_r\mathbf{x}\right)^* = \mathbf{x}^*\mathbf{M}_r\mathbf{x}$ and

$$\mathbf{x}^*\left(\frac{\mathbf{M} + \mathbf{M}^*}{2}\right)\mathbf{x}.$$

Hence, instead of working on $\mathbf{M}_r$, we can work on the symmetrized version.

We now decompose $\left(\mathbf{M}_r + \mathbf{M}_r^T\right)/2$ into the positive part and negative part $\mathbf{M}_+ + \mathbf{M}_-$, where $\mathbf{M}_+$ is positive semidefinite and $-\mathbf{M}_-$ is positive semidefinite (or 0 if $\left(\mathbf{M}_r + \mathbf{M}_r^T\right)/2$ is also positive semidefinite). This is possible since $\left(\mathbf{M}_r + \mathbf{M}_r^T\right)/2$ is real symmetric. We can carry out this decomposition by simply following its spectral decomposition. As pointed out before, both matrices have spectral norm less than κ.

Now we are in a position to consider the functional, which is our main purpose. The map

$$\phi : \mathbf{x} \mapsto \sqrt{\mathbf{x}^*\mathbf{M}_+\mathbf{x}}/\sqrt{n}$$

is K-Lipschitz with the coefficient $K = \sqrt{\kappa/n}$, with respect to the Euclidean norm. The map is also convex, by noting that

$$\phi : \mathbf{x} \mapsto \sqrt{\mathbf{x}^*\mathbf{M}_+\mathbf{x}}/\sqrt{n} = \left\| \mathbf{M}_+^{1/2}\mathbf{x}/\sqrt{n} \right\|_2.$$

All norms are convex. $\|\mathbf{A}\|_2 = \sqrt{\sum_{i,j} |A_{i,j}|^2}$ is the Euclidean norm. □

Theorem 4.15.4 (Lemma A.2 of El Karoui (2010) [230]). *Suppose the random vector* $\mathbf{z}$ *is a vector in* $\mathbb{R}^n$ *with i.i.d. entries of mean 0 and variance* σ^2. *The covariance matrix of* $\mathbf{z}$ *is* $\mathbf{\Sigma}$. *Suppose that the entries of* $\mathbf{z}$ *are bounded by 1. Let* $\mathbf{A}$ *be a symmetric matrix, with the largest singular value* $\sigma_1(\mathbf{A})$. *Set* $C_n = 128\exp(4\pi)\sigma_1(\mathbf{A})/n$. *Then, for all* $t/2 > C_n$, *we have*

$$\mathbb{P}\left(\left|\tfrac{1}{n}\mathbf{z}^T\mathbf{A}\mathbf{z} - \tfrac{1}{n}\sigma^2\,\mathrm{Tr}(\mathbf{A})\right| > t\right) \leqslant 8\exp(4\pi)e^{-n(t/2-C_n)^2/32/\left(1+2\sqrt{\sigma_1(\mathbf{\Sigma})}\right)^2/\sigma_1(\mathbf{A})}$$
$$+8\exp(4\pi)e^{-n/32/\left(1+2\sqrt{\sigma_1(\mathbf{\Sigma})}\right)^2/\sigma_1(\mathbf{A})}.$$

Talagrand's works [148,231] can be viewed as the infinite-dimensional analogue of Bernstein's inequality. Non-asymptotic bounds for statistics [136,232,233] are studied for model selection. For tutorial, we refer to [161, 234]. Bechar [235] establishes a new Bernstein-type inequality which controls quadratic forms of Gaussian random variables.

Theorem 4.15.5 (Bechar [235]). *Let* $\mathbf{a} = (a_i)_{i=1,\ldots,n}$ *and* $\mathbf{b} = (b_i)_{i=1,\ldots,n}$ *be two* n*-dimensional real vectors.* $\mathbf{z} = (z_i)_{i=1,\ldots,n}$ *is an* n*-dimensional standard Gaussian random vector, i.e.,* $z_i, i = 1,\ldots,n$ *are i.i.d. zero-mean Gaussian random variables with standard deviation 1. Set* $a_+ = \sup\left\{\sup_{i=1,\ldots,n}\{a_i\}, 0\right\}$, *and* $a_- = \sup\left\{\sup_{i=1,\ldots,n}\{-a_i\}, 0\right\}$. *Then the two concentration inequalities hold true for all* $t > 0$,

$$\mathbb{P}\left(\sum_{i=1}^n (a_i z_i^2 + b_i z_i) \geqslant \sum_{i=1}^n a_i^2 + 2\sqrt{\sum_{i=1}^n \left(a_i^2 + \tfrac{1}{2}b_i^2\right)}\cdot\sqrt{t} + 2a_+ t\right) \leqslant e^{-t}$$

$$\mathbb{P}\left(\sum_{i=1}^n (a_i z_i^2 + b_i z_i) \leqslant \sum_{i=1}^n a_i^2 - 2\sqrt{\sum_{i=1}^n \left(a_i^2 + \tfrac{1}{2}b_i^2\right)}\cdot\sqrt{t} - 2a_- t\right) \leqslant e^{-t}.$$

Theorem 4.15.6 (Real-valued quadratic forms—Bechar [235]). *Consider the random expression* $\mathbf{z}^T\mathbf{A}\mathbf{z} + \mathbf{b}^T\mathbf{z}$, *where* $\mathbf{A}$ *is* $n \times n$ *real square matrix,* $\mathbf{b}$ *is an* n*-dimensional real vector, and* $\mathbf{z} = (z_i)_{i=1,\ldots,n}$ *is an* n*-dimensional standard Gaussian random vector, i.e.,* $z_i, i = 1,\ldots,n$ *are i.i.d. zero-mean Gaussian* *random variables with standard deviation 1. Let us denote by* $\lambda_i, i = 1, \ldots, n$ *the eigenvalues of the symmetric matrix* $\frac{1}{2}\left(\mathbf{A} + \mathbf{A}^T\right)$, *and let us put* $\lambda_+ = \sup\left\{\sup_{i=1,\ldots,n}\{\lambda_i\}, 0\right\}$, *and* $\lambda_- = \sup\left\{\sup_{i=1,\ldots,n}\{-\lambda_i\}, 0\right\}$. *Then, the following two concentration results hold true for all* $t > 0$

$$\mathbb{P}\left(\mathbf{z}^T\mathbf{A}\mathbf{z} + \mathbf{b}^T\mathbf{z} \geqslant \mathrm{Tr}\left(\mathbf{A}\right) + 2\sqrt{\frac{1}{4}\|\mathbf{A} + \mathbf{A}^T\|^2 + \frac{1}{2}\|\mathbf{b}\|^2} \cdot \sqrt{t} + 2\lambda_+ t\right) \leqslant e^{-t}$$

$$\mathbb{P}\left(\mathbf{z}^T\mathbf{A}\mathbf{z} + \mathbf{b}^T\mathbf{z} \leqslant \mathrm{Tr}\left(\mathbf{A}\right) - 2\sqrt{\frac{1}{4}\|\mathbf{A} + \mathbf{A}^T\|^2 + \frac{1}{2}\|\mathbf{b}\|^2} \cdot \sqrt{t} - 2\lambda_- t\right) \leqslant e^{-t}. \tag{4.55}$$

Theorem 4.15.6 for quadratic forms of real-valued Gaussian random variables is extended to the complex form in the following theorem for Bernstein's inequality. Let $\mathbf{I}_n$ be the identity matrix of $n \times n$.

Theorem 4.15.7 (Complex-valued quadratic forms—Wang, So, Chang, Ma and Chi [236]). *Let* $\mathbf{z}$ *be a complex Gaussian with zero mean and covariance matrix* $\mathbf{I_n}$, *i.e.,* $\mathbf{z}_i \sim \mathcal{CN}\left(\mathbf{0}, \mathbf{I}_n\right)$, $\mathbf{Q}$ *is an* $n \times n$ *Hermitian matrix, and* $\mathbf{y} \in \mathbb{C}^n$. *Then, for any* $t > 0$, *we have*

$$\mathbb{P}\left(\mathbf{z}^H\mathbf{Q}\mathbf{z} + 2\,\mathrm{Re}\left(\mathbf{z}^H\mathbf{y}\right) \geqslant \mathrm{Tr}\left(\mathbf{Q}\right) - \sqrt{2t}\sqrt{\|\mathbf{Q}\|_F + 2\|\mathbf{y}\|^2} - t\lambda_+\left(\mathbf{Q}\right)\right) \geqslant 1 - e^{-t}, \tag{4.56}$$

with $\lambda_+\left(\mathbf{Q}\right) = \max\left\{\lambda_{\max}\left(-\mathbf{Q}\right), 0\right\}$.

Equation (4.56) is used to bound the probability that the quadratic form $\mathbf{z}^H\mathbf{Q}\mathbf{z} + 2\,\mathrm{Re}\left(\mathbf{z}^H\mathbf{y}\right)$ of complex Gaussian random variables deviates from its mean $\mathrm{Tr}\left(\mathbf{Q}\right)$.

Theorem 4.15.8 (Lopes, Jacob, Wainwright [237]). *Let* $\mathbf{A} \in \mathbb{R}^{n \times n}$ *be a positive semidefinite matrix* $\|\mathbf{A}\|_{op} > 0$, *and let Gaussian vectors* $\mathbf{z}$ *be* $\mathbf{z} \sim \mathcal{N}\left(0, \mathbf{I}_{n\times n}\right)$. *Then, for any* $t > 0$,

$$\mathbb{P}\left(\frac{\mathbf{z}^T\mathbf{A}\mathbf{z}}{\mathrm{Tr}\left(\mathbf{A}\right)} \geqslant \left(1 + t\sqrt{\frac{\|\mathbf{A}\|_{\mathrm{op}}}{\mathrm{Tr}\left(\mathbf{A}\right)}}\right)^2\right) \leqslant e^{-t^2/2}, \tag{4.57}$$

and for any $t \in \left(0, \sqrt{\frac{\mathrm{Tr}(\mathbf{A})}{\|\mathbf{A}\|_{op}}} - 1\right)$, *we have*

$$\mathbb{P}\left(\frac{\mathbf{z}^T\mathbf{A}\mathbf{z}}{\mathrm{Tr}\left(\mathbf{A}\right)} \leqslant \left(\sqrt{1 - \frac{\|\mathbf{A}\|_{\mathrm{op}}}{\mathrm{Tr}\left(\mathbf{A}\right)}} - t\sqrt{\frac{\|\mathbf{A}\|_{\mathrm{op}}}{\mathrm{Tr}\left(\mathbf{A}\right)}}\right)^2\right) \leqslant e^{-t^2/2}. \tag{4.58}$$

Here $||\mathbf{X}||_{\text{op}}$ denote the operator or matrix norm (largest singular value) of matrix $\mathbf{X}$. It is obvious that $\frac{\|\mathbf{A}\|_{\text{op}}}{\text{Tr}(\mathbf{A})}$ is less than or equal to 1. The error terms involve the operator norm as opposed to the Frobenius norm and hence are of independent interest.

Proof. We follow [237] for a proof. First $f(\mathbf{z}) = \sqrt{\mathbf{z}^T\mathbf{A}\mathbf{z}} = \left\|\mathbf{A}^{1/2}\mathbf{z}\right\|_2$ has Lipschitz constant $\sqrt{\|\mathbf{A}\|_{\text{op}}}$ with respect to the Euclidean norm on $\mathbb{R}^n$. By the Cirel'son-Ibragimov-Sudakov inequality for Lipschitz functions of Gaussian vectors [161], it follows that for any $s > 0$,

$$\mathbb{P}\left(f(\mathbf{z}) \leqslant \mathbb{E}\left[f(\mathbf{z})\right] - s\right) \leqslant \exp\left(-\frac{1}{2\|\mathbf{A}\|_{\text{op}}}s^2\right). \tag{4.59}$$

From the Poincare inequality for Gaussian measures [238], the variance of $f(\mathbf{z})$ is bounded above as

$$\text{var}\left[f(\mathbf{z})\right] \leqslant \|\mathbf{A}\|_{\text{op}}.$$

Since $\mathbb{E}\left[f(\mathbf{z})^2\right] = \text{Tr}\left(\mathbf{A}\right)$, the expectation of $f\left(\mathbf{z}\right)$ is lower bounded as

$$\mathbb{E}\left[f\left(\mathbf{z}\right)\right] \geqslant \sqrt{\text{Tr}\left(\mathbf{A}\right) - \|\mathbf{A}\|_{\text{op}}}.$$

Inserting this lower bound into the concentration inequality of (4.59) gives

$$\mathbb{P}\left(f(\mathbf{z}) \leqslant \sqrt{\text{Tr}\left(\mathbf{A}\right) - \|\mathbf{A}\|_{\text{op}}} - s\right) \leqslant \exp\left(-\frac{1}{2\|\mathbf{A}\|_{\text{op}}}s^2\right).$$

Let us turn to the bound (4.57). We start with the upper-tail version of (4.59), that is $\mathbb{P}\left(f(\mathbf{z}) \geqslant \mathbb{E}\left[f\left(\mathbf{z}\right)\right] - s\right) \leqslant \exp\left(-\frac{1}{2\|\mathbf{A}\|_{\text{op}}}s^2\right)$ for $s > 0$. By Jensen's inequality, it follows that

$$\mathbb{E}\left[f\left(\mathbf{z}\right)\right] = \mathbb{E}\left[\sqrt{\mathbf{z}^T\mathbf{A}\mathbf{z}}\right] \leqslant \sqrt{\mathbb{E}\left[\mathbf{z}^T\mathbf{A}\mathbf{z}\right]} = \sqrt{\text{Tr}\left(\mathbf{A}\right)},$$

from which we have $\mathbb{P}\left(f(\mathbf{z}) \geqslant \text{Tr}\left(\mathbf{A}\right) + s\right) \leqslant \exp\left(-\frac{1}{2\|\mathbf{A}\|_{\text{op}}}s^2\right)$, and setting $s^2 = t^2\|\mathbf{A}\|_{\text{op}}$ for $t > 0$ gives the result of (4.57).

□

In Sect. 10.18, we apply Theorem 4.15.8 for the two-sample test in high dimensions. We present small ball probability estimates for linear and quadratic functions of Gaussian random variables. C is a universal constant. The Frobenius norm is defined as $\|\mathbf{A}\|_F = \sqrt{\sum_{i,j} A_{i,j}^2} = \sqrt{\operatorname{Tr}(\mathbf{A}^2)}$. For a vector $\mathbf{x}$, $\|\mathbf{x}\|$ denotes the standard Euclidean norm—ℓ_2-norm.

Theorem 4.15.9 (Concentration of Gaussian random variables [239]). *Let* $a, b \in \mathbb{R}$. *Assume that* $\max\{|a|, |b|\} \geqslant \alpha > 0$. *Let* $X \sim \mathcal{N}(0,1)$. *Then, for* $t > 0$

$$\mathbb{P}(|a + bX| \leqslant t) \leqslant Ct/\alpha. \tag{4.60}$$

Theorem 4.15.10 (Concentration of Gaussian random vector [239]). *Let* $\mathbf{a} \in \mathbb{R}^n$, $\mathbf{A} \in \mathbb{R}^{n \times n}$. *Assume that*

$$\max\{\|\mathbf{a}\|_2, \|\mathbf{A}\|_F\} \geqslant \alpha > 0.$$

Let $\mathbf{z} \sim \mathcal{N}(0, \mathbf{I}_n)$. *Then, for* $t > 0$

$$\mathbb{P}(\|\mathbf{a} + \mathbf{A}\mathbf{z}\|_2 \leqslant t) \leqslant Ct\sqrt{n}/\alpha. \tag{4.61}$$

Theorem 4.15.11 ([239]). *Let* $\mathbf{a} \in \mathbb{R}^n$, $\mathbf{A} \in \mathbb{R}^{n \times n}$. *Let* $\mathbf{z} \sim \mathcal{N}(\boldsymbol{\mu}, \sigma^2 \mathbf{I}_n)$ *for some* $\boldsymbol{\mu} \in \mathbb{R}^n$. *Then, for* $t > 0$

$$\mathbb{P}\left(\|\mathbf{a} + \mathbf{A}\mathbf{z}\|_2 \leqslant t\sigma^2 \|\mathbf{A}\|_F\right) \leqslant Ct\sqrt{n}.$$

Theorem 4.15.12 (Concentration of quadrics in Gaussian random variables [239]). *Let* $a \in \mathbb{R}$, $\mathbf{b} \in \mathbb{R}^n$ *and let* $\mathbf{A} \in \mathbb{R}^{n \times n}$ *be a symmetric matrix. Assume that* $|a| \geqslant \alpha > 0$. *Let* $\mathbf{z} \sim \mathcal{N}(0, \mathbf{I}_n)$. *Then, for* $t > 0$

$$\mathbb{P}\left(\left|a + \mathbf{z}^T\mathbf{b} + \mathbf{z}^T\mathbf{A}\mathbf{z}\right| \leqslant t\right) \leqslant Ct^{1/6}\sqrt{n}/\alpha. \tag{4.62}$$

Example 4.15.13 (Hypothesis testing).

$$\begin{aligned} \mathcal{H}_0 &: \mathbf{y} = \mathbf{x} \\ \mathcal{H}_1 &: \mathbf{y} = \mathbf{x} + \mathbf{z} \end{aligned}$$

where $\mathbf{x}$ is independent of $\mathbf{z}$. See also Example 7.2.13. □

4.16 Distance Between a Random Vector and a Subspace

Let $\mathbf{P} = (p_{ij})_{1 \leqslant i \leqslant j \leqslant n}$ be the $n \times n$ orthogonal projection matrix on to $\mathbf{V}$. Obviously, we have

$$\mathrm{Tr}\left(\mathbf{P}^2\right) = \mathrm{Tr}\left(\mathbf{P}\right) = \sum_{i=1}^{n} p_{ii} = d$$

and $|p_{ij}| \leqslant 1$. Furthermore, the distance between a random vector $\mathbf{x}$ and a subspace $\mathbf{V}$ is defined by the orthogonal projection matrix $\mathbf{P}$ in a quadratic form

$$\begin{aligned} \mathrm{dist}\left(\mathbf{x}, \mathbf{V}\right)^2 &= \left(\mathbf{Px}\right)^* \left(\mathbf{Px}\right) = \mathbf{x}^*\mathbf{P}^*\mathbf{Px} = \mathbf{x}^*\mathbf{PPx} = \mathbf{x}^*\mathbf{P}^2\mathbf{x} = \mathbf{x}^*\mathbf{Px} \\ &= \sum_{1\leqslant i \neq j \leqslant n} p_{ij}\xi_i\xi_j^* = \sum_{i=1}^{n} p_{ii}|\xi_i|^2 + \sum_{1\leqslant i \neq j \leqslant n} p_{ij}\xi_i\xi_j^*. \end{aligned} \tag{4.63}$$

For instance, for a spectral decomposition of $\mathbf{A}$, $\mathbf{A} = \sum_{i=1}^{n} \lambda_i \mathbf{u}_i \mathbf{u}_i^H$, we can define $\mathbf{P} = \sum_{i=1}^{d} \mathbf{u}_i \mathbf{u}_i^H$, where $d \leq n$. For a random matrix $\mathbf{A}$, the projection matrix $\mathbf{P}$ is also a random matrix. What is the surprising is that the distance between a random vector and a subspace—the orthogonal projection of a random vector onto a large space—is strongly concentrated. This tool has a geometric flavor.

The distance $\mathrm{dist}\left(\mathbf{x}, \mathbf{V}\right)$ is a (scalar-valued) random variable. It is easy to show that $\mathbb{E}\,\mathrm{dist}\left(\mathbf{x}, \mathbf{V}\right)^2 = d$ so that it is indeed natural to expect that with high probability $\mathrm{dist}\left(\mathbf{x}, \mathbf{V}\right)$ is around $\sqrt{d}$.

We use a complex version of Talagrand's inequality, obtained by slightly modifying the proof [141].

Theorem 4.16.1. *Let* $\mathbf{D}$ *be the unit disk* $\{z \in \mathbb{C} : |z| \leqslant 1\}$. *For every product probability* $\mathbb{P}$ *on* $\mathbf{D}^n$, *every convex 1-Lipschitz function* $f : \mathbb{C}^n \to \mathbb{R}$ *and every* $t \geq 0$,

$$\mathbb{P}\left(|f - \mathbb{M}\left(f\right)| \geqslant t\right) \leqslant 4e^{-t^2/16},$$

where $\mathbb{M}(f)$ *denotes the median of* f.

The result still holds for the space $\mathbf{D}_1 \times \ldots \times \mathbf{D}_n$, where $\mathbf{D}_i$ are complex regions with diameter 2. An easy change of variable reveals the following generation of this inequality. If we consider the probability for a dilate $K \cdot \mathbf{D}^n$ of the unit disk for some $K > 0$, rather than $\mathbf{D}^n$ itself, then for every $t \geq 0$, we have instead

$$\mathbb{P}\left(|f - \mathbb{M}\left(f\right)| \geqslant t\right) \leqslant 4e^{-t^2/16K^2}. \tag{4.64}$$

Theorem 4.16.1 shows concentration around the median. In applications, it is more useful to have concentration around the mean. This can be done following the well known lemma [141,167], which shows that concentration around the median implies that the mean and the median are close.

Lemma 4.16.2. *Let* $\mathbf{X}$ *be a random variable such that for any* $t \geq 0$,

$$\mathbb{P}\left(|X - \mathbb{M}(X)| \geqslant t\right) \leqslant 4e^{-t^2}.$$

Then

$$|\mathbb{E}(X) - \mathbb{M}(X)| \leqslant 100.$$

The bound 100 is ad hoc and can be replaced by a much smaller constant.

Proof. Set $M = \mathbb{M}(X)$ and let $F(x)$ be the distribution function of X. We have

$$\mathbb{E}(X) = \sum_{i=1} \int_{M-i}^{M-i+1} x\partial F(x) \leqslant M + 4\sum_i |i| e^{-i^2} \leqslant M + 100.$$

We can prove the lower bound similarly. □

Now we are in a position to state a lemma and present its proof.

Theorem 4.16.3 (Projection Lemma, Lemma 68 of [240]). *Let* $\mathbf{x} = (\xi_1, \ldots, \xi_n) \in \mathbb{C}^n$ *be a random vector whose entries are independent with mean zero, variance 1, and are bounded in magnitude by K almost surely for some* $K \geqslant 10\left(\mathbb{E}|\xi|^4 + 1\right)$. *Let $\mathbf{V}$ be a subspace of dimension d and* $\text{dist}(\mathbf{x}, \mathbf{V})$ *be the orthogonal projection on $\mathbf{V}$. Then*

$$\mathbb{P}\left(\left|\text{dist}(\mathbf{x}, \mathbf{V}) - \sqrt{d}\right| > t\right) \leqslant 10\exp\left(-\frac{t^2}{10K^2}\right).$$

In particular, one has

$$\text{dist}(\mathbf{x}, \mathbf{V}) = \sqrt{d} + O(K\log n)$$

with overwhelming probability.

Proof. We follow [240, Appendix B] closely. See also [241, Appendix E]. This proof is a simple generalization of Tao and Vu in [242].

The standard approach of using Talagrand's inequality is to study the functional property of $f = \text{dist}(\mathbf{x}, \mathbf{V})$ as a function of vector $\mathbf{x} \in \mathbb{C}^n$. The functional map

$$\mathbf{x} \mapsto |\text{dist}(\mathbf{x}, \mathbf{V})|$$

is clearly convex and 1-Lipschitz. Applying (4.64), we have that

$$\mathbb{P}\left(|\text{dist}(\mathbf{x}, \mathbf{V}) - \mathbb{M}(\text{dist}(\mathbf{x}, \mathbf{V}))| \geqslant t\right) \leqslant 4e^{-t^2/16K^2}$$

for any $t > 0$.

To conclude the proof, it is sufficient to show the following

$$\left|\mathbb{M}(X) - \sqrt{d}\right| \leqslant 2K.$$

Consider the event $\mathcal{E}_+$ that $|\text{dist}(\mathbf{x}, \mathbf{V})| \geqslant \sqrt{d} + 2K$, which implies that

$$|\text{dist}(\mathbf{x}, \mathbf{V})|^2 \geqslant d + 4K\sqrt{d} + 4K^2.$$

From the definition of the orthogonal projection (4.63), we have

$$\mathbb{P}(\mathcal{E}_+) \leqslant \mathbb{P}\left(\sum_{i=1}^{n} p_{ii}|\xi_i|^2 \geqslant d + 2K\sqrt{d}\right) + \mathbb{P}\left(\sum_{1\leqslant i\neq j\leqslant n} p_{ij}\xi_i\xi_j^* \geqslant 2K\sqrt{d}\right).$$

Set $S_1 = \sum_{i=1}^{n} p_{ii}\left(|\xi_i|^2 - 1\right)$. Then it follows from Chebyshev's inequality that

$$\mathbb{P}\left(\sum_{i=1}^{n} p_{ii}|\xi_i|^2 \geqslant d + 2K\sqrt{d}\right) \leqslant \mathbb{P}\left(|S_1| \geqslant 2K\sqrt{d}\right) \leqslant \frac{\mathbb{E}\left(|S_1|^2\right)}{4dK^2}.$$

On the other hand, by the assumption of K,

$$\mathbb{E}\left(|S_1|^2\right) = \sum_{i=1}^{n} p_{ii}^2\mathbb{E}\left(|\xi_i|^2 - 1\right)^2 = \sum_{i=1}^{n} p_{ii}^2\left(\mathbb{E}\,|\xi_i|^4 - 1\right) \leqslant \sum_{i=1}^{n} p_{ii}^2 K = dK.$$

Thus we have

$$\mathbb{P}\left(|S_1| \geqslant 2K\sqrt{d}\right) \leqslant \frac{\mathbb{E}\left(|S_1|^2\right)}{4dK^2} \leqslant \frac{1}{K} \leqslant \frac{1}{10}.$$

Similarly, set $S_2 = \left|\sum_{1\leqslant i\neq j\leqslant n} p_{ij}\xi_i\xi_j^*\right|$. Then we have $\mathbb{E}\left(S_2^2\right) = \sum_{i\neq j} |p_{ij}|^2 \leqslant d$. Again, using Chebyshev's inequality, we have

$$\mathbb{P}\left(|S_2| \geqslant 2K\sqrt{d}\right) \leqslant \frac{d}{4dK^2} \leqslant \frac{1}{K} \leqslant \frac{1}{10}.$$

Combining the results, it follows that $\mathbb{P}(\mathcal{E}_+) \leqslant \frac{1}{5}$, and so $\mathbb{M}(\text{dist}(\mathbf{x}, \mathbf{V})) \leqslant \sqrt{d} + 2K$.

To prove the lower bound, let $\mathbb{E}_-$ be the event that $\text{dist}(\mathbf{x}, \mathbf{V}) \leqslant \sqrt{d} - 2K$ which implies that $\text{dist}(\mathbf{x}, \mathbf{V})^2 \leqslant d - 4K\sqrt{d} + K^2$. We thus have

$$\mathbb{P}(\mathcal{E}_+) \leqslant \mathbb{P}\left(\text{dist}(\mathbf{x}, \mathbf{V})^2 \leqslant d - 2K\sqrt{d}\right) \leqslant \mathbb{P}\left(S_1 \leqslant d - K\sqrt{d}\right) + \mathbb{P}\left(S_1 \geqslant K\sqrt{d}\right).$$

Both terms on the right-hand side can be bounded by $\frac{1}{5}$ by the same arguments as above. The proof is complete. □

Example. Similarity Based Hypothesis Detection

A subspace $\mathbf{V}$ is the subspace spanned by the first k eigenvectors (corresponding to the largest k eigenvalues).

4.17 Concentration of Random Matrices in the Stieltjes Transform Domain

As the Fourier transform is the tool of choice for a linear time-invariant system, the Stieltjes transform is the fundamental tool for studying the random matrix. For a random matrix $\mathbf{A}$ of $n \times n$, we define the Stieltjes transform as

$$m_{\mathbf{A}}(z) = \frac{1}{n}\operatorname{Tr}(\mathbf{A} - z\mathbf{I})^{-1}, \tag{4.65}$$

where $\mathbf{I}$ is the identity matrix of $n \times n$. Here z is a complex variable.

Consider $\mathbf{x}_i, i = 1, \ldots, N$ *independent* random (column) vectors in $\mathbb{R}^n$. $\mathbf{x}_i\mathbf{x}_i^T$ is a rank-one matrix of $n \times n$. We often consider the sample covariance matrix of $n \times n$ which is expressed as the sum of N rank-one matrices

$$\mathbf{S} = \sum_{i=1}^{N} \mathbf{x}_i\mathbf{x}_i^T = \mathbf{X}\mathbf{X}^T,$$

where the data matrix $\mathbf{X}$ of $N \times n$ is defined as

$$\mathbf{X} = \begin{bmatrix} \mathbf{x}_1 \\ \vdots \\ \mathbf{x}_N \end{bmatrix}.$$

Similar to the case of the Fourier transform, it is more convenient to study the Stieltjes transform of the sample covariance matrix $\mathbf{S}$ defined as

$$m_{\mathbf{S}}(z) = \frac{1}{n}\operatorname{Tr}(\mathbf{S} - z\mathbf{I})^{-1}.$$

It is remarkable that $m_{\mathbf{S}}(z)$ is strongly concentrated, as first shown in [229]. Let $\operatorname{Im}[z] = v$, we have [229]

$$\mathbb{P}\left(|m_{\mathbf{S}}(z) - \mathbb{E}m_{\mathbf{S}}(z)| > t\right) \leqslant 4\exp\left(-t^2n^2v^2/(16N)\right). \tag{4.66}$$

Note that (4.66) makes no assumption whatsoever about the structure of the vectors $\{\mathbf{x}_i\}_{i=1}^n$, other than the fact that they are *independent*.

Proof. We closely follow El Karou [229] for a proof. They use a sum of martingale difference, followed by Azuma's inequality[141, Lemma 4.1]. We define $\mathbf{S}_k = \mathbf{S} - \mathbf{x}_k\mathbf{x}_k^T$. Let $\mathcal{F}_i$ denote the filtration generated by random vectors $\{\mathbf{x}_l\}_{l=1}^i$. The first classical step is (from Bai [198, p. 649]) to express the random variable of interest as a sum of martingale differences:

$$m_{\mathbf{S}}(z) - \mathbb{E}m_{\mathbf{S}}(z) = \sum_{k=1}^n \mathbb{E}\left(m_{\mathbf{S}}(z)\,|\mathcal{F}_k\right) - \mathbb{E}\left(m_{\mathbf{S}}(z)\,|\mathcal{F}_{k-1}\right).$$

Note that

$$\mathbb{E}\left(\mathrm{Tr}\,(\mathbf{S}_k - z\mathbf{I})^{-1}\,|\mathcal{F}_k\right) = \mathbb{E}\left(\mathrm{Tr}\,(\mathbf{S}_k - z\mathbf{I})^{-1}\,|\mathcal{F}_{k-1}\right).$$

So we have that

$$\begin{aligned}
&\left|\mathbb{E}\left(m_{\mathbf{S}}(z)\,|\mathcal{F}_k\right) - \mathbb{E}\left(m_{\mathbf{S}}(z)\,|\mathcal{F}_{k-1}\right)\right| \\
&= \left|\mathbb{E}\left(m_{\mathbf{S}}(z)\,|\mathcal{F}_k\right) - \mathbb{E}\left(\frac{1}{n}\mathrm{Tr}\,(\mathbf{S}_k - z\mathbf{I})^{-1}\,|\mathcal{F}_k\right) + \mathbb{E}\left(\frac{1}{n}\mathrm{Tr}\,(\mathbf{S}_k - z\mathbf{I})^{-1}\,|\mathcal{F}_{k-1}\right)\right. \\
&\quad \left. - \mathbb{E}\left(m_{\mathbf{S}}(z)\,|\mathcal{F}_{k-1}\right)\right| \\
&\leqslant \left|\mathbb{E}\left(m_{\mathbf{S}}(z)\,|\mathcal{F}_k\right) - \mathbb{E}\left(\frac{1}{n}\mathrm{Tr}\,(\mathbf{S}_k - z\mathbf{I})^{-1}\,|\mathcal{F}_k\right)\right| + \left|\mathbb{E}\left(\frac{1}{n}\mathrm{Tr}\,(\mathbf{S}_k - z\mathbf{I})^{-1}\,|\mathcal{F}_{k-1}\right)\right. \\
&\quad \left. - \mathbb{E}\left(m_{\mathbf{S}}(z)\,|\mathcal{F}_{k-1}\right)\right| \\
&\leqslant \frac{2}{nv}.
\end{aligned}$$

The last inequality follows[243, Lemma 2.6]. As a result, the desired random variable $m_{\mathbf{S}}(z) - \mathbb{E}m_{\mathbf{S}}(z)$ is a sum of bounded martingale differences. The same would be true for both real and imaginary parts. For both of them, we apply Azuma's inequality[141, Lemma 4.1] to obtain that

$$\mathbb{P}\left(\left|\mathrm{Re}\left(m_{\mathbf{S}}(z) - \mathbb{E}m_{\mathbf{S}}(z)\right)\right| > t\right) \leqslant 2\exp\left(-t^2n^2v^2/\left(8N\right)\right),$$

and similarly for its imaginary part. We thus conclude that

$$\begin{aligned}
\mathbb{P}\left(\left|m_{\mathbf{S}}(z) - \mathbb{E}m_{\mathbf{S}}(z)\right| > t\right) &\leqslant \mathbb{P}\left(\left|\mathrm{Re}\left(m_{\mathbf{S}}(z) - \mathbb{E}m_{\mathbf{S}}(z)\right)\right| > t/\sqrt{2}\right) \\
&\quad + \mathbb{P}\left(\left|\mathrm{Im}\left(m_{\mathbf{S}}(z) - \mathbb{E}m_{\mathbf{S}}(z)\right)\right| > t/\sqrt{2}\right) \\
&\leqslant 4\exp\left(-t^2n^2v^2/\left(16N\right)\right).
\end{aligned}$$

□

The decay rate given by Azuma's inequality does not match the rate that appears in results concerning concentration behavior of linear spectral statistics (see Sect. 4.14). The rate given by Azuma's inequality is n, rather than $\sqrt{n}$. This decay rate is very important in practice. It is explicitly related to the detection probability and the false alarm, in a hypothesis testing problem.

The results using Azuma's inequality can handle many situations that are not covered by the current results on linear spectral statistics in Sect. 4.14. The "correct" rate can be recovered using ideas similar to Sect. 4.14. As a matter of fact, if we consider the Stieltjes transform of the measure that puts mass $1/n$ at each singular values of the sample covariance matrix $\mathbf{M} = \mathbf{X}^*\mathbf{X}/N$, it is an easy exercise to show that this function of $\mathbf{X}$ is K-Lipschitz with the Lipschitz coefficient $K = 1/\left(\sqrt{nN}v^2\right)$, with respect to the Euclidean norm (or Frobenius or Hilbert-Schmidt) norm.

Example 4.17.1 (Independent Random Vectors). Consider the hypothesis testing problem

$$\mathcal{H}_0 : \mathbf{y}_i = \mathbf{w}_i, i = 1, .., N$$
$$\mathcal{H}_1 : \mathbf{y}_i = \mathbf{x}_i + \mathbf{w}_i, i = 1, .., N$$

where N independent vectors are considered. Here w_i is a random noise vector and s_i is a random signal vector. Considering the sample covariance matrices, we have

$$\mathcal{H}_0 : \mathbf{S} = \sum_{i=1}^{N} \mathbf{y}_i \mathbf{y}_i^T = \sum_{i=1}^{N} \mathbf{w}_i \mathbf{w}_i^T = \mathbf{W}\mathbf{W}^T$$
$$\mathcal{H}_1 : \mathbf{S} = \sum_{i=1}^{N} \mathbf{y}_i \mathbf{y}_i^T = \sum_{i=1}^{N} (\mathbf{x}_i + \mathbf{w}_i)(\mathbf{x}_i + \mathbf{w}_i)^T = (\mathbf{X} + \mathbf{W})(\mathbf{X} + \mathbf{W})^T.$$

Taking the Stieltjes transform leads to

$$\mathcal{H}_0 : m_{\mathbf{S}}(z) = \frac{1}{n} \operatorname{Tr} \left(\mathbf{W}\mathbf{W}^T - z\mathbf{I}\right)^{-1}$$
$$\mathcal{H}_1 : m_{\mathbf{S}}(z) = \frac{1}{n} \operatorname{Tr} \left((\mathbf{X} + \mathbf{W})(\mathbf{X} + \mathbf{W})^T - z\mathbf{I}\right)^{-1}.$$

The Stieltjes transform is strongly concentrated, so

$$\mathcal{H}_0 : \mathbb{P}\left(\left|m_{\mathbf{S}}\left(z\right)-\mathbb{E}m_{\mathbf{S}}\left(z\right)\right|>t\right)\leqslant 4\exp\left(-t^2n^2v^2/\left(16N\right)\right),$$
$$\text{where } \mathbb{E}m_{\mathbf{S}}\left(z\right)=\frac{1}{n}\mathbb{E}\operatorname{Tr}\left(\mathbf{W}\mathbf{W}^T-z\mathbf{I}\right)^{-1}=\frac{1}{n}\operatorname{Tr}\mathbb{E}\left(\mathbf{W}\mathbf{W}^T-z\mathbf{I}\right)^{-1},$$
$$\mathcal{H}_1 : \mathbb{P}\left(\left|m_{\mathbf{S}}\left(z\right)-\mathbb{E}m_{\mathbf{S}}\left(z\right)\right|>t\right)\leqslant 4\exp\left(-t^2n^2v^2/\left(16N\right)\right),$$
$$\text{where } \mathbb{E}m_{\mathbf{S}}\left(z\right)=\frac{1}{n}\operatorname{Tr}\mathbb{E}\left(\left(\mathbf{X}+\mathbf{W}\right)\left(\mathbf{X}+\mathbf{W}\right)^T-z\mathbf{I}\right)^{-1}.$$

Note, both the expectation and the trace are linear functions so they commute—their order can be exchanged. Also note that

$$\left(\mathbf{A}+\mathbf{B}\right)^{-1}\neq\mathbf{A}^{-1}+\mathbf{B}^{-1}.$$

In fact

$$\mathbf{A}^{-1}-\mathbf{B}^{-1}=\mathbf{A}^{-1}\left(\mathbf{B}-\mathbf{A}\right)\mathbf{B}^{-1}.$$

Then we have

$$\left(\mathbf{W}\mathbf{W}^T-z\mathbf{I}\right)^{-1}-\left(\left(\mathbf{X}+\mathbf{W}\right)\left(\mathbf{X}+\mathbf{W}\right)^T-z\mathbf{I}\right)^{-1}$$
$$=\left(\mathbf{W}\mathbf{W}^T-z\mathbf{I}\right)^{-1}\left(\mathbf{X}\mathbf{X}^T+\mathbf{X}\mathbf{W}^T+\mathbf{W}\mathbf{X}^T\right)\left(\left(\mathbf{X}+\mathbf{W}\right)\left(\mathbf{X}+\mathbf{W}\right)^T-z\mathbf{I}\right)^{-1}.$$

Two relevant Taylor series are

$$\log\left(\mathbf{I}+\mathbf{A}\right)=\mathbf{A}-\frac{1}{2}\mathbf{A}^2+\frac{1}{3}\mathbf{A}^3-\frac{1}{4}\mathbf{A}^4+\cdots,\rho\left(\mathbf{A}\right)<1,$$
$$\left(\mathbf{I}-\mathbf{A}\right)^{-1}=\mathbf{A}-\frac{1}{2}\mathbf{A}^2+\frac{1}{3}\mathbf{A}^3-\frac{1}{4}\mathbf{A}^4+\cdots,\rho\left(\mathbf{A}\right)<1,$$

where $\rho\left(\mathbf{A}\right)$ is the (spectral) radius of convergence[20, p. 77]. The series for $\left(\mathbf{I}-\mathbf{A}\right)^{-1}$ is also called Neumann series [23, p. 7]. □

4.18 Concentration of von Neumann Entropy Functions

The von Neumann entropy [244] is a generalization of the classical entropy (Shannon entropy) to the field of quantum mechanics. The von Neumann entropy is one of the cornerstones of quantum information theory. It plays an essential role in the expressions for the best achievable rates of virtually every coding theorem. In particular, when proving the optimality of these expressions, it is the inequalities governing the relative magnitudes of the entropies of different subsystems which are important [245]. There are essentially two such inequalities known, the so-called basic inequalities known as *strong subadditivity* and *weak monotonicity*. To be precise, we are interested in only *linear* inequalities involving the entropies of various reduced states of a multiparty quantum state. We will demonstrate how the concentration inequalities are established for these basic inequalities. One motivation is for quantum information processing.

For any quantum state described by a Hermitian positive semi-definite matrix ρ, the von Neumann entropy of ρ is defined as

$$S(\rho) = -\operatorname{Tr}(\rho \log \rho) \tag{4.67}$$

A natural question occurs when the Hermitian positive semi-definite matrix ρ is a random matrix, rather than a deterministic matrix. For notation, here we prefer using bold upper-case symbols $\mathbf{X}, \mathbf{Y}, \mathbf{Z}$ to represent random matrices. If $\mathbf{X}$ is a symmetric (or Hermitian) matrix of $n \times n$ and f is a bounded measurable function, $f(\mathbf{X})$ is defined as the matrix with the same eigenvectors as $\mathbf{X}$ but with eigenvalues that are the images by f of those of $\mathbf{X}$; namely, if $\mathbf{e}$ is an eigenvector of $\mathbf{X}$ with eigenvalues λ, $\mathbf{X}\mathbf{e} = \lambda\mathbf{e}$, then we have $f(\mathbf{X})\,\mathbf{e} = f(\lambda)\,\mathbf{e}$. For the spectral decomposition $\mathbf{X} = \mathbf{U}\mathbf{D}\mathbf{U}^H$ with orthogonal (unitary) and $\mathbf{D} = \operatorname{diag}(\lambda_1, \ldots, \lambda_n)$ diagonal real, one has

$$f(\mathbf{X}) = \mathbf{U} f(\mathbf{D})\,\mathbf{U}^H$$

with $(f(\mathbf{D}))_{ii} = f(\lambda_i)\,, i = 1, \ldots, n$. We rewrite (4.67) as

$$S(\mathbf{X}) = -\operatorname{Tr}(\mathbf{X} \log \mathbf{X}) = -\sum_{i=1}^{n} \lambda_i(\mathbf{X}) \log \lambda_i(\mathbf{X}) \tag{4.68}$$

where $\lambda_i(\mathbf{X}), i = 1, \ldots, n$ are eigenvalues of $\mathbf{X}$. Recall from Corollary 4.6.4 or Lemma 4.3.1 that if $g : \mathbb{R} \to \mathbb{R}$ is a Lipschitz function with constant $|g|_{\mathcal{L}}$, the map $\mathbf{X} \in \mathbb{R}^{n^2} \to \sum\limits_{i=1}^{n} g(\lambda_k) \in \mathbb{R}$ is a Lipschitz function with constant $\sqrt{2}|g|_{\mathcal{L}}$. Observe from (4.68) that

$$g(t) = -t \log t, \quad t \in \mathbb{R}$$

is a Lipschitz function with the constant given by

$$|g|_{\mathcal{L}} = \sup_{s \neq t} \frac{|g(s) - g(t)|}{|s - t|} = \sup_{s \neq t} \frac{|s \log s - t \log t|}{|s - t|}. \tag{4.69}$$

Using Lemma 4.14.1, we have that

$$\mathbb{P}\left(\left|\frac{1}{n}\operatorname{Tr}\left(-\mathbf{X}\log\mathbf{X}\right)-\mathbb{E}\frac{1}{n}\operatorname{Tr}\left(-\mathbf{X}\log\mathbf{X}\right)\right|\geqslant t\right)\leqslant 2e^{-\frac{n^2t^2}{4c|g|_{\mathcal{L}}^2}}. \tag{4.70}$$

where $|g|_{\mathcal{L}}$ is given by (4.69).

Consider distinct quantum systems A and B. The joint system is described by a Hermitian positive semi-definite matrix ρ_{AB}. The individual systems are described by ρ_A and ρ_B, which are obtained from ρ_{AB} by taking partial trace. We simply use $S(A)$ to represent the entropy of System A, i.e., $S(\rho_A)$. In the following, the same convention applies to other joint or individual systems. It is well known that

$$|S(\rho_A)-S(\rho_B)|\leqslant S(\rho_{AB})\leqslant S(\rho_A)+S(\rho_B). \tag{4.71}$$

The second inequality above is called the subadditivity for the von Neumann entropy. The first one, called triangular inequality (also known as Araki-Lieb inequality [246]), is regarded as the quantum analog of the inequality

$$H(X)\leqslant H(X,Y)$$

for Shannon entropy $H(X)$ and $H(X,Y)$ the joint entropy of two random variables X and Y.

The *strong subadditivity* (SSA) of the von Neumann entropy proved by Lieb and Ruskai [247, 248] plays the same role as the basic inequalities for the classical entropy. For distinct quantum systems A, B, and C, strong subadditivity can be represented by the following two equivalent forms

$$\begin{aligned} &S(\rho_{AC})+S(\rho_{BC})-S(\rho_A)-S(\rho_B)\geqslant 0 \quad \text{(SSA)}\\ &S(\rho_{AB})+S(\rho_{BC})-S(\rho_B)-S(\rho_{ABC})\geqslant 0 \quad \text{(WM)} \end{aligned} \tag{4.72}$$

The expression on the left-hand side of the SSA inequality is known as the (quantum) conditional mutual information, and is commonly denoted as $I(A:B|C)$. Inequality (WM) is usually called *weak monotonicity*.

As pointed above, we are interested in only *linear* inequalities involving the entropies of various reduced states of a multiparty quantum state. Let us consider the (quantum) mutual information

$$\begin{aligned} I(A:B) &\triangleq S(\rho_A)+S(\rho_B)-S(\rho_{AB})\\ &=-\operatorname{Tr}(\rho_A\log\rho_A)-\operatorname{Tr}(\rho_A\log\rho_A)+\operatorname{Tr}(\rho_{AB}\log\rho_{AB}). \end{aligned} \tag{4.73}$$

where $\rho_A, \rho_B, \rho_{AB}$ are random matrices. Using the technique applied to treat the von Neumann entropy (4.67), we similarly establish the concentration inequalities like (4.70), by first evaluating the Lipschitz constant $|I(A:B)|_{\mathcal{L}}$ of the function $I(A:B)$ and $I(A:B|C)$. We can even extend to more general information inequalities [245, 249–252]. Recently, infinitely many new, constrained inequalities for the von Neumann entropy has been found by Cadney et al. [245].

Here, we make the explicit connection between concentration inequalities and these information inequalities. This new framework allows us to study the information stability when random states (thus random matrices), rather than deterministic states are considered. This direction needs more research.

4.19 Supremum of a Random Process

Sub-Gaussian random variables (Sect. 1.7) is a convenient and quite wide class, which includes as special cases the standard normal, Bernoulli, and all bounded random variables. Let $(Z_1, Z_2, \ldots)$ be (possibly dependable) mean-zero sub-Gaussian random variables, i.e., $\mathbb{E}[Z_i] = 0$, according to Lemma 1.7.1, there exists constants $\sigma_1, \sigma_2, \ldots$ such that

$$\mathbb{E}[\exp(tZ_i)] \leqslant \exp\left(\frac{t^2\sigma_i^2}{2}\right), \qquad t \in \mathbb{R}.$$

We further assume that the supremum of a random sequence is bounded, i.e., $v = \sup_i \sigma_i^2 < \infty$ and $\kappa = \frac{1}{v}\sum_i \sigma_i^2$.

Due to concentration of measure, *a Lipschitz function is nearly constant* [132, p. 17]. Even more important, the tails behave at worst like a scalar Gaussian random variable with absolutely controlled mean and variance.

Previously in this chapter, many functionals of random matrices, such as the largest eigenvalue and the trace, are shown to have a tail like

$$\mathbb{P}(|X| > t) \leqslant \exp\left(1 - t^2/C\right),$$

which, according to Lemma 1.7.1, implies that these functionals are sub-Gaussian random variables. We modify the arguments of [113] for our context.

Now let $\mathbf{X} = \mathrm{diag}(Z_1, Z_2, \ldots)$ be the random diagonal matrix with the Z_i on the diagonal. Since $\mathbb{E}[Z_i] = 0$, we have $\mathbb{E}\mathbf{X} = 0$. By the operator monotonicity of the matrix logarithm (Theorem 1.4.7), we have that

$$\log \mathbb{E}[\exp(t\mathbf{X})] \leqslant \mathrm{diag}\left(\frac{t^2\sigma_1^2}{2}, \frac{t^2\sigma_2^2}{2}, \ldots\right).$$

Due to (1.27): $\lambda_{\max}(\mathbf{A} + \mathbf{B}) \leqslant \lambda_{\max}(\mathbf{A}) + \lambda_{\max}(\mathbf{B})$, the largest eigenvalue has the following relation

$$\begin{aligned}\lambda_{\max}\left(\log \mathbb{E}\left[\exp\left(t\mathbf{X}\right)\right]\right) &\leqslant \lambda_{\max}\left(\mathrm{diag}\left(\tfrac{t^2\sigma_1^2}{2}, \tfrac{t^2\sigma_2^2}{2}, \ldots\right)\right)\\ &\leqslant \sup_i \frac{t^2\sigma_i^2}{2} = \frac{t^2}{2}v,\end{aligned}$$

and

$$\begin{aligned}\mathrm{Tr}\left(\log \mathbb{E}\left[\exp\left(t\mathbf{X}\right)\right]\right) &\leqslant \mathrm{Tr}\left(\mathrm{diag}\left(\tfrac{t^2\sigma_1^2}{2}, \tfrac{t^2\sigma_2^2}{2}, \ldots\right)\right)\\ &= \sum_i \frac{t^2\sigma_i^2}{2} = \frac{t^2}{2}\sum_i \sigma_i^2 = \frac{t^2 v\kappa}{2}.\end{aligned}$$

where we have used the property of the trace function (1.29): $\mathrm{Tr}\left(\mathbf{A}+\mathbf{B}\right) = \mathrm{Tr}\left(\mathbf{A}\right)+\mathrm{Tr}\left(\mathbf{B}\right)$.

By Theorem 2.16.3, we have

$$\mathbb{P}\left(\lambda_{\max}\left(\mathbf{X}\right) > \sqrt{2vt}\right) \leqslant \kappa \cdot t(e^t - t - 1)^{-1}.$$

Letting $t = 2\left(\tau + \log \kappa\right) > 2.6$ for $\tau > 0$ and interpreting $\lambda_{\max}\left(\mathbf{X}\right)$ as $\sup\limits_i Z_i$, finally we have that

$$\mathbb{P}\left(\sup_i Z_i > \sqrt{2\left(\sup_i \sigma_i^2\right)\left(\log \frac{\sum\limits_i \sigma_i^2}{\sup\limits_i \sigma_i^2} + \tau\right)}\right) \leqslant e^{-\tau}. \tag{4.74}$$

Consider the special case: $Z_i \sim \mathcal{N}\left(0,1\right)$ are N i.i.d. standard Gaussian random variables. Equation (4.74) says that the largest of the Z_i is $O(\log N + \tau)$ with probability at least $1 - e^{-\tau}$. This is known to be tight up to constants so the $\log N$ term cannot be removed. Besides, (4.74) can be applied to a countably infinite number of mean-zero Gaussian random variables $Z_i \sim \mathcal{N}\left(0, \sigma_i^2\right)$, or more generally, sub-Gaussian random variables, as long as the sum of the σ_i^2 is finite.

4.20 Further Comments

Reference [180] is the first paper to study the concentration of the spectral measure for large matrices.

Concentration of eigenvalues in kernel space [253–255]. We may use a kernel [159] to map the data to the high-dimensional feature space, even if the data samples are in low-dimensional space. We can exploit the high-dimensional space for concentration of eigenvalues in kernel space [254,255]. Concentration of random matrices in the kernel space is studied in [229,230,256,257].

For dependent random variables, we see [258]. Condition numbers of Gaussian random matrices is studied in [259]. Noncommutative Bennett and Rosenthal inequalities [260] is also relevant in this context. Concentration for noncommutative polynomials is studied in [190].

Chapter 5
Non-asymptotic, Local Theory of Random Matrices

Chapters 4 and 5 are the core of this book. Chapter 6 is included to form the comparison with this chapter. The development of the theory in this chapter will culminate in the sense of random matrices. The point of viewing this chapter as a novel statistical tool will have far-reaching impact on applications such as covariance matrix estimation, detection, compressed sensing, low-rank matrix recovery, etc. Two primary examples are: (1) approximation of covariance matrix; (2) restricted isometry property (see Chap. 7).

The non-asymptotic, local theory of random matrices is in its infancy. The goal of this chapter is to bring together the latest results to give a comprehensive account of this subject. No attempt is made to make the treatment exhaustive. However, for engineering problems, this treatment may contain the main relevant results in the literature.

The so-called geometric functional analysis studies high-dimensional sets and linear operators, combining ideas and methods from convex geometry, functional analysis and probability. While the complexity of a set may increase with the dimension, it is crucial to point out that passing to a high-dimensional setting may reveal properties of an object, which are obscure in low dimensions. For example, the average of a few random variables may exhibit a peculiar behavior, while the average of a large number of random variables will be close to a *constant* with high probability. This observation is especially relevant to big data [4]: one can do at a large scale that cannot be done at a smaller one, to extract new insights.

Another idea is probabilistic considerations in geometric problems. To prove the existence of a section of a convex body having a certain property, we can show that a random section possesses this property with positive probability. This powerful method allows to prove results in situations, where deterministic constructions are unknown, or unavailable.

In studying spectral properties of random matrices, the connection between the areas is interesting: the origins of the problems are purely probabilistic, while the methods draw from functional analysis and convexity.

R. Qiu and M. Wicks, *Cognitive Networked Sensing and Big Data*, DOI 10.1007/978-1-4614-4544-9_5,

5.1 Notation and Basics

We follow [261] for our notation here. In this chapter $|\mathbf{x}| = \|\mathbf{x}\|_2 = \left(\sum_{i=1}^{n} x_i^2\right)^{1/2}$ is the standard Euclidean norm. Sometimes by $\|\cdot\|$ we also denote the standard Euclidean norm. The vector is bold, lower case. The matrix is bold, uppercase. We work on $\mathbb{R}^n$, which is equipped with a Euclidean structure $\langle \cdot, \cdot \rangle$. We write B_2^n for the Euclidean unit ball and S^{n-1} for the unit sphere. We fix a coordinate system defined by an orthonormal basis $\{\mathbf{e}_1, \ldots, \mathbf{e}_n\}$. Volume ($n$-dimensional Lebesgue measure) and the cardinality of a finite set are also denoted by $|\cdot|$. We write ω_n for the volume of B_2^n.

Let K be a symmetric convex body in $\mathbb{R}^n$. The function

$$\|\mathbf{x}\|_K = \min\{t > 0 : \mathbf{x} \in tK\}$$

is a norm on $\mathbb{R}^n$. The normed space $(\mathbb{R}^n, \|\mathbf{x}\|_K)$ will be denoted by X_K. Conversely, if $X = (\mathbb{R}^n, \|\cdot\|_K)$ is a normed space, then the unit ball $K_X = \{\mathbf{x} \in \mathbb{R}^n : \|\mathbf{x}\| \leqslant 1\}$ of X is a symmetric convex body in $\mathbb{R}^n$.

The dual norm $\|\cdot\|_*$ of $\|\cdot\|$ is defined by $\|\mathbf{y}\|_* = \max\{|\langle \mathbf{x}, \mathbf{y} \rangle| : \|\mathbf{x}\| \leqslant 1\}$. From this definition it is clear that

$$|\langle \mathbf{x}, \mathbf{y} \rangle| \leqslant \|\mathbf{x}\| \, \|\mathbf{y}\|_*$$

for all $\mathbf{x}, \mathbf{y} \in \mathbb{R}^n$. If X^* is the dual space of X, then $K_* = K_X^o$ where

$$K^o = \{\mathbf{y} \in \mathbb{R}^n : \langle \mathbf{x}, \mathbf{y} \rangle \leqslant 1 \text{ for all } \mathbf{x} \in K\}$$

is the polar body K^o of K.

The Brunn-Minkowski inequality describes the effect of Minkowski addition to volumes: If A and B are two non-empty compact subsets of $\mathbb{R}^n$, then

$$|A + B|^{1/n} \geqslant |A|^{1/n} + |B|^{1/n}, \tag{5.1}$$

where $A + B = \{a + b : a \in A, b \in B\}$. It follows that, for every $\lambda \in (0, 1)$,

$$|\lambda A + (1 - \lambda) B|^{1/n} \geqslant \lambda |A|^{1/n} + (1 - \lambda) |B|^{1/n},$$

and, by the arithmetic-geometric means inequality,

$$|\lambda A + (1 - \lambda) B| \geqslant |A|^{\lambda} |B|^{(1-\lambda)}.$$

For a hypothesis testing

$$\mathcal{H}_0 : A$$
$$\mathcal{H}_1 : A + B$$

where A and B are two non-empty compact subsets of $\mathbb{R}^n$. It follows from (5.1) that: Claim $\mathcal{H}_0$ if

$$|A+B|^{1/n} - |B|^{1/n} \geqslant \gamma \geqslant |A|^{1/n},$$

where γ is the threshold. Often the set A is unknown in this situation.

The basic references on classical and asymptotic convex geometry are [152, 167, 262, 263].

Example 5.1.1 (A sequence of OFDM tones is modeled as a vector of random variables). Let $X_i, i = 1, \ldots, n$ be the random variables $X_i \in \mathbb{C}$. An arbitrary tone of the OFDM modulation waveform in the frequency domain may be modeled as X_i. For convenience, we form a vector of random variables $\mathbf{x} = (X_1, \ldots, X_n) \in \mathbb{C}^n$.

Due to the fading of the multipath channel and the radio resource allocation for all tones, the random variables X_i and X_j are not independent. The dependable random variables are difficult to deal with.

Each modulation waveform can be viewed as a random vector $\mathbf{x} \in \mathbb{C}^n$. We are interested in a sequence of independent copies $\mathbf{x}_1, \ldots, \mathbf{x}_N$ of the random vector $\mathbf{x}$.

□

5.2 Isotropic Convex Bodies

Lemma 5.2.1. *Let $\mathbf{x}, \mathbf{y}$ be independent isotropic random vectors in $\mathbb{R}^n$. Then $\mathbb{E}\,\|\mathbf{x}\|_2^2 = n$ and $\mathbb{E}\langle \mathbf{x}, \mathbf{y} \rangle^2 = n$.*

A *convex* body is a compact and full-dimensional set, $K \subseteq \mathbb{R}^n$ with $0 \in \text{int}\,(K)$. A convex body K is called *symmetric* if $\mathbf{x} \in K \Rightarrow -\mathbf{x} \Rightarrow K$. We say that K has a center of mass at the origin if

$$\int_K \langle \mathbf{x}, \mathbf{y} \rangle \, d\mathbf{x} = 0$$

for every $\mathbf{y} \in S^{n-1}$.

Definition 5.2.2 (Isotropic Position). Let K be a convex body in $\mathbb{R}^n$, and let $\mathbf{b}(K)$ denote its center of gravity. We say that K is in *isotropic position* if its center of gravity is in the origin, and for each i, j, $1 \le i \le j \le n$, we have

$$\frac{1}{\text{vol}\,(K)} \int_K \mathbf{x}_i \mathbf{x}_j d\mathbf{x} = \begin{cases} 1, \; i = j, \\ 0, \; i \neq j, \end{cases} \tag{5.2}$$

or equivalently, for every vector $\mathbf{y} \in \mathbb{R}^{\mathbf{n}}$,

$$\frac{1}{\mathrm{vol}\,(K)} \int_K \left(\mathbf{y}^T\mathbf{x}\right)^2 d\mathbf{x} = \frac{1}{\mathrm{vol}\,(K)} \int_K \langle \mathbf{y}, \mathbf{x} \rangle^2 d\mathbf{x} = \|\mathbf{y}\|^2. \tag{5.3}$$

By $\|\cdot\|$ we denote the standard Euclidean norm. Here $\mathbf{x}_i$ is the ith coordinate of $\mathbf{x}$. Our normalization is slightly different from [264]; their definition corresponds to applying a homothetical transformation to get $\mathrm{vol}(K) = 1$. The isotropic position has many interesting features. Among others, it minimizes $\frac{1}{\mathrm{vol}(K)} \int_K \|\mathbf{x}\|^2 d\mathbf{x}$ (see [264]).

If K is in isotropic position, then

$$\frac{1}{\mathrm{vol}\,(K)} \int_K \|\mathbf{x}\|^2 d\mathbf{x} = n,$$

from which it follows that "most" (i.e., all but a fraction of ϵ) of K is contained in a ball of radius $\sqrt{\frac{n}{\varepsilon}}$. Using a result of Borell [265], one can show that the radius of the ball could be replaced by $2\sqrt{2n}\log\left(1/\varepsilon\right)$. Also, if K is in isotropic position, it contains the unit ball [266, Lemma 5.1]. It is well known that for every convex body, there is an affine transformation to map it on a body in isotropic position, and this transformation is unique up to an isometry fixing the origin.

We have to allow an error $\varepsilon > 0$, and want to find an affine transformation bringing K into nearly isotropic position.

Definition 5.2.3 (Nearly Isotropic Position). We say that K is in ε-nearly isotropic position $(0 < \varepsilon \leq 1)$, if

$$\|\mathbf{b}\,(K)\| \leqslant \varepsilon,$$

and for every vector $\mathbf{y} \in \mathbb{R}^{\mathbf{n}}$,

$$(1-\varepsilon)\,\|\mathbf{y}\|^2 \leqslant \frac{1}{\mathrm{vol}\,(K)} \int_{K-b(K)} \left(\mathbf{y}^T\mathbf{x}\right)^2 d\mathbf{x} \leqslant (1+\varepsilon)\,\|\mathbf{y}\|^2. \tag{5.4}$$

Theorem 5.2.4 (Kannanand Lovász and Simonovits [266]). *Given* $0 < \delta, \varepsilon < 1$, *there exists a randomized algorithm finding an affine transformation* $\mathbf{A}$ *such that* $\mathbf{A}K$ *is in* ε*-nearly isotropic position with probability at least* $1-\delta$. *The number of oracle calls is*

$$O\left(\ln\left(\varepsilon\delta\right) n^5 \ln n\right).$$

Given a convex body $K \subseteq \mathbb{R}^n$ and a function $f : K \to \mathbb{R}^n$, we denote by $\mathbb{E}_K(f)$ the average of f over K, i.e.,

$$\mathbb{E}_K\left(f\right) = \frac{1}{\mathrm{vol}\,(K)} \int_K \|f\left(\mathbf{x}\right)\|^2 d\mathbf{x}.$$

We denote by $\mathbf{b} = \mathbf{b}(K) = \mathbb{E}_K(\mathbf{x})$ the center of gravity of K, and by $\mathbf{\Sigma}\,(K)$ the $n \times n$ matrix

$$\mathbf{\Sigma}(K) = \mathbb{E}_K\left((\mathbf{x}-\mathbf{b})(\mathbf{x}-\mathbf{b})^T\right).$$

The trace of $\mathbf{\Sigma}(K)$ is the average square distance of points of K from the center of gravity, which we also call the second moment of K. We recall from the definition of the isotropic position. The body $K \subseteq \mathbb{R}^n$ is isotropic position if and only if

$$\mathbf{b} = 0 \text{ and } \mathbf{\Sigma}(K) = \mathbf{I},$$

where $\mathbf{I}$ is the identity matrix. In this case, we have

$$\mathbb{E}_K(\mathbf{x}_i) = 0, \mathbb{E}_K\left(\mathbf{x}_i^2\right) = 1, \mathbb{E}_K(\mathbf{x}_i\mathbf{x}_j) = 0.$$

The second moment of K is n, and therefore all but a fraction of ε of its volume lies inside the ball $\sqrt{\frac{n}{\varepsilon}}\mathcal{B}$, where $\mathcal{B}$ is the unit ball.

If K is in isotropic position, then [266]

$$\sqrt{\frac{n+2}{n}}\mathcal{B} \subseteq K \subseteq \sqrt{n(n+2)}\mathcal{B}.$$

There is always a Euclidean structure, the canonical inner product denoted $\langle\cdot,\cdot\rangle$, on μ on $\mathbb{R}^n$ for which this measure is isotropic, i.e., for every $\mathbf{y} \in \mathbb{R}^n$,

$$\mathbb{E}\langle\mathbf{x},\mathbf{y}\rangle^2 = \int_{\mathbb{R}^n} \langle\mathbf{x},\mathbf{y}\rangle^2 d\mu(\mathbf{x}) = \|\mathbf{y}\|_2^2.$$

5.3 Log-Concave Random Vectors

We need to consider $\mathbf{x}$ that is an *isotropic, log-concave* random vectors in $\mathbb{R}^n$ (also a vector uniformly distributed in an isotropic convex body) [267]. A probability measure μ on $\mathbb{R}^n$ is said to be *log-concave* if for every compact sets A, B, and every $\lambda \in [0,1]$,

$$\mu(\lambda A + (1-\lambda)B) \geqslant \mu(A)^\lambda \mu(B)^{1-\lambda}.$$

In other words, an n-dimensional random vector is called *log-concave* if it has a log-concave distribution, i.e., for any compact nonempty sets $A, B \in \mathbb{R}^n$ and $\lambda \in (0,1)$,

$$\mathbb{P}(\mathbf{x} \in \lambda A + (1-\lambda)B) \geqslant \mathbb{P}(\mathbf{x} \in A)^\lambda \mathbb{P}(\mathbf{x} \in B)^{1-\lambda}.$$

According to Borell [268], a vector with full dimensional support is log-concave if and only has a density of the form e^{-f}, where $f : \mathbb{R}^n \to (-\infty, \infty)$ is a convex function. See [268, 269] for a general study of this class of measures.

It is known that any affine image, in particular any projection, of a log-concave random vector is log-concave. Moreover, if $\mathbf{x}$ and $\mathbf{y}$ are independent log-concave random vectors, then so is $\mathbf{x} + y$ (see [268, 270]).

One important and simple model of a centered log-concave random variable with variance 1 is the symmetric exponential random variable E which has density $f(\mathbf{t}) = \frac{1}{\sqrt{2}} \exp\left(-\sqrt{2}\|\mathbf{t}\|_2\right)$. In particular, for every $s > 0$ we have $\mathbb{P}\left(\|E\|_2 \geqslant s\right) \leqslant \exp\left(-s/\sqrt{2}\right)$. $|\mathbf{x}| = \|\mathbf{x}\|_2 = \left(\sum\limits_{i=1}^{n} x_i^2\right)^{1/2}$ is the standard Euclidean norm. Sometimes by $\|\cdot\|$ we denote also the standard Euclidean norm.

Log-concave measures are commonly encountered in convex geometry, since through the Brunn-Minkowski inequality, uniform distributions on convex bodies and their low-dimensional marginals are log-concave. The class of log-concave measures on $\mathbb{R}^n$ is the smallest class of probability measures that are closed under linear transformations and weak limits that contain uniform distributions on convex bodies. Vectors with logarithmically concave distributions are called log-concave.

The Euclidean norm of an n-dimensional log-concave random vector has moments of all orders [268]. A log-concave probability is supported on some convex subset of an affine subspace where it has a density. In particular when the support of the probability generates the whole space $\mathbb{R}^n$ (in which case we deal, in short, with full-dimensional probability) a characterization of Borell (see [268, 269]) states that the probability is absolutely continuous with respect to the Lebesgue measure and has a density which is log-concave. We say that a random vector is log-concave if its distribution is a log-concave measure.

The indicator function of a convex set and the density function of a Gaussian distribution are two canonical examples of log-concave functions [271, p. 43].

Every centered log-concave random variable Z, with variance 1 satisfies a sub-exponential inequality:

$$\text{for every } s > 0, \qquad \mathbb{P}\left(|Z| \geqslant s\right) \leqslant C \exp\left(-s/C\right),$$

where $C > 0$ is an absolute constant [268]. For a random variable Z we define the ψ_1-norm by

$$\|Z\|_{\psi_1} = \inf\left\{C > 0 : \mathbb{E}\exp\left(-|Z|/C\right) \leqslant 2\right\}$$

and we say that Z is ψ_1 with constant ψ, if $\|Z\|_{\psi_1} \leqslant \psi$.

A particular case of a log-concave probability measure is the normalized uniform (Lebesgue) measure on a convex body. Borell's inequality (see [267]) implies that the linear functionals $\mathbf{x} \mapsto \langle \mathbf{x}, \mathbf{y} \rangle$ satisfies Khintchine type inequalities with respect to log-concave probability measures. That is, if $p \geq 2$, then for every $\mathbf{y} \in \mathbb{R}^n$,

$$\left(\mathbb{E}|\langle \mathbf{x}, \mathbf{y} \rangle|^2\right)^{1/2} \leqslant \left(\mathbb{E}|\langle \mathbf{x}, \mathbf{y} \rangle|^p\right)^{1/p} \leqslant Cp\left(\mathbb{E}|\langle \mathbf{x}, \mathbf{y} \rangle|^2\right)^{1/2}. \tag{5.5}$$

Very recently, we have the following.

Theorem 5.3.1 (Paouris [272]). *There exists an absolute constant $c > 0$ such that if K is an isotropic convex body in $\mathbb{R}^n$, then*

$$\mathbb{P}\left(\left\{\mathbf{x} \in K : \|\mathbf{x}\|_2 \geqslant c\sqrt{n}L_K t\right\}\right) \leqslant e^{-t\sqrt{n}}$$

for every $t \geq 1$.

Theorem 5.3.2 (Paouris [272]). *These exists constants $c, C > 0$ such that for any isotropic, log-concave random vector $\mathbf{x}$ in $\mathbb{R}^n$, for any $p \leq c\sqrt{n}$,*

$$\left(\mathbb{E}\,\|\mathbf{x}\|_2^p\right)^{1/p} \leqslant C\left(\mathbb{E}\,\|\mathbf{x}\|_2^2\right)^{1/2}. \tag{5.6}$$

5.4 Rudelson's Theorem

A random vector $\mathbf{x} = (X_1, \ldots, X_n)$ is *isotropic* [93,264,273,274] if $\mathbb{E}X_i = 0$, and $\mathrm{Cov}\,(X_i, X_j) = \delta_{ij}$ for all $i, j \leq n$. Equivalently, an n-dimensional random vector with mean zero is *isotropic* if

$$\mathbb{E}\langle \mathbf{y}, \mathbf{x}\rangle^2 = \|\mathbf{y}\|_2^2,$$

for any $\mathbf{y} \in \mathbb{R}^n$. For any nondegenerate log-concave vector $\mathbf{x}$, there exists an affine transformation $\mathbf{T}$ such that $\mathbf{T}x$ is isotropic.

For $\mathbf{x} \in \mathbb{R}^n$, we define the Euclidean norm $\|\mathbf{x}\|_2$ as $\|\mathbf{x}\|_2 = \left(\sum\limits_{i=1}^{n} x_i^2\right)^{1/2}$. More generally, the l_p norm is defined as

$$\|\mathbf{x}\|_p = \left(\sum_{i=1}^{n} |x_i|^p\right)^{1/p}.$$

Consider N random points $\mathbf{y}_1, \ldots, \mathbf{y}_N$ independently, uniformly distributed in the body K and put

$$\hat{\boldsymbol{\Sigma}} = \frac{1}{N}\sum_{i=1}^{N} \mathbf{y}_i \otimes \mathbf{y}_i,$$

which is also sample covariance matrix. If N is sufficiently large, then with high probability

$$\left\|\hat{\boldsymbol{\Sigma}} - \boldsymbol{\Sigma}\right\|, \qquad \boldsymbol{\Sigma} = \frac{1}{\mathrm{vol}\,(K)}\int_K (\mathbf{y} \otimes \mathbf{y}),$$

will be small. Here $\boldsymbol{\Sigma}$ is also the true covariance matrix. Kannan et al. [266] proved that it is enough to take $N = c\frac{n^2}{\varepsilon}$ for some constant c. This result was greatly improved by Bourgain [273]. He has shown that one can take $N = C(\varepsilon)\, n\log^3 n$.

Since the situation is invariant under a linear transformation, we may assume that the body K is in the isotropic position. The the result of Bourgain may be reformulated as follows:

Theorem 5.4.1 (Bourgain [273]). *Let K be a convex body in $\mathbb{R}^n$ in the isotropic position. Fix $\varepsilon > 0$ and choose independently N random points $\mathbf{x}_1, \ldots, \mathbf{x}_N \in K$,*

$$N \geqslant C(\varepsilon)\, n\log^3 n.$$

Then with probability at least $1 - \varepsilon$ for any $\mathbf{y} \in \mathbb{R}^n$, one has

$$(1-\varepsilon)\left\|\mathbf{x}\right\|^2 < \frac{1}{N}\sum_{i=1}^{N}\left\langle \mathbf{x}_i, \mathbf{y}\right\rangle^2 < (1+\varepsilon)\left\|\mathbf{x}\right\|^2.$$

The work of Rudelson [93] is well-known. He has shown that this theorem follows from a general result about random vectors in $\mathbb{R}^n$. Let $\mathbf{y}$ be a random vector. Denote by $\mathbb{E}X$ the expectation of a random variable X. We say that $\mathbf{y}$ is the isotropic position if

$$\mathbb{E}\left(\mathbf{y} \otimes \mathbf{y}\right) = \mathbf{I}. \tag{5.7}$$

If $\mathbf{y}$ is uniformly distributed in a convex body K, then this is equivalent to the fact that K is in the isotropic position. The proof of Theorem 5.4.2 is taken from [93].

Theorem 5.4.2 (Rudelson [93]). *Let $\mathbf{y} \in \mathbb{R}^n$ be a random vector in the isotropic position. Let N be a natural number and let $\mathbf{y}_1, \ldots, \mathbf{y}_N$ be independent copies of $\mathbf{y}$. Then,*

$$\mathbb{E}\left\|\frac{1}{N}\sum_{i=1}^{N}\mathbf{y}_i \otimes \mathbf{y}_i - \mathbf{I}\right\| \leqslant C \cdot \frac{\sqrt{\log N}}{\sqrt{N}} \cdot \left(\mathbb{E}\left\|\mathbf{y}\right\|^{\log N}\right)^{1/\log N}, \tag{5.8}$$

provided that the last expression is smaller than 1.

Taking the trace of (5.7), we obtain that

$$\mathbb{E}\left\|\mathbf{y}\right\|^2 = n,$$

so to make the right hand side of (5.8) smaller than 1, we have to assume that

$$N \geqslant cn\log n.$$

Proof. The proof has two steps. The first step is relatively standard. First we introduce a Bernoulli random process and estimate the expectation of the norm in (5.8) by the expectation of its supremum. Then, we construct a majorizing measure to obtain a bound for the latest.

First, let be $\varepsilon_1, \dots, \varepsilon_N$ be independent Bernoulli variables taking values 1, -1 with probability 1/2 and let $\mathbf{y}_1, \dots, \mathbf{y}_N$, $\bar{\mathbf{y}}_1, \dots, \bar{\mathbf{y}}_N$ be independent copies of $\mathbf{y}$. Denote $\mathbb{E}_{\mathbf{y}}, \mathbb{E}_{\varepsilon}$ the expectation according to $\mathbf{y}$ and ε, respectively. Since $\mathbf{y}_i \otimes \mathbf{y}_i - \bar{\mathbf{y}}_i \otimes \bar{\mathbf{y}}_i$ is a symmetric random variable, we have

$$\mathbb{E}_{\mathbf{y}} \left\| \tfrac{1}{N} \sum_{i=1}^{N} \mathbf{y}_i \otimes \mathbf{y}_i - \mathbf{I} \right\| \leqslant \mathbb{E}_{\mathbf{y}} \mathbb{E}_{\bar{\mathbf{y}}} \left\| \tfrac{1}{N} \sum_{i=1}^{N} \mathbf{y}_i \otimes \mathbf{y}_i - \bar{\mathbf{y}}_i \otimes \bar{\mathbf{y}}_i \right\| =$$
$$\mathbb{E}_{\varepsilon} \mathbb{E}_{\mathbf{y}} \mathbb{E}_{\bar{\mathbf{y}}} \left\| \tfrac{1}{N} \sum_{i=1}^{N} \varepsilon_i \mathbf{y}_i \otimes \mathbf{y}_i - \bar{\mathbf{y}}_i \otimes \bar{\mathbf{y}}_i \right\| \leqslant 2 \mathbb{E}_{\mathbf{y}} \mathbb{E}_{\varepsilon} \left\| \tfrac{1}{N} \sum_{i=1}^{N} \varepsilon_i \mathbf{y}_i \otimes \mathbf{y}_i \right\|.$$

To estimate the last expectation, we need the following Lemma.

Lemma 5.4.3 (Rudelson [93]). *Let $\mathbf{y}_1, \dots, \mathbf{y}_N$ be vectors in $\mathbb{R}^n$ and $\varepsilon_1, \dots, \varepsilon_N$ be independent Bernoulli variables taking values 1, -1 with probability 1/2. Then*

$$\mathbb{E} \left\| \sum_{i=1}^{N} \varepsilon_i \mathbf{y}_i \otimes \mathbf{y}_i \right\| \leqslant C \sqrt{\log N} \cdot \max_{i=1,\dots,N} \|\mathbf{y}_i\| \cdot \left\| \sum_{i=1}^{N} \mathbf{y}_i \otimes \mathbf{y}_i \right\|^{1/2}.$$

The lemma was proven in [93]. Applying the Lemma, we get

$$\mathbb{E} \left\| \tfrac{1}{N} \sum_{i=1}^{N} \mathbf{y}_i \otimes \mathbf{y}_i - \mathbf{I} \right\| \leqslant$$
$$C \cdot \frac{\sqrt{\log N}}{\sqrt{N}} \cdot \left(\mathbb{E} \max_{i=1,\dots,N} \|\mathbf{y}_i\| \right)^{1/2} \cdot \left(\mathbb{E} \left\| \sum_{i=1}^{N} \mathbf{y}_i \otimes \mathbf{y}_i \right\| \right)^{1/2}. \tag{5.9}$$

We have

$$\left(\mathbb{E} \max_{i=1,\dots,N} \|\mathbf{y}_i\| \right)^{1/2} \leqslant \left(\mathbb{E} \left(\sum_{i=1}^{N} \|\mathbf{y}_i\|^{\log N} \right)^{2/\log N} \right)^{1/2}$$
$$\leqslant N^{1/\log N} \cdot \left(\mathbb{E} \|\mathbf{y}\|^{\log N} \right)^{1/\log N}.$$

Then, denoting

$$D = \mathbb{E} \left\| \frac{1}{N} \sum_{i=1}^{N} \mathbf{y}_i \otimes \mathbf{y}_i - \mathbf{I} \right\|,$$

we obtain through (5.9)

$$D \leqslant C \cdot \frac{\sqrt{\log N}}{\sqrt{N}} \cdot \left(\mathbb{E} \|\mathbf{y}\|^{\log N} \right)^{1/\log N} \cdot (D+1)^{1/2}.$$

If

$$C \cdot \frac{\sqrt{\log N}}{\sqrt{N}} \cdot \left(\mathbb{E}\|\mathbf{y}\|^{\log N}\right)^{1/\log N} \leqslant 1,$$

we arrive at

$$D \leqslant 2C \cdot \frac{\sqrt{\log N}}{\sqrt{N}} \cdot \left(\mathbb{E}\|\mathbf{y}\|^{\log N}\right)^{1/\log N},$$

which completes the proof of Theorem 5.4.2. □

Let us apply Theorem 5.4.2 to the problem of Kannan et al. [266].

Corollary 5.4.4 (Rudelson [93]). *Let $\varepsilon > 0$ and let K be an n-dimensional convex body in the isotropic position. Let*

$$N \geqslant C \cdot \frac{n}{\varepsilon^2} \cdot \log^2 \frac{n}{\varepsilon^2}$$

and let $\mathbf{y}_1, \ldots, \mathbf{y}_N$ be independent random vectors uniformly distributed in K. Then

$$\mathbb{E}\left\|\frac{1}{N}\sum_{i=1}^{N} \mathbf{y}_i \otimes \mathbf{y}_i - \mathbf{I}\right\| \leqslant \varepsilon.$$

Proof. It follows from a result of Alesker [275], that

$$\mathbb{E}\exp\left(\frac{\|\mathbf{y}\|^2}{c \cdot n}\right) \leqslant 2$$

for some absolute constant c. Then

$$\begin{aligned}\mathbb{E}\|\mathbf{y}\|^{\log N} &\leqslant \left(\mathbb{E}\exp\left(\tfrac{\|\mathbf{y}\|^2}{c\cdot n}\right)\right)^{1/2} \cdot \left(\mathbb{E}\left(\|\mathbf{y}\|^{2\log N} \cdot \exp\left(-\tfrac{\|\mathbf{y}\|^2}{c\cdot n}\right)\right)\right)^{1/2} \\ &\leqslant \sqrt{2} \cdot \left(\max_{t \geqslant 0} t^{\log N} \cdot e^{-\frac{t}{c\cdot n}}\right)^{1/2} \leqslant (Cn\log N)^{\frac{\log N}{2}}.\end{aligned}$$

Corollary 5.4.4 follows from this estimate and Theorem 5.4.2. By a Lemma of Borell [167, Appendix III], most of the volume of a convex body in the isotropic position is concerned within the Euclidean ball of radius $c\sqrt{n}$. So, it is might be of interest to consider a random vector uniformly distributed in the intersection of a convex body K and such a ball B_2^n. □

Corollary 5.4.5 (Rudelson [93]). *Let $\varepsilon, R > 0$ and let K be an n-dimensional convex body in the isotropic position. Suppose that $R \geqslant c\sqrt{\log 1/\varepsilon}$ and let*

$$N \geqslant C_0 \cdot \frac{R^2}{\varepsilon^2} \cdot \log n$$

and let $\mathbf{y}_1, \ldots, \mathbf{y}_N$ *be independent random vectors uniformly distributed in the intersection of* K *and the ball* $R\sqrt{n}B_2^n$, *i.e.,* $K \cap R\sqrt{n}B_2^n$. *Then*

$$\mathbb{E}\left\| \frac{1}{N}\sum_{i=1}^{N} \mathbf{y}_i \otimes \mathbf{y}_i - \mathbf{I} \right\| \leqslant \varepsilon.$$

See [93] for the proof.

5.5 Sample Covariance Matrices with Independent Rows

Singular values of matrices with independent rows (without assuming that the entries are independent) are treated here, with material taken from Mendelson and Pajor [274].

Let us first introduce a notion of isotropic position. Let $\mathbf{x}$ be a random vector selected randomly from a convex symmetric body in $\mathbb{R}^n$ which is in an isotropic position. By this we mean the following: let $\mathcal{K} \subset \mathbb{R}^n$ be a convex and symmetric set with a nonempty interior. We say that $\mathcal{K}$ is in an isotropic position if for any $\mathbf{y} \in \mathbb{R}^n$,

$$\frac{1}{\mathrm{vol}\,(\mathcal{K})} \int_{\mathcal{K}} |\langle \mathbf{y}, \mathbf{x} \rangle|^2 d\mathbf{x} = \|\mathbf{y}\|^2,$$

where the volume and the integral are with respect to the Lebesgue measure on $\mathbb{R}^n$ and $\langle \cdot, \cdot \rangle$ and $\|\cdot\|$ are, respectively, the scalar product and the norm in the Euclidean space l_2^n. In other words, if one considers the normalized volume measure on $\mathcal{K}$ and $\mathbf{x}$ is a random vector with that distribution, then a body is in an isotropic position if for any $\mathbf{y} \in \mathbb{R}^n$,

$$\mathbb{E}|\langle \mathbf{y}, \mathbf{x} \rangle|^2 = \|\mathbf{y}\|^2.$$

Let $\mathbf{x}$ be a random vector on $\mathbb{R}^n$ and consider $\{\mathbf{x}_i\}_{i=1}^N$ which are N independent random vectors distributed as $\mathbf{x}$. Consider the random operator $\mathbf{X} : \mathbb{R}^n \to \mathbb{R}^N$ defined by

$$\mathbf{X} = \begin{bmatrix} \mathbf{x}_1^T \\ \mathbf{x}_2^T \\ \vdots \\ \mathbf{x}_N^T \end{bmatrix}_{N \times n}$$

where $\{\mathbf{x}_i\}_{i=1}^N$ are independent random variables distributed according to the normalized volume measure on the body $\mathcal{K}$. A difficulty arises when the matrix $\mathbf{X}$ has dependent entries; in the standard setup in the theory of random matrices, one studies matrices with i.i.d. entries.

The method we are following from [274] is surprisingly simple. If $N \geq n$, the first n eigenvalues of $\mathbf{X}\mathbf{X}^* = \left(\langle \mathbf{x}_i, \mathbf{x}_j \rangle\right)_{i,j=1}^N$ are the same as the eigenvalues of $\mathbf{X}^*\mathbf{X} = \sum_{i=1}^N \mathbf{x}_i \otimes \mathbf{x}_i$. We will show that under very mild conditions on $\mathbf{x}$, with high probability,

$$\left\| \frac{1}{N} \sum_{i=1}^N \mathbf{x}_i \otimes \mathbf{x}_i - \mathbf{\Sigma} \right\|_{l_2^n \to l_2^n} \tag{5.10}$$

tends to 0 quickly as N tends to infinity, where $\mathbf{\Sigma} = \mathbb{E}\left(\mathbf{x} \otimes \mathbf{x}\right)$. In particular, with high probability, the eigenvalues of $\frac{1}{N} \sum_{i=1}^N \mathbf{x}_i \otimes \mathbf{x}_i$ are close to the eigenvalues of $\mathbf{\Sigma}$.

The general approximation question was motivated by an application in Complexity Theory, studied by Kannan, Lovasz and Simonovits [266], regarding algorithms that approximate the volume of convex bodies. Later, Bourgain and Rudelson obtain some results.

Theorem 5.5.1 (Bourgain [273]). *For every $\varepsilon > 0$ these exists a constant $c(\varepsilon)$ for which the following holds. If $\mathcal{K}$ is a convex symmetric body in $\mathbb{R}^n$ in isotropic position and $N \geqslant c(\varepsilon) n\log^3 n$, then with probability at least $1 - \varepsilon$, for any $\mathbf{y} \in S^{n-1}$,*

$$1 - \varepsilon \leqslant \frac{1}{N} \sum_{i=1}^N \langle \mathbf{x}_i, \mathbf{y} \rangle^2 = \frac{1}{N} \left\| \mathbf{\Sigma} \mathbf{y} \right\|^2 \leqslant 1 + \varepsilon.$$

Equivalently, this theorem says that $\frac{1}{N}\mathbf{\Sigma} : l_2^n \to l_2^N$ is a good embedding of l_2^n. When the random vector $\mathbf{x}$ has independent, standard Gaussian coordinates it is known that for any $\mathbf{y} \in S^{n-1}$,

$$1 - 2\sqrt{\frac{n}{N}} \leqslant \frac{1}{N} \sum_{i=1}^N \langle \mathbf{x}_i, \mathbf{y} \rangle^2 \leqslant 1 + 2\sqrt{\frac{n}{N}}$$

holds with high probability (see the survey [145, Theorem II.13]). In the Gaussian case, Theorem 5.5.1 is asymptotically optimal, up to a numerical constant.

Bourgain's result was improved by Rudelson [93], who removed one power of the logarithm while proving a more general statement.

Theorem 5.5.2 (Rudelson [93]). *There exists an absolute constant C for which the following holds. Let $\mathbf{y}$ be a random vector in $\mathbb{R}^n$ which satisfies that $\mathbb{E}\left(\mathbf{y} \otimes \mathbf{y}\right) = \mathbf{I}$. Then,*

$$\mathbb{E} \left\| \frac{1}{N} \sum_{i=1}^N \mathbf{x}_i \otimes \mathbf{x}_i - \mathbf{I} \right\| \leqslant C \sqrt{\frac{\log N}{N}} \left(\mathbb{E} \left\| \mathbf{y} \right\|^{\log N} \right)^{1/\log N}.$$

A proof of this theorem is given in Sect. 2.2.1.2, using the concentration of matrices.

For a vector-valued random variable $\mathbf{y}$ and $k \geq 1$, the ψ_k norm of $\mathbf{y}$ is

$$\|\mathbf{y}\|_{\psi_k} = \inf\left\{C > 0 : \mathbb{E}\exp\left(\frac{|\mathbf{y}|^k}{C^k}\right) \leqslant 2\right\}.$$

A standard argument [13] shows that if $\mathbf{y}$ has a bounded ψ_k norm, then the tail of $\mathbf{y}$ decays faster than $2\exp\left(-t^k / \|\mathbf{y}\|_{\psi_k}^2\right)$.

Assumption 5.5.3. Let $\mathbf{x}$ be a random vector in $\mathbb{R}^n$. We will assume that

1. There is some $\rho > 0$ such that for every $\mathbf{y} \in S^{n-1}$, $\left(\mathbb{E}|\langle \mathbf{x}, \mathbf{y}\rangle|^4\right)^{1/4} \leqslant \rho$.
2. Set $Z = \|\mathbf{x}\|$. Then $\|Z\|_{\psi_\alpha} \leqslant \infty$ for some α.

Assumption 5.5.3 implies that the average operator $\mathbf{\Sigma}$ satisfies that $\|\mathbf{\Sigma}\| \leq \rho^2$. Indeed,

$$\begin{aligned}\|\mathbf{\Sigma}\| &= \sup_{\mathbf{y}_1,\mathbf{y}_2\in S^{n-1}} \langle \mathbf{\Sigma}\mathbf{y}_1, \mathbf{y}_2\rangle = \sup_{\mathbf{y}_1,\mathbf{y}_2\in S^{n-1}} \mathbb{E}\langle \mathbf{x}, \mathbf{y}_1\rangle \langle \mathbf{x}, \mathbf{y}_2\rangle \\ &\leqslant \sup_{\mathbf{x}\in S^{n-1}} \mathbb{E}\langle \mathbf{x}, \mathbf{y}\rangle^2 \leqslant \rho^2.\end{aligned}$$

Before introducing the main result of [274], we give two results: a well known symmetrization theorem [13] and Rudelson [93]. A Rademacher random variable is a random variable taking values ± 1 with probability $1/2$.

Theorem 5.5.4 (Symmetrization Theorem [13]). *Let Z be a stochastic process indexed by a set F and let N be an integer. For every $i \leq N$, let $\mu_i : F \to \mathbb{R}$ be arbitrary functions and set $\{Z_i\}_{i=1}^N$ to be independent copies of Z. Under mild topological conditions on F and (μ_i) ensuring the measurability of the events below, for any $t > 0$,*

$$\beta_N(t)\,\mathbb{P}\left(\sup_{f\in F}\left|\sum_{i=1}^N Z_i(f)\right| > t\right) \leqslant 2\mathbb{P}\left(\sup_{f\in F}\left|\sum_{i=1}^N \varepsilon_i\left(Z_i(f) - \mu_i(f)\right)\right| > \frac{t}{2}\right),$$

where $\{\varepsilon_i\}_{i=1}^N$ are independent Rademacher random variables and

$$\beta_N(t) = \inf_{f\in F}\mathbb{P}\left(\left|\sum_{i=1}^N Z_i(f)\right| < \frac{t}{2}\right).$$

We express the operator norm of $\sum_{i=1}^{N}(\mathbf{x}_i \otimes \mathbf{x}_i - \mathbf{\Sigma})$ as the **supremum of an empirical process**. Indeed, let $\mathcal{Y}$ be the set of tensors $\mathbf{v} \otimes \mathbf{w}$, where $\mathbf{v}$ and $\mathbf{w}$ are vectors in the unit Euclidean ball. Then,

$$\sum_{i=1}^{N} (\mathbf{x} \otimes \mathbf{x} - \mathbf{\Sigma}) = \sup_{\mathbf{y} \in \mathcal{Y}} \langle \mathbf{x} \otimes \mathbf{x} - \mathbf{\Sigma}, \mathbf{y} \rangle .$$

Consider the process indexed by $\mathcal{Y}$ defined by $Z(\mathbf{y}) = \frac{1}{N} \sum_{i=1}^{N} \langle \mathbf{x}_i \otimes \mathbf{x}_i - \mathbf{\Sigma}, \mathbf{y} \rangle .$ Clearly, for every $\mathbf{y}$, $\mathbb{E} Z(\mathbf{y}) = 0$ (the expectation is a linear operator), and

$$\sup_{\mathbf{y} \in \mathcal{Y}} Z(\mathbf{y}) = \left\| \frac{1}{N} \sum_{i=1}^{N} (\mathbf{x}_i \otimes \mathbf{x}_i - \mathbf{\Sigma}) \right\| .$$

To apply Theorem 5.5.4, one has to estimate for any fixed $\mathbf{y} \in \mathcal{Y}$,

$$\mathbb{P}\left(\left| \frac{1}{N} \sum_{i=1}^{N} \langle \mathbf{x}_i \otimes \mathbf{x}_i - \mathbf{\Sigma}, \mathbf{y} \rangle \right| > Nt \right) .$$

For any fixed $\mathbf{y} \in \mathcal{Y}$, it follows that

$$\operatorname{var}\left(\frac{1}{N} \sum_{i=1}^{N} \langle \mathbf{x}_i \otimes \mathbf{x}_i - \mathbf{\Sigma}, \mathbf{y} \rangle \right) \leqslant \sup_{\mathbf{z} \in S^{n-1}} \mathbb{E} |\langle \mathbf{x}, \mathbf{z} \rangle|^4 \leqslant \rho^4 .$$

In particular, var $(Z(\mathbf{y}))\, \rho^4/N$, implying by Chebychev's inequality that

$$\beta_N(t) \geqslant 1 - \frac{\rho^4}{Nt^2} .$$

Corollary 5.5.5. *Let* $\mathbf{x}$ *be a random vector which satisfies Assumption 5.5.3 and let* $\mathbf{x}_1, \ldots, \mathbf{x}_N$ *be independent copies of* $\mathbf{x}$. *Then,*

$$\mathbb{P}\left(\left\| \sum_{i=1}^{N} \mathbf{x}_i \otimes \mathbf{x}_i - \mathbf{\Sigma} \right\| \geqslant tN \right) \leqslant 4\mathbb{P}\left(\left\| \sum_{i=1}^{N} \varepsilon_i \mathbf{x}_i \otimes \mathbf{x}_i \right\| > \frac{tN}{2} \right),$$

provided that $x \geqslant c\sqrt{\rho^4/N}$, *for some absolute constant* c.

Next, we need to estimate the norm of the symmetric random (vector-valued) variables $\sum_{i=1}^{N} \varepsilon_i \mathbf{x}_i \otimes \mathbf{x}_i$. We follow Rudelson [93], who builds on an inequality due to Lust-Piquard and Pisier [276].

Theorem 5.5.6 (Rudelson [93]). *There exists an absolute constant* c *such that for any integers* n *and* N, *and for any* $\mathbf{x}_1, \ldots, \mathbf{x}_N \in \mathbb{R}^n$ *and any* $k \geq 1$,

$$\left(\mathbb{E}\left\|\sum_{i=1}^{N}\varepsilon_i\mathbf{x}_i\otimes\mathbf{x}_i\right\|^k\right)^{1/k}\leqslant c\max\left\{\sqrt{\log n},\sqrt{k}\right\}\left\|\sum_{i=1}^{N}\mathbf{x}_i\otimes\mathbf{x}_i\right\|^{1/2}\max_{1\leqslant i\leqslant N}\|\mathbf{x}_i\|,$$

where $\{\varepsilon_i\}_{i=1}^N$ *are independent Rademacher random variables.*

This moment inequality immediately give a ψ_2 estimate on the random variable $\sum_{i=1}^{N}\varepsilon_i\mathbf{x}_i\otimes\mathbf{x}_i$.

Corollary 5.5.7. *These exists an absolute constant* c *such that for any integers* n *and* N, *and for any* $\mathbf{x}_1,\ldots,\mathbf{x}_N\in\mathbb{R}^n$ *and any* $t>0$,

$$\mathbb{P}\left(\left\|\sum_{i=1}^{N}\varepsilon_i\mathbf{x}_i\otimes\mathbf{x}_i\right\|\geq t\right)\leqslant 2\exp\left(-\frac{t^2}{\Delta^2}\right),$$

where $\Delta=c\sqrt{\log n}\left\|\sum_{i=1}^{N}\mathbf{x}_i\otimes\mathbf{x}_i\right\|^{1/2}\max_{1\leqslant i\leqslant N}\|\mathbf{x}_i\|.$

Finally, we are ready to present the main result of Mendelson and Pajor [274] and its proof.

Theorem 5.5.8 (Mendelson and Pajor [274]). *There exists an absolute constant* c *for which the following holds. Let* $\mathbf{x}$ *be a random vector in* $\mathbb{R}^n$ *which satisfies Assumption 5.5.3 and set* $Z=\|\mathbf{x}\|$. *For any integers* n *and* N *let*

$$A_{n,N}=\|Z\|_{\psi_\alpha}\frac{\sqrt{\log n}(\log N)^{1/\alpha}}{\sqrt{N}}\text{ and }B_{n,N}=\frac{\rho^2}{\sqrt{N}}+\|\mathbf{\Sigma}\|^{1/2}A_{n,N}.$$

Then, for any $t>0$,

$$\mathbb{P}\left(\left\|\sum_{i=1}^{N}(\mathbf{x}_i\otimes\mathbf{x}_i-\mathbf{\Sigma})\right\|\geqslant tN\right)\leqslant\exp\left[-\left(\frac{ct}{\max\left\{B_{n,N},A_{n,N}^2\right\}}\right)^{\beta}\right],$$

where $\beta=(1+2/\alpha)^{-1}$ *and* $\mathbf{\Sigma}=\mathbb{E}\left(\mathbf{x}\otimes\mathbf{x}\right)$.

Proof. First, recall that if $\mathbf{z}$ is a vector-valued random variable with a bounded ψ_α norm, and if $\mathbf{z}_1,\ldots,\mathbf{z}_N$ are N independent copies of $\mathbf{z}$, then

$$\left\|\max_{1\leqslant i\leqslant N}\mathbf{z}_i\right\|_{\psi_\alpha}\leqslant C\|\mathbf{z}\|_{\psi_\alpha}\log^{1/\alpha}N,$$

for an absolute constant C. Hence, for any k,

$$\left(\mathbb{E}\max_{1\leqslant i\leqslant N}|\mathbf{z}_i|^k\right)^{1/k}\leqslant Ck^{1/\alpha}\|\mathbf{z}\|_{\psi_\alpha}\log^{1/\alpha}N. \tag{5.11}$$

Consider the scalar-valued random variables

$$U=\left\|\frac{1}{N}\sum_{i=1}^{N}\varepsilon_i\mathbf{x}_i\otimes\mathbf{x}_i\right\|\text{ and }V=\left\|\frac{1}{N}\sum_{i=1}^{N}(\mathbf{x}_i\otimes\mathbf{x}_i-\mathbf{\Sigma})\right\|.$$

Combining Corollaries 5.5.5 and 5.5.7, we obtain

$$\begin{aligned}\mathbb{P}(V\geqslant t)&\leqslant 4\mathbb{P}(U\geqslant t/2)=4\mathbb{E}_{\mathbf{x}}\mathbb{P}_\varepsilon(U\geqslant t/2\,|\mathbf{x}_1,\ldots,\mathbf{x}_N)\\&\leqslant 8\mathbb{E}_{\mathbf{x}}\exp\left(-\tfrac{t^2N^2}{\Delta^2}\right).\end{aligned}$$

where $\Delta=c\sqrt{\log n}\left\|\sum_{i=1}^{N}\mathbf{x}_i\otimes\mathbf{x}_i\right\|^{1/2}\left\|\max_{1\leqslant i\leqslant N}\mathbf{x}_i\right\|$ for some constant c. Setting c_0 to be the constant in Corollary 5.5.5, then by Fubini's theorem and dividing the region of integration to $t\leqslant c_0\sqrt{\rho^4/N}$ (in this region there is no control on $\mathbb{P}(V\geqslant t)$) and $t>c_0\sqrt{\rho^4/N}$, it follows that the k-th order moments are

$$\begin{aligned}\mathbb{E}V^k&=\int_0^\infty kt^{k-1}\mathbb{P}(V\geqslant t)\,dt\\&=\int_0^{c_0\sqrt{\rho^4/N}}kt^{k-1}\mathbb{P}(V\geqslant t)\,dt+\int_0^\infty kt^{k-1}\mathbb{P}(V\geqslant t)\,dt\\&\leqslant\int_0^{c_0\sqrt{\rho^4/N}}kt^{k-1}\mathbb{P}(V\geqslant t)\,dt+8\mathbb{E}_{\mathbf{x}}\int_0^\infty kt^{k-1}\exp\left(-\tfrac{t^2N^2}{\Delta^2}\right)dt\\&\leqslant\left(c_0\sqrt{\rho^4/N}\right)^k+c^kk^{k/2}\mathbb{E}_{\mathbf{x}}\left(\tfrac{\Delta}{N}\right)^k\end{aligned}$$

for some new absolute constant c.

We can bound the second term by using

$$\begin{aligned}&c^k\left(\tfrac{k\log n}{N}\right)^{k/2}\mathbb{E}\left(\tfrac{1}{N}\left\|\sum_{i=1}^{N}\mathbf{x}_i\otimes\mathbf{x}_i\right\|^{k/2}\left\|\max_{1\leqslant i\leqslant N}\mathbf{x}_i\right\|^k\right)\\&\leqslant c^k\left(\tfrac{k\log n}{N}\right)^{k/2}\mathbb{E}\left(\left(\tfrac{1}{N}\left\|\sum_{i=1}^{N}\mathbf{x}_i\otimes\mathbf{x}_i-\mathbf{\Sigma}\right\|+\|\mathbf{\Sigma}\|\right)^{k/2}\max_{1\leqslant i\leqslant N}\|\mathbf{x}_i\|^k\right)\\&\leqslant c^k\left(\tfrac{k\log n}{N}\right)^{k/2}\left(\mathbb{E}\left(V+\|\mathbf{\Sigma}\|^k\right)^{1/2}\left(\mathbb{E}\max_{1\leqslant i\leqslant N}\|\mathbf{x}_i\|^{2k}\right)^{1/2}\right)\end{aligned}$$

for some new absolute constant c. Thus, setting $Z=\|\mathbf{x}\|$ and using Assumption 5.5.3 and (5.11), we obtain

$$\left(\mathbb{E}V^k\right)^{1/k}\leqslant c\frac{\rho^2}{\sqrt{N}}+ck^{\frac{1}{\alpha}+\frac{1}{2}}\left(\frac{\log n}{N}\right)^{1/2}\left(\log^{1/\alpha}N\right)\|Z\|_{\psi_\alpha}\left(\left(\mathbb{E}V^k\right)^{1/k}+\|\mathbf{\Sigma}\|\right)^{1/2},$$

for some new absolute constant c. Set $A_{n,N} = \left(\frac{\log n}{N}\right)^{1/2} \left(\log^{1/\alpha} N\right) \|Z\|_{\psi_\alpha}$ and $\beta = (1+2/\alpha)^{-1}$. Thus,

$$\left(\mathbb{E}V^k\right)^{1/k} \leqslant c\frac{\rho^2}{\sqrt{N}} + ck^{\frac{2}{\beta}} \|\mathbf{\Sigma}\|^{1/2} A_{n,N} + ck^{\frac{2}{\beta}} A_{n,N} \left(\mathbb{E}V^k\right)^{1/2k},$$

from which we have

$$\left(\mathbb{E}V^k\right)^{1/k} \leqslant ck^{\frac{1}{\beta}} \max\left\{\frac{\rho^2}{\sqrt{N}} + \|\mathbf{\Sigma}\|^{1/2} A_{n,N}, A_{n,N}^2\right\},$$

and thus,

$$\|V\|_{\psi_\beta} \leqslant c \max\left\{B_{n,N}, A_{n,N}^2\right\},$$

from which the estimator of the theorem follows by a standard argument. □

Consider the case that $\mathbf{x}$ is a bounded random vector. Thus, for any α, $\|Z\|_{\psi_\alpha} \leqslant \sup \|\mathbf{x}\| \equiv R$, and by taking $\alpha \to \infty$ one can set $\beta = 1$ and $A_{n,N} = R\sqrt{\frac{\log n}{N}}$. We obtain the following corollary.

Corollary 5.5.9 (Mendelson and Pajor [274]). *These exists an absolute constant c for which the following holds. Let $\mathbf{x}$ be a random vector in $\mathbb{R}^n$ supported in RB_2^n and satisfies Assumption 5.5.3. Then, for any $t > 0$*

$$\mathbb{P}\left(\left\|\sum_{i=1}^{N} \mathbf{x}_i \otimes \mathbf{x}_i - \mathbf{\Sigma}\right\| \geqslant tN\right) \leqslant \exp\left(-\frac{ct}{R^2} \min\left\{\frac{\sqrt{N}}{\sqrt{\log n}}, \frac{N}{\log n}\right\}\right).$$

The second case is when $\|Z\|_{\psi_\alpha} \leqslant c_1\sqrt{n}$, where $\mathbf{x}$ is a random vector associated with a convex body in an isotropic position.

Corollary 5.5.10 (Mendelson and Pajor [274]). *These exists an absolute constant c for which the following holds. Let $\mathbf{x}$ be a random vector in $\mathbb{R}^n$ that satisfies Assumption 5.5.3 with $\|Z\|_{\psi_\alpha} \leqslant c_1\sqrt{n}$. Then, for any $t > 0$*

$$\mathbb{P}\left(\left\|\sum_{i=1}^{N} \mathbf{x}_i \otimes \mathbf{x}_i - \mathbf{\Sigma}\right\| \geqslant tN\right)$$
$$\leqslant \exp\left(-c\left(x/\max\left\{\frac{\rho^2}{\sqrt{N}} + \rho c_1\sqrt{\frac{(n\log n)\log N}{N}}, c_1^2\frac{(n\log n)\log N}{N}\right\}\right)^{1/2}\right).$$

Let us consider two applications: the singular values and the integral operators. For the first one, the random vector $\mathbf{x}$ corresponds to the volume measure of some convex symmetric body in an isotropic position.

Corollary 5.5.11 (Mendelson and Pajor [274]). *These exist an absolute constant c_1, c_2, c_3, c_4 for which the following holds. Let $\mathcal{K} \subset \mathbb{R}^n$ be a symmetric convex body in an isotropic position, let $\mathbf{x}_1, \ldots, \mathbf{x}_N$ be N independent points sampled according to the normalized volume measure on $\mathcal{K}$, and set*

$$\mathbf{X} = \begin{bmatrix} \mathbf{x}_1^T \\ \mathbf{x}_2^T \\ \vdots \\ \mathbf{x}_N^T \end{bmatrix}$$

with non-zero singular values $\lambda_1, \ldots, \lambda_n$.

1. If $N \geqslant c_1 n\log^2 n$, then for every $t > 0$,

$$\mathbb{P}\left(\sqrt{1-t}\sqrt{N} \leqslant \lambda_i \leqslant \sqrt{1+t}\sqrt{N}\right) \geqslant 1 - \exp\left(-c_2 t^{1/2}\left(\frac{N}{(\log N)\,(n\log n)}\right)^{1/4}\right).$$

2. If $N > c_3 n$, then with probability at least $1/2$, $\lambda_1 \leqslant c_4\sqrt{N\log n}$.

Example 5.5.12 (Learning Integral Operators [274]). Let us apply the results above to the approximation of the integral operators. See also [277] for a learning theory from an approximation theory viewpoint. A compact integral operator with a symmetric kernel is approximated by the matrix of an empirical version of the operator [278]. What sample size yields given accuracy?

Let $\Omega \in \mathbb{R}^d$ and set ν to be a probability measure on Ω. Let t be a random variable on Ω distributed based on ν and consider $X(t) = \sum\limits_{i=1}^{\infty} \lambda_i \phi_i(x)\phi_i$, where $\{\phi_i\}_{i=1}^{\infty}$ is a complete basis in $\mathcal{L}(\Omega, \mu)$ and $\{\lambda_i\}_{i=1}^{\infty} \in l_1$.

Let $\mathcal{L}$ be a *bounded, positive-definite kernel* [89] on some probability space (Ω, μ). (See also Sect. 1.17 for the background on positive operators.) By Mercer's theorem, these is an orthonormal basis of $\mathcal{L}(\Omega, \mu)$, denoted by $\{\phi_i\}_{i=1}^{\infty}$ such that

$$\mathcal{L}\left(t, s\right) = \sum_{i=1}^{\infty} \lambda_i \phi_i(x)\phi_i\left(s\right).$$

Thus,

$$\left\langle X(t), X(s) \right\rangle = \mathcal{L}\left(t, s\right)$$

and the square of the singular values of the random matrix Σ are the eigenvalues of the Gram matrix (or empirical covariance matrix)

$$\mathbf{G} = \left(\left\langle X\left(t_i\right), X\left(t_j\right)\right\rangle\right)_{i,j=1}^{N}$$

where $t_1, \ldots, t_N$ are random variables distributed according to ν. It is natural to ask if the eigenvalues of this Gram matrix $\mathbf{G}$ converges in some sense to the eigenvalues of the integral operator

$$\mathcal{T}_L = \int \mathcal{L}(x, y)\, f(y) d\nu.$$

This question is useful to kernel-based learning [279]. It is not clear how the eigenvalues of the integral operator should be estimated from the given data in the form of the Gram matrix $\mathbf{G}$. Our results below enable us to do just that; Indeed, if $X(t) = \sum\limits_{i=1}^{\infty} \lambda_i \phi_i(x) \phi_i$, then

$$\mathbb{E}(X \otimes X) = \sum_{i=1}^{\infty} \lambda_i \langle \phi_i, \cdot \rangle \phi_i = \mathcal{T}_L.$$

Assume $\mathcal{L}$ is continuous and that Ω is compact. Thus, by Mercer's Theorem,

$$\mathcal{L}(x, y) = \sum_{i=1}^{\infty} \lambda_i \phi_i(x) \phi_i(y),$$

where $\{\lambda_i\}_{i=1}^{\infty}$ are the eigenvalues of $\mathcal{T}_L$, the integral operator associated with $\mathcal{L}$ and μ, and $\{\phi_i\}_{i=1}^{\infty}$ are complete bases in $\mathcal{L}(\mu)$. Also $\mathcal{T}_L$ is a trace-class operator since $\sum\limits_{i=1}^{\infty} \lambda_i = \int \mathcal{L}(x, x) d\mu(x)$.

To apply Theorem 5.5.8, we encounter a difficulty since $X(t)$ if of infinite dimensions. To overcome this, define

$$X_n(t) = \sum_{i=1}^{n} \lambda_i \phi_i(x) \phi_i,$$

where n is to be specified later. Mendelson and Pajor [274] shows that $X_n(t)$ is sufficiently close to $X(t)$; it follows that our problem is really finite dimensional.

A deviation inequality of (5.10) enables us to estimate with high probability the eigenvalues of the integral operators (infinite dimensions) using the eigenvalues of the Gram matrix or empirical covariance matrix (finite dimensions). This problem was also studied in [278], as pointed out above.

Learning with integral operators [127, 128] is relevant. □

Example 5.5.13 (Inverse Problems as Integral Operators [88]). The inverse problems may be formulated as integral operators [280], that in turn may be approximated by the empirical version of the integral operator, as shown in Example 5.5.12.

Integral operator are compact operators [89] in many natural topologies under very weak conditions on the kernel. Many examples are given in [281].

Once the connection between the integral operators and the concentration inequality is recognized, we can further extend this connection with electromagnetic inverse problems such as RF tomography [282, 283].

For relevant work, we refer to [127, 284–287]. □

5.6 Concentration for Isotropic, Log-Concave Random Vectors

The material here can be found in [288–291].

5.6.1 *Paouris' Concentration Inequality*

For our exposition, we closely follow [272], a breakthrough work. Let K be an isotropic convex body in $\mathbb{R}^n$. This implies that K has volume equal to 1, its center of mass is at the origin and its inertia matrix a multiple of the identity. Equivalently, there is a positive constant L_K, the isotropic constant of K, such that

$$\int_K \langle \mathbf{x}, \mathbf{y} \rangle^2 d\mathbf{x} = L_K^2 \tag{5.12}$$

for every $\mathbf{y} \in S^{n-1}$.

A major problem in Asymptotic Convex Geometry is whether there is an absolute constant $C > 0$ such that $L_K \le c$ for every n and every isotropic convex body of K in $\mathbb{R}^n$. The best known estimate is, due to Bourgain [292], $L_K^2 \leqslant c\sqrt[4]{n}\log n$, where c is an absolute constant. Klartag [293] has obtained an isomorphic answer to the question: For every symmetric convex body K in $\mathbb{R}^n$, there is a second symmetric convex body T in $\mathbb{R}^n$, whose Banach-Mazur distance from K is $O(\log n)$ and its isotropic constant is bounded by an absolute constant: $L_T \le c$.

The starting point of [272] is the following concentration estimate of Alesker [275]: There is an absolute constant $c > 0$ such that if K is an isotropic convex body in $\mathbb{R}^n$, then

$$\mathbb{P}\left(\left\{\mathbf{x} \in K : \|\mathbf{x}\|_2 \geqslant c\sqrt{n}L_K t\right\}\right) \leqslant 2e^{-t^2} \tag{5.13}$$

for every $t \ge 1$.

Bobkov and Nazrov [294, 295] have clarified the picture of volume distribution on isotropic, unconditional convex bodies. A symmetric convex body K is called *unconditional* if, for every choice of real numbers t_i, and every choice of $\varepsilon_i \in \{-1, 1\}, 1 \le i \le n$,

$$\|\varepsilon_1 t_1 \mathbf{e}_1 + \cdots + \varepsilon_n t_n \mathbf{e}_n\|_K = \|t_1 \mathbf{e}_1 + \cdots + t_n \mathbf{e}_n\|_K,$$

where $\|\cdot\|_K$ is the norm that corresponds to K and $\{\mathbf{e}_1, \ldots, \mathbf{e}_n\}$ is the standard orthonormal basis of $\mathbb{R}^n$. In particular, they obtained a striking strengthening of (5.13) in the case of 1-unconditional isotropic convex body: there is an absolute constant $c > 0$ such that if K is a 1-unconditional isotropic convex body in $\mathbb{R}^n$, then

$$\mathbb{P}\left(\left\{\mathbf{x} \in K : \|\mathbf{x}\|_2 \geqslant c\sqrt{n}t\right\}\right) \leqslant 2e^{-t\sqrt{n}} \tag{5.14}$$

for every $t \geq 1$. Note that $L_K \approx 1$ in the case of 1-unconditional convex bodies [264].

Paouris [272] obtained the following theorem in its full generality.

Theorem 5.6.1 (Paouris' Inequality for Isotropic Convex Body [272]). *There is an absolute constant $c > 0$ such that if K is an isotropic convex body in $\mathbb{R}^n$, then*

$$\mathbb{P}\left(\left\{\mathbf{x} \in K : \|\mathbf{x}\|_2 \geqslant c\sqrt{n}L_k t\right\}\right) \leqslant 2e^{-t\sqrt{n}}$$

for every $t \geq 1$.

Reference [291] gives a short proof of Paouris' inequality. Assume that $\mathbf{x}$ has a log-concave distribution (a typical example of such a distribution is a random vector uniformly distributed on a convex body). Assume further that it is centered and its covariance matrix is the identity such a random vector will be called isotropic. The tail behavior of the Euclidean norm $\|\mathbf{x}\|_2$ of an *isotropic, log-concave* random vector $\mathbf{x} \in \mathbb{R}^n$, states that for every $t > 1$,

$$\mathbb{P}\left(\|\mathbf{x}\|_2 \geqslant ct\sqrt{n}\right) \leqslant e^{-t\sqrt{n}}.$$

More precisely, we have that for any log-concave random vector $\mathbf{x}$ and any $p \geq 1$,

$$\left(\mathbb{E}\,\|\mathbf{x}\|_2^p\right)^{1/p} \sim \mathbb{E}\|\mathbf{x}\|_2 + \sup_{\mathbf{y} \in S^{n-1}} \left(\mathbb{E}|\langle \mathbf{y}, \mathbf{x}\rangle|^p\right)^{1/p}.$$

This result had a huge impact on the study of log-concave measures and has a lot of applications in that subject.

Let $\mathbf{x} \in \mathbb{R}^n$ be a random vector, denote the weak p-th moment of $\mathbf{x}$ by

$$\sigma_p(\mathbf{x}) = \sup_{\mathbf{y} \in S^{n-1}} \left(\mathbb{E}|\langle \mathbf{y}, \mathbf{x}\rangle|^p\right)^{1/p}.$$

Theorem 5.6.2 ([291]). *For any log-concave random vector $\mathbf{x} \in \mathbb{R}^n$ and any $p \geq 1$,*

$$\left(\mathbb{E}\,\|\mathbf{x}\|_2^p\right)^{1/p} \leqslant C\left(\mathbb{E}\|\mathbf{x}\|_2 + \sigma_p(\mathbf{x})\right),$$

where C is an absolute positive constant.

Theorem 5.6.1 allows us to prove the following in full generality.

Theorem 5.6.3 (Approximation of the identity operator [272]). *Let $\varepsilon \in (0,1)$. Assume that $n \geq n_0$ and let K be an isotropic convex body in $\mathbb{R}^n$. If $N \geqslant c(\varepsilon)\, n \log n$, where $c > 0$ is an absolute constant, and if $\mathbf{x}_1, \ldots, \mathbf{x}_N \in \mathbb{R}^n$ are independent random points uniformly distributed in K, then with probability greater than $1-\varepsilon$ we have*

$$(1-\varepsilon)\, L_K^2 \leqslant \frac{1}{N}\sum_{i=1}^{N} \langle \mathbf{x}_i, \mathbf{y}\rangle^2 \leqslant (1+\varepsilon)\, L_K^2,$$

for every $\mathbf{y} \in S^{n-1}$.

In the proof of Theorem 5.6.3, Paouris followed the argument of [296] that incorporates the concentration estimate of Theorem 5.6.1 into Rudelson's approach to the problem [93]. Theorem 5.4.2 from Rudelson [93] is used as a lemma.

We refer to the books [152, 167, 263] for basic facts from the Brunn-Minkowski and the asymptotic theory of finite dimensional normed spaces.

Aubrun [297] has proved that in the unconditional case, only $C(\epsilon)n$ random points are enough to obtain $(1+\varepsilon)$-approximation of the identity operator as in Theorem 5.6.3.

All the previous results remain valid if we replace Lebesgue measure on the isotropic convex body by an arbitrary isotropic, log-concave measure.

5.6.2 Non-increasing Rearrangement and Order Statistics

Log-concave vectors has recently received a lot of attention. Paouris' concentration of mass [272] says that, for any isotropic, log-concave vector $\mathbf{x}$ in $\mathbb{R}^n$,

$$\mathbb{P}\left(\|\mathbf{x}\|_2 \geqslant Ct\sqrt{n}\right) \leqslant e^{-t\sqrt{n}}. \tag{5.15}$$

where $|\mathbf{x}| = \|\mathbf{x}\|_2 = \left(\sum_{i=1}^{n} x_i^2\right)^{1/2}$. One wonders about the concentration of ℓ_p norm. For $p \in (1,2)$, this is an easy consequence of (5.15) and Hölder's inequality. For $p > 2$, new ideas are suggested by Latala [289], based on tail estimates of order statistics of $\mathbf{x}$.

Theorem 5.6.4 (Latala [289]). *For any $\delta > 0$ there exist constants $C_1(\delta), C_2(\delta) \leq C\delta^{-1/2}$ such that for any $p \geq 2+\delta$,*

$$\mathbb{P}\left(\|\mathbf{x}\|_p \geqslant t\right) \leqslant e^{-\frac{1}{C_1(\delta)}t} \textit{ for } t \geqslant C_1(\delta)\, pn^{1/p}$$

and

$$\left(\mathbb{E}\left\|\mathbf{x}\right\|_p^q\right)^{1/q} \leqslant C_2\left(\delta\right)\left(pn^{1/p}+q\right) \; for \; q \geqslant 2.$$

For an n-dimensional random vector $\mathbf{x}$, by $|X_1^*| \geqslant \ldots \geqslant |X_n^*|$ we denote the non-increasing rearrangement of $|X_1|,\ldots,|X_n|$. Random variable $X_1^*, 1 \leq k \leq n$, are called order statistics of X. In particular,

$$|X_1^*| = \max\left\{|X_1|,\ldots,|X_n|\right\}, and\, |X_n^*| = \min\left\{|X_1|,\ldots,|X_n|\right\}.$$

Random variables X_n^* are called order statistics of X.

By (5.15), we immediately have for isotropic, log-concave vectors $\mathbf{x} = (X_1,\ldots,X_n)$

$$\mathbb{P}\left(X_k^* \geqslant t\right) \leqslant e^{-\frac{1}{c}\sqrt{k}t} \tag{5.16}$$

for $t \geqslant \sqrt{Cn/k}$. The main result of [289] is to show (5.16) is valid for

$$t \geqslant C\log\left(en/k\right).$$

5.6.3 Sample Covariance Matrix

Taken from [298–300], the development here is motivated for the convergence of empirical (or sample) covariance matrix.

In the recent years a lot of work was done on the study of the empirical covariance matrix, and on understanding related random matrices with independent rows or columns. In particular, such matrices appear naturally in two important (and distinct) directions. That is (1) estimation of covariance matrices of high-dimensional distributions by empirical covariance matrices; and (2) the Restricted Isometry Property of sensing matrices defined in the Compressive Sensing theory. See elsewhere of the book for the background on RIP and covariance matrix estimation.

Let $\mathbf{x} \in \mathbb{R}^n$ be a centered vector whose covariance matrix is the identity matrix, and $\mathbf{x}_1,\ldots,\mathbf{x}_N \in \mathbb{R}^n$ are independent copies of $\mathbf{x}$. Let $\mathbf{A}$ be a random $n \times N$ matrix whose columns are $(\mathbf{x}_i)$. By $\lambda_{\min}$ (respectively $\lambda_{\max}$) we denote the smallest (respectively the largest) singular value of the empirical covariance matrix $\mathbf{AA}^T$. For Gaussian matrices, it is known that

$$1 - C\sqrt{\frac{n}{N}} \leqslant \frac{\lambda_{\min}}{N} \leqslant \frac{\lambda_{\max}}{N} \leqslant 1 + C\sqrt{\frac{n}{N}}. \tag{5.17}$$

with probability close to 1.

Let $\mathbf{x}_1, \ldots, \mathbf{x}_N \in \mathbb{R}^n$ be a sequence of random vectors on $\mathbb{R}^n$ (not necessarily identical). We say that is uniformly distributed if for some $\psi > 0$

$$\sup_{i \leqslant N} \sup_{\mathbf{y} \in S^{n-1}} \left\| \left| \langle \mathbf{x}_i, \mathbf{y} \rangle \right| \right\|_{\psi} \leqslant \psi, \tag{5.18}$$

for a random variable $Y \in \mathbb{R}$, $\|Y\|_{\psi_1} = \inf\left\{C > 0; \mathbb{E}\exp\left(|Y|/C\right) \leqslant 2\right\}$. We say it satisfies the boundedness condition with constant K (for some $K \geq 1$) if

$$\mathbb{P}\left(\max_{i \leqslant N} |X_i| / \sqrt{n} > K \max\left\{1, (N/n)^{1/4}\right\}\right) \leqslant e^{-\sqrt{n}}. \tag{5.19}$$

Theorem 5.6.5 (Adamczak et al. [299]). *Let N, n be positive integers and $\psi, K > 1$. Let $\mathbf{x}_1, \ldots, \mathbf{x}_N \in \mathbb{R}^n$ be independent random vectors satisfying* (5.18) *and* (5.19). *Then with probability at least $1 - 2e^{-c\sqrt{n}}$ one has*

$$\sup_{\mathbf{y} \in S^{n-1}} \left| \frac{1}{N} \sum_{i=1}^{N} \left(\left| \langle \mathbf{x}_i, \mathbf{y} \rangle \right|^2 - \mathbb{E}\left| \langle \mathbf{x}_i, \mathbf{y} \rangle \right|^2 \right) \right| \leqslant C(\psi + K)^2 \sqrt{\frac{n}{N}}.$$

Theorem 5.6.5 improves estimates obtained in [301] for log-concave, isotropic vectors. There, the result had a logarithmic factor. Theorem 5.6.5 removes this factor completely leading to the best possible estimate for an arbitrary N, that is to an estimate known for random matrices as in the Gaussian case.

As a consequence, we obtain in our setting, the quantitative version of Bai-Yin theorem [302] known for random matrices with i.i.d. entries.

Theorem 5.6.6 (Adamczak et al. [299]). *Let $\mathbf{A}$ be a random $n \times N$ matrix, whose columns $\mathbf{x}_1, \ldots, \mathbf{x}_N \in \mathbb{R}^n$ be isotropic random vectors satisfying Theorem 5.6.5. Then with probability at least $1 - 2e^{-c\sqrt{n}}$ one has*

$$1 - C(\psi + K)^2 \sqrt{\frac{n}{N}} \leqslant \frac{\lambda_{\min}}{N} \leqslant \frac{\lambda_{\max}}{N} \leqslant 1 + C(\psi + K)^2 \sqrt{\frac{n}{N}}. \tag{5.20}$$

The strength of the above results is that the conditions (5.18) and (5.19) are valid for many classes of distributions.

Example 5.6.7 (Uniformly distributed). Random vectors uniformly distributed on the Euclidean ball of radius $K\sqrt{n}$ clearly satisfy (5.19). They also satisfy (5.18) with $\psi = CK$. □

Lemma 5.6.8 (Lemma 3.1 of [301]). *Let $\mathbf{x}_1, \ldots, \mathbf{x}_N \in \mathbb{R}^n$ be i.i.d. random vectors, distributed according to an isotropic, log-concave probability measure on $\mathbb{R}^n$. There exists an absolute positive constant C_0 such that for any $N \leq e^{\sqrt{n}}$ and for every $K \geq 1$ one has*

$$\max_{i \leqslant N} |X_i| \leqslant C_0 K \sqrt{n}.$$

Proof. From [272] we have for every $i \leq N$

$$\mathbb{P}\left(|X_i| \geqslant Ct\sqrt{n}\right) \leqslant e^{-tc\sqrt{n}},$$

where C and c are absolute constants. The result follow by the union bound. □

Example 5.6.9 (Log-concave isotropic). Log-concave, isotropic random vectors in $\mathbb{R}^n$. Such vectors satisfy (5.18) and (5.19) for some absolute constants ψ and K. The boundedness condition follows from Lemma 5.6.8. A version of result with weaker was proven by Aubrun [297] in the case of isotropic, log-concave rand vectors under an additional assumption of unconditionality. □

Example 5.6.10 (Any isotropic random vector satisfying the Poincare inequality). Any isotropic random vectors $(\mathbf{x}_i)_{i \leqslant \mathrm{N}} \in \mathbb{R}^n$, satisfying the Poincare inequality with constant L, i.e., such that

$$var f\left(\mathbf{x}_i\right) \leqslant L^2 \mathbb{E}|\nabla f\left(\mathbf{x}_i\right)|^2$$

for all compactly supported smooth functions, satisfying (5.18) with $\psi = CL$ and (5.19) with $K = CL$.

The question from [303] regarding whether all log-concave, isotropic random vectors satisfy the Poincare inequality with an absolute constant is one of the major open problems in the theory of log-concave measures. □

5.7 Concentration Inequality for Small Ball Probability

We give a small ball probability inequality for isotropic log-concave probability measures, taking material from [79, 304]. There is an effort to replace the notion of independence by the "geometry" of convex bodies, since a log-concave measure should be considered as the measure-theoretic equivalent of a convex body. Most of these recent results make heavy use of tools from the asymptotic theory of finite-dimensional normed spaces.

Theorem 5.7.1 (Theorem 2.5 of Latala [79]). *Let* $\mathbf{A}$ *is an* $n \times n$ *matrix and let* $\mathbf{x} = (\xi_1, \ldots, \xi_n)$ *be a random vector, where* ξ_i *are independent sub-Gaussian random variables with* $\mathrm{var}\left(\xi_i\right) \geqslant 1$ *and sub-Gaussian bounded by* β*. Then, for any* $\mathbf{y} \in \mathbb{R}^n$*, one has*

$$\mathbb{P}\left(\|\mathbf{Ax} - \mathbf{y}\|_2 \leqslant \frac{\|\mathbf{A}\|_{\mathrm{HS}}}{2}\right) \leqslant 2\exp\left(-\frac{c_0}{\beta^4}\frac{\|\mathbf{A}\|_{\mathrm{HS}}}{\|\mathbf{A}\|_{\mathrm{op}}}\right),$$

where $c_0 > 0$ *is a universal constant.*

A standard calculation shows that in the situation of the theorem, for $\mathbf{A} = [a_{ij}]$ one has

$$\mathbb{E}|\mathbf{Ax} - \mathbf{y}|^2 \geqslant \mathbb{E}|\mathbf{A}\left(\mathbf{x} - \mathbb{E}\mathbf{x}\right)|^2 = \sum_{i,j} a_{ij}^2 \,\mathrm{Var}\left(\xi_i\right) \geqslant \|\mathbf{A}\|_F^2 .$$

Proof. Due to the highly accessible nature of this proof, we include the arguments taken from [79]. Let $\mathbf{x}' = \left(\xi_1', \ldots, \xi_n'\right)$ be the independent copy of $\mathbf{x}$ and set $\mathbf{z} = (Z_1, \ldots, Z_n) = \mathbf{x} - \mathbf{x}'$. Variables Z_i are independent **symmetric** with sub-Gaussian constants at most 2β. See also Sect. 1.11 for rademacher averages and symmetrization that will be used below.

Put

$$p := \mathbb{P}\left(|\mathbf{Ax} - \mathbf{y}| \leqslant \|\mathbf{A}\|_F/2\right).$$

Then

$$\begin{aligned} p^2 &= \mathbb{P}\left(|\mathbf{Ax} - \mathbf{y}| \leqslant \|\mathbf{A}\|_F/2, \mathbb{P}\left(|\mathbf{Ax}' - \mathbf{y}| \leqslant \|\mathbf{A}\|_F/2\right)\right) \\ &\leqslant \mathbb{P}\left(|\mathbf{Az}| \leqslant \|\mathbf{A}\|_F\right). \end{aligned}$$

Let $\mathbf{B} = \mathbf{AA}^T = (b_{ij})$. Then $b_{ii} = \sum\limits_j a_{ij}^2 \geqslant 0$ and

$$|\mathbf{Az}|^2 = \langle \mathbf{Bz}, \mathbf{z} \rangle = \sum_{i,j} b_{ij} Z_i Z_j = \sum_i b_{ii} Z_i^2 + 2 \sum_{i<j} b_{ij} Z_i Z_j.$$

Since $\mathrm{Var}\,(Z_i) = 2\,\mathrm{Var}\,(\xi_i) \geqslant 2$, so that

$$\sum_{i} b_{ii} \mathbb{E} Z_i^2 \geqslant 2\,\mathrm{Tr}\,(\mathbf{B}) = 2\,\|\mathbf{A}\|_F^2.$$

Thus

$$\begin{aligned} p^2 &\leqslant \mathbb{P}\left(|\mathbf{Az}|^2 \leqslant \|\mathbf{A}\|_F^2\right) \\ &\leqslant \mathbb{P}\left(2 \sum_{i<j} b_{ij} Z_i Z_j + \sum_i b_{ii}\left(Z_i^2 - \mathbb{E} Z_i^2\right) \leqslant -\|\mathbf{A}\|_F^2\right) \\ &\leqslant \mathbb{P}\left(\left|\sum_{i<j} b_{ij} Z_i Z_j\right| \geqslant \|\mathbf{A}\|_F^2/3\right) + \mathbb{P}\left(\sum_i b_{ii}\left(Z_i^2 - \mathbb{E} Z_i^2\right) \leqslant -\|\mathbf{A}\|_F^2/3\right) \end{aligned} \tag{5.21}$$

Note that we have $\|(b_{ij}\delta_{i\neq j})\|_F \leqslant \|\mathbf{B}\|_F = \left\|\mathbf{AA}^T\right\|_F \leqslant \|\mathbf{A}\|_{\mathrm{op}}\|\mathbf{A}\|_F$ and $\|(b_{ij}\delta_{i\neq j})\|_{op} \leqslant \|\mathbf{B}\|_{op} + \|(b_{ij}\delta_{i=j})\|_{op} \leqslant 2\|\mathbf{B}\|_{op} \leqslant 2\,\|\mathbf{A}\|_{\mathrm{op}}^2$. So by Lemma 1.12.1, we get

$$\mathbb{P}\left(\left|\sum_{i<j} b_{ij} Z_i Z_j\right| \geqslant \|\mathbf{A}\|_F^2/3\right) \leqslant 2\exp\left(-\left(C''\beta^4\right)^{-1}\left(\frac{\|\mathbf{A}\|_F}{\|\mathbf{A}\|_{\mathrm{op}}}\right)^2\right). \tag{5.22}$$

We have $\|Z_i\|_4 \leqslant 2\|\xi\|_4 \leqslant 2\beta\|g_i\|_4$, thus

$$\sum_i b_{ii}^2 \mathbb{E} Z_i^4 \leqslant 48\beta^4 \sum_i b_{ii}^2 \leqslant 48\beta^4 \|\mathbf{B}\|_F^2 \leqslant 48\beta^4 \|\mathbf{A}\|_{\text{op}}^2 \|\mathbf{A}\|_F^2 .$$

Therefore, by Lemma 1.12.2,

$$\mathbb{P}\left(\left|\sum_i b_{ii}\left(Z_i^2 - \mathbb{E}Z_i^2\right)\right| \geqslant \|\mathbf{A}\|_F^2 /3\right) \leqslant \exp\left(-\left(C'''\beta^4\right)^{-1}\left(\frac{\|\mathbf{A}\|_F}{\|\mathbf{A}\|_{\text{op}}}\right)^2\right). \tag{5.23}$$

Thus, combining (5.21)–(5.23), we have

$$p^2 \leqslant 4\exp\left(-\left(\max\left(C'', C'''\right)\beta^4\right)^{-1}\left(\frac{\|\mathbf{A}\|_F}{\|\mathbf{A}\|_{\text{op}}}\right)^2\right),$$

which completes the proof. □

Theorem 5.7.2 (Proposition 2.6 of Latala [79]). *Let* $\mathbf{A}$ *is a non-zero* $n \times n$ *matrix and let* $\mathbf{x} = (g_1, \ldots, g_n)$ *be a random vector, where* g_i *are independent standard Gaussian* $\mathcal{N}(0,1)$ *random variables. Then, for any* $t \in (0, c_1)$ *and any* $\mathbf{y} \in \mathbb{R}^n$, *one has*

$$\mathbb{P}\left(\|\mathbf{Ax} - \mathbf{y}\|_2 \leqslant t\|\mathbf{A}\|_{\text{HS}}\right) \leqslant t^{\left(c_2 \frac{\|\mathbf{A}\|_{\text{HS}}}{\|\mathbf{A}\|_{\text{op}}}\right)^2},$$

where $c_1, c_2 > 0$ *are universal constants.*

Theorem 5.7.3 (Paouris [304]). *Let* $\mathbf{x}$ *is an isotropic log-concave random vector in* $\mathbb{R}^n$, *which has sub-Gaussian constant* b. *Let* $\mathbf{A}$ *is a non-zero* $n \times n$ *matrix. Then, for any* $t \in (0, c_1)$ *and any* $\mathbf{y} \in \mathbb{R}^n$, *one has*

$$\mathbb{P}\left(\|\mathbf{Ax} - \mathbf{y}\|_2 \leqslant t\|\mathbf{A}\|_{\text{HS}}\right) \leqslant t^{\left(\frac{c_2}{b} \frac{\|\mathbf{A}\|_{\text{HS}}}{\|\mathbf{A}\|_{\text{op}}}\right)^2},$$

where $c_1, c_2 > 0$ *are universal constants.*

For a subset $A \in \mathbb{R}^n$ we denote the convex hull of A by $\operatorname{conv} A$ and the symmetric convex hull of A is denoted by $\operatorname{conv}(A \cup -A)$. By a symmetric body $B \in \mathbb{R}^n$ we mean a centrally symmetric compact subset $\mathbb{R}^n$ with nonempty interior, i.e., B is a convex body satisfying $B = -B$. Often, we identify such a symmetric convex body B with the n-dimensional Banach space $(\mathbb{R}^n, \|\cdot\|_B)$ for which B is the unit ball. B_2^n stands for the Euclidean unit ball in $\mathbb{R}^n$. The volume of a body $B \in \mathbb{R}^n$ is denoted by $|B|$.

In [79, Theorem 4.2], we get estimates of a similar type as in Theorem 5.7.1 for the probability that $\mathbf{A}x$ belongs to a general convex and symmetric set K rather than to a Euclidean ball.

Theorem 5.7.4 (Theorem 4.2 of Latala [79]). *Let* $\mathbf{A}$ *is a non-zero* $n \times n$ *matrix and let* $\mathbf{x} = (g_1, \ldots, g_n)$*, where* ξ_i *are independent sub-Gaussian random variables with* $\mathrm{Var}(\xi_i) \geqslant 1$ *and sub-Gaussian constants at most* β*. Let* $K \in \mathbb{R}^n$ *be a symmetric convex body satisfying* $B_2^n \subset K$*. Then*

$$\mathbb{P}\left(\mathbf{Ax} \in \alpha \|\mathbf{A}\|_{\mathrm{op}} \sqrt{n} K\right) \leqslant 3 \exp\left(-\frac{c}{2\beta^4} \frac{\|\mathbf{A}\|_F^2}{\|\mathbf{A}\|_{\mathrm{op}}^2}\right),$$

where $V_K = \left(|K| / |B_2^n|\right)^{1/n}$,

$$\alpha = 3\beta(4\pi V_K)^{(C/\eta)\ln\left(\|\mathbf{A}\|_F/(6\beta\sqrt{n})\right)},$$

and $\eta = \|\mathbf{A}\|_F^2 / (\beta^4 n)$*. Here* c_0 *is the constant from Theorem 5.7.1, and* C *is a universal constant.*

Example 5.7.5 (Hypothesis Testing). Consider the hypothesis testing

$$\begin{aligned} \mathcal{H}_0 &: \mathbf{y} = \mathbf{x} \\ \mathcal{H}_1 &: \mathbf{y} = \mathbf{s} + \mathbf{x} \end{aligned}$$

where $\mathbf{s} \in \mathbb{R}^n$ is the unknown signal vector and $\mathbf{x} \in \mathbb{R}^n$ is the noise vector $\mathbf{x} = (\xi_1, \ldots, \xi_n)$ where ξ_i are independent sub-random variables that satisfy the conditions of the above theorems. Let $\mathbf{B}$ be a non-zero $n \times n$ matrix. The algorithm is described by:

$$\begin{aligned} &\text{for} \quad \tfrac{\|\mathbf{y}-\mathbf{Bx}\|_2}{\|\mathbf{B}\|_F} \leqslant \tau_0, \quad \text{claim } \mathcal{H}_0. \\ &\text{for} \quad \tfrac{\|\mathbf{y}-\mathbf{Bx}\|_2}{\|\mathbf{B}\|_F} \geqslant \tau_1, \quad \text{claim } \mathcal{H}_1. \end{aligned}$$

We are interested in $\mathbb{P}\left(\frac{\|\mathbf{y}-\mathbf{Bx}\|_2}{\|\mathbf{B}\|_F} \leqslant \tau_1 \left|\mathcal{H}_1\right.\right)$ and $\mathbb{P}\left(\frac{\|\mathbf{y}-\mathbf{Bx}\|_2}{\|\mathbf{B}\|_F} \geqslant \tau_0 \left|\mathcal{H}0\right.\right)$, which may be handled by the above theorems. □

Example 5.7.6 (Hypothesis Testing for Compressed Sensed Data). In Sect. 10.16, the observation vector for compressed sensing is modeled in (10.54) and repeated here

$$\mathbf{y} = \mathbf{A}(\mathbf{s} + \mathbf{x}) \tag{5.24}$$

where $\mathbf{s} \in \mathbb{R}^n$ is an unknown vector in $\mathcal{S}$, $\mathbf{A} \in \mathbb{R}^{m \times n}$ is a random matrix with i.i.d. $\mathcal{N}(0,1)$ entries, and $\mathbf{x} \sim \mathcal{N}\left(\mathbf{0}, \sigma^2 \mathbf{I}_{n \times n}\right)$ denotes noise with known variance σ^2 that is independent of $\mathbf{A}$. The noise model (5.24) is different from a more commonly studied case

$$\mathbf{y}' = \mathbf{As} + \mathbf{x} \tag{5.25}$$

where $\mathbf{z} \sim \mathcal{N}\left(\mathbf{0}, \sigma^2 \mathbf{I}_{m \times m}\right)$.

Consider the hypothesis testing

$$\mathcal{H}_0 : \mathbf{y} = \mathbf{x}$$
$$\mathcal{H}_1 : \mathbf{y} = \mathbf{A}(\mathbf{s} + \mathbf{x}).$$

For $\mathbf{B} = \mathbf{A}^H$, we have

$$\frac{\|\mathbf{y} - \mathbf{B}\mathbf{x}'\|_2}{\|\mathbf{B}\|_F} = \frac{\|\mathbf{y} - \mathbf{B}\mathbf{A}\mathbf{x}\|_2}{\|\mathbf{B}\|_F} = \frac{\|\mathbf{y} - \mathbf{A}^H\mathbf{A}\mathbf{x}\|_2}{\|\mathbf{A}\|_F}.$$

The algorithm has become

$$\text{for} \quad \frac{\|\mathbf{y} - \mathbf{A}^H\mathbf{A}\mathbf{x}\|_2}{\|\mathbf{A}\|_F} < \tau_0, \quad \text{claim } \mathcal{H}_0.$$

$$\text{for} \quad \frac{\|\mathbf{y} - \mathbf{A}^H\mathbf{A}\mathbf{x}\|_2}{\|\mathbf{A}\|_F} > \tau_1, \quad \text{claim } \mathcal{H}_1.$$

We are interested $\mathbb{P}\left(\frac{\|\mathbf{y}-\mathbf{A}^H\mathbf{A}\mathbf{x}\|_2}{\|\mathbf{A}\|_F} \leqslant \tau_1 \left|\mathcal{H}_1\right.\right)$ and $\mathbb{P}\left(\frac{\|\mathbf{y}-\mathbf{A}^H\mathbf{A}\mathbf{x}\|_2}{\|\mathbf{A}\|_F} \geqslant \tau_0 \left|\mathcal{H}0\right.\right)$, which can be handled by above theorems. When using these theorems, we only need to replace $\mathbf{A}$ by $\mathbf{A}^H\mathbf{A}$. In a sense, $\mathbf{A}^H\mathbf{A}$ behaves like an identity matrix. □

For $N \geq n$ consider N random vectors $\mathbf{x}_i = (\xi_{1,i}, \xi_{2,i}, \ldots, \xi_{n,i}) \in \mathbb{R}^n, i = 1, \ldots, N$, where $\xi_{i,j}$ are independent sub-Gaussian random variables with $\text{Var}(\xi_i) \geqslant 1$ and sub-Gaussian constants at most β. Denote the matrix $[\xi_{i,j}]_{i \leqslant n, j \leqslant N}$ by $\mathbf{X}$. See Sect. 1.12 for the notation and relevant background.

5.8 Moment Estimates

For a random matrix $\mathbf{X} \in \mathbb{C}^{n_1 \times n_2}$, the singular values of the random matrix $\mathbf{X}$ are $\sigma_1 \geqslant \sigma_2 \geqslant \ldots \geqslant \sigma_n, n = \max\{n_1, n_2\}$. We can form a random vector $\boldsymbol{\sigma} = (\sigma_1, \sigma_2, \ldots, \sigma_n)^T$. For N independent copies $\mathbf{X}_1, \ldots, \mathbf{X}_N$ of the random matrix $\mathbf{X}$, we can form N independent copies $\boldsymbol{\sigma}_1, \ldots, \boldsymbol{\sigma}_N$ of the random vector $\boldsymbol{\sigma}$. So we can use the moments for random vectors to study these random matrices.

5.8.1 *Moments for Isotropic Log-Concave Random Vectors*

The aim here is to approximate the moments by the empirical averages with high probability. The material is taken from [267]. For every $1 \leq q \leq +\infty$,we define q^*

to be the conjugate of q, i.e., $1/q + 1/q^* = 1$. Let $\alpha > 0$ and let ν be a probability measure on (X, Ω). For a function $f : X \to \mathbb{R}$ define the ψ_α-norm by

$$\|f\|_{\psi_\alpha} = \inf\left\{\lambda > 0 \left| \int_X \exp\left(|f| / \lambda\right)^\alpha d\nu \leqslant 2 \right.\right\}.$$

Chebychev's inequality shows that the functions with bounded ψ_α-norm are strongly concentrated, namely $\nu\left\{x \left| |f(x)| > \lambda t\right.\right\} \leqslant C \exp\left(-t^\alpha\right)$. We denote by D the radius of the symmetric convex set K i.e., the smallest D such that $K \subset DB_2^n$, where B_2^n is the unit ball in $\mathbb{R}^n$.

Let $K \in \mathbb{R}^n$ be a convex symmetric body, and $\|\cdot\|_K$ the norm. The *modulus of convexity* of K is defined for any $\varepsilon \in (0, 2)$ by

$$\delta_K(\varepsilon) = \inf\left\{1 - \left\|\frac{\mathbf{x}+\mathbf{y}}{2}\right\|_K, \|\mathbf{x}\|_K = 1, \|\mathbf{y}\|_K = 1, \|\mathbf{x}-\mathbf{y}\|_K > \varepsilon\right\}. \quad (5.26)$$

We say that K has modulus of convexity of power type $q \geq 2$ if $\delta_K(\varepsilon) \geqslant c\varepsilon^q$ for every $\varepsilon \in (0, 2)$. This property is equivalent to the fact that the inequality

$$\left\|\frac{\mathbf{x}+\mathbf{y}}{2}\right\|_K^q + \frac{1}{\lambda^q}\left\|\frac{\mathbf{x}-\mathbf{y}}{2}\right\|_K^q \leqslant \frac{1}{2}\left(\|\mathbf{x}\|_K^q + \|\mathbf{y}\|_K^q\right),$$

for all $\mathbf{x}, \mathbf{y} \in \mathbb{R}^n$. Here $\lambda > 0$ is a constant depending only on c and q. We shall say that K has modulus of convexity of power type q with constant λ. Classical examples of convex bodies satisfying this property are unit balls of finite dimensional subspaces of L_q [305] or of non-commutative L_q-spaces (like Schatten trace class matrices [118]).

Given a random vector $\mathbf{x} \in \mathbb{R}^n$, let $\mathbf{x}_1, \ldots, \mathbf{x}_N$ be N independent copies of $\mathbf{x}$. Let $K \in \mathbb{R}^n$ be a convex symmetric body. Denote by

$$V_p(K) = \sup_{\mathbf{y} \in K}\left|\frac{1}{N}\sum_{i=1}^N |\langle \mathbf{x}_i, \mathbf{y}\rangle|^p - \mathbb{E}|\langle \mathbf{x}, \mathbf{y}\rangle|^p\right|$$

the maximal deviation of the empirical p-norm of $\mathbf{x}$ from the exact one. We want to bound $V_p(K)$ under minimum assumptions on the body K and random vector $\mathbf{x}$. We choose the size of the sample N such that this deviation is small with high probability.

To bound such random process, we must have some control of the random variable $\max_{1 \leqslant i \leqslant N} \|\mathbf{x}\|_2$, where $\|\cdot\|_2$ denotes the standard Euclidean norm. To this end we introduce the parameter

$$\kappa_{p,N}(\mathbf{x}) = \left(\mathbb{E}\max_{1 \leqslant i \leqslant N} \|\mathbf{x}\|_2^p\right)^{1/p}.$$

Theorem 5.8.1 (Guédon and Rudelson [267]). *Let $K \subset (\mathbb{R}^n, \langle\cdot,\cdot\rangle)$ be a symmetric convex body of radius D. Assume that K has modulus of convexity of power type q for some $q \geq 2$. Let $p \geq q$ and q^* be the conjugate of q.*

Let $\mathbf{x}$ be a random vector in $\mathbb{R}^n$, and let $\mathbf{x}_1, \ldots, \mathbf{x}_N$ be N independent copies of $\mathbf{x}$. Assume that

$$C_{p,\lambda} \frac{(\log N)^{2/q^*}}{N} (D \cdot \kappa_{p,N}(\mathbf{x}))^p \leqslant \beta^2 \cdot \sup_{\mathbf{y} \in K} \mathbb{E}|\langle \mathbf{x}, \mathbf{y}\rangle|^p$$

for some $\beta < 1$. Then

$$\mathbb{E}V_p(K) \leqslant 2\beta \cdot \sup_{\mathbf{a} \in K} \mathbb{E}|\langle \mathbf{x}, \mathbf{a}\rangle|^p.$$

The constant $C_{p,\lambda}$ in Theorem 5.8.1 depends on p and on the parameter λ in (5.26). That minimal assumptions on the vector $\mathbf{x}$ are enough to guarantee that $\mathbb{E}V_p(K)$ becomes small for large N. In most cases, $\kappa_{p,N}(\mathbf{x})$ may be bounded by a simple quantity:

$$\kappa_{p,N}(\mathbf{x}) \leqslant \left(\mathbb{E} \sum_{i=1}^N \|\mathbf{x}_i\|_2^p \right)^{1/p}.$$

Let us investigate the case of $\mathbf{x}$ being an isotropic, log-concave random vector in $\mathbb{R}^n$ (or also a vector uniformly distributed in an isotropic convex body). From (5.6), we have

$$\|\langle \cdot, \mathbf{y}\rangle\|_{\psi_1} \leqslant C\left(\mathbb{E}\langle \mathbf{x}, \mathbf{y}\rangle^2\right)^{1/2}.$$

From the sharp estimate of Theorem 5.3.2, we will deduce the following.

Theorem 5.8.2 (Guédon and Rudelson [267]). *Let $\mathbf{x}$ be an isotropic, log-concave random vector in $\mathbb{R}^n$, and let $\mathbf{x}_1, \ldots, \mathbf{x}_N$ be N independent copies of $\mathbf{x}$. If $N \leqslant e^{c\sqrt{n}}$, then for any $p \geq 2$,*

$$\kappa_{p,N}(\mathbf{x}) = \left(\mathbb{E}\max_{1 \leqslant i \leqslant N} \|\mathbf{x}\|_2^p\right)^{1/p} \leqslant \begin{cases} C\sqrt{n} & \text{if} \quad p \leqslant \log N \\ Cp\sqrt{n} & \text{if} \quad p \geqslant \log N. \end{cases}$$

Theorem 5.8.3 (Guédon and Rudelson [267]). *For any $\varepsilon \in (0,1)$ and $p \geq 2$ these exists $n_0(\varepsilon, p)$ such that for any $n \geq n_0$, the following holds: let $\mathbf{x}$ be an isotropic, log-concave random vector in $\mathbb{R}^n$, let $\mathbf{x}_1, \ldots, \mathbf{x}_N$ be N independent copies of $\mathbf{x}$, if*

$$N = \left\lfloor C_p n^{p/2} \log n / \varepsilon^2 \right\rfloor$$

then for any $t > \varepsilon$, with probability greater than $1 - C\exp\left(-\left(t/C_p'\varepsilon\right)^{1/p}\right)$, for any $\mathbf{y} \in \mathbb{R}^n$

$$(1-t)\,\mathbb{E}|\langle \mathbf{x},\mathbf{y}\rangle|^p \leqslant \frac{1}{N}\sum_{i=1}^{N}|\langle \mathbf{x}_i,\mathbf{y}\rangle|^p \leqslant (1+t)\,\mathbb{E}|\langle \mathbf{x},\mathbf{y}\rangle|^p.$$

The constants $C_p, C_p^{'} > 0$ are real numbers depending only on p.

Let us consider the classical case when $\mathbf{x}$ is a Gaussian random vector in $\mathbb{R}^n$. Let $\mathbf{x}_1,\ldots,\mathbf{x}_N$ be independent copies of $\mathbf{x}$. Let p^* denote the conjugate of p. For $\mathbf{t}=(t_1,\ldots,t_N)^T \in \mathbb{R}^N$, we have

$$\sup_{\mathbf{t}\in B_{p^*}^N}\sum_{i=1}^{N} t_i\left\langle \mathbf{x}_i,\mathbf{y}\right\rangle = \left(\sum_{i=1}^{N}|\langle \mathbf{x}_i,\mathbf{y}\rangle|^p\right)^{1/p},$$

where $\sum_{i=1}^{N} t_i\left\langle \mathbf{x}_i,\mathbf{y}\right\rangle$ is the Gaussian random variable.

Let Z and Y be Gaussian vectors in $\mathbb{R}^N$ and $\mathbb{R}^n$, respectively. Using Gordon's inequalities [306], it is easy to show that whenever $\mathbb{E}\|Z\|_p \geqslant \varepsilon^{-1}\mathbb{E}\|Y\|_2$ (i.e. for a universal constant c, we have $N \geqslant c^p p^{p/2} n^{p/2}/\varepsilon^p$)

$$\mathbb{E}\|Z\|_p - \mathbb{E}\|Y\|_2 \leqslant \mathbb{E}\inf_{\mathbf{y}\in S^{n-1}}\left(\frac{1}{N}\sum_{i=1}^{N}|\langle \mathbf{x}_i,\mathbf{y}\rangle|^p\right)^{1/p}$$

$$\leqslant \mathbb{E}\sup_{\mathbf{y}\in S^{n-1}}\left(\frac{1}{N}\sum_{i=1}^{N}|\langle \mathbf{x}_i,\mathbf{y}\rangle|^p\right)^{1/p} \leqslant \mathbb{E}\|Z\|_p + \mathbb{E}\|Y\|_2,$$

where $\left(\mathbb{E}\|Z\|_p + \mathbb{E}\|Y\|_2\right)/\left(\mathbb{E}\|Z\|_p - \mathbb{E}\|Y\|_2\right) \leqslant (1+\varepsilon)/(1-\varepsilon)$. It is therefore possible to get (with high probability with respect to the dimension n, see [307]) a family of N random vectors $\mathbf{x}_1,\ldots,\mathbf{x}_N$ such that for every $\mathbf{y}\in\mathbb{R}^n$

$$A\|\mathbf{y}\|_2 \leqslant \left(\frac{1}{N}\sum_{i=1}^{N}|\langle \mathbf{x}_i,\mathbf{y}\rangle|^p\right)^{1/p} \leqslant A\frac{1+\varepsilon}{1-\varepsilon}\|\mathbf{y}\|_2.$$

This argument significantly improves the bound on m in Theorem 5.8.3 for Gaussian random vectors.

Below we will be able to extend the estimate for the Gaussian random vector to random vector $\mathbf{x}$ satisfying the ψ_2-norm condition for linear functionals $\mathbf{y}\mapsto\langle \mathbf{x},\mathbf{y}\rangle$ with the same dependence on N. A random variable Z satisfies the ψ_2-norm condition if and only if for any $\lambda\in\mathbb{R}$

$$\mathbb{E}\exp(\lambda Z) \leqslant 2\exp\left(c\lambda^2\cdot\|Z\|_2^2\right).$$

Theorem 5.8.4 (Guédon and Rudelson [267]). *Let* $\mathbf{x}$ *be an isotropic random vector in* $\mathbb{R}^n$ *such that all functionals* $\mathbf{y} \mapsto \langle \mathbf{x}, \mathbf{y} \rangle$ *satisfy the* ψ_2*-norm condition. Let* $\mathbf{x}_1, \ldots, \mathbf{x}_N$ *be independent copies of the random vector* $\mathbf{x}$. *Then for every* $p \geq 2$ *and every* $N \geqslant n^{p/2}$

$$\sup_{\mathbf{y} \in B_2^n} \left(\frac{1}{N} \sum_{i=1}^{N} |\langle \mathbf{x}_i, \mathbf{y} \rangle|^p \right)^{1/p} \leqslant c\sqrt{p}.$$

5.8.2 Moments for Convex Measures

We follow [308] for our development. Related work is Chevet type inequality and norms of sub-matrices [267, 309].

Let $\mathbf{x} \in \mathbb{R}^n$ be a random vector in a finite dimensional Euclidean space E with Euclidean norm $\|\mathbf{x}\|$ and scalar product $< \cdot, \cdot >$. As above, for $p > 0$, we denote the weak p-th moment of $\mathbf{x}$ by

$$\sigma_p(\mathbf{x}) = \sup_{\mathbf{y} \in S^{n-1}} \left(\mathbb{E} \| \langle \mathbf{y}, \mathbf{x} \rangle \|^p \right)^{1/p}.$$

Clearly $\left(\mathbb{E} \|\mathbf{x}\|^p \right)^{1/p} \geqslant \sigma_p(\mathbf{x})$ and by Hölder's inequality, $\left(\mathbb{E} \|\mathbf{x}\|^p \right)^{1/p} \geqslant \mathbb{E} \|\mathbf{x}\|$. Sometimes we are interested in reversed inequalities of the form

$$\left(\mathbb{E} \|\mathbf{x}\|^p \right)^{1/p} \leqslant C_1 \mathbb{E} \|\mathbf{x}\| + C_2 \sigma_p(\mathbf{x}) \tag{5.27}$$

for $p \geq 1$ and constants C_1, C_2.

This is known for some classes of distributions and the question has been studied in a more general setting (see [288] and references there). Our objective here is to describe classes for which the relationship (5.27) is satisfied.

Let us recall some known results when (5.27) holds. It clearly holds for Gaussian vectors and it is not difficult to see that (5.27) is true for sub-Gaussian vectors.

Another example of such a class is the class of so-called log-concave vectors. It is known that for every log-concave random vector $\mathbf{x}$ in a finite dimensional Euclidean space and any $p > 0$,

$$\left(\mathbb{E} \|\mathbf{x}\|^p \right)^{1/p} \leqslant C \left(\mathbb{E} \|\mathbf{x}\| + \sigma_p(\mathbf{x}) \right),$$

where $C > 0$ is a universal constant.

Here we consider the class of complex measures introduced by Borell. Let $\kappa < 0$. A probability measure $\mathbb{P}$ on $\mathbb{R}^m$ is called κ-concave if for $0 < \theta < 1$ and for all compact subsets $A, B \in \mathbb{R}^m$ with positive measures one has

$$\mathbb{P}\left((1-\theta) A + \theta B \right) \geqslant \left((1-\theta) \mathbb{P}(A)^\kappa + \theta \mathbb{P}(B)^\kappa \right)^{1/\kappa}. \tag{5.28}$$

A random vector with a κ-concave distribution is called κ-concave. Note that a log-concave vector is also κ-concave for any $\kappa < 0$.

For $\kappa > -1$, a κ-concave vector satisfies (5.27) for all $0 < (1+\varepsilon)p < -1/\kappa$ with C_1 and C_2 depending only on ε.

Definition 5.8.5. Let $p > 0, m = \lceil p \rceil$, and $\lambda \geq 1$. We say that a random vector $\mathbf{x}$ in E satisfies the assumption $H(p, \lambda)$ if for every linear mapping $\mathbf{A} : E \to \mathbb{R}^m$ such that $\mathbf{y} = Ax$ is non-degenerate there is a gauge $|| \cdot ||$ on $\mathbb{R}^m$ such that $\mathbb{E} \|\mathbf{y}\| < \infty$ and

$$\left(\mathbb{E}\|\mathbf{y}\|^p\right)^{1/p} < \lambda \mathbb{E} \|\mathbf{y}\| . \tag{5.29}$$

For example, the standard Gaussian and Rademacher vectors satisfy the above condition. More generally, a sub-Gaussian random vector also satisfies the above condition.

Theorem 5.8.6 ([308]). *Let $p > 0$ and $\lambda \geq 1$. If a random vector $\mathbf{x}$ in a finite dimensional Euclidean space satisfies $H(p, \lambda)$, then*

$$\left(\mathbb{E}\|\mathbf{x}\|^p\right)^{1/p} < c\left(\lambda \mathbb{E} \|\mathbf{x}\| + \sigma_p(\mathbf{x})\right),$$

where c is a universal constant.

We can apply above results to the problem of the approximation of the covariance matrix by the empirical covariance matrix. For a random vector $\mathbf{x}$ the covariance matrix of $\mathbf{x}$ is given by $\mathbb{E}\mathbf{x}\mathbf{x}^T$. It is equal to the identity operator $\mathbf{I}$ if $\mathbf{x}$ is isotropic. The empirical covariance matrix of a sample of size N is defined by $\frac{1}{N}\sum\limits_{i=1}^{N} \mathbf{x}_i\mathbf{x}_i^T$, where $\mathbf{x}_1, \mathbf{x}_2, \ldots, \mathbf{x}_N$ are independent copies of $\mathbf{x}$. The main question is how small can be taken in order that these two matrices are close to each other in the operator norm.

It was proved there that for $N \geq n$ and log-concave n-dimensional vectors $\mathbf{x}_1, \mathbf{x}_2, \ldots, \mathbf{x}_N$ one has

$$\left\| \frac{1}{N}\sum_{i=1}^{N} \mathbf{x}_i\mathbf{x}_i^T - \mathbf{I} \right\|_{\text{op}} \leqslant C\sqrt{\frac{n}{N}}$$

with probability at least $1-2\exp\left(-c\sqrt{n}\right)$, where $\mathbf{I}$ is the identity matrix, and $\|\cdot\|_{\text{op}}$ is the operator norm and c, C are absolute positive constants.

In [310, Theorem 1.1], the following condition was introduced: an isotropic random vector $\mathbf{x} \in \mathbb{R}^n$ is said to satisfy the *strong regularity assumption* if for some $\eta, C > 0$ and every rank $r \leq n$ orthogonal projection $\mathbf{P}$, one has that for very $t > C$

$$\mathbb{P}\left(\|\mathbf{P}\mathbf{x}\|_2 \geqslant t\sqrt{r}\right) \leqslant C/\left(t^{2+2\eta}r^{1+\eta}\right).$$

In [308], it was shown that an isotropic $(-1/q)$ satisfies this assumption. For simplicity we give this (without proof) with $\eta = 1$.

Theorem 5.8.7 ([308]). *Let $n \geq 1, a > 0$ and $q = \max\{4, 2a\log n\}$. Let $\mathbf{x} \in \mathbb{R}^n$ be an isotropic $(-1/q)$ random vector. Then there is an absolute constant C such that for every rank r orthogonal projection $\mathbf{P}$ and every $t \geq C\exp(4/a)$, one has*

$$\mathbb{P}\left(\|\mathbf{P}\mathbf{x}\|_2 \geqslant t\sqrt{r}\right) \leqslant C\max\left\{(a\log a)^4, \exp(32/a)\right\} / \left(t^4r^2\right).$$

Theorem 1.1 from [310] and the above lemma immediately imply the following corollary on the approximation of the covariance matrix by the sample covariance matrix.

Corollary 5.8.8 ([308]). *Let $n \geq 1, a > 0$ and $q = \max\{4, 2a\log n\}$. Let $\mathbf{x}_1, \ldots, \mathbf{x}_N$ be independent $(-1/q)$-concave, isotropic random vector in $\mathbb{R}^n$. Then for every $\varepsilon \in (0,1)$ and every $N \geq C(\varepsilon)n$, one has*

$$\mathbb{E}\left\| \frac{1}{N}\sum_{i=1}^{N} \mathbf{x}_i \otimes \mathbf{x}_i^T - \mathbf{I}_{n\times n} \right\|_{\text{op}} \leqslant \varepsilon$$

where $C(\varepsilon, a)$ depends only on a and ε.

The following result was proved for small ball probability estimates.

Theorem 5.8.9 (Paouris [304]). *Let $\mathbf{x}$ be a centered log-concave random vector in a finite dimensional Euclidean space. For every $t \in (0, c')$ one has*

$$\mathbb{P}\left(\|\mathbf{x}\|_2 \leqslant t\left(\mathbb{E}\,\|\mathbf{x}\|_2^2\right)^{1/2} \right) \leqslant t^{c\left(\mathbb{E}\|\mathbf{x}\|_2^2\right)^{1/2}/\sigma_2(\mathbf{x})},$$

where $c, c' > 0$ are universal positive constants.

The following result generalizes the above result to the setting of convex distributions.

Theorem 5.8.10 (Paouris [304]). *Let $n \geq 1$ and $q > 1$. Let $\mathbf{x}$ be a centered n-dimensional $(-1/q)$-concave random vector. Assume $1 \leqslant p \leqslant \min\{q, n/2\}$. Then, for every $\varepsilon \in (0,1)$,*

$$\mathbb{P}\left(\|\mathbf{x}\|_2 \leqslant t\mathbb{E}\|\mathbf{x}\|_2\right) \leqslant \left(1 + \frac{c}{q-p}\right)(2c)^p \left(\frac{q^2}{(q-p)(q-1)}\right)^{3p} t^p,$$

whenever $\mathbb{E}\|\mathbf{x}\|_2 \geqslant 2C\sigma_p(\mathbf{x})$, where c, C are constants.

5.9 Law of Large Numbers for Matrix-Valued Random Variables

For $p \leq \infty$, the finite dimensional l_p spaces are denoted as l_p^n. Thus l_p^n is the Banach space $\left(\mathbb{R}^n, \|\cdot\|_p\right)$, where

$$\|\mathbf{x}\|_p = \left(\sum_{i=1}^{n} |x|_i^p\right)^{1/p}$$

for $p \leq \infty$, and $\|\mathbf{x}\|_\infty = \max_i |x_i|$. The closed unit ball of l_p is denoted by $B_p^n := \left\{\mathbf{x} : \ \|\mathbf{x}\|_p \leqslant 1\right\}$.

The canonical basis of $\mathbb{R}^n$ is denoted by $(\mathbf{e}_1, \ldots, \mathbf{e}_n)$. Let $\mathbf{x}, \mathbf{y} \in \mathbb{R}^n$. The canonical inner product is denoted by $\langle \mathbf{x}, \mathbf{y} \rangle := \mathbf{x}^T \mathbf{y}$. The tensor product (outer product) is defined as $\mathbf{x} \otimes \mathbf{y} = \mathbf{y}\mathbf{x}^T$; thus $\mathbf{z} = \langle \mathbf{x}, \mathbf{z} \rangle \mathbf{y}$ for all $\mathbf{z} \in \mathbb{R}^n$.

Let $\mathbf{A} = (A_{ij})$ be an $m \times n$ real matrix. The spectral norm of $\mathbf{A}$ is the operator norm $l_2 \to l_2$, defined as

$$\|\mathbf{A}\|_2 := \sup_{\mathbf{x} \in \mathbb{R}^n} \frac{\|\mathbf{A}\mathbf{x}\|_2}{\|\mathbf{x}\|_2} = \sigma_1(\mathbf{A}),$$

where $\sigma_1(\mathbf{A})$ is the largest singular value of $\mathbf{A}$. The Frobenius norm $\|\mathbf{A}\|_F$ is defined as

$$\|\mathbf{A}\|_F := \sum_{i,j} A_{ij}^2 = \sum_i \sigma_i(\mathbf{A})^2,$$

where $\sigma_i(\mathbf{A})$ are the singular values of $\mathbf{A}$.

C denotes positive absolute constants. The $a = O(b)$ notation means that $a \le Cb$ for some absolute constant C.

For the scalar random variables, the classical Law of Large Numbers says the following: let X be a bounded random variable and $X_1, \ldots, X_N$ be independent copies of X. Then

$$\mathbb{E}\left|\frac{1}{N}\sum_{i=1}^{N} X_i - \mathbb{E}X\right| = O\left(\frac{1}{\sqrt{N}}\right). \tag{5.30}$$

Furthermore, the large deviation theory allows one to estimate the probability that the empirical mean $\frac{1}{N}\sum\limits_{i=1}^{N} X_i$ stays close to the true mean $\mathbb{E}X$.

Matrix-valued versions of this inequality are harder to prove. The absolute value must be replaced by the matrix norm. So, instead of proving a large deviation estimate for a single random variable, we have to estimate the *supremum of a random process*. This requires deeper probabilistic techniques. The following theorem generalizes the main result of [93].

Theorem 5.9.1 (Rudelson and Vershynin [311]). *Let* $\mathbf{y}$ *be a random vector in* $\mathbb{R}^n$, *which is uniformly bounded almost everywhere:* $\|\mathbf{y}\|_2 \leqslant \alpha$. *Assume for normalization that* $\|\mathbf{y} \otimes \mathbf{y}\|_2 \leqslant 1$. *Let* $\mathbf{y}_1, \ldots, \mathbf{y}_N$ *be independent copies of* $\mathbf{y}$. *Let*

$$\sigma := C\sqrt{\frac{\log N}{N}} \cdot \alpha.$$

Then

1. If $\sigma < 1$, *then*

$$\mathbb{E}\left\|\frac{1}{N}\sum_{i=1}^{N} \mathbf{y}_i \otimes \mathbf{y} - \mathbb{E}(\mathbf{y} \otimes \mathbf{y})\right\|_2 \leqslant \sigma.$$

2. *For every* $t \in (0, 1)$,

$$\mathbb{P}\left\{\left\|\frac{1}{N}\sum_{i=1}^{N}\mathbf{y}_i \otimes \mathbf{y}_i - \mathbb{E}\left(\mathbf{y} \otimes \mathbf{y}\right)\right\|_2 > t\right\} \leqslant 2e^{-ct^2/\sigma^2}.$$

Theorem 5.9.1 generalizes Theorem 5.4.2. Part (1) is a law of large numbers, and part (2) is a large deviation estimate for matrix-valued random variables. The bounded assumption $\|\mathbf{y}\|_2 \leqslant \alpha$ can be too strong for some applications and can be relaxed to the moment assumption $\mathbb{E}\,\|\mathbf{y}\|_2^q \leqslant \alpha^q$, where $q = \log N$. The estimate in Theorem 5.9.1 is in general optimal (see [311]). Part 2 also holds under an assumption that the moments of $\|\mathbf{y}\|_2$ have a nice decay.

Proof. Following [311], we prove this theorem in two steps. First we use the standard symmetrization technique for random variables in Banach spaces, see e.g. [27, Sect. 6]. Then, we adapt the technique of [93] to obtain a bound on a symmetric random process. Note the expectation $\mathbb{E}(\cdot)$ and the average operation $\frac{1}{N}\sum_{i=1}^{N}\mathbf{x}_i \otimes \mathbf{x}_i$ are linear functionals.

Let $\varepsilon_1, \ldots, \varepsilon_N$ denote independent Bernoulli random variables taking values 1, -1 with probability 1/2. Let $\mathbf{y}_1, \ldots, \mathbf{y}_N, \bar{\mathbf{y}}_1, \ldots, \bar{\mathbf{y}}_N$ be independent copies of $\mathbf{y}$. We shall denote by $\mathbb{E}_{\mathbf{y}}, \mathbb{E}_{\bar{\mathbf{y}}}$, and $\mathbb{E}_{\varepsilon}$ the expectation according to $\mathbf{y}_i, \bar{\mathbf{y}}_i$, and ε_i, respectively.

Let $p \geq 1$. We shall estimate

$$E_p := \left(\mathbb{E}\left\|\frac{1}{N}\sum_{i=1}^{N}\mathbf{y}_i \otimes \mathbf{y}_i - \mathbb{E}\left(\mathbf{y} \otimes \mathbf{y}\right)\right\|_2^p\right)^{1/p}. \tag{5.31}$$

Note that

$$\mathbb{E}_{\mathbf{y}}\left(\mathbf{y} \otimes \mathbf{y}\right) = \mathbb{E}_{\bar{\mathbf{y}}}\left(\bar{\mathbf{y}} \otimes \bar{\mathbf{y}}\right) = \mathbb{E}_{\bar{\mathbf{y}}}\left(\frac{1}{N}\sum_{i=1}^{N}\bar{\mathbf{y}}_i \otimes \bar{\mathbf{y}}_i\right).$$

We put this into (5.31). Since $\mathbf{x} \mapsto \|\mathbf{x}\|_2^p$ is a convex function on $\mathbb{R}^n$, Jensen's inequality implies that

$$E_p \leqslant \left(\mathbb{E}_{\mathbf{y}}\mathbb{E}_{\bar{\mathbf{y}}}\left\|\frac{1}{N}\sum_{i=1}^{N}\mathbf{y}_i \otimes \mathbf{y}_i - \frac{1}{N}\sum_{i=1}^{N}\bar{\mathbf{y}}_i \otimes \bar{\mathbf{y}}_i\right\|_2^p\right)^{1/p}.$$

Since $\mathbf{y}_i \otimes \mathbf{y}_i - \bar{\mathbf{y}}_i \otimes \bar{\mathbf{y}}_i$ is a symmetric random variable, it is distributed identically with $\varepsilon_i\left(\mathbf{y}_i \otimes \mathbf{y}_i - \bar{\mathbf{y}}_i \otimes \bar{\mathbf{y}}_i\right)$. As a result, we have

$$E_p \leqslant \left(\mathbb{E}_{\mathbf{y}}\mathbb{E}_{\bar{\mathbf{y}}}\mathbb{E}_{\varepsilon}\left\|\frac{1}{N}\sum_{i=1}^{N}\varepsilon_i\left(\mathbf{y}_i \otimes \mathbf{y}_i - \bar{\mathbf{y}}_i \otimes \bar{\mathbf{y}}_i\right)\right\|_2^p\right)^{1/p}.$$

Denote

$$\mathbf{Y} = \frac{1}{N}\sum_{i=1}^{N} \varepsilon_i \mathbf{y}_i \otimes \mathbf{y}_i \text{ and } \bar{\mathbf{Y}} = \frac{1}{N}\sum_{i=1}^{N} \varepsilon_i \bar{\mathbf{y}}_i \otimes \bar{\mathbf{y}}_i.$$

Then

$$\left\|\mathbf{Y} - \bar{\mathbf{Y}}\right\|_2^p \leqslant \left(\|\mathbf{Y}\|_2 + \left\|\bar{\mathbf{Y}}\right\|_2\right)^p \leqslant 2^p \left(\|\mathbf{Y}\|_2^p + \left\|\bar{\mathbf{Y}}\right\|_2^p\right),$$

and $\mathbb{E}\,\|\mathbf{Y}\|_2^p = \mathbb{E}\left\|\bar{\mathbf{Y}}\right\|_2^p$. Thus we obtain

$$E_p \leqslant 2\left(\mathbb{E}_{\mathbf{y}}\mathbb{E}_{\varepsilon}\left\|\frac{1}{N}\sum_{i=1}^{N} \varepsilon_i\left(\mathbf{y}_i \otimes \mathbf{y}_i\right)\right\|_2^p\right)^{1/p}.$$

We shall estimate the last expectation using Lemma 5.4.3, which was a lemma from [93]. We need to consider the higher order moments:

Lemma 5.9.2 (Rudelson [93]). *Let $\mathbf{y}_1, \ldots, \mathbf{y}_N$ be vectors in $\mathbb{R}^k$ and $\varepsilon_1, \ldots, \varepsilon_N$ be independent Bernoulli variables taking values 1, −1 with probability 1/2. Then*

$$\left(\mathbb{E}\left\|\frac{1}{N}\sum_{i=1}^{N} \varepsilon_i \mathbf{y}_i \otimes \mathbf{y}_i\right\|_2^p\right)^p \leqslant C_0\sqrt{p + \log k} \cdot \max_{i=1,\ldots,N} \|\mathbf{y}_i\|_2 \cdot \left\|\sum_{i=1}^{N} \mathbf{y}_i \otimes \mathbf{y}_i\right\|_2^{1/2}.$$

Remark 5.9.3. We can consider the vectors $\mathbf{y}_1, \ldots, \mathbf{y}_N$ as vectors in their *linear span*, so we can always choose the dimension k of the ambient space at most N.

Combining Lemma 5.9.2 with Remark 5.9.3 and using Hölder's inequality, we obtain

$$E_p \leqslant 2C_0\frac{\sqrt{p + \log N}}{N} \cdot \alpha \cdot \left(\mathbb{E}\left\|\frac{1}{N}\sum_{i=1}^{N} \mathbf{y}_i \otimes \mathbf{y}_i\right\|_2^p\right)^{1/2p}. \tag{5.32}$$

By Minkowski's inequality we have

$$\begin{aligned}\left(\mathbb{E}\left\|\sum_{i=1}^{N} \mathbf{y}_i \otimes \mathbf{y}_i\right\|_2^p\right)^{1/p} &\leqslant N\left[\left(\mathbb{E}\left\|\frac{1}{N}\sum_{i=1}^{N} \mathbf{y}_i \otimes \mathbf{y}_i - \mathbb{E}\left(\mathbf{y} \otimes \mathbf{y}\right)\right\|_2^p\right)^{1/p} + \|\mathbb{E}\left(\mathbf{y} \otimes \mathbf{y}\right)\|_2\right] \\ &\leqslant N\left(E_p + 1\right).\end{aligned}$$

So we get

$$E_p \leqslant \frac{\sigma p^{1/2}}{2}\left(E_p + 1\right), \text{ where } \sigma = 4C_0\left(\frac{\log N}{N}\right)^{1/2}\alpha.$$

It follows that

$$\min\left(E_p, 1\right) \leqslant \sigma\sqrt{p}. \tag{5.33}$$

To prove part 1 of the theorem, note that $\sigma \leq 1$ by the assumption. We thus obtain $E_1 \leq \sigma$. This proves part 1.

To prove part 2, we consider $E_p = \left(\mathbb{E}Z^p\right)^{1/p}$, where

$$Z = \left\| \frac{1}{N}\sum_{i=1}^{N} \mathbf{y}_i \otimes \mathbf{y}_i - \mathbb{E}\left(\mathbf{y} \otimes \mathbf{y}\right) \right\|_2.$$

So (5.33) implies that

$$\left(\mathbb{E}\min\left(Z, 1\right)^p\right)^{1/p} \leqslant \min\left(E_p, 1\right) \leqslant \sigma\sqrt{p}. \tag{5.34}$$

We can express this moment bound as a tail probability estimate using the following standard lemma, see e.g. [27, Lemmas 3.7 and 4.10].

Lemma 5.9.4. *Let Z be a nonnegative random variable. Assume that there exists a constant $K > 0$ such that $\left(\mathbb{E}Z^p\right)^{1/p} \leqslant K\sqrt{p}$ for all $p \geq 1$. Then*

$$\mathbb{P}\left(Z > t\right) \leqslant 2\exp\left(-c_1 t^2/K^2\right) \textit{ for all } t > 0.$$

It thus follows from this and from (5.34) that

$$\mathbb{P}\left(\min\left(Z, 1\right) > t\right) \leqslant 2\exp\left(-c_1 t^2/K^2\right) \text{ for all } t > 0.$$

This completes the proof of the theorem. □

Example 5.9.5 (Bounded Random Vectors). In Theorem 5.9.1, we let $\mathbf{y}$ be a random vector in $\mathbb{R}^n$, which is uniformly bounded almost everywhere: $\left\|\mathbf{y}\right\|_2 \leqslant \alpha$. Since $\left\|\mathbf{A}\right\|_2 = \sigma_1\left(\mathbf{A}\right)$, where σ_1 is the largest singular value. This is very convenient to use in practice. In many problems, we have the prior knowledge that the random vectors are bounded as above. This bound constraint leads to more sharp inequalities. One task is how to formulate the problem using this additional bound constraint.

5.10 Low Rank Approximation

We assume that $\mathbf{A}$ has a small rank–or can be approximated by an (unknown) matrix of a small rank. We intend to find a low rank approximation of $\mathbf{A}$, from only a small random submatrix of $\mathbf{A}$.

Solving this problem is essential to development of fast Monte-Carlo algorithms for computations on large matrices. An extremely large matrix—say, of the order of $10^5 \times 10^5$—is impossible to upload into the random access memory (RAM) of a computer; it is instead stored in an external memory. On the other hand, sampling a submatrix of **A**, storing it in RAM and computing its small rank approximation is feasible.

The best fixed rank approximation to **A** is given by the partial sum of the Singular Value Decomposition (SVD)

$$\mathbf{A} = \sum_i \sigma_i \left(\mathbf{A}\right) \mathbf{u}_i \otimes \mathbf{v}_i$$

where $\sigma_i(\mathbf{A})$ are the nonincreasing and nonnegative sequence of the singular values of $\mathbf{A}$, and $\mathbf{u}_i$ and $\mathbf{v}_i$ are left and right singular vectors of $\mathbf{A}$, respectively. The best rank k approximation to $\mathbf{A}$ in both the spectral and Frobenius norms is thus $\mathbf{A}P_k$, where $\mathbf{P}_k$ is the orthogonal projection onto the top k left singular vectors of $\mathbf{A}$. In particular, for the spectral norm we have

$$\min_{\mathbf{B}:\mathrm{rank}(\mathbf{B})\leqslant k} \left\|\mathbf{A} - \mathbf{B}\right\|_2 = \left\|\mathbf{A} - \mathbf{A}\mathbf{P}_k\right\|_2 = \sigma_{k+1}\left(\mathbf{A}\right). \tag{5.35}$$

However, computing $\mathbf{P}_k$, which gives the first elements of the SVD of a $m \times n$ matrix $\mathbf{A}$ is often impossible in practice since (1) it would take many passes through $\mathbf{A}$, which is extremely slow for a matrix stored in an external memory; (2) this would take superlinear time in $m + n$. Instead, it was proposed in [312–316] to use the Monte-Carlo methodology: namely, appropriate the k-th partial sum of the SVD of $\mathbf{A}$ by the k-th partial sum of the SVD of a random submatrix of $\mathbf{A}$. Rudelson and Vershynin [311] have shown the following:

1. With almost linear sample complexity $O(r \log r)$, that is by sampling only $O(r \log r)$ random rows of $\mathbf{A}$, if $\mathbf{A}$ is approximiable by a rank r matrix;
2. In one pass through $\mathbf{A}$ if the matrix is stored row-by-row, and in two passes if its entries are stored in arbitrary order;
3. Using RAM space are stored and time $O(n + m)$ (and polynomial in r and k).

Theorem 5.10.1 (Rudelson and Vershynin [311]). *Let* $\mathbf{A}$ *be an* $m \times n$ *matrix with numerical rank* $r = \left\|\mathbf{A}\right\|_F^2 / \left\|\mathbf{A}\right\|_2^2$. *Let* $\varepsilon, \delta \in (0,1)$, *and let* $d \leq m$ *be an integer such that*

$$d \geqslant C\left(\frac{r}{\varepsilon^4\delta}\right)\log\left(\frac{r}{\varepsilon^4\delta}\right).$$

Consider a $d \times n$ *matrix* $\tilde{\mathbf{A}}$, *which consists of normalized rows of* $\mathbf{A}$ *picked independently with replacement, with probabilities proportional to the squares of their Euclidean lengths. Then, with probability at least* $1 - 2\exp(-c/\delta)$, *the following holds. For a positive integer* k, *let* $\mathbf{P}_k$ *be the orthogonal projection onto the top* k *left singular value vectors of* $\tilde{\mathbf{A}}$. *Then*

$$\left\|\mathbf{A} - \mathbf{A}\mathbf{P}_k\right\|_2 \leqslant \sigma_{k+1}\left(\mathbf{A}\right) + \varepsilon \left\|\mathbf{A}\right\|_2^2. \tag{5.36}$$

Here and in the sequel, $C, c, C_1, \ldots$ denote positive absolute constants.

We make the following remarks:

1. Optimality. The almost linear sample complexity $O(r \log r)$ achieved in Theorem 5.10.1 is optimal. The best previous result had $O(r^2)$ [314, 315]
2. Numerical rank. The numerical rank $r = \|\mathbf{A}\|_F^2 / \|\mathbf{A}\|_2^2$ is a relaxation of the exact notion of rank. Indeed, one always has $r(\mathbf{A})\text{rank}(\mathbf{A})$. The numerical rank is stable under small perturbation of the matrix, as opposed to the exact rank.
3. Law of large numbers for matrix-valued random variables. The new feature is a use of Rudelson's argument about random vectors in the isotropic position. See Sect. 5.4. It yields a law of large numbers for matrix-valued random variables. We apply it for independent copies of a rank one random matrix, which is given by a random row of the matrix $\mathbf{A}^T\mathbf{A}$—the sample covariance matrix.
4. Functional-analytic nature. A matrix is a linear operator between finite-dimensional normed spaces. It is natural to look for stable quantities tied to linear operators, which govern the picture. For example, operator (matrix) norms are stable quantities, while rank is not. The low rank approximation in Theorem 5.10.1 is only controlled by the numerical rank r. The dimension n does not play a separate role in these results.

Proof. By the homogeneity, we can assume $\|\mathbf{A}\|_2^2 = 1$. The following lemma from [314, 316] reduces Theorem 5.10.1 to a comparison of $\mathbf{A}$ and a sample $\tilde{\mathbf{A}}$ in the spectral norm.

Lemma 5.10.2 (Drineas and Kannan [314, 316]).

$$\|\mathbf{A} - \mathbf{A}\mathbf{P}_k\|_2^2 \leqslant \sigma_{k+1}(\mathbf{A})^2 + 2\left\|\mathbf{A}^T\mathbf{A} - \tilde{\mathbf{A}}^T\tilde{\mathbf{A}}\right\|_2.$$

Proof of the Lemma. We have

$$\begin{aligned}\|\mathbf{A}-\mathbf{A}\mathbf{P}_k\|_2^2 &= \sup_{\mathbf{x}\in\ker \mathbf{P}_k, \|\mathbf{x}\|_2=1} \|\mathbf{A}\mathbf{x}\|_2^2 = \sup_{\mathbf{x}\in\ker \mathbf{P}_k, \|\mathbf{x}\|_2=1} \langle \mathbf{A}^T\mathbf{A}\mathbf{x}, \mathbf{x}\rangle \\ &\leqslant \sup_{\mathbf{x}\in\ker \mathbf{P}_k, \|\mathbf{x}\|_2=1} \left\langle \left(\mathbf{A}^T\mathbf{A} - \tilde{\mathbf{A}}^T\tilde{\mathbf{A}}\right)\mathbf{x}, \mathbf{x}\right\rangle + \sup_{\mathbf{x}\in\ker \mathbf{P}_k, \|\mathbf{x}\|_2=1} \left\langle \tilde{\mathbf{A}}^T\tilde{\mathbf{A}}\mathbf{x}, \mathbf{x}\right\rangle \\ &= \left\|\mathbf{A}^T\mathbf{A} - \tilde{\mathbf{A}}^T\tilde{\mathbf{A}}\right\|_2 + \sigma_{k+1}\left(\tilde{\mathbf{A}}^T\tilde{\mathbf{A}}\right).\end{aligned}$$

$\tilde{\mathbf{A}}$ stands for the *kernel or null space* of matrix $\mathbf{A}$ [16].[1] By a result of perturbation theory, $\left|\sigma_{k+1}\left(\mathbf{A}^T\mathbf{A}\right) - \sigma_{k+1}\left(\tilde{\mathbf{A}}^T\tilde{\mathbf{A}}\right)\right| \leqslant \left\|\mathbf{A}^T\mathbf{A} - \tilde{\mathbf{A}}^T\tilde{\mathbf{A}}\right\|_2$. This proves the lemma.

Let $\mathbf{x}_1, \ldots, \mathbf{x}_m$ denote the rows of the matrix $\mathbf{A}$. Then

---

[1]Let $\mathcal{A}$ is a linear transformation from vector V to vector W. The subset in V

$$\ker(\mathcal{A}) = \{v \in V : \mathcal{A}(v) = 0 \in W\}$$

is a subspace of V, called the kernel or null space of A.

$$\mathbf{A}^T\mathbf{A} = \sum_{i=1}^{m} \mathbf{x}_i \otimes \mathbf{x}_i.$$

We shall regard the matrix $\mathbf{A}^T\mathbf{A}$ as the *true mean* of a bounded matrix valued random variable, while $\tilde{\mathbf{A}}^T\tilde{\mathbf{A}}$ will be its *empirical mean*; then the Law of Large Numbers for matrix valued random variables, Theorem 5.9.1, will be used. To this purpose, we define a random vector $\mathbf{y} \in \mathbb{R}^n$ as

$$\mathbb{P}\left(\mathbf{y} = \frac{\|\mathbf{A}\|_F}{\|\mathbf{x_i}\|_2}\mathbf{x_i}\right) = \frac{\|\mathbf{x_i}\|_2}{\|\mathbf{A}\|_F}.$$

Let $\mathbf{y}_1, \ldots, \mathbf{y}_N$ be independent copies of $\mathbf{y}$. Let the matrix $\tilde{\mathbf{A}}$ consist of rows $\frac{1}{\sqrt{N}}\mathbf{y_1}, \ldots, \frac{1}{\sqrt{N}}\mathbf{y_N}$. The normalization of $\tilde{\mathbf{A}}$ is different from the statement of Theorem 5.10.1: in the proof, it is convenient to multiply $\tilde{\mathbf{A}}$ by the factor $\frac{1}{\sqrt{N}}\|\mathbf{A}\|_F$. However the singular value vectors of $\tilde{\mathbf{A}}$ and thus $\mathbf{P}_k$ do not change. Then,

$$\mathbf{A}^T\mathbf{A} = \mathbb{E}\left(\mathbf{y} \otimes \mathbf{y}\right), \tilde{\mathbf{A}}^T\tilde{\mathbf{A}} = \frac{1}{\sqrt{N}}\sum_{i=1}^{N} \mathbf{y}_i \otimes \mathbf{y}_i, \alpha := \|\mathbf{y}\|_2 = \|\mathbf{A}\|_F = \sqrt{r}.$$

We can thus apply Theorem 5.9.1. Due to our assumption on N, we have

$$\sigma := 4C_0\left(\frac{\log N}{N} \cdot r\right)^{1/2} \leqslant \frac{1}{2}\varepsilon^2\delta^{1/2} < 1.$$

Thus Theorem 5.9.1 gives that, with probability at least $1 - 2\exp(-c/\delta)$, we have

$$\|\mathbf{A} - \mathbf{A}\mathbf{P}_k\|_2 \leqslant \sigma_{k+1}\left(\mathbf{A}\right) + \sqrt{2}\left\|\mathbf{A}^T\mathbf{A} - \tilde{\mathbf{A}}^T\tilde{\mathbf{A}}\right\|_2^{1/2} \leqslant \sigma_{k+1}\left(\mathbf{A}\right) + \varepsilon.$$

This proves Theorem 5.10.1. □

Let us comment on algorithmic aspects of Theorem 5.10.1. Finding a good low rank approximation to a matrix $\mathbf{A}$ comes down, due to Theorem 5.10.1, to sampling a random submatrix $\tilde{\mathbf{A}}$ and computing its SVD (actually, left singular vectors are needed). The algorithm works well if the numerical rank $r = \|\mathbf{A}\|_F^2 / \|\mathbf{A}\|_2^2$ of the matrix $\mathbf{A}$ is small. This is the case, in particular, when $\mathbf{A}$ is essentially a low rank matrix, since $r\left(\mathbf{A}\right) \leqslant \text{rank}\left(\mathbf{A}\right)$.

First, the algorithm samples $N = O(r\log r)$ random rows of $\mathbf{A}$. That is, it takes N independent samples of the random vector $\mathbf{y}$ whose law is

$$\mathbb{P}\left(\mathbf{y} = \frac{\mathbf{A}_i}{\|\mathbf{A}_i\|_2}\right) = \frac{\|\mathbf{A}_i\|_2^2}{\|\mathbf{A}\|_F^2}$$

where $\mathbf{A}_i$ is the i-th row of $\mathbf{A}$. This sampling can be done in one pass through the matrix $\mathbf{A}$ if the matrix is stored row-by-row, and in two passes if its entries are stored in arbitrary order [317, Sect. 5.1]. Second, the algorithm computes the SVD of the $N \times n$ matrix $\tilde{\mathbf{A}}$, which consists of the normalized sampled rows. This can be done in time $O(Nn)+$ the time needed to compute the SVD of a $N \times N$ matrix. The latter can be done by one of the known methods. This algorithm is takes significantly less time than computing SVD of the original $m \times n$ matrix $\mathbf{A}$. In particular, this algorithm is linear in the dimensions of the matrix (and polynomial in N).

5.11 Random Matrices with Independent Entries

The material here is taken from [72]. For a general random matrix $\mathbf{A}$ with independent centered entries bounded by 1, one can use Talagrand's concentration inequality for convex Lipschitz functions on the cube [142, 148, 231]. Since $\sigma_{\max}(\mathbf{A}) = \|\mathbf{A}\|$ (or $\sigma_1(\mathbf{A})$) is a convex function of $\mathbf{A}$. Talagrand's concentration inequality implies

$$\mathbb{P}\left(\left|\sigma_{\max}(\mathbf{A}) - \mathbb{M}\left(\sigma_{\max}(\mathbf{A})\right)\right| \geqslant t\right) \leqslant 2e^{-ct^2},$$

where $\mathbb{M}$ is the median. Although the precise value of the median may be unknown, integration of this inequality shows that

$$\left|\mathbb{E}\sigma_{\max}(\mathbf{A}) - \mathbb{M}\left(\sigma_{\max}(\mathbf{A})\right)\right| \leqslant C$$

Theorem 5.11.1 (Gordon's theorem for Gaussian matrices [72]). *Let $\mathbf{A}$ be an $N \times n$ matrix whose entries are independent standard normal random variables. Then,*

$$\sqrt{N} - \sqrt{n} \leqslant \mathbb{E}\sigma_{\min}(\mathbf{A}) \leqslant \mathbb{E}\sigma_{\max}(\mathbf{A}) \leqslant \sqrt{N} + \sqrt{n}.$$

Let f be a real valued Lipschitz function on $\mathbb{R}^n$ with Lipschitz constant K, i.e. $|f(\mathbf{x}) - f(\mathbf{y})| \leqslant K\|\mathbf{x} - \mathbf{y}\|_2$ for all $\mathbf{x}, y \in \mathbb{R}^n$ (such functions are also called K-Lipschitz).

Theorem 5.11.2 (Concentration in Gauss space [141]). *Let a real-valued function f is K-Lipschitz on $\mathbb{R}^n$. Let $\mathbf{x}$ be the standard normal random vector in $\mathbb{R}^n$. Then, for every $t \geq 0$, one has*

$$\mathbb{P}\left(f(\mathbf{x}) - \mathbb{E}f(\mathbf{x}) > t\right) \leqslant e^{-t^2/2K^2}.$$

Corollary 5.11.3 (Gaussian matrices, deviation; [145]). *Let $\mathbf{A}$ be an $N \times n$ matrix whose entries are independent standard normal random variables. Then, for every $t \geq 0$, with probability at least $1 - 2e^{-t^2/2}$, one has*

$$\sqrt{N} - \sqrt{n} - t \leqslant \sigma_{\min}(\mathbf{A}) \leqslant \sigma_{\max}(\mathbf{A}) \leqslant \sqrt{N} + \sqrt{n} + t.$$

Proof. $\sigma_{\min}(\mathbf{A})$ and $\sigma_{\max}(\mathbf{A})$ are 1-Lipschitz ($K = 1$) functions of matrices $\mathbf{A}$ considered as vectors in $\mathbb{R}^n$. The conclusion now follows from the estimates on the expectation (Theorem 5.11.1) and Gaussian concentration (Theorem 5.11.2). □

Lemma 5.11.4 (Approximate isometries [72]). *Consider a matrix* $\mathbf{X}$ *that satisfies*

$$\|\mathbf{X}^*\mathbf{X} - \mathbf{I}\| \leqslant \max\left(\delta, \delta^2\right) \tag{5.37}$$

for some $\delta > 0$. *Then*

$$1 - \delta \leqslant \sigma_{\min}(\mathbf{A}) \leqslant \sigma_{\max}(\mathbf{A}) \leqslant 1 + \delta. \tag{5.38}$$

Conversely, if $\mathbf{X}$ *satisfies* (5.38) *for some* $\delta > 0$, *then*

$$\|\mathbf{X}^*\mathbf{X} - \mathbf{I}\| \leqslant 3\max\left(\delta, \delta^2\right).$$

Often, we have $\delta = O\left(\sqrt{n/N}\right)$.

5.12 Random Matrices with Independent Rows

Independent rows are used to form a random matrix. In an abstract setting, an infinite-dimensional function (finite-dimensional vector) is regarded as a 'point' in some suitable space and an infinite-dimensional integral operator (finite-dimensional matrix) as a transformation of one 'point' to another. Since a point is conceptually simpler than a function, this view has the merit of removing some mathematical clutter from the problem, making it possible to see the salient issues more clearly [89, p. ix].

Traditionally, we require the entries of a random matrix are independent; Here, however, the requirements of independent rows are much more relaxed than independent entries. A row is a finite-dimensional vector (a 'point' in a finite-dimensional vector space).

The two proofs taken from [72] are used to illustrate the approach by showing how concentration inequalities are at the core of the proofs. In particular, the approach can handle the tough problem of matrix rows with heavy tails.

5.12.1 Independent Rows

$$\left\|\frac{1}{N}\mathbf{A}^*\mathbf{A} - \mathbf{I}\right\| \leqslant \max\left(\delta, \delta^2\right) \text{ where } \delta = C\sqrt{\frac{n}{N}} + \frac{t}{\sqrt{N}}. \tag{5.39}$$

Theorem 5.12.1 (Sub-Gaussian rows [72]). *Let* $\mathbf{A}$ *be an* $N \times n$ *matrix whose rows* $\mathbf{A}_i$ *are independent, sub-Gaussian, isotropic random vectors in* $\mathbb{R}^n$. *Then, for every* $t \geq 0$, *the following inequality holds with probability at least* $1 - 2n \cdot e^{-ct^2}$*:*

$$\sqrt{N} - C\sqrt{n} - t \leqslant \sigma_{\min}(\mathbf{A}) \leqslant \sigma_{\max}(\mathbf{A}) \leqslant \sqrt{N} + C\sqrt{n} + t. \tag{5.40}$$

Here $C = C_K, c = c_K > 0$ *depend only on the sub-Gaussian norm* $K = \max_i \|\mathbf{A}_i\|_{\psi_2}^2$ *of the rows.*

This result is a general version of Corollary 5.11.3; instead of independent Gaussian entries we allow independent sub-Gaussian entries such as Gaussian and Bernoulli. It also applies to some natural matrices whose entries are not independent.

Proof. The proof is taken from [72], and changed to our notation habits. The proof is a basic version of a *covering argument*, and it has three steps. The use of covering arguments in a similar context goes back to Milman's proof of Dvoretzky's theorem [318]. See e.g. [152,319] for an introduction. In the more narrow context of extremal singular values of random matrices, this type of argument appears recently e.g. in [301].

We need to control $\|\mathbf{Ax}\|_2$ for all vectors $\mathbf{x}$ on the unit sphere S^{n-1}. To this purpose, we *discretize the sphere* using the **net** $\mathcal{N}$ (called the approximation or sampling step), establish a tight control of $\|\mathbf{Ax}\|_2$ for every fixed vector $\mathbf{x} \in \mathcal{N}$ with high probability (the concentration step), and finish off by taking a union bound over all $\mathbf{x}$ in the net. The concentration step will be based on the deviation inequality for sub-exponential random variable, Corollary 1.9.4.

Step 1: Approximation. Recalling Lemma 5.11.4 for the matrix $\mathbf{B} = \mathbf{A}/\sqrt{N}$ we see that the claim of the theorem is equivalent to

$$\left\| \frac{1}{N}\mathbf{A}^*\mathbf{A} - \mathbf{I} \right\| \leqslant \max\left(\delta, \delta^2\right) \text{ where } \delta = C\sqrt{\frac{m}{N}}\frac{t}{\sqrt{N}}. \tag{5.41}$$

With the aid of Lemma 1.10.4, we can evaluate the norm in (5.41) on a $\frac{1}{4}$-net $\mathcal{N}$ of the unit sphere S^{n-1} :

$$\left\| \frac{1}{N}\mathbf{A}^*\mathbf{A} - \mathbf{I} \right\| \leqslant 2 \max_{\mathbf{x}\in\mathcal{N}} \left| \left\langle \left(\frac{1}{N}\mathbf{A}^*\mathbf{A} - \mathbf{I} \right) \mathbf{x}, \mathbf{x} \right\rangle \right| = 2 \max_{\mathbf{x}\in\mathcal{N}} \left| \frac{1}{N} \|\mathbf{Ax}\|_2^2 - 1 \right|.$$

To complete the proof, it is sufficient to show that, with the required probability,

$$\max_{\mathbf{x}\in\mathcal{N}} \left| \frac{1}{N} \|\mathbf{Ax}\|_2^2 - 1 \right| \leqslant \frac{\varepsilon}{2}.$$

By Lemma 1.10.2, we can choose the net $\mathcal{N}$ such that it has cardinality $|\mathcal{N}| \leqslant 9^n$.

Step 2: Concentration. Let us fix any vector $\mathbf{x} \in S^{n-1}$. We can rewrite $\|\mathbf{Ax}\|_2^2$ as a sum of independent (scalar) random variables

$$\|\mathbf{Ax}\|_2^2 = \sum_{i=1}^{N} \langle \mathbf{A}_i, \mathbf{x} \rangle^2 =: \sum_{i=1}^{N} Z_i^2 \tag{5.42}$$

where $\mathbf{A}_i$ denote the rows of the matrix $\mathbf{A}$. By assumption, $Z_i = \langle \mathbf{A}_i, \mathbf{x} \rangle$ are independent, sub-Gaussian random variables with $\mathbb{E}Z_i^2 = 1$ and $\|Z_i\|_{\psi_2} \leqslant K$. Thus, by Remark 1.9.5 and Lemma 1.9.1, $Z_i - 1$ are independent, centered sub-exponential random variables with $\|Z_i - 1\|_{\psi_1} \leqslant 2\|Z_i\|_{\psi_1} \leqslant 4\|Z_i\|_{\psi_2} \leqslant 4K$. Now we can use an exponential deviation inequality, Corollary 1.9.3, to control the sum (5.42). Since $K \geqslant \|Z_i\|_{\psi_2} \geqslant \frac{1}{\sqrt{2}}\left(\mathbb{E}Z_i^2\right)^{1/2} = \frac{1}{\sqrt{2}}$, this leads to

$$\begin{aligned}\mathbb{P}\left(\left|\frac{1}{N}\|\mathbf{Ax}\|_2^2 - 1\right| \geqslant \frac{\varepsilon}{2}\right) = \mathbb{P}\left(\left|\frac{1}{N}\sum_{i=1}^{N} Z_i^2 - 1\right| \geqslant \frac{\varepsilon}{2}\right) &\leqslant 2\exp\left[-\frac{c_1}{K^4}\min\left(\varepsilon^2, \varepsilon\right)N\right] \\ = 2\exp\left[-\frac{c_1}{K^4}\delta^2 N\right] &\leqslant 2\exp\left[-\frac{c_1}{K^4}\left(C^2 n + t^2\right)\right]\end{aligned} \tag{5.43}$$

where the last inequality follows by the definition of δ and using the inequality $(a+b)^2 \geqslant a^2 + b^2$ for $a, b \geq 0$.

Step 3: Union Bound. Taking the union bound over the elements of the net $\mathcal{N}$ with the cardinality $|\mathcal{N}| \leq 9^n$, together with (5.43), we have

$$\mathbb{P}\left(\max_{\mathbf{x}\in\mathcal{N}}\left|\frac{1}{N}\|\mathbf{Ax}\|_2^2 - 1\right| \geqslant \frac{\varepsilon}{2}\right) \leqslant 9^n \cdot 2\exp\left[-\frac{c_1}{K^4}\left(C^2 n + t^2\right)\right] \leqslant 2\exp\left(-\frac{c_1}{K^4}t^2\right)$$

where the second inequality follows for $C = C_K$ sufficiently large, e.g., $C = K^2\sqrt{\ln 9/c_1}$.

□

5.12.2 Heavy-Tailed Rows

Theorem 5.12.2 (Heavy-tailed rows [72]). *Let $\mathbf{A}$ be an $N \times n$ matrix whose rows $\mathbf{A}_i$ are independent random vectors in $\mathbb{R}^n$. Let m be a number such that $\|\mathbf{A}_i\|_2 \leqslant \sqrt{m}$ almost surely for all i. Then, for every $t \geq 0$, the following inequality holds with probability at least $1 - 2n \cdot e^{-ct^2}$:*

$$\sqrt{N} - t\sqrt{m} \leqslant \sigma_{\min}(\mathbf{A}) \leqslant \sigma_{\max}(\mathbf{A}) \leqslant \sqrt{N} + t\sqrt{m}. \tag{5.44}$$

Here c is an absolute constant.

Recall from Lemma 5.2.1 that $\mathbb{E}\|\mathbf{A}_i\|_2^2 = n$. This says that one would typically use Theorem 5.12.2 with $m = O(\sqrt{m})$. In this case the result has a form

$$\sqrt{N} - t\sqrt{n} \leqslant \sigma_{\min}(\mathbf{A}) \leqslant \sigma_{\max}(\mathbf{A}) \leqslant \sqrt{N} + t\sqrt{n},$$

probability at least $1 - 2n \cdot e^{-c't^2}$.

Proof. The proof is taken from [72] for the proof and changed to our notation habits. We shall use the non-commutative Bernstein's inequality (for a sum of independent random matrices).

Step 1: Reduction to a sum of independent random matrices. We first note that $m \ge n \ge 1$ since by Lemma 5.2.1 we have that $\mathbb{E}\left\|\mathbf{A}_i\right\|_2^2 = n$. Our argument here is parallel to Step 1 of Theorem 5.12.1. Recalling Lemma 5.11.4 for the matrix $\mathbf{B} = \mathbf{A}/\sqrt{N}$, we find that the desired inequality (5.44) is equivalent to

$$\left\|\frac{1}{N}\mathbf{A}^*\mathbf{A} - \mathbf{I}\right\| \leqslant \max\left(\delta, \delta^2\right) = \varepsilon, \quad \delta = t\sqrt{\frac{m}{N}}. \tag{5.45}$$

Here $\|\cdot\|$ is the operator (or spectral) norm. It is more convenient to express this random matrix as a sum of independent random matrices—the sum is a linear operator:

$$\frac{1}{N}\mathbf{A}^*\mathbf{A} - \mathbf{I} = \frac{1}{N}\sum_{i=1}^{N}\mathbf{A}_i \otimes \mathbf{A}_i - \mathbf{I} = \sum_{i=1}^{N}\mathbf{X}_i, \tag{5.46}$$

where $\mathbf{X}_i = \frac{1}{N}\left(\mathbf{A}_i \otimes \mathbf{A}_i - \mathbf{I}\right)$. Here $\mathbf{X}_i$ are independent centered $n \times n$ random matrices. Equation (5.46) is the standard form we have treated previously in this book.

Step 2: Estimating the mean, range, and variance. Now we are in a position to apply the non-commutative Bernstein inequality, for the sum $\sum\limits_{i=1}^{N}\mathbf{X}_i$. Since $\mathbf{A}_i$ are isotropic random vectors, we have $\mathbb{E}\mathbf{A}_i \otimes \mathbf{A}_i = \mathbf{I}$, which implies $\mathbb{E}\mathbf{X}_i = 0$, which is a required condition to use the non-commutative Bernstein inequality, Theorem 2.17.1.

We estimate the range of $\mathbf{X}_i$ using the assumption that $\left\|\mathbf{A}_i\right\|_2 \leqslant \sqrt{m}$ and $m \ge 1$:

$$\left\|\mathbf{X}_i\right\|_2 \leqslant \frac{1}{N}\left(\left\|\mathbf{A}_i \otimes \mathbf{A}_i\right\| + 1\right) = \frac{1}{N}\left(\left\|\mathbf{A}_i\right\|_2^2 + 1\right) \leqslant \frac{1}{N}\left(m+1\right) \leqslant \frac{2m}{N} = K.$$

Here $\|\cdot\|_2$ is the Euclidean norm. To estimate the total variance $\sum\limits_{i=1}^{N}\mathbb{E}\mathbf{X}_i^2$, we first need to compute

$$\mathbf{X}_i^2 = \frac{1}{N^2}\left(\left(\mathbf{A}_i \otimes \mathbf{A}_i\right)^2 - 2\frac{1}{N}\mathbf{A}_i \otimes \mathbf{A}_i + \mathbf{I}\right).$$

Then, using the isotropic vector assumption $\mathbb{E}\mathbf{A}_i \otimes \mathbf{A}_i = \mathbf{I}$, we have

$$\mathbb{E}\mathbf{X}_i^2 = \frac{1}{N^2}\left(\mathbb{E}\left(\mathbf{A}_i \otimes \mathbf{A}_i\right)^2 - \mathbf{I}\right). \tag{5.47}$$

Since

$$(\mathbf{A}_i \otimes \mathbf{A}_i)^2 = \|\mathbf{A}_i\|_2^2 \, \mathbf{A}_i \otimes \mathbf{A}_i$$

is a positive semi-definite matrix and $\|\mathbf{A}_i\|_2^2 \leqslant m$ by assumption, it follows that $\left\|\mathbb{E}(\mathbf{A}_i \otimes \mathbf{A}_i)^2\right\| \leqslant m \cdot \|\mathbb{E}\mathbf{A}_i \otimes \mathbf{A}_i\| = m$. Inserting this into (5.47), we have

$$\|\mathbb{E}\mathbf{X}_i^2\| \leqslant \frac{1}{N^2}(m+1) \leqslant \frac{2m}{N^2}.$$

where we used the assumption that $m \geq 1$. This leads to

$$\left\|\sum_{i=1}^{N} \mathbb{E}\mathbf{X}_i^2\right\| \leqslant N \cdot \max_i \|\mathbb{E}\mathbf{X}_i^2\| = \frac{2m}{N} = \sigma^2.$$

Step 3: Applying the non-commutative Bernstein's inequality. Applying the non-commutative Bernstein inequality, Theorem 2.17.1, and recalling the definitions of ε and δ in (5.45), we bound the probability in question as

$$\begin{aligned}\mathbb{P}\left(\left\|\tfrac{1}{N}\mathbf{A}^*\mathbf{A} - \mathbf{I}\right\| \geqslant \varepsilon\right) = \mathbb{P}\left(\left\|\sum_{i=1}^{N} \mathbf{X}_i\right\| \geqslant \varepsilon\right) &\leqslant 2n \cdot \exp\left[-c \min\left(\tfrac{\varepsilon^2}{\sigma^2}, \tfrac{\varepsilon}{K}\right)\right] \\ &\leqslant 2n \cdot \exp\left[-c \min\left(\varepsilon^2, \varepsilon\right) \cdot \tfrac{N}{2m}\right] \\ &= 2n \cdot \exp\left[-c \cdot \tfrac{\delta^2 N}{2m}\right] = 2n \cdot \exp\left[-ct^2/2\right].\end{aligned}$$

This completes the proof. □

Theorem 5.12.3 (Heavy-tailed rows, non-isotropic [72]). *Let* $\mathbf{A}$ *be an* $N \times n$ *matrix whose rows* $\mathbf{A}_i$ *are independent random vectors in* $\mathbb{R}^n$ *with the common second moment matrix* $\mathbf{\Sigma} = \mathbb{E}(\mathbf{x}_i \otimes \mathbf{x}_i)$. *Let* m *be a number such that* $\|\mathbf{A}_i\|_2 \leqslant \sqrt{m}$ *almost surely for all* i. *Then, for every* $t \geq 0$, *the following inequality holds with probability at least* $1 - n \cdot e^{-ct^2}$:

$$\left\|\frac{1}{N}\mathbf{A}^*\mathbf{A} - \mathbf{\Sigma}\right\| \leqslant \max\left(\|\mathbf{\Sigma}\|^{1/2}\delta, \delta^2\right) \text{ where } \delta = t\sqrt{\frac{m}{N}}. \tag{5.48}$$

Here c *is an absolute constant. In particular, this inequality gives*

$$\|\mathbf{A}\| \leqslant \|\mathbf{\Sigma}\|^{1/2}\sqrt{N} + t\sqrt{m}. \tag{5.49}$$

Proof. Since

$$\|\mathbf{\Sigma}\| = \|\mathbb{E}(\mathbf{A}_i \otimes \mathbf{A}_i)\| \leqslant \mathbb{E}\|\mathbf{A}_i \otimes \mathbf{A}_i\| = \mathbb{E}\|\mathbf{A}_i\|_2^2 \leqslant m,$$

we have $m \geqslant \|\Sigma\|$. Then (5.48) follows by a straightforward modification of the arguments of Theorem 5.12.2. Also, if (5.48) holds, then by triangle inequality

$$\begin{aligned}\tfrac{1}{N}\|\mathbf{A}\|_2^2 = \left\|\tfrac{1}{N}\mathbf{A}^*\mathbf{A}\right\| &\leqslant \|\Sigma\| + \left\|\tfrac{1}{N}\mathbf{A}^*\mathbf{A} - \Sigma\right\| \\ &\leqslant \|\Sigma\| + \|\Sigma\|^{1/2}\delta + \delta^2.\end{aligned}$$

Taking the square root and multiplying both sides by $\sqrt{N}$, we have (5.49). □

The *almost sure* boundedness in Theorem 5.12.2 may be too restrictive, and it can be relaxed to a bound in expectation.

Theorem 5.12.4 (Heavy-tailed rows; expected singular values [72]). *Let $\mathbf{A}$ be an $N \times n$ matrix whose rows $\mathbf{A}_i$ are independent, isotropic random vectors in $\mathbb{R}^n$. Let $m = \mathbb{E}\max_{i \leqslant N} \|\mathbf{A}_i\|_2^2$. Then*

$$\mathbb{E}\max_{i \leqslant N}\left|\sigma_i(\mathbf{A}) - \sqrt{N}\right| \leqslant C\sqrt{m \log \min(N, n)}$$

where C is an absolute constant.

The proof of this result is similar to that of Theorem 5.12.2, except that this time Rudelson's Lemma, Lemma 5.4.3, instead of matrix Bernstein's inequality. For details, we refer to [72].

Theorem 5.12.5 (Heavy-tailed rows, non-isotropic, expectation [72]). *Let $\mathbf{A}$ be an $N \times n$ matrix whose rows $\mathbf{A}_i$ are independent random vectors in $\mathbb{R}^n$ with the common second moment matrix $\Sigma = \mathbb{E}(\mathbf{x}_i \otimes \mathbf{x}_i)$. Let $m = \mathbb{E}\max_{i \leqslant N} \|\mathbf{A}_i\|_2^2$. Then*

$$\mathbb{E}\left\|\frac{1}{N}\mathbf{A}^*\mathbf{A} - \Sigma\right\| \leqslant \max\left(\|\Sigma\|^{1/2}\delta, \delta^2\right) \ \text{where } \delta = C\sqrt{\frac{m \log \min(N, n)}{N}}.$$

Here C is an absolute constant. In particular, this inequality gives

$$\left(\mathbb{E}\|\mathbf{A}\|^2\right)^{1/2} \leqslant \|\Sigma\|^{1/2}\sqrt{N} + C\sqrt{m \log \min(N, n)}.$$

Let us remark on non-identical second moment. The assumption that the rows $\mathbf{A}_i$ have a common second moment matrix Σ is not essential in Theorems 5.12.3 and 5.12.5. More general versions of these results can be formulated. For example, if $\mathbf{A}_i$ have arbitrary second moment matrices

$$\Sigma_i = \mathbb{E}(\mathbf{x}_i \otimes \mathbf{x}_i),$$

then the claim of Theorem 5.12.5 holds with $\Sigma = \frac{1}{N}\sum_{i=1}^{N}\Sigma_i$.

5.13 Covariance Matrix Estimation

Let $\mathbf{x}$ be a random vector in $\mathbb{R}^n$; for simplicity we assume that $\mathbf{x}$ is centered,[2] $\mathbb{E}\mathbf{x} = 0$. The covariance matrix of $\mathbf{x}$ is the $n \times n$ matrix

$$\mathbf{\Sigma} = \mathbb{E}\left(\mathbf{x} \otimes \mathbf{x}\right).$$

The simplest way to estimate $\mathbf{\Sigma}$ is to take some N independent samples $\mathbf{x}_i$ from the distribution and form the sample covariance matrix

$$\mathbf{\Sigma}_N = \frac{1}{N}\sum_{i=1}^{N} \mathbf{x}_i \otimes \mathbf{x}_i.$$

By the law of large numbers,

$$\mathbf{\Sigma}_N \to \mathbf{\Sigma} \text{ almost surely } N \to \infty.$$

So, taking sufficiently many samples, we are guaranteed to estimate the covariance matrix as well as we want. This, however, does not deal with the quantitative aspect of convergence: what is the minimal *sample size* N that guarantees this approximation with a given accuracy?

We can rewrite $\mathbf{\Sigma}_N$ as

$$\mathbf{\Sigma}_N = \frac{1}{N}\sum_{i=1}^{N} \mathbf{x}_i \otimes \mathbf{x}_i = \frac{1}{N}\mathbf{X}^*\mathbf{X},$$

where

$$\mathbf{X} = \begin{bmatrix} \mathbf{x}_1^T \\ \vdots \\ \mathbf{x}_N^T \end{bmatrix}.$$

The $\mathbf{X}$ is a $N \times n$ random matrix with independent rows $\mathbf{x}_i, i = 1, \ldots, N$, but usually not independent entries.

Theorem 5.13.1 (Covariance estimation for sub-Gaussian distributions [72]). *Consider a sub-Gaussian distribution in $\mathbb{R}^n$ with covariance matrix $\mathbf{\Sigma}$, and let $\varepsilon \in (0,1), t \geq 1$. Then, with probability at least $1 - 2e^{-t^2 n}$, one has*

$$\text{If } N \geqslant C(t/\varepsilon)^2 n \text{ then } \left\|\mathbf{\Sigma}_N - \mathbf{\Sigma}\right\| \leqslant \varepsilon. \tag{5.50}$$

Here $C = C_K$ depends only on the sub-Gaussian norm $K = \|\mathbf{x}\|_{\psi_2}$ of a random vector taken from this distribution.

[2] More generally, in this section we estimate the second moment matrix $\mathbb{E}\left(\mathbf{x} \otimes \mathbf{x}\right)$ of an arbitrary random vector $\mathbf{x}$ (not necessarily centered).

Proof. It follows from (5.39) that for every $s \geq 0$, with probability at least $1 - 2e^{-cs^2}$, we have $\|\mathbf{\Sigma}_N - \mathbf{\Sigma}\| \leqslant \max\left(\delta, \delta^2\right)$ where $\delta = C\sqrt{n/N} + s/\sqrt{N}$. The claim follows for $s = C^{'}t\sqrt{n}$ where $C^{'} = C^{'}_K$ is sufficiently large. □

For arbitrary *centered Gaussian distribution* in $\mathbb{R}^n$, (5.50) becomes

$$\text{If } N \geqslant C(t/\varepsilon)^2 n \text{ then } \|\mathbf{\Sigma}_N - \mathbf{\Sigma}\| \leqslant \varepsilon \|\mathbf{\Sigma}\| . \tag{5.51}$$

Here C is an absolute constant.

Theorem 5.12.3 gives a similar estimation result for arbitrary distribution, possibly heavy-tailed.

Theorem 5.13.2 (Covariance estimation for arbitrary distributions [72]). *Consider a distribution in $\mathbb{R}^n$ with covariance matrix $\mathbf{\Sigma}$, and supported in some centered Euclidean ball whose radius we denote $\sqrt{m}$. Let $\varepsilon \in (0,1)$, $t \geq 1$. Then, with probability at least $1 - n^{-t^2}$, one has*

$$\textit{If } N \geqslant C(t/\varepsilon)^2 \|\mathbf{\Sigma}\|^{-1} m \log n, \textit{ then } \|\mathbf{\Sigma}_N - \mathbf{\Sigma}\| \leqslant \varepsilon \|\mathbf{\Sigma}\| . \tag{5.52}$$

Here C is an absolute constant.

Proof. It follows from Theorem 5.12.3 that for every $s \geq 0$, with probability at least $1 - n \cdot e^{-cs^2}$, we have $\|\mathbf{\Sigma}_N - \mathbf{\Sigma}\| \leqslant \max\left(\|\mathbf{\Sigma}\|^{1/2}\delta, \delta^2\right)$ where $\delta = s\sqrt{m/N}$. Thus, if $N \geqslant C(s/\varepsilon)^2 \|\mathbf{\Sigma}\|^{-1} m \log n$, then $\|\mathbf{\Sigma}_N - \mathbf{\Sigma}\| \leqslant \varepsilon \|\mathbf{\Sigma}\|$. The claim follows for $s = C^{'}t\sqrt{\log n}$ where $C^{'}$ is a sufficiently large absolute constant. □

Theorem 5.52 is typically met with $m = O\left(\|\mathbf{\Sigma}\| n\right)$. For a random vector $\mathbf{x}$ chosen from the distribution at hand, the expected norm is

$$\mathbb{E}\|\mathbf{x}\|_2^2 = \mathrm{Tr}\left(\mathbf{\Sigma}\right) \leqslant n \|\mathbf{\Sigma}\| .$$

Recall that $\|\mathbf{\Sigma}\| = \sigma_1\left(\mathbf{\Sigma}\right)$ is the matrix norm which is also equal to the largest singular value. So, by Markov's inequality, most of the distribution is supported in a centered ball of radius $\sqrt{m}$ where $m = O\left(\|\mathbf{\Sigma}\| n\right)$. If all the distribution is supported there, i.e., if $\|\mathbf{x}\| = O\left(\|\mathbf{\Sigma}\| n\right)$ almost surely, then the claim of Theorem 5.52 holds with sample size

$$N \geqslant C(t/\varepsilon)^2 r \log n.$$

Let us consider low-rank estimation. In this case, the distribution in $\mathbb{R}^n$ lies close to a low dimensional subspace. As a result, a much smaller sample size N is sufficient for covariance estimation. The intrinsic dimension of the distribution can be expressed as the effective rank of the matrix $\mathbf{\Sigma}$, defined as

$$r\left(\mathbf{\Sigma}\right) = \frac{\mathrm{Tr}\left(\mathbf{\Sigma}\right)}{\|\mathbf{\Sigma}\|}.$$

One always has $r\left(\mathbf{\Sigma}\right) \leqslant \mathrm{rank}\left(\mathbf{\Sigma}\right) \leqslant n$, and this bound is sharp. For example, for an isotropic random vector $\mathbf{x}$ in $\mathbb{R}^n$, we have $\mathbf{\Sigma} = \mathbf{I}$ and $r\left(\mathbf{\Sigma}\right) = n$.

The effective rank $r = r(\mathbf{\Sigma})$ always controls the typical norm of $\mathbf{x}$, since $\mathbb{E}\|\mathbf{x}\|_2^2 = Tr(\mathbf{\Sigma}) = r\|\mathbf{\Sigma}\|$. It follows from Markov's inequality that most of the distribution is supported within a ball of radius $\sqrt{m}$ where $m = r\|\mathbf{\Sigma}\|$. Assume that all of the distribution is supported there, i.e., if $\|\mathbf{x}\| = O\left(\sqrt{r\|\mathbf{\Sigma}\|}\right)$ almost surely, then, the claim of Theorem 5.52 holds with sample size

$$N \geqslant C(t/\varepsilon)^2 r \log n.$$

The bounded assumption in Theorem 5.52 is necessary. For an isotropic distribution which is highly concentrated at the origin, the sample covariance matrix will likely equal 0. Still we can use a weaker assumption $\mathbb{E}\max_{i \leqslant N}\|\mathbf{x}_i\|_2^2 \leqslant m$ where $\mathbf{x}_i$ denote the sample points. In this case, the covariance estimation will be guaranteed in expectation rather than with high probability.

A different way to impose the bounded assumption is to reject any sample points $\mathbf{x}_i$ that fall outside the centered ball of radius $\sqrt{m}$. This is equivalent to sampling from the conditional distribution inside the ball. The conditional distribution satisfies the bounded requirement, so the results obtained above provide a good covariance estimation for it. In many cases, this estimate works even for the original distribution—that is, if only a small part of the distribution lies outside the ball of radius $\sqrt{m}$. For more details, refer to [320].

5.13.1 Estimating the Covariance of Random Matrices

The material here can be found in [321]. In recent years, interest in matrix valued random variables gained momentum. Many of the results dealing with real random variables and random vectors were extended to cover random matrices. Concentration inequalities like Bernstein, Hoeffding and others were obtained in the non-commutative setting. The methods used were mostly combination of methods from the real/vector case and some matrix inequalities like the Golden-Thompson inequality.

The method will work properly for a class of matrices satisfying a matrix strong regularity assumption which we denote by (MSR) and can be viewed as an analog to the property (SR) defined in [310]. For an $n \times n$ matrix, denote $\|\cdot\|_{\text{op}}$ by the operator norm of $\mathbf{A}$ on ℓ_2^n.

Definition 5.13.3 (Property (MSR)). Let $\mathbf{Y}$ be an $n \times n$ positive semi-definite random matrix such that $\mathbb{E}\mathbf{Y} = \mathbf{I}_{n\times n}$. We will say that $\mathbf{Y}$ satisfies (MSR) if for some $\eta > 0$ we have:

$$\mathbb{P}\left(\|\mathbf{PYP}\| \geqslant t\right) \leqslant \frac{c}{t^{1+\eta}} \quad \forall t \geqslant c \cdot \text{rank}\,(\mathbf{P})$$

where $\mathbf{P}$ is orthogonal projection of $\mathbb{R}^n$.

Theorem 5.13.4 (Youssef [321]). *Let* $\mathbf{X}$ *be an* $n \times n$ *positive semi-definite random matrix satisfying* $\mathbb{E}\mathbf{X} = \mathbf{I}_{n \times n}$ *and (MSR) for some* $\eta > 0$. *Then for every* $\varepsilon < 1$, *taking* $N = C_1(\eta)\frac{n}{\varepsilon^{2+2/\eta}}$ *we have*

$$\mathbb{E}\left\|\frac{1}{N}\sum_{i=1}^{N}\mathbf{X}_i - \mathbf{I}_{n \times n}\right\|_{\text{op}} \leqslant \varepsilon$$

where $\mathbf{X}_1, \ldots, \mathbf{X}_N$ *are independent copies of* $\mathbf{X}$.

We also introduce a regularity assumption on the moments which we denote by (MWR):

$$\exists p > 1 \text{ such that } \quad \mathbb{E}\langle \mathbf{X}\mathbf{z}, \mathbf{z}\rangle^p \leqslant C_p \qquad \forall \mathbf{z} \in S^{n-1}.$$

The proof of Theorem 5.13.4 is based on two theorems with the smallest and largest eigenvalues of $\frac{1}{N}\sum\limits_{i=1}^{N}\mathbf{X}_i$, which are of independent interest.

Theorem 5.13.5 (Youssef [321]). *Let* $\mathbf{X}_i$ $n \times n$ *independent, positive semi-definite random matrices satisfying* $\mathbb{E}\mathbf{X}_i = \mathbf{I}_{n \times n}$ *and (MWR). Let* $\varepsilon < 1$, *then for*

$$N \geqslant 16(16C_p)^{1/(p-1)}\frac{n}{\varepsilon^{\frac{2p-1}{p-1}}}$$

we have

$$\mathbb{E}\lambda_{\min}\left(\frac{1}{N}\sum_{i=1}^{N}\mathbf{X}_i\right) \geqslant 1 - \varepsilon.$$

Theorem 5.13.6 (Youssef [321]). *Let* $\mathbf{X}_i$ $n \times n$ *independent, positive semi-definite random matrices satisfying* $\mathbb{E}\mathbf{X}_i = \mathbf{I}_{n \times n}$ *and (MSR). Then for any* N *we have*

$$\mathbb{E}\lambda_{\max}\left(\sum_{i=1}^{N}\mathbf{X}_i\right) \leqslant C(\eta)(n + N).$$

Moreover, for $\varepsilon < 1$ *and* $N \geqslant C_2(\eta)\frac{n}{\varepsilon^{2+2/\eta}}$ *we have*

$$\mathbb{E}\lambda_{\max}\left(\frac{1}{N}\sum_{i=1}^{N}\mathbf{X}_i\right) \leqslant 1 + \varepsilon.$$

Consider the case of log-concave matrices is also covered in [321].

5.14 Concentration of Singular Values

We primarily follow Rudelson and Vershynin [322] and Vershynin [323] for our exposition. Relevant work also includes [72, 322, 324–333].

Let $\mathbf{A}$ be an $N \times n$ matrix whose entries with real independent, centered random variables with certain moment assumptions. Random matrix theory studies the distribution of the singular values $\sigma_k(\mathbf{A})$, which are the eigenvalues of $|\mathbf{A}| = \sqrt{\mathbf{A}^T\mathbf{A}}$ arranged in nonincreasing order. Of particular significance are the largest and the smallest random variables

$$\sigma_1(\mathbf{A}) = \sup_{\mathbf{x}:\|\mathbf{x}\|_2=1} \|\mathbf{Ax}\|_2, \qquad \sigma_n(\mathbf{A}) = \inf_{\mathbf{x}:\|\mathbf{x}\|_2=1} \|\mathbf{Ax}\|_2. \tag{5.53}$$

Here, we consider *sub-Gaussian random variables* ξ—those whose tails are dominated by that of the standard normal random variables. That is, a random variable is called sug-Gaussian if there exists $B > 0$ such that

$$\mathbb{P}(|\xi| > t) \leqslant 2e^{-t^2/B^2} \text{ for all } t > 0. \tag{5.54}$$

The minimal B in this inequality is called the sub-Gaussian moment of ξ. Inequality (5.54) is often equivalently formulated as the moment condition

$$(\mathbb{E}|\xi|^p)^{1/p} \leqslant CB\sqrt{p} \text{ for all } p \geqslant 1, \tag{5.55}$$

where C is an absolute constant. The class of sub-Gaussian random variables includes many random variables that arise naturally in applications, such as normal, symmetric ± 1 and general bounded random variables.

In this section, we study $N \times n$ real random matrices $\mathbf{A}$ whose entries are independent and identically distributed mean 0 sub-Gaussian random variables. The asymptotic behavior of the extremal singular values of $\mathbf{A}$ is well understood. If the entries have unit variance and the dimension n grows infinity while the aspect ratio n/N converges to a constant $\alpha \in (0, 1)$, then

$$\frac{\sigma_1(\mathbf{A})}{\sqrt{N}} \to 1 + \sqrt{\alpha}, \qquad \frac{\sigma_n(\mathbf{A})}{\sqrt{N}} \to 1 - \sqrt{\alpha}$$

almost surely. The result was proved in [302] for Gaussian matrices, and in [334] for matrices with independent and identically distributed entries with finite fourth moment. In other words, we have asymptotically

$$\sigma_1(\mathbf{A}) \sim \sqrt{N} + \sqrt{n}, \qquad \sigma_n(\mathbf{A}) \sim \sqrt{N} - \sqrt{n}. \tag{5.56}$$

Recently, consider efforts were made to understand non-asymptotic estimates similar to (5.56), which would hold for arbitrary fixed dimensions N and n.

Ledoux [149] is a survey on the largest singular value. A modern, non-asymptotic survey is given in [335], while a tutorial is given in [72]. The discussion in this section is on the smallest singular value, which is much harder to control.

5.14.1 Sharp Small Deviation

Let $\lambda_1(\mathbf{A})$ be the largest eigenvalue of the $n \times n$ random matrix $\mathbf{A}$. Following [336], we present

$$\mathbb{P}(\lambda_1(\mathbf{A}) \geqslant 2 + t) \leqslant Ce^{-cnt^{3/2}},$$

valid uniformly for all n and t. This inequality is sharp for "small deviation" and complements the usual "large deviation" inequality. Our motivation is to illustrate the simplest idea to get such a concentration inequality.

The Gaussian concentration is the easiest. It is straightforward consequence of the measure concentration phenomenon [145] that

$$\mathbb{P}(\lambda_1(\mathbf{A}) \geqslant \mathbb{M}\lambda_1(\mathbf{A}) + t) \leqslant e^{-nt^2}, \forall t > 0, \forall n \tag{5.57}$$

where $\mathbb{M}\lambda_1(\mathbf{A})$ stands for the median of $\lambda_1(\mathbf{A})$ with respect to the probability measure $\mathbb{P}$. One has the same upper bound estimate is the median $\mathbb{M}\lambda_1(\mathbf{A})$ is replaced by the expected value $\mathbb{E}\lambda_1(\mathbf{A})$, which is easier to compute.

The value of $\mathbb{M}\lambda_1(\mathbf{A})$ can be controlled: for example we have

$$\mathbb{M}\lambda_1(\mathbf{A}) \leqslant 2 + c/\sqrt{n}.$$

5.14.2 Sample Covariance Matrices

The entries of the random matrices will be (complex-valued) random variables ξ satisfying the following assumptions:

1. (A1) The distribution of ξ is symmetric; that is, ξ and $-\xi$ are identically distributed;
2. (A2) $\mathbb{E}|\xi|^{2k} \leqslant (C_0 k)^k$ for some constant $C_0 > 0$ (ξ has sub-Gaussian tails).

Also we assume that either $\mathbb{E}\xi^2 = \mathbb{E}\xi\bar{\xi} = 1$ or $\mathbb{E}\xi^2 = 0; \mathbb{E}\xi\bar{\xi} = 1$.

Theorem 5.14.1 ([324]).

$$\mathbb{P}\left\{\|\mathbf{A}\| \geqslant 2\sqrt{n}\,(1+t)\right\} \leqslant C\exp\left(-\frac{1}{C}nt^{3/2}\right),$$

$$\mathbb{P}\left\{\lambda_n\left(\mathbf{B}\right) \geqslant \left(\sqrt{n}+\sqrt{N}\right)^2 + nt\right\} \leqslant C\exp\left(-\frac{1}{C}Nt^{3/2}\right),$$
$$\mathbb{P}\left\{\lambda_1\left(\mathbf{B}\right) \leqslant \left(\sqrt{N}-\sqrt{n}\right)^2 - nt\right\} \leqslant \frac{C}{1-\sqrt{N/n}}\exp\left(-\frac{1}{C}Nt^{3/2}\right),$$

5.14.3 *Tall Matrices*

A result of [337] gives an optimal bound for tall matrices, those whose aspect ratio $\alpha = n/N$ satisfies $\alpha < \alpha_0$ for some sufficiently small constant α_0. Recalling (5.56), one should expect that tall matrices satisfy

$$\sigma_n\left(\mathbf{A}\right) \geqslant c\sqrt{N} \qquad \text{with high probability.} \tag{5.58}$$

It was indeed proven in [337] that for a tall ± 1 matrices one has

$$\mathbb{P}\left(\sigma_n\left(\mathbf{A}\right) \leqslant c\sqrt{N}\right) \leqslant e^{-cN} \tag{5.59}$$

where $\alpha_0 > 0$ and $c > 0$ are absolute constants.

5.14.4 *Almost Square Matrices*

As we move toward square matrices, making the aspect ratio $\alpha = n/N$ arbitrarily close to 1, the problem becomes harder. One still expect (5.58) to be true as long as $\alpha < 1$ is any constant. It was proved in [338] for arbitrary aspect ratios $\alpha < 1 - c/\log n$ and for general random matrices with independent sub-Gaussian entries. One has

$$\mathbb{P}\left(\sigma_n\left(\mathbf{A}\right) \leqslant c_\alpha\sqrt{N}\right) \leqslant e^{-cN}, \tag{5.60}$$

where $c_\alpha > 0$ depends on α and the maximal sub-Gaussian moment of the entries.

Later [339], the dependence of c_α on the aspect ratio in (5.60) was improved for random ± 1 matrices; however the probability estimates there was weaker than (5.60). An estimate for sub-Gaussian random matrices of all dimensions was obtained in a breakthrough work [340]. For any $\varepsilon \geqslant C/\sqrt{N}$, it was shown that

$$\mathbb{P}\left(\sigma_n\left(\mathbf{A}\right) \leqslant \varepsilon\left(1-\alpha\right)\left(\sqrt{N}-\sqrt{n}\right)\right) \leqslant \left(C\varepsilon\right)^{N-n} + e^{-cN}.$$

However, because of the factor $1 - \alpha$, this estimate is suboptimal and does not correspond to the expected asymptotic behavior (5.56).

5.14.5 Square Matrices

The extreme case for the problem of estimating the singular values is for the square matrices, where $N = n$. Equation (5.56) is useless for square matrices. However, for "almost" square matrices, those with constant defect $N - n = O(1)$ is of order $1/\sqrt{N}$, so (5.56) heuristically suggests that

$$\sigma_n(\mathbf{A}) \geqslant \frac{c}{\sqrt{N}} \qquad \text{with high probability.} \tag{5.61}$$

This conjecture was proved recently in [341] for all square sub-Gaussian matrices:

$$\mathbb{P}\left(\sigma_n(\mathbf{A}) \geqslant \frac{c}{\sqrt{N}}\right) \leqslant C\varepsilon + e^{-cN}. \tag{5.62}$$

5.14.6 Rectangular Matrices

Rudelson and Vershynin [322] proved the conjectural bound for $\sigma(\mathbf{A})$, which is valid for all sub-Gaussian matrices in all fixed dimensions N, n. The bound is optimal for matrices with all aspects we encountered above.

Theorem 5.14.2 (Rudelson and Vershynin [322]). *Let* $\mathbf{A}$ *be an* $N \times n$ *random matrix,* $N \geq n$*, whose elements are independent copies of a mean 0 sub-Gaussian random variable with unit variance. Then, for every* $t > 0$*, we have*

$$\mathbb{P}\left(\sigma_n(\mathbf{A}) \leqslant t\left(\sqrt{N} - \sqrt{n-1}\right)\right) \leqslant (Ct)^{N-n+1} + e^{-cN}, \tag{5.63}$$

where C, c *depend (polynomial) only on the sub-Gaussian moment* B.

For tall matrices, Theorem 5.14.2 clearly amounts to the known estimates (5.58) and (5.59). For square matrices $N = n$, the quantity $\sqrt{N} - \sqrt{N-1}$ is of order $1/\sqrt{N}$, so Theorem 5.14.2 amounts to the known estimates (5.61) and (5.62). Finally, for matrices that are arbitrarily close to square, Theorem 5.14.2 gives the new optimal estimate

$$\sigma_n(\mathbf{A}) \geqslant c\left(\sqrt{N} - \sqrt{n}\right) \qquad \text{with high probability.} \tag{5.64}$$

This is a version of (5.56), now valid for all fixed dimensions. This bound were explicitly conjectured in [342].

Theorem 5.14.2 seems to be new even for Gaussian matrices.

Vershynin [323] extends the argument of [322] for random matrices with bounded $(4+\varepsilon)$-th moment. It follows directly from the argument of [322] and [323, Theorem 1.1] (which is Eq. 5.69 below.)

Corollary 5.14.3 (Vershynin [323]). *Let $\varepsilon \in (0,1)$ and $N \geq n$ be positive integers. Let $\mathbf{A}$ be a random $N \times n$ matrix whose entries are i.i.d. random variables with mean 0, unit variance and $(4+\varepsilon)$-th moment bounded by α. Then, for every $\delta > 0$ there exists $t > 0$ and n_0 which depend only on ε, δ and α, and such that*

$$\mathbb{P}\left(\sigma_n\left(\mathbf{A}\right) \leqslant t\left(\sqrt{N} - \sqrt{n-1}\right)\right) \leqslant \delta, \textit{ for all } n \geqslant n_0.$$

After the paper of [323] was written, two important related results appeared on the universality of the smallest singular value in two extreme regimes–for almost square matrices [326] and for genuinely rectangular matrices [324]. The result of Tao and Vu [326] works for square and almost square matrices where the defect $N - n$ is constant. It is valid for matrices with i.i.d. entries with mean 0, unit variance, and bounded C-th moment where C is a sufficiently large absolute constant. The result says that the smallest singular value of such $N \times n$ matrices $\mathbf{A}$ is asymptotically the same as of the Gaussian matrix $\mathbf{G}$ of the same dimensions and with i.i.d. standard normal entries. Specifically,

$$\begin{aligned} \mathbb{P}\left(N\sigma_n(\mathbf{G})^2 \leqslant t - N^{-c}\right) - N^{-c} &\leqslant \mathbb{P}\left(N\sigma_n(\mathbf{A})^2 \leqslant t\right), \\ &\leqslant \mathbb{P}\left(N\sigma_n(\mathbf{G})^2 \leqslant t + N^{-c}\right) + N^{-c}. \end{aligned} \tag{5.65}$$

Another result was obtained by Feldheim and Sodin [324] for genuinely rectangular matrices, i.e. with aspect ratio N/n separated from 1 by a constant and with sub-Gaussian i.i.d. entries. In particular, they proved

$$\mathbb{P}\left(\sigma_n\left(\mathbf{A}\right) \leqslant \left(\sqrt{N} - \sqrt{n}\right)^2 - tN\right) \leqslant \frac{C}{1 - \sqrt{N/n}} e^{-cnt^{3/2}}. \tag{5.66}$$

Equations (5.63) and (5.66) complements each other—the former is multiplicative (and is valid for arbitrary dimensions) while the latter is additive (and is applicable for genuinely rectangular matrices.) Each of these two inequalities clearly has the regime where it is stronger.

The permanent of an $n \times n$ matrix $\mathbf{A}$ is defined as

$$\text{per}\left(\mathbf{A}\right) = \sum_{\pi \in \mathcal{S}_n} a_{1,\pi(1)} a_{2,\pi(2)} \cdots a_{n,\pi(n)},$$

where the summation is over all permutations of n elements. If $x_{i,j}$ are i.i.d. 0 mean variables with unit variance and $\mathbf{X}$ is an $n \times n$ matrix with entries $x_{i,j}$ then an easy computation shows that

$$\text{per}\left(\mathbf{A}\right) = \mathbb{E}\left(\det\left(\mathbf{A}_{1/2} \odot \mathbf{X}\right)\right)^2,$$

where for any two $n \times m$ matrices $\mathbf{A}, \mathbf{B}$, $\mathbf{D} = \mathbf{A} \odot \mathbf{B}$ denotes their Hadamard or Schur, product, i.e., the $n \times m$ matrix with entries $d_{i,j} = a_{i,j} \cdot b_{i,j}$, and where $A_{1/2}(i,j) = A(i,j)^{1/2}$. For a class of matrices that arise from δ, κ-strongly connected graph, i.e., graphs with good expansion properties, Rudelson and Zeitouni [343] showed the following: Let $\mathbf{G}$ be the $n \times n$ standard Gaussian matrix. The constants $C, C, c, c, \ldots$ depend only on δ, κ. For any $\tau \geq 1$ and any adjacency matrix $\mathbf{A}$ of a (δ, κ)-strongly connected graph,

$$\begin{aligned} &\mathbb{P}\left(\left|\log\det^2\left(\mathbf{A}_{1/2} \odot \mathbf{G}\right)\right| - \mathbb{E}\left|\log\det^2\left(\mathbf{A}_{1/2} \odot \mathbf{G}\right)\right| > C(\tau n \log n)^{1/3}\right) \\ &\leqslant \exp\left(-\tau\right) + \exp\left(-c\sqrt{n}/\log n\right). \end{aligned}$$

and

$$\mathbb{E}\left|\log\det^2\left(\mathbf{A}_{1/2} \odot \mathbf{G}\right)\right| \leqslant \log\operatorname{per}\left(\mathbf{A}\right) \leqslant \mathbb{E}\left|\log\det^2\left(\mathbf{A}_{1/2} \odot \mathbf{G}\right)\right| + C'\sqrt{n\log n}.$$

Further, we have

$$\mathbb{P}\left(\left|\log\frac{\det^2\left(\mathbf{A}_{1/2} \odot \mathbf{G}\right)}{\operatorname{per}\left(\mathbf{A}\right)}\right| > 2C'\sqrt{n\log n}\right) \leqslant \exp\left(-c\sqrt{n}/\log n\right).$$

For the smallest singular value, they showed that

$$\mathbb{P}\left(s_n\left(\mathbf{A} \odot \mathbf{G}\right) \leqslant ct/\sqrt{n}\right) \leqslant t + e^{-c'n},$$

and for any $n/2 < k < n - 4$

$$\mathbb{P}\left(s_k\left(\mathbf{A} \odot \mathbf{G}\right) \leqslant ct\frac{n-k}{\sqrt{n}}\right) \leqslant t^{(n-k)/4} + e^{-c'n}.$$

5.14.7 Products of Random and Deterministic Matrices

We study the $\mathbf{B} = \mathbf{\Gamma} A$, where $\mathbf{A}$ is a random matrix with independent 0 mean entries and $\mathbf{\Gamma}$ is a fixed matrix. Under the $(4+\varepsilon)$-th moment assumption on the entries of $\mathbf{A}$, it is shown in [323] that the spectral norm of such an $N \times n$ matrix $\mathbf{B}$ is bounded by $\sqrt{N} + \sqrt{n}$, which is sharp.

$\mathbf{B} = \mathbf{\Gamma} A$ can be equivalently regarded as sample covariance matrices of a wide class of random vectors—the linear transformation of vectors with independent entries.

Recall the spectral norm $\|\mathbf{W}\|_2$ is defined as the largest singular value of a matrix $\mathbf{W}$, which equals the largest eigenvalue of $|\mathbf{A}| = \sqrt{\mathbf{A}^T\mathbf{A}}$. Equivalently, the spectral norm can be defined as the $l_2 \to l_2$ operator norm:

$$\sigma_1\left(\mathbf{A}\right) = \sup_{\mathbf{x}:\|\mathbf{x}\|_2 = 1} \|\mathbf{A}\mathbf{x}\|_2,$$

where $\|\cdot\|_2$ denotes the Euclidean norm.

For random matrices with independent and identically distributed entries, the spectral norm is well studies. Let $\mathbf{B}$ be an $N \times n$ matrix whose entries are real independent and identically distributed random variables with mean 0, variance 1, and finite fourth moment. Estimates of the type

$$\sigma_1(\mathbf{B}) \sim \sqrt{N} + \sqrt{n}, \tag{5.67}$$

are known to hold (and are sharp) in both the limit regime for dimensions increasing to infinity, and the non-limit regime where the dimensions are fixed. The meaning of (5.67) is that, for a family of matrices as above whose aspect ratio N/n converges to a constant, the ratio $\sigma_1(\mathbf{B}) / \left(\sqrt{N} + \sqrt{n}\right)$ converges to 1 almost surely [344].

In the non-limit regime, i.e., for arbitrary dimensions n and N. Variants of (5.67) was proved by Seginer [107] and Latala [194]. If $\mathbf{B}$ is an $N \times n$ matrix whose entries are i.i.d. mean 0 random variables, then denoting the rows of $\mathbf{B}$ by $\mathbf{x}_i$ and the columns by $\mathbf{y}_j$, the result of Seginer [107] says that

$$\mathbb{E}\sigma_1(\mathbf{B}) \leqslant C \left(\mathbb{E}\max_i \|\mathbf{x}_i\|_2 + \mathbb{E}\max_i \|\mathbf{y}_i\|_2\right)$$

where C is an absolute constant. The estimate is sharp because $\sigma_1(\mathbf{B})$ is bounded below by the Euclidean norm of any row and any column of $\mathbf{B}$.

$$\max_i$$

If the entries of the matrix $\mathbf{B}$ are not necessarily identically distributed, the result of Latala [194] says that

$$\mathbb{E}\sigma_1(\mathbf{B}) \leqslant C \left(\mathbb{E}\max_i \|\mathbf{x}_i\|_2 + \mathbb{E}\max_i \|\mathbf{y}_i\|_2 + \left(\sum_{i,j} b_{ij}^4\right)^{1/4}\right),$$

where b_{ij} are entries of the matrix $\mathbf{B}$. In particular, if $\mathbf{B}$ is an $N \times n$ matrix whose entries are independent random variables with mean 0 and fourth moments bounded by 1, then one can deduce from either Seginer's or Latala's result that

$$\mathbb{E}\sigma_1(\mathbf{B}) \leqslant C \left(\sqrt{N} + \sqrt{n}\right). \tag{5.68}$$

This is a variant of (5.67) in the non-limit regime.

The fourth moment is known to be necessary. Consider again a family of matrices whose dimensions N and n increase to infinity, and whose aspect ratio N/n converges to a constant. If the entries are i.i.d. random variables with mean 0 and *infinite fourth moment*, then the upper limit of the ratio $\sigma_1(\mathbf{B}) / \left(\sqrt{N} + \sqrt{n}\right)$ is infinite almost surely [344].

The main result of [323] is an extension of the optimal bound (5.68) to the class of random matrices with non-independent entries, but which can be factored through a matrix with independent entries.

Theorem 5.14.4 (Vershynin [323]). *Let $\varepsilon \in (0,1)$ and let m, n, N be positive integers. Consider a random $m \times n$ matrix $\mathbf{B} = \mathbf{\Gamma} \boldsymbol{A}$, where $\mathbf{A}$ is an $N \times n$ random matrix whose entries are independent random variables with mean 0 and $(4+\varepsilon)$-th moment bounded by 1, and $\mathbf{\Gamma}$ is an $m \times N$ non-random matrix such that $\sigma_1(\mathbf{\Gamma}) \leqslant 1$. Then*

$$\mathbb{E}\sigma_1(\mathbf{B}) \leqslant C(\varepsilon)\left(\sqrt{m} + \sqrt{n}\right) \tag{5.69}$$

where $C(\varepsilon)$ is a function that depends only on ε.

Let us give some remarks on Eq. 5.69:

1. The conclusion is independent of N.
2. The proof of Eq. 5.69 gives the stronger estimate

$$\mathbb{E}\sigma_1(\mathbf{B}) \leqslant C(\varepsilon)\left(\sigma_1(\mathbf{\Gamma})\sqrt{m} + \|\mathbf{\Gamma}\|_{\mathrm{HS}}\right)$$

   which is valid for arbitrary (non-random) $m \times N$ matrix $\mathbf{\Gamma}$. Here $\|\cdot\|_{\mathrm{HS}}$ denotes the Hilbert-Schmidt norm or Frobenius norm. This result is independent of the dimensions of $\mathbf{\Gamma}$, therefore holds for an arbitrary linear operator $\mathbf{\Gamma}$ acting from the N-dimensional Euclidean space l_2^N to an arbitrary Hilbert space.
3. Equation 5.69 can be interpreted in terms of *sample covariance matrices* of random vectors in $\mathbb{R}^m$ of the form $\mathbf{\Gamma}\boldsymbol{x}$, where $\mathbf{x}$ is a random vector in $\mathbb{R}^n$ with independent entries. Let $\mathbf{A}$ be the random matrix whose columns are n independent samples of the random vector $\mathbf{x}$. Then $\mathbf{B} = \mathbf{\Gamma}\boldsymbol{A}$ is the matrix whose columns are n independent samples of the random vector $\mathbf{\Gamma}\boldsymbol{x}$. The sample covariance matrix of the random vector $\mathbf{\Gamma}\boldsymbol{x}$, is defined as $\mathbf{\Sigma} = \frac{1}{n}\mathbf{B}\mathbf{B}^T$. Equation 5.69 says that the largest eigenvalue of $\mathbf{\Sigma}$ is bounded by $C_1(\varepsilon)\left(1 + \frac{m}{n}\right)$, which is further bound by $C_2(\varepsilon)$ for the number of samples $n \geq m$ (and independently of the dimension N). This problem was studied [345, 346] in the asymptotic limit regime for $m = N$, where the result must of course dependent on N.

5.14.8 Random Determinant

Random determinant can be used the test metric for hypothesis testing, especially for extremely low signal to noise ratio. The variance of random determinant is much smaller than that of individual eigenvalues. We follow [347] for this exposition. Let $\mathbf{A}_n$ be an $n \times n$ random matrix whose entries $a_{ij}, 1 \leq i, j \leq n$, are independent real random variables of 0 mean and unit variance. We will refer to the entries a_{ij} as the atom variables. This shows that almost surely, $\log|\det(\mathbf{A}_n)|$ is $(1/2 + o(1))\, n \log n$ but does not provide any distributional information. For other models of random matrices, we refer to [348].

In [349], Goodman considered random Gaussian matrices where the atom variables are i.i.d. standard Gaussian variables. He noticed that in this case the determinant is a product of independent Chi-square variables. Therefore, its logarithm is the sum of independent variables and thus one expects a central limit theorem to hold. In fact, using properties of Chi square distribution, it is not very hard to prove

$$\frac{\log\left|\det\left(\mathbf{A}_n\right)\right| - \frac{1}{2}\log\left(n-1\right)!}{\sqrt{\frac{1}{2}\log n}} \to \mathcal{N}\left(0,1\right). \tag{5.70}$$

In [242], Tao and Vu proved that for Bernoulli random matrices, with probability tending to one (as n tents to infinity)

$$\sqrt{n!}\exp\left(-c\sqrt{n\log n}\right) \leqslant \left|\det\left(\mathbf{A}_n\right)\right| \leqslant \sqrt{n!}\omega\left(n\right) \tag{5.71}$$

for any function $\omega(n)$ tending to infinity with n. We say that a random variable ξ satisfies condition **C0** (with positive constants C_1, C_2) if

$$\mathbb{P}\left(\left|\xi\right| \geqslant t\right) \leqslant C_1\exp\left(-t^{C_2}\right) \tag{5.72}$$

for all $t > 0$. Nguyen and Vu [347] showed that the logarithm of $\left|\det\left(\mathbf{A}_n\right)\right|$ satisfies a central limit theorem. Assume that all atom variables a_{ij} satisfy condition **C0** with some positive constants C_1, C_2. Then

$$\sup_{t\in\mathbb{R}}\left|\mathbb{P}\left(\frac{\log\left|\det\left(\mathbf{A}_n\right)\right| - \frac{1}{2}\log\left(n-1\right)!}{\sqrt{\frac{1}{2}\log n}} \leqslant t\right) - \Phi(t)\right| \leqslant \log^{-1/3+o(1)}n. \tag{5.73}$$

Here $\Phi(t) = \mathbb{P}\left(\mathcal{N}\left(0,1\right) < t\right) = \frac{1}{\sqrt{2\pi}}\int_{-\infty}^{t}\exp\left(-x^2/2\right)dx$. An equivalent form is

$$\sup_{t\in\mathbb{R}}\left|\mathbb{P}\left(\frac{\log\det\left(\mathbf{A}_n^2\right) - \log\left(n-1\right)!}{\sqrt{2\log n}} \leqslant t\right) - \Phi(t)\right| \leqslant \log^{-1/3+o(1)}n. \tag{5.74}$$

For illustration, we see Fig. 5.1.

Example 5.14.5 (Hypothesis testing).

$$\begin{aligned}\mathcal{H}_0 &: \mathbf{R}_y = \mathbf{R}_x\\ \mathcal{H}_1 &: \mathbf{R}_y = \mathbf{R}_x + \mathbf{R}_s\end{aligned}$$

where $\mathbf{R}_x$ is an $n \times n$ random matrix whose entries are independent real random variables of 0 mean and unit variance, and $\mathbf{R}_s$ is an positive definite matrix of $n \times n$. Random determinant can be used the test metric for hypothesis testing, especially

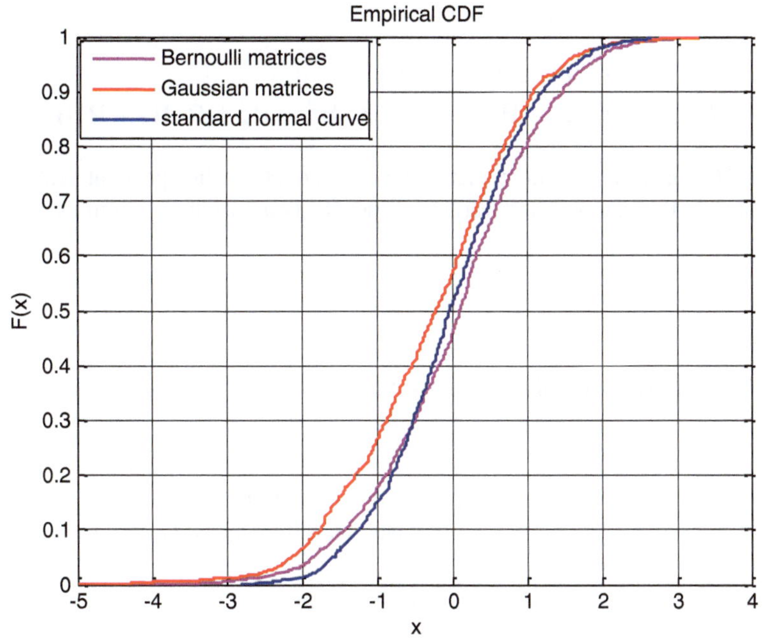

Fig. 5.1 The plot compares the distributions of $\left(\log\det\mathbf{A}^2 - \log(n-1)!\right)/\sqrt{2\log n}$ for random Bernoulli matrices, random Gaussian matrices, and $\mathcal{N}(0,1)$. We sampled 1,000 matrices of size 1,000 by 1,000 for each ensemble

for extremely low signal to noise ratio. The variance of random determinant is much smaller than that of individual eigenvalues. According to (5.73), the hypothesis test is

$$\begin{aligned}\mathcal{H}_0 &: \log|\det(\mathbf{R}_x)| \\ \mathcal{H}_1 &: \log|\det(\mathbf{R}_x+\mathbf{R}_s)|\end{aligned}$$

Let $\mathbf{A}, B$ be complex matrices of $n\times n$, and assume that $\mathbf{A}$ is positive definite. Then for $n\geq 2$ [18, p. 535]

$$\det\mathbf{A} \leqslant |\det\mathbf{A}| + |\det\mathbf{B}| \leqslant |\det(\mathbf{A}+\mathbf{B})| \tag{5.75}$$

It follows from (5.75) that

$$\begin{aligned}&\mathcal{H}_0 : \log|\det(\mathbf{R}_x)| \approx \frac{1}{2}\log(n-1)! \\ &\mathcal{H}_1 : \log|\det(\mathbf{R}_x+\mathbf{R}_s)| \geqslant \log(\det\mathbf{R}_s + |\det\mathbf{R}_x|) \geqslant \log|\det(\mathbf{R}_x)| \approx \frac{1}{2}\log(n-1)!\end{aligned}$$

where $\mathbf{R}_s$ is assumed to be positive definite and $\mathbf{R}_x$ is an arbitrary complex matrix. Our algorithm claims $\mathcal{H}_1$ if

$$\log|\det(\mathbf{R}_x+\mathbf{R}_s)| \geqslant \frac{1}{2}\log(n-1)!.$$

Equivalently, from (5.74), we can investigate

$$\begin{aligned}\mathcal{H}_0 &: \log\det\left(\mathbf{R}_x^2\right) \approx \log(n-1)!\\ \mathcal{H}_1 &: \log\det\left(\mathbf{R}_x + \mathbf{R}_s\right)^2 = \log\det\left(\mathbf{R}_x^2 + \mathbf{R}_s^2 + \mathbf{R}_x\mathbf{R}_s + \mathbf{R}_s\mathbf{R}_x\right)\end{aligned}$$

where $\mathbf{R}_x$ an $n \times n$ random matrix whose entries are independent real random variables of 0 mean and unit variance, and $\mathbf{R}_s$ is an arbitrary complex matrix of $n \times n$.

□

5.15 Invertibility of Random Matrix

We follow [350] for this exposition. Given an $n \times n$ random matrix $\mathbf{A}$, what is the probability that $\mathbf{A}$ is invertible, or at least "close" to being invertible? One natural way to measure this property is to estimate the following *small ball probability*

$$\mathbb{P}\left(s_n\left(\mathbf{A}\right) \leqslant t\right),$$

where

$$s_n\left(\mathbf{A}\right) \stackrel{\text{def}}{=} \inf_{\|\mathbf{x}\|_1=1} \|\mathbf{A}\mathbf{x}\|_2 = \frac{1}{\|\mathbf{A}^{-1}\|}.$$

In the case when the entries of $\mathbf{A}$ are i.i.d. random variables with appropriate moment assumption, the problem was studied in [239, 241, 341, 351, 352]. In particular, in [341] it is shown that if the above diagonal entries of $\mathbf{A}$ are continuous and satisfy certain regularity conditions, namely that the entries are i.i.d. subGaussian and satisfy certain smoothness conditions, then

$$\mathbb{P}\left(s_n\left(\mathbf{A}\right) \leqslant t\right) \leqslant C\sqrt{n}t + e^{-cn}. \tag{5.76}$$

where c, C depend on the moment of the entries.

Several cases of dependent entries have also been studied. A bound similar to (5.76) for the case when the rows are independent log-concave random vectors was obtained in [353, 354]. Another case of dependent entries is when the matrix is symmetric, which was studied in [355–360]. In particular, in [355], it is shown that if the above diagonal entries of $\mathbf{A}$ are continuous and satisfy certain regularity conditions, namely that the entries are i.i.d. subgaussian and satisfy certain smoothness conditions, then

$$\mathbb{P}\left[s_n\left(\mathbf{A}\right) \leqslant t\right] \leqslant C\sqrt{n}t.$$

The regularity assumptions were completely removed in [356] at the cost of a $n^{5/2}$ (independence of the entries in the non-symmetric part is still needed). On the other hand, in the discrete case, the result of [360] shows that if $\mathbf{A}$ is, say, symmetric whose above diagonal entries are i.i.d. Bernoulli random variables, then

$$\mathbb{P}\left[s_n\left(\mathbf{A}\right) = 0\right] \leqslant e^{-n^c},$$

where c is an absolute constant.

A more general case is the so called Smooth Analysis of random matrices, where now we replace the matrix $\mathbf{A}$ by $\mathbf{A} + \boldsymbol{\Gamma}$, where $\boldsymbol{\Gamma}$ being an arbitrary deterministic matrix. The first result in this direction can be found in [361], where it is shown that if $\mathbf{A}$ is a random matrix with i.i.d. standard normal entries, then

$$\mathbb{P}\left(s_n\left(\boldsymbol{\Gamma} + \mathbf{A}\right) \leqslant t\right) \leqslant C\sqrt{n}t. \tag{5.77}$$

Further development in this direction can be found in [362] estimates similar to (1.2) are given in the case when $\mathbf{A}$ is a Bernoulli random matrix, and in [356, 358, 359], where $\mathbf{A}$ is symmetric.

An alternative way to measure the invertibility of a random matrix $\mathbf{A}$ is to estimate $\det(\mathbf{A})$, which was studied in [242,363,364] (when the entries are discrete distributions). Here we show that if the diagonal entries of $\mathbf{A}$ are independent *continuous* random variables, we can easily get a small ball estimate for $\det(\boldsymbol{\Gamma}+\mathbf{A})$, where $\boldsymbol{\Gamma}$ being an arbitrary deterministic matrix.

Let $\mathbf{A}$ be an $n \times n$ random matrix, such that each diagonal entry $A_{i,i}$ is a continuous random variable, independent from all the other entries of $\mathbf{A}$. Friedland and Giladi [350] showed that for every $n \times n$ matrix $\boldsymbol{\Gamma}$ and every $t \geq 0$

$$\mathbb{P}\left(\left|\det\left(\mathbf{A} + \boldsymbol{\Gamma}\right)\right| \leqslant t\right) \leqslant 2\alpha n t,$$

where α is a uniform upper bound on the densities of $A_{i,i}$. Further, we have

$$\begin{aligned} &\mathbb{P}\left[\left\|\mathbf{A}\right\| \leqslant t\right] \leqslant 2\alpha n t, \\ &\mathbb{P}\left[s_n\left(\mathbf{A}\right) \leqslant t\right] \leqslant \left(2\alpha\right)^{n/(2n-1)}\left(\mathbb{E}\left\|\mathbf{A}\right\|\right)^{(n-1)/(2n-1)} t^{1/(2n-1)}. \end{aligned} \tag{5.78}$$

Equation (5.78) can be applied to the case when the random matrix $\mathbf{A}$ is symmetric, under very weak assumptions on the distributions and the moments of the entries and under no *independence* assumptions on the above diagonal entries. When $\mathbf{A}$ is symmetric, we have

$$\left\|\mathbf{A}\right\| = \sup_{\left\|\mathbf{x}\right\|_1 = 1} \left\langle \mathbf{A}\mathbf{x}, \mathbf{x}\right\rangle \geqslant \max_{1 \leqslant i \leqslant n} \left|A_{i,i}\right|.$$

Thus, in this case we get a far better small ball estimate for the norm

$$\mathbb{P}\left[\left\|\mathbf{A}\right\| \leqslant t\right] \leqslant \left(2\alpha t\right)^n.$$

Rudelson [76] gives an excellent self-contained lecture notes. We take some material from his notes to get a feel of the proof ingredients. His style is very elegant.

In the classic work on numerical inversion of large matrices, von Neumann and his associates used random matrices to test their algorithms, and they speculated [365, pp. 14, 477, 555] that

$$s_n(\mathbf{A}) \sim 1/\sqrt{n} \text{ with high probability.} \tag{5.79}$$

In a more precise form, this estimate was conjectured by Smale [366] and proved by Edelman[367] and Szarek[368] for random Gaussian matrices $\mathbf{A}$, i.e., those with i.i.d. standard normal entries. Edelman's theorem states that for every $t \in (0,1)$

$$\mathbb{P}\left(s_n(\mathbf{A}) \leqslant t/\sqrt{n}\right) \sim t. \tag{5.80}$$

In [341], the conjecture (5.79) is proved in full generality under the fourth moment assumption.

Theorem 5.15.1 (Invertibility: fourth moment [341]). *Let $\mathbf{A}$ be an $n \times n$ matrix whose entries are independent centered real random variables with variances at least 1 and fourth moments bounded by B. Then, for every $\delta > 0$ there exist $\varepsilon > 0$ and n_0 which depend (polynomially) only on δ and B, such that*

$$\mathbb{P}\left(s_n(\mathbf{A}) \leqslant t/\sqrt{n}\right) \leqslant \delta \text{ for all } n \geqslant n_0.$$

Spielman and Teng[369] conjectured that (5.80) should hold for the random sign matrices up to an exponentially small term that accounts for their singularity probability:

$$\mathbb{P}\left(s_n(\mathbf{A}) \leqslant t/\sqrt{n}\right) \leqslant \varepsilon + c^n.$$

Rudelson and Vershynin prove Spielman-Teng's conjecture up to a coefficient in front of t. Moreover, they show that this type of behavior is common for all matrices with subGaussian i.i.d. entries.

Theorem 5.15.2 (Invertibility: subGaussian [341]). *Let $\mathbf{A}$ be an $n \times n$ matrix whose entries are independent copies of a centered real subGaussian random variable. Then, for every $t \geq 0$, one has*

$$\mathbb{P}\left(s_n(\mathbf{A}) \leqslant t/\sqrt{n}\right) \leqslant C\varepsilon + c^n. \tag{5.81}$$

where $C > 0$ and $c \in (0,1)$.

5.16 Universality of Singular Values

Large complex system often exhibit remarkably simple universal patterns as the numbers of degrees of freedom increases [370]. The simplest example is the central limit theorem: the fluctuation of the sums of independent random scalars, irrespective of their distributions, follows the Gaussian distribution. The other cornerstone of probability theory is to treat the Poisson point process as the universal limit of many independent point-like evens in space or time. The mathematical assumption of independence is often too strong. What if independence is not realistic approximation and strong correlations need to be modelled? Is there a universality for strongly correlated models?

In a sensor network of time-evolving measurements consisting of many sensors—vector time series, it is probably realistic to assume that the measurements of sensors have strong correlations.

Let ξ be a real-valued or complex-valued random variable. Let $\mathbf{A}$ denote the $n \times n$ random matrix whose entries are i.i.d. copies of ξ. One of the two normalizations will be imposed on ξ:

- $\mathbb{R}$-normalization: ξ is real-valued with $\mathbb{E}\xi = 0$ and $\mathbb{E}\xi^2 = 1$.
- $\mathbb{C}$-normalization: ξ is complex-valued with $\mathbb{E}\xi = 0$, $\mathbb{E}\,\mathrm{Re}\,(\xi)^2 = \mathbb{E}\,\mathrm{Im}\,(\xi)^2 = \frac{1}{2}$, and $\mathbb{E}\,\mathrm{Re}\,(\xi)\,\mathrm{Im}\,(\xi) = 0$.

In both cases, ξ has mean 0 and variance 1.

Example 5.16.1 (Normalizations). A model example of a $\mathbb{R}$-normalized random variable is the real Gaussian $\mathcal{N}(0,1)$. Another $\mathbb{R}$-normalized random variable is Bernoulli, in which ξ equals +1 or -1 with an equal probability 1/2 of each.

A model example of $\mathbb{C}$-normalization is the complex Gaussian whose real and imaginary parts are i.i.d. copies of $\frac{1}{\sqrt{2}}\mathcal{N}(0,1)$. □

One frequently views $\sigma_n(\mathbf{A})^2$ as the eigenvalues of the sample covariance matrix $\mathbf{AA}^*$, where $*$ denotes the Hermitian (conjugate and transpose) of a matrix. It is more traditional to write down the limiting distributions in terms of σ^2. We study the "hard edge" of the spectrum, and specifically the least singular value $\sigma_n(\mathbf{A})$. This problem has a long history. It first appeared in the worked of von Neuman and Goldstein concerning numerical inversion of large matrices [371]. Later, Smale [366] made a specific conjecture about the magnitude of σ_n. Motivated by a question of Smale, Edelman [372] computed the distribution of $\sigma_n(\xi)$ for the real and complex Gaussian cases:

Theorem 5.16.2 (Limiting Distribution for Gaussian Models [372]). *For any fixed $t \geq 0$, we have, for real cases,*

$$\mathbb{P}\left(n\sigma_n(\mathbf{A})^2 \leqslant t\right) = \int_0^t \frac{1+\sqrt{x}}{2\sqrt{x}} e^{-(x/2+\sqrt{x})} dx + o(1) \tag{5.82}$$

as well as the exact (!) formula, for complex cases,

$$\mathbb{P}\left(n\sigma_n(\mathbf{A})^2 \leqslant t\right) = \int_0^t e^{-x}dx.$$

Both integrals can be computed explicitly. By exchange of variables, we have

$$\int_0^t \frac{1+\sqrt{x}}{2\sqrt{x}} e^{-(x/2+\sqrt{x})} dx = 1 - e^{-t/2-\sqrt{t}}. \tag{5.83}$$

Also, it is clear that

$$\int_0^t e^{-x}dx = 1 - e^{-t}. \tag{5.84}$$

The joint distribution of the bottom k singular values of real or complex $\mathbf{A}$ was computed in [373].

The error term $o(1)$ in (5.82) is not explicitly stated in [372], but Tao and Vu [326] gives the form of $O(n^{-c})$ for some absolute constant $c > 0$.

Under the assumption of bounded fourth moment $\mathbb{E}|\xi|^4 < \infty$, it was shown by Rudelson and Vershynin [341] that

$$\mathbb{P}\left(n\sigma_n(\mathbf{A})^2 \leqslant t\right) \leqslant f(t) + o(1)$$

for all fixed $t > 0$, where $g(t)$ goes to zero as $t \to 0$. Similarly, in [374] it was shown that

$$\mathbb{P}\left(n\sigma_n(\mathbf{A})^2 \geqslant t\right) \leqslant g(t) + o(1)$$

for all fixed $t > 0$, where $g(t)$ goes to zero as $t \to \infty$. Under stronger assumption that ξ is sub-Gaussian, the lower tail estimate was improved in [341] to

$$\mathbb{P}\left(n\sigma_n(\mathbf{A})^2 \leqslant t\right) \leqslant C\sqrt{t} + c^n \tag{5.85}$$

for some constant $C > 0$ and $0 < c < 1$ depending on the sub-Gaussian moments of ξ. At the other extreme, with no moment assumption on ξ, the bound

$$\mathbb{P}\left(n\sigma_n(\mathbf{A})^2 \leqslant n^{-1-\frac{5}{2}a-a^2}\right) \leqslant n^{-a+o(1)}$$

was shown for any fixed $a > 0$ in [362].

A common feature of the above mention results is that they give good upper and lower tail bounds on $n{\sigma_n}^2$, but not the distribution law. In fact, many papers [341,374,375] are partially motivated by the following conjecture of Spielman and Teng [369].

Conjecture 5.16.3 (Spielman and Teng [369]). Let ξ be the Bernoulli random variable. Then there is a constant $0 < c < 1$ such that for all $t \geq 0$

$$\mathbb{P}\left(n\sigma_n(\mathbf{A})^2 \leqslant t\right) \leqslant t + c^n. \tag{5.86}$$

A new method was introduced by Tao and Vu [326] to study small singular values. Their method is analytic in nature and enables us to prove the universality of the limiting distribution of $n{\sigma_n}^2$.

Theorem 5.16.4 (Universality for the least singular value [326]). *Let ξ be a (real or complex) random variable of mean 0 and variance 1. Suppose $\mathbb{E}|\xi|^{C_0} < \infty$ for some sufficiently large absolute constant C_0. Then, for all $t > 0$, we have,*

$$\mathbb{P}\left(n\sigma_n(\mathbf{A})^2 \leqslant t\right) = \int_0^t \frac{1+\sqrt{x}}{2\sqrt{x}} e^{-(x/2+\sqrt{x})} dx + o(n^{-c}) \tag{5.87}$$

if ξ is $\mathbb{R}$-normalized, and

$$\mathbb{P}\left(n\sigma_n(\mathbf{A})^2 \leqslant t\right) = \int_0^t e^{-x} dx + o(n^{-c})$$

if ξ is $\mathbb{C}$-normalized, where $c > 0$ is an absolute constant. The implied constants in the $O(\cdot)$ notation depends on $\mathbb{E}|\xi|^{C_0}$ but are uniform in t.

Very roughly, one can swap ξ with the appropriate Gaussian distribution $g_{\mathbb{R}}$ or $g_{\mathbb{C}}$, at which point one can basically apply Theorem 5.16.4 as a black box. In other words, the law of $n{\sigma_n}^2$ is *universal* with respect to the choice of ξ by a direct comparison to the Gaussian models. The exact formula $\int_0^t \frac{1+\sqrt{x}}{2\sqrt{x}} e^{-(x/2+\sqrt{x})} dx$ and $\int_0^t e^{-x} dx$ do not play any important role. This comparison (or coupling) approach is in the spirit of Lindeberg's proof [376] of the central limit theorem.

Tao and Vu's arguments are completely effective, and give an explicit value for C_0. For example, $C_0 = 10^4$ is certainly sufficient. Clearly, one can lower C_0 significantly.

Theorem 5.16.4 can be extended to rectangular random matrices of $(n - l) \times n$ dimensions (Fig. 5.2).

Theorem 5.16.5 (Universality for the least singular value of rectangular matrices [326]). *Let ξ be a (real or complex) random variable of mean 0 and variance 1. Suppose $\mathbb{E}|\xi|^{C_0} < \infty$ for some sufficiently large absolute constant C_0. Let l be a constant. Let*

$$X = n\sigma_{n-l}(\mathbf{A}(\xi))^2, X_{g_{\mathbb{R}}} = n\sigma_{n-l}(\mathbf{A}(g_{\mathbb{R}}))^2, X_{g_{\mathbb{C}}} = n\sigma_{n-l}(\mathbf{A}(g_{\mathbb{C}}))^2.$$

Then, there is a constant $c > 0$ such that for all $t \geq 0$, we have,

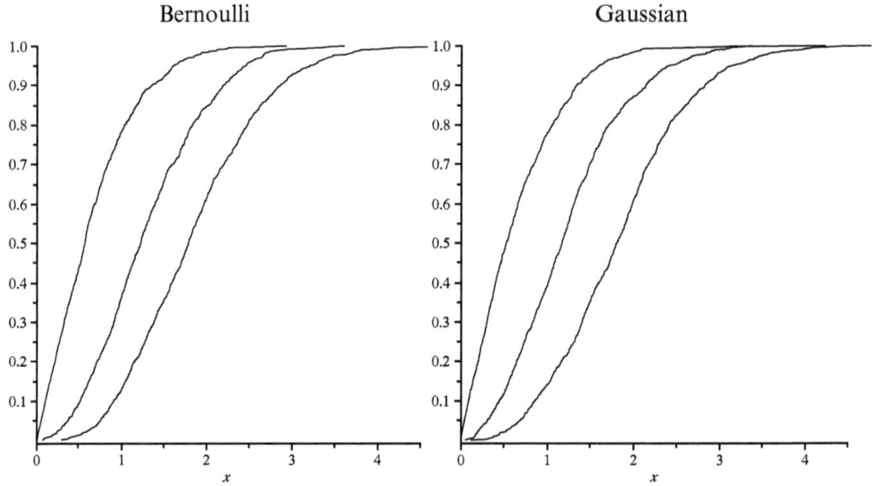

Fig. 5.2 Plotted above are the curves $\mathbb{P}\left(n\sigma_{n-l}(\mathbf{A}(\xi))^2 \leqslant x\right)$, for $l = 0, 1, 2$ based on data from 1,000 randomly generated matrices with $n = 100$. The curves on the *left* were generated with ξ being a random Bernoulli variable, taking the values +1 and -1 each with probability 1/2; The curves on the *right* were generated with ξ being a random Gaussian variable. In both cases, the curves from *left* to *right* correspond to the cases $l = 0, 1, 2$, respectively

$$\mathbb{P}\left(X \leqslant t - n^{-c}\right) - n^{-c} \leqslant \mathbb{P}\left(X_{\mathbf{g}\mathbb{R}} \leqslant t\right) \leqslant \mathbb{P}\left(X \leqslant t + n^{-c}\right) + n^{-c}$$

if ξ is $\mathbb{R}$-normalized, and

$$\mathbb{P}\left(X \leqslant t - n^{-c}\right) - n^{-c} \leqslant \mathbb{P}\left(X_{\mathbf{g}\mathbb{C}} \leqslant t\right) \leqslant \mathbb{P}\left(X \leqslant t + n^{-c}\right) + n^{-c}$$

if ξ is $\mathbb{C}$-normalized.

Theorem 5.16.4 can be extended to random matrices with independent (but not necessarily identical) entries.

Theorem 5.16.6 (Random matrices with independent entries [326]). *Let ξ_{ij} be a (real or complex) random variables with mean 0 and variance 1 ($\mathbb{R}$-normalized or $\mathbb{C}$-normalized). Suppose $\mathbb{E}|\xi|^{C_0} < C_1$ for some sufficiently large absolute constant C_0 and C_1. Then, for all $t > 0$, we have,*

$$\mathbb{P}\left(n\sigma_n(\mathbf{A})^2 \leqslant t\right) = \int_0^t \frac{1+\sqrt{x}}{2\sqrt{x}} e^{-(x/2+\sqrt{x})} dx + o(n^{-c}) \tag{5.88}$$

if ξ_{ij} are all $\mathbb{R}$-normalized, and

$$\mathbb{P}\left(n\sigma_n(\mathbf{A})^2 \leqslant t\right) = \int_0^t e^{-x} dx + o(n^{-c})$$

if ξ_{ij} are all $\mathbb{C}$-normalized, where $c > 0$ is an absolute constant. The implied constants in the $O(\cdot)$ notation depends on $\mathbb{E}|\xi|^{C_0}$ but are uniform in t.

Let us extend Theorem 5.16.4 for the condition number. Let $\mathbf{A}$ be an $n \times n$ matrix, its condition number $\kappa(\mathbf{X})$ is defined as

$$\kappa(\mathbf{A}) = \frac{\sigma_1(\mathbf{A})}{\sigma_n(\mathbf{A})}.$$

It is well known [2] that the largest singular value is concentrated strongly around $2\sqrt{n}$. Combining Theorem 5.16.4 with this fact, we have the following for the general setting.

Lemma 5.16.7 (Concentration of the largest singular value [326]). *Under the setting of Theorem 5.16.4, we have, with probability* $1 - \exp -n^{\Omega(1)}$, $\sigma_1(\mathbf{A}) = (2 + o(1))\sqrt{n}$.

Corollary 5.16.8 (Conditional number [326]). *Let* ξ_{ij} *be a (real or complex) random variables with mean 0 and variance 1 ($\mathbb{R}$-normalized or $\mathbb{C}$-normalized). Suppose* $\mathbb{E}|\xi|^{C_0} < C_1$ *for some sufficiently large absolute constant* C_0 *and* C_1*. Then, for all* $t > 0$*, we have,*

$$\mathbb{P}\left(\frac{1}{2n}\kappa(\mathbf{A}(\xi)) \geqslant t\right) = \int_0^t \frac{1+\sqrt{x}}{2\sqrt{x}} e^{-(x/2+\sqrt{x})} dx + o(n^{-c}) \tag{5.89}$$

if ξ_{ij} *are all $\mathbb{R}$-normalized, and*

$$\mathbb{P}\left(\frac{1}{2n}\kappa(\mathbf{A}(\xi)) \geqslant t\right) = \int_0^t e^{-x} dx + o(n^{-c})$$

if ξ_{ij} *are all $\mathbb{C}$-normalized, where* $c > 0$ *is an absolute constant. The implied constants in the* $O(\cdot)$ *notation depends on* $\mathbb{E}|\xi|^{C_0}$ *but are uniform in* t.

5.16.1 Random Matrix Plus Deterministic Matrix

Let ξ be a complex random variable with mean 0 and variance 1. Let $\mathbf{A}$ be the random matrix of size n whose entries are i.i.d. copies of ξ and $\mathbf{\Gamma}$ be a fixed matrix of the same size. Here we study the conditional number and least singular value of the matrix $\mathbf{B} = \mathbf{\Gamma} + \mathbf{A}$. This is called signal plus noise matrix model. It is interesting to find the "signal" matrix $\mathbf{\Gamma}$ does play a role on tail bounds for the least singular value of $\mathbf{\Gamma} + \mathbf{A}$.

Example 5.16.9 (Covariance Matrix). The conditional number is a random variable of interest to many applications [372, 377]. For example,

$$\hat{\mathbf{\Sigma}} = \mathbf{\Sigma} + \mathbf{Z} \quad \text{or} \quad \mathbf{Z} = \hat{\mathbf{\Sigma}} - \mathbf{\Sigma}$$

where $\mathbf{\Sigma}$ is the true covariance matrix (deterministic and assumed to be known) and $\hat{\mathbf{\Sigma}} = \frac{1}{n}\mathbf{X}\mathbf{X}^*$ is the sample (empirical) covariance matrix—random matrix; here $\mathbf{X}$ is the data matrix which is the only matrix available to the statistician. □

Example 5.16.10 (Hypothesis testing for two matrices).

$$\begin{aligned}\mathcal{H}_0 &: \mathbf{R}_y = \mathbf{R}_n\\ \mathcal{H}_1 &: \mathbf{R}_y = \mathbf{R}_x + \mathbf{R}_n\end{aligned}$$

where $\mathbf{R}_x$ is an arbitrary deterministic matrix and $\mathbf{R}_n$ is a random matrix. □

Let us consider the Gaussian case. Improving the results of Kostlan and Oceanu [366] and Edelman [372] computed the limiting distribution of $\sqrt{n}\sigma_n(\mathbf{A})$.

Theorem 5.16.11 (Gaussian random matrix with i.i.d. entries [372]). *There is a constant $C > 0$ such that the following holds. Let ξ be a real Gaussian random variable with mean 0 and variance 1, let $\mathbf{A}$ be the random matrix whose entries are i.i.d. copies of ξ, and let $\mathbf{\Gamma}$ be an arbitrary fixed matrix. Then, for any $t > 0$,*

$$\mathbb{P}(\sigma_n(\mathbf{A}) \leqslant t) \geqslant n^{1/2}t.$$

Considering the more general model $\mathbf{B} = \mathbf{\Gamma} + \mathbf{A}$, Sankar, Spielman and Teng proved [378].

Theorem 5.16.12 (Deterministic Matrix plus Gaussian random matrix [378]). *There is a constant $C > 0$ such that the following holds. Let ξ be a real Gaussian random variable with mean 0 and variance 1, let $\mathbf{A}$ be the random matrix whose entries are i.i.d. copies of ξ, and let $\mathbf{\Gamma}$ be an arbitrary fixed matrix. $\mathbf{B} = \mathbf{\Gamma} + \mathbf{A}$. Then, for any $t > 0$,*

$$\mathbb{P}(\sigma_n(\mathbf{B}) \leqslant t) \geqslant Cn^{1/2}t.$$

We say ξ is sub-Gaussian if there is a constant $\alpha > 0$ such that

$$\mathbb{P}(|\xi| \geqslant t) \leqslant 2e^{-t^2/\alpha^2}$$

for all $t > 0$. The smallest α is called the sub-Gaussian moment of ξ. For a more general sub-Gaussian random variable ξ, Rudelson and Vershynin proved [341] the following.

Theorem 5.16.13 (Sub-Gaussian random matrix with i.i.d. entries [341]). *Let ξ be a sub-Gaussian random variable with mean 0, variance 1 and sub-Gaussian moment α. Let c be an arbitrary positive constant. Let $\mathbf{A}$ be the random matrix whose entries are i.i.d. copies of ξ, Then, there is a constant $C > 0$ (depending on α) such that, for any $t \geq n^{-c}$,*

$$\mathbb{P}(\sigma_n(\mathbf{A}) \leqslant t) \geqslant Cn^{1/2}t.$$

For the general model $\mathbf{B} = \mathbf{\Gamma} + \mathbf{A}$, Tao and Vu proved [379]

Theorem 5.16.14 (A general model $\mathbf{B} = \mathbf{\Gamma} + \mathbf{A}$ [379]). *Let ξ be a random variable with non-zero variance. Then for any constants $C_1, C > 0$ there exists a constant $C_2 > 0$ (depending on C_1, C, ξ) such that the following holds. Let $\mathbf{A}$ be the random matrix whose entries are i.i.d. copies of ξ, and let $\mathbf{\Gamma}$ be any deterministic $n \times n$ matrix with norm $\|\mathbf{\Gamma}\| \leqslant n^C$. Then, we have*

$$\mathbb{P}\left(\sigma_n\left(\mathbf{\Gamma}+\mathbf{A}\right) \leqslant n^{-C_2}\right) \geqslant n^{-C_1}.$$

This theorem requires very little about the random variable ξ. It does not need to be sub-Gaussian nor even has bounded moments. All we ask is that the variable is bounded from zero, which basically means ξ is indeed "random". Thus, it guarantees the well-conditionness of $\mathbf{B} = \mathbf{\Gamma} + \mathbf{A}$ in a very general setting.

The weakness of this theorem is that the dependence of C_2 on C_1 and C, while explicit, is too generous. The work of [362] improved this dependence significantly.

Let us deal with the non-Gaussian random matrix.

Theorem 5.16.15 (The non-Gaussian random matrix [362]). *There are positive constants c_1 and c_2 such that the following holds. Let $\mathbf{A}$ be the $n \times n$ Bernoulli matrix with n even. For any $\alpha \geq n$, there is an $n \times n$ deterministic matrix $\mathbf{\Gamma}$ such that $\|\mathbf{\Gamma}\| = \alpha$ and*

$$\mathbb{P}\left(\sigma_n\left(\mathbf{\Gamma}+\mathbf{A}\right) \leqslant c_1\frac{n}{\alpha}\right) \geqslant c_2\frac{1}{\sqrt{n}}.$$

The main result of [362] is the following theorem

Theorem 5.16.16 (Bounded second moment on ξ—Tao and Vu [362]). *Let ξ be a random variable with mean 0 and bounded second moment, and let $\gamma \geq 1/2, C \geq 0$ be constants. There is a constant c depending on ξ, γ, C such that the following holds. Let $\mathbf{A}$ be the $n \times n$ matrix whose entries are i.i.d. copies of ξ, $\mathbf{\Gamma}$ be a deterministic matrix satisfying $\|\mathbf{\Gamma}\| \leqslant n^{\gamma}$. Then*

$$\mathbb{P}\left(\sigma_n\left(\mathbf{\Gamma}+\mathbf{A}\right) \leqslant n^{-(2C+1)\gamma}\right) \leqslant c\left(n^{-C+o(1)} + \mathbb{P}\left(\|\mathbf{A}\| \geqslant n^{\gamma}\right)\right).$$

This theorem only assumes bounded second moment on ξ. The assumption that the entries of $\mathbf{A}$ are i.i.d. is for convenience. A slightly weaker result would hold if one omit this assumption.

Let us deal with the condition number.

Theorem 5.16.17 (Conditional number—Tao and Vu [362]). *Let ξ be a random variable with mean 0 and bounded second moment, and let $\gamma \geq 1/2, C \geq 0$ be constants. There is a constant c depending on ξ, γ, C such that the following holds. Let $\mathbf{A}$ be the $n \times n$ matrix whose entries are i.i.d. copies of ξ, $\mathbf{\Gamma}$ be a deterministic matrix satisfying $\|\mathbf{\Gamma}\| \leqslant n^{\gamma}$. Then*

$$\mathbb{P}\left(\kappa\left(\mathbf{\Gamma}+\mathbf{A}\right) \geqslant 2n^{(2C+2)\gamma}\right) \leqslant c\left(n^{-C+o(1)} + \mathbb{P}\left(\|\mathbf{A}\| \geqslant n^{\gamma}\right)\right).$$

Proof. Since $\kappa(\mathbf{\Gamma}+\mathbf{A})=\sigma_1(\mathbf{\Gamma}+\mathbf{A})/\sigma_n(\mathbf{\Gamma}+\mathbf{A})$, it follows that if $\kappa(\mathbf{\Gamma}+\mathbf{A}) \geqslant 2n^{(2C+2)\gamma}$, then at least one of the two events $\sigma_n(\mathbf{\Gamma}+\mathbf{A}) \leqslant n^{-(2C+1)\gamma}$ and $\sigma_1(\mathbf{\Gamma}+\mathbf{A}) \geqslant 2n^{\gamma}$ holds. On the other hand, we have

$$\sigma_1(\mathbf{\Gamma}+\mathbf{A}) \leqslant \sigma_1(\mathbf{\Gamma}) + \sigma_1(\mathbf{A}) = \|\mathbf{\Gamma}\| + \|\mathbf{A}\| \leqslant n^{\gamma} + \|\mathbf{A}\|.$$

The claim follows. □

Let us consider several special cases and connect Theorem 5.16.16 with other existing results. First, we consider the sub-Gaussian case of ξ. Due to [341], one can have a strong bound on $\mathbb{P}(\|\mathbf{A}\| \geqslant n^{\gamma})$.

Theorem 5.16.18 (Tao and Vu [362]). *Let α be a positive constant. There are positive constants C_1, C_2 depending on α such that the following holds. Let ξ be a sub-Gaussian random variable with 0 mean, variance one and sub-Gaussian moment α, and $\mathbf{A}$ be a random matrix whose entries are i.i.d. copies of ξ. Then*

$$\mathbb{P}\left(\|\mathbf{A}\| \geqslant C_1\sqrt{n}\right) \leqslant e^{-C_2 n}.$$

If one replaces the sub-Gaussian condition by the weaker condition that ξ has fourth moment bounded α, then one has a weaker conclusion that

$$\mathbb{E}\|\mathbf{A}\| \leqslant C_1\sqrt{n}.$$

Combining Theorems 5.16.16 and 5.16.18, we have

Theorem 5.16.19 (Bounded second moment on ξ—Tao and Vu [362]). *Let C and γ be arbitrary positive constants. Let ξ be a sub-Gaussian random variable with mean 0 and variance 1. Let $\mathbf{A}$ be the $n \times n$ matrix whose entries are i.i.d. copies of ξ, $\mathbf{\Gamma}$ be a deterministic matrix satisfying $\|\mathbf{\Gamma}\| \leqslant n^{\gamma}$. Then*

$$\mathbb{P}\left(\sigma_n(\mathbf{\Gamma}+\mathbf{A}) \leqslant \left(\sqrt{n} + \|\mathbf{A}\|\right)^{-2C-1}\right) \leqslant n^{-C+o(1)}. \tag{5.90}$$

For $\|\mathbf{A}\| = O(\sqrt{n})$, (5.90) becomes

$$\mathbb{P}\left(\sigma_n(\mathbf{\Gamma}+\mathbf{A}) \leqslant n^{-C-1/2}\right) \leqslant n^{-C+o(1)}. \tag{5.91}$$

Up to a loss of magnitude $n^{o(1)}$, this matches Theorem 5.16.13, which treated the base case $\mathbf{\Gamma} = 0$.

If we assume bounded fourth moment instead of sub-Gaussian, we can use the second half of Theorem 5.16.18 to deduce the following, for $\|\mathbf{A}\| = O(\sqrt{n})$,

$$\mathbb{P}\left(\sigma_n(\mathbf{\Gamma}+\mathbf{A}) \leqslant \left(\sqrt{n} + \|\mathbf{A}\|\right)^{-1+o(1)}\right) = o(1). \tag{5.92}$$

In the case of $\|\mathbf{A}\| = O(\sqrt{n})$, this implies that almost surely $\sigma_n(\mathbf{\Gamma}+\mathbf{A}) \geqslant n^{-1/2+o(1)}$. For the special case of $\mathbf{\Gamma} = 0$, this matches [341, Theorem 1.1], up to the $o(1)$ term.

5.16.2 Universality for Covariance and Correlation Matrices

To state the results in[380–382], we need the following two conditions. Let $\mathbf{X} = (x_{ij})$ be a $M \times N$ data matrix with *independent* centered real valued entries with variance $1/M$:

$$x_{ij} = \frac{1}{\sqrt{M}}\xi_{ij}, \quad \mathbb{E}\xi_{ij} = 0, \quad \mathbb{E}\xi_{ij}^2 = 1. \tag{5.93}$$

Furthermore, the entries ξ_{ij} have a *sub-exponential* decay, i.e., there exists a constant

$$\mathbb{P}(|\xi_{ij}| > t) \leqslant \frac{1}{\kappa}\exp(-t^{\kappa}). \tag{5.94}$$

The sample covariance matrix corresponding to data matrix $\mathbf{X}$ is given by $\mathbf{S} = \mathbf{X}^H\mathbf{X}$. We are interested in the regime

$$d = d_N = N/M, \qquad \lim_{N\to\infty} d \neq 0, 1, \infty. \tag{5.95}$$

All the results here are also valid for complex valued entries with the moment condition (5.93) replaced with its complex valued analogue:

$$x_{ij} = \frac{1}{\sqrt{M}}\xi_{ij}, \quad \mathbb{E}\xi_{ij} = 0, \quad \mathbb{E}\xi_{ij}^2 = 0, \quad \mathbb{E}\left|\xi\right|_{ij}^2 = 1. \tag{5.96}$$

By the singular value decomposition of $\mathbf{X}$, there exist orthonormal bases

$$\mathbf{X} = \sum_{i=1}^{M} \sqrt{\lambda_i}\mathbf{u}_i\mathbf{v}_i^H = \sum_{i=1}^{M} \sqrt{\lambda_i}\mathbf{u}_i^H\mathbf{v}_i,$$

where $\lambda_1 \geqslant \lambda_2 \geqslant \cdots \geqslant \lambda_{\max\{M,N\}} \geqslant 0$, $\lambda_i = 0$ for $\lambda_i = 0, \min\{N,M\}+1 \leqslant i \leqslant \max\{N,M\}$.

Theorem 5.16.20 ([381]). *Let* $\mathbf{X}$ *with independent entries satisfying* (5.93) *and* (5.94)*. For any fixed* $k > 0$,

$$\left(\frac{M\lambda_1 - \left(\sqrt{N}+\sqrt{M}\right)^2}{\left(\sqrt{N}+\sqrt{M}\right)\left(\frac{1}{\sqrt{N}}+\frac{1}{\sqrt{M}}\right)^{1/3}}, \ldots, \frac{M\lambda_k - \left(\sqrt{N}+\sqrt{M}\right)^2}{\left(\sqrt{N}+\sqrt{M}\right)\left(\frac{1}{\sqrt{N}}+\frac{1}{\sqrt{M}}\right)^{1/3}}\right) \to TW_1, \tag{5.97}$$

where TW_1 *denotes the Tracy-Widom distribution. An analogous statement holds for the smallest eigenvalues.*

Let us study the correlation matrix. For a Euclidean vector $\mathbf{a} \in \mathbb{R}^n$, define the l_2 norm

$$\|\mathbf{a}\|_2 = \sqrt{\sum_{i=1}^{n} a_i^2}.$$

The matrix $\mathbf{X}^H\mathbf{X}$ is the usual covariance matrix. The j-th column of $\mathbf{X}$ is denoted by $\mathbf{x}_j$. Define the matrix $M \times N$ $\tilde{\mathbf{X}} = (\tilde{x}_{ij})$

$$\tilde{x}_{ij} = \frac{x_{ij}}{\|\mathbf{x}_j\|_2}. \tag{5.98}$$

Using the identity $\mathbb{E}\tilde{x}_{ij}^2 = \frac{1}{M}\mathbb{E}\sum_{i=1}^{M}\tilde{x}_{ij}^2$, we have

$$\mathbb{E}\tilde{x}_{ij}^2 = \frac{1}{M}.$$

Let $\tilde{\lambda}_i$ is the eigenvalues of the matrix $\tilde{\mathbf{X}}^H\tilde{\mathbf{X}}$, sorted decreasingly, similarly to λ_i, the eigenvalues of the matrix $\mathbf{X}^H\tilde{\mathbf{X}}$.

The key difficulty to be overcome is the *strong dependence* of the entries of the correlation matrix. The main result here states that, asymptotically, the k-point ($k \geq 1$) correlation functions of the extreme eigenvalues (at the both edges of the spectrum) of the correlation matrix $\tilde{\mathbf{X}}^H\tilde{\mathbf{X}}$ converge to those of Gaussian correlation matrix, i.e., Tracy-Widom law, and thus in particular, the largest and smallest eigenvalues of $\tilde{\mathbf{X}}^H\tilde{\mathbf{X}}$, after appropriate centering and rescaling, converge to the Tracy-Widom distribution.

Theorem 5.16.21 ([380]). *Let* $\mathbf{X}$ *with independent entries satisfying* (5.93), (5.94) *(or* (5.96) *for complex entries),* (5.95), *and* (5.98). *For any fixed* $k > 0$,

$$\left(\frac{M\tilde{\lambda}_1 - \left(\sqrt{N}+\sqrt{M}\right)^2}{\left(\sqrt{N}+\sqrt{M}\right)\left(\frac{1}{\sqrt{N}}+\frac{1}{\sqrt{M}}\right)^{1/3}}, \ldots, \frac{M\tilde{\lambda}_k - \left(\sqrt{N}+\sqrt{M}\right)^2}{\left(\sqrt{N}+\sqrt{M}\right)\left(\frac{1}{\sqrt{N}}+\frac{1}{\sqrt{M}}\right)^{1/3}}\right) \to \mathrm{TW}_1, \tag{5.99}$$

where TW_1 *denotes the Tracy-Widom distribution. An analogous statement holds for the* k*-smallest (non-trivial) eigenvalues.*

As a special case, we also obtain the Tracy-Widom law for the Gaussian correlation matrices. To reduce the variance of the test statistics, we prefer the sum of the functions of eigenvalues $\sum_{i=1}^{k} f\left(\tilde{\lambda}_i\right)$ for $k \leq \min\{M, N\}$ where $f : \mathbb{R} \to \mathbb{R}$ is a function (say convex and Lipschitz). Note that the eigenvalues $\tilde{\lambda}_i$ are highly correlated random variables. Often the sums of independent random variables are extensively studied. Their dependable counterparts are less studied: Stein's method [87, 258] can be used for this purpose.

Example 5.16.22 (Universality and Principal Component Analysis). Covariance matrices are ubiquitous in modern multivariate statistics where the advance of technology has leads to high dimensional data sets—Big Data. Correlation matrices are often preferred. The Principal Component Analysis (PCA) is not invariant to change of scale in the matrix entries. It is often recommended first to standardize the matrix entries and then perform PCA on the resulting correlation matrix [383]. Equivalently, one performs PCA on the sample correlation matrix.

The PCA based detection algorithms (hypothesis tests) are studied for spectrum sensing in cognitive radio [156, 157] where the signal to noise ratio is extremely low such as -20 dB. The kernel PCA is also studied in this context [384, 385].

Akin to the central limit theorem,*universality* [370, 380] refers to the phenomenon that the asymptotic distributions of various functionals of covariance/correlation matrices (such as eigenvalues, eigenvectors etc.) are identical to those of Gaussian covariance/correlation matrices. These results let us calculate the exact asymptotic distributions of various test statistics without restrictive distributional assumptions of the matrix entries. For example, one can perform various hypothesis tests under the assumption that the matrix entries are *not* normally distributed but use the same test statistic as in the Gaussian case.

For random vectors $\mathbf{a}_i, \mathbf{w}_i, \mathbf{y}_i \in \mathbb{C}^n$, our signal model is defined as

$$\mathbf{y}_i = \mathbf{a}_i + \mathbf{w}_i, \quad i = 1, \ldots, N$$

where $\mathbf{a}_i$ is the signal vector and $\mathbf{w}_i$ the random noise. Define the (random) sample covariance matrices as

$$\mathbf{Y} = \sum_{i=1}^{N} \mathbf{y}_i \mathbf{y}_i^H, \quad \mathbf{A} = \sum_{i=1}^{N} \mathbf{a}_i \mathbf{a}_i^H, \quad \mathbf{W} = \sum_{i=1}^{N} \mathbf{w}_i \mathbf{w}_i^H.$$

It follows that

$$\mathbf{Y} = \mathbf{A} + \mathbf{W}. \tag{5.100}$$

Often the entries of $\mathbf{W}$ are normally distributed (or Gaussian random variables). We are often interested in a more general matrix model

$$\mathbf{Y} = \mathbf{A} + \mathbf{W} + \mathbf{J} \tag{5.101}$$

where the entries of matrix $\mathbf{J}$ are not normally distributed (or non-Gaussian random variables). For example, when jamming signals are present in a communications or sensing system. We like to perform PCA on the matrix $\mathbf{Y}$. Often the rank of matrix $\mathbf{A}$ is much lower than that of $\mathbf{W}$. We can project the high dimensional matrix

$\mathbf{Y}$ into lower dimensions, hoping to expose more structures of $\mathbf{A}$. The Gaussian model of (5.100) is well studied. Universality implies that we are able to use the test statistic of (5.100) to study that of (5.101). □

We provide the analogous non-asymptotic bounds on the variance of eigenvalues for random covariance matrices, following [386]. Let X be a $m \times n$ (real or complex) random matrix, with $m > n$, such that its entries are independent, centered and have variance 1. The random covariance matrix (Wishart matrix) $\mathbf{S}$ is defined as

$$\mathbf{S} = \frac{1}{n}\mathbf{X}^H\mathbf{X}.$$

An important example is the case when all the entries of $\mathbf{X}$ are Gaussian. Then $\mathbf{S}$ belongs to the so-called Laguerre Unitary Ensemble (LUE) if the entries are complex and to the Laguerre Orthogonal Ensemble (LOE) if they are real. All the eigenvalues are nonnegative and will be sorted increasingly $0 \leqslant \lambda_1 \leqslant \cdots \leqslant \lambda_n$.

We say that $S_{m,n}$ satisfy condition ($C0$) if the real part ξ and imaginary part $\tilde{\xi}$ of $(S_{m,n})_{i,j}$ are independent and have an exponential decay: there are two positive constants β_1 and β_2 such that

$$\mathbb{P}\left(|\xi| \geqslant t^{\beta_1}\right) \leqslant e^{-t} \quad \text{and} \quad \mathbb{P}\left(\left|\tilde{\xi}\right| \geqslant t^{\beta_1}\right) \leqslant e^{-t}$$

for $t \geq \beta_2$.

We assume that (1)

$$1 < \alpha_1 \leqslant \frac{m}{n} \leqslant \alpha_2$$

where α_1 and α_2 are fixed constants and that $\mathbf{S}$ is a covariance matrix whose entries have an exponent decay (condition ($C0$)) and (2) have the same first four moments as those of a LUE matrix. The following summarizes a number of quantitative bounds.

Theorem 5.16.23 ([386]).

1. ***In the bulk of the spectrum.*** *Let $\eta \in (0, \frac{1}{2}]$. There is a constant $C > 0$ (depending on η, α_1, α_2) such that, for all covariance matrices $\mathbf{S}_{m,n}$, with $\eta n \leq i \leq (1-\eta)n$,*

$$\mathrm{Var}\left(\lambda_i\right) \leqslant C\frac{\log n}{n^2}.$$

2. ***Between the bulk and the edge of the spectrum.*** *There is a constant $\kappa > 0$ (depending on α_1, α_2) such that the following holds. For all $K > \kappa$, for all $\eta \in (0, \frac{1}{2}]$, there exists a constant $C > 0$ (depending on $K, \eta, \alpha_1, \alpha_2$) such that, for all covariance matrices $\mathbf{S}_{m,n}$, with $(1-\eta)n \leq i \leq n - K\log n$,*

$$\mathrm{Var}\left(\lambda_i\right) \leqslant C\frac{\log\left(n-i\right)}{n^{4/3}(n-i)^{2/3}}.$$

3. ***At the edge of the spectrum.*** *There exists a constant $C > 0$ (depending on α_1, α_2) such that, for all covariance matrices $\mathbf{S}_{m,n}$,*

$$\mathrm{Var}\left(\lambda_n\right) \leqslant C\frac{1}{n^{4/3}}.$$

5.17 Further Comments

This subject of this chapter is in its infancy. We tried to give a comprehensive review of this subject by emphasizing both techniques and results. The chief motivation is to make connections with the topics in other chapters. This line of work deserves further research.

The work [72] is the first tutorial treatment along this line of research. The two proofs taken from [72] form the backbone of this chapter. In the context of our book, this chapter is mainly a statistical tool for covariance matrix estimation—sample covariance matrix is a random matrix with independent rows. Chapter 6 is included here to highlight the contrast between two different statistical frameworks: non-asymptotic, local approaches and asymptotic, global approaches.

Chapter 6
Asymptotic, Global Theory of Random Matrices

The chapter contains standard results for asymptotic, global theory of random matrices. The goal is for readers to compare these results with results of non-asymptotic, local theory of random matrices (Chap. 5). A recent treatment of this subject is given by Qiu et al. [5].

The applications included in [5] are so rich; one wonder whether a parallel development can be done along the line of non-asymptotic, local theory of random matrices, Chap. 5. The connections with those applications are the chief reason why this chapter is included.

6.1 Large Random Matrices

Example 6.1.1 (Large random matrices). Consider n-dimensional random vectors $\mathbf{y}, \mathbf{x}, \mathbf{n} \in \mathbb{R}^n$

$$\mathbf{y} = \mathbf{x} + \mathbf{n}$$

where vector $\mathbf{x}$ is independent of $\mathbf{n}$. The components $x_1, \ldots, x_n$ of the random vector $\mathbf{x}$ are scalar valued random variables, and, in general, may be dependent random variables. For the random vector $\mathbf{n}$, this is similar. The true covariance matrix has the relation

$$\mathbf{R}_y = \mathbf{R}_x + \mathbf{R}_n,$$

due to the independence between $\mathbf{x}$ and $\mathbf{n}$.

Assume now there are N copies of random vector $\mathbf{y}$:

$$\mathbf{y}_i = \mathbf{x}_i + \mathbf{n}_i, \quad i = 1, 2, \ldots, N.$$

Let us consider the sample covariance matrix

R. Qiu and M. Wicks, *Cognitive Networked Sensing and Big Data*,
DOI 10.1007/978-1-4614-4544-9_6,

$$\frac{1}{N}\sum_{i=1}^{N}\mathbf{y}_{\mathrm{i}}\otimes\mathbf{y}_i^* = \frac{1}{N}\sum_{i=1}^{N}\mathbf{x}_{\mathrm{i}}\otimes\mathbf{x}_i^* + \frac{1}{N}\sum_{i=1}^{N}\mathbf{n}_{\mathrm{i}}\otimes\mathbf{n}_i^* + \text{junk}$$

where "junk" denotes two other terms. It is more convenient to consider the matrix form

$$\mathbf{Y} = \begin{bmatrix}\mathbf{y}_1^T\\ \vdots \\ \mathbf{y}_N^T\end{bmatrix}_{N\times n}, \quad \mathbf{X} = \begin{bmatrix}\mathbf{x}_1^T\\ \vdots \\ \mathbf{x}_N^T\end{bmatrix}_{N\times n}, \quad \mathbf{N} = \begin{bmatrix}\mathbf{n}_1^T\\ \vdots \\ \mathbf{n}_N^T\end{bmatrix}_{N\times n},$$

where all matrices are of size $n \times n$. Thus it follows that

$$\frac{1}{N}\mathbf{Y}^*\mathbf{Y} = \frac{1}{N}\mathbf{X}^*\mathbf{X} + \frac{1}{N}\mathbf{N}^*\mathbf{N} + \text{junk}. \tag{6.1}$$

A natural question arises from the above exercise: What happens if $\mathrm{n} \to \infty, \mathrm{N} \to \infty$, but $\frac{N}{n} \to \alpha$? □

The asymptotic regime

$$\mathrm{n} \to \infty, \mathrm{N} \to \infty, \text{ but } \frac{n}{N} \to \alpha? \tag{6.2}$$

calls for a global analysis that is completely different from that of non-asymptotic, local analysis of random matrices (Chap. 5). Stieltjes transform and free probability are two alternative frameworks (but highly related) used to conduct such an analysis. The former is an analogue of Fourier transform in a linear system. The latter is an analogue of independence for random variables.

Although this analysis is asymptotic, the result is very accurate even for small matrices whose size n is as small as less than five [5]. Therefore, the result is very relevant to practical limits. For example, we often consider $n = 100$, $N = 200$ with $\alpha = 0.5$.

In Sect. 1.6.3, a random matrix of arbitrary $N \times n$ is studied in the form of $\mathbb{E}\left[\mathrm{Tr}\left(\mathbf{A}^k\right)\right]$ and $\mathbb{E}\left[\mathrm{Tr}(\mathbf{B}^*\mathbf{B})^k\right]$ where $\mathbf{A}$ is a Gaussian random matrix (or Wigner matrix) and $\mathbf{B}$ is a Hermitian Gaussian random matrix.

6.2 The Limit Distribution Laws

The *empirical spectral distribution* (ESD) is defined as

$$\mu_{\mathbf{X}} = \frac{1}{n}\sum_{i=1}^{n}\delta_{\lambda_i(\mathbf{X})}, \tag{6.3}$$

of $\mathbf{X}$, where $\lambda_1(\mathbf{X}) \geqslant \cdots \geqslant \lambda_n(\mathbf{X})$ are the (necessarily real) eigenvalues of $\mathbf{X}$, counting multiplicity. The ESD is a probability measure, which can be viewed as a distribution of the normalized eigenvalues of $\mathbf{X}$.

Let $\mathbf{A}$ be the Wigner matrix defined in Sect. 1.6.3.

Theorem 6.2.1 (Semicircle Law). *Let* $\mathbf{A}$ *be the top left* $n \times n$ *minors of an infinite Wigner matrix* $(\xi_{ij})_{i,j\geqslant 1}$. *Then the ESDs* $\mu_{\mathbf{A}}$ *converge almost surely (and hence also in probability and in expectation) to Wigner semicircle distribution*

$$\mu_{sc} = \begin{cases} \frac{1}{2\pi}\left(4 - |x|^2\right)^{1/2} dx, & \text{if } |x| < 2, \\ 0, & \text{otherwise.} \end{cases}$$

Almost sure convergence (or with probability one) implies convergence in probability. Convergence in probability implies convergence in distribution (or expectation). The reverse is false in general.

When a sample covariance matrix $\mathbf{S} = \frac{1}{n}\mathbf{X}^*\mathbf{X}$ is considered, we will reach the so-called Marcenko-Pasture law, defined as

$$\mu_{MP} = \begin{cases} \frac{1}{2\pi x \alpha}\sqrt{(b-x)(x-a)}dx, & \text{if } a \leqslant x \leqslant b, \\ 0, & \text{otherwise,} \end{cases}$$

where α is defined in (6.2).

6.3 The Moment Method

Section 1.6.3 illustrates the power of Moment Method for the arbitrary matrix sizes in the Gaussian random matrix framework. Section 4.11 treats this topic in the non-asymptotic, local context.

The moment method is computationally intensive, but straightforward. Sometimes, we need to consider the asymptotic regime of (6.2), where moment method may not be feasible. In real-time computation, the computationally efficient techniques such as Stieltjes transform and free probability is attractive.

The basic starting point is the observation that the moments of the ESD $\mu_{\mathbf{A}}$ can be expressed as normalized traces of powers of $\mathbf{A}$ of $n \times n$

$$\int_{\mathbb{R}} x^k d\mu_{\mathbf{A}}(x) = \frac{1}{n}\mathrm{Tr}(\mathbf{A})^k. \tag{6.4}$$

where $\mathbf{A}$ is the Wigner matrix defined in Sect. 1.6.3. The matrix norm of $\mathbf{A}$ is typically of size $O(\sqrt{n})$, so it is natural to work with the normalized matrix $\frac{1}{\sqrt{n}}\mathbf{A}$. But for notation simplicity, we drop the normalization term.

Since expectation is linear, on taking expectation, it follows that

$$\int_{\mathbb{R}} x^k d\mathbb{E}\left[\mu_{\mathbf{A}}(x)\right] = \frac{1}{n}\mathbb{E}\left[\mathrm{Tr}(\mathbf{A})^k\right]. \tag{6.5}$$

The concentration of measure for the trace function $\mathrm{Tr}(\mathbf{A})^k$ is treated in Sect. 4.3. From this result, the $\mathbb{E}\left[\mu_{\mathbf{A}}(x)\right]$ are uniformly sub-Gaussian:

$$\mathbb{E}\left[\mu_{\mathbf{A}}\left\{|x| > t\right\}\right] \leqslant Ce^{-ct^2n^2}$$

for $t > C$, where C are absolute (so the decay improves quite rapidly with n). From this and the Carleman continuity theorem, Theorem 1.6.1, the circular law can be established, through computing the mean and variance of moments.

To prove the convergence in expectation to the semicircle law μ_{sc}, it suffices to show that

$$\frac{1}{n}\mathbb{E}\left[\mathrm{Tr}\left(\frac{1}{\sqrt{n}}\mathbf{A}\right)^k\right] = \int_{\mathbb{R}} x^k d\mu_{sc}(x) + o_k(1) \tag{6.6}$$

for $k = 1, 2, \ldots,$ where $o_k(1)$ is an expression that goes to zero as $n \to \infty$ for fixed k.

6.4 Stieltjes Transform

Equation (6.4) is the starting point for the moment method. The Stieltjes transform also proceeds from this fundamental identity

$$\int_{\mathbb{R}} \frac{1}{x - z} d\mu_{\mathbf{A}}(x) = \frac{1}{n}\mathrm{Tr}(\mathbf{A} - z\mathbf{I})^{-1} \tag{6.7}$$

for any complex z not in the support of $\mu_{\mathbf{A}}$. The expression in the left hand side is called Stieltjes transform of $\mathbf{A}$ or of $\mu_{\mathbf{A}}$, and denote it as $m_{\mathbf{A}}(z)$. The expression $(\mathbf{A} - z\mathbf{I})^{-1}$ is the *resolvent* of $\mathbf{A}$, and plays a significant role in the spectral theory of that matrix. Sometimes, we can consider the normalized version $\mathbf{M} = \frac{1}{\sqrt{n}}\mathbf{A}$ for an arbitrary matrix $\mathbf{A}$ of $n \times n$, since the matrix norm of $\mathbf{A}$ is typically of size $O(\sqrt{n})$. The Stieltjes transform, in analogy with Fourier transform for a linear time-invariant system, takes full advantage of specific linear-algebraic structure of this problem, and, in particular, of rich structure of resolvents. For example, one can use the Neumann series

$$(\mathbf{I} - \mathbf{X})^{-1} = \mathbf{I} + \mathbf{X} + \mathbf{X}^2 + \cdots + \mathbf{X}^k + \cdots.$$

One can further exploit the linearity of trace using:

$$\mathrm{Tr}(\mathbf{I} - \mathbf{X})^{-1} = \mathrm{Tr}\mathbf{I} + \mathrm{Tr}\mathbf{X} + \mathrm{Tr}\mathbf{X}^2 + \cdots + \mathrm{Tr}\mathbf{X}^k + \cdots.$$

The Stieltjes transform can be viewed as a generating function of the moments via the above Neumann series (an infinite Taylor series of matrices)

$$m_{\mathbf{M}}(z) = -\frac{1}{z} - \frac{1}{z^2}\frac{1}{n^{3/2}}\mathrm{Tr}\mathbf{M} - \frac{1}{z^3}\frac{1}{n^2}\mathrm{Tr}\mathbf{M} - \cdots,$$

valid for z sufficiently large. This is reminiscent of how the characteristic function $\mathbb{E}e^{jtX}$ of a scalar random variable can be viewed as a generating function of the moments $\mathbb{E}X^k$.

For fixed $z = a + jb$ away from the real axis, the Stieltjes transform $m_{\mathbf{M}_n}(z)$ is quite stable in n. When $\mathbf{M}_n$ is a Wigner matrix of size $n \times n$, using a standard concentration of measure result, such as McDiarmid's inequality, we conclude concentration of $m_{\mathbf{M}_n}(z)$ around its mean:

$$\mathbb{P}\left\{\left|m_{\mathbf{M}_n}(a+jb) - \mathbb{E}m_{\mathbf{M}_n}(a+jb)\right| > t/\sqrt{n}\right\} \leqslant Ce^{-ct^2} \tag{6.8}$$

for all $t > 0$ and some absolute constants $C, c > 0$. For details of derivations, we refer to [63, p. 146].

The concentration of measure says that $m_{\mathbf{M}_n}(z)$ is very close to its mean. It does not, however, tell much about what this mean *is*. We must exploit the linearity of trace (and expectation) such as

$$m_{\frac{1}{\sqrt{n}}\mathbf{A}_n}(z) = \frac{1}{n}\sum_{i=1}^{n}\mathbb{E}\left[\left(\frac{1}{\sqrt{n}}\mathbf{A}_n - z\mathbf{I}_n\right)^{-1}\right]_{ii}$$

where $[\mathbf{B}]_{ii}$ is the diagonal ii-th component of a matrix $\mathbf{B}$. Because $\mathbf{A}_n$ is Wigner matrix, on permuting the rows and columns, all of the random variables $\left[\left(\frac{1}{\sqrt{n}}\mathbf{A}_n - z\mathbf{I}_n\right)^{-1}\right]_{ii}$ have the same distribution. Thus we may simplify the above expression as

$$\mathbb{E}m_{\frac{1}{\sqrt{n}}\mathbf{A}_n}(z) = \mathbb{E}\left[\left(\frac{1}{\sqrt{n}}\mathbf{A}_n - z\mathbf{I}_n\right)^{-1}\right]_{nn}. \tag{6.9}$$

We only need to deal with the last entry of an inverse of a matrix.

Schur complement is very convenient. Let $\mathbf{B}_n$ be an $n \times n$ matrix, let $\mathbf{B}_{n-1}$ be the top left $n-1 \times n-1$ minor, let b_{nn} be the bottom right entry of $\mathbf{B}_n$ such that

$$\mathbf{B}_n = \begin{bmatrix} \mathbf{B}_{n-1} & \mathbf{x} \\ \mathbf{y}^* & b_{nn} \end{bmatrix}$$

where $\mathbf{x} \in \mathbb{C}^{n-1}$ is the right column of $\mathbf{B}_n$ with the bottom right entry b_{nn} removed, and $\mathbf{y}^* \in \left(\mathbb{C}^{n-1}\right)^*$ is the bottom row with the bottom right entry b_{nn} removed. Assume that $\mathbf{B}_n$ and $\mathbf{B}_{n-1}$ are both invertible, we have that

$$\left[\mathbf{B}_n^{-1}\right]_{nn} = \frac{1}{b_{nn} - \mathbf{y}^*\mathbf{B}_{n-1}^{-1}\mathbf{x}}. \tag{6.10}$$

This expression can be obtained as follows: Solve the equation $\mathbf{A}_n\mathbf{v} = \mathbf{e}_n$, where $\mathbf{e}_n$ is the n-th basis vector, using the method of Schur complements (or from first principles).

In our situations, the matrices $\frac{1}{\sqrt{n}}\mathbf{A}_n - z\mathbf{I}_n$ and $\frac{1}{\sqrt{n}}\mathbf{A}_{n-1} - z\mathbf{I}_{n-1}$ are automatically invertible. Inserting (6.10) into (6.9) (and recalling that we normalized the diagonal of $\mathbf{A}_n$ to vanish), it follows that

$$\mathbb{E}m_{\frac{1}{\sqrt{n}}\mathbf{A}_n}(z) = -\mathbb{E}\frac{1}{z + \frac{1}{n}\mathbf{x}^*\left(\frac{1}{\sqrt{n}}\mathbf{A}_{n-1} - z\mathbf{I}_{n-1}\right)^{-1}\mathbf{x} - \frac{1}{\sqrt{n}}\xi_{nn}}, \tag{6.11}$$

where $\mathbf{x} \in \mathbb{C}^{n-1}$ is the right column of $\mathbf{A}_n$ with the bottom right entry ξ_{nn} removed (the (ij)-th entry of $\mathbf{A}_n$ is a random variable ξ_{ij}). The beauty of (6.11) is to tie together the random matrix $\mathbf{A}_n$ of size $n \times n$ and the random matrix $\mathbf{A}_{n-1}$ of size $(n-1) \times (n-1)$.

Next, we need to understand the quadratic form $\mathbf{x}^*\left(\frac{1}{\sqrt{n}}\mathbf{A}_{n-1} - z\mathbf{I}_{n-1}\right)^{-1}\mathbf{x}$. We rewrite this as $\mathbf{x}^*\mathbf{R}\mathbf{x}$, where $\mathbf{R}$ is the resolvent matrix. This distribution of the random matrix $\mathbf{R}$ is understandably complicated. The core idea, however, is to exploit the observation that the $(n-1)$-dimensional vector $\mathbf{x}$ involves only these entries of $\mathbf{A}_n$ that do not lie in $\mathbf{A}_{n-1}$, so the random matrix $\mathbf{R}$ and the random vector $\mathbf{x}$ are *independent*. As a consequence of this key observation, we can use the randomness of $\mathbf{x}$ to do most of the work in understanding the quadratic form $\mathbf{x}^*\mathbf{R}\mathbf{x}$, without having to know much about $\mathbf{R}$ at all!

Concentration of quadratic forms like $\mathbf{x}^*\mathbf{R}\mathbf{x}$ has been studied in Chap. 5. It turns out that

$$\mathbb{P}\left\{\left|\mathbf{x}^*\mathbf{R}\mathbf{x} - \mathbb{E}\left(\mathbf{x}^*\mathbf{R}\mathbf{x}\right)\right| \geqslant t\sqrt{n}\right\} \leqslant Ce^{-ct^2}$$

for any determistic matrix $\mathbf{R}$ of operator norm $O(1)$. The expectation $\mathbb{E}\left(\mathbf{x}^*\mathbf{R}\mathbf{x}\right)$ is expressed as

$$\mathbb{E}\left(\mathbf{x}^*\mathbf{R}\mathbf{x}\right) = \sum_{i=1}^{n-1}\sum_{j=1}^{n-1}\mathbb{E}\bar{\xi}_{in}r_{ij}\xi_{jn}$$

where ξ_{in} are entries of $\mathbf{x}$, and r_{ij} are entries of $\mathbf{R}$. Since the ξ_{in} are iid with mean zero and variance one, the standard second moment computation shows that this expectation is less than the trace

$$\mathrm{Tr}\left(\mathbf{R}\right) = \sum_{i=1}^{n-1} r_{ii}$$

of $\mathbf{R}$. We have shown the concentration of the measure

$$\mathbb{P}\left\{\left|\mathbf{x}^*\mathbf{R}\mathbf{x} - \mathrm{Tr}\left(\mathbf{R}\right)\right| \geqslant t\sqrt{n}\right\} \leqslant Ce^{-ct^2} \tag{6.12}$$

for any deterministic matrix $\mathbf{R}$ of operator norm $O(1)$, and any $t > 0$. Informally, $\mathbf{x}^*\mathbf{R}\mathbf{x}$ is typically $\mathrm{Tr}\left(\mathbf{R}\right) + O\left(\sqrt{n}\right)$.

The bound (6.12) was shown for a deterministic matrix $\mathbf{R}$, but by using conditional expectation this works for any random matrix as long as the matrix $\mathbf{R}$ is independent of random vector $\mathbf{x}$. For our specific matrix

$$\mathbf{R} = \left(\frac{1}{\sqrt{n}}\mathbf{A}_{n-1} - z\mathbf{I}_{n-1}\right)^{-1},$$

we can apply conditional expectation. The trace of this matrix is nothing but the Stieltjes transform $m_{\frac{1}{\sqrt{n-1}}\mathbf{A}_{n-1}}(z)$ of matrix $\mathbf{A}_{n-1}$. Since the normalization factor is slightly off, we have

$$\mathrm{Tr}\left(\mathbf{R}\right) = n\frac{\sqrt{n}}{\sqrt{n-1}}m_{\frac{1}{\sqrt{n-1}}\mathbf{A}_{n-1}}\left(\frac{\sqrt{n}}{\sqrt{n-1}}z\right).$$

By some subtle arguments available at [63, p. 149], we have

$$\mathrm{Tr}\left(\mathbf{R}\right) = n\left(m_{\frac{1}{\sqrt{n-1}}\mathbf{A}_{n-1}}(z) + o(1)\right).$$

In particular, from (6.8) and (6.12), we see that

$$\mathbf{x}^*\mathbf{R}\mathbf{x} = n\left(\mathbb{E}m_{\frac{1}{\sqrt{n-1}}\mathbf{A}_{n-1}}(z) + o(1)\right)$$

with overwhelming probability. Putting this back to (6.11), we have the remarkable self-consistent equation

$$m_{\frac{1}{\sqrt{n}}\mathbf{A}_n}(z) = -\frac{1}{z + m_{\frac{1}{\sqrt{n}}\mathbf{A}_n}(z)} + o(1).$$

Following the arguments of [63, p. 150], we see that $m_{\frac{1}{\sqrt{n}}\mathbf{A}_n}(z)$ converges to a limit $m_{\mathbf{A}}(z)$, as $n \to \infty$. Thus we have shown that

$$m_{\mathbf{A}}(z) = -\frac{1}{z + m_{\mathbf{A}}(z)},$$

which has two solutions

$$m_{\mathbf{A}}(z) = -\frac{z \pm \sqrt{z^2 - 4}}{2}.$$

It is argued that the positive branch is the correct solution

$$m_{\mathbf{A}}(z) = -\frac{z + \sqrt{z^2 - 4}}{2},$$

through which we reach the semicircular law

$$\frac{1}{2\pi}\left(4 - x^2\right)_+^{1/2} dx = \mu_{sc}.$$

For details, we refer to [63, p. 151].

6.5 Free Probability

We take the liberty of freely drawing material from [63] for this brief exposition. We highlight concepts, basic properties, and practical material.

6.5.1 Concept

In the foundation of modern probability, as laid out by Kolmogorov, the basic objects of study are:

1. *Sample space* Ω, whose elements ω represent all the possible states.
2. One can select a $\sigma-$*algebra* of *events*, and assign *probabilities* to these events.
3. One builds (commutative) algebra of *random variables* X and one can assign *expectation*.

In *measure theory*, the underlying measure space Ω plays a prominent foundational role. In probability theory, in contrast, events and their probabilities are viewed as fundamental, with the sample space Ω being abstracted away as much as possible, and with the random variables and expectations being viewed as derived concepts.

If we take the above abstraction process one step further, we can view the *algebra of random variables* and their *expectations* as being the foundational concept, and ignoring both the presence of the original sample space, the algebra of events, or the probability of measure.

There are two reasons for considering the above foundational structures. First, it allows one to more easily take certain types of limits, such as: the large n limit $n \to \infty$ when considering $n \times n$ random matrices, because quantities build from the algebra of random variables and their expectations, the normalized moments of random matrices, which tend to be quite *stable* in the large n limit, even the sample space and event space varies with n.

Second, the abstract formalism allows one to generalize the classical commutative theory of probability to the more general theory of *non-commutative* probability which does not have the classical counterparts such as sample space or event space. Instead, this general theory is built upon a *non-commutative* algebra of random variables (or "observables") and their expectations (or "traces"). The more general formalism includes as special cases, classical probability and spectral theory (with matrices or operators taking the role of random variables and the trace taking the role of expectation). Random matrix theory is considered a natural blend of classical probability and spectrum theory whereas quantum mechanics with physical observables, takes the role of random variables and their expected values on a given state which is the expectation. In short, the idea is to make *algebra* the foundation of the theory, as opposed to other choices of foundations such as sets, measure, categories, etc. It is part of more general "non-commutative way of thinking.[1]"

6.5.2 Practical Significance

The significance of free probability to random matrix theory lies in the fundamental observation that random matrices that have independent entries in the classical sense, also tend to be independent[2] in the free probability sense, in the large n limit $n \to \infty$. Because of this observation, many tedious computations in random matrix theory, particularly those of an algebraic or enumerative combinatorial nature, can be performed more quickly and systematically by using the framework of free probability, which by design is optimized for algebraic tasks rather than analytical ones.

Free probability is an excellent tool for computing various expressions of interest in random matrix theory, such as asymptotic values of normalized moments in the large n limit $n \to \infty$. Questions like the *rate* of convergence cannot be answered by free probability that covers only the asymptotic regime in which n is sent to infinity. Tools such as concentration of measure (Chap. 5) can be combined with free probability to recover such rate information.

As an example, let us reconsider (6.1):

$$\frac{1}{N}\mathbf{Y}^*\mathbf{Y} = \frac{1}{N}(\mathbf{X}+\mathbf{N})^*(\mathbf{X}+\mathbf{N}) = \frac{1}{N}\mathbf{X}^*\mathbf{X} + \frac{1}{N}\mathbf{N}^*\mathbf{N} + \frac{1}{N}\mathbf{X}^*\mathbf{N} + \frac{1}{N}\mathbf{N}^*\mathbf{X}. \tag{6.13}$$

The rate of convergence how $\frac{1}{N}\mathbf{X}^*\mathbf{X}$ converges to its true covariance matrix $\mathbf{R}_x$ can be understood using concentration of measure (Chap. 5), by taking advantage of

[1]The foundational preference is a meta-mathematical one rather than a mathematical one.

[2]This is only possible because of the highly non-commutative nature of these matrices; this is not possible for non-trivial commuting independent random variables to be freely independent.

low rank structure of the $\mathbf{R}_x$. But the form of $\mathrm{Tr}(\mathbf{A}+\mathbf{B})^k$ can be easily handled by free probability.

If $\mathbf{A}, B$ are freely independent, and of expectation zero, then $\mathbb{E}[\mathbf{ABAB}]$ vanishes, but $\mathbb{E}[\mathbf{AABB}]$ instead factors as $\mathbb{E}\left[\mathbf{A}^2\right]\mathbb{E}\left[\mathbf{B}^2\right]$. Since

$$\mathbb{E}\mathrm{Tr}\left(\frac{1}{n}\left((\mathbf{A}+\mathbf{B})^*(\mathbf{A}+\mathbf{B})\right)^k\right)$$

can be calculated by free probability [387], (6.13) can be handled as a special case.

Qiu et al. [5] gives a comprehensive survey of free probability in wireless communications and signal processing. Couillet and Debbah [388] gives a deeper treatment of free probability in wireless communication. Tulino and Verdu [389] is the first book-form treatment of this topic in wireless communication.

6.5.3 Definitions and Basic Properties

In the classical (commutative) probability, two (bounded, real-valued) random variables X, Y are *independent* if one has

$$\mathbb{E}\left[f(X)\,g(Y)\right]=0$$

whenever $f, g : \mathbb{R} \to \mathbb{R}$ are well-behaved (such as polynomials) such that all of $\mathbb{E}[f(X)]$ and $\mathbb{E}[g(Y)]$ vanish. For two (bounded, Hermitian) *non-commutative* random variables X, Y, the classical notion no longer applies. We consider, as a substitute, the notion of being *freely independent* (or *free* for short), which means that

$$\mathbb{E}\left[f_1(X)\,g_1(Y)\cdots f_k(X)\,g_k(Y)\right]=0 \tag{6.14}$$

where $f_1, g_1, \ldots, f_k, g_k : \mathbb{R} \to \mathbb{R}$ are well-behaved functions such that $\mathbb{E}[f_1(X)]$, $\mathbb{E}[g_1(X)], \ldots, \mathbb{E}[f_k(X)], \mathbb{E}[g_k(X)]$ vanish.

Example 6.5.1 (Random matrix variables). Random matrix theory combines classical probability theory with finite-dimensional spectral theory, with the random variables of interest now being the random matrices $\mathbf{X}$, all of whose entries have all moments finite. The normalized trace τ is given by

$$\tau(\mathbf{X}) \triangleq \mathbb{E}\frac{1}{n}\mathrm{Tr}\mathbf{X}.$$

Thus one takes both the normalized matrix trace and the probabilistic expectation, in order to obtain a deterministic scalar. As seen before, the moment method for random matrices is based on the moments

$$\tau\left(\mathbf{X}^k\right)=\mathbb{E}\frac{1}{n}\mathrm{Tr}\mathbf{X}^k.$$

When $\mathbf{X}$ is a Gaussian random matrix, the exact expression for $\tau\left(\mathbf{X}^k\right) = \mathbb{E}\frac{1}{n}\mathrm{Tr}\mathbf{X}^k$ is available in a closed form (Sect. 1.6.3). When $\mathbf{Y}$ is a Hermitian Gaussian random matrix, the exact expression for $\tau\left((\mathbf{Y}^*\mathbf{Y})^k\right) = \mathbb{E}\frac{1}{n}\mathrm{Tr}(\mathbf{Y}^*\mathbf{Y})^k$ is also available in a closed form (Sect. 1.6.3). □

The expectation operator is a map that is linear. In fact, it is $*$-linear, which means that it is linear and also that $\mathbb{E}\left(\mathbf{X}^*\right) = \overline{\mathbb{E}\mathbf{X}}$, where the bar represents the complex conjugate. The analogue of the expectation operation for a deterministic matrix is the normalized trace $\tau\left(\mathbf{X}\right) = \frac{1}{n}\mathrm{Tr}\mathbf{X}$.

Definition 6.5.2 (Non-commutative probability space, preliminary definition). A non-commutative probability space $(\mathcal{A}, \tau)$ will consist of a (potentially non-commutative) $*$-algebra $\mathcal{A}$ of (potentially non-commutative) random variables (or observables) with identity 1, together with a trace $\tau : \mathcal{A} \to \mathbb{C}$, which is a $*$-linear functional that maps 1 to 1. This trace will be required to obey a number of additional axioms.

Axiom 6.5.3 (Non-negativity). *For any $\mathbf{X} \in \mathcal{A}$, we have $\tau\left(\mathbf{X}^*\mathbf{X}\right) \geqslant 0$. ($\mathbf{X}^*\mathbf{X}$ is Hermitian, and so its trace $\tau\left(\mathbf{X}^*\mathbf{X}\right)$ is necessarily a real number.)*

In the language of *Von Neumann algebras*, this axiom (together with the normalization $\tau\left(1\right) = 1$) is asserting that τ is a state. This axiom is the non-commutative analogue of the Kolmogorov axiom that all events have non-negative probability.

With this axiom, we can define a positive semi-definite inner product $\langle,\rangle_{L^2(\tau)}$ on $\mathcal{A}$ by

$$\langle\mathbf{X},\mathbf{Y}\rangle_{L^2(\tau)} \triangleq \tau\left(\mathbf{X}\mathbf{Y}\right).$$

This obeys the usual axioms of an inner product, except that it is only positive semi-definite rather than positive definite. One can impose positive definiteness by adding an axiom that the trace is *faithful*, which means that $\tau\left(\mathbf{X}^*\mathbf{X}\right) = 0$ if and only if $\mathbf{X} = 0$. However, this is not needed here.

Without the faithfulness, the norm can be defined using the positive semi-definite inner product

$$\|\mathbf{X}\|_{L^2(\tau)} \triangleq \left(\langle\mathbf{X},\mathbf{X}\rangle_{L^2(\tau)}\right)^{1/2} = \left(\tau\left(\mathbf{X}^*\mathbf{X}\right)\right)^{1/2}.$$

In particular, we have the *Cauchy-Schwartz inequality*

$$\left|\langle\mathbf{X},\mathbf{Y}\rangle_{L^2(\tau)}\right| \leqslant \|\mathbf{X}\|_{L^2(\tau)}\|\mathbf{Y}\|_{L^2(\tau)}.$$

This leads to an important monotonicity:

$$\left|\tau\left(\mathbf{X}^{2k-1}\right)\right|^{1/(2k-1)} \leqslant \left|\tau\left(\mathbf{X}^{2k}\right)\right|^{1/2k} \leqslant \left|\tau\left(\mathbf{X}^{2k+2}\right)\right|^{1/(2k+2)},$$

for any $k \geq 0$.

As a consequence, we can define the *spectral radius* $\rho(\mathbf{X})$ of a Hermitian matrix as

$$\rho(\mathbf{X}) = \lim_{k\to\infty} \left|\tau\left(\mathbf{X}^{2k}\right)\right|^{1/2k},$$

in which case we arrive at the inequality

$$\left|\tau\left(\mathbf{X}^{k}\right)\right| \leqslant (\rho(\mathbf{X}))^{k}$$

for $k = 0, 1, 2, \ldots$. We then say a Hermitian matrix is *bounded* if its spectral radius is finite. We can further obtain [9]

$$\|\mathbf{XY}\|_{L^2(\tau)} \leqslant \rho(\mathbf{X}) \|\mathbf{Y}\|_{L^2(\tau)}.$$

Proposition 6.5.4 (Boundedness [9]). *Let* $\mathbf{X}$ *be a bounded Hermitian matrix, and let* $P : \mathbb{C} \to \mathbb{C}$ *be a polynomial. Then*

$$|\tau(P(\mathbf{X}))| \leqslant \sup_{x\in[-\rho(\mathbf{X}),\rho(\mathbf{X})]} |P(x)|.$$

The spectral theorem is completely a single bounded Hermitian matrix in a non-commutative probability space. This can be extended to multiple commuting Hermitian elements. But, this is not true for multiple non-commuting elements.

We assume as a final (optional) axiom a weak form of commutativity in the trace.

Axiom 6.5.5 (Trace). *For any two elements* $\mathbf{X}, \mathbf{Y}$, *we have* $\tau(\mathbf{XY}) = \tau(\mathbf{YX})$.

From this axiom, we can cyclically permute products in a trace

$$\tau(\mathbf{XYZ}) = \tau(\mathbf{ZYX}).$$

Definition 6.5.6 (Non-commutative probability space, final definition [9]). A non-commutative probability space $(\mathcal{A}, \tau)$ consists of a $*$-algebra $\mathcal{A}$ with identity 1, together with a $*$-linear functional $\tau : \mathcal{A} \to \mathbb{C}$, that maps 1 to 1 and obeys the non-negativity axiom. If τ obeys the trace axiom, we say that the non-commutative probability space is *tracial*. If τ obeys the faithfulness axiom, we say that the non-commutative probability space is *faithful*.

6.5.4 Free Independence

We now come to the fundamental concept in free probability, namely the free independence.

Definition 6.5.7 (Free independence). A collection of $\mathbf{X}_1, \ldots, \mathbf{X}_k$ of random variables in a non-commutative probability space $(\mathcal{A}, \tau)$ is *freely independent* (or *free* for short) if one has

$$\tau\left\{\left[P_1\left(\mathbf{X}_{n,i_1}\right) - \tau\left(P_1\left(\mathbf{X}_{n,i_1}\right)\right)\right] \cdots \left[P_m\left(\mathbf{X}_{n,i_m}\right) - \tau\left(P_m\left(\mathbf{X}_{n,i_m}\right)\right)\right]\right\} = 0$$

whenever $P_1, \ldots, P_m$ are polynomials and $i_1, \ldots, i_m \in 1, \ldots, k$ are indices with no two adjacent i_j equal.

A sequence $\mathbf{X}_{n,1}, \ldots, \mathbf{X}_{n,k}$ of random variables in a non-commutative probability space $(\mathcal{A}, \tau)$ is *asymptotically freely independent* (or *asymptotically free* for short) if one has

$$\tau\left\{\left[P_1\left(\mathbf{X}_{n,i_1}\right) - \tau\left(P_1\left(\mathbf{X}_{n,i_1}\right)\right)\right] \cdots \left[P_m\left(\mathbf{X}_{n,i_m}\right) - \tau\left(P_m\left(\mathbf{X}_{n,i_m}\right)\right)\right]\right\} \to 0$$

as $n \to \infty$ whenever $P_1, \ldots, P_m$ are polynomials and $i_1, \ldots, i_m \in 1, \ldots, k$ are indices with no two adjacent i_j equal.

For classically independent commuting random (matrix valued) variables $\mathbf{X}, \mathbf{Y}$, knowledge of the individual moments $\tau\left(\mathbf{X}^k\right)$ and $\tau\left(\mathbf{Y}^k\right)$ give complete information on the joint moments $\tau\left(\mathbf{X}^k\mathbf{Y}^l\right) = \tau\left(\mathbf{X}^k\right)\tau\left(\mathbf{Y}^l\right)$. The same fact is true for freely independent random variables, though the situation is more complicated. In particular, we have that

$$\begin{aligned}
&\tau\left(\mathbf{XY}\right) = \tau\left(\mathbf{X}\right)\tau\left(\mathbf{Y}\right), \\
&\tau\left(\mathbf{XYX}\right) = \tau\left(\mathbf{Y}\right)\tau\left(\mathbf{X}^2\right), \\
&\tau\left(\mathbf{XYXY}\right) = \tau(\mathbf{X})^2\tau\left(\mathbf{Y}^2\right) + \tau\left(\mathbf{X}^2\right)\tau(\mathbf{Y})^2 - \tau(\mathbf{X})^2\tau(\mathbf{Y})^2.
\end{aligned} \tag{6.15}$$

For detailed derivations of the above formula, we refer to [9].

There is a fundamental connection between free probability and random matrices first observed by Voiculescu [66]: classically independent families of random matrices are *asymptotically free*!

Example 6.5.8 (Asymptotically free). As an illustration, let us reconsider (6.1):

$$\frac{1}{N}\mathbf{Y}^*\mathbf{Y} = \frac{1}{N}\left(\mathbf{X}+\mathbf{N}\right)^*\left(\mathbf{X}+\mathbf{N}\right) = \frac{1}{N}\mathbf{X}^*\mathbf{X} + \frac{1}{N}\mathbf{N}^*\mathbf{N} + \frac{1}{N}\mathbf{X}^*\mathbf{N} + \frac{1}{N}\mathbf{N}^*\mathbf{X}, \tag{6.16}$$

where random matrices $\mathbf{X}, N$ are classically independent. Thus, $\mathbf{X}, N$ are also asymptotically free. Taking the trace of (6.16), we have that

$$\frac{1}{N}\tau\left(\mathbf{Y}^*\mathbf{Y}\right) = \frac{1}{N}\tau\left(\mathbf{X}^*\mathbf{X}\right) + \frac{1}{N}\tau\left(\mathbf{N}^*\mathbf{N}\right) + \frac{1}{N}\tau\left(\mathbf{X}^*\mathbf{N}\right) + \frac{1}{N}\tau\left(\mathbf{N}^*\mathbf{X}\right).$$

Using (6.15), we have that

$$\tau\left(\mathbf{X}^*\mathbf{N}\right) = \tau\left(\mathbf{X}^*\right)\tau\left(\mathbf{N}\right) \text{ and } \tau\left(\mathbf{N}^*\mathbf{X}\right) = \tau\left(\mathbf{N}^*\right)\tau\left(\mathbf{X}\right),$$

which vanish since $\tau(\mathbf{N}) = \tau(\mathbf{N}^*) = 0$ for random matrices whose entries are zero-mean. Finally, we obtain that

$$\tau(\mathbf{Y}^*\mathbf{Y}) = \tau(\mathbf{X}^*\mathbf{X}) + \tau(\mathbf{N}^*\mathbf{N}). \tag{6.17}$$

□

The intuition here is that while a large random matrix $\mathbf{X}$ will certainly correlate with itself so that, Tr $(\mathbf{X}^*\mathbf{X})$ will be large, if we interpose an independent random matrix $\mathbf{N}$ of trace zero, the correlation is largely destroyed; for instance, Tr $(\mathbf{X}^*\mathbf{X}\mathbf{N})$ will be quite small.

We give a typical instance of this phenomenon here:

Proposition 6.5.9 (Asymptotic freeness of Wigner matrices [9]). *Let* $\mathbf{A}_{n,1}, \ldots, \mathbf{A}_{n,k}$ *be a collection of independent $n \times n$ Wigner matrices, where the coefficients all have uniformly bounded m-th moments for each m. Then the random variables* $\mathbf{A}_{n,1}, \ldots, \mathbf{A}_{n,k}$ *are* ***asymptotically free***.

A Wigner matrix is called Hermitian random Gaussian matrices. In Sect. 1.6.3. We consider

$$\mathbf{A}_{n,i} = \mathbf{U}_i^*\mathbf{D}_i\mathbf{U}_i$$

where $\mathbf{D}_i$ are deterministic Hermitian matrices with uniformly bounded eigenvalues, and the $\mathbf{U}_i$ are iid unitary matrices drawn from Haar measure on the unitary group $\mathcal{U}(n)$. One can also show that the $\mathbf{A}_{n,i}$ are asymptotically free.

6.5.5 Free Convolution

When two classically independent random variables X, Y are added up, the distribution μ_{X+Y} of the sum $X + Y$ is the convolution $\mu_{X+Y} = \mu_X \otimes \mu_Y$ of the distributions μ_X and μ_Y. This convolution can be computed by means of the characteristic function

$$F_X = \tau\left(e^{jtX}\right) = \int_{\mathbb{R}} e^{jtx} d\mu_X(x)$$

using the simple formula

$$\tau\left(e^{jt(X+Y)}\right) = \tau\left(e^{jtX}\right)\tau\left(e^{jtY}\right).$$

There is an analogous theory when summing two freely independent (Hermitian) non-commutative random variables $\mathbf{A}, \mathbf{B}$; the distribution $\mu_{\mathbf{A}+\mathbf{B}}$ turns out to be a certain combination $\mu_{\mathbf{A}} \boxplus \mu_{\mathbf{B}}$, known as *free convolution* of $\mu_{\mathbf{A}}$ and $\mu_{\mathbf{B}}$. The Stieltjes transform, rather than the characteristic function, is the correct tool to use

Table 6.1 Common random matrices and their moments (The entries of $\mathbf{W}$ are i.i.d. with zero mean and variance $\frac{1}{N}$; $\mathbf{W}$ is square $N \times N$, unless otherwise specified. $\operatorname{tr}(\mathbf{H}) \triangleq \lim_{N\to\infty} \frac{1}{N} \operatorname{Tr}(\mathbf{H})$)

Convergence laws	Definitions	Moments
Full-circle law	$\mathbf{W}$ square $N \times N$	
Semi-circle law	$\mathbf{K} = \frac{\mathbf{W}+\mathbf{W}^H}{\sqrt{2}}$	$\operatorname{tr}\left(\mathbf{K}^{2m}\right) = \frac{1}{m+1}\binom{2m}{m}$
Quarter circle law	$\mathbf{Q} = \sqrt{\mathbf{W}\mathbf{W}^H}$	$\operatorname{tr}\left(\mathbf{Q}^{m}\right) = \frac{2^{2m}}{\pi m}\frac{1}{\left(\frac{m}{2}+1\right)}\binom{m-1}{\frac{m-1}{2}}\ \forall \mathrm{m\ \ odd}$
	$\mathbf{Q}^2$	
Deformed quarter circle law	$\mathbf{R} = \sqrt{\mathbf{W}^H\mathbf{W}}$, $\mathbf{W} \in \mathbb{C}^{N\times\beta N}$	
	$\mathbf{R}^2$	$\operatorname{tr}\left(\mathbf{R}^{2m}\right) = \frac{1}{m}\sum_{i=1}^{m}\binom{m}{i}\binom{m}{i-1}\beta^i$
Haar distribution	$\mathbf{T} = \mathbf{W}\left(\mathbf{W}^H\mathbf{W}\right)^{-\frac{1}{2}}$	
Inverse semi-circle law	$\mathbf{Y} = \mathbf{T} + \mathbf{T}^H$	

$$m_{\mathbf{A}}(z) = \tau\left(\left(\mathbf{A} - z\right)^{-1}\right) = \int_{\mathbb{R}} \frac{1}{x-z} d\mu_X(x),$$

which has been discussed earlier.

6.6 Tables for Stieltjes, R- and S-Transforms

Table 6.1 gives common random matrices and their moments. Table 6.2 gives definition of commonly encountered random matrices for convergence laws, Table 6.3 gives the comparison of Stieltjes, R- and S-Transforms.

Let the random matrix $\mathbf{W}$ be square $N \times N$ with i.i.d. entries with zero mean and variance $\frac{1}{N}$. Let Ω be the set containing eigenvalues of $\mathbf{W}$. The empirical distribution of the eigenvalues

$$P_{\mathbf{H}}(z) \triangleq \frac{1}{N}\left|\{\lambda \in \Omega : \operatorname{Re}\lambda < \operatorname{Re} z \text{ and } \operatorname{Im}\lambda < \operatorname{Im} z\}\right|$$

converges a non-random distribution functions as $N \to \infty$. Table 6.2 lists commonly used random matrices and their density functions.

Table 6.1 compiles some moments for commonly encountered matrices from [390]. Calculating eigenvalues λ_k of a matrix $\mathbf{X}$ is not a linear operation. Calculation of the moments of the eigenvalue distribution is, however, conveniently done using a normalized trace since

Table 6.2 Definition of commonly encountered random matrices for convergence laws (The entries of $\mathbf{W}$ are i.i.d. with zero mean and variance $\frac{1}{N}$; $\mathbf{W}$ is square $N \times N$, unless otherwise specified)

Convergence laws	Definitions	Density functions
Full-circle law	$\mathbf{W}$ square $N \times N$	$p_{\mathbf{W}}(z) = \begin{cases} \frac{1}{\pi} & \|z\| < 1 \\ 0 & \text{elsewhere} \end{cases}$
Semi-circle law	$\mathbf{K} = \frac{\mathbf{W}+\mathbf{W}^H}{\sqrt{2}}$	$p_{\mathbf{K}}(z) = \begin{cases} \frac{1}{2\pi}\sqrt{4-x^2} & \|x\| < 2 \\ 0 & \text{elsewhere} \end{cases}$
Quarter circle law	$\mathbf{Q} = \sqrt{\mathbf{W}\mathbf{W}^H}$	$p_{\mathbf{Q}}(z) = \begin{cases} \frac{1}{\pi}\sqrt{4-x^2} & 0 \le x \le 2 \\ 0 & \text{elsewhere} \end{cases}$
	$\mathbf{Q}^2$	$p_{\mathbf{Q}^2}(z) = \begin{cases} \frac{1}{2\pi}\sqrt{\frac{4-x}{x}} & 0 \le x \le 4 \\ 0 & \text{elsewhere} \end{cases}$
Deformed quarter circle law	$\mathbf{R} = \sqrt{\mathbf{W}^H\mathbf{W}}$, $\mathbf{W} \in \mathbb{C}^{N\times\beta N}$	$p_{\mathbf{R}}(z) = \begin{cases} \frac{\sqrt{4\beta-(x^2-1-\beta)^2}}{\pi x} & a \le x \le b \\ (1-\sqrt{\beta})^+\delta(x) & \text{elsewhere} \end{cases}$ $a = \left\|1-\sqrt{\beta}\right\|, b = 1+\sqrt{\beta}$
	$\mathbf{R}^2$	$p_{\mathbf{R}^2}(z) = \begin{cases} \frac{\sqrt{4\beta-(x-1-\beta)^2}}{2\pi x} & a^2 \le x \le b^2 \\ (1-\sqrt{\beta})^+\delta(x) & \text{elsewhere} \end{cases}$
Haar distribution	$\mathbf{T} = \mathbf{W}(\mathbf{W}^H\mathbf{W})^{-\frac{1}{2}}$	$p_{\mathbf{T}}(z) = \frac{1}{2\pi}\delta(\|z\|-1)$
Inverse semi-circle law	$\mathbf{Y} = \mathbf{T} + \mathbf{T}^H$	$p_{\mathbf{Y}}(z) = \begin{cases} \frac{1}{\pi}\frac{1}{\sqrt{4-x^2}} & \|x\| < 2 \\ 0 & \text{elsewhere} \end{cases}$

$$\frac{1}{N}\sum_{k=1}^{N}\lambda_k^m = \frac{1}{N}\mathrm{Tr}\left(\mathbf{X}^m\right).$$

Thus, in the large matrix limit, we define $\mathrm{tr}(\mathbf{X})$ as

$$\mathrm{tr}(\mathbf{X}) \triangleq \lim_{N\to\infty}\frac{1}{N}\mathrm{Tr}(\mathbf{X}).$$

Table 6.2 is made self-contained and only some remarks are made here. For Haar distribution, all eigenvalues lie on the complex unit circle since the matrix $\mathbf{T}$ is unitary. The essential nature is that the eigenvalues are uniformly distributed. Haar distribution demands for Gaussian distributed entries in the random matrix $\mathbf{W}$. This condition does not seem to be necessary, but allowing for any complex distribution with zero mean and finite variance is not sufficient.

Table 6.3[3] lists some transforms (Stieltjes, R-, S-transforms) and their properties. The Stieltjes Transform is more fundamental since both R-transform and S-transform can be expressed in terms of the Stieltjes transform.

[3] This table is primarily compiled from [390].

Table 6.3 Table of Stieltjes, R- and S-transforms

Stieltjes transform	R-transform	S-transform
$G(z) \triangleq \int \frac{1}{x-z} dP(x)$, Im$z$>0, Im$G(z) \geq 0$	$R(z) \triangleq G^{-1}(-z) - z^{-1}$	$S(z) \triangleq \frac{1+z}{z} \Upsilon^{-1}(z)$, $\Upsilon(z) \triangleq -z^{-1} G^{-1}(z^{-1}) - 1$
$G_{\alpha\mathbf{I}}(z) = \frac{1}{\alpha - z}$	$R_{\alpha\mathbf{I}}(z) = \alpha$	$S_{\alpha\mathbf{I}}(z) = \frac{1}{\alpha}$,
$G_{\mathbf{K}}(z) = \frac{z}{2}\sqrt{1-\frac{4}{z^2}} - \frac{z}{2}$	$R_{\mathbf{K}}(z) = z$	$S_{\mathbf{K}}(z)$ =undefined
$G_{\mathbf{Q}}(z) = \sqrt{1-\frac{4}{z^2}}\left(\frac{z}{2} - \arcsin\frac{2}{z}\right) - \frac{z}{2} - \frac{1}{2\pi}$	$R_{\mathbf{Q}^2}(z) = \frac{1}{1-z}$	$S_{\mathbf{Q}^2}(z) = \frac{1}{1+z}$
$G_{\mathbf{Q}^2}(z) = \frac{1}{2}\sqrt{1-\frac{4}{z}} - \frac{1}{2}$	$R_{\mathbf{R}^2}(z) = \frac{\beta}{1-z}$	$S_{\mathbf{R}^2}(z) = \frac{1}{\beta + z}$
$G_{\mathbf{R}^2}(z) = \sqrt{\frac{(1-\beta)^2}{4z^2} - \frac{1+\beta}{2z} + \frac{1}{4}} - \frac{1}{2} - \frac{(1-\beta)}{2z}$	$R_{\mathbf{Y}}(z) = \frac{-1+\sqrt{1+4z^2}}{z}$	$S_{\mathbf{Y}}(z)$=undefined
$G_{\mathbf{Y}}(z) = \frac{-\text{sign}(\text{Re } z)}{\sqrt{z^2-4}}$	$R_{\alpha\mathbf{X}}(z) = \alpha R_{\mathbf{X}}(\alpha z)$	$S_{\mathbf{AB}}(z) = S_{\mathbf{A}}(z) S_{\mathbf{B}}(z)$
$G_{\lambda^2}(z) = \frac{G_\lambda(\sqrt{z}) - G_\lambda(-\sqrt{z})}{2\sqrt{z}}$	$\lim_{z \to \infty} R(z) = \int x dP(x)$	
$G_{\mathbf{XX}^H}(z) = \beta G_{\mathbf{X}^H\mathbf{X}}(z) + \frac{\beta - 1}{z}$, $\mathbf{X} \in \mathbb{C}^{N \times \beta N}$	$R_{\mathbf{A}+\mathbf{B}}(z) = R_{\mathbf{A}}(z) R_{\mathbf{B}}(z)$	
	$G_{\mathbf{A}+\mathbf{B}}\left(R_{\mathbf{A}+\mathbf{B}}(-z) - z^{-1}\right) = z$	
$G_{\mathbf{X}+\mathbf{WYW}^H}(z) = G_{\mathbf{X}}\left(z - \beta \int \frac{y dP_{\mathbf{Y}}(x)}{1 + y G_{\mathbf{X}+\mathbf{WYW}^H}(z)}\right)$ Im z>0, $\mathbf{X}, \mathbf{Y}, \mathbf{W}$ jointly independent. $G_{\mathbf{WW}^H}(z) = \int_0^1 u(x,z) dx$, $u(x,z) = \left[-z + \int_0^\beta \frac{w(x,y) dy}{1 + \int_0^1 u(x',z) w(x',y) dx'}\right]^{-1}$, $x \in [0,1]$		

Part II
Applications

Chapter 7
Compressed Sensing and Sparse Recovery

The central mathematical tool for algorithm analysis and development is the concentration of measure for random matrices. This chapter is motivated to provide applications examples for the theory developed in Part I. We emphasize the central role of random matrices.

Compressed sensing is a recent revolution. It is built upon the observation that sparsity plays a central role in the structure of a vector. The unexpected message here is that for a sparse signal, the relevant "information" is much less that what we thought previously. As a result, to recover the sparse signal, the required samples are much less than what is required by the traditional Shannon's sampling theorem.

7.1 Compressed Sensing

The compressed sensing problem deals with how to recover sparse vectors from highly incomplete information using efficient algorithms. To formulate the procedure, a complex vector $\mathbf{x} \in \mathbb{C}^N$ is called s-sparse if

$$\|\mathbf{x}\|_0 := |\{\ell : x_\ell \neq 0\}| = \#\{\ell : x_\ell \neq 0\} \leqslant s.$$

where $\|\mathbf{x}\|_0$ denotes the ℓ_0-norm of the vector $\mathbf{x}$. The ℓ_0-norm represents the total number of how many non-zero components there are in the vector $\mathbf{x}$. The ℓ_p-norm for a real number p is defined as

$$\|\mathbf{x}\|_p = \left(\sum_{i=1}^{N} |x_i|^p \right)^{1/p}, \qquad 1 \leq p < \infty.$$

Given a rectangular complex matrix $\mathbf{\Phi} \in \mathbb{C}^{n \times N}$, called the measurement matrix, the task is to reconstruct the complex vector $\mathbf{x} \in \mathbb{C}^N$ from the *linear* measurements

$$\mathbf{y} = \mathbf{\Phi} \boldsymbol{x}.$$

R. Qiu and M. Wicks, *Cognitive Networked Sensing and Big Data*,
DOI 10.1007/978-1-4614-4544-9_7,

We are interested in the case $n \ll N$, so that this system is under-determined. It is well known that without additional structure constraints on the complex vector $\mathbf{x}$, this linear system problem has no solution. The central message of the compressed sensing is the *surprising discovery* that under the additional structure constraint that the complex vector $\mathbf{x} \in \mathbb{C}^N$ is s-sparse, then the situation changes.

The naive approach for reconstruction, namely, ℓ_0-minimization,

$$\text{minimize} \quad \|\mathbf{z}\|_0 \qquad \text{subject to} \quad \mathbf{\Phi z} = \mathbf{y}, \tag{7.1}$$

which is NP-hard in general. There are several well-known tractable alternatives—for instance, ℓ_1-minimization

$$\text{minimize} \quad \|\mathbf{z}\|_1 \qquad \text{subject to} \quad \mathbf{\Phi z} = \mathbf{y}, \tag{7.2}$$

where $\|\mathbf{z}\|_1 = |z_1|+|z_2|+\cdots+|z_N|$ for $\mathbf{z} = (z_1, z_2, \ldots, z_N) \in \mathbb{C}^N$. Equation (7.2) is a convex optimization problem and may be solved efficiently by tools such as CVX.

The restricted isometry property (RIP) streamlines the analysis of recovery algorithms. The restricted isometry property (RIP) also offers a very elegant way to analyze ℓ_1-minimization and greedy algorithms.

To guarantee recoverability of the sparse vector x in (7.1) by means of ℓ_1-minimization and greedy algorithms, it suffices to establish the restricted isometry property (RIP) of the so-called measurement matrix $\mathbf{\Phi}$: For a matrix $\mathbf{\Phi} \in \mathbb{C}^{n \times N}$ and sparsity $s < N$, the restricted isometry constant δ_s is defined as the *smallest* positive number such that

$$(1-\delta_s)\|\mathbf{x}\|_2^2 \leqslant \|\mathbf{\Phi x}\|_2^2 \leqslant (1+\delta_s)\|\mathbf{x}\|_2^2 \quad \text{for all } \mathbf{x} \in \mathbb{C}^N \text{ with } \|\mathbf{x}\|_0 \leq s. \tag{7.3}$$

In words, the statement (7.3) requires that all column submatrices of $\mathbf{\Phi}$ with at most s columns are well-conditioned. Informally, $\mathbf{\Phi}$ is said to satisfy the RIP with order s when the level δ_s is "small".

A number of recovery algorithms (Table 7.1) are provably effective for sparse recovery if the matrix $\mathbf{\Phi}$ satisfies the RIP. More precisely, suppose that the matrix $\mathbf{\Phi}$ obeys (7.3) with

$$\delta_{\kappa s} < \delta^\star \tag{7.4}$$

for suitable constants $\kappa \geq 1$ and $\delta^\star$, then many algorithms precisely recover any s-sparse vectors $\mathbf{x}$ from the measurements $\mathbf{y} = \mathbf{\Phi x}$. Moreover, if $\mathbf{x}$ can be well approximated by an s sparse vector, then for noisy observations

$$\mathbf{y} = \mathbf{\Phi x} + \mathbf{e}, \qquad \|\mathbf{e}\|_2 \leqslant \alpha, \tag{7.5}$$

these algorithms return a reconstruction $\tilde{\mathbf{x}}$ that satisfies an error bound of the form

$$\|\mathbf{x} - \tilde{\mathbf{x}}\|_2 \leqslant C_1 \frac{1}{\sqrt{s}} \sigma_s(\mathbf{x})_1 + C_2 \alpha, \tag{7.6}$$

Table 7.1 Values of the constants [391] κ and $\delta^\star$ in (7.4) that guarantee success for various recovery algorithms

Algorithm	κ	$\delta^\star$	References
ℓ_1-minimization (7.2)	2	$\frac{3}{4+\sqrt{6}} \approx 0.4652$	[392–395]
CoSaMP	4	$\sqrt{\frac{2}{5+\sqrt{73}}} \approx 0.3843$	[396, 397]
Iterative hard thresholding	3	$1/2$	[395, 398]
Hard thresholding pursuit	3	$1/\sqrt{3} \approx 0.5774$	[399]

where

$$\sigma_s(\mathbf{x})_1 = \inf_{\|\mathbf{z}\|_0 \leqslant s} \|\mathbf{x} - \mathbf{z}\|_1$$

denotes the error of best s-term approximation in ℓ_1 and $C_1, C_2 > 0$ are constants.

For illustration, we include Table 7.1 (from [391]) which lists available values for the constants κ and δ_s in (7.4) that guarantee (7.6) for several algorithms along with respective references.

Remarkably, all optimal measurement matrices known so far are random matrices. For example, a Bernoulli random matrix $\mathbf{\Phi} \in \mathbb{R}^{n \times N}$ has entries

$$\phi_{jk} = \varepsilon_{jk}/\sqrt{n}, \qquad 1 \le j \le n, 1 \le k \le N,$$

where ε_{ik} are independent, symmetric $\{-1, 1\}$-valued random variables. Its restricted isometry constant satisfies

$$\delta_s \leqslant \delta$$

with probability at least $1 - \eta$ provided that

$$n \geqslant C\delta^{-2} \left(s \ln (eN/s) + \ln (1/\eta) \right),$$

where C is an absolute constant.

On the other hand, Gaussian random matrices, that is, matrices that have independent, normally distributed entries with mean zero and variance one, have been shown [400, 406, 407] to have restricted isometry constants of $\frac{1}{\sqrt{n}}\mathbf{\Phi}$ satisfy $\delta_s \le \delta$ with high probability provided that

$$n \geqslant C\delta^{-2} s \log (N/s).$$

That is, the number n of Gaussian measurements required to reconstruct an s-sparse signal of length N is linear in the sparsity and logarithmic in the ambient dimension. It follows [391] from lower estimates of Gelfand widths that this bound on the required samples is optimal [408–410], that is, the log-factor must be present.

More structured measurement matrices are considered for practical considerations in Sect. 7.3.

Table 7.2 List of measurement matrices [391] that have been proven to be RIP, scaling of sparsity s in the number of measurements n, and the respective Shannon entropy of the (random) matrix

$n \times N$ measurement matrix	Shannon entropy	RIP regime	References
Gaussian	$nN\frac{1}{2}\log(2\pi e)$	$s \leqslant Cn/\log N$	[400–402]
Rademacher entries	nN	$s \leqslant Cn/\log N$	[400]
Partial Fourier matrix	$N\log_2 N - n\log_2 n$ $-(N-n)\log_2(N-n)$	$s \leqslant Cn/\log^4 N$	[402, 403]
Partial circulant Rademacher	N	$s \leqslant Cn^{2/3}/\log^{2/3} N$	[403]
Gabor, Rademacher window	n	$s \leqslant Cn^{2/3}/\log^2 n$	[404]
Gabor, alltop window	0	$s \leqslant C\sqrt{n}$	[405]

In Table 7.2 we list the Shannon entropy (in bits) of various random matrices along with the available RIP estimates.

7.2 Johnson–Lindenstrauss Lemma and Restricted Isometry Property

The ℓ_p^N-norm of a vector $\mathbf{x} = (x_1, \ldots, x_N)^T \in \mathbb{R}^N$ is defined by

$$\|\mathbf{x}\|_{\ell_2} = \|\mathbf{x}\|_{\ell_2^N} = \begin{cases} \left(\sum\limits_{i=1}^{N} |x_i|^p\right)^{1/p}, & 0 < p < \infty, \\ \max\limits_{i=1,\ldots,N} |x_i|, & p = \infty. \end{cases} \tag{7.7}$$

We are given a set $\mathcal{A}$ of points in $\mathbb{R}^N$ with N typically large. We would like to embed these points into a lower-dimensional Euclidean space $\mathbb{R}^n$ which approximately preserving the relative distance between any two of these points.

Lemma 7.2.1 (Johnson–Lindenstrauss [411]). *Let $\varepsilon \in (0,1)$ be given. For every set $\mathcal{A}$ of k points in $\mathbb{R}^N$, if n is a positive integer such that $n \geqslant n_0 = O\left(\ln k/\varepsilon^2\right)$, there exists a Lipschitz mapping $f : \mathbb{R}^N \to \mathbb{R}^n$ such that*

$$(1-\varepsilon)\|\mathbf{x}-\mathbf{y}\|_{\ell_2^N}^2 \leqslant \|f(\mathbf{x}) - f(\mathbf{y})\|_{\ell_2^n} \leqslant (1+\varepsilon)\|\mathbf{x}-\mathbf{y}\|_{\ell_2^N}^2$$

for $\mathbf{x}, y \in \mathcal{A}$.

The Johnson–Lindenstrauss (JL) lemma states [412] that any set of k points in high dimensional Euclidean space can be embedded into $O(\log(k)/\varepsilon^2)$ dimensions, without distorting the distance between any two points by more than a factor between $1-\varepsilon$ and $1+\varepsilon$. As a consequence, the JL lemma has become a valuable tool for dimensionality reduction.

In its original form, the Johnson–Lindenstrauss lemma reads as follows.

Lemma 7.2.2 (Johnson–Lindenstrauss [411]). *Let $\varepsilon \in (0,1)$ be given. Let $\mathbf{x}_1, \ldots, \mathbf{x}_k \in \mathbb{R}^N$ be arbitrary points. Let $m = O\left(\log k/\varepsilon^2\right)$ be a natural number. Then there exists a Lipschitz map $f : \mathbb{R}^N \to \mathbb{R}^m$ such that*

$$(1-\varepsilon)\left\|\mathbf{x}_i - \mathbf{x}_j\right\|_2^2 \leqslant \left\|f\left(\mathbf{x}_i\right) - f\left(\mathbf{x}_j\right)\right\|_2^2 \leqslant (1+\varepsilon)\left\|\mathbf{x}_i - \mathbf{x}_j\right\|_2^2 \qquad (7.8)$$

for all $i, j \in \{1, 2, \ldots, k\}$. Here $\|\cdot\|_2$ stands for the Euclidean norm in $\mathbb{R}^N$ or $\mathbb{R}^m$, respectively.

It is known that this setting of m is nearly tight; Alon [413] showed that one must have

$$m = \Omega\left(\varepsilon^{-2}\log(k)\log\left(1/\varepsilon\right)\right).$$

In most of these frameworks, the map f under consideration is a *linear map* represented by an $m \times N$ matrix $\mathbf{\Phi}$. In this case, one can consider the set of differences $E = \{\mathbf{x}_i - \mathbf{x}_j\}$; To prove the theorem, one then needs to show that

$$(1-\varepsilon)\left\|\mathbf{y}\right\|_2^2 \leqslant \left\|\mathbf{\Phi y}\right\|_2^2 \leqslant (1+\varepsilon)\left\|\mathbf{y}\right\|_2^2, \quad \text{for all } \mathbf{y} \in E. \qquad (7.9)$$

When $\mathbf{\Phi}$ is a random matrix, the proof that $\mathbf{\Phi}$ satisfies the JL lemma with high probability boils down to showing a concentration inequality of the typev

$$\mathbb{P}\left((1-\varepsilon)\left\|\mathbf{x}\right\|_2^2 \leqslant \left\|\mathbf{\Phi x}\right\|_2^2 \leqslant (1+\varepsilon)\left\|\mathbf{x}\right\|_2^2\right) \geqslant 1 - 2\exp\left(-c_0\varepsilon^2 m\right), \qquad (7.10)$$

for an arbitrary fixed $\mathbf{x} \in \mathbb{R}^N$, where c_0 is an absolute constant in the optimal case, and in addition possibly dependent on N in almost-optimal sense as e.g. in [414]. In order to reduce storage space and implementation time of such embeddings, the design of structured random JL embeddings has been an active area of research in recent years [412]. Of particular importance in this context is whether fast (i.e. $O(N\log(N))$) multiplication algorithms are available for the resulting matrices.

The map f is a linear mapping represented by an $n \times N$ matrix $\mathbf{\Phi}$ whose entries are randomly drawn from certain probability distributions. A concise description of this evolution is provided in [415]. As observed by Achlioptas in [416], the mapping $f : \mathbb{R}^N \to \mathbb{R}^m$ may be realized by a random matrix, where each component is selected independently at random with a fixed distribution. This decreases the time for evaluation of the function $f(\mathbf{x})$ essentially. An important breakthrough was achieved by Ailon and Chazelle in [417, 418].

Here let us show how to prove the JL lemma using such random matrices, following [400]. One first shows that for any $\mathbf{x} \in \mathbb{R}^N$, the random variable

$$\mathbb{E}\left(\left\|\mathbf{\Phi x}\right\|_{\ell_2^n}^2\right) = \left\|\mathbf{x}\right\|_{\ell_2^N}^2. \qquad (7.11)$$

Next, one must show that for any $\mathbf{x} \in \mathbb{R}^N$, the random variable $\|\mathbf{\Phi x}\|_{\ell_2^n}^2$ is sharply concentrated a round its expected value (concentration of measure), thanks to the moment conditions; that is,

$$\mathbb{P}\left(\left|\|\mathbf{\Phi x}\|_{\ell_2^n}^2 - \|\mathbf{x}\|_{\ell_2^N}^2\right| \geqslant t\,\|\mathbf{x}\|_{\ell_2^N}^2\right) \leqslant 2e^{-ng(t)}, \quad 0 < t < 1, \tag{7.12}$$

where the probability is taken over all $n \times N$ matrices $\mathbf{\Phi}$ and $g(t)$ is a constant depending only on t such that for all $t \in (0, 1)$, $g(t) > 0$.

Finally, one uses the union bound to the set of differences between all possible pairs of points in $\mathcal{A}$.

$$\phi_{ij} \sim \begin{cases} +1/\sqrt{n} & \text{with probability } 1/2, \\ -1/\sqrt{n} & \text{with probability } 1/2, \end{cases} \tag{7.13}$$

or related distributions such as

$$\phi_{ij} = \begin{cases} +\sqrt{3/n} & \text{with probability } 1/6, \\ 0 & \text{with probability } 2/3, \\ +\sqrt{3/n} & \text{with probability } 1/6. \end{cases} \tag{7.14}$$

Perhaps, the most prominent example is the $n \times N$ random matrices $\mathbf{\Phi}$ whose entries $\phi_{i,j}$, are independent realizations of Gaussian random variables

$$\phi_{ij} \sim \mathcal{N}\left(0, \frac{1}{n}\right). \tag{7.15}$$

The verification of (7.12) with $g(t) = t^2/4 - t^3/6$ is elementary using Chernoff inequalities and a comparison of the moments of a Bernouli random variable to those of a Gaussian random variable.

In [415], it is shown that we can also use matrices whose entries are independent realizations of ± 1 Bernoulli random variables.

Let $(\Omega, \mathcal{F}, \mu)$ be the probability space where μ a probability measure and let Z be a random variable on Ω. Given n and N, we can generate random matrices by choosing the entries ϕ_{ij} as independent realizations of Z. This gives the random matrices $\mathbf{\Phi}(\omega), \omega \in \Omega^{nN}$. Given any set of indices T with the number of elements of the set (cardinality) $|T| \leq k$, denote by $\mathcal{X}_T$ the set of all vectors in $\mathbb{R}^N$ that are zero outside of T.

Theorem 7.2.3 (Baraniuk, Davenport, DeVore, Wakin [400]). *Let $\mathbf{\Phi}(\omega), \omega \in \Omega^{nN}$, be a random matrix of size $n \times N$ drawn according to any distribution that satisfies the concentration inequality* (7.12). *Then, for any set T with the cardinality $|T| = k < n$ and any $0 < \delta < 1$, we have*

$$(1-\delta)\,\|\mathbf{x}\|_{\ell_2^N} \leqslant \|\mathbf{\Phi x}\|_{\ell_2^n} \leqslant (1+\delta)\,\|\mathbf{x}\|_{\ell_2^N}, \quad \textit{for all } \mathbf{x} \in \mathcal{X}_T \tag{7.16}$$

with probability

$$\geqslant 1 - 2(12/\delta)^k e^{-g(\delta/2)n}. \tag{7.17}$$

Proof. 1. *Nets of points.* It is sufficient to prove (7.16) for the case of $\|\mathbf{x}\|_{\ell_2^N} = 1$, due to the linearity of $\boldsymbol{\Phi}$. We choose a finite set of points $\mathcal{A}_T$ such that $\mathcal{A}_T \subseteq \mathcal{X}_T$, $\|\mathbf{a}\|_{\ell_2^N} = 1$ for all $\mathbf{a} \in \mathcal{A}_T$, and for all $\mathbf{x} \in \mathcal{X}_T$ with $\|\mathbf{x}\|_{\ell_2^N} = 1$ we have

$$\min_{\mathbf{a} \in \mathcal{A}_T} \|\mathbf{x} - \mathbf{a}\|_{\ell_2^N} \leqslant \delta/4. \tag{7.18}$$

It is well known from covering numbers (see, e.g., [419]) that we can choose such a set with the cardinality $|\mathcal{A}_T| \leqslant (12/\delta)^k$.

2. *Concentration of measure through the union bound.* We use the union bound to apply (7.12) to this set of points $t = \delta/2$. It thus follows that, with probability at least the right-hand side of (7.17), we have

$$(1 - \delta/2) \|\mathbf{a}\|_{\ell_2^N}^2 \leqslant \|\boldsymbol{\Phi}\mathbf{a}\|_{\ell_2^n} \leqslant (1 + \delta/2) \|\mathbf{a}\|_{\ell_2^N}^2, \quad \text{for all } \mathbf{a} \in \mathcal{A}_T \tag{7.19}$$

3. *Extension to all possible k-dimensional signals.* We define α as the smallest number such that

$$\|\boldsymbol{\Phi}\mathbf{x}\|_{\ell_2^n} \leqslant (1 + \alpha) \|\mathbf{x}\|_{\ell_2^N}^2, \quad \text{for all } \mathbf{x} \in \mathcal{X}_T. \tag{7.20}$$

The goal is to show that $\alpha \leq \delta$. To do this, we recall from (7.18) that for any $\mathbf{x} \in \mathcal{X}_T$ we can pick up a point (vector) $\mathbf{a} \in \mathcal{A}_T$ such that $\|\mathbf{x} - \mathbf{a}\|_{\ell_2^N} \leqslant \delta/4$. In this case we have

$$\|\boldsymbol{\Phi}\mathbf{x}\|_{\ell_2^n} \leqslant \|\boldsymbol{\Phi}\mathbf{a}\|_{\ell_2^n} + \|\boldsymbol{\Phi}(\mathbf{x} - \mathbf{a})\|_{\ell_2^n} \leqslant (1 + \delta/2) + (1 + \alpha)\delta/4. \tag{7.21}$$

Since by definition α is the smallest number for which (7.20) holds, we obtain $\alpha \leqslant \delta/2 + (1 + \alpha)\delta/4$. Solving for α gives $\alpha \leqslant 3\delta/4/(1 - \delta/4) \leqslant \delta$, as desired. So we have shown the upper inequality in (7.16). The lower inequality follows from this since

$$\|\boldsymbol{\Phi}\mathbf{x}\|_{\ell_2^n} \geqslant \|\boldsymbol{\Phi}\mathbf{a}\|_{\ell_2^n} - \|\boldsymbol{\Phi}(\mathbf{x} - \mathbf{a})\|_{\ell_2^n} \geqslant (1 - \delta/2) - (1 + \delta)\delta/4 \geqslant 1 - \delta,$$

which complements the claim. □

Now we can apply Theorem 7.2.3 to obtain the so-called *restricted isometry property* (RIP) in compressive sensing (CS). Given a matrix $\boldsymbol{\Phi}$ and any set T of column indices, we denote by $\boldsymbol{\Phi}_T$ the $n \times |T|$ matrix composed of these columns. Similarly, for $\mathbf{x} \in \mathbb{R}^N$, we denote by $\mathbf{x}_T$ the vector obtained by retaining only the entries in $\mathbf{x}$ corresponding to the column indices T. We say that a matrix $\boldsymbol{\Phi}$ satisfies the restricted isometry property of order k is there exists a $\delta_k \in (0, 1)$ such that

$$(1-\delta_k)\|\mathbf{x}_T\|_{\ell_2^N}^2 \leqslant \|\mathbf{\Phi}_T\mathbf{x}_T\|_{\ell_2^N}^2 \leqslant (1+\delta_k)\|\mathbf{x}_T\|_{\ell_2^N}^2 \tag{7.22}$$

holds for all sets T with $|T| \leq k$. The condition (7.22) is equivalent to requiring that the Grammian matrix $\mathbf{\Phi}_T^T\mathbf{\Phi}_T$ has all of its eigenvalues in $[1-\delta_k, 1+\delta_k]$. (Here $\mathbf{\Phi}_T^T$ is the transpose of $\mathbf{\Phi}_T$.)

The **similarity** between the expressions in (7.9) and (7.22) suggests a connection between the JL lemma and the Restricted Isometry Property. A first result in this direction was established in [400], wherein it was shown that random matrices satisfying a concentration inequality of type (7.10) (and hence the JL Lemma) satisfy the RIP of optimal order. More precisely, the authors prove the following theorem.

Theorem 7.2.4 (Baraniuk, Davenport, DeVore, Wakin [400]). *Suppose that* $n, N,$ *and* $0 < \delta < 1$ *are given. For* $\mathbf{x} \in \mathbb{R}^N$, *if the probability distribution generating the* $n \times N$ *matrices* $\mathbf{\Phi}(\omega), \omega \in \mathbb{R}^{nN}$, *satisfies the concentration inequalities* (7.12)

$$\mathbb{P}\left(\left|\|\mathbf{\Phi x}\|_{\ell_2^n}^2 - \|\mathbf{x}\|_{\ell_2^n}^2\right| \geqslant t\|\mathbf{x}\|_{\ell_2^n}^2\right) \leqslant 2e^{-ng(t)}, \quad 0<t<1, \tag{7.23}$$

then, there exist constant $c_1, c_2 > 0$ *depending only on* δ *such that the restricted isometry property* (7.22) *holds for* $\mathbf{\Phi}(\omega)$ *with the prescribed* δ *and any* $k \leqslant c_1 n/\log(N/k)$ *with probability at least* $1-2e^{-c_2 n}$.

Proof. From Theorem 7.2.3, we know that for each of the k-dimensional spaces $\mathcal{X}_T$, the matrix $\mathbf{\Phi}(\omega)$ will fail to satisfy (7.23) with probability at most

$$2(12/\delta)^k e^{-g(\delta/2)n}. \tag{7.24}$$

There are $\binom{N}{k} \leqslant (2N/k)^k$ such subspaces. Thus, (7.23) will fail to hold with probability (7.23) at most

$$2(2N/k)^k(12/\delta)^k e^{-g(\delta/2)n} = 2\exp\left[-g(\delta/2)n + k(\log(eN/k) + \log(12/\delta))\right]. \tag{7.25}$$

Thus, for a fixed $c_1 > 0$, whenever $k \leqslant c_1 n/\log(N/k)$, we will have that the exponent in the exponential on the right side of (7.25) is at most $-c_2 n$ if

$$c_2 \leqslant g(\delta/2) - c_1\left[1 + (1+\log(12/\delta))/\log(N/k)\right].$$

As a result, we can always choose $c_1 > 0$ sufficiently small to ensure that $c_2 > 0$. This prove that with probability at least $1-2e^{-c_2 n}$, the matrix $\mathbf{\Phi}(\omega)$ will satisfy (7.16) for each $\mathbf{x}$. From this one can easily obtain the theorem. □

The JL lemma implies the Restricted Isometry Property. Theorem 7.2.5 below is a converse result to Theorem 7.2.4: They show that RIP matrices, with *randomized column signs*, provide Johnson–Lindenstrauss embeddings that are optimal up to logarithmic factors in the ambient dimension. In particular, RIP matrices of optimal order provide Johnson–Lindenstrauss embeddings of optimal order as such, up to a logarithmic factor in N (see Theorem 7.2.5). Note that without randomization, such a converse is impossible as vectors in the null space of the fixed parent matrix are always mapped to zero.

For a vector $\mathbf{x} \in \mathbb{R}^N$, we denote $\mathbf{D_x} = (D_{i,j}) \in \mathbb{R}^{N \times N}$ the diagonal matrix satisfying $D_{j,j} = x_j$.

Theorem 7.2.5 (Theorem 3.1 of Krahmer and Ward [412]). *Fix* $\eta > 0$ *and* $\varepsilon \in (0,1)$, *and consider a finite set* $E \in \mathbb{R}^N$ *if cardinality* $|E| = p$. *Set* $k \geqslant 40 \log \frac{4p}{\eta}$, *and suppose that* $\mathbf{\Phi} \in \mathbb{R}^{m \times N}$ *satisfies the Restricted Isometry Property of order* k *and level* $\delta \leqslant \varepsilon/4$. *Let* $\boldsymbol{\xi} \in \mathbb{R}^N$ *be a Rademacher sequence, i.e., uniformly distributed on* $\{-1,1\}^N$. *Then with probability exceeding* $1-\eta$,

$$(1-\varepsilon)\left\|\mathbf{x}\right\|_2^2 \leqslant \left\|\mathbf{\Phi D_{\boldsymbol{\xi}} x}\right\|_2^2 \leqslant (1+\varepsilon)\left\|\mathbf{x}\right\|_2^2 \tag{7.26}$$

uniformly for all $\mathbf{x} \in E$.

The proof of Theorem 7.2.5 follows from the use of three ingredients: (1) Concentration of measure result: Hoeffding's inequality; (2) Concentration of measure result: Theorem 1.5.5; (3) RIP matrices: Theorem 7.2.6.

Theorem 7.2.6 (Proposition 2.5 of Rauhut [30]). *Suppose that* $\mathbf{\Phi} \in \mathbb{R}^{m \times N}$ *has the Restricted Isometry Property of order* $2s$ *and level* δ. *Then for any two disjoint subsets* $\mathcal{J}, \mathcal{L} \subset \{1,\ldots,N\}$ *of size* $|\mathcal{J}| \leqslant s, |\mathcal{L}| \leqslant s$,

$$\left\|\mathbf{\Phi}_{(\mathcal{J})}^H \mathbf{\Phi}_{(\mathcal{L})}\right\| \leqslant \delta.$$

Now we closely follow [29] to develop some useful techniques and at the same time gives a shorter proof of Johnson–Lindenstrauss lemma. To prove Lemma 7.2.2, it actually suffices to prove the following.

Lemma 7.2.7 (Nelson [29]). *For any* $0 < \varepsilon, \delta < 1/2$ *and positive integer* d, *there exists a distribution* $\mathcal{D}$ *over* $\mathbb{R}^{m \times N}$ *for* $m = O\left(\varepsilon^{-2}\log\left(1/\delta\right)\right)$ *such that for any* $\mathbf{x} \in \mathbb{R}^N$ *with* $\left\|\mathbf{x}\right\|_2 = 1$,

$$\mathbb{P}\left(\left|\left\|\mathbf{Ax}\right\|_2^2 - 1\right| > \varepsilon\right) < \delta.$$

Let $\mathbf{x}_1, \ldots, \mathbf{x}_n \in \mathbb{R}^N$ be arbitrary points. Lemma 7.2.7 implies Lemma 7.2.2, since we can set $\delta = 1/n^2$ then apply a union bound on the vectors $(\mathbf{x}_i - \mathbf{x}_j)/\left\|\mathbf{x}_i - \mathbf{x}_j\right\|_2$ for all $i < j$. The first proof of Lemma 7.2.2 was given by Johnson and Lindenstrauss [411]. Later, proofs are given when $\mathcal{D}$ can be taken as a distribution over matrices with independent Gaussian or Bernoulli entries, or even more generally, $\Omega\left(\log\left(1/\delta\right)\right)$-wise independent entries which each have mean zero, variance $1/m$, and a subGaussian tail. See [416, 420–426].

For $\mathbf{A} \in \mathbb{R}^{n\times n}$, we define the Frobenius norm as $\|\mathbf{A}\|_F = \sqrt{\sum_{i,j} A_{i,j}^2} = \sqrt{\mathrm{Tr}\left(\mathbf{A}^2\right)}$. We also define $\|\mathbf{A}\|_{2\to 2} = \sup_{\|\mathbf{x}\|_2=1} \|\mathbf{A}\mathbf{x}\|_2$, which is also equal to the largest magnitude of an eigenvalue of $\mathbf{A}$ when $\mathbf{A}$ has all real eigenvalues (e.g., it is symmetric). We need the Hanson-Wright inequality [61]. We follow [29] for the proof since its approach is modern.

Theorem 7.2.8 (Hanson-Wright inequality [29,61]). *For $\mathbf{A} \in \mathbb{R}^{n\times n}$ symmetric and $\mathbf{x} \in \mathbb{R}^n$ with the x_i independent having subGaussian entries of mean zero and variance one,*

$$\mathbb{P}\left(\left|\mathbf{x}^T\mathbf{A}\mathbf{x} - \mathrm{Tr}\left(\mathbf{A}^2\right)\right| > t\right) \leqslant C\exp\left(-\min\left\{C't^2/\left\|\mathbf{A}\right\|_F^2, C't/\|\mathbf{A}\|_{2\to 2}\right\}\right),$$

where $C, C' > 0$ are universal constants. Also, this holds even if the x_i are only $\Omega\left(1+\min\left\{t^2/\left\|\mathbf{A}\right\|_F^2, t/\|\mathbf{A}\|_{2\to 2}\right\}\right)$-wise independent.

Proof. By Markov's inequality,

$$\mathbb{P}\left(\left|\mathbf{x}^T\mathbf{A}\mathbf{x} - \mathrm{Tr}\left(\mathbf{A}\right)\right| > t\right) \leqslant t^{\alpha}\cdot\mathbb{E}\left[\left|\mathbf{x}^T\mathbf{A}\mathbf{x} - \mathrm{Tr}\left(\mathbf{A}\right)\right|^{\alpha}\right]$$

for any $\alpha > 0$. We apply Theorem 7.2.12 with $\alpha = \min\left\{t^2/\left(256\cdot\left\|\mathbf{A}\right\|_F^2\right), t/\left(256\cdot\|\mathbf{A}\|_{2\to 2}\right)\right\}$. □

The following theorem implies Lemma 7.2.7.

Theorem 7.2.9 (Sub-Gaussian matrix [29]). *For $N > 0$ an integer and any $0 < \varepsilon, \delta < 1/2$, let $\mathbf{A}$ be an $m \times N$ random matrix with subGaussian entries of mean zero and variance $1/m$ for $m = \Omega\left(\varepsilon^{-2}\log\left(1/\delta\right)\right)$. Then for any $\mathbf{x} \in \mathbb{R}^N$ with $\|\mathbf{x}\|_2 = 1$,*

$$\mathbb{P}\left(\left|\|\mathbf{A}\mathbf{x}\|_2^2 - 1\right| > \varepsilon\right) < \delta.$$

Proof. Observe that

$$\|\mathbf{A}\mathbf{x}\|_2^2 = \frac{1}{m}\cdot\sum_{i=1}^{m}\left(\sum_{(j,k)\in[N]\times[N]} x_j x_k z_{i,j} z_{i,k}\right), \tag{7.27}$$

where $\mathbf{z}$ is an mN-dimensional vector formed by concatenating the rows of $\sqrt{m}\cdot\mathbf{A}$, and $[N] = \{1,\ldots,N\}$. We use the block-diagonal matrix $\mathbf{T} \in \mathbb{R}^{mN\times mN}$ with m blocks, where each block is the $N \times N$ rank-one matrix $\mathbf{x}\mathbf{x}^T/m$. Now we get our desired quadratic form $\|\mathbf{A}\mathbf{x}\|_2^2 = \mathbf{z}^T\mathbf{T}\mathbf{z}$. Besides, $\mathrm{Tr}\left(\mathbf{T}\right) = \|\mathbf{x}\|_2^2 = 1$. Next, we want to argue that the quadratic form $\mathbf{z}^T\mathbf{T}\mathbf{z}$ has a concentration around $\mathrm{Tr}\left(\mathbf{T}\right)$, for which we can use Theorem 7.2.8. In particular, we have that

$$\mathbb{P}\left(\left|\|\mathbf{Ax}\|_2^2 - 1\right| > \varepsilon\right) = \mathbb{P}\left(\left|\mathbf{z}^T\mathbf{Tz} - \mathrm{Tr}\left(\mathbf{T}\right)\right| > \varepsilon\right) \leqslant C\exp\left(-\min\left\{C'\varepsilon^2/\left\|\mathbf{T}\right\|_F^2, C'\varepsilon/\|\mathbf{T}\|_{2\to 2}\right\}\right).$$

Direct computation yields $\|\mathbf{T}\|_F^2 = 1/m \cdot \|\mathbf{x}\|_2^4 = 1/m$. Also, $\mathbf{x}$ is the only eigenvector of the rank one matrix $\mathbf{xx}^T/m$ with non-zero eigenvalues, and furthermore its eigenvalue is $\|\mathbf{x}\|_2^2/m = 1/m$. Thus, we have the induced matrix norm $\|\mathbf{A}\|_{2\to 2} = 1/m$. Plugging these in gives error probability δ for $m = \Omega\left(\varepsilon^{-2}\log\left(1/\delta\right)\right)$. □

Lemma 7.2.10 (Khintchine inequality [29]). *For* $\mathbf{a} \in \mathbb{R}^n$, $\mathbf{x} \in \{-1, 1\}^n$ *uniform, and* $k \geq 2$ *an even integer,* $\mathbb{E}\left[\left(\mathbf{a}^T\mathbf{x}\right)^k\right] \leqslant \|\mathbf{a}\|_2^k \cdot k^{k/2}$.

Lemma 7.2.11 (Moments [29]). *If* X, Y *are independent with* $\mathbb{E}[Y] = 0$ *and* $k \geq 2$, *then* $\mathbb{E}\left[|X|^k\right] \leqslant \mathbb{E}\left[|X - Y|^k\right]$.

We here give a proof of Theorem 7.2.8 in the case that $\mathbf{x}$ is uniform in $\{-1, 1\}^n$.

Theorem 7.2.12 (Concentration for quadratic form [29]). *For* $\mathbf{A} \in \mathbb{R}^{n\times n}$ *symmetric and* $\mathbf{x} \in \mathbb{R}^n$ *with the* x_i *independent having subGaussian entries of mean zero and variance one,*

$$\mathbb{E}\left[\left|\mathbf{x}^T\mathbf{Ax} - \mathrm{Tr}\left(\mathbf{A}\right)\right|^k\right] \leqslant C^k \cdot \max\left\{\sqrt{k}\left\|\mathbf{A}\right\|_F^2, k\|\mathbf{A}\|_{2\to 2}\right\}^k$$

where $C > 0$ *is a universal constant.*

Proof. The proof is taken from [29] and we only slightly change some wording and notation for the sake of our habits. Without loss of generality we can assume $\mathrm{Tr}\left(\mathbf{A}\right) = 0$. The reason is that if we consider $\mathbf{A}' = \mathbf{A} - \left(\mathrm{Tr}\left(\mathbf{A}\right)/n\right) \cdot \mathbf{I}$, then $\mathbf{x}^T\mathbf{Ax} - \mathrm{Tr}\left(\mathbf{A}\right) = \mathbf{x}^T\mathbf{A}'\mathbf{x}$, and we obtain $\|\mathbf{A}'\|_F \leqslant \|\mathbf{A}\|_F$, and $\|\mathbf{A}'\|_{2\to 2} \leqslant 2\|\mathbf{A}\|_{2\to 2}$. We use the induction method. We consider k a power of 2. For $k = 2$, $\mathbb{E}\left[\left(\mathbf{x}^T\mathbf{Ax}\right)^2\right] = 4\sum\limits_{i<j} A_{i,j}^2$, and $\|\mathbf{A}\|_F^2 = \sum\limits_i A_{i,i}^2 + 2\sum\limits_{i<j} A_{i,j}^2$. Thus $\mathbb{E}\left[\left(\mathbf{x}^T\mathbf{Ax}\right)^2\right] \leqslant 2\left\|\mathbf{A}\right\|_F^2$.

Next, we assume the statement of our theorem for $k/2$ and prove it for the hypothesis of k. Lemma 7.2.11 is used to establish

$$\mathbb{E}\left[\left(\mathbf{x}^T\mathbf{Ax}\right)^k\right] \leqslant \mathbb{E}\left[\left|\mathbf{x}^T\mathbf{Ax} - \mathbf{y}^T\mathbf{Ay}\right|^k\right] = \mathbb{E}\left[\left|\left(\mathbf{x}+\mathbf{y}\right)^T\mathbf{A}\left(\mathbf{x}-\mathbf{y}\right)\right|^k\right],$$

where $\mathbf{y} \in \{-1, 1\}^n$ is random and independent of $\mathbf{x}$.

If we swap x_i with y_i, then $\mathbf{x}+y$ remains constant as does $|x_i - y_i|$ and that $x_i - y_i$ is replaced with its negation. See Sect. 1.11 for this symmetrization. The advantage of this approach is that we can conditional expectation and apply sophisticated methods to estimate the moments of the Rademacher series. Let us do averaging over all such swaps. Let $\xi_i = \left(\left(\mathbf{x}+\mathbf{y}\right)^T\mathbf{A}\right)_i$, and $\eta_i = x_i - y_i$. Let z_i the indicator random variable: z_i is 1 if we did not swap and -1 if we did. Let $\mathbf{z} = (z_1, \ldots, z_n)$.

Then we have $(\mathbf{x}+\mathbf{y})^T\mathbf{A}(\mathbf{x}-\mathbf{y}) = \sum_i \xi_i\eta_i z_i$. Averaging over all swaps,

$$\mathbb{E}_{\mathbf{z}}\left[\left|(\mathbf{x}+\mathbf{y})^T\mathbf{A}(\mathbf{x}-\mathbf{y})\right|^k\right] = \sum_i \xi_i\eta_i z_i \leqslant \left(\sum_i \xi_i^2\eta_i^2\right)^{k/2} \cdot k^{k/2} \leqslant 2^k k^{k/2} \cdot \left(\sum_i \xi_i^2\right)^{k/2}.$$

The first inequality is by Lemma 7.2.10, and the second uses that $|\eta_i| \leqslant 2$. Note that

$$\sum_i \xi_i^2 = \|\mathbf{A}(\mathbf{x}+\mathbf{y})\|_2^2 \leqslant 2\|\mathbf{Ax}\|_2^2 + 2\|\mathbf{Ay}\|_2^2,$$

and thus

$$\mathbb{E}\left[\left(\mathbf{x}^T\mathbf{Ax}\right)^k\right] \leqslant 2^k k^{k/2} \cdot \mathbb{E}\left[\left(2\|\mathbf{Ax}\|_2^2 + 2\|\mathbf{Ay}\|_2^2\right)^{k/2}\right] \leqslant 4^k k^{k/2} \cdot \mathbb{E}\left[\left(\|\mathbf{Ax}\|_2^2\right)^{k/2}\right],$$

with the final inequality using Minkowski's inequality (namely that $|\mathbb{E}|X+Y|^p|^{1/p} \leqslant |\mathbb{E}|X|^p + \mathbb{E}|Y|^p|^{1/p}$ for any random variables X, Y and any $1 \leq p\infty$). Next let us deal with $\|\mathbf{Ax}\|_2^2 = \langle \mathbf{Ax}, \mathbf{Ax}\rangle = \mathbf{x}^T\mathbf{A}^2\mathbf{x}$. Let $\mathbf{B} = \mathbf{A}^2 - \mathrm{Tr}\left(\mathbf{A}^2\right)\mathbf{I}/n$. Then $\mathrm{Tr}\left(\mathbf{B}\right) = 0$. Also $\|\mathbf{B}\|_F \leqslant \|\mathbf{A}\|_F\|\mathbf{A}\|_{2\to 2}$, and $\|\mathbf{B}\|_{2\to 2} \leqslant \|\mathbf{A}\|_{2\to 2}^2$. The former holds since

$$\|\mathbf{B}\|_F^2 \leqslant \left(\sum_i \lambda_i^4\right) - \left(\sum_i \lambda_i^2\right)^2 /n \leqslant \sum_i \lambda_i^4 \leqslant \|\mathbf{A}\|_F \|\mathbf{A}\|_{2\to 2}^2.$$

The latter is valid since the eigenvalues of $\mathbf{B}$ are $\lambda_i^2 - \left(\sum_{j=1}^n \lambda_j^2\right)/n$ for each $i \in [n]$. The largest eigenvalue of $\mathbf{B}$ is thus at most that of $\mathbf{A}^2$, and since $\lambda_i^2 \geq 0$, the smallest eigenvalue of $\mathbf{B}$ cannot be smaller than $-\|\mathbf{A}\|_{2\to 2}^2$.

Then we have that

$$\mathbb{E}\left[\left(\|\mathbf{Ax}\|_2^2\right)^{k/2}\right] = \mathbb{E}\left[\left|\|\mathbf{A}\|_F^2 + \mathbf{x}^T\mathbf{Bx}\right|^{k/2}\right] \leqslant 2^k \max\left\{\|\mathbf{A}\|_F^k, \mathbb{E}\left[\left|\mathbf{x}^T\mathbf{Bx}\right|^{k/2}\right]\right\}.$$

Hence using the inductive hypothesis on $\mathbf{B}$ we have that

$$\begin{aligned}
\mathbb{E}\left[\left|\mathbf{x}^T\mathbf{Ax}\right|^k\right] &\leqslant 8^k \max\left\{\sqrt{k}\|\mathbf{A}\|_F, C^{k/2}k^{3/4}\|\mathbf{B}\|_F, C^{k/2}k\sqrt{\|\mathbf{B}\|_{2\to 2}}\right\}^k \\
&\leqslant 8^k C^{k/2} \max\left\{\sqrt{k}\|\mathbf{A}\|_F, k^{3/4}\sqrt{\|\mathbf{A}\|_F\|\mathbf{A}\|_{2\to 2}}, k\|\mathbf{A}\|_{2\to 2}\right\}^k \\
&= 8^k C^{k/2} \max\left\{\sqrt{k}\|\mathbf{A}\|_F, k\|\mathbf{A}\|_{2\to 2}\right\}^k,
\end{aligned}$$

where the final equality follows since the middle term above is the geometric mean of the other two, and thus is dominated by at least one of them. This proves our hypothesis as long as $C \geq 64$.

To prove our statement for general k, set $k' = 2^{\lceil \log_2 k \rceil}$. Then using the power mean inequality and our results for k' a power of 2,

$$\mathbb{E}\left[\left|\mathbf{x}^T\mathbf{A}\mathbf{x}\right|^k\right] \leqslant \left(\mathbb{E}\left[\left|\mathbf{x}^T\mathbf{A}\mathbf{x}\right|^{k'}\right]\right)^{k/k'} \leqslant 128^k \max\left\{\sqrt{k}\|\mathbf{A}\|_F, k\|\mathbf{A}\|_{2\to 2}\right\}^k.$$

□

Example 7.2.13 (Concentration using for quadratic form).

$$\begin{aligned} \mathcal{H}_0 &: \mathbf{y} = \mathbf{x}, \qquad \mathbf{x} \in \mathbb{R}^n \\ \mathcal{H}_1 &: \mathbf{y} = \mathbf{x} + \mathbf{z}, \quad \mathbf{z} \in \mathbb{R}^n \end{aligned}$$

where $\mathbf{x} = (x_1, \ldots, x_n)^T$ and x_i are independent, subGaussian random variables with zero mean and variance one and $\mathbf{z}$ is independent of $\mathbf{x}$. For $\mathcal{H}_0$, it follows from Hanson-Wright inequality that

$$\mathbb{E}\left[\left|\mathbf{x}^T\mathbf{A}\mathbf{x} - \mathrm{Tr}\left(\mathbf{A}\right)\right|^k\right] \leqslant C^k \cdot \max\left\{\sqrt{k}\,\|\mathbf{A}\|_F^2, k\|\mathbf{A}\|_{2\to 2}\right\}^k$$

where C is a universal constant. The algorithm claims $\mathcal{H}_1$ if the test metric

$$\mathbb{E}\left[\left|\mathbf{y}^T\mathbf{A}\mathbf{y} - \mathrm{Tr}\left(\mathbf{A}\right)\right|^k\right] > \gamma$$

where γ is the threshold. A good estimate of the threshold is

$$\gamma_0 = C^k \cdot \max\left\{\sqrt{k}\,\|\mathbf{A}\|_F^2, k\|\mathbf{A}\|_{2\to 2}\right\}^k.$$

□

7.3 Structured Random Matrices

As pointed out above, remarkably, all optimal measurement matrices known so far are random matrices. In practice, structure is an additional requirement on the measurement matrix $\mathbf{\Phi}$. Indeed, certain applications impose constraints on the matrix $\mathbf{\Phi}$ and recovery algorithms can be accelerated when fast matrix vector multiplication routines are available for $\mathbf{\Phi}$.

7.3.1 *Partial Random Fourier Matrices*

Partial random Fourier matrices [406, 427] $\mathbf{\Phi} \in \mathbb{C}^{m \times n}$ arise as random row submatrices of the discrete Fourier matrix and their restricted isometry constants satisfy $\delta_s \leqslant \delta$ with high probability provided that

$$m \geqslant C\delta^{-2} s\log^3 s \log n.$$

7.4 Johnson–Lindenstrauss Lemma for Circulant Matrices

Beyond Nyquist: Efficient sampling of sparse bandlimited signals [28] Johnson–Lindenstrauss notes[29]

A variant of the Johnson–Lindenstrauss lemma for circulant matrices [428]

Johnson–Lindenstrauss lemma for circulant matrices [429]

7.5 Composition of a Random Matrix and a Deterministic Dictionary

The theory of compressed sensing has been developed for classes of signals that have a very sparse representation in an orthonormal basis. This is a rather stringent restriction. Indeed, allowing the signal to be sparse with respect to a redundant dictionary adds a lot of flexibility and significantly extends the range of applicability. Already the use of two orthonormal basis instead of just one dramatically increases the class of signals that can be modelled in this way [430].

Throughout this section, $\|\mathbf{x}\|$ denotes the standard Euclidean norm. Signals $\mathbf{y}$ are not sparse in an orthonormal basis but rather in a redundant dictionary dictionary $\mathbf{\Phi} \in \mathbb{R}^{d \times K}$ with $K > d$. Now $\mathbf{y} = \mathbf{\Phi x}$, where $\mathbf{x}$ has only few non-zero components. The goal is to reconstruct $\mathbf{y}$ from few measurements. Given a suitable measurement matrix $\mathbf{A} \in \mathbb{R}^{n \times d}$, we want to recover $\mathbf{y}$ from the measurements $\mathbf{s} = \mathbf{Ay} = \mathbf{A\Phi x}$. The key idea then is to use the sparse representation in $\mathbf{\Phi}$ to drive the reconstruction procedure, i.e., try to identify the sparse coefficient sequence x and from that reconstruct $\mathbf{y}$. Clearly, we may represent $\mathbf{s} = \mathbf{\Psi x}$ with

$$\mathbf{\Psi} = \mathbf{A\Phi} \in \mathbb{R}^{n \times K}.$$

In Table 7.3, two greedy algorithms are listed.

We will assume that $\mathbf{A}$ is an $n \times N$ random matrix that satisfies

$$\mathbb{P}\left(\left|\|\mathbf{Av}\|^2 - \|\mathbf{z}\|^2\right| \geqslant t\|\mathbf{v}\|^2\right) \leqslant 2e^{-cnt^2/2}, \quad t \in (0, 1/3) \tag{7.28}$$

Table 7.3 Greedy algorithms. Goal: reconstruct $\mathbf{x}$ from $\mathbf{s} = \boldsymbol{\Psi}\mathbf{x}$. Columns of $\boldsymbol{\Psi}$ are denoted by $\boldsymbol{\psi}_i$, and $\boldsymbol{\Psi}_\Lambda^\dagger$ is the pseudo-inverse of $\boldsymbol{\Psi}_\Lambda$

Orthogonal matching pursuits	Thresholding
Initialize: $z = 0, \mathbf{r} = \mathbf{s}, \Lambda = \emptyset$ find: $k = \arg\max_i \lvert\langle \mathbf{r}, \boldsymbol{\psi}_i\rangle\rvert$ update: $\Lambda = \Lambda \cup \{k\}, \quad \mathbf{r} = \mathbf{s} - \boldsymbol{\Psi}_\Lambda \boldsymbol{\Psi}_\Lambda^\dagger \mathbf{s}$ iterate until stopping criterion is attained. output: $\mathbf{x} = \boldsymbol{\Psi}_\Lambda^\dagger \mathbf{s}$.	find: Λ that contains the indices corresponding to the S largest values of $\lvert\langle \mathbf{s}, \boldsymbol{\psi}_i\rangle\rvert$ output: $\mathbf{x} = \boldsymbol{\Psi}_\Lambda^\dagger \mathbf{s}$.

for all $\mathbf{v} \in \mathbb{R}^d$ and some constant $c > 0$. Let us list some examples of random matrices that satisfy the above condition: (1) Gaussian ensemble; (2) Bernoulli ensemble; (3) Isotropic subGaussian ensembles; (4) Basis transformation. See [430].

Using the concentration inequality (7.28), we can now investigate the isometry constants for a matrix of the type $\mathbf{A}\boldsymbol{\Phi}$, where $\mathbf{A}$ is an $n \times d$ random measurement matrix and $\boldsymbol{\Phi}$ is a $d \times K$ deterministic dictionary, [430] follows the approach taken in [400], which was inspired by proofs for the Johnson–Lindenstrauss lemma [416]. See Sect. 7.2.

A matrix, which is a composition of a random matrix of certain type and a deterministic dictionary, has small restricted isometry constants.

Theorem 7.5.1 (Lemma 2.1 of Rauhut, Schnass, and Vandergheynst [430]). *Let* $\mathbf{A}$ *be a random matrix of size* $n \times d$ *drawn from a distribution that satisfies the concentration inequality* (7.28). *Extract from the* $d \times K$ *deterministic dictionary* $\boldsymbol{\Phi}$ *any sub-dictionary* $\boldsymbol{\Phi}_\Lambda$ *of size* S, *in* $\mathbb{R}^d$ *i.e.,* $|\Lambda| = S$ *with (local) isometry constant* $\delta_\Lambda = \delta_\Lambda(\boldsymbol{\Phi})$. *For* $0 < \delta < 1$, *we have set*

$$\nu := \delta_\Lambda + \delta + \delta_\Lambda \delta. \tag{7.29}$$

Then

$$(1 - \nu)\,\|\mathbf{x}\|^2 \leqslant \|\mathbf{A}\boldsymbol{\Phi}_\Lambda \mathbf{x}\|_2^2 \leqslant \|\mathbf{x}\|^2 (1 + \nu) \tag{7.30}$$

with probability exceeding

$$1 - 2(1 + 12/\delta)^S e^{-c\delta^2 n/9}.$$

The key ingredient for the proof Theorem 7.5.1 is a finite ε-covering (a set of points[1]) of the unit sphere, which is included below for convenience.

We denote the unit Euclidean ball by $\mathcal{B}_2^n = \{\mathbf{x} \in \mathbb{R}^n : \|\mathbf{x}\| \leqslant 1\}$ and the unit sphere $\mathcal{S}^{n-1} = \{\mathbf{x} \in \mathbb{R}^n : \|\mathbf{x}\| = 1\}$, respectively. For a finite set, the cardinality

[1]A point in a vector space is a vector.

of A is denoted by $|A|$, and for a set $A \in \mathbb{R}^n$, conv A denotes the convex hull of A. The following fact is well-known and standard: see, e.g, [152, Lemma 4.10] and [407, Lemma 2.2].

Theorem 7.5.2 (ε-Cover). *Let $n \geq 1$ and $\varepsilon > 0$. There exists an ε-cover $\Gamma \subset \mathcal{B}_2^n$ of the unit Euclidean ball $\mathcal{B}_2^n$ with respect to the Euclidean metric such that $\mathcal{B}_2^n \subset (1-\varepsilon)^{-1} \operatorname{conv} \Gamma$ and*

$$|\Gamma| \leqslant (1 + 2/\varepsilon)^n.$$

Similarly, there exists $\Gamma' \subset \mathcal{S}^{n-1}$ which is an ε-cover of the sphere $\mathcal{S}^{n-1}$ and

$$|\Gamma'| \leqslant (1 + 2/\varepsilon)^n.$$

Proof of Theorem 7.5.1. First we choose a finite ε-covering of the unit sphere in $\mathbb{R}^S$, i.e., a set of points $\mathcal{Q}$, with $\|\mathbf{q}\| = 1$ for all $\mathbf{q} \in \mathcal{Q}$, such that for all $\mathbf{q} \in \mathcal{Q}$, such that for all $\|\mathbf{x}\| = 1$

$$\min_{\mathbf{q} \in \mathcal{Q}} \|\mathbf{x} - \mathbf{q}\| \leqslant \varepsilon$$

for some $\varepsilon \in (0, 1)$. According to Theorem 7.5.2, there exists such a $\mathcal{Q}$ with $|\mathcal{Q}| \leqslant (1 + 2/\varepsilon)^S$. Applying the measure concentration in (7.29) with $t < 1/3$ to all the points $\mathbf{\Phi}_\Lambda \mathbf{q}$ and taking the union bound we obtain

$$(1-t) \|\mathbf{\Phi}_\Lambda \mathbf{q}\|^2 \leqslant \|\mathbf{A}\mathbf{\Phi}_\Lambda \mathbf{q}\|^2 \leqslant (1+t) \|\mathbf{\Phi}_\Lambda \mathbf{q}\|^2 \quad \text{for all} \quad \mathbf{q} \in \mathcal{Q}. \tag{7.31}$$

with probability larger than

$$1 - 2\left(1 + \frac{2}{\varepsilon}\right)^S e^{-cnt^2}.$$

Define ν as the *smallest* number such that

$$\|\mathbf{A}\mathbf{\Phi}_\Lambda \mathbf{x}\|^2 \leqslant (1+\nu) \|\mathbf{x}\|^2 \quad \text{for all } x \text{ supported on } \Lambda. \tag{7.32}$$

Next we estimate ν in terms of ε, t. Since for all $\mathbf{x}$ with $\|\mathbf{x}\| = 1$ we can choose a point $\mathbf{q}$ such that $\|\mathbf{x} - \mathbf{q}\| \leqslant \varepsilon$ and obtain

$$\begin{aligned}
\|\mathbf{A}\mathbf{\Phi}_\Lambda \mathbf{x}\| &\leqslant \|\mathbf{A}\mathbf{\Phi}_\Lambda \mathbf{q}\| + \|\mathbf{A}\mathbf{\Phi}_\Lambda (\mathbf{x} - \mathbf{q})\| \\
&\leqslant (1+t)^{1/2} \|\mathbf{\Phi}_\Lambda \mathbf{q}\| + \|\mathbf{A}\mathbf{\Phi}_\Lambda (\mathbf{x} - \mathbf{q})\| \\
&\leqslant (1+t)^{1/2} (1+\delta_\Lambda)^{1/2} + (1+\nu)^{1/2} \varepsilon.
\end{aligned}$$

Since ν is the smallest possible constant for which (7.32) holds it also has to satisfy

$$\sqrt{1+\nu} \leqslant \sqrt{1+t}\sqrt{1+\delta_\Lambda} + (1+\nu)\,\varepsilon.$$

Simplifying the above equation gives

$$1+\nu \leqslant \frac{1+\varepsilon}{(1-t)^2}\,(1+\delta_\Lambda)\,.$$

Now we choose $\varepsilon = \delta/6$, and $t = \delta/3 < 1/3$. Then

$$\frac{1+t}{(1-\varepsilon)} = \frac{1+\delta/3}{(1-\delta/6)} = \frac{1+\delta/3}{1-\delta/3+\delta^2/36} < \frac{1+\delta/3}{1-\delta/3} = 1+\frac{2\delta/3}{1-\delta/3} < 1+\delta.$$

Thus,

$$\nu < \delta + \delta_\Lambda\,(1+\delta)\,.$$

To get a lower bound we operate in a similar fashion,

$$\begin{aligned}\|\mathbf{A}\mathbf{\Phi}_\Lambda\mathbf{x}\| &\geqslant \|\mathbf{A}\mathbf{\Phi}_\Lambda\mathbf{q}\| - \|\mathbf{A}\mathbf{\Phi}_\Lambda\,(\mathbf{x}-\mathbf{q})\| \\ &\geqslant (1+t)^{1/2}\,\|\mathbf{\Phi}_\Lambda\mathbf{q}\| - \|\mathbf{A}\mathbf{\Phi}_\Lambda\,(\mathbf{x}-\mathbf{q})\| \\ &\geqslant (1-t)^{1/2}(1-\delta_\Lambda)^{1/2} - (1+\nu)^{1/2}\varepsilon.\end{aligned}$$

Now square both sides and observe that $\nu < 1$ (otherwise we have nothing to show). Then we finally arrive at

$$\begin{aligned}\|\mathbf{A}\mathbf{\Phi}_\Lambda\mathbf{x}\|^2 &\geqslant \left((1-t)^{1/2}(1-\delta_\Lambda)^{1/2} - \varepsilon\sqrt{2}\right)^2 \\ &\geqslant (1-t)\,(1-\delta_\Lambda) - 2\sqrt{2}\varepsilon(1-t)^{1/2}(1-\delta_\Lambda)^{1/2} + 2\varepsilon^2 \\ &\geqslant 1-\delta_\Lambda - t - 2\sqrt{2}\varepsilon \geqslant 1-\delta_\Lambda-\delta \geqslant 1-\nu.\end{aligned}$$

This completes the proof. □

Based on the previous theorem it is easy to derive an estimation of the global restricted isometry constants of the composed matrix $\mathbf{\Psi} = \mathbf{A}\mathbf{\Phi}$.

Theorem 7.5.3 (Theorem 2.2 of Rauhut, Schnass, and Vandergheynst [430]). *Let $\mathbf{\Phi} \in \mathbb{R}^{d\times K}$ be a deterministic dictionary of size K in $\mathbb{R}^d$ with restricted isometry constant $\delta_S\,(\mathbf{\Phi})\,, S \in \mathbb{N}$. Let $\mathbf{A} \in \mathbb{R}^{n\times d}$ be a random matrix satisfying* (7.28) *and assume*

$$n \geqslant C\delta^{-2}\,(S\log{(K/S)} + \log{(2e\,(1+12/\delta))} + t) \tag{7.33}$$

for some $\delta \in (0,1)$ and $t > 0$. Then with probability at least $1-e^{-t}$ the composed matrix $\mathbf{\Psi} = \mathbf{A}\mathbf{\Phi}$ has restricted isometry constant

$$\delta_S(\mathbf{A\Phi}) \leqslant \delta_S(\mathbf{\Phi}) + \delta(1 + \delta_S(\mathbf{\Phi})). \tag{7.34}$$

The constant satisfies $C \leq 9/c$.

Proof. Using Theorem 7.5.1 we can estimate the probability that a sub-dictionary

$$\mathbf{\Psi}_S = (\mathbf{A\Phi})_S = \mathbf{A\Phi}_S, S = \{1, \ldots, K\}$$

does not hold for (local) isometry constants $\delta_S(\mathbf{A\Phi}) \leqslant \delta_S(\mathbf{\Phi}) + \delta(1 + \delta_S(\mathbf{\Phi}))$ by the probability

$$\mathbb{P}(\delta_S(\mathbf{A\Phi}) > \delta_S(\mathbf{\Phi}) + \delta(1 + \delta_S(\mathbf{\Phi}))) \leqslant 2\left(1 + \frac{12}{\delta}\right)^S \exp\left(-c\delta^2 n/9\right).$$

By taking the union bound over all $\binom{K}{S}$ possible sub-dictionaries of size S we can estimate the probability of $\delta_S(\mathbf{A\Phi}) = \sup\limits_{\Lambda=\{1,\ldots,K\},|\Lambda|=S} \delta_\Lambda(\mathbf{A\Phi})$ not satisfying (7.34) by

$$\mathbb{P}(\delta_S(\mathbf{A\Phi}) > \delta_S(\mathbf{\Phi}) + \delta(1 + \delta_S(\mathbf{\Phi}))) \leqslant 2\binom{K}{S}\left(1 + \frac{12}{\delta}\right)^S \exp\left(-c\delta^2 n/9\right).$$

Using Stirling's formula $\binom{K}{S} \leqslant (eK/S)^S$ and demanding that the above term is less than e^{-t} completes the proof. □

It is interesting to observe the stability of inner products under multiplication with a random matrix $\mathbf{A}$, i.e.,

$$\langle \mathbf{Ax}, \mathbf{Ay} \rangle \approx \langle \mathbf{x}, \mathbf{y} \rangle.$$

Theorem 7.5.4 (Lemma 3.1 of Rauhut, Schnass, and Vandergheynst [430]). *Let* $\mathbf{x}, \mathbf{y} \in \mathbb{R}^N$ *with* $\|\mathbf{x}\|_2, \|\mathbf{y}\|_2 \leqslant 1$. *Assume that* $\mathbf{A}$ *is an* $n \times N$ *random matrix with independent* $\mathcal{N}(0, 1/n)$ *entries (independent of* $\mathbf{x}, y$*). Then for all* $t > 0$

$$\mathbb{P}(|\langle \mathbf{Ax}, \mathbf{Ay} \rangle - \langle \mathbf{x}, \mathbf{y} \rangle| \geqslant t) \leqslant 2\exp\left(-n\frac{t^2}{C_1 + C_2 t}\right), \tag{7.35}$$

with $C_1 = \frac{4e}{\sqrt{6\pi}} \approx 2.5044$, *and* $C_2 = e\sqrt{2} \approx 3.8442$.

The analogue statement holds for a random matrix $\mathbf{A}$ *whose entries are independent* $\pm 1/\sqrt{n}$ *Bernoulli random variables.*

Taking $\mathbf{x} = \mathbf{y}$ in the theorem provides the concentration inequality (7.28) for Gaussian and Bernoulli matrices. Due to the elementary nature of the method used in the proof [430], we include the proof here as an example below.

We need Theorem 1.3.4 due to Bennett [12, Eq. (7)].

Example 7.5.5 (Concentration of measure for inner products). Observe that

$$\langle \mathbf{Ax}, \mathbf{Ay} \rangle = \frac{1}{n} \sum_{\ell=1}^{n} \sum_{k=1}^{n} \sum_{j=1}^{n} g_{\ell k} g_{\ell j} x_k y_k$$

where $g_{\ell k}, \ell = 1, \ldots, n, k = 1, \ldots, N$ are independent standard Gaussian random variables. We consider the random variable

$$Y = \sum_{k=1}^{N} \sum_{j=1}^{N} g_k g_j x_k y_j$$

where again $g_k, k = 1, \ldots, N$ are independent standard Gaussian random variables. Now we can write

$$\langle \mathbf{Ax}, \mathbf{Ay} \rangle = \frac{1}{n} \sum_{\ell=1}^{n} Y_\ell$$

where Y_ℓ are independent copies of Y.

The expectation of Y is easily expressed as

$$\mathbb{E}Y = \sum_{k=1}^{N} x_k y_k = \langle \mathbf{x}, \mathbf{y} \rangle.$$

Hence, also $\mathbb{E}\left[\langle \mathbf{Ax}, \mathbf{Ay} \rangle\right] = \langle \mathbf{x}, \mathbf{y} \rangle$. Let

$$Z = Y - \mathbb{E}Y = \sum_{k \neq j} g_j g_k x_j y_k + \sum_{k} \left(g_k^2 - 1\right) x_k y_k.$$

The new random variable Z is known as Gaussian chaos of order 2. Note that $\mathbb{E}Z = 0$.

To apply Theorem 1.3.4, we need to show the moment bound (1.24) for the random variable Z. A general bound for Gaussian chaos [27, p. 65] gives

$$\mathbb{E}|Z|^p \leqslant (p-1)^p \left(\mathbb{E}|Z|^2\right)^{p/2}. \tag{7.36}$$

for $p \geq 2$. Using Stirling's formula, $p! = \sqrt{2\pi p} p^p e^{-p} e^{R_p}$, $\frac{1}{12p+1} \leqslant R_p \leqslant \frac{1}{12p}$, we have that, for $p \geq 3$,

$$\begin{aligned}\mathbb{E}|Z|^p &= p!\frac{(p-1)^p}{e^{R_p}\sqrt{2\pi p}e^{-p}p^p}\left(\mathbb{E}|Z|^2\right)^{p/2}\\&= \left(1-\frac{1}{p}\right)^p\frac{e^2p!}{e^{R_p}\sqrt{2\pi p}}\left(e^2\mathbb{E}|Z|^2\right)^{(p-2)/2}\mathbb{E}|Z|^2\\&\leqslant \frac{e}{e^{R_p}\sqrt{2\pi p}}p!\left(e^2\mathbb{E}|Z|^2\right)^{(p-2)/2}\mathbb{E}|Z|^2\\&\leqslant p!\left(e\left(\mathbb{E}|Z|^2\right)^{1/2}\right)^{(p-2)}\frac{e}{\sqrt{6\pi}}\mathbb{E}|Z|^2.\end{aligned}$$

Compare the above with the moment bound (1.24) holds for all $p \geq 3$ with

$$M = e\left(\mathbb{E}|Z|^2\right)^{1/2}, \sigma^2 = \frac{2e}{\sqrt{6\pi}}\mathbb{E}|Z|^2,$$

and by direct inspection we see that it also holds for $p = 2$.

Now we need to determine $\mathbb{E}|Z|^2$. Using the independence of the Gaussian random variables g_k, we obtain

$$\begin{aligned}\mathbb{E}|Z|^2 &= \mathbb{E}\left[\sum_{j\neq k}\sum_{j'\neq k'} g_j g_k g_{j'} g_{k'} x_j x_k x_{j'} x_{k'} + 2\sum_{j\neq k}\sum_{k'} g_j g_k \left(g_{k'}^2 - 1\right) x_j x_k x_{j'} x_{k'}\right.\\&\left.+\sum_{k}\sum_{k'}\left(g_k^2-1\right)\left(g_{k'}^2-1\right) x_k y_k x_{k'} y_{k'}\right]\end{aligned}$$

Further

$$\begin{aligned}\mathbb{E}|Z|^2 &= \sum_{j\neq k}\mathbb{E}\left[g_j^2\right]\mathbb{E}\left[g_k^2\right] x_j y_j x_k y_k + \sum_{k\neq j}\mathbb{E}\left[g_j^2\right]\mathbb{E}\left[g_k^2\right] x_j^2 y_k^2\\&+\sum_k \mathbb{E}\left[\left(g_k^2-1\right)^2\right] x_k^2 y_k^2.\end{aligned} \tag{7.37}$$

Finally,

$$\begin{aligned}\mathbb{E}|Z|^2 &= \sum_{k\neq j} x_j y_j x_k y_k + \sum_{k\neq j} x_j^2 y_k^2 + 2\sum_k x_k^2 y_k^2\\&= \sum_{j,k} x_j y_j x_k y_k + \sum_{j,k} x_j^2 y_k^2\\&= \langle \mathbf{x}, \mathbf{y}\rangle^2 + \|\mathbf{x}\|_2^2\,\|\mathbf{y}\|_2^2 \leqslant 2\end{aligned} \tag{7.38}$$

since by assumption $\|\mathbf{x}\|_2, \|\mathbf{y}\|_2 \leqslant 1$. Denoting by $Z_\ell, \ell = 1, \ldots, n$ independent copies of Z, Theorem 1.3.4 gives

$$\mathbb{P}\left(|\langle \mathbf{Ax}, \mathbf{Ay}\rangle - \langle \mathbf{x}, \mathbf{y}\rangle| \geqslant t\right) = \mathbb{P}\left(\left|\sum_{\ell=1}^{n} Z_\ell\right| \geqslant nt\right)$$
$$\leqslant 2\exp\left(-\frac{1}{2}\frac{n^2t^2}{n\sigma^2 + nMt}\right) = 2\exp\left(-n\frac{t^2}{C_1 + C_2 t}\right),$$

with $C_1 = \frac{2e}{\sqrt{6\pi}}\mathbb{E}|Z|^2 \leqslant \frac{4e}{\sqrt{6\pi}} \approx 2.5044$ and $C_2 = e\sqrt{2} \approx 3.8442$.

For the case of Bernoulli random matrices, the derivation is completely analogue. We just have to replace the standard Gaussian random variables g_k by $\varepsilon_k = \pm 1$ Bernoulli random variables. In particular, the estimate (7.36) for the chaos variable Z is still valid, see [27, p. 105]. Furthermore, for Bernoulli variables $\varepsilon_k = \pm 1$ we clearly have $\varepsilon_k^2 = 1$. Hence, going through the estimate above we see that in (7.37) the last term is actually zero, so the final bound in (7.38) is still valid. □

Now we are in a position to investigate recovery from random measurements by thresholding, using Theorem 7.5.4. Thresholding works by comparing inner products of the signal with the atoms of the dictionary.

Example 7.5.6 (Recovery from random measurements by thresholding [430]). Let $\mathbf{A}$ be an $n \times K$ random matrix satisfying one of the two probability models of Theorem 7.5.4. We know thresholding will succeed if we have

$$\min_{i\in\Lambda} |\langle \mathbf{Ay}, \mathbf{Az}_i\rangle| > \max_{k\in\bar{\Lambda}} |\langle \mathbf{Ay}, \mathbf{Az}_k\rangle|.$$

We need to estimate the probability that the above inequality is violated

$$\mathbb{P}\left(\min_{i\in\Lambda} |\langle \mathbf{Ay}, \mathbf{Az}_i\rangle| \leqslant \max_{k\in\bar{\Lambda}} |\langle \mathbf{Ay}, \mathbf{Az}_k\rangle|\right)$$
$$\leqslant \mathbb{P}\left(\min_{i\in\Lambda} |\langle \mathbf{Ay}, \mathbf{Az}_i\rangle| \leqslant \min_{i\in\Lambda} |\langle \mathbf{y}, \mathbf{z}_i\rangle| - \tfrac{\varepsilon}{2}\right) + \mathbb{P}\left(\max_{k\in\bar{\Lambda}} |\langle \mathbf{Ay}, \mathbf{Az}_k\rangle| \geqslant \max_{k\in\bar{\Lambda}} |\langle \mathbf{y}, \mathbf{z}_k\rangle| + \tfrac{\varepsilon}{2}\right).$$

The probability of the good components having responses lower than the threshold can be further estimated as

$$\mathbb{P}\left(\min_{i\in\Lambda} |\langle \mathbf{Ay}, \mathbf{Az}_i\rangle| \leqslant \min_{i\in\Lambda} |\langle \mathbf{y}, \mathbf{z}_i\rangle| - \tfrac{\varepsilon}{2}\right)$$
$$\leqslant \mathbb{P}\left(\bigcup_{i\in\Lambda}\left\{|\langle \mathbf{Ay}, \mathbf{Az}_i\rangle| \leqslant \min_{i\in\Lambda} |\langle \mathbf{y}, \mathbf{z}_i\rangle| - \tfrac{\varepsilon}{2}\right\}\right)$$
$$\leqslant \sum_{i\in\Lambda} \mathbb{P}\left(|\langle \mathbf{y}, \mathbf{z}_i\rangle - \langle \mathbf{Ay}, \mathbf{Az}_i\rangle| \geqslant \tfrac{\varepsilon}{2}\right)$$
$$\leqslant 2|\Lambda| \exp\left(-n\frac{t^2/4}{C_1 + C_2 t/2}\right).$$

Similarly, we can bound the probability of the bad components being higher than the threshold,

$$\begin{aligned}
&\mathbb{P}\left(\max_{k\in\bar{\Lambda}}|\langle \mathbf{Ay},\mathbf{Az}_k\rangle| \geqslant \max_{k\in\bar{\Lambda}}|\langle \mathbf{y},\mathbf{z}_k\rangle| + \tfrac{\varepsilon}{2}\right)\\
&\quad\leqslant \mathbb{P}\left(\bigcup_{k\in\bar{\Lambda}}\left\{|\langle \mathbf{Ay},\mathbf{Az}_k\rangle| \geqslant \max_{k\in\bar{\Lambda}}|\langle \mathbf{y},\mathbf{z}_k\rangle| + \tfrac{\varepsilon}{2}\right\}\right)\\
&\quad\leqslant \sum_{k\in\bar{\Lambda}}\mathbb{P}\left(|\langle \mathbf{Ay},\mathbf{Az}_k\rangle - \langle \mathbf{y},\mathbf{z}_k\rangle| \geqslant \tfrac{\varepsilon}{2}\right)\\
&\quad\leqslant 2\left|\bar{\Lambda}\right|\exp\left(-n\tfrac{t^2/4}{C_1+C_2t/2}\right).
\end{aligned}$$

Combining the these two estimates we obtain that the probability of success for thresholding is exceeding

$$1 - 2K\exp\left(-n\frac{t^2/4}{C_1+C_2t/2}\right).$$

Theorem 7.5.4 finally follows from requiring this probability to be higher than $1 - e^{-t}$ and solving for n. □

We summarize the result in the following theorem.

Theorem 7.5.7 (Theorem 3.2 of Rauhut, Schnass, and Vandergheynst [430]). *Let $\mathbf{\Psi}$ be a $d \times K$ dictionary. Assume that the support of $\mathbf{x}$ for a signal $\mathbf{y} = \mathbf{\Phi x}$, normalized to have $\|\mathbf{y}\|_2 = 1$, could be recovered by thresholding with a margin ε, i.e.*

$$\min_{i\in\Lambda}|\langle \mathbf{Ay},\mathbf{Az}_i\rangle| > \max_{k\in\bar{\Lambda}}|\langle \mathbf{Ay},\mathbf{Az}_k\rangle| + \varepsilon.$$

Let $\mathbf{A}$ be an $n \times d$ random matrix satisfying one of the two probability models of Theorem 7.5.4. Then, with probability exceeding $1 - e^{-t}$, the support and thus the signal can be reconstructed via thresholding from the n-dimensional measurement vector $\mathbf{w} = \mathbf{Ay} = \mathbf{A\Phi x}$ as long as

$$n \geqslant C(\varepsilon)\left(\log(2K) + t\right).$$

where $C(\varepsilon) = 4C_1\varepsilon^{-2} + 2C_2\varepsilon^{-1}$ and C_1, C_2 are constants from Theorem 7.5.4.

7.6 Restricted Isometry Property for Partial Random Circulant Matrices

Circular matrices are connected to circular convolution, defined for two vectors $\mathbf{x},\mathbf{z} \in \mathbb{C}^n$ by

$$(\mathbf{z} \otimes \mathbf{x})_j := \sum_{k=1}^{n} z_{j\ominus k}x_k, \quad j = 1,\ldots,n,$$

where

$$j \ominus k = j - k \bmod n$$

is the cyclic subtraction. The circular matrix $\mathbf{H} = \mathbf{H}_{\mathbf{z}} \in \mathbb{C}^{n \times n}$ associated with $\mathbf{z}$ is given by

$$\mathbf{Hx} = \mathbf{z} \otimes \mathbf{x}$$

and has entries $H_{jk} = z_{j \ominus k}$. Given a vector $\mathbf{z} = (z_0, \ldots, z_{n-1})^T \in \mathbb{C}^n$, we introduce the circulant matrix

$$\mathbf{H}_{\mathbf{z}} = \begin{bmatrix} z_0 & z_{n-1} & \cdots & z_1 \\ z_1 & z_0 & \cdots & z_2 \\ \vdots & \vdots & & \vdots \\ z_{n-1} & z_{n-2} & \cdots & z_0 \end{bmatrix} \in \mathbb{C}^{n \times n}.$$

Square matrices are not very interesting for compressed sensing, so we our attention to a row submatrix of $\mathbf{H}$. Consider an *arbitrary* index set $\Omega \subset \{0, 1, \ldots, n-1\}$ whose cardinality $|\Omega| = m$. We define the operator $\mathbf{R}_\Omega : \mathbb{C}^n \to \mathbb{C}^m$ that restricts a vector $\mathbf{x} \in \mathbb{C}^n$ to its entries in Ω. Let $\boldsymbol{\varepsilon} = \{\varepsilon_i\}_{i=1}^n$ be a Rademacher vector of length n, i.e., a random vector with independent entries distributed according to $\mathbb{P}(\varepsilon_i = \pm 1) = 1/2$. The associated partial random circulant matrix is given by

$$\mathbf{\Phi} = \frac{1}{\sqrt{m}} \mathbf{R}_\Omega \mathbf{H}_{\boldsymbol{\varepsilon}} \in \mathbb{R}^{m \times n} \tag{7.39}$$

and acts on complex vectors $\mathbf{x} \in \mathbb{C}^n$ via

$$\mathbf{\Phi x} = \frac{1}{\sqrt{m}} \mathbf{R}_\Omega \mathbf{H}_\varepsilon \mathbf{x} = \frac{1}{\sqrt{m}} \mathbf{R}_\Omega (\boldsymbol{\varepsilon} \otimes \mathbf{x}). \tag{7.40}$$

In other words, $\mathbf{\Phi x}$ is a circular matrix generated by a Rademacher vector, where the rows outside Ω are removed.

Example 7.6.1 (Expectation of the restricted isometry constant [403]). We study the expectation of the restricted isometry constant, $\mathbb{E}[\delta_s]$. The goal is to convert $\mathbb{E}[\delta_s]$ to another form that is easier to bound. Let T denote the set of all s-sparse signals in the Euclidean unit ball:

$$T = \{\mathbf{x} \in \mathbb{R}^n : \|\mathbf{x}\|_0 \leqslant s, \|\mathbf{x}\|_2 \leqslant 1\}. \tag{7.41}$$

Define a function $|||\cdot|||$ on Hermitian $n \times n$ matrices via the expression

$$|||\mathbf{A}||| = \sup_{\mathbf{x} \in T} \left| \mathbf{x}^T \mathbf{A} \mathbf{x} \right|.$$

Now consider

$$|||\mathbf{\Phi}^H\mathbf{\Phi} - \mathbf{I}||| = \sup_{\mathbf{x}\in T} \left| \left\langle \left(\mathbf{\Phi}^H\mathbf{\Phi} - \mathbf{I} \right) \mathbf{x}, \mathbf{x} \right\rangle \right| = \sup_{\mathbf{x}\in T} \left| \|\mathbf{\Phi}\mathbf{x}\|_2^2 - \|\mathbf{x}\|_2^2 \right| = \delta_s. \quad (7.42)$$

Let $\mathbf{S}$ be the cyclic shift down operator on column vectors in $\mathbb{R}^n$. Applying the power $\mathbf{S}^k$ to $\mathbf{x}$ will cycle $\mathbf{x}$ downward by k coordinates:

$$\left(\mathbf{S}^k\mathbf{x}\right)_\ell = x_{\ell\ominus k},$$

where $\ominus$ is subtraction modulo n. Note that $\left(\mathbf{S}^k\right)^H = \mathbf{S}^{-k} = \mathbf{S}^{n-k}$. Then we rewrite $\mathbf{\Psi}$ as a random sum of shift operators,

$$\mathbf{\Phi} = \frac{1}{\sqrt{m}} \sum_{k=1}^{n} \varepsilon_k \mathbf{R}_\Omega \mathbf{S}^k.$$

It follows that

$$\mathbf{\Phi}^H\mathbf{\Phi} - \mathbf{I} = \frac{1}{m} \sum_{k\neq\ell}^{n} \varepsilon_k\varepsilon_\ell \mathbf{S}^{-k}\mathbf{R}_\Omega^H\mathbf{R}_\Omega\mathbf{S}^\ell = \frac{1}{m} \sum_{k\neq\ell}^{n} \varepsilon_k\varepsilon_\ell \mathbf{S}^{-k}\mathbf{P}_\Omega\mathbf{S}^\ell. \quad (7.43)$$

where $\mathbf{P}_\Omega = \mathbf{R}_\Omega^H\mathbf{R}_\Omega$ is the $n \times n$ diagonal projector onto the coordinates in Ω. Applying $\mathbf{P}_\Omega$ to the vector $\mathbf{x}$ preserves the values of $\mathbf{x}$ on the set Ω while setting the values outside of Ω to zero. Combining (7.42) and (7.43), we get the final form

$$\delta_s = \sup_{\mathbf{x}\in T} |G_\mathbf{x}| \quad \text{where} \quad G_\mathbf{x} = \frac{1}{m} \sum_{k\neq\ell}^{n} \varepsilon_k\varepsilon_\ell \mathbf{x}^H\mathbf{S}^{-k}\mathbf{P}_\Omega\mathbf{S}^\ell\mathbf{x}. \quad (7.44)$$

We may regard the restricted isometry constant as the supremum of a random process indexed by the set T defined above. The expected supremum of this process can be bounded using some sophisticated techniques like Rademacher chaos, covering number estimates, and chaining [27, 81]. □

The next example is to re-express the random process $G_\mathbf{x}$ (defined in (7.44)) in the Fourier domain. This is a key tool. [403] uses a version of the classical Dudley inequality—Sect. 3.5—for Rademacher chaos that bounds the expectation of its supremum by the maximum of two entropy integrals that involve covering numbers with respect to two different metrics. Then they use elementary ideas from Fourier analysis to provide bounds for these metrics. This reduction allows them to exploit covering number estimates from the RIP analysis for partial Fourier matrices [30, 402] to complete the argument.

Example 7.6.2 (Fourier representation of the random process [403]). This approach studied here is elementary and is of independent interest. Let $\mathbf{F}$ be the $n \times n$ discrete Fourier transform matrix whose entries are given by the expression

$$F(\omega, \ell) = e^{-i2\pi\omega\ell/n}, 0 \leqslant \omega, \ell \leqslant n-1.$$

Here $\mathbf{F}$ is unnormalized. The hat symbol denotes the Fourier transform of a vector: $\hat{\mathbf{x}} = \mathbf{F}\mathbf{x}$. Use the property of Fourier transform: a shift in the time domain followed by a Fourier transform may be written as a Fourier transform followed by a frequency modulation

$$\mathbf{F}\mathbf{S}^k = \mathbf{M}^k\mathbf{F},$$

where $\mathbf{M}$ is the diagonal matrix with entries $\mathbf{M}(\omega, \omega) = e^{-i2\pi\omega/n}$ for $0 \leqslant \omega \leqslant n-1$. Now we are ready to handle $G_{\mathbf{x}}$. The random process $G_{\mathbf{x}}$ has the Fourier transform representation

$$G_{\mathbf{x}} = \frac{1}{m}\sum_{k \neq \ell}^{n} \varepsilon_k \varepsilon_\ell \hat{\mathbf{x}}^H \mathbf{M}^{-k} \hat{\mathbf{P}}_\Omega \mathbf{M}^\ell \hat{\mathbf{x}}, \tag{7.45}$$

where $\hat{\mathbf{P}}_\Omega = \frac{1}{n}\mathbf{F}\mathbf{P}_\Omega\mathbf{F}^{-1}$. The matrix $\hat{\mathbf{P}}_\Omega$ has several nice properties:

1. $\hat{\mathbf{P}}_\Omega$ is *circulant* and conjugate symmetric.
2. Along the diagonal $\hat{\mathbf{P}}_\Omega(\omega, \omega) = m/n^2$, and off the diagonal $\left|\hat{\mathbf{P}}_\Omega(\omega, \omega)\right| \leqslant m/n^2$.
3. Since the rows and columns of $\hat{\mathbf{P}}_\Omega$ are circular shifts of one another,

$$\sum_\omega \left|\hat{\mathbf{P}}_\Omega(\omega, \xi)\right|^2 = \sum_\xi \left|\hat{\mathbf{P}}_\Omega(\omega, \xi)\right|^2 = \left\|\hat{\mathbf{P}}_\Omega\right\|_F^2 / n = m/n^3.$$

4. $\hat{\mathbf{P}}_\Omega$ has exactly m nonzero eigenvalues $\lambda_i, i = 1, \ldots, m$, each of which is equal to $\lambda_i = 1/n$. As such, $\hat{\mathbf{P}}_\Omega$ has spectral norm $\left\|\hat{\mathbf{P}}_\Omega\right\| = 1/n$ and Frobenius norm $\left\|\hat{\mathbf{P}}_\Omega\right\|_F^2 = m/n^2$.

These properties immediately follow from the fact that $\mathbf{P}_\Omega = \mathbf{R}_\Omega^H \mathbf{R}_\Omega$ is a diagonal matrix with 0–1 entries. The matrix $\hat{\mathbf{P}}_\Omega$ inherits conjugate symmetry from $\mathbf{P}_\Omega$. $\hat{\mathbf{P}}_\Omega$ is circulant since it is diagonalized by the Fourier transform. Since we form $\hat{\mathbf{P}}_\Omega$ by applying a similar transform to $\mathbf{P}_\Omega$, they have the same eigenvalue modulo the scalar factor $1/n$. We can further rewrite the random process (7.45) in terms of the form (7.46) we desire for sophisticated techniques. □

Example 7.6.3 (Integrability of chaos processes [403]). Let $\boldsymbol{\varepsilon} = (\varepsilon_1, \ldots, \varepsilon_n)^T$. The process (7.45) can be written as a quadratic form

$$G_{\mathbf{x}} = \langle \boldsymbol{\varepsilon}, \mathbf{Z}_{\mathbf{x}} \boldsymbol{\varepsilon} \rangle \quad \text{where} \quad \mathbf{x} \in T. \tag{7.46}$$

The matrix $\mathbf{Z}_{\mathbf{x}}$ has entries

$$Z_{\mathbf{x}}(k,\ell) = \begin{cases} \frac{1}{m}\hat{\mathbf{x}}^H \mathbf{M}^{-k}\hat{\mathbf{P}}_\Omega \mathbf{M}^\ell \hat{\mathbf{x}}, & k \neq \ell \\ 0, & k = \ell \end{cases}.$$

A short calculation verifies that this matrix can be expressed compactly as

$$\mathbf{Z}_{\mathbf{x}} = \frac{1}{m}\left(\mathbf{F}^H \hat{\mathbf{X}}^H \hat{\mathbf{P}}_\Omega \hat{\mathbf{X}}\mathbf{F}\right) - \mathrm{diag}\left(\mathbf{F}^H \hat{\mathbf{X}}^H \hat{\mathbf{P}}_\Omega \hat{\mathbf{X}}\mathbf{F}\right), \tag{7.47}$$

where $\hat{\mathbf{X}} = \mathrm{diag}(\hat{\mathbf{x}})$ is the diagonal matrix constructed from the vector $\hat{\mathbf{x}}$. The term *homogeneous second-order chaos* is used to refer to a random process $G_{\mathbf{x}}$ of the form (7.46) where each matrix $\mathbf{Z}_{\mathbf{x}}$ is conjugate symmetric and hollow, i.e., has zeros on the diagonal. To bound the expected supremum of the random process $G_{\mathbf{x}}$ over the set T, we apply a version of Dudleys inequality that is specialized to this setting. Define two pseudo-metrics on the index set T:

$$d_1(\mathbf{x},\mathbf{y}) = \|\mathbf{Z}_{\mathbf{x}} - \mathbf{Z}_{\mathbf{y}}\| \quad \text{and} \quad d_2(\mathbf{x},\mathbf{y}) = \|\mathbf{Z}_{\mathbf{x}} - \mathbf{Z}_{\mathbf{y}}\|_F.$$

Let $N(T, d_i, r)$ denote the minimum number of balls of radius r in the metric d_1, d_2 that we need to cover the set T.

Proposition 7.6.4 (Dudley's inequality for chaos). *Suppose that $G_{\mathbf{x}}$ is a homogeneous second-order chaos process indexed by a set T. Fix a point $\mathbf{x}_0 \in T$. There exists a universal constant C such that*

$$\mathbb{E}\sup_{\mathbf{x}\in T}|G_{\mathbf{x}} - G_{\mathbf{x}_0}| \leqslant C\max\left\{\int_0^\infty \log N(T,d_1,r)\,dr,\ \int_0^\infty \sqrt{\log N(T,d_2,r)}dr\right\}. \tag{7.48}$$

Our statement of the proposition follows [403] and looks different from the versions presented in the literature [27, Theorem 11.22] and [81, Theorem 2.5.2]. For details about how to bound these integrals, we see [403]. □

Immediately following Example 7.6.3, we can get the following theorem.

Theorem 7.6.5 (Theorem 1.1 of Rauhut, Romberg, and Tropp [403]). *Let Ω be an arbitrary subset of $\{0,1,\ldots,n-1\}$ with cardinality $|\Omega| = m$. Let $\mathbf{\Psi}$ be the corresponding partial random circulant matrix* (7.39) *generated by a Rademacher sequence, and let δ_s denote the s-th restricted isometry constant. Then,*

$$\mathbb{E}[\delta_s] \leqslant C_1\max\left\{\frac{s^{3/2}}{m}\log^{3/2} n,\ \sqrt{\frac{s}{m}}\log s\log n\right\} \tag{7.49}$$

where $C_1 > 0$ is a universal constant.

In particular, (7.49) implies that for a given $\delta \in (0,1)$, we have $\mathbb{E}[\delta_s] \leqslant \delta$ provided

$$m \geqslant C_2 \max\left\{\delta^{-1}s^{3/2}\log^{3/2}n, \quad \delta^{-2}s\log^2 n\log^2 s, \right\}, \tag{7.50}$$

where $C_2 > 0$ is another universal constant.

Theorem 7.6.5 also says that partial random circulant matrices $\boldsymbol{\Phi}$ obey the RIP (7.3) *in expectation.* The following theorem tell us that the random variable δ_s does not deviate much from its expectation.

Theorem 7.6.6 (Theorem 1.2 of Rauhut, Romberg, and Tropp [403]). *Let the random variable δ_s defined as in Theorem 7.6.5. Then for $0 \leq t \leq 1$*

$$\mathbb{P}\left(\delta_s \geqslant \mathbb{E}\left[\delta_s\right] + t\right) \leqslant e^{-t^2/\sigma^2} \quad \textit{where} \quad \sigma^2 = C_3 \frac{s}{m} \log^2 s \log^2 n,$$

for a universal constant $C_3 > 0$.

Proof. Since the techniques used here for the proof are interesting, we present the detailed derivations, closely following [403]. We require Theorem 1.5.6, which is Theorem 17 in [62].

Let $\mathcal{F}$ denote a collection of $n \times n$ symmetric matrices $\mathbf{Z}$, and $\varepsilon_1, \ldots, \varepsilon_n$ are i.i.d. Rademacher variables. Assume that $\mathbf{Z}$ has zero diagonal, that is, $Z(i,i) = 0, i = 1, \ldots, n.$ for each $\mathbf{Z} \in \mathcal{F}$. We are interested in the concentration variable

$$Y = \sup_{\mathbf{Z} \in \mathcal{F}} \sum_{k=1}^{n} \sum_{\ell=1}^{n} \varepsilon_k \varepsilon_\ell Z\left(k, \ell\right).$$

Define two variance parameters

$$U = \sup_{\mathbf{Z} \in \mathcal{F}} \|\mathbf{Z}\| \quad \text{and} \quad V^2 = \mathbb{E} \sup_{\mathbf{Z} \in \mathcal{F}} \sum_{k=1}^{n} \left| \sum_{\ell=1}^{n} \varepsilon_\ell Z\left(k, \ell\right) \right|^2.$$

Proposition 7.6.7 (Tail Bound for Chaos). *under the preceding assumptions, for all $t \geq 0$,*

$$\mathbb{P}\left(Y \geqslant \mathbb{E}\left[Y\right] + t\right) \leqslant \exp\left(-\frac{t^2}{32V^2 + 65Ut/3}\right) \tag{7.51}$$

Putting together (7.44), (7.46), and (7.47), we have

$$\delta_s = \sup_{\mathbf{x} \in T} \left|G_{\mathbf{x}}\right| = \sup_{\mathbf{x} \in T} \sum_{k=1}^{n} \sum_{\ell=1}^{n} \varepsilon_k \varepsilon_\ell Z_{\mathbf{x}}\left(k, \ell\right)$$

where the matrix $\mathbf{Z}_{\mathbf{x}}$ has the expression

$$\mathbf{Z}_{\mathbf{x}} = \mathbf{A}_{\mathbf{x}} - \operatorname{diag}\left(\mathbf{A}_{\mathbf{x}}\right) \quad \text{for} \quad \mathbf{A}_{\mathbf{x}} = \frac{1}{m} \mathbf{F}^H \hat{\mathbf{X}}^H \hat{\mathbf{P}}_\Omega \hat{\mathbf{X}} \mathbf{F}.$$

As a result, (7.51) applies to the random variable δ_s.

To bound the other parameter V^2, we use the following "vector version" of the Dudley inequality.

Proposition 7.6.8 ([403]). *Consider the vector-valued random process*

$$\mathbf{h_x} = \mathbf{Z_x}\boldsymbol{\varepsilon} \quad \textit{for} \quad \mathbf{x} \in T.$$

The pseudo-metric is defined as

$$d_2(\mathbf{x}, \mathbf{y}) = \|\mathbf{Z_x} - \mathbf{Z_y}\|_F.$$

For a point $\mathbf{x}_0 \in T$. *There exists a universal constant* $C > 0$ *such that*

$$\left(\mathbb{E} \sup_{\mathbf{x} \in T} \|\mathbf{h_x} - \mathbf{h}_{x_0}\|_2^2\right)^{1/2} \leqslant C \int_0^{\infty} \sqrt{N(T, d_2, r)}dr. \tag{7.52}$$

With $\mathbf{x}_0 = \mathbf{0}$, the left-hand side of (7.52) is exactly V. The rest is straightforward. See [403] for details. □

Theorem 7.6.9 (Theorem 1.1 of Krahmer, Mendelson, and Rauhut [31]). *Let* $\mathbf{\Phi} \in \mathbb{R}^{m \times n}$ *be a draw of partial random circulant matrix generated by a Rademacher vector* $\boldsymbol{\varepsilon}$. *If*

$$m \geqslant c\delta^{-2}s\left(\log^2 s\right)\left(\log^2 n\right), \tag{7.53}$$

then with probability at least $1 - n^{-(\log n)(\log^2 s)}$, *the restricted isometry constant of* $\mathbf{\Phi}$ *satisfies* $\delta_s \leq \delta$, *in other words,*

$$\mathbb{P}(\delta_s \geqslant \delta) \leqslant n^{-(\log n)(\log^2 s)}.$$

The constant $c > 0$ *is universal.*

Combining Theorem 7.6.9 with the work [412] on the relation between the restricted isometry property and the Johnson–Lindenstrauss lemma, we obtain the following. See [428, 429] for previous work.

Theorem 7.6.10 (Theorem 1.2 of Krahmer, Mendelson, and Rauhut [31]). *Fix* $\eta, \delta \in (0,1)$, *and consider a finite set* $E \in \mathbb{R}^n$ *of cardinality* $|E| = p$. *Choose*

$$m \geqslant C_1\delta^{-2}\log(C_2 p)\left(\log\log(C_2 p)\right)^2(\log n)^2,$$

where the constants C_1, C_2 *depend only on* η. *Let* $\mathbf{\Phi} \in \mathbb{R}^{m \times n}$ *be a partial circulant matrix generated by a Rademacher vector* $\boldsymbol{\epsilon}$. *Furthermore, let* $\boldsymbol{\epsilon}' \in \mathbb{R}^n$ *be a Rademacher vector independent of* $\boldsymbol{\epsilon}$ *and set* $\mathbf{D}_{\boldsymbol{\epsilon}'}$ *to be the diagonal matrix with diagonal* $\boldsymbol{\epsilon}'$. *Then with probability exceeding* $1 - \eta$, *for every* $\mathbf{x} \in E$,

$$(1-\delta)\left\|\mathbf{x}\right\|_2^2 \leqslant \left\|\mathbf{\Phi}\mathbf{D}_{\boldsymbol{\epsilon}'}\mathbf{x}\right\|_2^2 \leqslant (1+\delta)\left\|\mathbf{x}\right\|_2^2.$$

Now we present a result that is more general than Theorem 7.6.9. The L-sub-Gaussian random variables are defined in Sect. 1.8.

Theorem 7.6.11 (Theorem 4.1 of Krahmer, Mendelson, and Rauhut [31]). *Let $\boldsymbol{\xi} = \{\xi_i\}_{i=1}^n$ be a random vector with independent mean-zero, variance one, L-sub-Gaussian entries. If, for $s \leq n$ and $\eta, \delta \in (0,1)$,*

$$m \geqslant c\delta^{-2} s \max\left\{(\log s)^2 (\log n)^2, \log(1/\eta)\right\} \tag{7.54}$$

then with probability at least $1-\eta$, the restricted isometry constant of the partial random circulant matrix $\mathbf{\Phi} \in \mathbb{R}^{m\times n}$ generated by $\boldsymbol{\xi}$ satisfies $\delta_s \leqslant \delta$. The constant $c > 0$ depends only on L.

Here, we only introduce the proof ingredient. Let $\mathbf{V}_{\mathbf{x}}\mathbf{z} = \frac{1}{\sqrt{m}}\mathbf{P}_\Omega(\mathbf{x}\otimes\mathbf{z})$, where the projection operator $\mathbf{P}_\Omega : \mathbb{C}^n \to \mathbb{C}^m$ is given by the positive-definite (sample covariance) matrix

$$\mathbf{P}_\Omega = \mathbf{R}_\Omega^H \mathbf{R}_\Omega,$$

that is,

$$(\mathbf{P}_\Omega\mathbf{x})_\ell = x_\ell \quad \text{for} \quad \ell \in \Omega \quad \text{and} \quad (\mathbf{P}_\Omega\mathbf{x})_\ell = 0 \quad \text{for} \quad \ell \notin \Omega.$$

Define the unit ball with sparsity constrain

$$\mathcal{T}_s = \left\{\mathbf{x} \in \mathbb{C}^n : \left\|\mathbf{x}\right\|_2^2 \leqslant 1, \left\|\mathbf{x}\right\|_0 \leqslant s\right\}.$$

Let the operator norm be defined as $\left\|\mathbf{A}\right\| = \sup\limits_{\left\|\mathbf{x}\right\|_2=1} \left\|\mathbf{A}\mathbf{x}\right\|_2$. The restricted isometry constant of $\mathbf{\Phi}$ is expressed as

$$\delta_s = \sup_{\mathbf{x}\in\mathcal{T}_s}\left|\left\|\mathbf{R}_\Omega(\boldsymbol{\xi}\otimes\mathbf{x})\right\|^2 - \left\|\mathbf{x}\right\|_2^2\right| = \sup_{\mathbf{x}\in\mathcal{T}_s}\left|\left\|\mathbf{P}_\Omega(\mathbf{x}\otimes\boldsymbol{\xi})\right\|^2 - \left\|\mathbf{x}\right\|_2^2\right| = \sup_{\mathbf{x}\in\mathcal{T}_s}\left|\left\|\mathbf{V}_{\mathbf{x}}\boldsymbol{\xi}\right\|^2 - \left\|\mathbf{x}\right\|_2^2\right|.$$

The δ_s is indexed by the *set* $\mathcal{T}_s$ of vectors. Since $|\Omega| = m$, it follows that

$$\mathbb{E}\left\|\mathbf{V}_{\mathbf{x}}\xi\right\|^2 = \frac{1}{m}\sum_{\ell\in\Omega}\mathbb{E}\sum_{k,l=1}^{n}\xi_j\bar{\xi}_k x_{\ell\ominus j}\bar{x}_{\ell\ominus j} = \frac{1}{m}\sum_{\ell\in\Omega}\sum_{k,l=1}^{n}\left|x_{\ell\ominus j}\right|^2 = \left\|\mathbf{x}\right\|_2^2$$

and hence

$$\delta_s = \sup_{\mathbf{x}\in\mathcal{T}_s}\left|\left\|\mathbf{V}_{\mathbf{x}}\xi\right\|_2^2 - \mathbb{E}\left\|\mathbf{V}_{\mathbf{x}}\xi\right\|_2^2\right|,$$

which is the process studied in Theorem 7.8.3.

The proof of Theorem 7.6.11 requires a Fourier domain description of $\mathbf{\Phi}$. Let the matrix $\mathbf{F}$ by the unnormalized Fourier transform with elements $F_{jk} = e^{i2\pi jk/n}$. By convolution theorem, for every $1 \le j \le n$,

$$\mathbf{F}(\mathbf{x} \otimes \mathbf{y})_j = \mathbf{F}(\mathbf{x})_j \cdot \mathbf{F}(\mathbf{y})_j .$$

So, we have

$$\mathbf{V}_{\mathbf{x}}\boldsymbol{\xi} = \frac{1}{\sqrt{m}}\mathbf{P}_{\Omega}\mathbf{F}^{-1}\hat{\mathbf{X}}\mathbf{F}\boldsymbol{\xi},$$

where $\hat{\mathbf{X}}$ is the diagonal matrix, whose diagonal is the Fourier transform $\mathbf{F}x$. In short,

$$\mathbf{V}_{\mathbf{x}} = \frac{1}{\sqrt{m}}\hat{\mathbf{P}}_{\Omega}\hat{\mathbf{X}}\mathbf{F},$$

where $\hat{\mathbf{P}}_{\Omega} = \mathbf{P}_{\Omega}\mathbf{F}^{-1}$. For details of the proof, we refer to [31]. The proof ingredient is novel. The approach of suprema of chaos processes is *indexed by a set of matrices*, which is based on a chaining method due to Talagrand. See Sect. 7.8.

7.7 Restricted Isometry Property for Time-Frequency Structured Random Matrices

Applications of random Gabor synthesis matrices include operator identification (channel estimation in wireless communications), radar and sonar [405, 431, 431]. The restricted isometry property for time-frequency structured random matrices is treated in [31, 391, 404, 432, 433].

Here we follow [31] to highlight the novel approach. The translation and modulation operators on $\mathbb{C}^m$ are defined by $(\mathbf{T}\mathbf{y})_j = e^{i2\pi j/m}y_j = \omega^j y_j$, where $\omega = e^{i2\pi/m}$ and $\ominus$ again denotes cyclic subtraction, this time modulo m. Observe that

$$\left(\mathbf{T}^k\mathbf{y}\right)_j = y_{j\ominus k} \quad \text{and} \quad \left(\mathbf{M}^\ell\mathbf{y}\right)_j = e^{i2\pi j\ell/m}y_j = \omega^{\ell j}y_j. \tag{7.55}$$

The time-frequency shifts are given by

$$\mathbf{\Pi}(k,\ell) = \mathbf{M}^\ell\mathbf{T}^k,$$

where $(k,\ell) \in \mathbb{Z}_m^2 = \mathbb{Z}_m \times \mathbb{Z}_m = \{\{0,\ldots,m-1\}\{0,\ldots,m-1\}\}$. For $\mathbf{y} \in \mathbb{C}^m \backslash \{0\}$, the system $\left\{\mathbf{\Pi}(k,\ell)\mathbf{y} : (k,\ell) \in \mathbb{Z}_m^2\right\}$, is called a Gabor system [434,

435]. The $m \times m^2$ matrix $\mathbf{\Psi}_{\mathbf{y}}$ whose columns are vectors $\mathbf{\Pi}(k,\ell)\,\mathbf{y}$, $(k,\ell) \in \mathbb{Z}_m^2$ is called a Gabor synthesis matrix,

$$\mathbf{\Psi}_{\mathbf{y}} = \mathbf{M}^{\ell}\mathbf{T}^{k}\mathbf{y} \in \mathbb{C}^{m\times m^2}, \quad (k,\ell) \in \mathbb{Z}_m^2. \tag{7.56}$$

The operators $\mathbf{\Pi}(k,\ell) = \mathbf{M}^{\ell}\mathbf{T}^{k}$ are called time-frequency shifts and the system $\mathbf{\Pi}(k,\ell)$ of all time-frequency shifts forms a basis of the matrix space $\mathbb{C}^{m\times m}$ [436, 437]. $\mathbf{\Psi}_{\mathbf{y}}$ allows for fast matrix vector multiplication algorithms based on the FFT.

Theorem 7.7.1 (Theorem 1.3 of Krahmer, Mendelson, and Rauhut [31]). *Let ε be a Rademacher vector and consider the Gabor synthesis matrix $\mathbf{\Psi}_{\mathbf{y}} \in \mathbb{C}^{m\times m^2}$ defined in* (7.56) *generated by $\mathbf{y} = \frac{1}{\sqrt{m}}\varepsilon$. If*

$$m \geqslant c\delta^{-2}s(\log s)^2(\log m)^2,$$

then with probability at least $1 - m^{-(\log m)\cdot(\log^2 s)}$, the restricted isometry constant of $\mathbf{\Psi}_{\mathbf{y}}$ satisfies $\delta_s \leq \delta$.

Now we consider $\boldsymbol{\epsilon} \in \mathbb{C}^n$ to be a Rademacher or Steinhaus sequence, that is, a vector of independent random variables taking values $+1$ and -1 with equal probability, respectively, taking values uniformly distributed on the complex torus $S^1 = \{z \in \mathbb{C} : |z| = 1\}$. The normalized window is

$$\mathbf{g} = \frac{1}{\sqrt{n}}\boldsymbol{\epsilon}.$$

Theorem 7.7.2 (Pfander and Rauhut [404]). *Let $\mathbf{\Psi}_{\mathbf{g}} \in \mathbb{C}^{n\times n^2}$ be a draw of the random Gabor synthesis with normalized Steinhaus or Rademacher generating vector.*

1. *The expectation of the restricted isometry constant δ_s of $\mathbf{\Psi}_{\mathbf{g}}$, $s \leq n$, satisfies*

$$\mathbb{E}\delta_s \leqslant \max\left\{C_1\sqrt{\frac{s^{3/2}}{n}}\sqrt{\log n}\log s,\quad C_2\frac{s^{3/2}\log^{3/2}n}{n}\right\}, \tag{7.57}$$

   where C_1, C_2 are universal constants.
2. *For $0 \leq t \leq 1$, we have*

$$\mathbb{P}(\delta_s \geqslant \mathbb{E}\delta + t) \leqslant e^{-t^2/\sigma^2}, \qquad \text{where } \sigma^2 = \frac{C_3 s^{3/2}(\log n)\log^2 s}{n}, \tag{7.58}$$

   where $C_3 > 0$ is a universal constant.

With slight variations of the proof one can show similar statements for normalized Gaussian or subGaussian random windows $\mathbf{g}$.

Example 7.7.3 (Wireless communications and radar [404]). A common finite-dimensional model for the channel operator, which combines digital (discrete) to analog conversion, the analog channel, and the analog to digital conversion. It is given by

$$\mathbf{H} = \sum_{(k,\ell)\in\mathbb{Z}_n\times\mathbb{Z}_n} x_{(k,\ell)}\mathbf{\Pi}\left(k,\ell\right).$$

Time-shifts delay is due to the multipath propagation, and the frequency-shifts are due to the Doppler effects caused by moving transmitter, receiver and scatterers. Physical considerations often suggest that $\mathbf{x}$ be rather sparse as, indeed, the number of present scatterers can be assumed to be small in most cases. The same model is used in sonar and radar.

Given a single input-output pair $(\mathbf{g}, \mathbf{Hg})$, our task is to find the sparse coefficient vector $\mathbf{x}$. In other words, we need to find $\mathbf{H} \in \mathbb{C}^{n\times n}$, or equivalently $\mathbf{x}$, from its action $\mathbf{y} = \mathbf{Hz}$ on a single vector $\mathbf{z}$. Writing

$$\mathbf{y} = \mathbf{Hg} = \sum_{(k,\ell)\in\mathbb{Z}_n\times\mathbb{Z}_n} x_{(k,\ell)}\mathbf{\Pi}\left(k,\ell\right)\mathbf{g} = \mathbf{\Psi}_{\mathbf{g}}\mathbf{x}, \tag{7.59}$$

with *unknown but sparse* $\mathbf{x}$, we arrive at a compressed sensing problem. In this setup, we clearly have the freedom to choose the vector $\mathbf{g}$, and we may choose it as a random Rademacher or Steinhaus sequence. Then, the restricted isometry property of $\mathbf{\Psi}_{\mathbf{g}}$, as shown in Theorem 7.7.2, ensures recovery of sufficiently sparse $\mathbf{x}$, and thus of the associated operator $\mathbf{H}$.

Recovery of the sparse vector $\mathbf{x}$ in (7.59) can be also interpreted as finding a sparse time-frequency representation of a given $\mathbf{y}$ with respect to the window $\mathbf{g}$. □

Let us use one example to highlight the approach that is used to prove Theorem 7.7.2.

Example 7.7.4 (Expectation of the restricted isometry constant [404]). We first rewrite the restricted isometry constants δ_s. Let the set of s-sparse vectors with unit ℓ_2-norm be defined as

$$T = T_s = \left\{\mathbf{x}\in\mathbb{C}^{n^2} : \quad \left\|\mathbf{x}\right\|_2 = 1, \left\|\mathbf{x}\right\|_0 \leqslant s\right\}.$$

We express δ_s as the following semi-norm on Hermitian matrices

$$\delta_s = \sup_{\mathbf{x}\in T_s}\left|\mathbf{x}^H\left(\mathbf{\Psi}^H\mathbf{\Psi} - \mathbf{I}\right)\mathbf{x}\right| \tag{7.60}$$

where $\mathbf{I}$ is the identity matrix and $\mathbf{\Psi} = \mathbf{\Psi}_{\mathbf{g}}$. The Gabor synthesis matrix $\mathbf{\Psi}_{\mathbf{g}}$ has the form

$$\Psi_{\mathbf{g}} = \sum_{k=0}^{n-1} g_k \mathbf{A}_k$$

with

$$\mathbf{A}_0 = \left(\mathbf{I}|\mathbf{M}|\mathbf{M}^2|\cdots|\mathbf{M}^{n-1}\right), \quad \mathbf{A}_1 = \left(\mathbf{I}|\mathbf{MT}|\mathbf{M}^2\mathbf{T}|\cdots|\mathbf{M}^{n-1}\mathbf{T}^k\right),$$

and so on. In short, for $k \in \mathbb{Z}_n$,

$$\mathbf{A}_k = \left(\mathbf{T}^k|\mathbf{MT}^k|\mathbf{M}^2\mathbf{T}^k|\cdots|\mathbf{M}^{n-1}\mathbf{T}^k\right).$$

With

$$\sum_{k=0}^{n-1} \mathbf{A}_k^H \mathbf{A}_k = n\mathbf{I},$$

it follows that

$$\Psi^H \Psi - \mathbf{I} = -\mathbf{I} + \frac{1}{n}\sum_{k=0}^{n-1}\sum_{k'=0}^{n-1} \overline{\epsilon_{k'}}\epsilon_k \mathbf{A}_{k'}^H \mathbf{A}_k = \frac{1}{n}\sum_{k=0}^{n-1}\sum_{k'=0}^{n-1} \overline{\epsilon_{k'}}\epsilon_k \mathbf{W}_{k',k},$$

where

$$\mathbf{W}_{k',k} = \begin{cases} \mathbf{A}_{k'}^H \mathbf{A}_k, & k' \neq k, \\ 0, & k' = k. \end{cases}$$

We use the matrix $\mathbf{B}(\mathbf{x}) \in \mathbb{C}^{n \times n}, \mathbf{x} \in T_s$, given by matrix entries

$$B(\mathbf{x})_{k,k'} = \mathbf{x}^H \mathbf{A}_{k'}^H \mathbf{A}_k \mathbf{x}.$$

Then we have

$$n\mathbb{E}\delta_s = \mathbb{E}\sup_{\mathbf{x}\in T_s} |Z_\mathbf{x}| = \mathbb{E}\sup_{\mathbf{x}\in T_s} |Z_\mathbf{x} - Z_\mathbf{0}|, \tag{7.61}$$

where

$$Z_\mathbf{x} = \sum_{k' \neq k} \bar{\varepsilon}_{k'}\varepsilon_k \mathbf{x}^H \mathbf{A}_{k'}^H \mathbf{A}_k \mathbf{x} = \boldsymbol{\epsilon}^H \mathbf{B}(\mathbf{x})\,\boldsymbol{\epsilon}, \tag{7.62}$$

with $\mathbf{x} \in T_s = \{\mathbf{x} \in \mathbb{C}^{n\times n}: \quad \|\mathbf{x}\|_2 = 1, \|\mathbf{x}\|_0 \leqslant s\}$. □

A process of the type (7.62) is called Rademacher or Steinhaus chaos process of order 2. In order to bound such a process, [404] uses the following Theorem, see for example, [27, Theorem 11.22] or [81, Theorem 2.5.2], where it is stated for Gaussian processes and in terms of majorizing measure (generic chaining) conditions. The formulation below requires the operator norm $\|\mathbf{A}\|_{2\to 2} = \max_{\|\mathbf{x}\|_2=1} \|\mathbf{Ax}\|_2$ and the Frobenius norm $\|\mathbf{A}\|_F = \mathrm{Tr}\left(\mathbf{A}^H\mathbf{A}\right)^{1/2} = \left(\sum_{i,j} |A_{i,j}|^2\right)^{1/2}$, where $\mathrm{Tr}\left(\mathbf{A}\right)$ denotes the trace of a matrix $\mathbf{A}$. $\|\mathbf{z}\|_2$ denotes the ℓ_2-norm of the vector $\mathbf{z}$.

Theorem 7.7.5 (Pfander and Rauhut [404]). *Let $\epsilon = (\epsilon_1, \ldots, \epsilon_n)^T$ be a Rademacher or Steinhaus sequence, and let*

$$Z_{\mathbf{x}} = \sum_{k' \neq k} \bar{\varepsilon}_{k'}\varepsilon_k \mathbf{x}^H \mathbf{A}_{k'}^H \mathbf{A}_k \mathbf{x} = \boldsymbol{\epsilon}^H \mathbf{B}\left(\mathbf{x}\right) \boldsymbol{\epsilon}$$

be an associated chaos process of order 2, indexed by $\mathbf{x} \in T$, where assume that $\mathbf{B}(x)$ is Hermitian with zero diagonal, that is, $B(\mathbf{x})_{k,k} = 0$ and $B(\mathbf{x})_{k',k} = \overline{B(\mathbf{x})_{k,k'}}$. We define two (pseudo-)metrics on the set T,

$$d_1\left(\mathbf{x}, \mathbf{y}\right) = \|\mathbf{B}(\mathbf{x}) - \mathbf{B}(\mathbf{y})\|_{2\to 2},$$
$$d_2\left(\mathbf{x}, \mathbf{y}\right) = \|\mathbf{B}(\mathbf{x}) - \mathbf{B}(\mathbf{y})\|_F.$$

Let $N\left(T, d_i, r\right)$ be the minimum number of balls of radius r in the metric d_i needed to cover the set T. Then, these exists a universal constant $C > 0$ such that, for an arbitrary $\mathbf{x}_0 \in T$,

$$\mathbb{E} \sup_{\mathbf{x}\in T} |Z_{\mathbf{x}} - Z_{\mathbf{x}_0}| \leqslant C \max\left\{\int_0^\infty \log N\left(T, d_1, r\right) dr, \quad \int_0^\infty \log N\left(T, d_2, r\right) dr\right\}. \tag{7.63}$$

The proof ingredients include: (1) decoupling [56, Theorem 3.1.1]; (2) the contraction principle [27, Theorem 4.4]. For a Rademacher sequence, the result is stated in [403, Proposition 2.2].

The following result is a slight variant of Theorem 17 in [62], which in turn is an improved version of a striking result due to Talagrand [231].

Theorem 7.7.6 (Pfander and Rauhut [404]). *Let the set of matrices $\mathcal{B} = \{\mathbf{B}\left(\mathbf{x}\right)\} \in \mathbb{C}^{n\times n}, \mathbf{x} \in T$, where T is the set of vectors. Let $\boldsymbol{\varepsilon} = \left(\varepsilon_1, \ldots, \varepsilon_n\right)^T$ be a sequence of i.i.d. Rademacher or Steinhaus random variables. Assume that the matrix $\mathbf{B}(\mathbf{x})$ has zero diagonal, i.e., $B_{i,i}(\mathbf{x}) = 0$ for all $\mathbf{x} \in T$. Let Y be the random variable*

$$Y = \sup_{\mathbf{x}\in T} \left|\boldsymbol{\varepsilon}^H \mathbf{B}\left(\mathbf{x}\right) \boldsymbol{\varepsilon}\right| = \left|\sum_{k=1}^{n-1}\sum_{k'=1}^{n-1} \overline{\varepsilon_{k'}}\varepsilon_k B(\mathbf{x})_{k',k}\right|^2.$$

Define U and V as

$$U = \sup_{\mathbf{x} \in T} \|\mathbf{B}(\mathbf{x})\|_{2 \to 2}$$

and

$$V = \mathbb{E} \sup_{\mathbf{x} \in T} \|\mathbf{B}(\mathbf{x}) \boldsymbol{\varepsilon}\|_2^2 = \mathbb{E} \sup_{\mathbf{x} \in T} \sum_{k'=1}^{n} \left| \sum_{k=1}^{n} \varepsilon_k B(\mathbf{x})_{k',k} \right|^2.$$

Then, for $t \geq 0$,

$$\mathbb{P}(Y \geqslant \mathbb{E}[Y] + t) \leqslant \exp\left(-\frac{t^2}{32V + 65Ut/3}\right).$$

7.8 Suprema of Chaos Processes

The approach of suprema of chaos processes is *indexed by a set of matrices*, which is based on a chaining method.

Both for partial random matrices and for time-frequency structured random matrices generated by Rademacher vectors, the restricted isometry constants δ_s can be expressed as a (scalared) random variable X of the form

$$X = \sup_{\mathbf{A} \in \mathcal{A}} \left| \|\mathbf{A}\boldsymbol{\epsilon}\|_2^2 - \mathbb{E}\|\mathbf{A}\boldsymbol{\epsilon}\|_2^2 \right|, \tag{7.64}$$

where $\mathcal{A}$ is a set of matrices and $\boldsymbol{\epsilon}$ is a Rademacher vector. By expanding the ℓ_2-norms, we rewrite (7.64) as

$$X = \sup_{\mathbf{A} \in \mathcal{A}} \left| \sum_{i \neq j} \epsilon_i \epsilon_j \left(\mathbf{A}^H \mathbf{A}\right)_{i,j} \right|, \tag{7.65}$$

which is a homogeneous chaos process of order 2 indexed by the positive semidefinite matrices $\mathbf{A}^H \mathbf{A}$. Talagrand [81] considers general homogenous chaos process of the form

$$X = \sup_{\mathbf{B} \in \mathcal{B}} \left| \sum_{i \neq j} \epsilon_i \epsilon_j B_{i,j} \right|, \tag{7.66}$$

where $\mathcal{B} \subset \mathbb{C}^{n \times n}$ is a set of (not necessarily positive semidefinite) matrices. He derives the bound

$$\mathbb{E}Y \leqslant C_1 \gamma_2 (\mathcal{B}, \|\cdot\|_F) + C_2 \gamma_2 (\mathcal{B}, \|\cdot\|_{2 \to 2}). \tag{7.67}$$

where the Talagrand's functional γ_α is defined below.

The core of the generic chaining methodology is based on the following definition:

Definition 7.8.1 (Talagrand [81]). For a metric space (T, d), an admissible sequence of T is a collection of subsets of T, $\{T_s : s \geqslant 0\}$, such that for every $s \geq 1$, $|T_s| \leqslant 2^{2^s}$ and $|T_0| = 1$. For $\beta \geq 1$, define the γ_β functional by

$$\gamma_\beta (T, d) = \inf \sup_{t \in T} \sum_{s=0}^{\infty} 2^{s/\beta} d (t, T_s),$$

where the infimum is taken with respect to all admissible sequences of T.

We need some new notation to proceed. For a set of matrices $\mathcal{A}$, the radius of the set $\mathcal{A}$ in the Frobenius norm

$$\|\mathbf{A}\|_F = \sqrt{\operatorname{Tr}(\mathbf{A}^H \mathbf{A})}$$

is denoted by $d_F(\mathcal{A})$. Similarly, the radius of the set $\mathcal{A}$ in the operator norm

$$\|\mathbf{A}\|_{2\to 2} = \sup_{\|\mathbf{x}\|_2 \leqslant 1} \|\mathbf{A}\mathbf{x}\|_2$$

is denoted by $d_{2\to 2}(\mathcal{A})$. That is,

$$d_F(\mathcal{A}) = \sup_{\mathbf{A} \in \mathcal{A}} \|\mathbf{A}\|_F, \qquad d_{2\to 2}(\mathcal{A}) = \sup_{\mathbf{A} \in \mathcal{A}} \|\mathbf{A}\|_{2\to 2}.$$

A metric space is a set T where a notion of distance d (called a metric) between elements of the set is defined. We denote the metric space by (T, d). For a metric space (T, d) and $r > 0$, the covering number $N(T, d, r)$ is the minimum number of open balls of radius r in the metric space (T, d) to cover. Talagrand's functionals γ_α can be bounded in terms of such covering numbers by the well known Dudley integral (see, e.g., [81]). A more specific formulation for the γ_2-functional of a set of matrices $\mathcal{A}$ equipped with the operator norm is

$$\gamma_2 (\mathcal{A}, \|\cdot\|_{2\to 2}) \leqslant c \int_0^{d_{2\to 2}(\mathcal{A})} \sqrt{\log N (\mathcal{A}, \|\cdot\|_{2\to 2}, r)} dr. \tag{7.68}$$

This type of entropy integral was suggested by Dudley [83] to bound the supremum of Gaussian processes.

Under mild measurability assumptions, if $\{G_t : t \in T\}$ is a centered Gaussian process by a set T, then

$$c_1 \gamma_2 (T, d) \leqslant \mathbb{E} \sup_{t \in T} G_t \leqslant c_2 \gamma_2 (T, d), \tag{7.69}$$

where c_1 and c_2 are absolute constants, and for every $s, t \in T$,

$$d^2(s,t) = \mathbb{E}|G_s - G_t|^2.$$

The upper bound is due to Fernique [84] while the lower bound is due to Talagrand's majorizing measures theorem [81, 85].

With the notions above, we are ready to stage the main results.

Theorem 7.8.2 (Theorem 1.4 of Krahmer, Mendelson, and Rauhut [31]). *Let $\mathcal{A} \in \mathbb{R}^{m\times n}$ be a symmetric set of matrices, $\mathcal{A} = -\mathcal{A}$. Let ϵ be a Rademacher vector of length n. Then*

$$\mathbb{E}\sup_{\mathbf{A}\in\mathcal{A}} \left| \|\mathbf{A}\epsilon\|_2^2 - \mathbb{E}\|\mathbf{A}\epsilon\|_2^2 \right| \leqslant C_1\left(d_F(\mathcal{A})\,\gamma_2\left(\mathcal{A}, \|\cdot\|_{2\to2}\right) + \gamma_2\left(\mathcal{A}, \|\cdot\|_{2\to2}\right)^2 \right) =: C_2 E \tag{7.70}$$

Furthermore, for $t \geqslant 0$,

$$\mathbb{P}\left(\sup_{\mathbf{A}\in\mathcal{A}} \left| \|\mathbf{A}\epsilon\|_2^2 - \mathbb{E}\|\mathbf{A}\epsilon\|_2^2 \right| \geqslant C_2 E + t \right) \leqslant 2\exp\left(-C_3 \min\left\{ \frac{t^2}{V^2}, \frac{t}{U} \right\} \right), \tag{7.71}$$

where

$$V = d_{2\to2}(\mathcal{A})\left[\gamma_2\left(\mathcal{A}, \|\cdot\|_{2\to2}\right) + d_F(\mathcal{A})\right] \quad \textit{and} \quad U = d_{2\to2}^2(\mathcal{A}).$$

The constants C_1, C_2, C_3 are universal.

The symmetry assumption $\mathcal{A} = -\mathcal{A}$ was made for the sake of simplicity. The following more general theorem does not use this assumption.

One proof ingredient is the well-known bound relating strong and weak moments for L-sub-Gaussian random vectors, see Sect. 1.8.

Theorem 7.8.3 (Theorem 3.1 of Krahmer, Mendelson, and Rauhut [31]). *Let $\mathcal{A}$ be a set of matrices, and let $\boldsymbol{\xi}$ be a random vector whose entries ξ_i are independent, mean-zero, variance one, and L-sub-Gaussian random variables. Set*

$$E = \gamma_2\left(\mathcal{A}, \|\cdot\|_{2\to2}\right)\left[\gamma_2\left(\mathcal{A}, \|\cdot\|_{2\to2}\right) + d_F(\mathcal{A})\right] + d_F(\mathcal{A})\, d_{2\to2}(\mathcal{A}),$$
$$V = d_{2\to2}(\mathcal{A})\left[\gamma_2\left(\mathcal{A}, \|\cdot\|_{2\to2}\right) + d_F(\mathcal{A})\right], \quad \textit{and} \quad U = d_{2\to2}^2(\mathcal{A}).$$

Then, for $t > 0$,

$$\mathbb{P}\left(\sup_{\mathbf{A}\in\mathcal{A}} \left| \|\mathbf{A}\boldsymbol{\xi}\|_2^2 - \mathbb{E}\|\mathbf{A}\boldsymbol{\xi}\|_2^2 \right| \geqslant c_1 E + t \right) \leqslant 2\exp\left(-c_3 \min\left\{ \frac{t^2}{V^2}, \frac{t}{U} \right\} \right). \tag{7.72}$$

The constants c_1, c_2 depend only on L.

Some notation is needed. We write $x \lesssim y$ if there is an absolute constant c for which $x \lesssim cy$. $x \sim y$ means that $c_1 x \leq y \leq c_2 y$ for absolute constants c_1, c_2. If the constants depend on some parameter u, we write $x \lesssim_u y$. The L_p-norm of a random variable X, or its p-th moment, is given by

$$\|X\|_{L_p} = \left(\mathbb{E}|X|^p\right)^{1/p}.$$

Theorem 7.8.4 (Lemma 3.3 of Krahmer, Mendelson, and Rauhut [31]). *Let $\mathcal{A}$ be a set of matrices, let $\boldsymbol{\xi} = (\xi_1, \ldots, \xi_n)$ be an L-subGaussian random vector, and let $\boldsymbol{\xi}'$ be an independent copy of $\boldsymbol{\xi}$. Then, for every $p \geq 1$,*

$$\left\| \sup_{\mathbf{A} \in \mathcal{A}} \left\langle \mathbf{A}\boldsymbol{\xi}, \mathbf{A}\boldsymbol{\xi}' \right\rangle \right\|_{L_p} \lesssim_L \gamma_2\left(\mathcal{A}, \|\cdot\|_{2\to 2}\right) \left\| \sup_{\mathbf{A} \in \mathcal{A}} \|\mathbf{A}\boldsymbol{\xi}\|_2 \right\|_{L_p} + \sup_{\mathbf{A} \in \mathcal{A}} \left\| \left\langle \mathbf{A}\boldsymbol{\xi}, \mathbf{A}\boldsymbol{\xi}' \right\rangle \right\|_{L_p}.$$

The proof in [31] follows a chaining argument. We also refer to Sect. 1.5 for decoupling from dependance to independence.

Theorem 7.8.5 (Theorem 3.4 of Krahmer, Mendelson, and Rauhut [31]). *Let $L \geq 1$ and $\boldsymbol{\xi} = (\xi_1, \ldots, \xi_n)$, where $\xi_i, i = 1, \ldots, n$ are independent mean-zero, variance one, L-subGaussian random variables, and let $\mathcal{A}$ be a set of matrices. Then, for every $p \geq 1$,*

$$\begin{aligned}
\left\| \sup_{\mathbf{A} \in \mathcal{A}} \|\mathbf{A}\xi\|_2 \right\|_{L_p} &\lesssim_L \gamma_2\left(\mathcal{A}, \|\cdot\|_{2\to 2}\right) + d_F\left(\mathcal{A}\right) + \sqrt{p}\, d_{2\to 2}\left(\mathcal{A}\right), \\
\sup_{\mathbf{A} \in \mathcal{A}} \left| \|\mathbf{A}\xi\|_2^2 - \mathbb{E}\|\mathbf{A}\xi\|_2^2 \right| &\lesssim_L \gamma_2\left(\mathcal{A}, \|\cdot\|_{2\to 2}\right)\left[\gamma_2\left(\mathcal{A}, \|\cdot\|_{2\to 2}\right) + d_F\left(\mathcal{A}\right)\right] \\
&\quad + \sqrt{p}\, d_{2\to 2}\left(\mathcal{A}\right)\left[\gamma_2\left(\mathcal{A}, \|\cdot\|_{2\to 2}\right) + d_F\left(\mathcal{A}\right)\right] + p\, d_{2\to 2}^2\left(\mathcal{A}\right).
\end{aligned}$$

7.9 Concentration for Random Toeplitz Matrix

Unstructured random matrices [415] are studied for concentration of measure. In practical applications, measurement matrices possess a certain structure[438–440]. Toeplitz matrices arise from the convolution process. For a linear time-invariant (LTI) system with system impulse response $\mathbf{h} = \{h_k\}_{k=1}^N$. Let $\mathbf{x} = \{x_k\}_{k=1}^{N+M-1}$ be the applied input discrete-time waveform. Suppose the x_k and h_k are zero-padded from both sides. The output waveform is

$$y_k = \sum_{j=1}^{N} a_j x_{k-j}. \tag{7.73}$$

Keeping only M consecutive observations of the output waveform, $\mathbf{y} = \{y_k\}_{k=N+1}^{N+M}$, we rewrite (7.73) as

$$\mathbf{y} = \mathbf{X}\mathbf{h}, \tag{7.74}$$

where

$$\mathbf{X} = \begin{bmatrix} x_N & x_{N-1} & \cdots & x_1 \\ x_{N+1} & x_N & \cdots & x_2 \\ \vdots & \vdots & \ddots & \vdots \\ x_{N+M-1} & x_{N+M-2} & \cdots & X_M \end{bmatrix} \tag{7.75}$$

is an $M \times N$ Toeplitz matrix. Here we consider the entries $\mathbf{x} = \{x_i\}_{i=1}^{N+M-1}$ drawn from an i.i.d. Gaussian random sequences. To state the result, we need the eigenvalues of the covariance matrix of the vector $\mathbf{h}$ defined as

$$\mathbf{R} = \begin{bmatrix} A(0) & A(1) & \cdots & A(M-1) \\ A(1) & A(0) & \cdots & A(M-2) \\ \vdots & \vdots & \ddots & \vdots \\ A(M-1) & A(M-2) & \cdots & A(0) \end{bmatrix} \tag{7.76}$$

where

$$A(\tau) = \sum_{i=1}^{N-\tau} h_i h_{i+\tau}, \quad \tau = 0, 1, \ldots, M-1$$

is the un-normalized sample autocorrelation function of $\mathbf{h} \in \mathbb{R}^N$. Let $\|\mathbf{a}\|_2$ be the Euclidean norm of the vector $\mathbf{a}$.

Theorem 7.9.1 (Sanandaji, Vincent, and Wakin [439]). *Let* h $\in \mathbb{R}^N$ *be fixed. Define two quantities*

$$\rho(\mathbf{h}) = \frac{\max_i \lambda_i(\mathbf{R})}{\|\mathbf{h}\|_2^2} \quad \text{and} \quad \mu(\mathbf{h}) = \frac{\sum_{i=1}^{M} \lambda_i^2(\mathbf{R})}{M\|\mathbf{h}\|_2^4},$$

where $\lambda_i(\mathbf{R})$ *is the* i*-th eigenvalue of* $\mathbf{R}$. *Let* $\mathbf{y} = \mathbf{X}\mathbf{h}$, *where* $\mathbf{X}$ *is a random Toeplitz matrix (defined in (7.75)) with i.i.d. Gaussian entries having zero-mean and unit variance. Noting that* $\mathbb{E}\left[\|\mathbf{y}\|_2^2\right] = M\|\mathbf{h}\|_2^2$, *then for any* $t \in (0,1)$, *the upper tail probability bound is*

$$\mathbb{P}\left\{\|\mathbf{y}\|_2^2 - M\|\mathbf{h}\|_2^2 \geqslant tM\|\mathbf{h}\|_2^2\right\} \leqslant \exp\left(-\frac{M}{8\rho(\mathbf{h})}t^2\right) \tag{7.77}$$

and the lower tail probability bound is

$$\mathbb{P}\left\{\|\mathbf{y}\|_2^2 - M\|\mathbf{h}\|_2^2 \leqslant -tM\|\mathbf{h}\|_2^2\right\} \leqslant \exp\left(-\frac{M}{8\mu(\mathbf{h})}t^2\right). \tag{7.78}$$

Allowing $\mathbf{X}$ to have $M \times N$ i.i.d. Gaussian entries with zero mean and unit variance (thus no Toeplitz structure) will give the concentration bound [415]

$$\mathbb{P}\left\{\|\mathbf{y}\|_2^2 - M\|\mathbf{h}\|_2^2 \geqslant tM\|\mathbf{h}\|_2^2\right\} \leqslant 2\exp\left(-\frac{M}{4}t^2\right).$$

Thus, to achieve the same probability bound for Toeplitz matrices requires choosing M larger by a factor of $2\rho(\mathbf{h})$ or $2\mu(\mathbf{h})$.

See [441], for the spectral norm of a random Toeplitz matrix. See also [442].

7.10 Deterministic Sensing Matrices

The central goal of compressed sensing is to capture attributes of a signal using very few measurements. In most work to date, this broader objective is exemplified by the important special case in which a k-sparse vector $\mathbf{x} \in \mathbb{R}^n$ (with n large) is to be reconstructed from a small number N of linear measurements with $k < N < n$. In this problem, measurement data constitute a vector $\frac{1}{\sqrt{N}}\Phi\mathbf{x}$, where Φ is an $N \times n$ matrix called the sensing matrix.

The two fundamental questions in compressed sensing are: how to construct suitable sensing matrices Φ, and how to recover $\mathbf{x}$ from $\frac{1}{\sqrt{N}}\Phi\mathbf{x}$ efficiently. In [443] the authors constructed a large class of deterministic sensing matrices that satisfy a statistical restricted isometry property. Because we will be interested in expected-case performance only, we need not impose RIP; we shall instead work with the weaker Statistical Restricted Isometry Property.

Definition 7.10.1 (Statistical restricted isometry property). An $N \times n$ (sensing) matrix Φ is said to be a (k, δ, ε)-statistical restricted isometry property matrix if, for k-sparse vectors $\mathbf{x} \in \mathbb{R}^n$, the inequalities

$$(1-\delta)\|\mathbf{x}\|_2^2 \leqslant \left\|\frac{1}{\sqrt{N}}\Phi\mathbf{x}\right\|^2 \leqslant (1+\delta)\|\mathbf{x}\|_2^2,$$

hold with probability exceeding $1-\varepsilon$ (with respect to a uniform distribution of the vectors $\mathbf{x}$ among all k-sparse vectors in $\mathbb{R}^n$ of the same norm).

Norms without subscript denote ℓ_2-norms. Discrete chirp sensing matrices are studied in [443]. The proof of unique reconstruction [443] uses a version of the classical McDiarmid concentration inequality.

Chapter 8
Matrix Completion and Low-Rank Matrix Recovery

This chapter is a natural development following Chap. 7. In other words, Chaps. 7 and 8 may be viewed as two parallel developments. In Chap. 7, compressed sensing exploits the sparsity structure in a vector, while low-rank matrix recovery—Chap. 8—exploits the low-rank structure of a matrix: sparse in the vector composed of singular values. The theory ultimately traces back to concentration of measure due to high dimensions.

8.1 Low Rank Matrix Recovery

Sparsity recovery and compressed sensing are interchangeable terms. This sparsity concept can be extended to the matrix case: sparsity recovery of the vector of singular values. We follow [444] for this exposition.

The observed data $\mathbf{y}$ is modeled as

$$\mathbf{y} = \mathcal{A}(\mathbf{M}) + \mathbf{z}, \tag{8.1}$$

where $\mathbf{M}$ is an unknown $n_1 \times n_2$ matrix, $\mathcal{A} : \mathbb{R}^{n_1 \times n_2} \mapsto \mathbb{R}^m$ is a *linear* mapping, and $\mathbf{z}$ is an m-dimensional noise term. For example, $\mathbf{z}$ is a Gaussian vector with i.i.d. $\mathcal{N}\left(0, \sigma^2\right)$ entries, written as $\mathbf{z} \sim \mathcal{N}\left(0, \sigma^2 \mathbf{I}\right)$ where the covariance matrix is essentially the identify matrix. The goal is to recover a good approximation of $\mathbf{M}$ while requiring as few measurements as possible.

For some sequences of matrices $\mathbf{A}_i$ and with the standard inner product $\langle \mathbf{A}, \mathbf{X} \rangle = \operatorname{Tr}\left(\mathbf{A}^* \mathbf{X}\right)$ where $\mathbf{A}^*$ is the adjoint of $\mathbf{A}$. Each $\mathbf{A}_i$ is similar to a compressed sensing matrix. We has the intuition of forming a large matrix

R. Qiu and M. Wicks, *Cognitive Networked Sensing and Big Data*,
DOI 10.1007/978-1-4614-4544-9_8,

$$\mathcal{A}(\mathbf{X}) = \begin{bmatrix} \mathrm{vec}(\mathbf{A}_1) \\ \mathrm{vec}(\mathbf{A}_2) \\ \vdots \\ \mathrm{vec}(\mathbf{A}_m) \end{bmatrix} \mathrm{vec}(\mathbf{X}) \tag{8.2}$$

where vec $(\mathbf{X})$ is a long vector obtained by stacking the columns of matrix $\mathbf{X}$.

The matrix version of the restricted isometry property (RIP) is an integral tool in proving theoretical results. For each integer $r = 1, 2, \ldots, n$, the isometry constant δ_r of $\mathcal{A}$ is the *smallest* value such that

$$(1-\delta_r)\|\mathbf{X}\|_F^2 \leqslant \|\mathcal{A}(\mathbf{X})\|_{\ell_2}^2 \leqslant (1+\delta_r)\|\mathbf{X}\|_F^2 \tag{8.3}$$

holds for all matrices $\mathbf{X}$ of rank at most r.

We present only two algorithms. First, it is direct to have the optimization problem

$$\begin{aligned} \text{minimize} \quad & \|\mathbf{X}\|_* \\ \text{subject to} \quad & \|\mathcal{A}^*(\mathbf{v})\| \leqslant \gamma \\ & \mathbf{v} = \mathbf{y} - \mathcal{A}(\mathbf{X}) \end{aligned} \tag{8.4}$$

where $\|\cdot\|$ is the operator norm and $\|\cdot\|_*$ is its dual, i.e., the nuclear norm. The nuclear norm of a matrix $\mathbf{X}$ is the sum of the singular values of $\mathbf{X}$ and the operator norm is its largest singular value. $\mathbf{A}^*$ is the adjoint of $\mathbf{A}$. $\|\mathbf{X}\|_F$ is the Frobenius norm (the ℓ_2-norm of the vector of singular values).

Suppose $\mathbf{z}$ is a Gaussian vector with i.i.d. $\mathcal{N}(0,\sigma^2)$ entries, and let $n = \max\{n_1, n_2\}$. Then if $C_0 > 4\sqrt{(1+\delta_1)\log 12}$

$$\|\mathcal{A}^*(\mathbf{z})\| \leqslant C_0\sqrt{n}\sigma, s \tag{8.5}$$

with probability at least $1 - 2e^{-cn}$ for a fixed numerical constant $c > 0$. The scalar δ_1 is the restricted isometry constant at rank $r = 1$.

We can reformulate (8.4) as a semi-definite program (SDP)

$$\begin{aligned} \text{minimize} \quad & \mathrm{Tr}(\mathbf{W}_1)/2 + \mathrm{Tr}(\mathbf{W}_2)/2 \\ \text{subject to} \quad & \begin{bmatrix} \mathbf{W}_1 & \mathbf{X} \\ \mathbf{X}^* & \mathbf{W}_2 \end{bmatrix} \geqslant 0 \end{aligned} \tag{8.6}$$

with optimization variables $\mathbf{X}, \mathbf{W}_1, \mathbf{W}_2 \in \mathbb{R}^{n\times n}$. We say a matrix $\mathbf{Q} \geq \mathbf{0}$ if $\mathbf{Q}$ is positive semi-definite (all its eigenvalues are nonnegative).

Second, the constraint $\|\mathcal{A}^*(\mathbf{v})\| \leqslant \gamma$ is an SDP constraint since it can be expressed as the linear matrix inequality (LMI)

$$\begin{bmatrix} \gamma \mathbf{I}_n & \mathcal{A}^* (\mathbf{v}) \\ [\mathcal{A}^* (\mathbf{v})]^* & \gamma \mathbf{I}_n \end{bmatrix} \geqslant 0.$$

As a result, (8.4) can be reformulated as the SDP

$$\begin{aligned} \text{minimize} \quad & \operatorname{Tr}(\mathbf{W}_1)/2 + \operatorname{Tr}(\mathbf{W}_2)/2 \\ \text{subject to} \quad & \begin{bmatrix} \mathbf{W}_1 & \mathbf{X} & 0 & 0 \\ \mathbf{X}^* & \mathbf{W}_2 & 0 & 0 \\ 0 & 0 & \gamma \mathbf{I}_n & \mathcal{A}^*(\mathbf{v}) \\ 0 & 0 & [\mathcal{A}^*(\mathbf{v})]^* & \gamma \mathbf{I}_n \end{bmatrix} \geqslant 0 \\ & \mathbf{v} = \mathbf{y} - \mathcal{A}(\mathbf{X}), \end{aligned} \tag{8.7}$$

with optimization variables $\mathbf{X}, \mathbf{W}_1, \mathbf{W}_2 \in \mathbb{R}^{n \times n}$.

8.2 Matrix Restricted Isometry Property

Non-Asymptotic Theory of Random Matrices Lecture 6: Norm of a Random Matrix [445]

The matrix $\mathbf{X}^*$ is the adjoint of $\mathbf{X}$, and for the *linear* operator $\mathcal{A} : \mathbb{R}^{n_1 \times n_2} \mapsto \mathbb{R}^m$, $\mathcal{A}^* : \mathbb{R}^m \mapsto \mathbb{R}^{n_1 \times n_2}$ is the adjoint operator. Specifically, if $[\mathcal{A}(\mathbf{X})]_i = \langle \mathbf{A}_i, \mathbf{X} \rangle$ for all matrices $\mathbf{X} \in \mathbb{R}^{n_1 \times n_2}$, then

$$\mathcal{A}^*(\mathbf{X}) = \sum_{i=1}^{m} v_i \mathbf{A}_i$$

for all vectors $\mathbf{v} = (v_1, \ldots, v_m)^T \in \mathbb{R}^m$.

The matrix version of the restricted isometry property (RIP) is an integral tool in proving theoretical results. For each integer $r = 1, 2, \ldots, n$, the isometry constant δ_r of $\mathcal{A}$ is the *smallest* value such that

$$(1 - \delta_r) \|\mathbf{X}\|_F^2 \leqslant \|\mathcal{A}(\mathbf{X})\|_{\ell_2}^2 \leqslant (1 + \delta_r) \|\mathbf{X}\|_F^2 \tag{8.8}$$

holds for all matrices $\mathbf{X}$ of rank at most r. We say that $\mathcal{A}$ satisfies the RIP at rank r if δ_r (or δ_{4r}) is bounded by a sufficiently small constant between 0 and 1.

Which linear maps $\mathcal{A}$ satisfy the RIP? One example is the Gaussian measurement ensemble. $\mathcal{A}$ is a Gaussian measurement ensemble if each 'row' $\mathbf{a}_i, 1 \leqslant i \leqslant m$, contains i.i.d. $\mathcal{N}(0, 1/m)$ entries (and the $\mathbf{a}_i$'s are independent from each other). We have selected the variance of the entries to be $1/m$ so that for a fixed matrix $\mathbf{X}$, $\mathbb{E} \|\mathcal{A}(\mathbf{X})\|_{\ell_2}^2 = \|\mathbf{X}\|_F^2$.

Theorem 8.2.1 (Recht et al. [446]). *Fix* $0 \leq \delta < 1$ *and let* $\mathcal{A}$ *is a random measurement ensemble obeying the following conditions: for any given* $\mathbf{X} \in \mathbb{R}^{n_1 \times n_2}$ *and fixed* $0 < t < 1$,

$$\mathbb{P}\left(\left| \|\mathcal{A}(\mathbf{X})\|_{\ell_2}^2 - \|\mathbf{X}\|_F^2 \right| > t\|\mathbf{X}\|_F^2 \right) \leqslant C\exp(-cm) \tag{8.9}$$

for fixed constants $C, t > 0$ *(which may depend on t). Then, if* $m \geq Dnr$, $\mathcal{A}$ *satisfies the RIP with isometry constant* $\delta_r \leq \delta$ *with probability exceeding* $1 - Ce^{-dm}$ *for fixed constants* $D, d > 0$.

If $\mathcal{A}$ is a Gaussian random measurement ensemble, $\|\mathcal{A}(\mathbf{X})\|_{\ell_2}^2$ is distributed as $m^{-1}\|\mathbf{X}\|_F^2$ times a chi-squared random variable with m degrees of freedom, and we have

$$\mathbb{P}\left(\left| \|\mathcal{A}(\mathbf{X})\|_{\ell_2}^2 - \|\mathbf{X}\|_F^2 \right| > t\|\mathbf{X}\|_F^2 \right) \leqslant 2\exp\left(\frac{m}{2}\left(t^2/2 - t^3/3\right) \right). \tag{8.10}$$

Similarly, $\mathcal{A}$ satisfies (8.10) in the case when each entry of each 'row' $\mathbf{a}_i$ has i.i.d. entries that are equally likely to take the values $+1/\sqrt{m}$ or $-1/\sqrt{m}$ [446], or if $\mathcal{A}$ is a random projection [416, 446]. Finally, $\mathcal{A}$ satisfies (8.9) if the "rows" $\mathbf{a}_i$ contain sub-Gaussian entries [447].

In Theorem 8.2.1, the degrees of freedom of $n_1 \times n_2$ matrix of rank r is $r(n_1 + n_2 - r)$.

8.3 Recovery Error Bounds

Given the observed vector $\mathbf{y}$, the optimal solution to (8.4) is our estimator $\hat{\mathbf{M}}(\mathbf{y})$.

For the data vector and the linear model

$$\mathbf{y} = \mathbf{A}\mathbf{x} + \mathbf{z} \tag{8.11}$$

where $\mathbf{A} \in \mathbb{R}^{m \times n}$ and the z_i's are i.i.d. $\mathcal{N}(0, \sigma^2)$. Let $\lambda_i(\mathbf{A}^T\mathbf{A})$ be the eigenvalues of the matrix $\mathbf{A}^T\mathbf{A}$. Then [448, p. 403]

$$\inf_{\hat{\mathbf{x}}} \sup_{\mathbf{x} \in \mathbb{R}^n} \mathbb{E}\|\hat{\mathbf{x}} - \mathbf{x}\|_{\ell_2}^2 = \sigma^2 \operatorname{Tr}\left(\mathbf{A}^T\mathbf{A}\right)^{-1} = \sum_{i=1}^{n} \frac{\sigma^2}{\lambda_i(\mathbf{A}^T\mathbf{A})} \tag{8.12}$$

where $\hat{\mathbf{x}}$ is estimate of $\mathbf{x}$.

Suppose that the measurement operator is fixed and satisfies the RIP, and that the noise vector $\mathbf{z} = (z_1, \ldots, z_n)^T \sim \mathcal{N}(0, \sigma^2\mathbf{I}_n)$. Then any estimator $\hat{\mathbf{M}}(\mathbf{y})$ obeys [444, 4.2.8]

$$\sup_{\mathbf{M}:\operatorname{rank}(\mathbf{M}) \leqslant r} \mathbb{E}\left\|\hat{\mathbf{M}}(\mathbf{y}) - \mathbf{M}\right\|_F \geqslant \frac{1}{1+\delta_r} nr\sigma^2. \tag{8.13}$$

Further, we have [444, 4.2.9]

$$\sup_{\mathbf{M}:\text{rank}(\mathbf{M})\leqslant r} \mathbb{P}\left(\left\|\hat{\mathbf{M}}(\mathbf{y})-\mathbf{M}\right\|_F^2 \geqslant \frac{1}{2(1+\delta_r)} nr\sigma^2\right) \geqslant 1-e^{-nr/16}. \quad (8.14)$$

8.4 Low Rank Matrix Recovery for Hypothesis Detection

Example 8.4.1 (Different convergence rates of sums of random matrices). Consider n-dimensional random vectors $\mathbf{y}, \mathbf{x}, \mathbf{n} \in \mathbb{R}^n$

$$\mathbf{y} = \mathbf{x} + \mathbf{n}$$

where vector $\mathbf{x}$ is independent of $\mathbf{n}$. The components $x_1, \ldots, x_n$ of the random vector $\mathbf{x}$ are scalar valued random variables, and, in general, may be dependent random variables. For the random vector $\mathbf{n}$, this is similar. The true covariance matrix has the relation

$$\mathbf{R}_y = \mathbf{R}_x + \mathbf{R}_n,$$

due to the independence between $\mathbf{x}$ and $\mathbf{n}$.

Assume now there are N copies of random vector $\mathbf{y}$:

$$\mathbf{y}_i = \mathbf{x}_i + \mathbf{n}_i, \quad i = 1, 2, \ldots, N.$$

Assume that $\mathbf{x}_i$ are ***dependent*** random vectors, while $\mathbf{n}_i$ are ***independent*** random vectors.

Let us consider the sample covariance matrix

$$\frac{1}{N}\sum_{i=1}^{N} \mathbf{y}_i \otimes \mathbf{y}_i^* = \frac{1}{N}\sum_{i=1}^{N} \mathbf{x}_i \otimes \mathbf{x}_i^* + \frac{1}{N}\sum_{i=1}^{N} \mathbf{n}_i \otimes \mathbf{n}_i^* + \text{junk}$$

where "junk" denotes another two terms. When the sample size N increases, concentration of measure phenomenon occurs. It is remarkable that, on the right hand side, the convergence rate of the first sample covariance matrix $\frac{1}{N}\sum_{i=1}^{N} \mathbf{x}_i \otimes \mathbf{x}_i^*$ and that of the second sample covariance matrix $\frac{1}{N}\sum_{i=1}^{N} \mathbf{n}_i \otimes \mathbf{n}_i^*$ are different! If we further assume $\mathbf{R}_x$ is of low rank, $\hat{\mathbf{R}}_x = \frac{1}{N}\sum_{i=1}^{N} \mathbf{x}_i \otimes \mathbf{x}_i^*$ converges to its true value $\mathbf{R}_x$; $\hat{\mathbf{R}}_n = \frac{1}{N}\sum_{i=1}^{N} \mathbf{n}_i \otimes \mathbf{n}_i^*$ converges to its true value $\mathbf{R}_n$. Their convergence rates, however, are different. □

Example 8.4.1 illustrates the fundamental role of low rank matrix recovery in the framework of signal plus noise. We can take advantage of the faster convergence of the low rank structure of signal, since, for a given recovery accuracy, the required samples N for low rank matrix recovery is $O(r\log(n))$, where r is the rank of the signal covariance matrix. Results such as Rudelson's theorem in Sect. 5.4 are the basis for understanding such a convergence rate.

8.5 High-Dimensional Statistics

High-dimensional statistics is concerned with models in which the ambient dimension d of the problem may be of the same order as—or substantially larger than—the sample size n. It is so called in the "large d, small n" regime.

The rapid development of data collection technology is a major driving force: it allows for more observations (larger n) and also for more variables to be measured (larger d). Examples are ubiquitous throughout science and engineering, including gene array data, medical imaging, remote sensing, and astronomical data. Terabytes of data are produced.

In the absence of additional structure, it is often impossible to obtain consistent estimators unless the ratio d/n converges to zero. On the other hand, the advent of the big data age requires solving inference problems with

$$d \gg n,$$

so that consistency is not possible without imposing additional structure. Typical values of n and d include: $n =$ 100–2,500 and $d =$ 100–20,000.

There are several lines of work within high-dimensional statistics, all of which are based on low-dimensional constraint on the model space, and then studying the behavior of different estimators. Examples [155, 449, 450] include

- Linear regression with sparse constraints
- Multivariate or multi-task forms of regression
- System identification for autoregressive processes
- Estimation of structured covariance or inverse covariance matrices
- Graphic model selection
- Sparse principal component analysis
- Low rank matrix recovery from random projections
- Matrix decomposition problems
- Estimation of sparse additive non-parametric models
- Collaborative filtering

On the computation side, many well-known estimators are based on a convex optimization problem formed by the sum of a loss function with a weighted regularizer. Examples of convex programs include

- ℓ_1-regularized quadratic programs (also known as the Lasso) for sparse linear regression
- Second-order cone program (SOCP) for the group Lasso
- Semidefinite programming relaxation (SDP) for various problems include sparse PCA and low-rank matrix estimation.

8.6 Matrix Compressed Sensing

Section 3.8 is the foundation for this section.

For a vector $\mathbf{a}$, the ℓ_p-norm is denoted by $||\mathbf{a}||_p$. $||\mathbf{a}||_2$ represents the Euclidean norm. For pairs of matrices $\mathbf{A}, \mathbf{B} \in \mathbb{R}^{m_1 \times m_2}$, we define the trace inner product of two matrices $\mathbf{A}, \mathbf{B}$ as

$$\langle \mathbf{A}, \mathbf{B} \rangle = \mathrm{Tr}\left(\mathbf{A}^T \mathbf{B}\right).$$

The Frobenius or trace norm is defined as $\|\mathbf{A}\|_F = \sqrt{\sum_{i=1}^{m_1} \sum_{j=1}^{m_2} |a_{ij}|^2}$. The element-wise ℓ_1-norm $\|\mathbf{A}\|_1$ is defined as $\|\mathbf{A}\|_1 = \sum_{i=1}^{m_1} \sum_{j=1}^{m_2} |a_{ij}|$.

8.6.1 *Observation Model*

A *linear* observation model [155] is defined as

$$Y_i = \langle \mathbf{X}_i, \mathbf{A} \rangle + Z_i, \qquad i = 1, 2, \ldots, N, \tag{8.15}$$

which is specified by a sequence of random matrices $\mathbf{X}_i$, and observation noise Z_i. Of course, Y_i, Z_i and the matrix inner products $\varphi_i = \langle \mathbf{X}_i, \mathbf{A} \rangle$ are scalar-valued random variables, but not necessarily Gaussian. After defining the vectors

$$\mathbf{y} = [Y_1, \ldots, Y_N]^T, \mathbf{z} = [Z_1, \ldots, Z_N]^T, \boldsymbol{\varphi} = [\varphi_1, \ldots, \varphi_N]^T,$$

we rewrite (8.15) as

$$\mathbf{y} = \boldsymbol{\varphi} + \mathbf{z}. \tag{8.16}$$

In order to highlight the $\boldsymbol{\varphi}$ as a linear functional of the matrix $\mathbf{A}$, we use

$$\mathbf{y} = \boldsymbol{\varphi}(\mathbf{A}) + \mathbf{z}. \tag{8.17}$$

The vector $\boldsymbol{\varphi}(\mathbf{A})$ is viewed as a (high but finite-dimensional) random Gaussian operator mapping $\mathbb{R}^{m_1 \times m_2}$ to $\mathbb{R}^N$. In a typical matrix compressed sensing [451], the observation matrix $\mathbf{X}_i \in \mathbb{R}^{m_1 \times m_2}$ has i.i.d. zero-mean, unit-variance Gaussian $\mathcal{N}(0,1)$ entries. In a more general observation model [155], the entries of $\mathbf{X}_i$ are allowed to have general Gaussian dependencies.

8.6.2 Nuclear Norm Regularization

For a rectangular matrix $\mathbf{B} \in \mathbb{R}^{m_1 \times m_2}$, the nuclear or trace norm is defined as

$$\|\mathbf{B}\|_* = \sum_{i=1}^{\min\{m_1,m_2\}} \sigma_i(\mathbf{B}),$$

which is the sum of its singular values that are sorted in non-increasing order. The maximum singular value is $\sigma_{\max} = \sigma_1$ and the minimum singular value $\sigma_{\min} = \sigma_{\min\{m_1,m_2\}}$. The operator norm is defined as $\|\mathbf{A}\|_{op} = \sigma_1(\mathbf{A})$. Given a collection of observation $(Y_i, \mathbf{X}_i) \in \mathbb{R} \times \mathbb{R}^{m_1 \times m_2}$, the problem at hand is to estimate the unknown matrix $\mathbf{A}^\star \in \mathcal{S}$, where $\mathcal{S}$ is a general convex subset of $\mathbb{R}^{m_1 \times m_2}$. The problem may be formulated as an optimization problem

$$\hat{\mathbf{A}} \in \arg\min_{\mathbf{A} \in \mathcal{S}} \left\{ \frac{1}{2N} \|\mathbf{y} - \varphi(\mathbf{A})\|_2^2 + \gamma_N \|\mathbf{A}\|_* \right\}. \tag{8.18}$$

Equation (8.18) is a semidefinite program (SDP) convex optimization problem [48], which can be solved efficiently using standard software packages.

A natural question arises: How accurate will the solution $\hat{\mathbf{A}}$ in (8.18) be when compared with the true unknown $\mathbf{A}^\star$?

8.6.3 Restricted Strong Convexity

The key condition for us to control the matrix error $\mathbf{A}^\star - \hat{\mathbf{A}}$ between $\hat{\mathbf{A}}$, the SDP solution (8.18), and the unknown matrix $\mathbf{A}^\star$ is the so-called *restricted strong convexity*. This condition guarantees that the quadratic loss function in the SDP (8.18) is strictly convex over a restricted set of directions. Let the set $\mathcal{C} \subseteq \mathbb{R} \times \mathbb{R}^{m_1 \times m_2}$ denote the restricted directions. We say the random operator φ satisfies restricted strong convexity over the set $\mathcal{C}$ if there exists some $\kappa(\varphi) > 0$ such that

$$\frac{1}{2N} \|\varphi(\boldsymbol{\Delta})\|_2^2 \geqslant \kappa(\varphi) \|\boldsymbol{\Delta}\|_F^2 \quad \text{for all} \quad \boldsymbol{\Delta} \in \mathcal{C}. \tag{8.19}$$

Recall that N is the number of observations defined in (8.15).

Let r an integer $r \leqslant m = \min\{m_1, m_2\}$ and $\delta \geq 0$ be a tolerance parameter. The set $C(r;\delta)$ defines a set whose conditions are too technical to state here.

Another ingredient is the choice of the regularization parameter γ_N used in solving the SDP (8.18).

8.6.4 Error Bounds for Low-Rank Matrix Recovery

Now we are ready to state our main results.

Theorem 8.6.1 (Exact low-rank matrix recovery [155]). *Suppose* $\mathbf{A}^\star \in S$ *has rank* r*, and the random operator* φ *satisfies the restricted strong convexity with respect to the set* $C(r;0)$*. Then, as long as* $\gamma_N \geqslant 2\left\|\sum_{i=1}^{N} \varepsilon_i \mathbf{X}_i\right\|_{op}/N$*, where* $\{\varepsilon_i\}_{i=1}^N$ *are random variables, any optimal solution* $\hat{\mathbf{A}}$ *to the SDP* (8.18) *satisfies the bound*

$$\left\|\hat{\mathbf{A}} - \mathbf{A}^\star\right\|_F \leqslant \frac{32\gamma_N\sqrt{r}}{\kappa(\varphi)}. \tag{8.20}$$

Theorem 8.6.1 is a deterministic statement on the SDP error.

Sometimes, the unknown matrix $\mathbf{A}^\star$ is nearly low rank: Its singular value sequence $\{\sigma_i(\mathbf{A}^\star)\}_{i=1}^N$ decays quickly enough. For a parameter $q \in (0,1)$ and a positive radius R_q, we define the ball set

$$\mathbb{B}(R_q) = \left\{\mathbf{A} \in \mathbb{R}^{m_1 \times m_2} : \sum_{i=1}^{\min\{m_1,m_2\}} |\sigma_i(\mathbf{A})|^q \leqslant R_q\right\}. \tag{8.21}$$

When $q = 0$, the set $\mathbb{B}(R_0)$ corresponds to the set of matrices with rank at most R_0.

Theorem 8.6.2 (Nearly low-rank matrix recovery [155]). *Suppose that* $\mathbf{A} \in \mathbb{B}(R_q) \cap S$*, the regularization parameter is lower bounded as* $\gamma_N \geqslant 2\left\|\sum_{i=1}^{N} \varepsilon_i \mathbf{X}_i\right\|_{op}/N$*, where* $\{\varepsilon_i\}_{i=1}^N$ *are random variables, and the random operator* φ *satisfies the restricted strong convexity with parameter* $\kappa(\varphi) \in (0,1)$ *over the set* $C(R_q/\gamma_N{}^q;\delta)$*. Then, any solution* $\hat{\mathbf{A}}$ *to the SDP* (8.18) *satisfies the bound*

$$\left\|\hat{\mathbf{A}} - \mathbf{A}^\star\right\|_F \leqslant \max\left\{\delta, 32\sqrt{R_q}\left(\frac{\gamma_N}{\kappa(\varphi)}\right)^{1-q/2}\right\}. \tag{8.22}$$

The error (8.22) reduces to the exact rank case (8.20) when $q = 0$ and $\delta = 0$.

Example 8.6.3 (Matrix compressed sensing with dependent sampling [155]). A standard matrix compressed sensing has the form

$$Y_i = \langle \mathbf{X}_i, \mathbf{A} \rangle + Z_i, \qquad i = 1, 2, \ldots, N, \tag{8.23}$$

where the observation matrix $\mathbf{X}_i \in \mathbb{R}^{m_1 \times m_2}$ has i.i.d. standard Gaussian $\mathcal{N}(0,1)$ entries. Equation (8.23) is an instance of (8.15). Here, we study a more general observation model, in which the entries of $\mathbf{X}_i$ are allowed to have general Gaussian *dependence*.

Equation (8.17) involves a random Gaussian operator mapping $\mathbb{R}^{m_1 \times m_2}$ to $\mathbb{R}^N$.

We repeat some definitions in Sect. 3.8 for convenience. For a matrix $\mathbf{A} \in \mathbb{R}^{m_1 \times m_2}$, we use vector $\text{vec}(\mathbf{A}) \in \mathbb{R}^M, M = m_1 m_2$. Given a symmetric positive definite matrix $\mathbf{\Sigma} \in \mathbb{R}^{M \times M}$, we say that the random matrix $\mathbf{X}_i$ is sampled from the $\mathbf{\Sigma}$-ensemble if

$$\text{vec}(\mathbf{X}_i) \sim \mathcal{N}(0, \mathbf{\Sigma}).$$

We define the quantity

$$\rho^2(\mathbf{\Sigma}) = \sup_{\|\mathbf{u}\|_1 = 1, \|\mathbf{v}\|_1 = 1} \text{var}\left(\mathbf{u}^T \mathbf{X} \mathbf{v}\right),$$

where the random matrix $\mathbf{X} \in \mathbb{R}^{m_1 \times m_2}$ is sampled from the $\mathbf{\Sigma}$-ensemble. For the special case (white Gaussian random vector) $\mathbf{\Sigma} = \mathbf{I}$, we have $\rho^2(\mathbf{\Sigma}) = 1$.

The noise vector $\boldsymbol{\epsilon} \in \mathbb{R}^N$ satisfies the bound $\|\boldsymbol{\epsilon}\|_2 \leqslant 2\nu\sqrt{N}$ for some constant ν. This assumption holds for any bounded noise, and also holds with high probability for any random noise vector with sub-Gaussian entries with parameter ν. The simplest case is that of Gaussian noise $\mathcal{N}(0, \nu^2)$.

Suppose that the matrices $\{\mathbf{X}_i\}_{i=1}^N$ are drawn i.i.d. from the $\mathbf{\Sigma}$-ensemble, and that the unknown matrix $\mathbf{A}^\star \in \mathbb{B}(R_q) \cap \mathcal{S}$ for some $q \in (0,1]$. Then there are universal constant c_0, c_1, c_2 such that a sample size

$$N > c_1 \rho^2(\mathbf{\Sigma}) R_q^{1-q/2} (m_1 + m_2),$$

any solution $\hat{\mathbf{A}}$ to the SDP (8.18) with regularization parameter

$$\gamma_N = c_0 \rho^2(\mathbf{\Sigma}) \nu \sqrt{\frac{m_1 + m_2}{N}}$$

satisfies the bound

$$\mathbb{P}\left(\left\| \hat{\mathbf{A}} - \mathbf{A}^\star \right\|_F^2 \geqslant c_2 R_q \left(\frac{m_1 + m_2}{N} \left(\nu^2 \vee 1\right) \left(\rho^2(\mathbf{\Sigma}) / \sigma_{\min}^2(\mathbf{\Sigma})\right) \right)^{1-q/2} \right)$$
$$\leqslant c_3 \exp\left(-c_4 (m_1 + m_2)\right). \tag{8.24}$$

For the special case of $q = 0$ and $\hat{\mathbf{A}}$ of rank r, we have

$$\mathbb{P}\left(\left\| \hat{\mathbf{A}} - \mathbf{A}^\star \right\|_F^2 \geqslant c_2 \frac{\rho^2(\mathbf{\Sigma}) \nu^2}{\sigma_{\min}^2(\mathbf{\Sigma})} \frac{r(m_1 + m_2)}{N} \right) \leqslant c_3 \exp\left(-c_4 (m_1 + m_2)\right). \tag{8.25}$$

In other words, we have

$$\left\|\hat{\mathbf{A}}-\mathbf{A}^{\star}\right\|_F^2 \leqslant c_2 \frac{\rho^2(\mathbf{\Sigma})\,\nu^2}{\sigma_{\min}^2(\mathbf{\Sigma})} \frac{r(m_1+m_2)}{N}$$

with probability at least $1-c_3\exp\left(-c_4\left(m_1+m_2\right)\right)$.

The central challenge to prove (8.25) is to use Theorem 3.8.5. □

Example 8.6.4 (Low-rank multivariate regression [155]). Consider the observation pairs linked by the vector pairs

$$\mathbf{y}_i = \mathbf{A}^{\star}\mathbf{z}_i + \mathbf{w}_i, \quad i=1,2,\ldots,n,$$

where $\mathbf{w}_i \sim \mathcal{N}\left(0,\nu^2\mathbf{I}_{m_1\times m_2}\right)$ is observation noise vector. We assume that the covariates $\mathbf{z}_i$ are random, i.e., $\mathbf{z}_i \sim \mathcal{N}(0,\mathbf{\Sigma})$, i.i.d. for some m_2-dimensional covariance matrix $\mathbf{\Sigma}>0$.

Consider $\mathbf{A}^{\star}\in\mathbb{B}(R_q)\cap\mathcal{S}$. There are universal constants c_1,c_2,c_3 such that if we solve the SDP (8.18) with regularization parameter

$$\gamma_N = 10\frac{\nu}{m_1}\sqrt{\sigma_{\max}(\mathbf{\Sigma})}\sqrt{\frac{m_1+m_2}{n}},$$

we have

$$\mathbb{P}\left(\left\|\hat{\mathbf{A}}-\mathbf{A}^{\star}\right\|_F^2 \geqslant c_1\left(\frac{\nu^2\sigma_{\max}^2(\mathbf{\Sigma})}{\sigma_{\min}^2(\mathbf{\Sigma})}\right)^{1-q/2} R_q\left(\frac{m_1+m_2}{n}\right)^{1-q/2}\right) \leqslant c_2\exp\left(-c_3\left(m_1+m_2\right)\right). \tag{8.26}$$

When $\mathbf{\Sigma}=\mathbf{I}_{m_2\times m_2}$, there is a constant c_1' such that

$$\mathbb{P}\left(\left\|\hat{\mathbf{A}}-\mathbf{A}^{\star}\right\|_F^2 \geqslant c_1'\nu^{2-q}R_q\left(\frac{m_1+m_2}{n}\right)^{1-q/2}\right) \leqslant c_2\exp\left(-c_3\left(m_1+m_2\right)\right).$$

When $\mathbf{A}^{\star}$ is exactly low rank—that is $q=0$ and $r=R_0$—this simplifies further to

$$\mathbb{P}\left(\left\|\hat{\mathbf{A}}-\mathbf{A}^{\star}\right\|_F^2 \geqslant {c_1}'\nu^2 r\left(\frac{m_1+m_2}{n}\right)\right) \leqslant c_2\exp\left(-c_3\left(m_1+m_2\right)\right).$$

In other words, we have

$$\left\|\hat{\mathbf{A}}-\mathbf{A}^{\star}\right\|_F^2 \leqslant {c_1}'\nu^2 r\left(\frac{m_1+m_2}{n}\right)$$

with probability at least $1-c_2\exp\left(-c_3\left(m_1+m_2\right)\right)$. □

Example 8.6.5 (Vector autoregressive processes [155]). A vector autoregressive (VAR) process [452] in m-dimension is a stochastic process $\{\mathbf{z}_t\}_{t=1}^n$ specified by an initialization $\mathbf{z}_1 \in \mathbb{R}^m$, followed by the recursion

$$\mathbf{z}_{t+1} = \mathbf{A}^\star \mathbf{z}_t + \mathbf{w}_t, \qquad t = 1, 2, .., n. \tag{8.27}$$

In this recursion, the sequence $\mathbf{w}_t \in \mathbb{R}^m$ consists of i.i.d. samples of innovation noise. We assume that each vector $\mathbf{w}_t \in \mathbb{R}^m$ is zero-mean and has a covariance matrix $\mathbf{C} > 0$, so that the process $\{\mathbf{z}_t\}_{t=1}^n$ is zero-mean and has a covariance matrix $\boldsymbol{\Sigma}$ defined by the discrete-time Ricatti equation

$$\boldsymbol{\Sigma} = \mathbf{A}^\star \boldsymbol{\Sigma} (\mathbf{A}^\star)^T + \mathbf{C}.$$

The goal of the problem is to estimate the unknown matrix $\mathbf{A}^\star \in \mathbb{R}^{m \times m}$ on the basis of a sequence of vector samples $\{\mathbf{z}_t\}_{t=1}^n$.

It is natural to expect that the system is controlled primarily by a low-dimensional subset of variables, implying that $\mathbf{A}^\star$ is of low rank. Besides, $\mathbf{A}^\star$ is a Hankel matrix.

Since $\mathbf{z}_t = [Z_{t1} \cdots Z_{tm}]^T$ is m-dimension column vector, the sample size of scalar random variables is $N = nm$. Letting $k = 1, 2, \ldots, m$ index the dimension, we have

$$Z_{(t+1)k} = \left\langle \mathbf{e}_k \mathbf{z}_t^T, \mathbf{A}^\star \right\rangle + W_{tk}. \tag{8.28}$$

We re-index the collection of $N = nm$ observations via the mapping

$$(t, k) \mapsto i = (t-1)k + k.$$

After doing this, the autoregressive problem can be written in the form of (8.15) with $Y_i = Z_{(t+1)k}$ and observation matrix $\mathbf{X}_i = \mathbf{e}_k \mathbf{z}_t^T$.

Suppose that we are given n samples $\{\mathbf{z}_t\}_{t=1}^n$ from a m-dimensional autoregressive process (8.27) that is stationary, based on a system matrix that is stable ($\|\mathbf{A}^\star\|_{\text{op}} \leqslant \alpha \leqslant 1$) and approximately low-rank ($\mathbf{A}^\star \in \mathbb{B}(R_q) \cap \mathcal{S}$). Then, there are universal constants c_1, c_2, c_3 such that if we solve the SDP (8.18) with regularization parameter

$$\gamma_N = \frac{2c_0 \|\boldsymbol{\Sigma}\|_{\text{op}}}{m(1-\alpha)} \sqrt{\frac{m}{n}},$$

then any solution $\hat{\mathbf{A}}$ satisfies

$$\mathbb{P}\left(\left\| \hat{\mathbf{A}} - \mathbf{A}^\star \right\|_F^2 \geqslant c_1 \left(\frac{\sigma_{\max}^2(\boldsymbol{\Sigma})}{\sigma_{\min}^2(\boldsymbol{\Sigma})} \right)^{1-q/2} R_q \left(\frac{m}{n} \right)^{1-q/2} \right) \leqslant c_2 \exp(-c_3 m). \tag{8.29}$$

To prove (8.29), we need the following results (8.30) and (8.31). We need the notation

$$\mathbf{X} = \begin{bmatrix} \mathbf{z}_1^T \\ \mathbf{z}_2^T \\ \vdots \\ \mathbf{z}_n^T \end{bmatrix} \in \mathbb{R}^{n \times m} \quad \text{and} \quad \mathbf{Y} = \begin{bmatrix} \mathbf{z}_2^T \\ \mathbf{z}_3^T \\ \vdots \\ \mathbf{z}_{n+1}^T \end{bmatrix} \in \mathbb{R}^{n \times m}.$$

Let $\mathbf{W}$ be a matrix where each row is sampled i.i.d. from the $\mathcal{N}(0, \mathbf{C})$ distribution corresponding to the innovation noise driving the VAR process. With this notation, and the relation $N = nm$, the SDP objective function (8.18) is written as

$$\frac{1}{m}\left\{\frac{1}{2n}\left\|\mathbf{Y} - \mathbf{X}\mathbf{A}^T\right\|_F^2 + \gamma_n \left\|\mathbf{A}\right\|_*\right\},$$

where $\gamma_n = \gamma_N m$.

The eigenspectrum of the matrix of the matrix $\mathbf{X}^T\mathbf{X}/n$ is well controlled in terms of the stationary covariance matrix: in particular, as long as $n > c_3 m$, we have

$$\begin{aligned} &\mathbb{P}\left(\sigma_{\max}\left(\mathbf{X}^T\mathbf{X}/n\right) \geqslant \frac{24\sigma_{\max}(\mathbf{\Sigma})}{1-\alpha}\right) \leqslant 2c_1 \exp\left(-c_2 m\right), \quad \text{and} \\ &\mathbb{P}\left(\sigma_{\min}\left(\mathbf{X}^T\mathbf{X}/n\right) \leqslant 0.25\sigma_{\min}(\mathbf{\Sigma})\right) \leqslant 2c_1 \exp\left(-c_2 m\right). \end{aligned} \tag{8.30}$$

There exists constants $c_i > 0$, independent of $n, m, \mathbf{\Sigma}$, etc. such that

$$\mathbb{P}\left(\frac{1}{n}\left\|\mathbf{X}^T\mathbf{W}\right\|_{op} \geqslant \frac{c_0\left\|\mathbf{\Sigma}\right\|_{op}}{1-\alpha}\sqrt{\frac{m}{n}}\right) \leqslant c_2 \exp\left(-c_3 m\right). \tag{8.31}$$

□

8.7 Linear Regression

Consider a standard linear regression model

$$\mathbf{y} = \mathbf{X}\boldsymbol{\beta} + \mathbf{w} \tag{8.32}$$

where $\mathbf{y} \in \mathbb{R}^n$ is an observation vector, $\mathbf{X} \in \mathbb{R}^{n \times d}$ is a design matrix, $\boldsymbol{\beta}$ is the unknown regression vector, and $\mathbf{w} \in \mathbb{R}^d$ is additive Gaussian noise, i.e., $\mathbf{w} \sim \mathcal{N}\left(0, \sigma^2 \mathbf{I}_{d \times d}\right)$, where $\mathbf{I}_{n \times n}$ is the $n \times n$ identity matrix. As pointed out above, the consistent estimation of $\boldsymbol{\beta}$ is impossible unless we impose some additional structure on the unknown vector $\boldsymbol{\beta}$. We consider sparsity constraint here: $\boldsymbol{\beta}$ has exactly $s \ll d$ non-zero entries.

The notions of sparsity can be defined more precisely in terms of the ℓ_p-balls[1] for $p \in (0, 1]$, defined as [135]

$$\mathbb{B}_p\left(R_p\right) = \left\{ \boldsymbol{\beta} \in \mathbb{R}^d : \quad \|\boldsymbol{\beta}\|_p^p = \sum_{i=1}^{d} |\boldsymbol{\beta}_i|^p \leqslant R_p \right\}, \tag{8.33}$$

In the limiting case of $p = 0$, we have the ℓ_0-ball

$$\mathbb{B}_0\left(s\right) = \left\{ \boldsymbol{\beta} \in \mathbb{R}^d : \quad \sum_{i=1}^{d} I\left[\boldsymbol{\beta}_i \neq 0\right] \leqslant s \right\}, \tag{8.34}$$

where I is the indicator function and $\boldsymbol{\beta}$ has exactly $s \ll d$ non-zero entries.

The unknown vector $\boldsymbol{\beta}$ can be computed by solving the convex optimization problem

$$\begin{array}{ll} \text{minize} & \|\mathbf{y} - \mathbf{X}\boldsymbol{\beta}\|_2^2 \\ \text{subject to} & \left\| \|\boldsymbol{\beta}\|_p^p \leqslant R_p \right\| . \end{array} \tag{8.35}$$

Sometimes we are interested in the vector $\mathbf{v}$ whose sparsity is bounded by $2s$ and ℓ_2-norm is bounded by R

$$\mathbb{S}\left(s, R\right) = \left\{ \mathbf{v} \in \mathbb{R}^d : \quad \|\mathbf{v}\|_0 \leqslant 2s, \quad \|\mathbf{v}\|_2 \leqslant R \right\}.$$

The following random variable Z_n is of independent interest, defined as

$$Z_n = \sup_{\mathbf{v} \in \mathbb{S}(s,R)} \frac{1}{n} \left| \mathbf{w}^T \mathbf{X} \mathbf{v} \right|, \tag{8.36}$$

where $\mathbf{X}, \mathbf{w}$ are given in (8.32). Let us show how to bound the random variable Z_n. The approach used by Raskutti et al. [135] is emphasized and followed closely here.

For a given $\varepsilon \in (0, 1)$ to be chosen, we need to bound the minimal cardinality of a set that covers $\mathbb{S}\left(s, R\right)$ up to $(R\varepsilon)$-accuracy in ℓ-norm. We claim that we may find such a covering set $\left\{ \mathbf{v}^1, \ldots, \mathbf{v}^N \right\} \subset \mathbb{S}\left(s, R\right)$ with cardinality $N = N(s, R, \varepsilon)$ that is upper bounded by

$$\log N(s, R, \varepsilon) \leqslant \log \binom{d}{2s} + 2s \log\left(1/\varepsilon\right).$$

[1] Strictly speaking, these sets are not "balls" when $p < 1$, since they fail to be convex.

To establish the claim, we note that there are $\binom{d}{2s}$ subsets of size $2s$ within $\{1,2,\ldots,d\}$. Also, for any $2s$-sized subset, there are an $(R\varepsilon)$-covering in ℓ_2-norm of the ball $\mathbb{B}_2(R)$ with at most $2^{2s\log(1/\varepsilon)}$ elements (e.g., [154]).

As a result, for each vector $\mathbf{v} \in \mathbb{S}(s,R)$, we may find some $\mathbf{v}^k$ such that $\left\|\mathbf{v}-\mathbf{v}^k\right\|_2 \leqslant R\varepsilon$. By triangle inequality, we obtain

$$\begin{aligned}\tfrac{1}{n}\left|\mathbf{w}^T\mathbf{X}\mathbf{v}\right| &\leqslant \tfrac{1}{n}\left|\mathbf{w}^T\mathbf{X}\mathbf{v}^k\right| + \tfrac{1}{n}\left|\mathbf{w}^T\mathbf{X}\left(\mathbf{v}-\mathbf{v}^k\right)\right| \\ &\leqslant \tfrac{1}{n}\left|\mathbf{w}^T\mathbf{X}\mathbf{v}^k\right| + \tfrac{1}{n}\|\mathbf{w}\|_2\left\|\mathbf{X}\left(\mathbf{v}-\mathbf{v}^k\right)\right\|_2.\end{aligned}$$

Now we make explicit assumption that

$$\frac{1}{\sqrt{n}}\frac{\|\mathbf{X}\mathbf{v}\|_2}{\|\mathbf{v}\|_2} \leqslant \kappa, \quad \text{for all} \quad \mathbf{v} \in \mathbb{B}_0(2s).$$

With this assumption, it follows that

$$\left\|\mathbf{X}\left(\mathbf{v}-\mathbf{v}^k\right)\right\|_2/\sqrt{n} \leqslant \kappa R\left\|\mathbf{v}-\mathbf{v}^k\right\|_2 \leqslant \kappa\varepsilon.$$

Since the vector $\mathbf{w} \in \mathbb{R}^n$ is Gaussian, the variate $\|\mathbf{w}\|_2^2/\sigma^2$ is the χ^2 distribution with n degrees of freedom, we have $\frac{1}{\sqrt{n}}\|\mathbf{w}\|_2 \leqslant 2\sigma$ with probability at least $1 - c_1\exp(-c_2 n)$, where c_1, c_2 are two numerical constants, using standard tail bounds (see Sect. 3.2). Putting all the pieces together, we obtain

$$\frac{1}{n}\left|\mathbf{w}^T\mathbf{X}\mathbf{v}\right| \leqslant \frac{1}{n}\left|\mathbf{w}^T\mathbf{X}\mathbf{v}^k\right| + 2\kappa\sigma R\varepsilon$$

with high probability. Taking the supremum over $\mathbf{v}$ on both sides gives

$$Z_n \leqslant \max_{k=1,2,\ldots,N}\frac{1}{n}\left|\mathbf{w}^T\mathbf{X}\mathbf{v}^k\right| + 2\kappa\sigma R\varepsilon.$$

It remains to bound the finite maximum over the covering set. First, we see that each variate $\frac{1}{n}\mathbf{w}^T\mathbf{X}\mathbf{v}^k$ is zero-mean Gaussian with variance $\sigma^2\left\|\mathbf{X}\mathbf{v}^i\right\|_2^2/n^2$. Now by standard Gaussian tail bounds, we conclude that

$$\begin{aligned}Z_n &\leqslant \sigma R\kappa\sqrt{3\log N(s,R,\varepsilon)}/\sqrt{n} + 2\kappa\sigma R\varepsilon \\ &= \sigma R\kappa\left[\sqrt{3\log N(s,R,\varepsilon)}/\sqrt{n} + 2\varepsilon\right].\end{aligned} \tag{8.37}$$

with probability greater than $1 - c_1\exp(-c_2\log N(s,R,\varepsilon))$.

Finally, suppose $\varepsilon = \sqrt{s\log(d/2s)}/\sqrt{n}$. With this choice and assuming that $n \leq d$, we have

$$\frac{\log N(s,R,\varepsilon)}{n} \leqslant \frac{\log\binom{d}{2s}}{n} + \frac{s\log\left(\frac{n}{s\log(d/2s)}\right)}{n}$$
$$\leqslant \frac{\log\binom{d}{2s}}{n} + \frac{s\log(d/s)}{n}$$
$$\leqslant \frac{2s+2s\log(d/s)}{n} + \frac{s\log(d/s)}{n},$$

where the final line uses standard bounds on binomial coefficients. Since $d/s \geq 2$ by assumption, we conclude that our choice of ε guarantees that

$$\frac{\log N(s,R,\varepsilon)}{n} \leqslant 5s\log(d/s).$$

Inserting these relations into (8.37), we conclude that

$$Z_n \leqslant 6\sigma R\kappa\left[\frac{s\log(d/s)}{n}\right].$$

Since $\log N(s,R,\varepsilon) \geqslant s\log(d-2s)$, this event occurs with probability at least $1-c_1\exp(-c_2\min\{n, s\log(d-s)\})$. We summarize the results in this theorem.

Theorem 8.7.1 ([135]). *If the ℓ_2-norm of random matrix $\mathbf{X}$ is bounded by $\frac{1}{\sqrt{n}}\frac{\|\mathbf{Xv}\|_2}{\|\mathbf{v}\|_2} \leqslant \kappa$ for all $\mathbf{v}$ with at most $2s$ non-zeros, i.e., $\mathbf{v} \in \mathbb{B}_0(2s)$, and $\mathbf{w} \in \mathbb{R}^d$ is additive Gaussian noise, i.e., $\mathbf{w} \sim \mathcal{N}(0, \sigma^2\mathbf{I}_{n\times n})$, then for any radius $R > 0$, we have*

$$\sup_{\|\mathbf{v}\|_0 \leqslant 2s,\ \|\mathbf{v}\|_2 \leqslant R} \frac{1}{n}\left|\mathbf{w}^T\mathbf{Xv}\right| \leqslant 6\sigma\kappa R\left[\frac{s\log(d/s)}{n}\right],$$

with probability at least $1-c_1\exp(-c_2\min\{n, s\log(d-s)\})$.

Let us apply Theorem 8.7.1. Let $\boldsymbol{\beta}^\star$ be a feasible solution of (8.35). We have

$$\|\mathbf{y}-\mathbf{X}\boldsymbol{\beta}\|_2^2 \leqslant \|\mathbf{y}-X\boldsymbol{\beta}^\star\|_2^2.$$

Define the error vector $\mathbf{e} = (\boldsymbol{\beta}-\boldsymbol{\beta}^\star)$. After some algebra, we obtain

$$\frac{1}{n}\|\mathbf{Xv}\|_2^2 \leqslant \frac{2}{n}\left|\mathbf{w}^T\mathbf{Xe}\right|$$

the right-hand side of which is exactly the expression required by Theorem 8.7.1, if we identify $\mathbf{v} = \mathbf{e}$.

8.8 Multi-task Matrix Regression

We are given a collection of d_2 regression problems in $\mathbb{R}^{d_1}$, each of the form

$$\mathbf{y}_i = \mathbf{X}\boldsymbol{\beta}_i + \mathbf{w}_i, \quad i = 1, 2, \ldots, d_2,$$

where $\boldsymbol{\beta}_i \in \mathbb{R}^{d_1}$ is an unknown regression vector, $\mathbf{w}_i \in \mathbb{R}^n$ is the observation noise, and $\mathbf{X} \in \mathbb{R}^{n \times d_1}$ is the design (random) matrix. In a convenient matrix form, we have

$$\mathbf{Y} = \mathbf{X}\mathbf{B} + \mathbf{W} \tag{8.38}$$

where $\mathbf{Y} = [\mathbf{y}_1, \ldots, \mathbf{y}_{d_2}]$ and $\mathbf{W} = [\mathbf{w}_1, \ldots, \mathbf{w}_{d_2}]$ are both matrices in $\mathbb{R}^{n \times d_2}$ and $\mathbf{B} = [\boldsymbol{\beta}_1, \ldots, \boldsymbol{\beta}_{d_2}] \in \mathbb{R}^{d_1 \times d_2}$ is a matrix of regression vectors. In multi-task learning, each column of $\mathbf{B}$ is called a task and each row of $\mathbf{B}$ is a feature.

A special structure has the form of low-rank plus sparse decomposition

$$\mathbf{B} = \boldsymbol{\Theta} + \boldsymbol{\Gamma}$$

where $\boldsymbol{\Theta}$ is of low rank and $\boldsymbol{\Gamma}$ is sparse, with a small number of non-zero entries. For example, $\boldsymbol{\Gamma}$ is row-sparse, with a small number of non-zero rows. It follows from (8.38) that

$$\mathbf{Y} = \mathbf{X}(\boldsymbol{\Theta} + \boldsymbol{\Gamma}) + \mathbf{W} \tag{8.39}$$

In the following examples, the entries of $\mathbf{W}$ are assumed to be i.i.d. zero-mean Gaussian with variance ν^2, i.e. $W_{ij} \sim \mathcal{N}(0, \nu^2)$.

Example 8.8.1 (Concentration for product of two random matrices [453]). Consider the product of two random matrices defined above

$$\mathbf{Z} = \mathbf{X}^T\mathbf{W} \in \mathbb{R}^{d_1 \times d_2}.$$

It can be shown that the matrix $\mathbf{Z}$ has independent columns, with each column $\mathbf{z}_j \sim \mathcal{N}(0, \nu^2\mathbf{X}^T\mathbf{X}/n)$. Let $\sigma_{\max}$ be the maximum eigenvalue of matrix $\mathbf{X}$. Since $\left\|\mathbf{X}^T\mathbf{X}\right\|_{op} \leqslant \sigma_{\max}^2$, known results on the singular values of Gaussian random matrices [145] imply that

$$\mathbb{P}\left(\left\|\mathbf{X}^T\mathbf{W}\right\|_{op} \geqslant \frac{4(d_1 + d_2)\nu\sigma_{\max}}{\sqrt{n}}\right) \leqslant 2\exp\left(-c(d_1 + d_2)\right).$$

Let $\mathbf{x}_j$ be the j-th column of matrix $\mathbf{X}$. Let $\kappa_{\max} = \max_{j=1,\ldots,d_1} \left\|\mathbf{x}_j\right\|_2$ be the maximum ℓ_2-norm over columns. Since the ℓ_2-norm of the columns of $\mathbf{X}$ are bounded by $\kappa_{\max}$, the entries of $\mathbf{X}^T\mathbf{W}$ are i.i.d. and Gaussian with variance at most $(\nu\kappa_{\max})^2/n$. As a result, the standard Gaussian tail bound combined with union bound gives

$$\mathbb{P}\left(\left\|\mathbf{X}^T\mathbf{W}\right\|_\infty \geqslant \frac{4\nu\sigma_{\max}}{\sqrt{n}}\log\left(d_1d_2\right)\right) \leqslant \exp\left(-\log\left(d_1d_2\right)\right),$$

where $\|\mathbf{A}\|_\infty$ for matrix $\mathbf{A}$ with the (i,j)-element a_{ij} is defined as

$$\|\mathbf{A}\|_\infty = \max_{i=1,\ldots,d_1}\max_{j=1,\ldots,d_2}|a_{ij}|. \qquad \square$$

Example 8.8.2 (Concentration for the columns of a random matrix [453]). As defined above, $\mathbf{w}_i$ is the k-th column of the matrix $\mathbf{W}$. The function

$$\mathbf{w}_k \to \|\mathbf{w}_k\|_2$$

is Lipschitz. By concentration of measure for Gaussian Lipschitz functions [141], we have that for all $t > 0$

$$\mathbb{P}\left(\|\mathbf{w}_k\|_2 \geqslant \mathbb{E}\|\mathbf{w}_k\|_2 + t\right) \leqslant \exp\left(-\frac{t^2d_1d_2}{2\nu^2}\right).$$

Using the Gaussianity of $\mathbf{w}_i$, we have

$$\mathbb{E}\|\mathbf{w}_k\|_2 \leqslant \frac{\nu}{\sqrt{d_1d_2}}\sqrt{d_1} = \frac{\nu}{\sqrt{d_2}}.$$

Applying union bound over all d_2 columns, we conclude that

$$\mathbb{P}\left(\max_{k=1,2,\ldots,d_2}\|\mathbf{w}_k\|_2 \geqslant \frac{\nu}{\sqrt{d_2}} + t\right) \leqslant \exp\left(-\frac{t^2d_1d_2}{2\nu^2} + \log d_2\right).$$

That is, with probability greater than $1 - \exp\left(-\frac{t^2d_1d_2}{2\nu^2} + \log d_2\right)$, we have $\max\limits_k \|\mathbf{w}_k\|_2 \leqslant \frac{\nu}{\sqrt{d_2}} + t$. Setting $t = 4\nu\sqrt{\frac{\log d_2}{d_1d_2}}$ gives

$$\mathbb{P}\left(\max_{k=1,2,\ldots,d_2}\|\mathbf{w}_k\|_2 \geqslant \frac{\nu}{\sqrt{d_2}} + 4\nu\sqrt{\frac{\log d_2}{d_1d_2}}\right) \leqslant \exp\left(-3\log d_2\right). \qquad \square$$

Example 8.8.3 (Concentration for trace inner product of two matrices [453]). We study the function defined as

$$Z(s) = \sup_{\|\mathbf{\Delta}\|_1 \leqslant \sqrt{s},\ \|\mathbf{\Delta}\|_F \leqslant 1} |\langle \mathbf{W}, \mathbf{\Delta}\rangle|.$$

Viewed as a function of matrix $\mathbf{W}$, the random variable $Z(s)$ is a Lipschitz function with constant $\frac{\nu}{\sqrt{d_1 d_2}}$. Using Talagrand's concentration inequality, we obtain

$$\mathbb{P}\left(Z\left(s\right) \geqslant \mathbb{E}\left[Z\left(s\right)\right] + t\right) \leqslant \exp\left(-\frac{t^2 d_1 d_2}{2\nu^2}\right).$$

Setting $t^2 = \frac{4s\nu^2}{d_1 d_2} \log\left(\frac{d_1 d_2}{s}\right)$, we have

$$Z\left(s\right) \leqslant \mathbb{E}\left[Z\left(s\right)\right] + \frac{2s\nu}{d_1 d_2} \log\left(\frac{d_1 d_2}{s}\right)$$

with probability greater than at least

$$1 - \exp\left(-2s \log\left(\frac{d_1 d_2}{s}\right)\right).$$

It remains to upper bound the expected value $\mathbb{E}\left[Z\left(s\right)\right]$. In order to do so, we use [153, Theorem 5.1(ii)] with $(q_0, q_1) = (1, 2), n = d_1 d_2$, and $t = \sqrt{s}$, thereby obtaining

$$\mathbb{E}\left[Z\left(s\right)\right] \leqslant c' \frac{\nu}{d_1 d_2} \sqrt{s} \sqrt{2 + \log\left(\frac{2 d_1 d_2}{s}\right)} \leqslant c \frac{\nu}{d_1 d_2} \sqrt{s \log\left(\frac{2 d_1 d_2}{s}\right)}.$$

Define the notation

$$\left\|\mathbf{A}\right\|_{2,1} = \sum_{i=1}^{d_2} \left\|\mathbf{a}_k\right\|_2$$

where $\mathbf{a}_k$ is the k-th column of matrix $\mathbf{A} \in \mathbb{R}^{d_1 \times d_2}$. We can study the function

$$\tilde{Z}\left(s\right) = \sup_{\left\|\boldsymbol{\Delta}\right\|_{2,1} \leqslant \sqrt{s}, \quad \left\|\boldsymbol{\Delta}\right\|_F \leqslant 1} \left|\left\langle \mathbf{W}, \boldsymbol{\Delta} \right\rangle\right|$$

which is Lipschitz with constant $\frac{\nu}{\sqrt{d_1 d_2}}$. Similar to above, we can use the standard approach: (1) derive concentration of measure for Gaussian Lipschitz functions; (2) upper bound the expectation. For details, we see [453]. □

8.9 Matrix Completion

This section is taken from Recht [103] for low rank matrix recovery, primarily due to his highly accessible presentation.

The nuclear norm $\|\mathbf{X}\|_*$ of a matrix $\mathbf{X}$ is equal to the sum of its singular values $\sum_i \sigma_i(\mathbf{X})$, and is the best convex lower bound of the rank function that is NP-hard. The intuition behind this heuristic is that while the rank function counts the number of nonvanishing singular values, the nuclear norm sums their amplitudes, much like how the ℓ_1 norm is a useful surrogate for counting the number of nonzeros in a vector. Besides, the nuclear norm can be minimized subject to equality constraints via semidefinite programming (SDP).[2]

Let us review some matrix preliminaries and also fix the notation. Matrices are bold capital, vectors are bold lower case and scalars or entries are not bold. For example, $\mathbf{X}$ is a matrix, $\mathbf{X}_{ij}$ its (i,j)-th entry. Likewise, $\mathbf{x}$ is a vector, and x_i its i-th component. If $\mathbf{u}_k \in \mathbb{R}^n$ for $1 \leqslant k \leqslant d$ is a collection of vectors, $[\mathbf{u}_1, \ldots, \mathbf{u}_d]$ will denote the $n \times d$ matrix whose k-th column is $\mathbf{u}_k$. $\mathbf{e}_k$ will denote the k-th standard basis vector in $\mathbb{R}^d$, equal to 1 in component k and 0 everywhere else. $\mathbf{X}^*$ and $\mathbf{x}^*$ denote the transpose of matrices $\mathbf{X}$ and $\mathbf{x}$.

The spectral norm of a matrix is denoted $\|\mathbf{X}\|$. The Euclidean inner product between two matrices is $\langle \mathbf{X}, \mathbf{Y} \rangle = \mathrm{Tr}(\mathbf{X}^*\mathbf{Y})$, and the corresponding Euclidean norm, called the Frobenius or Hilbert-Schmidt norm, is denoted $\|\mathbf{X}\|_F$. That is, $\|\mathbf{X}\|_F = \langle \mathbf{X}, \mathbf{X} \rangle^{\frac{1}{2}}$. Or

$$\|\mathbf{X}\|_F^2 = \langle \mathbf{X}, \mathbf{X} \rangle = \mathrm{Tr}\left(\mathbf{X}^T\mathbf{X}\right), \tag{8.40}$$

which is a linear operator of $\mathbf{X}^T\mathbf{X}$ since the trace function is linear. The nuclear norm of a matrix is $\|\mathbf{X}\|_*$. The maximum entry of $\mathbf{X}$ (in absolute value) is denoted by $\|\mathbf{X}\|_\infty = \max_{ij} |X_{ij}|$, where of course $|\cdot|$ is the absolute value. For vectors, the only norm applied is the Euclidean ℓ_2 norm, simply denoted $\|\mathbf{x}\|$.

Linear transformations that act on matrices will be denoted by calligraphic letters. In particular, the identity operator is $\mathcal{I}$. The spectral norm (the top singular value) of such an operator is $\|\mathcal{A}\| = \sup_{\mathbf{X}:\|\mathbf{X}\|_F \leqslant 1} \|\mathcal{A}(\mathbf{X})\|_F$. Subspaces are also denoted by calligraphic letters.

8.9.1 Orthogonal Decomposition and Orthogonal Projection

We suggest the audience to review [454, Chap. 5] for background. We only review the key definitions needed later. For a set of vectors $\mathcal{S} = \{\mathbf{v}_1, \ldots, \mathbf{v}_r\}$, the subspace

$$\mathrm{span}(\mathcal{S}) = \{\alpha_1\mathbf{v}_1 + \alpha_2\mathbf{v}_2 + \cdots + \alpha_r\mathbf{v}_r\}$$

[2]The SDP is of course the convex optimization. It is a common practice that once a problem is recast in terms of a convex optimization problem, then the problem may be solved, using many general-purpose solvers such as CVX.

generated by forming *all linear combinations* of vectors from $\mathcal{S}$ is called the *space spanned by* $\mathcal{S}$. For a subset $\mathcal{M}$ of an inner-product space $\mathcal{V}$, the *orthogonal complement* $\mathcal{M}^{\perp}$ of $\mathcal{M}$ is defined to be the set of all vectors in $\mathcal{V}$ that are orthogonal to very vector in $\mathcal{M}$. That is,

$$\mathcal{M}^{\perp} = \{\mathbf{x} \in \mathcal{V} : \langle \mathbf{m}, \mathbf{x} \rangle = 0 \text{ for all } \mathbf{m} \in \mathcal{M}\}.$$

Let $\mathbf{u}_k$ (respectively $\mathbf{v}_k$) denote the k-th column of $\mathbf{U}$ (respectively $\mathbf{V}$). Set

$$\mathcal{U} \equiv \text{span}(\mathbf{u}_1, \ldots, \mathbf{u}_r), \text{ and } \mathcal{V} \equiv \text{span}(\mathbf{v}_1, \ldots, \mathbf{v}_r).$$

Also, assume, without loss of generality, that $n_1 \le n_2$. It is useful to introduce orthogonal decomposition

$$\mathbb{R}^{n_1 \times n_2} = \mathcal{T} \oplus \mathcal{T}^{\perp}$$

where $\mathcal{T}$ is the linear space spanned by elements of the form $\mathbf{u}_k\mathbf{y}^*$ and $\mathbf{x}\mathbf{v}_k^*, 1 \leqslant k \leqslant r$, where $\mathbf{x}$ and $\mathbf{y}$ are arbitrary, and $\mathcal{T}^{\perp}$ is its orthogonal complement. $\mathcal{T}^{\perp}$ is the subspace of matrices spanned by the family $(\mathbf{x}\mathbf{y}^*)$, where $\mathbf{x}$ (respectively $\mathbf{y}$) is any vector orthogonal to $\mathcal{U}$ (respectively $\mathcal{V}$).

The orthogonal projection $\mathcal{P}_{\mathcal{T}}$ of a matrix $\mathbf{Z}$ onto subspace $\mathcal{T}$ is defined as

$$\mathcal{P}_{\mathcal{T}}(\mathbf{Z}) = \mathbf{P}_{\mathcal{U}}\mathbf{Z} + \mathbf{Z}\mathbf{P}_{\mathcal{V}} - \mathbf{P}_{\mathcal{U}}\mathbf{Z}\mathbf{P}_{\mathcal{V}}, \tag{8.41}$$

where $\mathbf{P}_{\mathcal{U}}$ and $\mathbf{P}_{\mathcal{V}}$ are the orthogonal projections onto $\mathcal{U}$ and $\mathcal{V}$, respectively. While $\mathbf{P}_{\mathcal{U}}$ and $\mathbf{P}_{\mathcal{V}}$ are matrices, $\mathcal{P}_{\mathcal{T}}$ is a linear operator that maps a matrix to another matrix. The orthogonal projection of a matrix $\mathbf{Z}$ onto $\mathcal{T}^{\perp}$ is written as

$$\mathcal{P}_{\mathcal{T}^{\perp}}(\mathbf{Z}) = (\mathcal{I} - \mathcal{P}_{\mathcal{T}})(\mathbf{Z}) = (\mathbf{I}_{n_1} - \mathbf{P}_{\mathcal{U}})\mathbf{Z}(\mathbf{I}_{n_2} - \mathbf{P}_{\mathcal{V}})$$

where $\mathbf{I}_d$ denotes the $d \times d$ identity matrix. It follows from the definition

$$\mathcal{P}_{\mathcal{T}}(\mathbf{e}_a\mathbf{e}_b^*) = (\mathbf{P}_{\mathcal{U}}\mathbf{e}_a)\mathbf{e}_b^* + \mathbf{e}_a(\mathbf{P}_{\mathcal{V}}\mathbf{e}_b)^* - (\mathbf{P}_{\mathcal{U}}\mathbf{e}_a)(\mathbf{P}_{\mathcal{V}}\mathbf{e}_b)^*.$$

With the aid of (8.40), the Frobenius norm of $\mathcal{P}_{\mathcal{T}^{\perp}}(\mathbf{e}_a\mathbf{e}_b^*)$ is given as[3]

$$\begin{aligned}\|\mathcal{P}_{\mathcal{T}}(\mathbf{e}_a\mathbf{e}_b^*)\|_F^2 &= \langle \mathcal{P}_{\mathcal{T}}(\mathbf{e}_a\mathbf{e}_b^*), \mathcal{P}_{\mathcal{T}}(\mathbf{e}_a\mathbf{e}_b^*)\rangle = \|\mathbf{P}_{\mathcal{U}}\mathbf{e}_a\|^2 + \|\mathbf{P}_{\mathcal{V}}\mathbf{e}_b\|^2 \\ &- \|\mathbf{P}_{\mathcal{U}}\mathbf{e}_d\|^2\|\mathbf{P}_{\mathcal{V}}\mathbf{e}_b\|^2.\end{aligned} \tag{8.42}$$

In order to upper bound (8.42), we are motivated to define a scalar $\mu(\mathcal{W})$, called the coherence of a subspace $\mathcal{W}$, such that

[3]This equation in the original paper [104] has a typo and is corrected here.

$$\|\mathbf{P}_{\mathcal{U}}\mathbf{e}_a\|^2 \leqslant \mu(\mathcal{U})\, r/n_1, \qquad \|\mathbf{P}_{\mathcal{V}}\mathbf{e}_b\|^2 \leqslant \mu(\mathcal{V})\, r/n_2, \tag{8.43}$$

With the help of (8.43), the Frobenius norm is upper bounded by

$$\|\mathcal{P}_T(\mathbf{e}_a\mathbf{e}_b^*)\|_F^2 \leqslant \max\{\mu(\mathcal{U}), \mu(\mathcal{V})\}\, r\frac{n_1+n_2}{n_1 n_2} \leqslant \mu_0 r \frac{n_1+n_2}{n_1 n_2}, \tag{8.44}$$

which will be used frequently.

For a subspace $\mathcal{W}$, let us formally define its *coherence*, which plays a central role in the statement of the final theorem on matrix completion.

Definition 8.9.1 (Coherence of a subspace). Let $\mathcal{W}$ be a subspace of dimension r and $\mathbf{P}_{\mathcal{W}}$ be the orthogonal projection of a matrix onto $\mathcal{W}$. Then, the coherence of $\mathcal{W}$ (using the standard basis $(\mathbf{e}_i)$) is defined to be

$$\mu(\mathcal{W}) \equiv \frac{n}{r} \max_{1 \leqslant i \leqslant n} \|\mathbf{P}_{\mathcal{W}}\mathbf{e}_i\|^2.$$

For any subspace, the smallest value which $\mu(\mathcal{W})$ can be is 1, achieved, for example, if $\mathcal{W}$ is spanned by vectors whose entries all have magnitude $1/\sqrt{n}$. The largest value for $\mu(\mathcal{W})$, on the other hand, is n/r which would correspond to any subspace that contains a standard basis element. If a matrix has row and column spaces with low coherence, then each entry can be expected to provide about the same amount of information.

8.9.2 *Matrix Completion*

The main contribution of Recht [104] is an analysis of uniformly sampled sets via the study of a sampling with replacement model. In particular, Recht analyze the situation where each entry index is sampled independently from uniform distribution on $\{1, \ldots, n_1\} \times \{1, \ldots, n_2\}$.

Proposition 8.9.2 (Sampling with replacement [104]). *The probability that the nuclear norm heuristic fails when the set of observed entries is sampled uniformly from the collection of sets of size N is less than or equal to the probability that the heuristic fails when N entries are sampled independently with replacement.*

Theorem 2.2.17 is repeated here for convenience.

Theorem 8.9.3 (Noncommutative Bernstein Inequality [104]). *Let $\mathbf{X}_1, \ldots, \mathbf{X}_n$ be independent zero-mean random matrices of dimension $d_1 \times d_2$. Suppose $\rho_k^2 = \max\{\|\mathbb{E}(\mathbf{X}_k\mathbf{X}_k^*)\|, \|\mathbb{E}(\mathbf{X}_k^*\mathbf{X}_k)\|\}$ and $\|\mathbf{X}_k\| \leqslant M$ almost surely for all k. Then, for any $\tau > 0$,*

$$\mathbb{P}\left[\left\|\sum_{k=1}^{L}\mathbf{X}_k\right\| > \tau\right] \leqslant (d_1+d_2)\exp\left(\frac{-\tau^2/2}{\sum\limits_{k=1}^{L}\rho_k^2 + M\tau/3}\right). \tag{8.45}$$

Theorem 8.9.4 (Matrix Completion Recht [104]). *Let* $\mathbf{M}$ *be an* $n_1 \times n_2$ *matrix of rank* r *with singular value decomposition* $\mathbf{U\Sigma V}^H$. *Without loss of generality, impose the convention* $n_1 \le n_2$, $\mathbf{\Sigma} \in \mathbb{R}^{r\times r}, \mathbf{U} \in \mathbb{R}^{n_1\times r}, \mathbf{U} \in \mathbb{R}^{r\times n_2}$. *Assume that*

A0 *The row and column spaces have coherences bounded above by some positive* μ_0.

A1 *The matrix* $\mathbf{UV}^H$ *has a maximum entry bounded by* $\mu_1\sqrt{r/(n_1 n_2)}$ *in absolute value for some positive* μ_1.

Suppose m *entries of* $\mathbf{M}$ *are observed with locations sampled uniformly at random. Then if*

$$m \geqslant 32\max\left\{\mu_1^2, \mu_0\right\} r\,(n_1+n_2)\,\beta\log^2(2n_2)$$

for some $\beta > 1$, *the minimizer to the problem*

$$\begin{array}{ll} \text{minimize} & \|\mathbf{X}\|_* \\ \text{subject to } X_{ij} = M_{ij} & (i,j)\in\Omega. \end{array} \tag{8.46}$$

is unique and equal to $\mathbf{M}$ *with probability at least* $1 - 6\log(n_2)(n_1+n_2)^{2-2\beta} - n_2^{2-2\sqrt{\beta}}$.

The proof is very short and straightforward. It only uses basic matrix analysis, elementary large deviation bounds and a noncommutative version of Bernsterin's inequality (See Theorem 8.9.3).

Recovering low-Rank matrices is studied by Gross [102].

Example 8.9.5 (A secure communications protocol that is robust to sparse errors [455]). We want to securely transmit a binary message across a communications channel. Our theory shows that decoding the message via deconvolution also makes this secure scheme perfectly robust to *sparse* corruptions such as erasures or malicious interference.

We model the binary message as a sign vector $\mathbf{m}_0 \in \{\pm 1\}^d$. Choose a random basis $\mathbf{Q} \in \mathbb{O}_d$. The transmitter sends the scrambled message $\mathbf{s}_0 = \mathbf{Qm}_0$ across the channel, where it is corrupted by an unknown sparse vector $\mathbf{c}_0 \in \mathbb{R}^d$. The receiver must determine the original message given only the corrupted signal

$$\mathbf{z}_0 = \mathbf{s}_0 + \mathbf{c}_0 = \mathbf{Qm}_0 + \mathbf{c}_0$$

and knowledge of the scrambling matrix $\mathbf{Q}$.

The signal model is perfectly suited to the deconvolution recipe of [455, Sect. 1.2]. The ℓ_1 and ℓ_∞ are natural complexity measures for the structured signals $\mathbf{c}_0$ and $\mathbf{m}_0$. Since the message $\mathbf{m}_0$ is a sign vector, we also have the side information $\|\mathbf{m}_0\|_{\ell_\infty} = 1$. Our receiver then recovers the message with the convex deconvolution method

$$\begin{array}{ll} \text{minimize} & \|\mathbf{c}\|_{\ell_1} \\ \text{subject to } \|\mathbf{m}\|_{\ell_\infty} = 1 & \text{and} \quad \mathbf{Q}\mathbf{m} + \mathbf{c} = \mathbf{z}_0, \end{array} \tag{8.47}$$

where the decision variables are $\mathbf{c}, \mathbf{m} \in \mathbb{R}^d$. For example, $d = 100$. This method succeeds if $(\mathbf{c}_0, \mathbf{m}_0)$ is the unique optimal point of (8.47). □

Example 8.9.6 (Low-rank matrix recovery with generic sparse corruptions [455]). Consider the matrix observation

$$\mathbf{Z}_0 = \mathbf{X}_0 + \mathcal{R}(\mathbf{Y}_0) \in \mathbb{R}^{n \times n},$$

where $\mathbf{X}_0$ has low rank, $\mathbf{Y}_0$ is sparse, and $\mathcal{R}$ is a random rotation on $\mathbb{R}^{n \times n}$. For example $n = 35$. We aim to discovery the matrix $\mathbf{X}_0$ given the corrupted observation $\mathbf{Z}_0$ and the basis $\mathcal{R}$.

The Schatten 1-norm $\|\cdot\|_{S_1}$ serves as a natural complexity measure for the low-rank structure of $\mathbf{X}_0$, and the matrix ℓ_1 norm $\|\cdot\|_{\ell_1}$ is appropriate for the sparse structure of $\mathbf{Y}_0$. We further assume the side information $\alpha = \|\mathbf{Y}_0\|_{\ell_1}$. We then solve

$$\begin{array}{l} \text{minimize } \|\mathbf{X}\|_{S_1} \\ \text{subject to } \|\mathbf{Y}\|_{\ell_1} \leqslant \alpha \quad \text{and} \quad \mathbf{X} + \mathcal{R}(\mathbf{Y}) = \mathbf{Z}_0. \end{array} \tag{8.48}$$

This convex deconvolution method succeeds if $(\mathbf{X}_0, \mathbf{Y}_0)$ is the unique solution to (8.48). This problem is related to latent variable selection and robust principal component analysis [456]. □

8.10 Von Neumann Entropy Penalization and Low-Rank Matrix Estimation

Following [457], we study a problem of estimating a Hermitian nonnegatively definite matrix $\mathbf{R}$ of unit trace, e.g., a density matrix of a quantum system and a covariance matrix of a measured data. Our estimation is based on n i.i.d. measurements

$$(\mathbf{X}_1, Y_1), \ldots, (\mathbf{X}_n, Y_n), \tag{8.49}$$

where

$$Y_i = \operatorname{Tr}(\mathbf{R}\mathbf{X}_i) + W_i, \; i = 1, \ldots, n, \tag{8.50}$$

Here, $\mathbf{X}_i$, $i = 1, \ldots, n$ are random i.i.d. Hermitian matrices (or matrix-valued random variables) and W_i i.i.d. (scalar-valued) random variables with $\mathbb{E}\left(W_i | \mathbf{X}_i\right) = 0$. We consider the estimator

$$\hat{\mathbf{R}}^{\varepsilon} = \arg\min_{\mathbf{S} \in \mathbb{S}^{m \times m}} \left[\frac{1}{n} \sum_{i=1}^{n} \left(Y_i - \mathrm{Tr}\left(\mathbf{S}\mathbf{X}_i\right)\right)^2 + \varepsilon \, \mathrm{Tr}\left(\mathbf{S} \log \mathbf{S}\right) \right], \tag{8.51}$$

where $\mathbb{S}^{m \times m}$ is the set of all nonnegatively definite Hermitian $m \times m$ matrices of trace 1. The goal is to derive oracle inequalities showing how the estimation error depends on the accuracy of approximation of the unknown state $\mathbf{R}$ by low-rank matrices.

8.10.1 System Model and Formalism

Let $\mathbb{M}^{m \times m}$ be the set of all $m \times m$ matrices with complex entries. $\mathrm{Tr}(\mathbf{S})$ denotes the trace of $\mathbf{S} \in \mathbb{M}^{m \times m}$, and $\mathbf{S}^*$ denotes its adjoint matrix. Let $\mathbb{H}^{m \times m}$ be the set of all $m \times m$ Hermitian matrices with complex entries, and let

$$\mathbb{S}^{m \times m} \equiv \left\{ \mathbf{S} \in \mathbb{H}^{m \times m}\left(\mathbb{C}\right) : \mathbf{S} \geqslant 0, \mathrm{Tr}\left(\mathbf{S}\right) = 1 \right\}$$

be the set of all nonnegatively definite Hermitian matrices of trace 1. The matrices of $\mathbb{S}^{m \times m}$ can be interpreted, for instance, as *density matrices*, describing the states of a quantum system; or *covariance matrices*, describing the states of the observed phenomenon.

Let $\mathbf{X} \in \mathbb{H}^{m \times m}\left(\mathbb{C}\right)$ be a matrix (an observable) with spectral representation

$$\mathbf{X} = \sum_{i=1}^{m} \lambda_i \mathbf{P}_i, \tag{8.52}$$

where λ_i are the eigenvalues of $\mathbf{X}$ and $\mathbf{P}_i$ are its spectral projectors. Then, a matrix-valued measurement of $\mathbf{X}$ in a state of $\mathbf{R} \in \mathbf{S} \in \mathbb{M}^{\mathbf{m} \times \mathbf{m}}$ would result in outcomes λ_i with probabilities $\lambda_i = \mathrm{Tr}\left(\mathbf{R}\mathbf{P}_i\right)$ and its expectation is $\mathbb{E}_{\mathbf{R}}\mathbf{X} = \mathrm{Tr}\left(\mathbf{R}\mathbf{X}\right)$.

Let $\mathbf{X}_1, \ldots, \mathbf{X}_n \in \mathbb{H}^{m \times m}\left(\mathbb{C}\right)$ be given matrices (observables), and let $\mathbf{R} \in \mathbb{S}^{m \times m}$ be an *unknown* state of the system. A statistical problem is to estimate the unknown $\mathbf{R}$, based on the matrix-valued observations $\left(\mathbf{X}_1, Y_1\right), \ldots, \left(\mathbf{X}_n, Y_n\right)$, where $Y_1, \ldots, Y_n$ are outcomes of matrix-valued measurements of the observables $\mathbf{X}_1, \ldots, \mathbf{X}_n$ for the system identically prepared n times in the state $\mathbf{R}$. In other words, the unknown state $\mathbf{R}$ of the system is to be "learned" from a set of n linear measurements in a number of "directions" $\mathbf{X}_1, \ldots, \mathbf{X}_n$.

It is assumed that the *matrix-valued design variables* $\mathbf{X}_1, \ldots, \mathbf{X}_n$ are also random; specifically, they are i.i.d. Hermitian $m \times m$ matrices with distribution Π.

In this case, the observations $(\mathbf{X}_1, Y_1), \ldots, (\mathbf{X}_n, Y_n)$ are i.i.d., and they satisfy the following model:

$$Y_i = \operatorname{Tr}(\mathbf{R}\mathbf{X}_i) + W_i, \quad i = 1, \ldots, n, \tag{8.53}$$

where $W_i, i{=}1, \ldots, n$ are i.i.d. (scalar-valued) random variables with $\mathbb{E}(W_i|\, \mathbf{X}_i) = 0,\ i = 1, \ldots, n$.

8.10.2 Sampling from an Orthogonal Basis

The linear space of matrices $\mathbb{M}^{m\times m}(\mathbb{C})$ can be equipped with the Hilbert-Schmidt inner product,

$$\langle \mathbf{A}, \mathbf{B}\rangle = \operatorname{Tr}(\mathbf{A}\mathbf{B}^*).$$

Let $\mathbf{E}_i, i = 1, \ldots, m^2$, be an **orthonormal basis** of $\mathbb{M}^{m\times m}(\mathbb{C})$ consisting of Hermitian matrices $\mathbf{E}_i$. Let $\mathbf{X}_i, i = 1, \ldots, n$, be i.i.d. matrix-valued random variables sampled from a distribution Π on the set $\mathbf{E}_i, i = 1, \ldots, m^2$. We will refer to this model as *sampling from an orthonormal basis*.

Most often, we will use the **uniform distribution** Π that assigns probability $\frac{1}{m^2}$ to each basis matrix $\mathbf{E}_i$. In this case,

$$\mathbb{E}|\langle \mathbf{A}, \mathbf{X}\rangle|^2 = \frac{1}{m^2}\|\mathbf{A}\|_2^2,$$

where $\|\cdot\|_2 = \langle\cdot,\cdot\rangle^{1/2}$ is the Hilbert-Schmidt (or Frobenius) norm.

Example 8.10.1 (Matrix Completion). Let $\{\mathbf{e}_i : i = 1, \ldots, m\}$ the canonical basis of $\mathbb{C}^m$, where $\mathbf{e}_i$ are m-dimensional vectors. We first define

$$\begin{aligned} &\mathbf{E}_{ii} = \mathbf{e}_i \otimes \mathbf{e}_i, i = 1, \ldots, m \\ &\mathbf{E}_{ij} = \frac{1}{\sqrt{2}}(\mathbf{e}_i \otimes \mathbf{e}_j + \mathbf{e}_j \otimes \mathbf{e}_i), \quad \mathbf{E}_{ji} = \frac{1}{\sqrt{2}}(\mathbf{e}_i \otimes \mathbf{e}_j - \mathbf{e}_j \otimes \mathbf{e}_i), \end{aligned} \tag{8.54}$$

for $i < j,\ \ i, j = 1, \ldots, m$. Here $\otimes$ denotes the tensor (or Kronecker) product of vectors or matrices [16]. Then, the set of Hermitian matrices $\{\mathbf{E}_{ij} : 1 \leqslant i, j \leqslant m\}$ forms an orthogonal basis of $\mathbb{H}^{m\times m}(\mathbb{C})$. For $i < j$, the Fourier coefficients of a Hermitian matrix $\mathbf{R}$ in this basis are equal to the real and imaginary parts of the entries $\mathbf{R}_{ij}, i < j$ of matrix $\mathbf{R}$ multiplied by $\sqrt{2}$; for $i = j$, they are just the diagonal entries of $\mathbf{R}$ that are real.

If now Π is the uniform distribution in this basis, then $\mathbb{E}|\langle \mathbf{A}, \mathbf{X}\rangle|^2 = \frac{1}{m^2}\|\mathbf{A}\|_2^2$. Sampling from this distribution is equivalent to sampling at random real and imaginary parts of the entries of matrix $\mathbf{R}$. □

Example 8.10.2 (Sub-Gaussian design). A scalar-valued random variable X is called sub-Gaussian with parameter σ, if and only if, for all $\lambda \in \mathbb{R}$,

$$\mathbb{E}e^{\lambda X} \leqslant e^{\lambda^2\sigma^2/2}.$$

The inner product $\langle \mathbf{A}, \mathbf{X} \rangle$ is a sub-Gaussian scalar-valued random variables for each $\mathbf{A} \in \mathbb{H}^{m\times m}(\mathbb{C})$. This is an important model, closely related to randomized designs in compressed sensing, for which one can use powerful tools developed in the high-dimensional probability.

Let us consider two examples: Gaussian design and Rademacher design.

1. *Gaussian design*: $\mathbf{X}$ is a symmetric random matrix with real entries such that $\{X_{ij} : 1 \leqslant i \leqslant j \leqslant m\}$ are independent, centered normal random variables with

$$\mathbb{E}X_{ij}^2 = 1, i = 1, \ldots, m, \text{ and } \mathbb{E}X_{ij}^2 = \tfrac{1}{2}, i < j.$$

2. *Rademacher design*:

$$X_{ii} = \varepsilon_{ii}, i = 1, .., m, \text{ and } X_{ij} = \tfrac{1}{2}\varepsilon_{ij}, i < j,$$

   where $\varepsilon_{ij} : 1 \leq i \leq j \leq m$ are i.i.d. Rademacher random variables: random variables taking values $+1$ or -1 with probability $1/2$ each.

   In both cases, we have

$$\mathbb{E}|\langle \mathbf{A}, \mathbf{X} \rangle|^2 = \frac{1}{m^2}\left\| \mathbf{A} \right\|_2^2, \mathbf{A} \in \mathbb{H}^{m\times m}(\mathbb{C}),$$

(such matrix-valued random variables are called *isotropic*) and $\langle \mathbf{A}, \mathbf{X} \rangle$ is a sub-Gaussian random variable whose sub-Gaussian parameter is equal to $\left\| \mathbf{A} \right\|_2^2$ (up to a constant). □

8.10.3 Low-Rank Matrix Estimation

We deal with random sampling from an orthonormal basis and sub-Gaussian isotropic design such as Gaussian or Rademacher, as mentioned above. Assume, for simplicity, the noise W_i is a sequence of i.i.d. $\mathbb{N}(0, \sigma_w^2)$ random variables independent of $\mathbf{X}_1, \ldots, \mathbf{X}_n \in \mathbb{H}^{m\times m}(\mathbb{C})$ (a Gaussian noise).

We write

$$f(\mathbf{S}) = \sum_{i=1}^{m} f(\lambda_i)(\boldsymbol{\phi}_i \otimes \boldsymbol{\phi}_i)$$

for any Hermitian matrix $\mathbf{S}$ with spectral representation $\mathbf{S} = \sum_{i=1}^{m} \lambda (\phi_i \otimes \phi_i)$ and any function f defined on a set that contains the spectrum of $\mathbf{S}$. See Sect. 1.4.13.

Let us consider the case of sampling from an orthonormal basis $\{\mathbf{E}_1, \ldots, \mathbf{E}_{m^2}\}$ of $\mathbb{H}^{m\times m}(\mathbb{C})$ (that consists of Hermitian matrices). Let us call the distribution Π in $\{\mathbf{E}_1, \ldots, \mathbf{E}_{m^2}\}$ *nearly uniform* if and only there exist constants c_1, c_2 such that

$$\max_{1\leqslant i\leqslant m^2} \Pi(\{\mathbf{E}_i\}) \leqslant c_1 \text{ and } \frac{1}{m^2}\|\mathbf{A}\|_{L_2(\Pi)}^2 \geqslant c_2 \frac{1}{m^2}\|\mathbf{A}\|_2^2, \ \mathbf{A} \in \mathbb{H}^{m\times m}(\mathbb{C}).$$

Clearly the matrix completion design (Example 8.10.1) is a special case of sampling from such nearly uniform distributions.

We study the following estimator of the unknown state $\mathbf{R}$ defined a solution of a penalized empirical risk minimization problem:

$$\hat{\mathbf{R}}^{\varepsilon} = \arg\min_{\mathbf{S}\in\mathbb{S}^{m\times m}} \left[\frac{1}{n}\sum_{i=1}^{n} (Y_i - \mathrm{Tr}(\mathbf{S}\mathbf{X}_i))^2 + \varepsilon \,\mathrm{Tr}(\mathbf{S}\log\mathbf{S}) \right], \tag{8.55}$$

where ε is a regularized parameter. The penalty term is based on the function $\mathrm{Tr}(\mathbf{S}\log\mathbf{S}) = -\mathcal{E}(\mathbf{S})$, where $\mathcal{E}(\mathbf{S})$ is the *von Neumann entropy* of the state $\mathbf{S}$. Thus the method here is based on a trade-off between fitting the model by the least square in the class of all density matrices and maximizing the entropy of the state.

The optimization of (8.55) is convex: this is based on convexity of the penalty term that follows from the concavity of von Neumann entropy; see [458].

It is shown that the solution $\hat{\mathbf{R}}^{\varepsilon}$ of (8.55) is **always** a full rank matrix; see the proof of Proposition 3 of [457]. Nevertheless, when the target matrix $\mathbf{R}$ is nearly low rank, $\hat{\mathbf{R}}^{\varepsilon}$ is also well approximated by low rank matrices and the error $\left\|\mathbf{R} - \hat{\mathbf{R}}^{\varepsilon}\right\|_{L_2(\Pi)}^2$ can be controlled in terms of the "approximate rank" of $\mathbf{R}$.

Let $t > 0$ be fixed, and hence $t_m \equiv t + \log(2m)$, and $\tau_n \equiv t + \log\log_2(2n)$. To simplify the bounds, assume that $\log\log_2(2n) \leqslant \log(2m)$ (so, $\tau_n \leq \tau_m$), that $n \geqslant m t_m \log^2 m$, and finally, that $\sigma_w \geqslant \frac{1}{\sqrt{m}}$. The last condition just means that the variance of the noise is not "too small" which allows one to suppress "exponential tail term" in Bernstein-type inequalities used in the derivation of the bounds. Recall that $\mathbf{R} \in \mathbb{S}^{m\times m}$.

We state two theorems without proof.

Theorem 8.10.3 (Sampling from a nearly uniform distribution-Koltchinskii [457]). *Suppose* $\mathbf{X}$ *is sampled from a nearly uniform distribution* Π. *Then, these exists a constant* $C > 0$ *such that, for all* $\varepsilon \in [0,1]$, *with probability at least* $1 - e^{-t}$,

$$\left\|\hat{\mathbf{R}}^{\varepsilon} - \mathbf{R}\right\|_{L_2(\Pi)}^2 \leqslant C\left(\varepsilon\left(\|\log p\| \wedge \log\left(\frac{m}{\varepsilon}\right)\right) \vee \sigma_w \sqrt{\frac{m t_m}{nm}}\right). \tag{8.56}$$

In addition, for all sufficiently large $D > 0$*, these exists a constant* $C > 0$ *such that, for all* $\varepsilon \equiv D\sigma_w\sqrt{\frac{t_m}{mn}}$*, with probability at least* $1 - e^{-t}$*,*

$$\left\|\hat{\mathbf{R}}^\varepsilon - \mathbf{R}\right\|_{L_2(\Pi)}^2 \leqslant \inf_{\mathbf{S}\in\mathbb{S}^{m\times m}} \left[2\left\|\mathbf{S} - \mathbf{R}\right\|_{L_2(\Pi)}^2 + C\sigma_w^2 \vee m^{-1}\frac{\text{rank}\,(\mathbf{S})\, mt_m \log^2 (mn)}{n}\right]. \tag{8.57}$$

where $a \vee b = \max\{a, b\}$, $a \wedge b = \min\{a, b\}$.

Theorem 8.10.4 (Sub-Gaussian isotropic matrix—Koltchinskii [457]). *Suppose* $\mathbf{X}$ *is a sub-Gaussian isotropic matrix. There exist constants* $C > 0; c > 0$ *such that the following hold. Under the assumptions that* $\tau_n \leqslant cn$ *and* $t_m \leq n$*, for all* $\varepsilon \in [0, 1]$*, with probability at least* $1 - e^{-t}$

$$\left\|\hat{\mathbf{R}}^\varepsilon - \mathbf{R}\right\|_{L_2(\Pi)}^2 \leqslant C\left(\varepsilon\left(\left\|\log p\right\| \wedge \log\frac{m}{\varepsilon}\right) \vee \sigma_w\sqrt{\frac{mt_m}{n}} \vee \left(\sigma_w \vee \sqrt{m}\right)\frac{\sqrt{m}\left(\tau_n \log n \vee t_m\right)}{n}\right). \tag{8.58}$$

Moreover, there exists a constant $c > 0$ *and, for all sufficiently large* $D > 0$*, a constant* $C > 0$ *such that, for* $\varepsilon \equiv D\sigma_w\sqrt{\frac{mt_m}{n}}$*, with probability at least* $1 - e^{-t}$*,*

$$\left\|\hat{\mathbf{R}}^\varepsilon - \mathbf{R}\right\|_{L_2(\Pi)}^2 \leqslant \inf_{\mathbf{S}\in\mathbb{S}^{m\times m}} \left[2\left\|\mathbf{S} - \mathbf{R}\right\|_{L_2(\Pi)}^2 + C\left(\frac{\sigma_w^2\,\text{rank}\,(\mathbf{S})\, mt_m \log^2 (mn)}{n} \vee \frac{m\left(\tau_n \log n \vee t_m\right)}{n}\right)\right]. \tag{8.59}$$

8.10.4 Tools for Low-Rank Matrix Estimation

Let us present three tools that have been used for low-rank matrix estimation, since they are of general interest. We must bear in mind that random matrices are noncommutative, which is fundamentally different form the scalar-valued random variables.

Noncommutative Kullback-Liebler and other distances. We use noncommutative extensions of classical distances between probability distributions such as Kullback-Liebler and Hellinger distances. We use the symmetrized Kullback-Liebler distance between two states $\mathbf{S}_1, \mathbf{S}_1 \in \mathbb{S}^{m\times m}$ defined as

$$\begin{aligned} K(\mathbf{S}_1;\mathbf{S}_2) &= \mathbb{E}_{\mathbf{S}_1}(\log \mathbf{S}_1 - \log \mathbf{S}_2) + \mathbb{E}_{\mathbf{S}_2}(\log \mathbf{S}_2 - \log \mathbf{S}_1) \\ &= \mathrm{Tr}\left[(\mathbf{S}_1 - \mathbf{S}_2)(\log \mathbf{S}_1 - \log \mathbf{S}_2)\right]. \end{aligned}$$

Empirical processes bounds. Let $\mathbf{X}_1, \ldots, \mathbf{X}_n$ be i.i.d. matrix-valued random variables with common distribution. If the class of measurable functions $\mathbb{F}$ is uniformly bounded by a number Θ, then the famous Talagrand concentration inequality implies that, for all $t > 0$, with probability at least $1 - e^{-t}$,

$$\begin{aligned} &\sup_{f \in \mathcal{F}} \left| \frac{1}{n} \sum_{i=1}^{n} f(\mathbf{X}_i) - \mathbb{E}f(\mathbf{X}) \right| \\ &\leqslant 2 \left[\mathbb{E} \sup_{f \in \mathcal{F}} \left| \frac{1}{n} \sum_{i=1}^{n} f(\mathbf{X}_i) - \mathbb{E}f(\mathbf{X}) \right| + \sigma \sqrt{\frac{t}{n}} + \Theta \frac{t}{n} \right], \end{aligned}$$

where $\sigma^2 = \sup_{f \in \mathcal{F}} \mathrm{Var}(f(\mathbf{X}))$.

Noncommutative Bernstein-type Inequalities. Let $\mathbf{X}_1, \ldots, \mathbf{X}_n$ be i.i.d. Hermitian matrix-valued random variables with $\mathbb{E}\mathbf{X} = 0$ and $\sigma_X^2 = \left\|\mathbb{E}\mathbf{X}^2\right\|$. We need to study the partial sum of $\mathbf{X}_1 + \cdots + \mathbf{X}_n = \sum_{i=1}^{n} \mathbf{X}_i$. Chapter 2 gives an exhaustive treatment of this subject.

8.11 Sum of a Large Number of Convex Component Functions

The sums of random matrices can be understood, with the help of concentration of measure. A matrix may be viewed as a vector in n-dimensional space. In this section, we make the connection between the sum of random matrices and the optimization problem. We draw material from [459] for the background of incremental methods. Consider the sum of a large number of component functions $\sum_{i=1}^{N} f_i(\mathbf{x})$. The number of components N is very large. We can further consider the optimization problem

$$\begin{aligned} &\text{minimize} \sum_{i=1}^{N} f_i(\mathbf{x}) \\ &\text{subject to } \mathbf{x} \in \mathcal{X}, \end{aligned} \tag{8.60}$$

where $f_i : \mathbb{R}^n \to \mathbb{R}, i = 1, \ldots, N$, and $\mathcal{X} \in \mathbb{R}^n$. The standard Euclidean norm is defined as $\|\mathbf{x}\|_2 = \left(\mathbf{x}^T\mathbf{x}\right)^{1/2}$. There is an incentive to use incremental methods that operate on a single component $f_i(\mathbf{x})$ at each iteration, rather than on the entire cost function. If each incremental iteration tends to make reasonable progress in some "average" sense, then, depending on the value of N, an incremental method may significantly outperform (by orders of magnitude) its nonincremental counterparts. This framework provides flexibility in exploiting the special structure of f_i, including randomization in the selection of components. It is suitable for large-scale, distributed optimization—such as in Big Data [5].

Incremental subgradient methods apply to the case where the component functions f_i are convex and nondifferentiable at some points

$$\mathbf{x}_k = P_{\mathcal{X}}\left(\mathbf{x_k} - \alpha_k \nabla f_{i_k}\left(\mathbf{x}_k\right)\right)$$

where α_k is a positive stepsize, $P_{\mathcal{X}}$ denotes the projection on $\mathcal{X}$, and i_k is the index of the cost component that is iterated on.

An extension of the incremental approach to proximal algorithms is considered by Sra et al. [459] in a unified algorithmic framework that includes incremental gradient, subgradient, and proximal methods and their combinations, and highlights their common structure and behavior. The only further restriction on (8.60) is that $f_i(\mathbf{x})$ is a real-valued *convex* function. Fortunately, the convex function class includes many eigenvalue functions of $n \times n$ matrices such as the largest eigenvalue $\lambda_{\max}$ the smallest eigenvalue $\lambda_{\min}$, the first K largest eigenvalues $\sum\limits_{i=1}^{K} \lambda_i$, and the last K largest eigenvalues $\sum\limits_{i=1}^{K} \lambda_{n-i+1}$. When $K = n$, the sum is replace with the linear trace function. As a result, Chaps. 4 and 5 are relevant along the line of concentration of measure.

Some examples are given to use (8.60).

Example 8.11.1 (Sample covariance matrix estimation and geometric functional analysis). For N independent samples $\mathbf{x}_i, i = 1, \ldots, N$ of a random vector $\mathbf{x} \in \mathbb{R}^n$, a sample covariance matrix of $n \times n$ is obtained as

$$\hat{\mathbf{R}}_x = \sum_{i=1}^{N} \mathbf{x}_i \otimes \mathbf{x}_i,$$

which implies that

$$f_i\left(\mathbf{x}\right) = \mathbf{x}_i \otimes \mathbf{x}_i,$$

where $\otimes$ is the outer product of two matrices. In fact, $\mathbf{x}_i \otimes \mathbf{x}_i$ is a rank-one, positive matrix and a basic building block for low-rank matrix recovery. The so-called geometric functional analysis (Chap. 5) is explicitly connected with convex optimization through (8.60). This deep connection may be fruitful in the future.

For example, consider a connection with the approximation of a convex body by another one having a small number of contact points [460]. Let $\mathcal{K}$ be a convex body in $\mathbb{R}^n$ such that the ellipsoid of minimum volume containing it [93] is the standard Euclidean ball $\mathcal{B}_2^n$. Then by the theorem of John, there exists $N \leqslant (n+3)\,n/2$

points $\mathbf{z}_1, \ldots, \mathbf{z}_N \in \mathcal{K}$, $\|\mathbf{z}_i\|_2 = 1$ and N positive numbers $c_1, \ldots, c_N$ satisfying the following system of equations

$$\mathbf{I} = \sum_{i=1}^{N} c_i \mathbf{z}_i \otimes \mathbf{z}_i,$$

$$\mathbf{0} = \sum_{i=1}^{N} c_i \mathbf{z}_i.$$

It is established by Rudelson [93] (Theorem 5.4.2) that $N = O(n \log(n))$ is sufficient for a good approximation. By choosing

$$f_i(\mathbf{x}) = c_i \mathbf{x}_i \otimes \mathbf{x}_i,$$

in (8.60), we are able to calculate $c_1, \ldots, c_N$, by solving a convex optimization problem. Recall that a random vector $\mathbf{y}$ is in the isotropic position if

$$\mathbf{R}_y \triangleq \mathbb{E}\left[\mathbf{y}_i \otimes \mathbf{y}_i\right] = \mathbf{I},$$

where the true covariance matrix is defined as $\mathbf{R}_y$.

For a high dimensional convex body [266] $\mathcal{K} \in \mathbb{R}^n$, our algorithm formulated in terms of (8.60) brings the body into isotropic positions. □

Example 8.11.2 (Least squares and inference). An important class is the cost function $\sum_{i=1}^{m} f_i(\mathbf{x})$, where $f_i(\mathbf{x})$ is the error between some data and the output a parametric model, with $\mathbf{x}$ being the vector of parameters. (The standard Euclidean norm is defined as $\|\mathbf{x}\|_2 = \left(\mathbf{x}^T \mathbf{x}\right)^{1/2}$.) An example is linear least-squares problem, where f_i has a quadratic structure

$$\sum_{i=1}^{N} \left(\mathbf{a}_i^T \mathbf{x} - \mathbf{b}_i\right) + \gamma \|\mathbf{x} - \bar{\boldsymbol{x}}\|_2^2, \text{ s.t. } \mathbf{x} \in \mathbb{R}^n,$$

where $\bar{\boldsymbol{x}}$ is given, or nondifferentiable, as in the l_1-regulation

$$\sum_{i=1}^{N} \left(\mathbf{a}_i^T \mathbf{x} - \mathbf{b}_i\right)^2 + \gamma \sum_{j=1}^{n} |x_j|, \text{ s.t. } (x_1, \ldots, x_n) \in \mathbb{R}^n,$$

More generally, nonlinear least squares may be used

$$f_i(\mathbf{x}) = \left(h_i(\mathbf{x})\right)^2,$$

where $h_i(\mathbf{x})$ represents the difference between the i-th measurement (out of N) from a physical system and the output a parametric model whose parameter vector is $\mathbf{x}$. Another is the choice

$$f_i(\mathbf{x}) = g\left(\mathbf{a}_i^T\mathbf{x} - \mathbf{b}_i\right),$$

where g is a convex function. Still another example is maximum likelihood estimation, where f_i is a log-likelihood function of the form

$$f_i(\mathbf{x}) = -\log P_Y(\mathbf{y}_i; \mathbf{x}),$$

where $\mathbf{y}_1, \ldots, \mathbf{y}_N$ represent values of independent samples of a random vector whose distribution $P_Y(\cdot; \mathbf{x})$ depends on an unknown parameter vector $\mathbf{x} \in \mathbb{R}^n$ that one wishes to estimate. Related contexts include "incomplete" data cases, where the expectation-maximization (EM approach) is used. □

Example 8.11.3 (Minimization of an expected value: stochastic programming). Consider the minimization of an expected value

$$\begin{aligned}&\text{minimize } \mathbb{E}\left[F(\mathbf{x}, \mathbf{w})\right]\\&\text{subject to } \mathbf{x} \in \mathbb{R}^n,\end{aligned}$$

where $\mathbf{w}$ is a random variable taking a finite but very large number of values $\mathbf{w}_i, i = 1, \ldots, N$, with corresponding probabilities π_i. Then the cost function consists of the sum of the N random functions $\pi_i F(\mathbf{x}, \mathbf{w}_i)$. □

Example 8.11.4 (Distributed incremental optimization in sensor networks). Consider a network of N sensors where data are collected and used to solve some inference problem involving a parameter vector $\mathbf{x}$. If $f_i(\mathbf{x})$ represents an error penality for the data collected by the i-th sensor, then the inference problem is of the form (8.60). One approach is to use the centralized approach: to collect all the data at a fusion center. The preferable alternative is to adopt the distributed approach: to save data communication overhead and/or take advantage of parallelism in computation. In the age of Big Data, this distributed alternative is almost mandatory due to the need for storing massive amount of data.

In such an approach, the current iterate $\mathbf{x}_k$ is passed from one sensor to another, with each sensor i performing an incremental iteration improving just its local computation function f_i, and the entire cost function need be known at any one location! See [461, 462] for details. □

8.12 Phase Retrieval via Matrix Completion

Our interest in the problem of spectral factorization and phase retrieval is motivated by the pioneering work of [463, 464], where this problem at hand is connected with the recently developed machinery—matrix completion, e.g., see [102, 465–472] for the most cited papers. This connection has been first made in [473], followed by Candes et al. [463, 464]. The first experimental demonstration is made, via a LED source with 620 nm central wavelength, in [474], following a much simplified version of the approach described in [464]. Here, we mainly want to explore the mathematical techniques, rather than specific applications. We heavily rely on [463, 464] for our exposition.

8.12.1 Methodology

Let x_n be a finite-length real-valued sequence, and r_n its autocorrelation, that is,

$$r_n \triangleq \sum_k x_k x_{k-n} = (x_k * x_{n-k}), \tag{8.61}$$

where $*$ represents the discrete convolution of two finite-length sequences. The goal of spectral factorization is to recover x_n from r_n. It is more explicit in the discrete Fourier domain, that is,

$$R\left(e^{j\omega}\right) = X\left(e^{j\omega}\right) X^*\left(e^{j\omega}\right) = \left|X\left(e^{j\omega}\right)\right|^2, \tag{8.62}$$

where

$$X\left(e^{j\omega}\right) = \frac{1}{\sqrt{n}} \sum_{0 \leqslant k \leqslant n} x[k] e^{-j2\pi k/n}, \qquad \omega \in \Omega,$$

$$R\left(e^{j\omega}\right) = \frac{1}{\sqrt{n}} \sum_{0 \leqslant k \leqslant n} r[k] e^{-j2\pi k/n}, \qquad \omega \in \Omega.$$

are the discrete Fourier Transform of x_n and r_n, respectively. The task of spectral factorization is equivalent to recovering the missing phase information of $X\left(e^{j\omega}\right)$ from its squared magnitude $\left|X\left(e^{j\omega}\right)\right|^2$ [475]. This problem is often called phase retrieval in the literature [476]. Spectral factorization and phase retrieval have been extensively studied, e.g. see [476, 477] for comprehensive surveys.

Let the unknown $\mathbf{x}$ and the observed vector $\mathbf{b}$ be collected as

$$\mathbf{x} = \begin{bmatrix} x_1 \\ x_2 \\ \vdots \\ x_N \end{bmatrix} \text{ and } \mathbf{b} = \begin{bmatrix} b_1 \\ b_2 \\ \vdots \\ b_N \end{bmatrix}.$$

Suppose $\mathbf{x} \in \mathbb{C}^N$ about which we have *quadratic measurements* of the form

$$b_i = \left|\langle \mathbf{z}_i, \mathbf{x} \rangle\right|^2, \quad i = 1, 2, \ldots, N. \tag{8.63}$$

where $\langle \mathbf{c}, \mathbf{d} \rangle$ is the (scalar-valued) inner product of finite-dimensional column vectors $\mathbf{c}, \mathbf{d}$. In other words, we are given information about the squared modus of the inner product between the signal and some vectors $\mathbf{z}_i$. Our task is to find the unknown vector $\mathbf{x}$. The most important observation is to linearize the nonlinear quadratic measurements. It is well known that this can be done by *interpreting the quadric measurements as linear measurements about the unknown rank-one matrix* $\mathbf{X} = \mathbf{x}\mathbf{x}^*$. By using this "trick", we can solve a linear problem of unknown, matrix-valued random variable $\mathbf{X}$. It is remarkable that one additional dimension (from one to two dimensions) can make such a decisive difference. This trick has been systemically exploited in the context of matrix completion.

As a result of this trick, we have

$$\begin{aligned}
|\langle \mathbf{z}_i, \mathbf{x}\rangle|^2 &= \mathrm{Tr}\left(|\langle \mathbf{z}_i, \mathbf{x}\rangle|^2\right) && \text{(trace property of a scalar)}\\
&= \mathrm{Tr}\left((\mathbf{z}_i^*\mathbf{x})(\mathbf{z}_i^*\mathbf{x})^*\right) && \text{(definitions of } \langle\cdot,\cdot\rangle \text{ and } |\cdot|^2)\\
&= \mathrm{Tr}\left(\mathbf{z}_i^*\mathbf{x}\mathbf{x}^*\mathbf{z}_i\right) && \text{(property of transpose (Hermitian))}\\
&= \mathrm{Tr}\left(\mathbf{z}_i^*\mathbf{X}\mathbf{z}_i\right) && \text{(KEY: identifying } \mathbf{X} = \mathbf{x}\mathbf{x}^*)\\
&= \mathrm{Tr}\left(\mathbf{z}_i\mathbf{z}_i^*\mathbf{X}\right) && \text{(cyclical property of trace)}\\
&= \mathrm{Tr}\left(\mathbf{A}_i\mathbf{X}\right) && \text{(identifying } \mathbf{A}_i = \mathbf{z}_i\mathbf{z}_i^*).
\end{aligned} \tag{8.64}$$

The first equality follows from the fact that a trace of a scalar equals the scalar itself, that is, $\mathrm{Tr}(\alpha) = \alpha$. The second equality follows from the definition of the inner product $\langle \mathbf{c}, \mathbf{d}\rangle = \mathbf{c}^*\mathbf{d}$ and the definition of the squared modus of a scalar, that is for any complex scalar α, $|\alpha|^2 = \alpha\alpha^* = \alpha^*\alpha$. The third equality from the property of Hermitian (transpose and conjugation), that is $(\mathbf{AB})^* = \mathbf{B}^*\mathbf{A}^*$. The fourth step is critical, by identifying the rank-one matrix $\mathbf{X} = \mathbf{x}\mathbf{x}^*$. Note that $\mathbf{A}^*\mathbf{A}$ and $\mathbf{A}\mathbf{A}^*$ are always positive semidefinite, $\mathbf{A}^*\mathbf{A}, \mathbf{A}\mathbf{A}^* \geq 0$, for any matrix $\mathbf{A}$. The fifth equality follows from the famous cyclical property [17, p. 31]

$$\mathrm{Tr}(\mathbf{ABC}) = \mathrm{Tr}(\mathbf{CAB}) = \mathrm{Tr}(\mathbf{BCA}) \neq \mathrm{Tr}(\mathbf{ACB}).$$

Note that trace is a linear operator. The last step is reached by identifying another rank-one matrix $\mathbf{A}_i = \mathbf{z}_i\mathbf{z}_i^*$.

Since trace is a linear operator [17, p. 30], the phase retrieval problem, by combining (8.63) and (8.64), comes down to a linear problem of unknown, rank-one matrix $\mathbf{X}$. Let $\mathcal{A}$ be the linear operator that mappes positive semidefinite matrices into $\{\mathrm{Tr}(\mathbf{A}_i\mathbf{X}) : i = 1, \ldots, N\}$. In other words, we are given N observed pairs $\{y_i, b_i : i = 1, \ldots, N\}$, where $y_i = \mathrm{Tr}(\mathbf{A}_i\mathbf{X})$ is a scalar-valued random variable. Thus, the phase retrievable problem is equivalent to

$$\begin{array}{ll} \text{find} & \mathbf{X} \\ \text{subject to} & \mathcal{A}(\mathbf{X}) = \mathbf{b} \\ & \mathbf{X} \geqslant 0 \\ & \mathrm{rank}(\mathbf{X}) = 1 \end{array} \quad \Leftrightarrow \quad \begin{array}{ll} \text{minimize} & \mathrm{rank}(\mathbf{X}) \\ \text{subject to} & \mathcal{A}(\mathbf{X}) = \mathbf{b} \\ & \mathbf{X} \geqslant 0. \end{array} \tag{8.65}$$

After solving the lef-hand side of (8.65), we can factorize the rank-one solution $\mathbf{X}$ as $\mathbf{x}\mathbf{x}^*$. The equivalence between the left and right-hand side of (8.65) is straightforward, since, by definition, these exists one rank-one solution. This problem is a standard rank minimizing problem over an affine slice of the positive semidefinite cone. Thus, it can be solved using the recently developed machinery—low rank *matrix completion* or *matrix recovery*.

8.12.2 Matrix Recovery via Convex Programming

The rank minimization problem (8.65) is NP-hard. It is well known that the trace norm is replaced with a convex surrogate for the rank function [478, 479]. This techniques gives the familiar semi-definite programming (SDP)

$$\begin{aligned} &\text{minimize } \operatorname{Tr}(\mathbf{X}) \\ &\text{subject to } \mathcal{A}(\mathbf{X}) = \mathbf{b} \\ &\qquad\quad \mathbf{X} \geqslant 0, \end{aligned} \tag{8.66}$$

where $\mathcal{A}$ is a linear operator. This problem is convex and there exists a wide array of general purpose solvers. As far as we are concerned, the problem is solved once the problem can be formulated in terms of convex optimization. For example, in [473], (8.66) was solved by using the solver SDPT3 [480] with the interface provided by the package CVX [481]. In [474], the singular value thresholding (SVT) method [466] was used. In [463], all algorithms were implemented in MATLAB using TFOCS [482].

The trace norm promotes low-rank solution. This is the reason why it is used so often as a *convex proxy* for the rank. We can solve a sequence of weighted trace-norm problem, a technique which provides even more accurate solutions [483,484].

Choose $\varepsilon > 0$; start with $\mathbf{W}_0 = \mathbf{I}$ and for $k = 0, 1, \ldots$, inductively define $\mathbf{X}_k$ as the optimal solution to

$$\begin{aligned} &\text{minimize } \operatorname{Tr}(\mathbf{W}_k\mathbf{X}) \\ &\text{subject to } \mathcal{A}(\mathbf{X}) = \mathbf{b} \\ &\qquad\quad \mathbf{X} \geqslant 0. \end{aligned} \tag{8.67}$$

and update the 'reweight matrix' as

$$\mathbf{W}_k = (\mathbf{X}_k + \varepsilon\mathbf{I})^{-1}.$$

The algorithm terminates on convergence or when the iteration attains a specific maximum number of iterations k_{max}. The reweighting scheme [484, 485] can be viewed as attempting to solve

$$\begin{aligned} &\text{minimize } f(\mathbf{X}) = \log(\det(\mathbf{X}_k + \varepsilon\mathbf{I})) \\ &\text{subject to } \mathcal{A}(\mathbf{X}) = \mathbf{b} \\ &\qquad\quad \mathbf{X} \geqslant 0. \end{aligned} \tag{8.68}$$

by minimizing the tangent approximation to f at each iterate.

The noisy case can be solved. For more details, see [463].

In Sect. 8.10, following [457], we have studied a problem of estimating a Hermitian nonnegatively definite matrix $\mathbf{R}$ of unit trace, e.g., a density matrix of a quantum system and a covariance matrix of a measured data. Our estimation is based on n i.i.d. measurements

$$(\mathbf{X}_1, Y_1), \ldots, (\mathbf{X}_n, Y_n), \tag{8.69}$$

where

$$Y_i = \mathrm{Tr}(\mathbf{R}\mathbf{X}_i) + W_i, \; i = 1, \ldots, n. \tag{8.70}$$

By identifying $\mathbf{R} = \mathbf{X}$, $\mathbf{X}_i = \mathbf{A}_i$, and $Y_i = b_i$, our phase retrieval problem is equivalent to this problem of (8.69).

8.12.3 Phase Space Tomography

We closely follow [474] for our development. Let us consider a quasi-monochromatic light [486, Sect. 4.3.1] represented by a statistically stationary ensemble of analytic signal $V(\mathbf{r}, t)$. For any wide-sense stationary (WSS) random process, its the 'ensemble cross-correlation function' $\Gamma(\mathbf{r}_1, \mathbf{r}_2; t_1, t_2)$ is independent of the origin of time, and may be replace by the corresponding temporal cross-correlation function. This function depends on the two time arguments only through their difference $\tau = t_2 - t_1$. Thus

$$\begin{aligned}\langle \Gamma(\mathbf{r}_1, \mathbf{r}_2; \tau) \rangle &= \mathbb{E}\left[V^*(\mathbf{r}_1, t) V(\mathbf{r}_2, t+\tau)\right] = \langle V^*(\mathbf{r}_1, t) V(\mathbf{r}_2, t+\tau) \rangle \\ &= \lim_{T \to \infty} \tfrac{1}{2T} \int_{-T}^{T} V^*(\mathbf{r}_1, t) V(\mathbf{r}_2, t+\tau)\, dt.\end{aligned}$$

where $\langle \cdot \rangle$ represents the expectation value over a statistical ensemble of realizations of the random fields. The cross-correlation function $\Gamma(\mathbf{r}_1, \mathbf{r}_2; \tau)$ is known as the mutual coherence function and is the central quantity of the elementary theory of optical coherence. We set $\tau = 0$, $\Gamma(\mathbf{r}_1, \mathbf{r}_2; 0)$ is just the mutual intensity $J(\mathbf{r}_1, \mathbf{r}_2)$. The Fourier transform of $\Gamma(\mathbf{r}_1, \mathbf{r}_2; \tau)$ with respect to the delay τ is given by

$$W(\mathbf{r}_1, \mathbf{r}_2; \omega) = \int \Gamma(\mathbf{r}_1, \mathbf{r}_2; \tau) e^{-j\omega\tau} d\tau.$$

To make the formulation more transparent, we neglect the time or temporal frequency dependence by restricting our discussions to quasi-monochromatic illumination and to one-dimensional description, although these restrictions are not necessary for the following. The measurable quantity of the classical field is the intensity. The simplified quantity is called the mutual intensity and is given by

$$J(x_1, x_2) = \langle V(x_1) V^*(x_2) \rangle .$$

The measurable quantity of the classical field after propagation over distance z is [474, 486]

$$I(x_0; z) = \iint dx_1 dx_2 J(x_1, x_2) \exp\left(-\frac{j\pi}{\lambda z}\left(x_1^2 - x_2^2\right)\right) \exp\left(j2\pi \frac{x_1 - x_2}{\lambda} x_0\right). \tag{8.71}$$

It is more insightful to express this in operator form as

$$I = \mathrm{Tr}\left(P_{x_0} J\right),$$

where P_x is the free-space propagation operator that combines both the quadratic phase and Fourier transform operations in (8.71). Here x_0 is the lateral coordinate of the observation plane. Note that P is an infinite-dimensional operator. In practice, we can consider the discrete, finite-dimensional operator (matrix) to avoid some subtlety. By changing variables $x = \frac{x_1 + x_2}{2}$ and $\Delta x = x_1 - x_2$ and Fourier transforming the mutual intensity with respect to x, we obtain the Ambiguity Function

$$A(\mu, \Delta x) = \int u(x + \Delta x/2) u(x - \Delta x/2) \exp(-j2\pi\mu x)\, dx.$$

We can rewrite (8.71) as

$$\bar{I}(\mu; z) = A(\mu, \lambda z \mu),$$

where $\bar{I}(\mu; z)$ is the Fourier transform of the vector of measured intensities with respect to x_0. Thus, radial slices of the Ambiguity Function may be obtained from Fourier transforming the vectors of intensities measured at corresponding propagation distances z. From the Ambiguity Function, the mutual intensity $J(x, \Delta x)$ can be recovered by an additional inverse Fourier transform, subject to sufficient sampling,

$$J(x, \Delta x) = \int A(y, \Delta x) \exp(j2\pi x y)\, dy.$$

Let us formulate the problem into a linear model. The measured intensity data is first arranged in the Ambiguity Function space. The mutual density J is defined as the unknown to solve for. To relate the unknowns (mutual density J) to the measurements (Ambiguity Function A), the center-difference coordination-transform is first applied, which can be expressed as a linear transform $\mathcal{L}$ upon the mutual intensity J; then this process is followed by Fourier transform $\mathcal{F}$, and adding measurement noise. Formally, we have

$$A = \mathcal{F} \cdot \mathcal{L} \cdot J + e.$$

The propagation operator for mutual intensity P_x is unitary and Hermitian, since it preserves energy. The goal of low rank matrix recovery is to minimize the rank (effective number of coherent modes). We formulate the physically meaningful belief: the significant coherent modes is very few (most eigenvalues of these modes are either very small or zero).

Mathematically, if we define all the eigenvalues λ_i and the estimated mutual intensity as $\hat{J}$, the problem can be formulated as

$$\begin{aligned} &\text{minimize rank}\left(\hat{J}\right) \\ &\text{subject to } A = \mathcal{F} \cdot \mathcal{L} \cdot J, \\ &\qquad \lambda_i \geqslant 0 \quad \text{and} \quad \sum_i \lambda_i = 1. \end{aligned} \tag{8.72}$$

Direct rank minimization is NP-hard. We solve instead a proxy problem: with the rank with the "nuclear norm". The nuclear norm of a matrix is defined as the sum of singular values of the matrix. The corresponding problem is stated as

$$\begin{aligned} &\text{minimize } \left\|\hat{J}\right\|_* \\ &\text{subject to } A = F \cdot T \cdot J, \\ &\qquad \lambda_i \geqslant 0 \text{ and } \sum_i \lambda_i = 1. \end{aligned} \tag{8.73}$$

This problem is a convex optimization problem, which can be solved using general purpose solvers. In [474], the singular value thresholding (SVT) method [466] was used.

8.12.4 Self-Coherent RF Tomography

We follow our paper [487] about a novel single-step approach for self-coherent tomography for the exposition. Phase retrieval is implicitly executed.

8.12.4.1 System Model

In self-coherent tomography, we only know amplitude-only total fields and the full-data incident fields. The system model in 2-D near field configuration of self-coherent tomography can be described as follows. There are N_t transmitter sensors on the source domain with locations $\mathbf{l}_{n_t}^t$, $n_t = 1, 2, \ldots, N_t$. There are N_r receiver sensors on the measurement domain with locations $\mathbf{l}_{n_r}^r$, $n_r = 1, 2, \ldots, N_r$. The target domain Ω is discretized into a total number of N_d subareas with the center of the subarea located at $\mathbf{l}_{n_d}^d$, $n_d = 1, 2, \ldots, N_d$. The corresponding target scattering strength is τ_{n_d}, $n_d = 1, 2, \ldots, N_d$. If the n_t^{th} sensor sounds the target domain and the n_r^{th} sensor receives the amplitude-only total field, the full-data measurement equation is shown as,

$$E_{\text{tot}}\left(\mathbf{l}_{n_t}^t \to \mathbf{l}_{n_r}^r\right) = E_{\text{inc}}\left(\mathbf{l}_{n_t}^t \to \mathbf{l}_{n_r}^r\right) + E_{\text{scatter}}\left(\mathbf{l}_{n_t}^t \to \Omega \to \mathbf{l}_{n_r}^r\right) \tag{8.74}$$

where $\left|E_{\text{tot}}\left(\mathbf{l}_{n_t}^t \to \mathbf{l}_{n_r}^r\right)\right|$ is the amplitude-only total field measured by the n_r^{th} receiver sensor due to the sounding signal from the n_t^{th} transmitter sensor; $E_{\text{inc}}\left(\mathbf{l}_{n_t}^t \to \mathbf{l}_{n_r}^r\right)$ is the incident field directly from the n_t^{th} transmitter sensor to the n_r^{th} receiver sensor; $E_{\text{scatter}}\left(\mathbf{l}_{n_t}^t \to \Omega \to \mathbf{l}_{n_r}^r\right)$ is the scattered field from the target domain which can be expressed as,

$$E_{\text{scatter}}\left(\mathbf{l}_{n_t}^t \to \Omega \to \mathbf{l}_{n_r}^r\right) = \sum_{n_d=1}^{N_d} G\left(\mathbf{l}_{n_d}^d \to \mathbf{l}_{n_r}^r\right) E_{\text{tot}}\left(\mathbf{l}_{n_t}^t \to \mathbf{l}_{n_d}^d\right) \tau_{n_d} \tag{8.75}$$

In Eq. (8.75), $G\left(\mathbf{l}_{n_d}^d \to \mathbf{l}_{n_r}^r\right)$ is the wave propagation Green's function from location $\mathbf{l}_{n_d}^d$ to location $\mathbf{l}_{n_r}^r$ and $E_{\text{tot}}\left(\mathbf{l}_{n_t}^t \to \mathbf{l}_{n_d}^d\right)$ is the total field in the target subarea $\mathbf{l}_{n_d}^d$ caused by the sounding signal from the n_t^{th} transmitter sensor which can be represented as the state equation shown as

$$\begin{aligned} E_{\text{tot}}\left(\mathbf{l}_{n_t}^t \to \mathbf{l}_{n_d}^d\right) = E_{\text{inc}}\left(\mathbf{l}_{n_t}^t \to \mathbf{l}_{n_d}^d\right) + \sum_{n_{d'}=1, n_{d'} \neq n_d}^{N_d} G\left(\mathbf{l}_{n_{d'}}^d \to \mathbf{l}_{n_d}^d\right) \\ E_{\text{tot}}\left(\mathbf{l}_{n_t}^t \to \mathbf{l}_{n_{d'}}^d\right) \tau_{n_{d'}} \end{aligned} \tag{8.76}$$

Hence, the goal of self-coherent tomography is to recover τ_{n_d}, $n_d = 1, 2, \ldots, N_d$ and image the target domain Ω based on Eqs. (8.74)–(8.76).

Define $\mathbf{e}_{\text{tot},m} \in R^{N_t N_r \times 1}$ as

$$\mathbf{e}_{\text{tot},m} = \begin{bmatrix} \left|E_{\text{tot}}\left(\mathbf{l}_1^t \to \mathbf{l}_1^r\right)\right| \\ \left|E_{\text{tot}}\left(\mathbf{l}_1^t \to \mathbf{l}_2^r\right)\right| \\ \vdots \\ \left|E_{\text{tot}}\left(\mathbf{l}_1^t \to \mathbf{l}_{N_r}^r\right)\right| \\ \left|E_{\text{tot}}\left(\mathbf{l}_2^t \to \mathbf{l}_1^r\right)\right| \\ \vdots \\ \left|E_{\text{tot}}\left(\mathbf{l}_{N_t}^t \to \mathbf{l}_{N_r}^r\right)\right| \end{bmatrix}. \tag{8.77}$$

Define $\mathbf{e}_{\text{inc},m} \in C^{N_t N_r \times 1}$ as

$$\mathbf{e}_{\text{inc},m} = \begin{bmatrix} E_{\text{inc}}\left(\mathbf{l}_1^t \to \mathbf{l}_1^r\right) \\ E_{\text{inc}}\left(\mathbf{l}_1^t \to \mathbf{l}_2^r\right) \\ \vdots \\ E_{\text{inc}}\left(\mathbf{l}_1^t \to \mathbf{l}_{N_r}^r\right) \\ E_{\text{inc}}\left(\mathbf{l}_2^t \to \mathbf{l}_1^r\right) \\ \vdots \\ E_{\text{inc}}\left(\mathbf{l}_{N_t}^t \to \mathbf{l}_{N_r}^r\right) \end{bmatrix}. \tag{8.78}$$

Define $\mathbf{e}_{\text{scatter},m} \in C^{N_t N_r \times 1}$ as

$$\mathbf{e}_{\text{scatter},m} = \begin{bmatrix} \mathbf{e}_{\text{scatter},m,1} \\ \mathbf{e}_{\text{scatter},m,2} \\ \vdots \\ \mathbf{e}_{\text{scatter},m,N_t} \end{bmatrix} \tag{8.79}$$

where based on Eq. (8.75) $\mathbf{e}_{\text{scatter},m,n_t} \in C^{N_r \times 1}$ is described as,

$$\mathbf{e}_{\text{scatter},m,n_t} = \mathbf{G}_m \mathbf{E}_{\text{tot},s,n_t} \boldsymbol{\tau} \tag{8.80}$$

where $\mathbf{G}_m \in C^{N_r \times N_d}$ is defined as

$$\mathbf{G}_m = \begin{bmatrix} G\left(\mathbf{l}_1^d \to \mathbf{l}_1^r\right) & G\left(\mathbf{l}_2^d \to \mathbf{l}_1^r\right) & \cdots & G\left(\mathbf{l}_{N_d}^d \to \mathbf{l}_1^r\right) \\ G\left(\mathbf{l}_1^d \to \mathbf{l}_2^r\right) & G\left(\mathbf{l}_2^d \to \mathbf{l}_2^r\right) & \cdots & G\left(\mathbf{l}_{N_d}^d \to \mathbf{l}_2^r\right) \\ \vdots & \vdots & \vdots & \vdots \\ G\left(\mathbf{l}_1^d \to \mathbf{l}_{N_r}^r\right) & G\left(\mathbf{l}_2^d \to \mathbf{l}_{N_r}^r\right) & \cdots & G\left(\mathbf{l}_{N_d}^d \to \mathbf{l}_{N_r}^r\right) \end{bmatrix} \tag{8.81}$$

and $\boldsymbol{\tau}$ is

$$\boldsymbol{\tau} = \begin{bmatrix} \tau_1 \\ \tau_2 \\ \vdots \\ \tau_{N_d} \end{bmatrix}. \tag{8.82}$$

Besides, $\mathbf{E}_{\text{tot},s,n_t} = \text{diag}(\mathbf{e}_{\text{tot},s,n_t})$ and $\mathbf{e}_{\text{tot},s,n_t} \in C^{N_d \times 1}$ can be expressed as based on Eq. (8.76),

$$\mathbf{e}_{\text{tot},s,n_t} = \left(\mathbf{I} - \mathbf{G}_s \text{diag}(\boldsymbol{\tau})\right)^{-1} \mathbf{e}_{\text{inc},s,n_t} \tag{8.83}$$

where $\mathbf{G}_s \in C^{N_d \times N_d}$ is defined as,

$$\mathbf{G}_s = \begin{bmatrix} 0 & G\left(\mathbf{l}_2^d \to \mathbf{l}_1^d\right) & \cdots & G\left(\mathbf{l}_{N_d}^d \to \mathbf{l}_1^d\right) \\ G\left(\mathbf{l}_1^d \to \mathbf{l}_2^d\right) & 0 & \cdots & G\left(\mathbf{l}_{N_d}^d \to \mathbf{l}_2^d\right) \\ \vdots & \vdots & \vdots & \vdots \\ G\left(\mathbf{l}_1^d \to \mathbf{l}_{N_d}^d\right) & G\left(\mathbf{l}_2^d \to \mathbf{l}_{N_d}^d\right) & \cdots & 0 \end{bmatrix} \tag{8.84}$$

and $\mathbf{e}_{\text{inc},s,n_t} \in C^{N_d \times 1}$ is

$$\mathbf{e}_{\text{inc},s,n_t} = \begin{bmatrix} E_{\text{inc}}\left(\mathbf{l}_{n_t}^t \to \mathbf{l}_1^d\right) \\ E_{\text{inc}}\left(\mathbf{l}_{n_t}^t \to \mathbf{l}_2^d\right) \\ \vdots \\ E_{\text{inc}}\left(\mathbf{l}_{n_t}^t \to \mathbf{l}_{N_d}^d\right) \end{bmatrix}. \tag{8.85}$$

Define $\mathbf{E}_{\text{tot},s} \in C^{N_t N_d \times N_d}$ as

$$\mathbf{E}_{\text{tot},s} = \begin{bmatrix} \text{diag}((\mathbf{I} - \mathbf{G}_s \text{diag}(\tau))^{-1} \mathbf{e}_{\text{inc},s,1}) \\ \text{diag}((\mathbf{I} - \mathbf{G}_s \text{diag}(\tau))^{-1} \mathbf{e}_{\text{inc},s,2}) \\ \vdots \\ \text{diag}((\mathbf{I} - \mathbf{G}_s \text{diag}(\tau))^{-1} \mathbf{e}_{\text{inc},s,N_t}) \end{bmatrix}. \tag{8.86}$$

Define $\mathbf{B}_m \in C^{N_t N_r \times N_d}$ as

$$\mathbf{B}_m = \begin{bmatrix} \mathbf{G}_m & 0 & \cdots & 0 \\ 0 & \mathbf{G}_m & \cdots & 0 \\ \vdots & \vdots & \vdots & \vdots \\ 0 & 0 & \cdots & \mathbf{G}_m \end{bmatrix} \mathbf{E}_{\text{tot},s}. \tag{8.87}$$

From Eqs. (8.77) to (8.87), we can safely express $\mathbf{e}_{\text{tot},m}$ as

$$\mathbf{e}_{\text{tot},m} = \left|\mathbf{e}_{\text{inc},m} + \mathbf{B}_m \boldsymbol{\tau}\right|. \tag{8.88}$$

8.12.4.2 Mathematical Background

In a linear model for phase retrieval problem $\mathbf{y} = \mathbf{A}\mathbf{x}$ where $\mathbf{y} \in C^{M \times 1}$, $\mathbf{A} \in C^{M \times m}$, and $\mathbf{x} \in C^{m \times 1}$, only the squared magnitude of the output $\mathbf{y}$ is observed,

$$o_i = |y_i|^2 = |\mathbf{a}_i \mathbf{x}|^2, \; i = 1, 2, \ldots, M \tag{8.89}$$

where

$$\mathbf{A} = [\mathbf{a}_1^H \, \mathbf{a}_2^H \, \ldots \, \mathbf{a}_M^H]^H \tag{8.90}$$

$$\mathbf{y} = [y_1^H \, y_2^H \, \ldots \, y_M^H]^H \tag{8.91}$$

and

$$\mathbf{o} = [o_1^T \, o_2^T \, \ldots \, o_M^T]^T. \tag{8.92}$$

where H is Hermitian operator and T is transpose operator.

We assume $\{o_i, \mathbf{a}_i\}_{i=1}^M$ are known and seek $\mathbf{x}$ which is called the generalized phase retrieval problem. Derivation from Eq. (8.89) to get,

$$\begin{aligned} o_i &= \mathbf{a}_i \mathbf{x} (\mathbf{a}_i \mathbf{x})^H \\ &= \mathbf{a}_i \mathbf{x} \mathbf{x}^H \mathbf{a}_i^H \\ &= \text{trace}(\mathbf{a}_i^H \mathbf{a}_i \mathbf{x} \mathbf{x}^H) \end{aligned} \tag{8.93}$$

where trace returns the trace value of matrix. Define $\mathbf{A}_i = \mathbf{a}_i^H \mathbf{a}_i$ and $\mathbf{X} = \mathbf{x}\mathbf{x}^H$. Both $\mathbf{A}_i$ and $\mathbf{X}$ are rank-1 positive semidefinite matrices. Then,

$$o_i = \text{trace}(\mathbf{A}_i \mathbf{X}) \tag{8.94}$$

which is called semidefinite relaxation.

In order to seek $\mathbf{x}$, we can first obtain the rank-1 positive semidefinite matrix $\mathbf{X}$ which can be the solution to the following optimization problem

$$\begin{array}{l} \text{minimize} \\ \text{rank}(\mathbf{X}) \\ \text{subject to} \\ o_i = \text{trace}(\mathbf{A}_i \mathbf{X}),\ i = 1, 2, \ldots, M \\ \mathbf{X} \geq 0 \end{array} . \tag{8.95}$$

However, the rank function is not a convex function and the optimization problem (8.95) is not a convex optimization problem. Hence, the rank function is relaxed to the trace function or the nuclear norm function which is a convex function. The optimization problem (8.95) can be relaxed to an SDP,

$$\begin{array}{l} \text{minimize} \\ \text{trace}(\mathbf{X}) \\ \text{subject to} \\ o_i = \text{trace}(\mathbf{A}_i \mathbf{X}),\ i = 1, 2, \ldots, M \\ \mathbf{X} \geq 0 \end{array} \tag{8.96}$$

which can be solved by CVX which is a Matlab-based modeling system for convex optimization [488]. If the solution $\mathbf{X}$ to the optimization problem (8.96) is a rank-1 matrix, then the optimal solution $\mathbf{x}$ to the original phase retrieval problem is achieved by eigen-decomposition of $\mathbf{X}$. However, there is still a phase ambiguity problem. When the number of measurements M are fewer than necessary for a unique solution, additional assumptions are needed to select one of the solutions [489]. Motivated by compressive sensing, if we would like to seek the sparse vector $\mathbf{x}$, the objective function in SDP (8.96) can be replaced by $\text{trace}(\mathbf{X}) + \delta \|\mathbf{X}\|_1$ where $\|\cdot\|_1$ returns the l_1 norm of matrix and δ is a design parameter [489].

8.12.4.3 The Solution to Self-Coherent Tomography

Here, the solution to the linearized self-coherent tomography will be given first. Then, a novel single-step approach based on Born iterative method will be proposed to deal with self-coherent tomography with consideration of mutual multi-scattering. Distorted wave born approximation (DWBA) is used here to linearize self-coherent tomography. Specifically speaking, all the scattering within the target domain will be ignored in DWBA [490, 491]. Hence, $E_{\text{tot}}\left(\mathbf{l}_{n_t}^t \to \mathbf{l}_{n_d}^d\right)$ in Eq. (8.76) is reduced to $E_{\text{tot}}\left(\mathbf{l}_{n_t}^t \to \mathbf{l}_{n_d}^d\right) = E_{\text{inc}}\left(\mathbf{l}_{n_t}^t \to \mathbf{l}_{n_d}^d\right)$ and $\mathbf{B}_m$ in Eq. (8.87) is simplified as,

$$\mathbf{B}_m = \begin{bmatrix} \mathbf{G}_m & 0 & \cdots & 0 \\ 0 & \mathbf{G}_m & \cdots & 0 \\ \vdots & \vdots & \vdots & \vdots \\ 0 & 0 & \cdots & \mathbf{G}_m \end{bmatrix} \begin{bmatrix} \text{diag}(\mathbf{e}_{\text{inc},s,1}) \\ \text{diag}(\mathbf{e}_{\text{inc},s,2}) \\ \vdots \\ \text{diag}(\mathbf{e}_{\text{inc},s,N_t}) \end{bmatrix}. \tag{8.97}$$

In this way, $\mathbf{B}_m$ is independent of $\boldsymbol{\tau}$ and can be calculated through Green's function. The goal of the linearized self-coherent tomography is to recover $\boldsymbol{\tau}$ given $\mathbf{e}_{\text{tot},m}$, $\mathbf{e}_{\text{inc},m}$, and $\mathbf{B}_m$ based on Eq. (8.88).

Let $\mathbf{o} = \mathbf{e}_{\text{tot},m}$; $\mathbf{c} = \mathbf{e}_{\text{inc},m}$; $\mathbf{A} = \mathbf{B}_m$; and $\mathbf{x} = \boldsymbol{\tau}$. Equation (8.88) is equivalent to

$$\mathbf{o} = |\mathbf{c} + \mathbf{A}\mathbf{x}| \tag{8.98}$$

and

$$\begin{aligned} o_i &= |c_i + \mathbf{a}_i\mathbf{x}| \\ &= \text{trace}(\mathbf{A}_i\mathbf{X}) + |c_i|^2 + (\mathbf{a}_i\mathbf{x})\, c_i^* + (\mathbf{a}_i\mathbf{x})^*\, c_i \end{aligned} \tag{8.99}$$

where $*$ returns the conjugate value of the complex number. There are two unknown variables $\mathbf{X}$ and $\mathbf{x}$ in Eq. (8.99) which is different from Eq. (8.94) where there is only one unknown variable $\mathbf{X}$. In order to solve a set of non-linear equations in Eq. (8.98) to get $\mathbf{x}$, the following SDP is proposed,

$$
\begin{aligned}
&\text{minimize} \\
&\text{trace}(\mathbf{X}) + \delta \left\|\mathbf{x}\right\|_2 \\
&\text{subject to} \\
&o_i = \text{trace}(\mathbf{A}_i\mathbf{X}) + \left|c_i\right|^2 + (\mathbf{a}_i\mathbf{x})\, c_i^* + (\mathbf{a}_i\mathbf{x})^*\, c_i \\
&i = 1, 2, \ldots, N_t N_r \\
&\begin{bmatrix} \mathbf{X} & \mathbf{x} \\ \mathbf{x}^H & 1 \end{bmatrix} \geq 0;\ \mathbf{X} \geq 0
\end{aligned}
\tag{8.100}
$$

where $\left\|\cdot\right\|_2$ returns the l_2 norm of vector and δ is a design parameter. The optimization solution $\mathbf{x}$ can be achieved without phase ambiguity. Furthermore, if we know additional prior information about $\mathbf{x}$, for example, the bound of the real or imaginary part of each entry in $\mathbf{x}$, this prior information can be put into the optimization problem (8.100) as linear constraints,

$$
\begin{aligned}
&\text{minimize} \\
&\text{trace}(\mathbf{X}) + \delta \left\|\mathbf{x}\right\|_2 \\
&\text{subject to} \\
&o_i = \text{trace}(\mathbf{A}_i\mathbf{X}) + \left|c_i\right|^2 + (\mathbf{a}_i\mathbf{x})\, c_i^* + (\mathbf{a}_i\mathbf{x})^*\, c_i \\
&i = 1, 2, \ldots, N_t N_r \\
&\mathbf{b}_{\text{real}}^{\text{lower}} \leq \text{real}\,(\mathbf{x}) \leq \mathbf{b}_{\text{real}}^{\text{upper}} \\
&\mathbf{b}_{\text{imag}}^{\text{lower}} \leq \text{imag}\,(\mathbf{x}) \leq \mathbf{b}_{\text{imag}}^{\text{upper}} \\
&\begin{bmatrix} \mathbf{X} & \mathbf{x} \\ \mathbf{x}^H & 1 \end{bmatrix} \geq 0;\ \mathbf{X} \geq 0
\end{aligned}
\tag{8.101}
$$

where real returns the real part of the complex number and imag returns the imaginary part of the complex number. $\mathbf{b}_{\text{real}}^{\text{lower}}$ and $\mathbf{b}_{\text{real}}^{\text{upper}}$ are the lower and upper bounds of the real part of $\mathbf{x}$, respectively. Similarly, $\mathbf{b}_{\text{real}}^{\text{lower}}$ and $\mathbf{b}_{\text{real}}^{\text{upper}}$ are the lower and upper bounds of the imaginary part of $\mathbf{x}$, respectively.

If mutual multi-scattering is considered, we have to solve Eq. (8.88) to obtain $\boldsymbol{\tau}$, i.e., $\mathbf{x}$. The novel single-step approach based on Born iterative method will be proposed as follows:

1. Set $\boldsymbol{\tau}^{(0)}$ to be zero; $t = -1$;
2. $t = t + 1$; get $\mathbf{B}_m^{(t)}$ based on Eqs. (8.87) and (8.86) using $\boldsymbol{\tau}^{(t)}$;
3. Solve the inverse problem in Eq. (8.98) by the following SDP using $\mathbf{B}_m^{(t)}$ to get $\boldsymbol{\tau}^{(t+1)}$

$$\begin{aligned}
&\text{minimize} \\
&\text{trace}(\mathbf{X}) + \delta_1 \left\|\mathbf{x}\right\|_2 + \delta_2 \left\|\mathbf{o} - \mathbf{u}\right\|_2 \\
&\text{subject to} \\
&u_i = \text{trace}(\mathbf{A}_i\mathbf{X}) + \left|c_i\right|^2 + (\mathbf{a}_i\mathbf{x})\, c_i^* + (\mathbf{a}_i\mathbf{x})^*\, c_i \\
&i = 1, 2, \ldots, N_t N_r \\
&\mathbf{b}_{\text{real}}^{\text{lower}} \leq \text{real}\,(\mathbf{x}) \leq \mathbf{b}_{\text{real}}^{\text{upper}} \\
&\mathbf{b}_{\text{imag}}^{\text{lower}} \leq \text{imag}\,(\mathbf{x}) \leq \mathbf{b}_{\text{imag}}^{\text{upper}} \\
&\begin{bmatrix} \mathbf{X} & \mathbf{x} \\ \mathbf{x}^H & 1 \end{bmatrix} \geq 0;\ \mathbf{X} \geq 0
\end{aligned} \tag{8.102}$$

where the definitions of $\mathbf{o}$ and $\mathbf{u}$ can be referred to Eq. (8.92);

4. If $\boldsymbol{\tau}$ converges, the approach is stopped; otherwise the approach goes to step 2.

8.13 Further Comments

Noisy low-rank matrix completion with general sampling distribution was studied by Klopp [492]. Concentration-based guarantees are studied by Foygel et al. [493], Foygel and Srebro [494], and Koltchinskii and Rangel [495]. The paper [496] introduces a penalized matrix estimation procedure aiming at solutions which are sparse and low-rank at the same time.

Related work to phase retrieval includes [463, 464, 473, 474, 497–502]. In particular, robust phase retrieval for sparse signals [503].

Chapter 9
Covariance Matrix Estimation in High Dimensions

Statistical structures start with covariance matrices. In practice, we must estimate the covariance matrix from the big data. One may think this chapter should be more basic than Chaps. 7 and 8—thus should be treated earlier chapters. Recent work on compressed sensing and low-rank matrix recovery supports the idea that sparsity can be exploited for statistical estimation, too. The treatment of this subject is very superficial, due to the limited space. This chapter is mainly developed to support the detection theory in Chap. 10.

9.1 Big Picture: Sense, Communicate, Compute, and Control

The nonasymptotic point of view [108] may turn out to be relevant when the number of observations is large. It is to fit large complex sets of data that one needs to deal with possibly huge collections of models at different scales. This approach allows the collections of models together with their dimensions to vary freely, letting the dimensions be possibly of the same order of magnitude as the number of observations. Concentration inequalities are the probabilistic tools that we need to develop a nonasymptotic theory.

A hybrid, large-scale cognitive radio network (CRN) testbed consisting of 100 hybrid nodes: 84 USRP2 nodes and 16 WARP nodes, as shown in Fig. 9.1, is deployed at Tennessee Technological University. In each node, non-contiguous orthogonal frequency division multiplexing (NC-OFDM) waveforms are agile and programmable, as shown in Fig. 9.2, due to the use of software defined radios; such waveforms are ideal for the convergence of communications and sensing. The network can work in two different modes: sense and communicate. They can even work in a hybrid mode: communicating while sensing. From sensing point of view, this network is an active wireless sensor network. Consequentially, many analytical tools can be borrowed from wireless sensor network; on the other hand, there is a fundamental difference between oursensing problems and the traditional

R. Qiu and M. Wicks, *Cognitive Networked Sensing and Big Data*, DOI 10.1007/978-1-4614-4544-9_9,

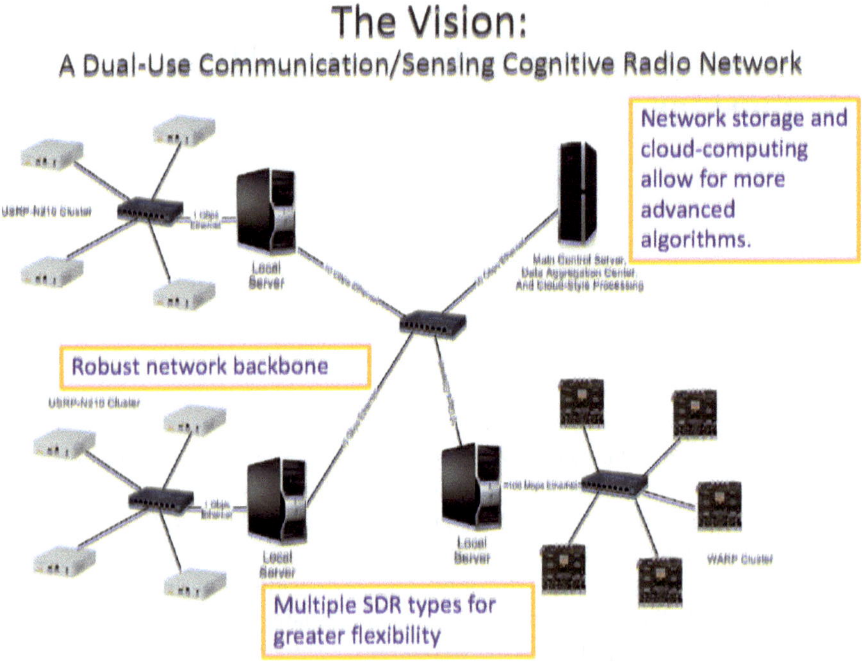

Fig. 9.1 A large-scale cognitive radio network is deployed at Tennessee Technological University, as an experimental testbed on campus. A hybrid network consisting of 80 USRP2 nodes and 16 WARP nodes. The ultimate goal is to demonstrate the big picture: sense, communicate, compute, and control

wireless sensor network. The main difference derives from the nature of the SDR and dynamical spectrum access (DSA) for a cognitive radio. The large-scale CRN testbed has received little attention in the literature.

With the vision of the big picture: sense, communicate, compute and control, we deal with the Big Data. A fundamental problem is to determine what information needs to be stored locally and what information to be communicated in a real-time manner. The communications data rates ultimately determine how the computing is distributed among the whole network. It is impossible to solve this problem analytically, since the answer depends on applications. The expertise of this network will enables us to develop better ways to approach this problem. At this point, through a heuristic approach, we assume that only the covariance matrix of the data is measured at each node and will be communicated in real time. More specifically, at each USRP2 or WARP node, only the covariance matrix of the data are communicated across the network in real time, at a data rate of say 1 Mbs. These (sensing) nodes record the data much faster than the communications speed. For example, a data rate of 20 Mbps can be supported by using USRP2 nodes.

The problem is sometimes called wireless distributed computing. Our view emphasizes the convergence of sensing and communications. Distributed (parallel) computing is needed to support all kinds of applications in mind, with a purpose of control across the network.

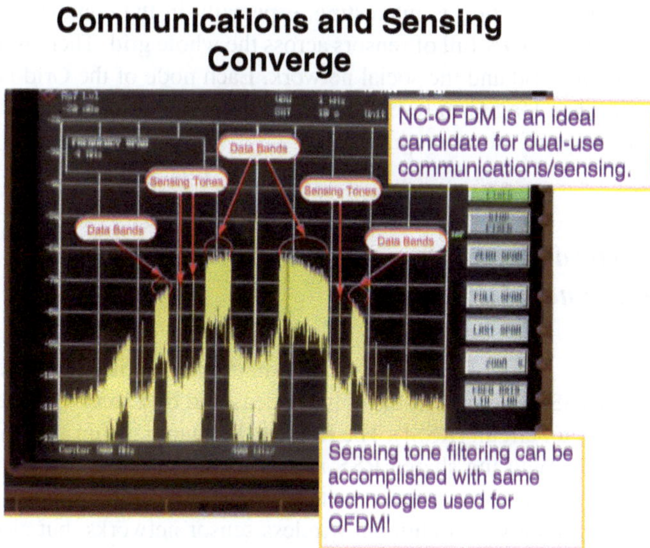

Fig. 9.2 The non-contiguous orthogonal frequency division multiplexing (NC-OFDM) waveforms are suitable for both communications and sensing. The agile, programmable waveforms are made available by software defined radios (SDR)

To support the above vision, we distill the following mathematical problems:

1. High dimensional data processing. We focus on high-dimensional data processing.One can infer dependent structures among variables by estimating the associated covariance matrices. Sample covariance matrices are most commonly used.
2. Data fusing. A sample covariance matrix is a random matrix. As a result, a sum of random matrices is a fundamental mathematical problem.
3. Estimation and detection. Intrusion/activity detection can be enabled. Estimation of network parameters is possible.
4. Machine learning. Machine learning algorithms can be distributed across the network.

These mathematical problems are of special interest, in the context of social networks. The data and/or the estimated information can be shared within the social networks. The concept is very remote at the writing of this monograph, it is our belief that the rich information contained in the radio waveforms will make a difference when integrated into the social networks. The applications are almost of no limit.

Cameras capture the information of optical fields (signals), while the SDR nodes sense the environment using the radio frequency (RF). A multi-spectral approach consists of sensors of a broad electromagnetic wave spectrum, even another physical signal: acoustic sensors.

The vision of this section is interesting, especially in the context of the Smart Grid that is a huge network full of sensors across the whole grid. There is an analogy between the Smart Grid and the social network. Each node of the Grid is an agent. This connection is a long term research topic. The in-depth treatment of this topic is beyond this scope of this monograph.

9.1.1 Received Signal Strength (RSS) and Applications to Anomaly Detection

Sensing across a network of mobiles (such as smart phones) is emerging. Received signal strength (RSS) is defined as the voltage measured by a receiver's received signal strength indicator circuit (RSSI). The RSS can be shared within a social network, such as Facebook. With the shared RSS across such a network, we can sense the radio environment. Let us use an example to illustrate this concept. This concept can be implemented not only in traditional wireless sensor networks, but also wireless communications network. For example, Wi-Fi nodes and cognitive radio nodes can be used to form such a "sensor" network. The big picture is "sense, communicate, compute and control."

An real-world application of using RSS, Chen, Wiesel and Hero [504] demonstrates the proposed robust covariance estimator in a real application: activity/intrusion detection using an active wireless sensor network. They show that the measured data exhibit strong non-Gaussian behavior.

The experiment was set up on an Mica2 sensor network platform, which consists of 14 sensor nodes randomly deployed inside and outside a laboratory at the University of Michigan. Wireless sensors communicated with each other *asynchronously* by **broadcasting an RF signal every 0.5 seconds**. The received signal strength was recorded for each pair of transmitting and receiving nodes. There were pairs of RSSI measurements over a 30-min period, and samples were acquired every 0.5 s. During the experiment period, persons walked into and out of the lab at random times, causing anomaly patterns in the RSSI measurements. Finally, for ground truth, a Web camera was employed to record the actual activity.

9.1.2 NC-OFDM Waveforms and Applications to Anomaly Detection

The OFDM modulation waveforms can be measured for spectrum sensing in a cognitive radio network. Then these waveforms data can be stored locally for further processing. The first step is to estimate covariance from these stored data. Our sampling rate is about 20 mega samples per second (Msps), in contrast with 2 samples per second in Sect. 9.1.1 for received signal strength indicator circuit (RSSI). The difference is seven orders of magnitude. This fundamental difference asks for a different approach. This difference is one basic motivation for writing this book.

For more details on the network testbed, see Chap. 13.

9.2 Covariance Matrix Estimation

Estimating a covariance matrix (or a dispersion matrix) is a fundamental problem in statistical signal processing. Many techniques for detection and estimation rely on accurate estimation of the true covariance. In recent years, estimating a high dimensional $p \times p$ covariance matrix under small sample size n has attracted considerable attention. In these large p, small n problems, the classical sample covariance suffers from a systematically distorted eigenstructure [383], and improved estimators are required.

9.2.1 Classical Covariance Estimation

Consider a random vector

$$\mathbf{x} = (X_1, X_2, \ldots, X_p)^H,$$

where H denotes the Hermitian of a matrix. Let $\mathbf{x}_1, \ldots, \mathbf{x}_n$ be independent random vectors that follow the same distribution as $\mathbf{x}$. For simplicity, we assume that the distribution has zero mean: $\mathbb{E}\mathbf{x} = \mathbf{0}$. The covariance matrix $\mathbf{\Sigma}$ is the $p \times p$ matrix that tabulates the second-order statistics of the distribution:

$$\mathbf{\Sigma} = \mathbb{E}\left(\mathbf{x}\mathbf{x}^H\right). \tag{9.1}$$

The classical estimator for the covariance matrix is the sample covariance matrix

$$\hat{\mathbf{\Sigma}}_n = \frac{1}{n}\sum_{i=1}^{n} \mathbf{x}_i \mathbf{x}_i^H. \tag{9.2}$$

The sample covariance matrix is an unbiased estimator of the covariance matrix: $\mathbb{E}\hat{\mathbf{\Sigma}} = \mathbf{\Sigma}$.

Given a tolerance $\varepsilon \in (0,1)$, we can study how many samples n are typically required to provide an estimate with relative error ε in the spectral norm:

$$\mathbb{E}\left\|\hat{\mathbf{\Sigma}} - \mathbf{\Sigma}\right\| \leqslant \varepsilon \left\|\mathbf{\Sigma}\right\|. \tag{9.3}$$

where $\|\mathbf{A}\|$ is the l_2 norm. The symbol $\|\cdot\|_q$ refers to the Schatten q-norm of a matrix.:

$$\|\mathbf{A}\|_q \triangleq \left[\mathrm{Tr}\left|\mathbf{A}\right|^q\right]^{1/q}$$

where $|\mathbf{A}| = \left(\mathbf{A}^H\mathbf{A}\right)^{1/2}$. This type of spectral-norm error bound defined in (9.3) is quite powerful. It limits the magnitude of the estimator error for each entry of the covariance matrix; it even controls the error in estimating the eigenvalues of the covariance using the eigenvalues of the sample covariance.

Unfortunately, the error bound (9.3) for the sample covariance estimator demands a lot of samples. Typical positive results state that the sample covariance matrix estimator is precise when the number of samples is proportional to the number of variables, provided that the distribution decays fast enough. For example, assuming that $\mathbf{x}$ follows a normal distribution:

$$n \geqslant C\varepsilon^{-2}p \Rightarrow \left\|\hat{\mathbf{\Sigma}} - \mathbf{\Sigma}\right\| \leqslant \varepsilon \left\|\mathbf{\Sigma}\right\| \text{ with high probability,} \tag{9.4}$$

where C is an absolute constant.

We are often interested in the largest and smallest eigenvalues of the empirical covariance matrix of sub-Gaussian random vectors: sums of random vector outer products. We present a version of [113]. This result (with non-explicit constants) was originally obtained by Litvak et al. [338] and Vershynin [72].

Theorem 9.2.1 (Sums of random vector outer products [72, 113, 338]). *Let* $\mathbf{x}_1, \ldots, \mathbf{x}_N$ *be random vectors in* $\mathbb{R}^n$ *such that, for some* $\gamma \geq 0$,

$$\begin{aligned} &\mathbb{E}\left[\mathbf{x}_i\mathbf{x}_i^T \left|\mathbf{x}_1, \ldots, \mathbf{x}_{i-1}\right.\right] = \mathbf{I} \quad \textit{and} \\ &\mathbb{E}\left[\exp\left(\boldsymbol{\alpha}\mathbf{x}_i^T\right) \left|\mathbf{x}_1, \ldots, \mathbf{x}_{i-1}\right.\right] \leqslant \exp\left(\left\|\boldsymbol{\alpha}\right\|^2 \gamma/2\right) \textit{ for all } \boldsymbol{\alpha} \in \mathbb{R}^n \end{aligned}$$

for all $i = 1, \ldots, N$, *almost surely. For all* $\varepsilon \in (0, 1/2)$ *and* $\delta \in (0, 1)$,

$$\mathbb{P}\left[\lambda_{\max}\left(\frac{1}{N}\sum_{i=1}^N \mathbf{x}_i\mathbf{x}_i^T\right) > 1 + \frac{1}{1-2\varepsilon}\cdot C_{\varepsilon,\delta,N} \quad \text{or} \quad \lambda_{\min}\left(\frac{1}{N}\sum_{i=1}^N \mathbf{x}_i\mathbf{x}_i^T\right) < 1 - \frac{1}{1-2\varepsilon}\cdot C_{\varepsilon,\delta,N}\right] \leqslant \delta$$

where

$$C_{\varepsilon,\delta,N} = \gamma\cdot\left(\sqrt{\frac{32\left(N\log\left(1+2/\varepsilon\right)\right) + \log\left(2/\delta\right)}{N}} + \frac{2\left(N\log\left(1+2/\varepsilon\right) + \log\left(2/\delta\right)\right)}{N}\right).$$

The sub-Gaussian property most readily lends itself to bounds on linear combinations of sub-Gaussian random variables. However, the outer products are in certain quadratic combinations. We bootstrap from the bound for linear combinations to bound the moment generating function of the quadratic combinations. From there, we get the desired tail bound.

For a (scalar-valued) non-negative random variable W. For any $\beta \in \mathbb{R}$, we have

$$\mathbb{E}\left[\exp\left(\beta W\right)\right] - \beta\mathbb{E}\left[W\right] - 1 = \beta\int_0^\infty \left(\exp\left(\beta t\right) - 1\right)\cdot\mathbb{P}\left[W > t\right]\cdot dt. \tag{9.5}$$

The claim follows using integration-by-parts.

Theorem 9.2.2 (Sums of random vector outer products (quadratic form) [113]). *Let $\mathbf{x}_1, \ldots, \mathbf{x}_N$ be random vectors in $\mathbb{R}^n$ such that, for some $\gamma \geq 0$,*

$$\begin{aligned}&\mathbb{E}\left[\mathbf{x}_i\mathbf{x}_i^T \left|\mathbf{x}_1, \ldots, \mathbf{x}_{i-1}\right.\right] = \mathbf{I} \quad \textit{and}\\&\mathbb{E}\left[\exp\left(\boldsymbol{\alpha}\mathbf{x}_i^T\right) \left|\mathbf{x}_1, \ldots, \mathbf{x}_{i-1}\right.\right] \leqslant \exp\left(\left\|\boldsymbol{\alpha}\right\|^2 \gamma/2\right) \textit{ for all } \boldsymbol{\alpha} \in \mathbb{R}^n\end{aligned}$$

for all $i = 1, \ldots, N$, almost surely. For all $\boldsymbol{\alpha} \in \mathbb{R}^n$ such that $\|\boldsymbol{\alpha}\| = 1$ and all $\delta \in (0,1)$,

$$\begin{aligned}&\mathbb{P}\left[\boldsymbol{\alpha}^T\left(\frac{1}{N}\sum_{i=1}^N \mathbf{x}_i\mathbf{x}_i^T\right)\boldsymbol{\alpha} > 1 + \sqrt{\frac{32\gamma^2\log(1/\delta)}{N}} + \frac{2\gamma\log(1/\delta)}{N}\right] \leqslant \delta \textit{ and}\\&\mathbb{P}\left[\boldsymbol{\alpha}^T\left(\frac{1}{N}\sum_{i=1}^N \mathbf{x}_i\mathbf{x}_i^T\right)\boldsymbol{\alpha} < 1 - \sqrt{\frac{32\gamma^2\log(1/\delta)}{N}}\right] \leqslant \delta.\end{aligned}$$

We see [113] for a proof, based on (9.5). With this theorem, we can bound the smallest and largest eigenvalues of the empirical covariance matrix, when we apply the bound for the Rayleigh quotient (quadratic form) in the above theorem, together with a covering argument from Pisier [152].

9.2.2 Masked Sample Covariance Matrix

One way to circumvent the problem of covariance estimation in large p, small n is to assume that the covariance matrix is nearly sparse and to focus on estimating only the significant entries [130, 505]. A formalism called masked covariance estimation is introduced here. This approach uses a mask, constructed *a prior*, to specify the importance we place on each entry of the covariance matrix. By re-weighting the sample covariance matrix estimate using a mask, we can reduce the error that arises from imprecise estimates of covariances that are small or zero. The mask matrix formalism was first introduced by Levina and Vershynin [505].

Modern applications often involve a small number of samples and a large number of variables. The paucity of data make it impossible to obtain an accurate estimator of a general covariance matrix. As a sequence, we must frame additional model assumptions and develop estimators that exploit this extra structure. A number of papers have focused on the situation where the covariance matrix is sparse or nearly so. Thus we limit our attention to the significant entries of the covariance matrix and thereby perform more accurate estimation with fewer samples.

Our analysis follows closely that of [130], using matrix concentration inequalities that are suitable for studying a sum of independent random matrices (Sect. 2.2). Indeed, matrix concentration inequalities can be viewed as a far-reaching extensions of the classical inequalities for a sum of scalar random variables (Sect. 1.4.10). Matrix concentration inequalities sometimes allow us to replace devilishly hard calculations with simple arithmetic. These inequalities streamline the analysis of random matrices. We believe that the simplicity of the arguments and the strength of the conclusions make a compelling case for the value of these methods. We hope matrix concentration inequalities will find a place in the toolkit of researchers working on multivariate problems in statistics.

In the regime n $\ll$ p, were we have very few samples, we cannot hope to achieve an estimate like (9.3) for a general covariance matrix. Instead, we must instate additional assumptions and incorporate this prior information to construct a regularized estimator.

One way to formalize this idea is to construct a symmetric $p \times p$ matrix $\mathbf{M}$ with real entries, which we call the mask matrix. In the simplest case, the mask matrix has 0–1 values that indicate which entries of the covariance matrix we attend to. A unit entry $m_{ij} = 1$ means that we estimate the interaction between the ith and jth variables, while a zero entry $m_{ij} = 0$ means that we ignore their interaction when making estimation. More generally, we allow the entries of the mask matrix to vary over the interval $[0, 1]$, in which case the relative values of m_{ij} is proportional to the importance of estimating the (i, j) entry of the covariance matrix.

Given a mask $\mathbf{M}$, we define the masked sample covariance matrix estimator

$$\mathbf{M} \odot \hat{\mathbf{\Sigma}},$$

where the symbol $\odot$ denotes the component-wise (i.e., Schur or Hadamard) product. The following expression bounds the root-mean-square spectral-norm error that this estimator incurs:

$$\left[\mathbb{E}\left\|\mathbf{M} \odot \hat{\mathbf{\Sigma}} - \mathbf{\Sigma}\right\|^2\right]^{1/2} \leqslant \underbrace{\left[\mathbb{E}\left\|\mathbf{M} \odot \hat{\mathbf{\Sigma}} - \mathbf{M} \odot \mathbf{\Sigma}\right\|^2\right]^{1/2}}_{\text{variance}} + \underbrace{\left[\mathbb{E}\|\mathbf{M} \odot \mathbf{\Sigma} - \mathbf{\Sigma}\|^2\right]^{1/2}}_{\text{bias}}. \tag{9.6}$$

This bound is analogous to the classical bias-variance decomposition for the mean-squared-error (MSE) of a point estimator. To obtain an effective estimator, we must design a mask that controls both the bias and the variance in (9.6). We cannot neglect too many components of the covariance matrix, or else the bias in the masked estimator may compromise its accuracy. On the other hand, each additional component we add in our estimator contributes to the size of the variance term. In the case where the covariance matrix is sparse, it is natural to strike a balance between these two effects by refusing to estimate entries of the covariance matrix that we know *a prior* to be small or zero.

For a stationary random process, the covariance matrix is Toeplitz. A Toeplitz matrix or diagonal-constant matrix, named after Otto Toeplitz, is a matrix in which each descending diagonal from left to right is constant. For instance, the following matrix is a Toeplitz matrix

$$\mathbf{\Sigma}_n = \left(\gamma_{i-j}\right)_{1 \leqslant i,j \leqslant n}, \tag{9.7}$$

Example 9.2.3 (The Banded Estimator of a Decaying Matrix). Let us consider the example where entries of the covariance matrix $\boldsymbol{\Sigma}$ decay away from the diagonal.

Suppose that, for a fixed parameter $\alpha > 1$,

$$\left|(\boldsymbol{\Sigma})_{ij}\right| \leqslant |i - j + 1|^{-\alpha} \text{for each pair } (i, j) \text{ of indices.}$$

This type of property may hold for a random process whose correlation are localized in time. Related structure arises from random fields that have short spatial correlation scales.

A simple (suboptimal) approach to this covariance estimation problem is to focus on a band of entries near the diagonal. Suppose that the bandwidth $B = 2b+1$ for a nonnegative integer b. For example, a mask with bandwidth $B = 3$ for an ensemble of $p = 5$ variables takes the form

$$\mathbf{M}_{band} = \begin{bmatrix} 1 & 1 & & & \\ 1 & 1 & 1 & & \\ & 1 & 1 & 1 & \\ & & 1 & 1 & 1 \\ & & & 1 & 1 \end{bmatrix}.$$

In this setting, it is easy to compute the bias term in (9.6). Indeed,

$$\left|(\mathbf{M} \odot \boldsymbol{\Sigma} - \boldsymbol{\Sigma})_{ij}\right| \leqslant \begin{cases} |i - j + 1|^{-\alpha}, & |i - j| > b \\ 0, & \text{otherwise.} \end{cases}$$

Gershgorin's theorem [187, Sect. 6.1] implies that the spectral norm of a symmetric matrix is dominated by the maximum l_1 norm of a column, so

$$\|\mathbf{M} \odot \boldsymbol{\Sigma} - \boldsymbol{\Sigma}\| \leqslant 2 \sum_{k>b} (k+1)^{-\alpha} \leqslant \frac{2}{\alpha - 1} (b+1)^{1-\alpha}.$$

The second inequality follows when we compare with the sum with an integral. A similar calculation shows

$$\|\boldsymbol{\Sigma}\| \leqslant 1 + 2(\alpha - 1)^{-1}.$$

Assuming the covariance matrix really does have constant spectral norm, it follows that

$$\|\mathbf{M} \odot \boldsymbol{\Sigma} - \boldsymbol{\Sigma}\| \leqslant B^{1-\alpha} \|\boldsymbol{\Sigma}\|.$$

□

9.2.2.1 Masked Covariance Estimation for Multivariate Normal Distributions

The main result of using masked covariance estimation is presented here. The norm $\|\cdot\|_\infty$ returns the maximum absolute entry of a vector, but we use a separate notation $\|\cdot\|_{max}$ for the maximum absolute entry of a matrix. We also require the norm

$$\|\mathbf{A}\|_{1\to 2}^2 \triangleq \max_j \left(\sum_i |a_{ij}|^2 \right)^{1/2}.$$

The notation reflects the fact that this is the natural norm for linear maps from l_1 into l_2.

Theorem 9.2.4 (Chen and Tropp [130]). *Fix a $p \times p$ symmetric mask matrix* $\mathbf{M}$, *where $p \geq 3$. Suppose that* $\mathbf{x}$ *is a Gaussian random vector in $\mathbb{R}^p$ with mean zero. Define the covariance matrix* $\boldsymbol{\Sigma}$ *and* $\hat{\boldsymbol{\Sigma}}$ *in* (9.1) *and* (9.2). *Then the variance of the masked sample covariance estimator satisfies*

$$\left[\mathbb{E}\left\|\mathbf{M}\odot\hat{\boldsymbol{\Sigma}}-\mathbf{M}\odot\boldsymbol{\Sigma}\right\|^2\right]^{1/2} \leqslant C\left[\left(\frac{\|\boldsymbol{\Sigma}\|_{\max}}{\|\boldsymbol{\Sigma}\|}\frac{\|\boldsymbol{\Sigma}\|_{1\to 2}^2\log p}{n}\right)^{1/2}+\frac{\|\boldsymbol{\Sigma}\|_{\max}}{\|\boldsymbol{\Sigma}\|}\frac{\|\mathbf{M}\|\log p\cdot\log(np)}{n}\right]\|\boldsymbol{\Sigma}\|. \tag{9.8}$$

9.2.2.2 Two Complexity Metrics of a Mask Design

In this subsection, we use masks that take 0–1 values to gain intuition. Our analysis uses two separate metrics that quantify the complexity of the mask. The first complexity is the square of the maximum column norm:

$$\|\mathbf{M}\|_{1\to 2}^2 \triangleq \max_j \left(\sum_i |m_{ij}|^2 \right)^{1/2}.$$

Roughly, the bracket counts the number of interactions we want to estimate that involve the variable j, and the maximum computes a bound over all p variables. This metric is "local" in nature. The second complexity metric is the spectral norm $\|\mathbf{M}\|$ of the mask matrix, which provides a more "global" view of the complexity of the interactions that we estimate.

Let us use some examples to illustrate. First, suppose we estimate the entire covariance matrix so the mask is the matrix of ones:

$$\mathbf{M} = \text{matrix of ones} \Rightarrow \|\boldsymbol{\Sigma}\|_{1\to 2}^2 = p \text{ and } \|\mathbf{M}\| = p.$$

Next, consider the mask that arises from the banded estimator in Example 9.2.3:

$$\mathbf{M} = 0 - 1 \text{ matrix, bandwidth B} \Rightarrow \|\boldsymbol{\Sigma}\|_{1\to 2}^2 \leqslant \text{B and } \|\mathbf{M}\| \leqslant B,$$

since there are at most B ones in each row and column. When $B \ll p$, the banded matrix asks us to estimate fewer interactions than the full mask, so we expect the estimation problem to be much easier.

9.2.2.3 Covariance Matrix for Wide-Sense Stationary (WSS)

A random process is wide-sense stationary (WSS) if its mean is constant for all time indices (i.e., independent of time) and its autocorrelation depends on only the time index difference. WSS discrete random process $x[n]$ is statistically characterized by a constant mean

$$\bar{x}[n] = \bar{x},$$

and an autocorrelation sequence

$$r_{xx}[m] = \mathbb{E}\left\{x[n+m]x^*[n]\right\},$$

where $*$ denotes the complex conjugate. The terms "correlation" and "covariance" are often used synonymously in the literature, but formally identical only for zero-mean processes. The covariance matrix

$$\mathbf{R}_M = \begin{bmatrix} r_{xx}[0] & r_{xx}^*[1] & \cdots & r_{xx}^*[M] \\ r_{xx}[1] & r_{xx}[0] & \cdots & r_{xx}^*[M-1] \\ \vdots & \vdots & \ddots & \vdots \\ r_{xx}[M] & r_{xx}[M-1] & \cdots & r_{xx}[0] \end{bmatrix}$$

is a Hermitian Toeplitz autocorrelation matrix of order M, and, therefore, has dimension $(M+1) \times (M+1)$. Then, the quadratic form

$$\mathbf{a}^H \mathbf{R}_{xx} \mathbf{a} = \sum_{m=0}^{M} \sum_{n=0}^{M} a[m] a^*[n] r_{xx}[m-n] \geqslant 0 \tag{9.9}$$

must be positive semi-definite (or non-negative) for any arbitrary $M \times 1$ vector $\mathbf{a}$ if $r_{xx}[m]$ is a valid autocorrelation sequence. From (9.9), it follows that the covariance matrix $\mathbf{R}_M$ is positive semi-definite, implying all its eigenvalues must be non-negative:

$$\lambda_i\left(\mathbf{R}_M\right) \geq 0, \quad i = 1, \ldots, M.$$

This property is fundamental for the covariance matrix estimation.

9.2.2.4 Signal Plus Noise Model

As mentioned above, in the case where the covariance matrix is sparse, it is natural to strike a balance between these two effects by refusing to estimate entries of the covariance matrix that we know *a prior* to be small or zero. Let us illustrate this point by using an example that is crucial to high-dimensional data processing.

Example 9.2.5 (A Sum of Sinusoids in White Gaussian Noise [506]). Let us sample the continuous-time signal at an sampling interval T_s. If there are L real sinusoids

$$x[n] = \sum_{l=1}^{L} A_l \sin\left(2\pi f_l n T_s + \theta_l\right),$$

each of which has a phase that is uniformly distributed on the interval 0 to 2π, independent of the other phases, then the mean of the L sinusoids is zero and the autocorrelation sequence is

$$r_{xx}[m] = \sum_{l=1}^{L} \frac{A_l^2}{2} \cos\left(2\pi f_l m T_s\right).$$

If the process consists of L complex sinusoids

$$x[n] = \sum_{l=1}^{L} A_l \exp\left[j\left(2\pi f_l n T_s + \theta_l\right)\right],$$

then the autocorrelation sequence is

$$r_{xx}[m] = \sum_{l=1}^{L} A_l^2 \exp\left(j 2\pi f_l m T_s\right).$$

A white Gaussian noise is uncorrelated with itself for all lags, except at $m = 0$, for which the variance is σ^2. The autocorrelation sequence is

$$r_{ww}[m] = \sigma^2 \delta[m],$$

which is a constant for all frequencies, justifying the name *white noise*. The covariance matrix is

$$\mathbf{R}_{ww} = \sigma^2 \mathbf{I} = \sigma^2 \begin{bmatrix} 1 & & 0 \\ & \ddots & \\ 0 & & 1 \end{bmatrix}. \tag{9.10}$$

If an independent white noise process $w[n]$ is added to the complex sinusoids with random phases, then the combined process

$$y[n] = x[n] + w[n]$$

will have an autocorrelation sequence

$$\begin{aligned} r_{yy}[m] &= r_{xx}[m] + r_{ww}[m] \\ &= \sum_{l=1}^{L} A_l^2 \exp\left(j2\pi f_l m T_s\right) + \sigma^2 \delta[m]. \end{aligned} \tag{9.11}$$

Equation (9.11) can be rewritten as

$$\mathbf{R}_{yy} = \mathbf{R}_{xx} + \mathbf{R}_{ww} = \sum_{l=1}^{L} A_l^2 \mathbf{v}_M\left(f_l\right) \mathbf{v}_M^H\left(f_l\right) + \sigma^2 \mathbf{I}, \tag{9.12}$$

where $\mathbf{I}$ is an $(M+1) \times (M+1)$ identity matrix and

$$\mathbf{v}_M\left(f_l\right) = \begin{bmatrix} 1 \\ \exp\left(j2\pi f_l T_s\right) \\ \vdots \\ \exp\left(j2\pi f_l m T_s\right) \end{bmatrix}$$

is a complex sinusoidal vector at frequency f_l.

The impact of additive white Gaussian noise $w[n]$ on the signal is, according to (9.12), through only diagonals since $\mathbf{R}_{ww} = \sigma^2 \mathbf{I}$. This is an ideal model that is not valid for the "large p, small n" problem: p variables and n data samples—in this case the sample covariance matrix $\hat{\mathbf{R}}_{ww}$, the most commonly encountered estimate of $\mathbf{R}_{ww}$, is a positive semi-definite random matrix, that is a dense matrix of full rank. In other words, $\hat{\mathbf{R}}_{ww}$ is far away from the ideal covariance matrix $\sigma^2 \mathbf{I}$. This observation has far-reaching impact since the ideal covariance matrix $\mathbf{R}_{ww} = \sigma^2 \mathbf{I}$ is a sparse matrix but the sample covariance matrix $\hat{\mathbf{R}}_{ww}$ is not sparse at all. □

Example 9.2.6 (Tridiagonal Toeplitz Matrix [506]). An $n \times n$ tridiagonal Toeplitz matrix $\mathbf{T}$ has the form

$$\mathbf{T} = \begin{bmatrix} b & a & 0 & \cdots & 0 \\ a & b & a & \ddots & \vdots \\ 0 & a & \ddots & \ddots & 0 \\ \vdots & \ddots & \ddots & \ddots & a \\ 0 & \cdots & 0 & a & b \end{bmatrix},$$

where a and b are constants. The eigenvalues of $\mathbf{T}$ are

$$\lambda_k = a + 2b\cos\left(\frac{k\pi}{n} + 1\right), \quad k = 1, \ldots, n,$$

and the corresponding eigenvectors are

$$\mathbf{v}_k = \sqrt{\frac{2}{n+1}} \begin{pmatrix} \sin\left(\frac{k\pi}{n+1}\right) \\ \vdots \\ \sin\left(\frac{kn\pi}{n+1}\right) \end{pmatrix}.$$

If an additive Gaussian white noise is added to the signal whose covariance matrix is $\mathbf{T}$, then the resultant noisy signal has a covariance matrix

$$\mathbf{R}_{yy} = \mathbf{R}_{xx} + \mathbf{R}_{ww} = \mathbf{T} + \sigma^2\mathbf{I} =$$

$$= \begin{bmatrix} b & a & 0 & \cdots & 0 \\ a & b & a & \ddots & \vdots \\ 0 & a & \ddots & \ddots & 0 \\ \vdots & \ddots & \ddots & \ddots & a \\ 0 & \cdots & 0 & a & b \end{bmatrix} + \sigma^2 \begin{bmatrix} 1 & 0 & 0 & \cdots & 0 \\ 0 & 1 & 0 & \ddots & \vdots \\ 0 & 0 & \ddots & \ddots & 0 \\ \vdots & \ddots & \ddots & \ddots & 0 \\ 0 & \cdots & 0 & 0 & 1 \end{bmatrix}$$

$$= \begin{bmatrix} b+\sigma^2 & a & 0 & \cdots & 0 \\ a & b+\sigma^2 & a & \ddots & \vdots \\ 0 & a & \ddots & \ddots & 0 \\ \vdots & \ddots & \ddots & \ddots & a \\ 0 & \cdots & 0 & a & b+\sigma^2 \end{bmatrix}.$$

Only diagonal entries are affected by the ideal covariance matrix of the noise. □

A better model for modeling the noise is

$$\hat{\mathbf{R}}_{ww} = \begin{bmatrix} 1 & \rho & \rho & \cdots & \rho \\ \rho & 1 & \rho & \ddots & \vdots \\ \rho & \rho & \ddots & \ddots & \rho \\ \vdots & \ddots & \ddots & \ddots & \rho \\ \rho & \cdots & \rho & \rho & 1 \end{bmatrix},$$

where the correlation coefficient $\rho \leq 1$ is typically small, for example, $\rho = 0.01$. A general model for the Gaussian noise is

$$\hat{\mathbf{R}}_{ww} = \sigma^2 \begin{bmatrix} 1+\rho_{11} & \rho_{12} & \rho_{13} & \cdots & \rho_{1M} \\ \rho_{21} & 1+\rho_{22} & \rho_{23} & \cdots & \rho_{2M} \\ \rho_{31} & \rho_{32} & 1+\rho_{33} & \cdots & \rho_{3M} \\ \vdots & \vdots & \vdots & \vdots & \vdots \\ \rho_{M1} & \rho_{M2} & \rho_{M3} & \cdots & 1+\rho_{MM} \end{bmatrix},$$

where $\rho_{ij} \leq 1, i, j = 1, \ldots, M$ are random variables of the same order, e.g. 0.01. The accumulation effect of the weak random variables ρ_{ij} will have a decisive influence on the performance of the covariance estimation. This covariance matrix is dense and of full rank but positive semi-definite. It is difficult to enforce the Toeplitz structure on the estimated covariance matrix.

When a random noise vector is added to a random signal vector

$$\mathbf{y} = \mathbf{x} + \mathbf{w},$$

where it is assumed that the random signal and the random noise are independent, it follows [507] that

$$\mathbf{R}_{yy} = \mathbf{R}_{xx} + \mathbf{R}_{ww} \tag{9.13}$$

The difficulty arises from the fact that $\mathbf{R}_{xx}$ is unknown. Our task at hand is to separate the two matrices—a matrix separation problem. Our problems will be greatly simplified if some special structures of these three matrices can be exploited!!! Two special structures are important: (1) $\mathbf{R}_{xx}$ is of low rank; (2) $\mathbf{R}_{xx}$ is sparse.

In a real world, we are given the date to estimate the covariance matrix of the noisy signal $\mathbf{R}_{yy}$

$$\hat{\mathbf{R}}_{yy} = \hat{\mathbf{R}}_{xx} + \hat{\mathbf{R}}_{ww}, \tag{9.14}$$

where we have used the assumption that the random signal and the random noise are independent—which is reasonable for most covariance matrix estimators in mind.

Two estimators $\hat{\mathbf{R}}_{xx}, \hat{\mathbf{R}}_{ww}$ are required. It is very critical to remember that their difficulties are fundamentally different; two different estimators must be used. The basic reason is that the signal subspace and the noise subspace are different—even though we cannot always separate the two subspaces using tools such as singular value decomposition or eigenvalue decomposition.

Example 9.2.7 (Sample Covariance Matrix). The classical estimator for the covariance matrix is the sample covariance matrix defined in (9.2) and repeated here for convenient:

$$\hat{\mathbf{\Sigma}} = \frac{1}{n} \sum_{i=1}^{n} \mathbf{x}_i \mathbf{x}_i^H. \tag{9.15}$$

Using (9.15) for $\mathbf{y} = \mathbf{x} + \mathbf{w}$, it follows that

$$\begin{aligned}
\hat{\mathbf{R}}_{yy} &= \tfrac{1}{n}\sum_{i=1}^{n} \mathbf{y}_i\mathbf{y}_i^H = \tfrac{1}{n}\sum_{i=1}^{n} (\mathbf{x}_i + \mathbf{w}_i)(\mathbf{x}_i + \mathbf{w}_i)^H \\
&= \tfrac{1}{n}\sum_{i=1}^{n} \mathbf{x}_i\mathbf{x}_i^H + \tfrac{1}{n}\sum_{i=1}^{n} \mathbf{w}_i\mathbf{w}_i^H + \underbrace{\frac{1}{n}\sum_{i=1}^{n} \mathbf{x}_i\mathbf{w}_i^H}_{\to 0, n\to\infty} + \underbrace{\frac{1}{n}\sum_{i=1}^{n} \mathbf{w}_i\mathbf{x}_i^H}_{\to 0, n\to\infty} \\
&\quad \text{(zero mean random vectors)} \\
&= \tfrac{1}{n}\sum_{i=1}^{n} \mathbf{x}_i\mathbf{x}_i^H + \tfrac{1}{n}\sum_{i=1}^{n} \mathbf{w}_i\mathbf{w}_i^H, n \to \infty \\
&= \hat{\mathbf{R}}_{xx} + \hat{\mathbf{R}}_{ww}
\end{aligned} \tag{9.16}$$

Our ideal equation is the following:

$$\mathbf{R}_{yy} = \mathbf{R}_{xx} + \mathbf{R}_{ww}. \tag{9.17}$$

In what conditions does (9.16) approximate (9.17) with high accuracy? The asymptotic process in the derivation of (9.16) hides the difficulty of data processing. We need to make the asymptotic process explicit via feasible algorithms. In other words, we require a non-asymptotic theory for high-dimensional processing. Indeed, n approaches a very large value but finite! (say $n = N = 10^5$) For $\varepsilon \in (0, 1)$, we require that

$$\left\| \hat{\mathbf{R}}_{xx} - \mathbf{R}_{xx} \right\| \leqslant \varepsilon \left\| \mathbf{R}_{xx} \right\| \text{ and } \left\| \hat{\mathbf{R}}_{ww} - \mathbf{R}_{ww} \right\| \leqslant \varepsilon \left\| \mathbf{R}_{ww} \right\|.$$

To achieve the same accuracy ε, the sample size $n = N_x$ required for the signal covariance estimator $\hat{\mathbf{R}}_{xx}$ is much less than $n = N_w$ required for the noise covariance estimator $\hat{\mathbf{R}}_{ww}$. This observation is very critical in data processing.

For a given n, how close does $\hat{\mathbf{R}}_{xx}$ become to $\mathbf{R}_{xx}$? For a given n, how close does $\hat{\mathbf{R}}_{ww}$ become to $\mathbf{R}_{ww}$?

Let $\mathbf{A}, \mathbf{B} \in \mathbb{C}^{N\times N}$ be Hermitian matrices. Then [16]

$$\lambda_i(\mathbf{A}) + \lambda_N(\mathbf{B}) \leqslant \lambda_i(\mathbf{A} + \mathbf{B}) \leqslant \lambda_i(\mathbf{A}) + \lambda_1(\mathbf{B}). \tag{9.18}$$

Using (9.18), we have

$$\lambda_i\left(\hat{\mathbf{R}}_{xx}\right) + \lambda_M\left(\hat{\mathbf{R}}_{ww}\right) \leqslant \lambda_i\left(\hat{\mathbf{R}}_{yy}\right) = \lambda_i\left(\hat{\mathbf{R}}_{xx} + \hat{\mathbf{R}}_{ww}\right) \leqslant \lambda_i\left(\hat{\mathbf{R}}_{xx}\right) + \lambda_1\left(\hat{\mathbf{R}}_{ww}\right) \tag{9.19}$$

for $i = 1, \ldots, M$. All these covariance matrices are the positive semi-definite matrices that have non-negative eigenvalues. If $\mathbf{R}_{xx}$ is of low rank, for $q \ll M$, only the first q dominant eigenvalues are of interest. Using (9.19) q times and summing both sides of these q inequalities yield

$$\sum_{i=1}^{q}\lambda_i\left(\hat{\mathbf{R}}_{xx}\right)+q\lambda_M\left(\hat{\mathbf{R}}_{ww}\right)\leqslant\sum_{i=1}^{q}\lambda_i\left(\hat{\mathbf{R}}_{xx}+\hat{\mathbf{R}}_{ww}\right)\leqslant\sum_{i=1}^{q}\lambda_i\left(\hat{\mathbf{R}}_{xx}\right)+q\lambda_1\left(\hat{\mathbf{R}}_{ww}\right). \tag{9.20}$$

For $q = 1$, we have

$$\lambda_1\left(\hat{\mathbf{R}}_{xx}\right)+\lambda_M\left(\hat{\mathbf{R}}_{ww}\right)\leqslant\lambda_1\left(\hat{\mathbf{R}}_{xx}+\hat{\mathbf{R}}_{ww}\right)\leqslant\lambda_1\left(\hat{\mathbf{R}}_{xx}\right)+\lambda_1\left(\hat{\mathbf{R}}_{ww}\right),$$

For $q = M$, it follows that

$$\begin{aligned}\operatorname{Tr}\left(\hat{\mathbf{R}}_{xx}\right)+M\lambda_M\left(\hat{\mathbf{R}}_{ww}\right)&\leqslant\operatorname{Tr}\left(\hat{\mathbf{R}}_{yy}\right)=\operatorname{Tr}\left(\hat{\mathbf{R}}_{xx}+\hat{\mathbf{R}}_{ww}\right)\\&\leqslant\operatorname{Tr}\left(\hat{\mathbf{R}}_{xx}\right)+M\lambda_1\left(\hat{\mathbf{R}}_{ww}\right)\end{aligned} \tag{9.21}$$

where we have used the standard linear algebra identity:

$$\operatorname{Tr}\left(\mathbf{A}\right)=\sum_{i=1}^{n}\lambda_i\left(\mathbf{A}\right).$$

More generally

$$\operatorname{Tr}\left(\mathbf{A}^k\right)=\sum_{i=1}^{n}\lambda_i^k\left(\mathbf{A}\right),\mathbf{A}\in\mathbb{C}^{n\times n},k\in\mathbb{N}.$$

In particular, if $k = 2, 4, \ldots$ is an even integer, then $\operatorname{Tr}\left(\mathbf{A}^k\right)^{1/k}$ is just the l^k norm of these eigenvalues, and we have [9, p. 115]

$$\|\mathbf{A}\|_{op}^k\leqslant\operatorname{Tr}\left(\mathbf{A}^k\right)\leqslant n\,\|\mathbf{A}\|_{op}^k,$$

where $\|\cdot\|_{op}$ is the operator norm.

All eigenvalues we deal with here are non-negative since the sample covariance matrix defined in (9.15) is non-negative. The eigenvalues, their sum, and the trace of a random matrix are scalar-valued random variables. The expectation $\mathbb{E}$ of these scalar-valued random variables can be considered. Since expectation and trace are both linear, they commute [38, 91]:

$$\mathbb{E}\operatorname{Tr}\left(\mathbf{A}\right)=\operatorname{Tr}\left(\mathbb{E}\mathbf{A}\right). \tag{9.22}$$

Taking the expectation of it, we have

$$\begin{aligned}\mathbb{E}\operatorname{Tr}\left(\hat{\mathbf{R}}_{xx}\right)+M\mathbb{E}\lambda_M\left(\hat{\mathbf{R}}_{ww}\right)&\leqslant\mathbb{E}\operatorname{Tr}\left(\hat{\mathbf{R}}_{yy}\right)=\mathbb{E}\operatorname{Tr}\left(\hat{\mathbf{R}}_{xx}+\hat{\mathbf{R}}_{ww}\right)\\&\leqslant\mathbb{E}\operatorname{Tr}\left(\hat{\mathbf{R}}_{xx}\right)+M\mathbb{E}\lambda_1\left(\hat{\mathbf{R}}_{ww}\right)\end{aligned} \tag{9.23}$$

and, with the aid of (9.22),

$$\begin{aligned}&\operatorname{Tr}\left(\mathbb{E}\hat{\mathbf{R}}_{xx}\right)+M\mathbb{E}\lambda_M\left(\hat{\mathbf{R}}_{ww}\right)\leqslant \operatorname{Tr}\left(\mathbb{E}\hat{\mathbf{R}}_{yy}\right)\\ &=\operatorname{Tr}\left(\mathbb{E}\hat{\mathbf{R}}_{xx}+\mathbb{E}\hat{\mathbf{R}}_{ww}\right)\leqslant \operatorname{Tr}\left(\mathbb{E}\hat{\mathbf{R}}_{xx}\right)+M\mathbb{E}\lambda_1\left(\hat{\mathbf{R}}_{ww}\right).\end{aligned} \tag{9.24}$$

Obviously, $\mathbb{E}\lambda_M\left(\hat{\mathbf{R}}_{ww}\right)$ and $\mathbb{E}\lambda_1\left(\hat{\mathbf{R}}_{ww}\right)$ are non-negative scalar values, since $\lambda_i\left(\hat{\mathbf{R}}_{ww}\right)\geqslant 0, i=1,\ldots,M$.

We are really concerned with

$$\left|\lambda_1\left(\sum_{k=1}^{K}\mathbf{R}_{yy,k}\right)-\lambda_1\left(\sum_{k=1}^{K}\hat{\mathbf{R}}_{yy,k}\right)\right|\leqslant \varepsilon\lambda_1\left(\sum_{k=1}^{K}\mathbf{R}_{yy,k}\right). \tag{9.25}$$

Sample covariance matrices are random matrices. In analogy with a sum of independent scalar-valued random variables, we can consider a sum of matrix-valued random variables. Instead of considering the sum, we can consider the expectation. □

9.2.3 *Covariance Matrix Estimation for Stationary Time Series*

We follow [508]. For a stationary random process, the covariance matrix is Toeplitz. A Toeplitz matrix or diagonal-constant matrix, named after Otto Toeplitz, is a matrix in which each descending diagonal from left to right is constant. For instance, the following matrix is a Toeplitz matrix

$$\mathbf{\Sigma}_n=\left(\gamma_{i-j}\right)_{1\leqslant i,j\leqslant n}, \tag{9.26}$$

A thresholded covariance matrix estimator can better characterize sparsity if the true covariance matrix is sparse. Toeplitzs connection of eigenvalues of matrices and Fourier transforms of their entries is used. The thresholded sample covariance matrix is defined as

$$\hat{\mathbf{\Sigma}}_{T,A_T}=\left(\gamma_{s-t}\mathbf{1}_{|\gamma_{s-t}|\geqslant A_T}\right)_{1\leqslant s,t\leqslant T}$$

for $A_T=2c\sqrt{\log T/T}$ where c is a constant. The diagonal elements are never thresholded. The thresholded estimate may not be positive definite.

In the context of time series, the observations have an intrinsic temporal order and we expect that observations are weakly dependent if they are far apart, so banding seems to be natural. However, if there are many zeros or very weak correlations within the band, the banding method does not automatically generate a sparse estimate.

9.3 Covariance Matrix Estimation

$\|\cdot\|$ is the operator norm and $\|\cdot\|_2$ the Euclidean norm in $\mathbb{R}^n$. For N copies of a random vector $\mathbf{x}$, the sample covariance matrix is defined as

$$\hat{\mathbf{\Sigma}}_N = \frac{1}{N}\sum_{i=1}^{N} \mathbf{x}_i \otimes \mathbf{x}_i.$$

Theorem 9.3.1 ([310]). *Consider independent, isotropic random vectors $\mathbf{x}_i$ valued in $\mathbb{R}^n$. Assume that $\mathbf{x}_i$ satisfy the strong regularity assumption: for some $C_0, \eta > 0$, one has*

$$\mathbb{P}\left\{\|\mathbf{P}\mathbf{x}_i\|_2^2 > t\right\} \leqslant C_0 t^{-1-\eta} \textit{ for } t > C_0 \operatorname{rank}(\mathbf{P}) \tag{9.27}$$

for every orthogonal projection $\mathbf{P}$ in $\mathbb{R}^n$. Then, for $\varepsilon \in (0,1)$ and for

$$N \geqslant C\varepsilon^{-2-2/\eta} \cdot n$$

one has

$$\mathbb{E}\left\|\frac{1}{N}\sum_{i=1}^{N} \mathbf{x}_i \otimes \mathbf{x}_i - \mathbf{I}\right\| \leqslant \varepsilon. \tag{9.28}$$

Here, $C = 512(48C_0)^{2+2/\eta}(6+6/\eta)^{1+4/\eta}$.

Corollary 9.3.2 (Covariance estimation [310]). *Consider a random vector $\mathbf{x}$ valued in $\mathbb{R}^n$ with covariance matrix $\mathbf{\Sigma}$. Assume that: for some $C_0, \eta > 0$, the isotropic random vector $\mathbf{z} = \mathbf{\Sigma}^{-1/2}\mathbf{x}$ satisfies*

$$\mathbb{P}\left\{\|\mathbf{P}\mathbf{x}_i\|_2^2 > t\right\} \leqslant C_0 t^{-1-\eta} \textit{ for } t > C_0 \operatorname{rank}(\mathbf{P}) \tag{9.29}$$

for every orthogonal projection $\mathbf{P}$ in $\mathbb{R}^n$. Then, for $\varepsilon \in (0,1)$ and for

$$N \geqslant C\varepsilon^{-2-2/\eta} \cdot n$$

the sample covariance matrix $\hat{\mathbf{\Sigma}}_N$ obtained from N independent copies of $\mathbf{x}$ satisfies

$$\mathbb{E}\left\|\hat{\mathbf{\Sigma}}_N - \mathbf{\Sigma}\right\| \leqslant \varepsilon \left\|\mathbf{\Sigma}\right\|. \tag{9.30}$$

Theorem 9.3.1 says that, for sufficiently large N, all eigenvalues of the sample covariance matrix $\hat{\mathbf{\Sigma}}_N$ are concentrated near 1. This following corollary extends to a result that holds for all N.

Corollary 9.3.3 (Extreme eigenvalues [310]). *Let n, N be arbitrary positive integers, suppose $\mathbf{x}_i$ are N independent, isotropic random vectors satisfying* (9.27), *and let $y = n/N$. Then the sample covariance matrix $\hat{\boldsymbol{\Sigma}}_N = \frac{1}{N}\sum_{i=1}^{N} \mathbf{x}_i \otimes \mathbf{x}_i$ satisfies*

$$1 - C_1 y^c \leqslant \mathbb{E}\lambda_{\min}\left(\hat{\boldsymbol{\Sigma}}_N\right) \leqslant \mathbb{E}\lambda_{\max}\left(\hat{\boldsymbol{\Sigma}}_N\right) \leqslant 1 + C_1\left(y + y^c\right). \tag{9.31}$$

Here $c = \frac{\eta}{2\eta+2}$, $C_1 = 512(16C_0)^{2+2/\eta}(6+6/\eta)^{1+4/\eta}$, and $\lambda_{\min}\left(\hat{\boldsymbol{\Sigma}}_N\right)$, $\lambda_{\max}\left(\hat{\boldsymbol{\Sigma}}_N\right)$ denote the smallest and the largest eigenvalues of $\hat{\boldsymbol{\Sigma}}_N$, respectively.

It is sufficient to assume $2 + \eta$ moments for one-dimensional marginals rather than for marginals in all dimensions. This is only slightly stronger than the isotropy assumption, which fixes the second moments of one-dimensional marginals.

Corollary 9.3.4 (Smallest Eigenalue [310]). *Consider N independent isotropic random vectors $\mathbf{x}_i$ valued in $\mathbb{R}^n$. Assume that $\mathbf{x}_i$ satisfy the weak regularity assumption: for some $C_0, \eta > 0$,*

$$\sup_{\|\mathbf{y}\|_2 \leqslant 1} \left|\langle \mathbf{x}_i, \mathbf{y}\rangle\right|^{2+\eta} \leqslant C_0 \tag{9.32}$$

Then, for $\varepsilon > 0$ and for

$$N \geqslant C\varepsilon^{-2-2/\eta} \cdot n,$$

the minimum eigenvalue of the sample covariance matrix $\hat{\boldsymbol{\Sigma}}_N$ satisfies

$$\mathbb{E}\lambda_{\min}\left(\hat{\boldsymbol{\Sigma}}_N\right) \geqslant 1 - \varepsilon.$$

Here $C = 40(10C_0)^{2/\eta}$.

9.4 Partial Estimation of Covariance Matrix

Theorem 9.4.1 (Estimation of Hadamard products [505]). *Let M be an arbitrary fixed symmetric $n \times n$ matrix. Then*

$$\mathbb{E}\left\|\mathrm{M} \cdot \hat{\boldsymbol{\Sigma}}_N - \mathrm{M} \cdot \boldsymbol{\Sigma}\right\| \leqslant C\log^3(2n)\left(\frac{\|\mathbf{M}\|_{1,2}}{\sqrt{N}} + \frac{\|\mathbf{M}\|}{N}\right)\|\boldsymbol{\Sigma}\|. \tag{9.33}$$

Here M does not depend on $\hat{\boldsymbol{\Sigma}}_N$ and $\boldsymbol{\Sigma}$.

Corollary 9.4.2 (Partial estimation [505]). *Let* M *be an arbitrary fixed symmetric* $n \times n$ *matrix such that all of the entries are equal to 0 or 1, and there are at most* k *nonzero entries in each column. Then*

$$\mathbb{E}\left\|\mathbf{M}\cdot\hat{\mathbf{\Sigma}}_N - \mathbf{M}\cdot\mathbf{\Sigma}\right\| \leqslant C\log^3(2n)\left(\frac{\sqrt{k}}{\sqrt{N}}+\frac{k}{N}\right)\|\mathbf{\Sigma}\|. \tag{9.34}$$

Proof. We note that $\|\mathbf{M}\|_{1,2} \leqslant \sqrt{k}$ and $\|\mathbf{M}\| \leqslant k$ and apply Theorem 9.4.1. □

Corollary 9.4.2 implies that for every $\varepsilon \in (0,1)$, the sample size

$$N \geqslant 4C^2\varepsilon^{-2}k\log^6(2n) \ \text{ suffices for } \mathbb{E}\left\|\mathbf{M}\cdot\hat{\mathbf{\Sigma}}_N - \mathbf{M}\cdot\mathbf{\Sigma}\right\| \leqslant \varepsilon\|\mathbf{\Sigma}\|. \tag{9.35}$$

For sparse matrices $\mathbf{M}$ with $k \ll n$, this makes partial estimation possible with $N \ll n$ observations. Therefore, (9.35) is a satisfactory "sparse" version of the classical bound such as given in Corollary 9.29.

Identifying the non-zero entries of $\mathbf{\Sigma}$ by thresholding. If we assume that all non-zero entries in $\mathbf{\Sigma}$ are bounded away from zero by a margin of $h > 0$, then a sample size of

$$N \gtrsim h^{-2}\log(2n)$$

would assure that all their locations are estimated correctly with probability approaching 1. With this assumption, we could derive a bound for the thresholded estimator.

Example 9.4.3 (Thresholded estimator). An $n \times n$ tridiagonal Toeplitz matrix has the form

$$\mathbf{\Sigma} = \begin{bmatrix} b & a & 0 & \cdots & 0 \\ a & b & a & \ddots & 0 \\ 0 & a & \ddots & \ddots & 0 \\ \vdots & \vdots & \ddots & \ddots & a \\ 0 & 0 & \cdots & a & b \end{bmatrix}_{n\times n}.$$

$$\mathbf{R}_{ww} = \begin{bmatrix} 1 & 0 & 0 & \cdots & 0 \\ 0 & 1 & 0 & \ddots & 0 \\ 0 & 0 & \ddots & \ddots & 0 \\ \vdots & \vdots & \ddots & \ddots & 0 \\ 0 & 0 & \cdots & 0 & 1 \end{bmatrix}_{n\times n} + \begin{bmatrix} h & h & h & \cdots & h \\ h & h & h & \ddots & h \\ h & h & \ddots & \ddots & h \\ \vdots & \vdots & \ddots & \ddots & h \\ h & h & \cdots & h & h \end{bmatrix}_{n\times n}.$$

All non-zero entries in $\mathbf{\Sigma}$ are bounded away from zero by a margin of $h > 0$. □

9.5 Covariance Matrix Estimation in Infinite-Dimensional Data

In the context of kernel principal component analysis of high (or infinite) dimensional data, covariance matrix estimation is relevant to Big Data. Let $||\mathbf{A}||_2$ denote the spectral norm of matrix $\mathbf{A}$. If $\mathbf{A}$ is symmetric, then $||\mathbf{A}||_2 = \max\{\lambda_{\max}(\mathbf{A}), -\lambda_{\min}(\mathbf{A})\}$, where $\lambda_{\max}(\mathbf{A})$ and $\lambda_{\min}(\mathbf{A})$ are, respectively, the largest and smallest eigenvalue of $\mathbf{A}$.

Example 9.5.1 (Infinite-dimensional data [113]). Let $\mathbf{x}_1, \ldots, \mathbf{x}_N$ be i.i.d. random vectors with their true covariance matrix $\mathbf{\Sigma} = \left[\mathbf{x}_i\mathbf{x}_i^T\right]$, $\mathbf{K} = \mathbb{E}\left[\mathbf{x}_i\mathbf{x}_i^T\mathbf{x}_i\mathbf{x}_i^T\right]$, and $\|\mathbf{x}\|_2 \leqslant \alpha$ almost surely for some $\alpha > 0$. Define random matrices $\mathbf{X}_i = \mathbf{x}_i\mathbf{x}_i^T - \mathbf{\Sigma}$ and the sample covariance matrix (a random matrix) $\hat{\mathbf{\Sigma}} = \frac{1}{N}\sum_{i=1}^{N}\mathbf{x}_i\mathbf{x}_i^T$. We have $\lambda_{\max}(\mathbf{X}_i) \leqslant \alpha^2 - \lambda_{\min}(\mathbf{X}_i)$. Also,

$$\lambda_{\max}\left(\frac{1}{N}\sum_{i=1}^{N}\mathbf{X}_i^2\right) = \lambda_{\max}\left(\mathbf{K} - \mathbf{\Sigma}^2\right)$$

and

$$\mathbb{E}\left[\mathrm{Tr}\left(\frac{1}{N}\sum_{i=1}^{N}\mathbf{X}_i^2\right)\right] = \mathrm{Tr}\left(\mathbf{K} - \mathbf{\Sigma}^2\right).$$

By using Theorem 2.16.4, we have that

$$\begin{aligned}\mathbb{P}\left(\lambda_{\max}\left(\hat{\mathbf{\Sigma}}-\mathbf{\Sigma}\right) > \sqrt{\frac{2t\lambda_{\max}\left(\mathbf{K}-\mathbf{\Sigma}^2\right)}{N}} + \frac{\left(\alpha^2-\lambda_{\min}\left(\mathbf{X}_i\right)\right)t}{3N}\right) \\ \leqslant \frac{\mathrm{Tr}\left(\mathbf{K}-\mathbf{\Sigma}^2\right)}{\lambda_{\max}\left(\mathbf{K}-\mathbf{\Sigma}^2\right)} \cdot t\left(e^t-t-1\right)^{-1}.\end{aligned}$$

Since $\lambda_{\max}(-\mathbf{X}_i) = \lambda_{\max}\left(\mathbf{\Sigma} - \mathbf{x}_i\mathbf{x}_i^T\right) \leqslant \lambda_{\max}(\mathbf{\Sigma})$, by using Theorem 2.16.4, we thus have

$$\mathbb{P}\left(\lambda_{\max}\left(\mathbf{\Sigma}-\hat{\mathbf{\Sigma}}\right) > \sqrt{\frac{2t\lambda_{\max}\left(\mathbf{K}-\mathbf{\Sigma}^2\right)}{N}} + \frac{\lambda_{\max}(\mathbf{\Sigma})\,t}{3N}\right) \leqslant \frac{\mathrm{Tr}\left(\mathbf{K}-\mathbf{\Sigma}^2\right)}{\lambda_{\max}\left(\mathbf{K}-\mathbf{\Sigma}^2\right)} \cdot t\left(e^t-t-1\right)^{-1}.$$

Combing the above two inequalities, finally we have that

$$\mathbb{P}\left(\left\|\boldsymbol{\Sigma} - \hat{\boldsymbol{\Sigma}}\right\|_2 > \sqrt{\frac{2t\lambda_{\max}\left(\mathbf{K}-\boldsymbol{\Sigma}^2\right)}{N}} + \frac{\max\left\{\alpha^2-\lambda_{\min}\left(\mathbf{X}_i\right), \lambda_{\max}\left(\boldsymbol{\Sigma}\right)\right\}t}{3N}\right)$$
$$\leqslant \frac{\mathrm{Tr}\left(\mathbf{K}-\boldsymbol{\Sigma}^2\right)}{\lambda_{\max}\left(\mathbf{K}-\boldsymbol{\Sigma}^2\right)} \cdot 2t\left(e^t-t-1\right)^{-1}.$$

The relevant notion of intrinsic dimension is $\frac{\mathrm{Tr}\left(\mathbf{K}-\boldsymbol{\Sigma}^2\right)}{\lambda_{\max}\left(\mathbf{K}-\boldsymbol{\Sigma}^2\right)}$, which can be finite even when the random vectors $\mathbf{x}_i$ take on values in an infinite dimensional Hilbert space.

□

9.6 Matrix Model of Signal Plus Noise $\mathbf{Y} = \mathbf{S} + \mathbf{X}$

We follow [509, 510] for our exposition. Consider a matrix model of signal plus noise

$$\mathbf{Y} = \mathbf{S} + \mathbf{X} \tag{9.36}$$

where $\mathbf{S} \in \mathbb{R}^{n\times n}$ is a *deterministic* matrix ("signal") and $\mathbf{X} \in \mathbb{R}^{n\times n}$ is a centered Gaussian matrix ("noise") whose entries are independent with variance σ^2. Our goal is to study the non-asymptotic upper and lower bounds on the accuracy of approximation which involves explicitly the singular values of $\mathbf{S}$. Our work is motivated for high-dimensional setting, in particular low-rank matrix recovery.

The Schatten-p norm is defined as

$$\|\mathbf{A}\|_{S_p} = \left(\sum_{i=1}^{n}\lambda_i^p\right)^{1/p} \quad \text{for} \quad 1 \leqslant p \leqslant \infty, \quad \text{and} \quad \|\mathbf{A}\|_\infty = \|\mathbf{A}\|_{\mathrm{op}} = \lambda_1,$$

for a matrix $\mathbf{A} \in \mathbb{R}^{n\times n}$. When $p = \infty$, we obtain the operator norm $\|\mathbf{A}\|_{\mathrm{op}}$, which is the largest singular value. When $p = 2$, we obtain the commonly called Hilbert-Schmidt norm or Frobenius norm $\|\mathbf{A}\|_{S_2} = \|\mathbf{A}\|_F = \|\mathbf{A}\|_2$. When $p = 1$, $\|\mathbf{A}\|_{S_1}$ denotes the nuclear norm.

Let the projection matrix $\mathbf{P}_r$ be the rank-r projection, which maximizes the Hilbert-Schmidt norm

$$\|\mathbf{P}_r\mathbf{A}\|_F = \|\mathbf{P}_r\mathbf{A}\|_2 = \|\mathbf{P}_r\mathbf{A}\|_{S_2}.$$

Let $\mathbb{O}_{n\times n,r}$ be the set of all orthogonal rank-r projections into subspaces of $\mathbb{R}^n$ so that we can say $\mathbf{P}_r \in \mathbb{O}_{n\times n,r}$. For any $\mathbf{A} \in \mathbb{R}^{n\times n}$, its singular values $\lambda_1, \ldots, \lambda_n$ are ordered in decreasing magnitude. In terms of singular values, we have

$$\|\mathbf{A}\|_F^2 = \sum_{i=1}^{n} \lambda_i^2, \quad \|\mathbf{P}_r\mathbf{A}\|_F^2 = \sum_{i=1}^{r} \lambda_i^2.$$

Let us review some basic properties of orthogonal projections. By definition we have

$$\mathbf{P}_r = \mathbf{P}_r^T \quad \text{and} \quad \mathbf{P}_r = \mathbf{P}_r\mathbf{P}_r, \quad \mathbf{P}_r \in \mathbb{S}_{n,r}.$$

Every orthogonal projection $\mathbf{P}_r$ is positive-semidefinite. Let $\mathbf{I}_{r\times r}$ be the identity matrix of $r \times r$.For $\mathbf{P}_r^{(1)}, \mathbf{P}_r^{(2)} \in \mathbb{S}_{n,r}$ with eigendecomposition

$$\mathbf{P}_r^{(1)} = \mathbf{U}\mathbf{I}_{r\times r}\mathbf{U}^T \quad \text{and} \quad \mathbf{P}_r^{(2)} = \tilde{\mathbf{U}}\mathbf{I}_{r\times r}\tilde{\mathbf{U}}^T,$$

we have

$$\mathrm{Tr}\left(\mathbf{P}_r^{(1)}\mathbf{P}_r^{(2)}\right) = \mathrm{Tr}\left(\mathbf{U}\mathbf{I}_{r\times r}\mathbf{U}^T\tilde{\mathbf{U}}\mathbf{I}_{r\times r}\tilde{\mathbf{U}}^T\right) = \mathrm{Tr}\left(\tilde{\mathbf{U}}^T\mathbf{U}\mathbf{I}_{r\times r}\mathbf{U}^T\tilde{\mathbf{U}}\mathbf{I}_{r\times r}\right).$$

The matrix $\mathbf{\Pi} = \tilde{\mathbf{U}}^T\mathbf{U}\mathbf{I}_{r\times r}\mathbf{U}^T\tilde{\mathbf{U}}$ is also an orthogonal projection. Since $\mathbf{\Pi}$ is positive semidefinite, the diagonal entries of $\mathbf{\Pi}$ are nonnegative. It follows that

$$\mathrm{Tr}\left(\mathbf{P}_r^{(1)}\mathbf{P}_r^{(2)}\right) = \mathrm{Tr}\left(\mathbf{\Pi}\mathbf{I}_{r\times r}\right) = \sum_{i=1}^{r} \Pi_{ii} \geqslant 0.$$

We conclude that

$$\left\|\mathbf{P}_r^{(2)} - \mathbf{P}_r^{(1)}\right\|_F = \left\|\left(\mathbf{I}_{r\times r} - \mathbf{P}_r^{(1)}\right) - \left(\mathbf{I}_{r\times r} - \mathbf{P}_r^{(2)}\right)\right\| \leqslant \sqrt{2r\left(n-r\right)}.$$

Let S^{n-1} be the Euclidean sphere in n-dimensional space. Finally, by the symmetry of $\mathbf{P}_r^{(2)} - \mathbf{P}_r^{(1)}$, we obtain

$$\begin{aligned}
\|\mathbf{P}_r^{(2)} - \mathbf{P}_r^{(1)}\|_{S_\infty} = \|\mathbf{P}_r^{(2)} - \mathbf{P}_r^{(1)}\|_{\mathrm{op}} = \lambda_1\left(\mathbf{P}_r^{(2)} - \mathbf{P}_r^{(1)}\right) &= \sup_{\mathbf{x}\in S^{n-1}} \left|\mathbf{x}^T\left(\mathbf{P}_r^{(2)} - \mathbf{P}_r^{(1)}\right)\mathbf{x}\right| \\
&= \sup_{\mathbf{x}\in S^{n-1}} \left|\underbrace{\mathbf{x}^T\mathbf{P}_r^{(2)}\mathbf{x}}_{\in[0,1]} - \underbrace{\mathbf{x}^T\mathbf{P}_r^{(1)}\mathbf{x}}_{\in[0,1]}\right| \leqslant 1.
\end{aligned}$$

The largest singular value (the operator norm) for the difference of two projection matrices is bounded by 1.

For (9.36), it is useful to bound the trace $\mathrm{Tr}\left(\mathbf{X}^T\left(\mathbf{P}_r^{(2)} - \mathbf{P}_r^{(1)}\right)\mathbf{S}\right)$ and $\left\|\tilde{\mathbf{P}}_r\mathbf{Y}\right\|_F^2 - \|\mathbf{P}_r\mathbf{Y}\|_F^2$. Motivated for this purpose, we consider the trace

$\mathrm{Tr}\left(\mathbf{A}^T\left(\mathbf{P}_r^{(2)} - \mathbf{P}_r^{(1)}\right)\mathbf{B}\right)$, for two arbitrary rank-r projections $\mathbf{P}_r^{(1)}, \mathbf{P}_r^{(2)} \in \mathbb{S}_{n,r}$, and arbitrary two matrices $\mathbf{A}, \mathbf{B} \in \mathbb{R}^{n\times n}$.

First, we observe that

$$\mathbf{P}_r^{(2)} - \mathbf{P}_r^{(1)} = \mathbf{P}_r^{(2)} - \mathbf{P}_r^{(2)}\mathbf{P}_r^{(1)} + \mathbf{P}_r^{(2)}\mathbf{P}_r^{(1)} - \mathbf{P}_r^{(1)} = \mathbf{P}_r^{(2)}\left(\mathbf{I} - \mathbf{P}_r^{(1)}\right) + \left(\mathbf{P}_r^{(2)} - \mathbf{I}\right)\mathbf{P}_r^{(1)}.$$

According to the proof of Proposition 8.1 of Rohde [509], we have

$$\left\|\left(\mathbf{I} - \mathbf{P}_r^{(2)}\right)\mathbf{P}_r^{(1)}\right\|_F = \left\|\mathbf{P}_r^{(2)}\left(\mathbf{I} - \mathbf{P}_r^{(1)}\right)\right\|_F = \frac{1}{\sqrt{2}}\left\|\mathbf{P}_r^{(2)} - \mathbf{P}_r^{(1)}\right\|_F.$$

By the Cauchy-Schwarz inequality we obtain

$$\begin{aligned}
\mathrm{Tr}\left(\mathbf{A}^T\left(\mathbf{P}_r^{(2)} - \mathbf{P}_r^{(1)}\right)\mathbf{B}\right) &= \mathrm{Tr}\left(\mathbf{A}^T\mathbf{P}_r^{(2)}\left(\mathbf{I} - \mathbf{P}_r^{(1)}\right)\mathbf{B}\right) + \mathrm{Tr}\left(\mathbf{A}^T\left(\mathbf{I} - \mathbf{P}_r^{(2)}\right)\mathbf{P}_r^{(1)}\mathbf{B}\right) \\
&\leqslant \left\|\mathbf{B}\mathbf{A}^T\mathbf{P}_r^{(2)}\right\|_F \cdot \left\|\left(\mathbf{I} - \mathbf{P}_r^{(1)}\right)\mathbf{B}\mathbf{A}^T\right\|_F \cdot \left\|\mathbf{P}_r^{(2)}\left(\mathbf{I} - \mathbf{P}_r^{(1)}\right)\right\|_F \\
&\quad + \left\|\mathbf{P}_r^{(1)}\mathbf{B}\mathbf{A}^T\right\|_F \cdot \left\|\mathbf{B}\mathbf{A}^T\left(\mathbf{I} - \mathbf{P}_r^{(2)}\right)\right\|_F \cdot \left\|\left(\mathbf{P}_r^{(2)} - \mathbf{I}\right)\mathbf{P}_r^{(1)}\right\|_F \\
&\leqslant \frac{1}{\sqrt{2}}\sqrt{r(n-r)} \cdot \sqrt{\|\mathbf{B}\mathbf{A}^T\mathbf{A}\mathbf{B}^T\|_{S_\infty}} \cdot \left\|\mathbf{P}_r^{(2)} - \mathbf{P}_r^{(1)}\right\|_F \\
&\quad + \frac{1}{\sqrt{2}}\sqrt{r(n-r)} \cdot \sqrt{\|\mathbf{B}\mathbf{A}^T\mathbf{A}\mathbf{B}^T\|_{S_\infty}} \cdot \left\|\mathbf{P}_r^{(2)} - \mathbf{P}_r^{(1)}\right\|_F \\
&\leqslant \sqrt{2r(n-r)}\lambda_1(\mathbf{A})\lambda_1(\mathbf{B})\left\|\mathbf{P}_r^{(2)} - \mathbf{P}_r^{(1)}\right\|_F.
\end{aligned}$$

Note $\|\cdot\|_{S_\infty}$ is the largest singular value $\lambda_1(\cdot)$. Thus,

$$\mathrm{Tr}\left(\mathbf{A}^T\left(\mathbf{P}_r^{(2)} - \mathbf{P}_r^{(1)}\right)\mathbf{B}\right) \leqslant \sqrt{2r(n-r)}\lambda_1(\mathbf{A})\lambda_1(\mathbf{B})\left\|\mathbf{P}_r^{(2)} - \mathbf{P}_r^{(1)}\right\|_F. \tag{9.37}$$

The inequality (9.37) is optimal to the effect for the following case: When, for $n \geq 2r$, there are r orthonormal vectors $\mathbf{u}_1, \ldots, \mathbf{u}_r$ and $\tilde{\mathbf{u}}_1, \ldots, \tilde{\mathbf{u}}_r$ such that we can form two arbitrary rank-r projection matrices

$$\mathbf{P}_r^{(1)} = \sum_{i=1}^{r}\mathbf{u}_i\mathbf{u}_i^T, \quad \text{and} \quad \mathbf{P}_r^{(2)} = \sum_{i=1}^{r}\left(\sqrt{1-\alpha^2}\mathbf{u}_i + \alpha\tilde{\mathbf{u}}_i\right)\left(\sqrt{1-\alpha^2}\mathbf{u}_i + \alpha\tilde{\mathbf{u}}_i\right)^T,$$

and

$$\mathbf{A} = \mu\mathbf{I}, \quad \mathbf{B} = \nu\left(\mathbf{P}_r^{(1)} - \mathbf{P}_r^{(2)}\right).$$

In fact, for the case, the left-hand side of the inequality (9.37) attains the upper bound for any real numbers $0 \leqslant \alpha \leqslant 1$, and $\mu, \nu > 0$.

For the considered case, let us explicitly evaluate the left-hand side of the inequality (9.37)

$$\begin{aligned}
\operatorname{Tr}\left(\mathbf{A}^T\left(\mathbf{P}_r^{(2)}-\mathbf{P}_r^{(1)}\right)\mathbf{B}\right) &= \mu\nu \operatorname{Tr}\left(\begin{array}{l}\left(\mathbf{P}_r^{(2)}-\sum\limits_{i=1}^{r}\left(\sqrt{1-\alpha^2}\mathbf{u}_i+\alpha\tilde{\mathbf{u}}_i\right)\left(\sqrt{1-\alpha^2}\mathbf{u}_i+\alpha\tilde{\mathbf{u}}_i\right)^T\right)\\ \times\left(\mathbf{P}_r^{(1)}-\sum\limits_{i=1}^{r}\left(\sqrt{1-\alpha^2}\mathbf{u}_i+\alpha\tilde{\mathbf{u}}_i\right)\left(\sqrt{1-\alpha^2}\mathbf{u}_i+\alpha\tilde{\mathbf{u}}_i\right)^T\right)\end{array}\right)\\
&= \mu\nu\left(2r - 2\operatorname{Tr}\left(\mathbf{P}_r^{(2)}\mathbf{P}_r^{(1)}\right)\right)\\
&= \mu\nu\left(2r - 2r\left(1-\alpha^2\right)\right)\\
&= \sqrt{2}\mu\nu\alpha^2\sqrt{2r}.
\end{aligned}$$

We have used $\left\|\mathbf{P}_r^{(1)} - \mathbf{P}_r^{(2)}\right\|_F = \alpha\sqrt{2r}$ and $\lambda_1\left(\mathbf{P}_r^{(1)} - \mathbf{P}_r^{(2)}\right) = \alpha$. Let us establish them now. The first one is simple since

$$\left\|\mathbf{P}_r^{(1)} - \mathbf{P}_r^{(2)}\right\|_F = \sqrt{\operatorname{Tr}\left(\left(\mathbf{P}_r^{(1)} - \mathbf{P}_r^{(2)}\right)\left(\mathbf{P}_r^{(1)} - \mathbf{P}_r^{(2)}\right)\right)} = \alpha\sqrt{2r}.$$

To prove the second one, we can check that α and $-\alpha$ are the only non-zero eigenvalues of the difference matrix $\mathbf{P}_r^{(1)} - \mathbf{P}_r^{(2)}$ and their eigen spaces are given by

$$\begin{aligned}
&\mathbb{W}_\alpha = \operatorname{span}\left\{\sqrt{\tfrac{1+\alpha}{2}}\mathbf{u}_i - \sqrt{\tfrac{1-\alpha}{2}}\tilde{\mathbf{u}}_i, \quad i = 1,\ldots,r\right\}\\
\text{and}\quad &\mathbb{W}_{-\alpha} = \operatorname{span}\left\{\sqrt{\tfrac{1-\alpha}{2}}\mathbf{u}_i + \sqrt{\tfrac{1+\alpha}{2}}\tilde{\mathbf{u}}_i, \quad i = 1,\ldots,r\right\}.
\end{aligned}$$

Since $\mathbf{P}_r^{(1)} - \mathbf{P}_r^{(2)}$ is symmetric, it follows that

$$\lambda_1\left(\mathbf{P}_r^{(1)} - \mathbf{P}_r^{(2)}\right) = \max\left(\lambda_{\max}\left(\mathbf{P}_r^{(1)} - \mathbf{P}_r^{(2)}\right), \left|\lambda_{\min}\left(\mathbf{P}_r^{(1)} - \mathbf{P}_r^{(2)}\right)\right|\right) = \alpha.$$

Now we are in a position to consider the model (9.36) using the process

$$Z = \left\|\tilde{\mathbf{P}}_r\mathbf{Y}\right\|_F^2 - \left\|\mathbf{P}_r\mathbf{Y}\right\|_F^2 = \left\|\tilde{\mathbf{P}}_r\left(\mathbf{S}+\mathbf{X}\right)\right\|_F^2 - \left\|\mathbf{P}_r\left(\mathbf{S}+\mathbf{X}\right)\right\|_F^2,$$

for a pair of rank-r projections. Recall that the projection matrix $\mathbf{P}_r$ be the rank-r projection, which *maximizes* the Hilbert-Schmidt norm

$$\left\|\mathbf{P}_r\mathbf{A}\right\|_F = \left\|\mathbf{P}_r\mathbf{A}\right\|_2 = \left\|\mathbf{P}_r\mathbf{A}\right\|_{S_2}.$$

On the other hand, $\tilde{\mathbf{P}}_r \in \mathbb{O}_{n\times n,r}$ is an orthogonal rank-r projections into subspaces of $\mathbb{R}^n$. Obviously, Z is a functional of the projection matrix $\tilde{\mathbf{P}}_r$. The supremum is denoted by

$$Z_{\tilde{\mathbf{P}}_r} = \sup_{\tilde{\mathbf{P}}_r \in \mathbb{O}_{n\times n,r}} Z_{\tilde{\mathbf{P}}_r},$$

where $\tilde{\mathbf{P}}_r$ is a location of the supremum. In general, $\tilde{\mathbf{P}}_r$ is not unique. In addition, the differences $\left\|\tilde{\mathbf{P}}_r\mathbf{Y}\right\|_F^2 - \|\mathbf{P}_r\mathbf{Y}\|_F^2$ are usually not centered.

Theorem 9.6.1 (Upper bound for Gaussian matrices [510]). *Let the distribution of X_{ij} be centered Gaussian with variance σ^2 and* $\mathrm{rank}(\mathbf{X}) \geqslant r$. *Then, for $r \leq n-r$, the following bound holds*

$$\mathbb{E}Z \lesssim \sigma^2 rn \left(\min\left(\frac{\lambda_1^2}{\lambda_r^2}, 1+\frac{\lambda_1}{\sigma\sqrt{n}}\right) + \min\left(\left(\frac{\frac{1}{r}\sum_{i=r+1}^{2r}\lambda_i^2}{\lambda_r^2}\right)^{1/2} \cdot \frac{\lambda_1}{\sigma\sqrt{n}}, \frac{\lambda_1^2}{\lambda_r^2-\lambda_{r+1}^2}\right)\right),$$

where $\frac{\lambda_1^2}{\lambda_r^2-\lambda_{r+1}^2}$ is set to infinity, if $\lambda_r = \lambda_{r+1}$

Theorem 9.6.1 is valid for the Gaussian entries, while the following theorem is more general to i.i.d. entries with finite fourth moment.

Theorem 9.6.2 (Universal upper bound [510]). *Assume that the i.i.d. entries X_{ij} of the random matrix $\mathbf{X}$ has finite variance σ^2 and finite fourth moment m_4. Then, we have*

$$\mathbb{E}Z \lesssim r(n-r)\min(\mathrm{I},\mathrm{II},\mathrm{III}), \tag{9.38}$$

where

$$\begin{aligned}
\mathrm{I} &= \sigma^2 + \sqrt{m_4} + \tfrac{\lambda_1}{\sqrt{n}}\left(\sigma + m_4^{1/4}\right),\\
II &= \begin{cases} \frac{\lambda_1^2}{\lambda_r^2-\lambda_{r+1}^2}\left(\sigma^2+\sqrt{m_4}\right) & \text{if } \lambda_r > \lambda_{r+1},\\ \infty & \text{if } \lambda_r = \lambda_{r+1},\end{cases}\\
III &= \begin{cases} \frac{\lambda_1^2}{\lambda_r^2}\left(\sigma^2+\sqrt{m_4}\right) + \sqrt{\frac{\lambda_1^2\sum_{i=r+1}^{2r}\lambda_i^2}{r(n-r)\lambda_r^2}}\left(\sigma+m_4^{1/4}\right) & \text{if } \lambda_r > 0,\\ \infty & \text{if } \lambda_r = 0.\end{cases}
\end{aligned} \tag{9.39}$$

Let us consider some examples for the model $\mathbf{Y} = \mathbf{S} + \mathbf{X}$ as defined in (9.36). Let $\lambda_1 \geqslant \lambda_2 \geqslant \ldots \geqslant \lambda_n$ and $\hat{\lambda}_1 \geqslant \hat{\lambda}_2 \geqslant \ldots \geqslant \hat{\lambda}_n$ denote the singular values of $\mathbf{S}$ and $\mathbf{Y} = \mathbf{S} + \mathbf{X}$, respectively. Recall that $\left\|\hat{\mathbf{P}}_r\mathbf{Y}\right\|_F^2 = \sum_{i=1}^{r}\hat{\lambda}_i^2$ and $\|\mathbf{P}_r\mathbf{S}\|_F^2 = \sum_{i=1}^{r}\lambda_i^2$ with the rank-r projections $\hat{\mathbf{P}}_r$ and $\mathbf{P}_r$. The hat version standards for the non-centered Gaussian random matrix $\mathbf{Y}$ and the version without hat stands for the deterministic matrix $\mathbf{S}$.

Example 9.6.3 (The largest singular value [509]).

Consider estimating λ_1^2, the largest eigenvalue of $\mathbf{S}^T\mathbf{S}$, based on the observation $\mathbf{Y} = \mathbf{S} + \mathbf{X}$ defined by (9.36). The maximum eigenvalue of $\mathbf{Y}^T\mathbf{Y}$ is positively biased as an estimate for, since

$$\mathbb{E}\hat{\lambda}_1^2 = \mathbb{E}\left\|\hat{\mathbf{P}}_1\mathbf{Y}\right\|_F^2 \geqslant \mathbb{E}\left\|\mathbf{P}_1\mathbf{Y}\right\|_F^2 = \lambda_1^2 + \sigma^2 n.$$

It is natural to consider $\hat{s} = \hat{\lambda}_1^2 - \sigma^2 n$ as an estimator for λ_1^2. However, the analysis in [509] reveals that

$$\mathbb{E}\hat{s} - \lambda_1^2 = \mathbb{E}\hat{\lambda}_1^2 - \sigma^2 n - \lambda_1^2$$

is strictly positive and bounded away from zero, uniformly over $\mathbf{S} \in \mathbb{R}^{n\times n}$. In fact,

$$\mathbb{E}\hat{s} - \lambda_1^2 \in \left[c_1\sigma^2 n, c_2\left(\sigma^2 n + \sigma\sqrt{n}\lambda_1\left(\mathbf{S}\right)\right)\right]$$

for some universal constants $c_1, c_2 > 0$, which do not depend on n, σ^2 and $\mathbf{S}$. □

Example 9.6.4 (Quadratic functional of low-rank matrices [509]).

One natural candidate for estimating $\|\mathbf{S}\|_F^2$, based on the observation $\mathbf{Y} = \mathbf{S} + \mathbf{X}$ defined by (9.36), is the unbiased estimator $\|\mathbf{Y}\|_F^2 - \sigma^2 n^2$. Simple calculation gives

$$\mathrm{Var}\left(\|\mathbf{Y}\|_F^2 - \sigma^2 n^2\right) = 2\sigma^4 n^2 + 4\sigma^2\|\mathbf{S}\|_F^2. \tag{9.40}$$

The disadvantage of this estimator is its large variance for large values of n : it depends quadratically on the dimension. If $r = \mathrm{rank}\left(\mathbf{S}\right) < n$, the matrix $\mathbf{S}$ can be fully characterized by $(2n - r)r$ parameters as it can be seen by the singular value decomposition. In other words, if $r \ll n$, the intrinsic dimension of the problem is of the order rn rather than n^2. For every matrix with $r = \mathrm{rank}\left(\mathbf{S}\right) = r$, we have

$$\|\mathbf{S}\|_F^2 = \|\mathbf{P}_r\mathbf{S}\|_F^2.$$

Elementary analysis shows that $\|\mathbf{P}_r\mathbf{S}\|_F^2 - \sigma^2 rn$ unbiasedly estimates $\|\mathbf{S}\|_F^2$, and

$$\mathrm{Var}\left(\|\mathbf{P}_r\mathbf{S}\|_F^2 - \sigma^2 rn\right) = 2\sigma^4 rn + 4\sigma^2\|\mathbf{S}\|_F^2. \tag{9.41}$$

Further, it follows that

$$\mathbb{E}\Big(\|\mathbf{P}_r\mathbf{S}\|_F^2 - \sigma^2 rn - \|\mathbf{S}\|_F^2 - 2\sigma\,\mathrm{Tr}\left(\mathbf{X}^T\mathbf{S}\right)\Big)^2 = 2\sigma^4 rn,$$

that is, $\sigma^{-1}\left(\|\mathbf{P}_r\mathbf{S}\|_F^2 - \sigma^2 rn - \|\mathbf{S}\|_F^2\right)$ is approximately centered Gaussian with variance $4\|\mathbf{S}\|_F^2$ if $\sigma^2 rn = o(1)$ in an asymptotic framework, and $4\|\mathbf{S}\|_F^2$ is the asymptotic efficiency lower bound [134]. The statistics $\|\mathbf{P}_r\mathbf{S}\|_F^2 - \sigma^2 rn$, however, cannot be used for estimator since $\mathbf{P}_r = \mathbf{P}_r(\mathbf{S})$ depends on $\mathbf{S}$ itself and is unknown a prior. The analysis of [509] argues that empirical low-rank projections $\left\|\hat{\mathbf{P}}_r\mathbf{Y}\right\|_F^2 - \sigma^2 rn$ cannot be successfully used for efficient estimation of $\|\mathbf{S}\|_F^2$, even if the rank $(\mathbf{S}) < n$ is explicitly known beforehand.

□

9.7 Robust Covariance Estimation

Following [453], we introduce a robust covariance matrix estimation. For $i = 1, 2, \ldots, N$, let $\mathbf{x}_i \in \mathbb{R}^n$ be samples from a zero-mean distribution with *unknown* covariance matrix $\mathbf{R}_x$, which is positive definite. Suppose that the data associated with some subsets $\mathcal{S}$ of individuals is arbitrarily corrupted. This adversarial corruption can be modeled as

$$\mathbf{y}_i = \mathbf{x}_i + \mathbf{v}_i, \quad i = 1, \ldots, N,$$

where $\mathbf{v}_i \in \mathbb{R}^n$ is a vector supported on the set $\mathcal{S}$. Let

$$\hat{\mathbf{R}}_y = \frac{1}{N}\sum_{i=1}^{N} \mathbf{y}_i\mathbf{y}_i^T$$

be the sample covariance matrix of the corrupted samples. We define

$$\tilde{\mathbf{R}}_x = \frac{1}{N}\sum_{i=1}^{N} \mathbf{x}_i\mathbf{x}_i^T - \mathbf{R}_x,$$

which is a type of re-centered Wishart noise. After some algebra, we have

$$\hat{\mathbf{R}}_y = \mathbf{R}_x + \tilde{\mathbf{R}}_x + \boldsymbol{\Delta} \tag{9.42}$$

where

$$\boldsymbol{\Delta} = \frac{1}{N}\sum_{i=1}^{N} \mathbf{v}_i\mathbf{v}_i^T + \frac{1}{N}\sum_{i=1}^{N}\left(\mathbf{x}_i\mathbf{v}_i^T + \mathbf{v}_i\mathbf{x}_i^T\right).$$

Let us demonstrate how to use concentration of measure (see Sects. 3.7 and 3.9) in this context. Let us assume that $\mathbf{R}_x$ has rank at most r. We can write

$$\tilde{\mathbf{R}}_x = \mathbf{Q}\left\{\frac{1}{N}\sum_{i=1}^{N}\mathbf{z}_i\mathbf{z}_i^T - \mathbf{I}_{r\times r}\right\}\mathbf{Q}^T,$$

where $\mathbf{R}_x = \mathbf{Q}\mathbf{Q}^T$, and $\mathbf{z}_i \sim \mathcal{N}\left(0, \mathbf{I}_{r\times r}\right)$ is standard Gaussian in dimension r. As a result, by known results on singular values of Wishart matrices [145], we have [453]

$$\frac{\left\|\tilde{\mathbf{R}}_x\right\|_{op}}{\left\|\mathbf{R}_x\right\|_{op}} \leqslant 4\sqrt{\frac{r}{N}}, \tag{9.43}$$

with probability greater than $1 - 2\exp\left(-c_1 r\right)$.

Chapter 10
Detection in High Dimensions

This chapter is the core of Part II: Applications.

Detection in high dimensions is fundamentally different from the traditional detection theory. Concentration of measure plays a central role due to the high dimensions. We exploit the bless of dimensions.

10.1 OFDM Radar

We propose to study the weak signal detection under the framework of sums of random matrices. This matrix setting is natural for many radar problems, such as orthogonal frequency division multiplexing (OFDM) radar and distributed aperture. Each subcarrier (or antenna sensor) can be modeled as a random matrix, via, e.g., sample covariance matrix estimated using the time sequence of the data. Often the data sequence is very extremely long. One fundamental problem is to break the long data record into shorter data segments. Each short data segment is sufficiently long to estimate the sample covariance matrix of the underlying distribution. If we have 128 subcarriers and 100 short data segments, we will have 12,800 random matrices at our disposal. The most natural approach for data fusion is to sum up the 12,800 random matrices.

The random matrix (here sample covariance matrix) is the basic information block in our proposed formalism. In this novel formalism, *we take the number of observations as it is* and try to evaluate the effect of all the influential parameters. The number of observations, large but finite-dimensional, is taken as it is—and treated as "given". From this given number of observations, we want algorithms to achieve the performance as good as they can. We desire to estimate the covariance matrix using a smaller number of observations; this way, a larger number of covariance matrices can be obtained. In our proposed formalism, low rank matrix recovery (or matrix completion) plays a fundamental role.

R. Qiu and M. Wicks, *Cognitive Networked Sensing and Big Data*,
DOI 10.1007/978-1-4614-4544-9_10,

10.2 Principal Component Analysis

Principal component analysis (PCA) [208] is a classical method for reducing the dimension of data, say, from high-dimensional subset of $\mathbb{R}^n$ down to some subsets of $\mathbb{R}^d$, with $d \ll n$. PCA operates by projecting the data onto the d directions of maximal variance, as captured by eigenvectors of the $n \times n$ true covariance matrix $\mathbf{\Sigma}$. See Sect. 3.6 for background and notation on induced operator norms. See the PhD dissertation [511] for a treatment of high-dimensional principal component analysis. We freely take material from [511] in this section to give some background on PCA and its SDP formulation.

PCA as subspace of maximal variance. Consider a collection of data points $\mathbf{x}_i, i = 1, \ldots, N$ in $\mathbb{R}^n$, drawn i.i.d. from a distribution $\mathbb{P}$. We denote the expectation with respect to this distribution by $\mathbb{E}$. Assume that the distribution is centered, i.e., $\mathbb{E}\mathbf{x} = 0$, and that $\mathbb{E}\left\|\mathbf{x}\right\|_2^2 < \infty$. We collect $\{\mathbf{x}_i\}_{i=1}^N$ in a matrix $\mathbf{X} \in \mathbb{R}^{N \times n}$. Thus, $\mathbf{x}_i$ represents the i-th row of $\mathbf{X}$. Let $\mathbf{\Sigma}$ and $\hat{\mathbf{\Sigma}} = \hat{\mathbf{\Sigma}}_N$ denote the true covariance matrix and the sample covariance matrix, respectively. We have

$$\mathbf{\Sigma} := \mathbb{E}\mathbf{x}\mathbf{x}^T, \quad \hat{\mathbf{\Sigma}} := \frac{1}{N}\mathbf{X}^T\mathbf{X} = \frac{1}{N}\sum_{i=1}^{N}\mathbf{x}_i\mathbf{x}_i^T. \tag{10.1}$$

The first *principal component* of the distribution $\mathbb{P}$ is a vector $\mathbf{z}^\star \in \mathbb{R}^n$ satisfying

$$\mathbf{z}^\star \in \arg\max_{\|\mathbf{z}\|_2=1} \mathbb{E}\left(\mathbf{z}^T\mathbf{x}\right)^2, \tag{10.2}$$

that is, $\mathbf{z}^\star$ is a direction that the projection of the distribution along which has maximal variance. Noting that $\mathbb{E}\left(\mathbf{z}^T\mathbf{x}\right)^2 = \mathbb{E}\left(\mathbf{z}^T\mathbf{x}\right)\left(\mathbf{z}^T\mathbf{x}\right) = \mathbf{z}^T\left(\mathbb{E}\mathbf{x}\mathbf{x}^T\right)\mathbf{z}$, we obtain

$$\mathbf{z}^\star \in \arg\max_{\|\mathbf{z}\|_2=1} \mathbf{z}^T\mathbf{\Sigma}\mathbf{z}. \tag{10.3}$$

By a well-known result in linear analysis, called Rayleigh-Ritz or Courant-Fischer theorem [23], (10.3) is the variational characterization of maximal eigenvectors of $\mathbf{\Sigma}$.

The second principal component is obtained by removing the contribution form the first principal component and applying the same procedure; that is, obtaining the first principal component of $\mathbf{x} - \left(\left(\mathbf{z}^\star\right)^T\mathbf{x}\right)\mathbf{z}^\star$. The subsequent principal components are obtained recursively until all the variance in $\mathbf{x}$ is explained, i.e., the remainder is zero. In case of ambiguity, one chooses a direction orthogonal to all the previous components. Thus, principal components form an orthonormal basis for the eigenspace of $\mathbf{\Sigma}$ corresponding to nonzero eigenvalues.

SDP formulation. Let us derive a SDP equivalent to (10.3). Using the cyclic property of the trace, we have $\mathrm{Tr}\left(\mathbf{z}^T\boldsymbol{\Sigma}\mathbf{z}\right) = \mathrm{Tr}\left(\boldsymbol{\Sigma}\mathbf{z}\mathbf{z}^T\right)$. For a matrix $\mathbf{Z} \in \mathbb{R}^{n\times n}, \mathbf{Z} \geqslant 0$ and $\mathrm{rank}\left(\mathbf{Z}\right) = 1$ is equivalent to $\mathbf{Z} = \mathbf{z}\mathbf{z}^T$ for some $\mathbf{z} \in \mathbb{R}^n$. Imposing the additional condition $\mathrm{Tr}\left(\mathbf{Z}\right) = 1$ is equivalent to the additional constraint $\|\mathbf{z}\|_2 = 1$. Now after dropping the $\mathrm{rank}\left(\mathbf{Z}\right) = 1$, we obtain a relaxation of (10.3)

$$\mathbf{Z}^\star \in \arg \max_{\mathbf{Z}\geqslant 0, \mathrm{Tr}(\mathbf{Z})} \mathrm{Tr}\left(\boldsymbol{\Sigma}\mathbf{Z}\right). \tag{10.4}$$

It turns out that this relaxation is in fact exact! That is,

Lemma 10.2.1. *There is always a rank one solution* $\mathbf{Z} = \mathbf{z}^\star\left(\mathbf{z}^\star\right)^T$ *of* (10.4) *where* $\mathbf{z}^\star = \vartheta_{\max}\left(\boldsymbol{\Sigma}\right)$.

Any member of the set of eigenvectors of $\mathbf{A}$ associated with an eigenvalue is denoted as $\vartheta\left(\mathbf{A}\right)$. Similarly, $\vartheta_{max}\left(\mathbf{A}\right)$ represents any eigenvector associated with the maximal eigenvalue (occasionally referred to as a"maximal eigenvector").

Proof. It is enough to show that all $\mathbf{Z}$ feasible for (10.4), one has $\mathrm{Tr}\left(\boldsymbol{\Sigma}\mathbf{Z}\right) \leqslant \lambda_{\max}\left(\boldsymbol{\Sigma}\right)$. Using eigenvalue decomposition of $\mathbf{Z} = \sum_{i=1}^{n} \lambda_i \mathbf{u}_i \mathbf{u}_i^T$, this is equivalent to $\sum_{i=1}^{n} \lambda_i \mathbf{u}_i \mathbf{u}_i^T \leqslant \lambda_{\max}\left(\boldsymbol{\Sigma}\right)$. But this is true, by (10.3) and $\sum_{i=1}^{n} \lambda_i = 1$. □

As the optimization problem in (10.4) is over the cone of semidefinite matrices ($\mathbf{Z} \geq 0$) with an objective and extra constraints which are *linear* in matrix $\mathbf{Z}$, the optimization problem (10.4) is a textbook example of a SDP [48]. The SDPs belong to the class of conic programs for which fast methods of solution are currently available [512]. Software tools such as CVX can be used to solve (10.4).

Noisy Samples. In practice, one does not access to the true covariance matrix $\boldsymbol{\Sigma}$, but instead must rely on a "noisy" version of the form

$$\hat{\boldsymbol{\Sigma}} = \boldsymbol{\Sigma} + \boldsymbol{\Delta} \tag{10.5}$$

where $\boldsymbol{\Delta} = \boldsymbol{\Delta}_N$ denotes a *random noisy matrix*, typically arising from having only a finite number N of samples.

A natural question is under what conditions the sample eigenvectors based on $\hat{\boldsymbol{\Sigma}}$ are consistent estimators of their true analogues $\boldsymbol{\Sigma}$. In the classical theory of PCA, the model dimension n is viewed as fixed, asymptotic statements are established as N goes to infinity, $N \to \infty$. However, such "fixed n, large N" scaling may be inappropriate for today's big data applications, where the model dimension n is comparable or even larger than the number of observations N, or $n \leq N$.

10.2.1 PCA Inconsistency in High-Dimensional Setting

We briefly study some inconsistency results for PCA, in the high-dimensional setting where $(N, n) \to \infty$. We observe data points $\{\mathbf{x}_i\}_{i=1}^N$ i.i.d. from a distribution with true covariance matrix $\mathbf{\Sigma} := \mathbb{E}\mathbf{x}_i{\mathbf{x}_i}^T$. The single spiked covariance model assumes the following structure on $\mathbf{\Sigma}$

$$\mathbf{\Sigma} = \beta \mathbf{z}^\star (\mathbf{z}^\star)^T + \mathbf{I}_{n \times n} \tag{10.6}$$

where $\beta > 0$ is some positive constant, measuring signal-to-noise ratio (SNR). The eigenvalues of $\mathbf{\Sigma}$ are all equal to 1 except for the largest one which is $1+\beta$. $\mathbf{z}^\star$ is the leading principal component for $\mathbf{\Sigma}$. One then forms the sample covariance matrix $\hat{\mathbf{\Sigma}}$ and obtains its maximal eigenvector $\hat{\mathbf{z}}$, hoping that $\hat{\mathbf{z}}$ is a consistent estimate of $\mathbf{z}^\star$.

This unfortunately does not happen unless $n/N \to 0$ as shown by Paul and Johnston [513] among others. See also [203]. As $(N, n) \to \infty$, $n/N \to \alpha$, asymptotically, the following phase transition occurs:

$$\langle \hat{\mathbf{z}}, \mathbf{z}^\star \rangle_2 \to \begin{cases} 0, & \beta \leqslant \sqrt{\alpha} \\ \frac{1-\alpha/\beta^2}{1+\alpha/\beta^2}, & \beta > \sqrt{\alpha}. \end{cases} \tag{10.7}$$

Note that $\langle \hat{\mathbf{z}}, \mathbf{z}^\star \rangle_2$ measures cosine of the angle between $\hat{\mathbf{z}}$ and $\mathbf{z}^\star$ and is related to the projection of 2-distance between the corresponding 1-dimensional subspaces.

Nether case in (10.7) show consistency, i.e., $\langle \hat{\mathbf{z}}, \mathbf{z}^\star \rangle_2 \to 1$. This has led to research on additional structure/constraints that one may impose on $\mathbf{z}^\star$ to allow for consistent estimation.

10.3 Space-Time Coding Combined with CS

$$\mathbf{y} = \mathbf{A}\mathbf{x} + \mathbf{z}$$

$$\mathbf{y} = \mathbf{H}\mathbf{x} + \mathbf{z}$$

where $\mathbf{H}$ is flat fading channel MIMO information theory

10.4 Sparse Principal Components

We follow [514, 515]. Let $\mathbf{x}_1, \ldots, \mathbf{x}_n$ be n i.i.d. realizations of a random variable $\mathbf{x}$ in $\mathbb{R}^N$. Our task is to test whether the sphericity hypothesis is true, i.e., that the distribution of $\mathbf{x}$ is invariant by rotation in $\mathbb{R}^N$. For a Gaussian distribution, this is equivalent to testing if the covariance matrix of $\mathbf{x}$ is of the form $\sigma^2\mathbf{I}_N$ for some known $\sigma^2 > 0$, where $\mathbf{I}_N$ is identity matrix.

Without loss of generality, we may assume $\sigma^2 = 1$, so that the covariance matrix is the identity in $\mathbb{R}^N$ under the null hypothesis. For alternative hypotheses, there exists a privileged direction, along which $\mathbf{x}$ has more variance. Here we consider the case where the privileged direction is sparse. The covariance matrix is a sparse rank one matrix perturbation of the identity matrix $\mathbf{I}_N$. Formally, let $v \in \mathbb{R}^N$ be such that $||v||_2 = 1, ||v||_0 \leq k$, and $\theta > 0$. The hypothesis problem is

$$\begin{aligned} \mathcal{H}_0 &: \mathbf{x} \sim \mathcal{N}(0, \mathbf{I}) \\ \mathcal{H}_1 &: \mathbf{x} \sim \mathcal{N}\left(0, \mathbf{I} + \theta \mathbf{v}\mathbf{v}^T\right). \end{aligned} \tag{10.8}$$

This problem is considered in [157] where $\mathbf{v}$ is the so-called feature. Later a perturbation of several rank one matrices is considered in [158, 159]. This idea is carried out in a Kernel space [385]. What is new in this section is to include the sparsity of $\mathbf{v}$. The model under $\mathcal{H}_1$ is a generalization of the spiked covariance model since it allows $\mathbf{v}$ to be k-sparse on the unit Euclidean sphere. The statement of $\mathcal{H}_1$ is invariant under the rotation on the k relevant variables.

Denote $\boldsymbol{\Sigma}$ the covariance matrix of $\mathbf{x}$. We most often use the empirical covariance matrix $\hat{\boldsymbol{\Sigma}}$ defined by

$$\hat{\boldsymbol{\Sigma}} = \frac{1}{n}\sum_{i=1}^{n} \mathbf{x}_i\mathbf{x}_i^T = \frac{1}{n}\mathbf{X}\mathbf{X}^T$$

where

$$\mathbf{X} = \begin{bmatrix} x_{11} & x_{12} & \cdots & x_{1n} \\ x_{21} & x_{22} & \cdots & x_{2n} \\ \vdots & \vdots & \vdots & \vdots \\ x_{N1} & x_{N2} & \cdots & x_{Nn} \end{bmatrix}_{N \times n}$$

where $\mathbf{X}$ is an random matrix of $N \times n$. Here $\hat{\boldsymbol{\Sigma}}$ is the maximum likelihood estimation in the Gaussian case, when the mean is zero. Often, $\hat{\boldsymbol{\Sigma}}$ is the only data provided to the statistician.

10.5 Information Plus Noise Model Using Sums of Random Vectors

A ubiquitous model is information plus noise. We consider the matrix setting:

$$\mathbf{y}_k = \mathbf{x}_k + \mathbf{z}_k, \quad k = 1, 2, \ldots, n$$

where $\mathbf{x}_k$ represents the information and $\mathbf{z}_k$ the noise. Often, we have n independent copies of $\mathbf{Y}$ at our disposal for high-dimensional data processing. It is natural to consider the matrix concentration of measure:

$$\mathbf{y}_1 + \cdots + \mathbf{y}_n = (\mathbf{x}_1 + \cdots + \mathbf{x}_n) + (\mathbf{z}_1 + \cdots + \mathbf{z}_n),$$

where these matrices $\mathbf{y}_k, \mathbf{x}_k, \mathbf{z}_k$ may be independent or dependent. The power of expectation is due to the fact that expectation is valid for both independent and dependent (matrix-valued) random variables. Expectation is also linear, which is fundamentally useful. The linearity of expectation implies that

$$\begin{aligned}\mathbb{E}\left(\mathbf{y}_1 + \cdots + \mathbf{y}_n\right) &= \mathbb{E}\mathbf{y}_1 + \cdots + \mathbb{E}\mathbf{y}_n \\ &= \mathbb{E}\left(\mathbf{x}_1 + \cdots + \mathbf{x}_n\right) + \mathbb{E}\left(\mathbf{z}_1 + \cdots + \mathbf{z}_n\right) \\ &= \mathbb{E}\mathbf{x}_1 + \cdots + \mathbb{E}\mathbf{x}_n + \mathbb{E}\mathbf{z}_1 + \cdots + \mathbb{E}\mathbf{z}_n.\end{aligned}$$

Consider a hypothesis testing problem in the setting of sums of random matrices:

$$\begin{aligned}&\mathcal{H}_0 : \rho = \mathbf{z}_1^{'} + \cdots + \mathbf{z}_n^{'} \\ &\mathcal{H}_1 : \sigma = \left(\mathbf{x}_1 + \cdots + \mathbf{x}_n\right) + \left(\mathbf{z}_1 + \cdots + \mathbf{z}_n\right).\end{aligned}$$

10.6 Information Plus Noise Model Using Sums of Random Matrices

A ubiquitous model is information plus noise. We consider the matrix setting:

$$\mathbf{Y}_k = \mathbf{X}_k + \mathbf{Z}_k, \quad k = 1, 2, \ldots, n$$

where $\mathbf{X}_k$ represents the information and $\mathbf{Z}_k$ the noise. Often, we have n independent copies of $\mathbf{Y}$ at our disposal for high-dimensional data processing. It is natural to consider the matrix concentration of measure:

$$\mathbf{Y}_1 + \cdots + \mathbf{Y}_n = \left(\mathbf{X}_1 + \cdots + \mathbf{X}_n\right) + \left(\mathbf{Z}_1 + \cdots + \mathbf{Z}_n\right),$$

where these matrices $\mathbf{Y}_k, \mathbf{X}_k, \mathbf{Z}_k$ may be independent or dependent. The power of expectation is due to the fact that expectation is valid for both independent and dependent (matrix-valued) random variables. Expectation is also linear, which is fundamentally useful. The linearity of expectation implies that

$$\begin{aligned}\mathbb{E}\left(\mathbf{Y}_1 + \cdots + \mathbf{Y}_n\right) &= \mathbb{E}\mathbf{Y}_1 + \cdots + \mathbb{E}\mathbf{Y}_n \\ &= \mathbb{E}\left(\mathbf{X}_1 + \cdots + \mathbf{X}_n\right) + \mathbb{E}\left(\mathbf{Z}_1 + \cdots + \mathbf{Z}_n\right) \\ &= \mathbb{E}\mathbf{X}_1 + \cdots + \mathbb{E}\mathbf{X}_n + \mathbb{E}\mathbf{Z}_1 + \cdots + \mathbb{E}\mathbf{Z}_n.\end{aligned}$$

Consider a hypothesis testing problem in the setting of sums of random matrices:

$$\begin{aligned}&\mathcal{H}_0 : \rho = \mathbf{Z}_1^{'} + \cdots + \mathbf{Z}_n^{'} \\ &\mathcal{H}_1 : \sigma = \left(\mathbf{X}_1 + \cdots + \mathbf{X}_n\right) + \left(\mathbf{Z}_1 + \cdots + \mathbf{Z}_n\right).\end{aligned}$$

Trace is linear, so

$$\mathcal{H}_1: \begin{aligned} \operatorname{Tr} \boldsymbol{\sigma} &= \operatorname{Tr}\left(\mathbf{X}_1 + \cdots + \mathbf{X}_n\right) + \operatorname{Tr}\left(\mathbf{Z}_1 + \cdots + \mathbf{Z}_n\right) \\ &= \left(\operatorname{Tr} \mathbf{X}_1 + \cdots + \operatorname{Tr} \mathbf{X}_n\right) + \left(\operatorname{Tr} \mathbf{Z}_1 + \cdots + \operatorname{Tr} \mathbf{Z}_n\right) \end{aligned}$$

It is natural to consider $\boldsymbol{\sigma} - \boldsymbol{\rho}$, as in quantum information processing. We are naturally led to the tail bounds of $\boldsymbol{\sigma} - \boldsymbol{\rho}$. It follows that

$$\boldsymbol{\sigma} - \boldsymbol{\rho} = \left(\mathbf{X}_1 + \cdots + \mathbf{X}_n\right) + \left(\mathbf{Z}_1 + \cdots + \mathbf{Z}_n\right) - \left(\mathbf{Z}_1^{'} + \cdots + \mathbf{Z}_n^{'}\right). \tag{10.9}$$

We assume that the eigenvalues, singular values and diagonal entries of Hermitian matrices are arranged in decreasing order. Thus, $\lambda_1 = \lambda_{\max}$, and $\lambda_n = \lambda_{\min}$.

Theorem 10.6.1 (Eigenvalues of Sums of Two Matrices [16]). *Let* $\mathbf{A}, \mathbf{B}$ *are* $n \times n$ *Hermitian matrices. Then*

$$\lambda_i\left(\mathbf{A}\right) + \lambda_n\left(\mathbf{B}\right) \leqslant \lambda_i\left(\mathbf{A} + \mathbf{B}\right) \leqslant \lambda_i\left(\mathbf{A}\right) + \lambda_1\left(\mathbf{B}\right).$$

In particular,

$$\begin{aligned} \lambda_1\left(\mathbf{A}\right) + \lambda_n\left(\mathbf{B}\right) \leqslant \lambda_1\left(\mathbf{A} + \mathbf{B}\right) \leqslant \lambda_1\left(\mathbf{A}\right) + \lambda_1\left(\mathbf{B}\right) \\ \lambda_n\left(\mathbf{A}\right) + \lambda_n\left(\mathbf{B}\right) \leqslant \lambda_n\left(\mathbf{A} + \mathbf{B}\right) \leqslant \lambda_n\left(\mathbf{A}\right) + \lambda_1\left(\mathbf{B}\right). \end{aligned}$$

It is natural to consider the maximum eigenvalue of $\boldsymbol{\sigma} - \boldsymbol{\rho}$ and the minimum eigenvalue of $\boldsymbol{\sigma} - \boldsymbol{\rho}$. The use of Theorem 10.6.1 in (10.9) leads to the upper bound

$$\begin{aligned} &\lambda_{\max}\left[\left(\mathbf{X}_1 + \cdots + \mathbf{X}_n\right) + \left(\mathbf{Z}_1 + \cdots + \mathbf{Z}_n\right) + \left(-\mathbf{Z}_1^{'} - \cdots - \mathbf{Z}_n^{'}\right)\right] \\ &\leqslant \lambda_{\max}\left[\left(\mathbf{X}_1 + \cdots + \mathbf{X}_n\right) + \left(\mathbf{Z}_1 + \cdots + \mathbf{Z}_n\right)\right] + \lambda_{\max}\left[-\left(\mathbf{Z}_1^{'} + \cdots + \mathbf{Z}_n^{'}\right)\right] \\ &= \lambda_{\max}\left[\left(\mathbf{X}_1 + \cdots + \mathbf{X}_n\right) + \left(\mathbf{Z}_1 + \cdots + \mathbf{Z}_n\right)\right] - \lambda_{\min}\left(\mathbf{Z}_1^{'} + \cdots + \mathbf{Z}_n^{'}\right) \\ &\leqslant \lambda_{\max}\left[\left(\mathbf{X}_1 + \cdots + \mathbf{X}_n\right)\right] + \lambda_{\max}\left[\left(\mathbf{Z}_1 + \cdots + \mathbf{Z}_n\right)\right] - \lambda_{\min}\left(\mathbf{Z}_1^{'} + \cdots + \mathbf{Z}_n^{'}\right). \end{aligned} \tag{10.10}$$

The third line of (10.10) follows from the fact that [53, p. 13]

$$\lambda_{\min}\left(\mathbf{A}\right) = -\lambda_{\max}\left(-\mathbf{A}\right),$$

where $\mathbf{A}$ is a Hermitian matrix. In the fourth line, we have made the assumption that the sum of information matrices $\left(\mathbf{X}_1 + \cdots + \mathbf{X}_n\right)$ and noise matrices $\left(\mathbf{Z}_1 + \cdots + \mathbf{Z}_n\right)$ are independent from each. Similarly, we have the lower bound

$$\begin{aligned}
&\lambda_{\min}\left[(\mathbf{X}_1+\cdots+\mathbf{X}_n)+(\mathbf{Z}_1+\cdots+\mathbf{Z}_n)+\left(-\mathbf{Z}_1^{'}-\cdots-\mathbf{Z}_n^{'}\right)\right]\\
&\geqslant \lambda_{\min}\left[(\mathbf{X}_1+\cdots+\mathbf{X}_n)+(\mathbf{Z}_1+\cdots+\mathbf{Z}_n)\right]+\lambda_{\min}\left[-\left(\mathbf{Z}_1^{'}+\cdots+\mathbf{Z}_n^{'}\right)\right]\\
&= \lambda_{\min}\left[(\mathbf{X}_1+\cdots+\mathbf{X}_n)+(\mathbf{Z}_1+\cdots+\mathbf{Z}_n)\right]-\lambda_{\max}\left(\mathbf{Z}_1^{'}+\cdots+\mathbf{Z}_n^{'}\right).\\
&\geqslant \lambda_{\min}\left[(\mathbf{X}_1+\cdots+\mathbf{X}_n)\right]+\lambda_{\min}\left[(\mathbf{Z}_1+\cdots+\mathbf{Z}_n)\right]-\lambda_{\max}\left(\mathbf{Z}_1^{'}+\cdots+\mathbf{Z}_n^{'}\right)
\end{aligned} \tag{10.11}$$

10.7 Matrix Hypothesis Testing

Let us consider the matrix hypothesis testing

$$\begin{aligned}
&\mathcal{H}_0 : \mathbf{N}\\
&\mathcal{H}_1 : \mathbf{Y} = \sqrt{SNR}\cdot\mathbf{X}+\mathbf{N}
\end{aligned} \tag{10.12}$$

where SNR represents the signal-to-noise ratio, and $\mathbf{X}$ and $\mathbf{N}$ are two random matrices of $m \times n$. We assume that $\mathbf{X}$ is independent of $\mathbf{N}$. The problem of (10.12) is equivalent to the following:

$$\begin{aligned}
&\mathcal{H}_0 : \mathbf{N}\mathbf{N}^H\\
&\mathcal{H}_1 : \mathbf{Y}\mathbf{Y}^H = SNR\cdot\mathbf{X}\mathbf{X}^H+\mathbf{N}\mathbf{N}^H+\sqrt{SNR}\cdot\left(\mathbf{X}\mathbf{N}^H+\mathbf{N}\mathbf{X}^H\right)
\end{aligned} \tag{10.13}$$

One metric of our interest is the covariance matrix with its trace

$$f(SNR,\mathbf{X}) = \mathrm{Tr}\left(\left(\mathbb{E}\left(\mathbf{Y}\mathbf{Y}^H\right)-\mathbb{E}\left(\mathbf{N}\mathbf{N}^H\right)\right)^2\right). \tag{10.14}$$

This function is not only positive but also linear (trace function is linear). When only N independent realizations are available, we can replace the expectation with its average form

$$\hat{f}(SNR,\mathbf{X}) = \mathrm{Tr}\left(\left(\frac{1}{N}\sum_{i=1}^{N}\mathbf{Y}_i\mathbf{Y}_i^H-\frac{1}{N}\sum_{i=1}^{N}\mathbf{N}_i\mathbf{N}_i^H\right)^2\right). \tag{10.15}$$

Hypothesis $\mathcal{H}_0$ can be viewed as the extreme case of $SNR = 0$. It can be shown that $\hat{f}(SNR,\mathbf{X})$ is a Lipschitz continuous function of SNR and $\mathbf{X}$. $\hat{f}(SNR,\mathbf{X})$ is a trace functional of $\mathbf{X}$. It is known that the trace functional is strongly concentrated [180]. It has the form as follows

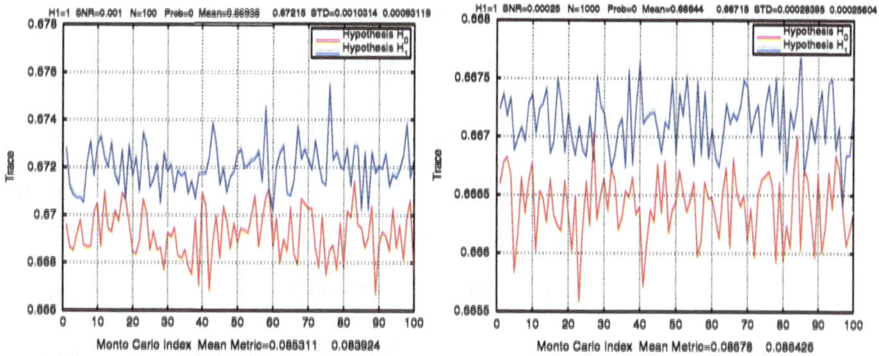

Fig. 10.1 Random matrix detection: (**a**) SNR = −30 dB, N = 100; (**b**) SNR = −36 dB, N = 1,000

$$\mathbb{P}\left(\left|\hat{f}(SNR, \mathbf{X}) - \mathbb{E}\hat{f}(SNR, \mathbf{X})\right| > t\right) \leqslant Ce^{-n^2t^2/c} \tag{10.16}$$

where C, c are two absolute constants independent of dimension n.

Figure 10.1 illustrates the concentration of the trace function $\hat{f}(SNR, \mathbf{X})$ around the mean of hypothesis $\mathcal{H}_0$ and that of hypothesis $\mathcal{H}_1$, respectively. We use the following: the entries of $\mathbf{X}, \mathbf{N}$ are zero-mean, Gaussian with variance 1, m = 200, and n = 100. We plot the function $\hat{f}(SNR, \mathbf{X})$ for $K = 100$ Monte Carlo simulations since $\hat{f}(SNR, \mathbf{X})$ is a scalar valued (always positive) random variable. It is interesting to observe that the two hypotheses are more separated even if the second case (Fig. 10.1b) has a lower SNR—6 dB lower. The reason is that we have used $N = 1{,}000$ realizations (measurements) of random matrices, while in the first case only $N = 100$ is used. As claimed in (10.16), the fluctuations $\hat{f}(SNR, \mathbf{X}) - \mathbb{E}\hat{f}(SNR, \mathbf{X})$ is strongly concentrated around its expectation.

10.8 Random Matrix Detection

When A, B are Hermitian, it is fundamental to realize that the $\mathbf{TAT}^*$ and $\mathbf{TBT}^*$ are two *commutative* matrices. Obviously $\mathbf{TAT}^*$ and $\mathbf{TBT}^*$ are Hermitian since $\mathbf{A}^* = \mathbf{A}, \mathbf{B}^* = \mathbf{B}$. Using the fact that for any two complex matrices $\mathbf{C}, \mathbf{D}$. $(\mathbf{CD})^* = \mathbf{D}^*\mathbf{C}^*$, we get

$$\begin{aligned}(\mathbf{TBT}^*)(\mathbf{TAT}^*) &= \mathbf{TBT}^*\mathbf{TAT}^* = (\mathbf{AT}^*)^*(\mathbf{TBT}^*\mathbf{T})^* \\ &= \mathbf{TA}^*(\mathbf{T}^*\mathbf{T})^*(\mathbf{TB})^* = \mathbf{TA}^*\mathbf{T}^*\mathbf{TB}^*\mathbf{T}^* = (\mathbf{TA}^*\mathbf{T}^*)(\mathbf{TBT}^*),\end{aligned} \tag{10.17}$$

which says that $\mathbf{TAT}^*\mathbf{TBT}^* = \mathbf{TBT}^*\mathbf{TAT}^*$, verifying the claim.

For commutative matrices $\mathbf{C}, \mathbf{D}$, $\mathbf{C} \leq \mathbf{D}$ is equivalent to $e^{\mathbf{C}} \leq e^{\mathbf{D}}$.

The matrix exponential has the property [20, p. 235] that

$$e^{\mathbf{A}+\mathbf{B}} = e^{\mathbf{A}}e^{\mathbf{B}}$$

if and only if two matrices $\mathbf{A}, \mathbf{B}$ are commutative: $\mathbf{AB} = \mathbf{BA}$. Thus, it follows that

$$e^{\mathbf{TAT}^*} < e^{\mathbf{TBT}^*}, \tag{10.18}$$

when $\mathbf{A}, \mathbf{B}$ are Hermitian and $\mathbf{T}^*\mathbf{T} > 0$. We have that

$$e^{\mathbf{TAT}^*+\mathbf{TBT}^*} = e^{\mathbf{TAT}^*}e^{\mathbf{TBT}^*}. \tag{10.19}$$

Let $\mathcal{A}$ be the C^*-algebra. $\mathcal{A}_s$ is the set of all the self-adjoint (Hermitian) matrices. So $\mathcal{A}_s$ is the self-adjoint part of $\mathcal{A}$. Let us recall a lemma that has been proven in Sect. 2.2.4. We repeat the lemma and its proof here for convenience.

Lemma 10.8.1 (Large deviations and Bernstein trick). *For a matrix-valued random variable* $\mathbf{A}$, $\mathbf{B} \in \mathcal{A}_s$, *and* $\mathbf{T} \in \mathcal{A}$ *such that* $\mathbf{T}^*\mathbf{T} > 0$

$$\mathbb{P}\{\mathbf{A} \not\leqslant \mathbf{B}\} \leqslant \operatorname{Tr}\left[\mathbb{E}\mathrm{e}^{\mathbf{TAT}^*-\mathbf{TBT}^*}\right] = \mathbb{E}\left[\operatorname{Tr}\mathrm{e}^{\mathbf{TAT}^*-\mathbf{TBT}^*}\right]. \tag{10.20}$$

Proof. We directly calculate

$$\begin{aligned}
\mathbb{P}\left(\mathbf{A} \nleq \mathbf{B}\right) &= \mathbb{P}\left(\mathbf{A} - \mathbf{B} \nleq 0\right) \\
&= \mathbb{P}\left(\mathbf{TAT}^* - \mathbf{TBT}^* \nleq 0\right) \\
&= \mathbb{P}\left[\mathrm{e}^{\mathbf{TAT}^*-\mathbf{TBT}^*} \nleq \mathbf{I}\right] \\
&\leqslant \operatorname{Tr}\left[\mathbb{E}\mathrm{e}^{\mathbf{TAT}^*-\mathbf{TBT}^*}\right].
\end{aligned} \tag{10.21}$$

Here, the second line is because the mapping $\mathbf{X} \mapsto \mathbf{TXT}^*$ is bijective and preserves the order. As shown above in (10.17), when A, B are Hermitian, the $\mathbf{TAT}^*$ and $\mathbf{TBT}^*$ are two *commutative* matrices. For commutative matrices $\mathbf{C}, \mathbf{D}$, $\mathbf{C} \leq \mathbf{D}$ is equivalent to $e^{\mathbf{C}} \leq e^{\mathbf{D}}$, from which the third line follows. The last line follows from Chebyshev's inequality (2.2.11). □

The closed form of $\mathbb{E}\left[\operatorname{Tr}\left(e^{\mathbf{X}}\right)\right]$ is available in Sect. 1.6.4.

The famous Golden-Thompson inequality is recalled here

$$\operatorname{Tr}\left(e^{\mathbf{A}+\mathbf{B}}\right) \leqslant \operatorname{Tr}\left(e^{\mathbf{A}} \cdot e^{\mathbf{B}}\right), \tag{10.22}$$

where $\mathbf{A}, \mathbf{B}$ are arbitrary Hermitian matrices. This inequality is very tight (almost sharp).

$$\operatorname{Tr}(\mathbf{AB}) \leqslant \|\mathbf{A}\|_{\mathrm{op}} \operatorname{Tr}(\mathbf{B})$$

The random matrix based hypothesis testing problem is formulated as follows:

$$\begin{aligned}\mathcal{H}_0 &: \mathbf{A}, \mathbf{A} \geqslant 0 \\ \mathcal{H}_1 &: \mathbf{A}+\mathbf{B}, \mathbf{A} \geqslant 0, \mathbf{B} \geqslant 0\end{aligned} \tag{10.23}$$

where $\mathbf{A}$ and $\mathbf{B}$ are random matrices. One cannot help with the temptation of using

$$\mathbb{P}\left(\mathbf{A}+\mathbf{B} > \mathbf{A}\right) = \mathbb{P}\left(e^{\mathbf{A}+\mathbf{B}} > e^{\mathbf{A}}\right),$$

which is false. It is true only when $\mathbf{A}$ and $\mathbf{B}$ commute, i.e., $\mathbf{AB} = \mathbf{BA}$. In fact, in general, we have that

$$\mathbb{P}\left(\mathbf{A}+\mathbf{B} > \mathbf{A}\right) \neq \mathbb{P}\left(e^{\mathbf{A}+\mathbf{B}} > e^{\mathbf{A}}\right).$$

However, the $\mathbf{TAT}^*$ and $\mathbf{TBT}^*$ are two *commutative* matrices, when A, B are Hermitian.

Let us consider another formulation

$$\begin{aligned}&\mathcal{H}_0 : \mathbf{X}, \\ &\mathcal{H}_1 : \mathbf{C}+\mathbf{X}, \ \mathbf{C} \text{ is fixed.}\end{aligned}$$

where $\mathbf{X}$ is a Hermitian random matrix and $\mathbf{C}$ is a *fixed* Hermitian matrix. In particular, $\mathbf{X}$ can be a Hermitian Gaussian random matrix that is treated in Sect. 1.6.4. This formulation is related to a simple but powerful corollary of Lieb's theorem, which is Corollary 1.4.18 that has been shown previously. This result connects expectation with the trace exponential.

Corollary 10.8.2. *Let* $\mathbf{H}$ *be a fixed Hermitian matrix, and let* $\mathbf{X}$ *be a random Hermitian matrix. Then*

$$\mathbb{E}\,\mathrm{Tr}\exp\left(\mathbf{H}+\mathbf{X}\right) \leqslant \mathrm{Tr}\exp\left(\mathbf{H}+\log\left(\mathbb{E}e^{\mathbf{X}}\right)\right). \tag{10.24}$$

We claim $\mathcal{H}_1$ if the decision metric $\mathbb{E}\,\mathrm{Tr}\exp\left(\mathbf{C}+\mathbf{X}\right)$, is greater than some positive threshold t we can freely set. We can compare this expression with the left-hand-side of (10.24), by replacing $\mathbf{C}$ with $\mathbf{H}$. The probability of detection for this algorithm is

$$\mathbb{P}\left(\mathbb{E}\,\mathrm{Tr}\exp\left(\mathbf{C}+\mathbf{X}\right) > t\right),$$

which is upper bounded by

$$\mathbb{P}\left(\mathrm{Tr}\exp\left(\mathbf{C}+\log\left(\mathbb{E}e^{\mathbf{X}}\right)\right) > t\right),$$

due to (10.24). As a result, $\mathbb{E}e^{\mathbf{X}}$, the expectation of the exponential of the random matrix $\mathbf{X}$, plays a basic role in this hypothesis testing problem.

On the other hand, it follows that

$$\mathbb{E}\left[\operatorname{Tr}\left(e^{\mathbf{C}+\mathbf{X}}\right)\right] \leqslant \mathbb{E}\left[\operatorname{Tr}\left(e^{\mathbf{C}} \cdot e^{\mathbf{X}}\right)\right] \leqslant \mathbb{E}\left[\operatorname{Tr}\left(e^{\mathbf{C}}\right) \operatorname{Tr}\left(e^{\mathbf{X}}\right)\right]$$
$$= \operatorname{Tr}\left(e^{\mathbf{C}}\right) \mathbb{E}\left(\operatorname{Tr}\left(e^{\mathbf{X}}\right)\right). \tag{10.25}$$

The first inequality follows from the Golden-Thompson inequality (10.22). The second inequality follows from $\operatorname{Tr}(\mathbf{AB}) \leqslant \operatorname{Tr}(\mathbf{A}) \operatorname{Tr}(\mathbf{B})$ when $\mathbf{A} \geq 0$ and $\mathbf{B} \geq 0$ are of the same size [16, Theorem 6.5]. Note that the all the eigenvalues of an exponential matrix are nonnegative. The final step follows from the fact that $\mathbf{C}$ is fixed. It follows from (10.25) that

$$\mathbb{P}\left(\mathbb{E}\left[\operatorname{Tr}\left(e^{\mathbf{C}+\mathbf{X}}\right)\right] > t\right) \leqslant \mathbb{P}\left(\operatorname{Tr}\left(e^{\mathbf{C}}\right) \mathbb{E}\left(\operatorname{Tr}\left(e^{\mathbf{X}}\right)\right) > t\right),$$

which is the final upper bound of interest. The $\mathbb{E}\left(\operatorname{Tr}\left(e^{\mathbf{X}}\right)\right)$ plays a basic role. Fortunately, for Hermitian Gaussian random matrices $\mathbf{X}$, the closed form expression of $\mathbb{E}\left(\operatorname{Tr}\left(e^{\mathbf{X}}\right)\right)$ is obtained in Sect. 1.6.4.

Example 10.8.3 (Commutative property of $\mathbf{TAT}^*$ *and* $\mathbf{TBT}^*$*).* EXPM(X) is the matrix exponential of X. EXPM is computed using a scaling and squaring algorithm with a Pade approximation, while EXP(X) is the exponential of the elements of X. For example EXP(0) gives a matrix whose entries are all ones, while the EXPM(X) is the unit matrix whose diagonal elements are ones and all non-diagonal elements are zeros. It is very critical to realize that EXPM(X) should be used to calculate the matrix exponential of X, rather than EXP(X).

Without loss of generality, we set T = randn(m,n) where randn(m,n) gives a random matrix of $m \times n$ whose entries are normally distributed pseudorandom numbers, since we only need to require $\mathbf{T}^*\mathbf{T} > 0$. For two Hermitian matrices $\mathbf{A}, \mathbf{B}$, the MATLAB expression

```
expm(T*A*T' - T*B*T')*expm( T*B*T') - expm(T*A*T')
```

gives zeros, while

```
expm(A - B)*expm( B) - expm(A)
```

gives non-zeros. In Latex, we have that

$$e^{\mathbf{TAT}^*-\mathbf{TBT}^*} e^{\mathbf{TBT}^*} - e^{\mathbf{TAT}^*} = 0.$$

Since $e^{\mathbf{TAT}^*-\mathbf{TBT}^*} e^{\mathbf{TBT}^*} = e^{\mathbf{TAT}^*}$. This demonstrates the fundamental role of the commutative property of $\mathbf{TAT}^*$ and $\mathbf{TBT}^*$:

$$\mathbf{TAT}^*\mathbf{TBT}^* = \mathbf{TBT}^*\mathbf{TAT}^*.$$

On the other hand, since A, B are not commutative matrices:

$$\mathbf{AB} \neq \mathbf{BA},$$

the expression $e^{\mathbf{A}-\mathbf{B}}e^{\mathbf{B}} \neq e^{\mathbf{A}}$. Note that the inverse matrix exponential is defined as

$$\left(e^{\mathbf{A}}\right)^{-1} = e^{-\mathbf{A}}.$$

Also the zero matrix $\mathbf{0}$ has $e^{\mathbf{0}} = \mathbf{I}$. □

The product rule of matrix expectation

$$\mathbb{E}(\mathbf{XY}) = \mathbb{E}(\mathbf{X})\,\mathbb{E}(\mathbf{Y})$$

play a basic rule in random matrices analysis. Often, we can only observe the product of two matrix-valued random variables X, Y. We can thus readily calculate the expectation of the product $\mathbf{X}$ and $\mathbf{Y}$. Assume we want to "deconvolve" the role of $\mathbf{X}$ to obtain the expectation of $\mathbf{Y}$. This can be done using

$$\mathbb{E}(\mathbf{XY})\left(\mathbb{E}(\mathbf{X})\right)^{-1} = \mathbb{E}(\mathbf{Y}),$$

assuming $\left(\mathbb{E}(\mathbf{X})\right)^{-1}$ exists.[1]

Example 10.8.4 (The product rule of matrix expectation: $\mathbb{E}(\mathbf{XY}) = \mathbb{E}(\mathbf{X})\,\mathbb{E}(\mathbf{Y})$*).* In particular, we consider the matrix exponentials

$$\mathbf{X} = e^{\mathbf{TAT}^* - \mathbf{TBT}^*} \qquad \mathbf{Y} = e^{\mathbf{TBT}^*},$$

where $\mathbf{T}, \mathbf{A}, \mathbf{B}$ are assumed the same as Example 10.8.3. Then, we have that

$$\begin{aligned}\mathbb{E}\left(e^{\mathbf{TAT}^* - \mathbf{TBT}^*}\right)\mathbb{E}\left(e^{\mathbf{TBT}^*}\right) = \mathbb{E}(\mathbf{XY}) &= \mathbb{E}\left(e^{\mathbf{TAT}^* - \mathbf{TBT}^*}e^{\mathbf{TBT}^*}\right)\\ &= \mathbb{E}\left(e^{\mathbf{TAT}^* - \mathbf{TBT}^* + \mathbf{TBT}^*}\right)\\ &= \mathbb{E}\left(e^{\mathbf{TAT}^*}\right).\end{aligned}$$

The first line uses the product rule of matrix expectation. The second line follows from (10.19). In MATLAB simulation, the expectation will be implemented using N independent Monto Carlo simulations and replaced with the average of the N random matrices

$$\frac{1}{N}\sum_{i=1}^{N}\mathbf{Z}_i.$$

□

[1]This is not guaranteed. The probability that the random matrix $\mathbf{X}$ is singular is studied in the literature [241,242].

Our hypothesis testing problem of (10.23) can be reformulated in terms of

$$\begin{aligned}\mathcal{H}_0 &: e^{\mathbf{TAT}^*}\\ \mathcal{H}_1 &: e^{\mathbf{TAT}^*+\mathbf{TBT}^*}\end{aligned} \tag{10.26}$$

where $\mathbf{TT}^* > 0$, and $\mathbf{A}, \mathbf{B}$ are two Hermitian random matrices. Usually, here $\mathbf{A}$ is a Gaussian random matrix representing the noise. Using Monto Carlo simulations, one often has the knowledge of $\mathbb{E}\left(e^{\mathbf{TAT}^*}\right)$ and $\mathbb{E}\left(e^{\mathbf{TAT}^*+\mathbf{TBT}^*}\right)$. Using the arguments similar to Example 10.8.4, we have that

$$\begin{aligned}\mathbb{E}\left(e^{\mathbf{TAT}^*+\mathbf{TBT}^*}\right)\mathbb{E}\left(e^{-\mathbf{TAT}^*}\right) &= \mathbb{E}\left(e^{\mathbf{TAT}^*+\mathbf{TBT}^*}e^{-\mathbf{TAT}^*}\right)\\ &= \mathbb{E}\left(e^{\mathbf{TAT}^*+\mathbf{TBT}^*-\mathbf{TAT}^*}\right)\\ &= \mathbb{E}\left(e^{\mathbf{TBT}^*}\right).\end{aligned}$$

If $\mathbf{B} = \mathbf{0}$, then $e^{\mathbf{TBT}^*} = \mathbf{I}$ and thus $\mathbb{E}\left(e^{\mathbf{TBT}^*}\right) = \mathbf{I}$, since $e^{\mathbf{0}} = \mathbf{I}$. If $\mathbf{B}$ is very weak but $\mathbf{B} \neq \mathbf{0}$, we often encounter $\mathbf{B}$ as a random matrix. Note that

$$\log \mathbb{E}e^{\mathbf{TBT}^*} = \mathbf{0}$$

if $\mathbf{B} = \mathbf{0}$, where log represents the matrix logarithm (MATLAB function LOGM(A) for matrix A). One metric to describe the difference away from zero (hypothesis $\mathcal{H}_0$) is to use the matrix norm

$$\left\|\mathbb{E}\left(e^{\mathbf{TAT}^*+\mathbf{TBT}^*}\right)\mathbb{E}\left(e^{-\mathbf{TAT}^*}\right)\right\|_{op} = \lambda_{\max}\left(\mathbb{E}\left(e^{\mathbf{TAT}^*+\mathbf{TBT}^*}\right)\mathbb{E}\left(e^{-\mathbf{TAT}^*}\right)\right).$$

Another metric is use the trace

$$\mathrm{Tr}\left(\mathbb{E}\left(e^{\mathbf{TAT}^*+\mathbf{TBT}^*}\right)\mathbb{E}\left(e^{-\mathbf{TAT}^*}\right)\right).$$

Dyson's expansion [23, p. 311]

Claim $\mathcal{H}_1$ if

$$\mathrm{Tr}\left(\mathbb{E}\left(e^{\mathbf{TAT}^*+\mathbf{TBT}^*}\right)\mathbb{E}\left(e^{-\mathbf{TAT}^*}\right)\right) > \gamma.$$

γ is set using the typical value of $\mathcal{H}_0$.

$$e^{\mathbf{A}+\mathbf{B}} - e^{\mathbf{A}} = \int_0^1 e^{(1-t)\mathbf{A}}\mathbf{B}e^{t(\mathbf{A}+\mathbf{B})}dt$$

can be used to study the difference $e^{\mathbf{A}+\mathbf{B}} - e^{\mathbf{A}}$. Our goal is to understand the perturbation of very weak $\mathbf{B}$.

10.9 Sphericity Test with Sparse Alternative

Let $\mathbf{x}_1, \ldots, \mathbf{x}_N$ be N i.i.d. realizations of a random variable $\mathbf{x}$ in $\mathbb{R}^n$. Our goal is to test the sphericity hypothesis, i.e., that the distribution of $\mathbf{x}$ is invariant by rotation in $\mathbb{R}^n$. For a Gaussian distribution, this is equivalent to testing if the covariance matrix of $\mathbf{x}$ is of the form $\sigma^2 \mathbf{I}_n$ for some known $\sigma^2 > 0$.

Without loss of generality, we assume $\sigma^2 = 1$ so that the covariance matrix is the identity matrix in $\mathbb{R}^n$. under the null hypothesis. Possible alternative hypotheses include the idea that there exists a *privileged direction*, along which $\mathbf{x}$ has more variance. In the spirit of sparse PCA [516, 517], we focus on the case where the *privileged direction is sparse*. The alternative hypothesis has the covariance matrix that is a sparse, rank 1 perturbation of the identity matrix $\mathbf{I}_n$. Formally, let $\mathbf{v} \in \mathbb{R}^n$ be such that $\|\mathbf{v}\|_2 = 1$, $\|\mathbf{v}\|_0 \leqslant k$, and $\theta > 0$. Here, for any $p \geq 1$, we denote by $\|\mathbf{v}\|_p$ the l_p norm of a vector and by extension, we denote $\|\mathbf{v}\|_0$ by its l_0 norm, that is its number of non-zero elements.

The hypotheses testing problem is written as

$$\begin{aligned} \mathcal{H}_0 &: \mathbf{x} \sim \mathcal{N}(0, \mathbf{I}_n) \\ \mathcal{H}_1 &: \mathbf{x} \sim \mathcal{N}\left(0, \mathbf{I}_n + \theta \mathbf{v}\mathbf{v}^T\right). \end{aligned}$$

The model under $\mathcal{H}_1$ is a generalization of the spiked covariance model since it allows $\mathbf{v}$ to be k-sparse on the unit Euclidean sphere. In particular, the statement of $\mathcal{H}_1$ is invariant under the k relevant variables.

Denote $\boldsymbol{\Sigma}$ the covariance matrix of $\mathbf{x}$. A most commonly used statistic is the empirical (or sample) covariance matrix $\hat{\boldsymbol{\Sigma}}$ defined by

$$\hat{\boldsymbol{\Sigma}} = \frac{1}{N} \sum_{i=1}^{N} \mathbf{x}_i \mathbf{x}_i^T.$$

It is an unbiased estimator for the covariance matrix of $\mathbf{x}$, the maximum likelihood estimator in the Gaussian case, when the mean is known to be 0. $\hat{\boldsymbol{\Sigma}}$ *is often the only data provided to the statistician.*

We say that a test discriminates between $\mathcal{H}_0$ and $\mathcal{H}_1$ with probability $1 - \delta$ if the type I and type II errors both have a probability smaller than δ. Our objective is to find a statistic $\varphi\left(\hat{\boldsymbol{\Sigma}}\right)$ and thresholds $\tau_0 < \tau_1$, depending on (n, N, k, δ) such that

$$\begin{aligned} \mathbb{P}_{\mathcal{H}_0}\left(\varphi\left(\hat{\boldsymbol{\Sigma}}\right) > \tau_0\right) &\leqslant \delta \\ \mathbb{P}_{\mathcal{H}_1}\left(\varphi\left(\hat{\boldsymbol{\Sigma}}\right) < \tau_1\right) &\leqslant \delta. \end{aligned}$$

Taking $\tau \in [\tau_0, \tau_1]$ allows us to control the type I and type II errors of the test

$$\psi\left(\hat{\mathbf{\Sigma}}\right) = \mathbf{1}\left\{\varphi\left(\hat{\mathbf{\Sigma}}\right) > \tau\right\},$$

where $\mathbf{1}\{\}$ denotes the indicator function. As desired, this test has the property to discriminate between the hypotheses with probability $1 - \delta$.

10.10 Connection with Random Matrix Theory

The sample covariance matrix $\hat{\mathbf{\Sigma}}$ has been studied extensively [5,388]. Convergence of the empirical covariance matrix to the true covariance matrix in spectral norm has received attention [518–520] under various elementwise sparsity and using thresholding methods. Our assumption allows for relevant variables to produce arbitrary small entries and thus we cannot use such results. A natural statistic would be, for example, using the largest eigenvalue of the covariance matrix.

10.10.1 Spectral Methods

For any unit vector, we have

$$\lambda_{\max}\left(\mathbf{I}_n\right) = 1 \text{ and } \lambda_{\max}\left(\mathbf{I}_n + \theta\mathbf{v}\mathbf{v}^T\right) = 1 + \theta.$$

In high dimension setting, where n may grow with N, the behavior of $\lambda_{\max}\left(\hat{\mathbf{\Sigma}}\right)$ is different. If $n/N \to \alpha > 0$, Geman [521] showed that, in accordance with the Marcenko-Pastur distribution, we have

$$\lambda_{\max}\left(\hat{\mathbf{\Sigma}}\right) \to \left(1 + \sqrt{\alpha}\right)^2 > 1,$$

where the convergence holds almost surely [198, 383]. Yin et al. [344] established that $\mathbb{E}\left(\mathbf{x}\right) = 0$ and $\mathbb{E}\left(\mathbf{x}^4\right) < \infty$ is a necessary and sufficient condition for this almost sure convergence to hold. As $\hat{\mathbf{\Sigma}} \geqslant 0$, its number of positive eigenvalues is equal to its rank (which is smaller than N), and we have

$$\lambda_{\max}\left(\hat{\mathbf{\Sigma}}\right) \geqslant \frac{1}{\operatorname{rank}\left(\hat{\mathbf{\Sigma}}\right)}\sum_{i=1}^{n}\lambda_i\left(\hat{\mathbf{\Sigma}}\right) \geqslant \frac{1}{N}\operatorname{Tr}\left(\hat{\mathbf{\Sigma}}\right) \geqslant \frac{n}{N}\frac{\operatorname{Tr}\left(\hat{\mathbf{\Sigma}}\right)}{Nn}.$$

As the sum of nN squared norms of independent standard Gaussian vectors, $\operatorname{Tr}\left(\hat{\mathbf{\Sigma}}\right) \sim \chi^2_{nN}$, hence almost surely, for $n/N \to \infty$, we have $\lambda_{\max}\left(\hat{\mathbf{\Sigma}}\right) \to \infty$ under the null hypothesis.

These two results indicate that the largest eigenvalue will not be able to discriminate between the two hypotheses unless $\theta > Cn/N$ for some positive constant C. In a "large n/small N" scenario, this corresponds to a very strong signal indeed.

10.10.2 Low Rank Perturbation of Wishart Matrices

When adding a finite rank perturbation to a Wishart matrix, a phase transition [522] arises already in the moderate dimensional regime where $n/N \to \alpha \in (0, 1)$. A very general class of random matrices exhibit similar behavior, under finite rank perturbation, as shown by Tao [523]. These results are extended to more general distributions in [524]. The analysis of [514] indicates that detection using the largest eigenvalue is impossible already for moderate dimensions, without further assumptions. Nevertheless, resorting to the sparsity assumption allows us to bypass this intrinsic limitation of using the largest eigenvalue as a test statistic.

10.10.3 Sparse Eigenvalues

To exploit the sparsity assumption, we use the fact that only a small submatrix of the empirical covariance matrix will be affected by the perturbation. Let $\mathbf{A}$ be a $n \times n$ matrix and fix $k < n$. We define the k-sparse largest eigenvalue by

$$\lambda_{\max}^{k}(\mathbf{A}) = \max_{|\mathcal{S}|=k} \lambda_{\max}^{k}(\mathbf{A}_{\mathcal{S}}).$$

For a set $\mathcal{S}$, we denote by $|\mathcal{S}|$ the cardinality of $\mathcal{S}$. We have the same equalities as for regular eigenvalues

$$\lambda_{\max}^{k}(\mathbf{I}_n) = 1 \text{ and } \lambda_{\max}^{k}\left(\mathbf{I}_n + \theta \mathbf{v}\mathbf{v}^T\right) = 1 + \theta.$$

The k-sparse largest eigenvalue behaves differently under the two hypotheses as soon as there is a $k \times k$ matrix with a significantly higher largest eigenvalue.

10.11 Sparse Principal Component Detection

The test statistic $\varphi\left(\hat{\mathbf{\Sigma}}\right) = \lambda_{\max}^{k}\left(\hat{\mathbf{\Sigma}}\right)$ can be equivalently defined as

$$\lambda_{\max}^{k}(\mathbf{A}) = \max_{\|\mathbf{x}\|_2=1, \|\mathbf{x}\|_0 \leqslant k} \mathbf{x}^T \mathbf{A} \mathbf{x} \tag{10.27}$$

for any $\mathbf{A} \succeq 0$.

10.11.1 Concentration Inequalities for the k-Sparse Largest Eigenvalue

Finding the optimal detection thresholds comes down to the concentration inequalities of the test statistic $\lambda_{\max}^{k}\left(\hat{\boldsymbol{\Sigma}}\right)$ both under the null and the alternative hypotheses. The concentration of measure phenomenon plays a fundamental role in this framework.

Consider $\mathcal{H}_1$ first. There is a unit vector with sparsity k, such that $\mathbf{x} \sim \mathcal{N}\left(0, \mathbf{I}_n + \theta \mathbf{v}\mathbf{v}^T\right)$. By definition of $\hat{\boldsymbol{\Sigma}}$, it follows that

$$\lambda_{\max}^{k}\left(\hat{\boldsymbol{\Sigma}}\right) \geqslant \mathbf{v}^T\hat{\boldsymbol{\Sigma}} = \frac{1}{N}\sum_{i=1}^{N}\left(\mathbf{x}_i^T\mathbf{v}\right)^2.$$

This problem involves only linear functionals. Since $\mathbf{x} \sim \mathcal{N}\left(0, \mathbf{I}_n + \theta \mathbf{v}\mathbf{v}^T\right)$, we have $\mathbf{x}_i^T\mathbf{v} \sim \mathcal{N}\left(0, 1+\theta\right)$.

Define the new random variable

$$Y = \frac{1}{N}\sum_{i=1}^{N}\left[\frac{1}{1+\theta}\left(\mathbf{x}_i^T\mathbf{v}\right)^2 - 1\right],$$

which has a χ^2 distribution. Using Laurent and Massart [134, Lemma 1] on concentration of the χ^2 distribution—see Lemma 3.2.1 for this and its proof, we get for any $t > 0$, that

$$\mathbb{P}\left(Y \leqslant -2\sqrt{t/N}\right) \leqslant e^{-t}.$$

Hence, taking $t = \log\left(1/\delta\right)$, we have $Y \geqslant -2\sqrt{\log\left(1/\delta\right)/N}$ with probability $1-\delta$. Therefore, under $\mathcal{H}_1$, we have with probability $1-\delta$

$$\lambda_{\max}^{k}\left(\hat{\boldsymbol{\Sigma}}\right) \geqslant 1 + \theta - 2\left(1+\theta\right)\sqrt{\frac{\log\left(1/\delta\right)}{N}}. \tag{10.28}$$

We now establish the following: Under $\mathcal{H}_0$, with probability $1-\delta$

$$\lambda_{\max}^{k}\left(\hat{\boldsymbol{\Sigma}}\right) \leqslant 1 + 4\sqrt{\frac{k\log\left(9en/k\right) + \log\left(1/\delta\right)}{N}} + 4\frac{k\log\left(9en/k\right) + \log\left(1/\delta\right)}{N}. \tag{10.29}$$

We adapt a technique from [72, Lemma 3]. Let $\mathbf{A}$ be a symmetric $n \times n$ matrix, and let $\mathcal{N}_\epsilon$ be an ϵ-net of sphere S^{n-1} for some $\epsilon \in [0,1]$. For ϵ-net, please refer to Sect. 1.10. Then,

$$\lambda_{\max}\left(\mathbf{A}\right) = \sup_{\mathbf{x}\in S^{n-1}}\left|\left\langle \mathbf{A}\mathbf{x}, \mathbf{x}\right\rangle\right| \leqslant \left(1-2\varepsilon\right)^{-1}\sup_{\mathbf{x}\in\mathcal{N}_\varepsilon}\left|\left\langle \mathbf{A}\mathbf{x}, \mathbf{x}\right\rangle\right|.$$

Using a 1/4-net over the unit sphere of $\mathbb{R}^k$, there exists a subset $\mathcal{N}_k$ of the unit sphere of $\mathbb{R}^k$, with cardinality smaller than 9^k, such that for any $\mathbf{A} \in \mathbf{S}_k^+$

$$\lambda_{\max}(\mathbf{A}) \leqslant 2 \max_{\mathbf{x} \in \mathcal{N}_\varepsilon} \mathbf{x}^T \mathbf{A} \mathbf{x}.$$

Under $\mathcal{H}_0$, we have

$$\lambda_{\max}^k\left(\hat{\mathbf{\Sigma}}\right) = 1 + \max_{|S|=k} \left\{ \lambda_{\max}\left(\hat{\mathbf{\Sigma}}_S\right) - 1 \right\},$$

where the maximum in the right-hand side is taken over all subsets of $\{1, \ldots, n\}$ that have cardinality k. See [514] for details.

10.11.2 Hypothesis Testing with $\lambda_{\max}^k$

Using above results, we have

$$\mathbb{P}_{\mathcal{H}_0}\left(\varphi\left(\hat{\mathbf{\Sigma}}\right) > \tau_0\right) \leqslant \delta$$

$$\mathbb{P}_{\mathcal{H}_1}\left(\varphi\left(\hat{\mathbf{\Sigma}}\right) < \tau_1\right) \leqslant \delta,$$

where

$$\tau_0 = 1 + 4\sqrt{\frac{k \log(9en/k) + \log(1/\delta)}{N}} + 4\frac{k \log(9en/k) + \log(1/\delta)}{N}$$

$$\tau_1 = 1 + \theta - 2(1+\theta)\sqrt{\frac{\log(1/\delta)}{N}}.$$

When $\tau_1 > \tau_0$, we take $\tau \in [\tau_0, \tau_1]$ and define the following test

$$\psi\left(\hat{\mathbf{\Sigma}}\right) = \mathbf{1}\left\{\varphi\left(\hat{\mathbf{\Sigma}}\right) > \tau\right\}.$$

It follows from the previous subsection that it discriminates between $\mathcal{H}_0$ and $\mathcal{H}_1$ with probability $1 - \delta$.

It remains to find for which values of θ, the condition $\tau_1 > \tau_0$. It corresponds to our minimum detection threshold.

Theorem 10.11.1 (Berthet and Rigollet [514]). *Assume that k, n, N and δ are such that $\bar{\theta} \leq 1$, where*

$$\bar{\theta} = 4\sqrt{\frac{k \log(9en/k) + \log(1/\delta)}{N}} + 4\frac{k \log(9en/k) + \log(1/\delta)}{N} + 4\sqrt{\frac{\log(1/\delta)}{N}}.$$

Then, for any $\theta > \bar{\theta}$ and for any $\tau \in [\tau_0, \tau_1]$, the test $\psi\left(\hat{\mathbf{\Sigma}}\right) = \mathbf{1}\left\{\varphi\left(\hat{\mathbf{\Sigma}}\right) > \tau\right\}$ discriminates between $\mathcal{H}_0$ and $\mathcal{H}_1$ with probability $1 - \delta$.

Considering the asymptotic regimes, for large n, N, k, taking $\delta = n^{-\beta}$ with $\beta > 0$, gives a sequence of tests ψ_N that discriminate between $\mathcal{H}_0$ and $\mathcal{H}_1$ with probability converging to 1, for any fixed $\theta > 0$, as soon as

$$\frac{k \log(n)}{N} \to 0.$$

Theorem 10.11.1 gives the upper bound. The lower bound for the probability of error is also found in [514]. We observe a gap between the upper and lower bound, with a term in $\log(n/k)$ in the upper bound, and one $\log(n/k^2)$ in the lower bound. However, by considering some regimes for n, N and k, it disappears. Indeed, as soon as $n \geq k^{2+\epsilon}$, for some $\epsilon > 0$, upper bound and lower bounds match up to constants, and the detection rate for the sparse eigenvalue is optimal in a minimax sense. Under this assumption, detection becomes impossible if

$$\theta < C\sqrt{\frac{k \log(n/k)}{N}},$$

for a small enough constant $C > 0$.

10.12 Semidefinite Methods for Sparse Principal Component Testing

10.12.1 Semidefinite Relaxation for $\lambda_{\max}^k$

Computing $\lambda_{\max}^k$ is a NP-hard problem. We need a relaxation to solve this problem. Semidefinite programming (SDP) is the matrix equivalent of linear programing. Define the Euclidean scalar product in $\mathbf{S}_d^+$ by $\langle \mathbf{A}, \mathbf{B} \rangle = \mathrm{Tr}(\mathbf{AB})$. A semidefinite program can be written in the canonical form:

$$\begin{aligned} \text{SDP} = \text{maximize} \quad & \mathrm{Tr}(\mathbf{CX}) \\ \text{subject to} \quad & \mathrm{Tr}(\mathbf{A}_i\mathbf{X}) \leqslant \mathbf{b}_i, \quad i \in \{1, \ldots, m\} \\ & \mathbf{X} \geqslant 0 \end{aligned} \tag{10.30}$$

A major breakthrough for sparse PCA was achieved in [516], who introduced a SDP relaxation for $\lambda_{\max}^k$, but tightness of this relaxation is, to this day, unknown. Making the change of variables $\mathbf{X} = \mathbf{x}\mathbf{x}^T$, in (10.27) yields

$$\begin{aligned} \lambda_{\max}^k(\mathbf{A}) = \text{maximize} \quad & \mathrm{Tr}(\mathbf{AX}) \\ \text{subject to} \quad & \mathrm{Tr}(\mathbf{X}) = 1, \quad \|\mathbf{X}\|_0 \leqslant k^2 \\ & \mathbf{X} \geqslant 0, \\ & rank(\mathbf{X}) = 1. \end{aligned}$$

This problem contains two sources of non-convexity: the l_0 norm constraint and the rank constraint. We make two relaxations in order to have a convex feasible set. First, for a semidefinite matrix $\mathbf{X}$, with trace 1, and sparsity k^2, the Cauchy-Schwartz inequality yields $||\mathbf{X}||_1 \leq k$, which is substituted to the cardinality constraint in this relaxation. Simply dropping the rank constraint leads to the following relaxation of our original problem:

$$\begin{aligned}\mathrm{SDP}_k(\mathbf{A}) = &\text{maximize } \mathrm{Tr}\left(\mathbf{AX}\right)\\ &\text{subject to } \mathrm{Tr}\left(\mathbf{X}\right) = 1, \quad \|\mathbf{X}\|_1 \leqslant k\\ &\mathbf{X} \geqslant 0.\end{aligned} \tag{10.31}$$

This optimization problem is convex since it consists in minimizing a linear objective over a convex set. It is a standard exercise to prove that it can be expressed in the canonical form (10.30). As such, standard convex optimization algorithms can be used to solve this problem efficiently. A relaxation of the original problem, for any $\mathbf{A} \geq 0$, it holds

$$\lambda_{\max}^k\left(\mathbf{A}\right) \leqslant \mathrm{SDP}_k\left(\mathbf{A}\right). \tag{10.32}$$

Since we have proved in Sect. 10.11.1 that $\lambda_{\max}^k\left(\hat{\mathbf{\Sigma}}\right)$ takes large values under $\mathcal{H}_1$, this inequality says that using $\mathrm{SDP}_k\left(\mathbf{A}\right)$ as a test statistic will be to our advantage under $\mathcal{H}_1$. Of course, we have to prove this stays small under $\mathcal{H}_0$. This can be obtained using the dual formulation of the SDP.

Lemma 10.12.1 (Bach et al. [525]). *For a given* $\mathbf{A} \geq 0$*, we have by duality*

$$\mathrm{SDP}_k\left(\mathbf{A}\right) = \min_{\mathbf{U} \in \mathbf{S}_p} \left\{\lambda_{\max}\left(\mathbf{A}+\mathbf{U}\right)\right\} + k\|\mathbf{U}\|_\infty. \tag{10.33}$$

Together with (10.32), Lemma 10.12.1 implies that for any $z \geq 0$ and any matrix $\mathbf{U}$ such that $\|\mathbf{U}\|_\infty \leqslant z$, it holds

$$\lambda_{\max}{}^k\left(\mathbf{A}\right) \leqslant \mathrm{SDP}_k\left(\mathbf{A}\right) \leqslant \lambda_{\max}\left(\mathbf{A}+\mathbf{U}\right) + kz. \tag{10.34}$$

A direct consequence of (10.34) is that the functional $\lambda_{\max}^k\left(\mathbf{A}\right)$ is robust to perturbations by matrices that have small $||\cdot||_\infty$-norm. Formally, let $\mathbf{A} \geq 0$ be such that its largest eigenvector has l_0 norm bounded by k. Then, for any matrix $\mathbf{W}$, (10.34) gives

$$\lambda_{\max}^k\left(\mathbf{A}+\mathbf{W}\right) \leqslant \lambda_{\max}\left(\left(\mathbf{A}+\mathbf{W}\right)-\mathbf{W}\right) + k\|\mathbf{W}\|_\infty = \lambda_{\max}^k\left(\mathbf{A}\right) + k\|\mathbf{W}\|_\infty.$$

10.12.2 High Probability Bounds for Convex Relaxation

The $\mathrm{SDP}_k\left(\hat{\boldsymbol{\Sigma}}\right)$ and other computationally efficient variants can be used as test statistics for our detection problem. Recall that $\mathrm{SDP}_k\left(\hat{\boldsymbol{\Sigma}}\right) \geqslant \lambda_{\max}^k\left(\hat{\boldsymbol{\Sigma}}\right)$. In view of (10.32), the following follows directly from (10.28): Under $\mathcal{H}_1$, we have, with high probability $1-\delta$,

$$\mathrm{SDP}_k\left(\hat{\boldsymbol{\Sigma}}\right) \geqslant 1+\theta-2\left(1+\theta\right)\sqrt{\frac{\log\left(1/\delta\right)}{N}}.$$

Similarly, we can obtain (see [514]): Under $\mathcal{H}_0$, we have, with high probability $1-\delta$,

$$\mathrm{SDP}_k\left(\hat{\boldsymbol{\Sigma}}\right) \leqslant 1+2\sqrt{\frac{k^2\log\left(4n^2/\delta\right)}{N}}+2\frac{k\log\left(4n^2/\delta\right)}{N}+2\sqrt{\frac{\log\left(2n/\delta\right)}{N}}+2\frac{\log\left(2n/\delta\right)}{N}.$$

10.12.3 Hypothesis Testing with Convex Methods

The results of the previous subsection can be written as

$$\begin{aligned}
&\mathbb{P}_{\mathcal{H}_0}\left(\mathrm{SDP}_k\left(\hat{\boldsymbol{\Sigma}}\right) > \hat{\tau}_0\right) \leqslant \delta \\
&\mathbb{P}_{\mathcal{H}_1}\left(\mathrm{SDP}_k\left(\hat{\boldsymbol{\Sigma}}\right) < \hat{\tau}_1\right) \leqslant \delta,
\end{aligned}$$

where $\hat{\tau}_0$ and $\hat{\tau}_1$ are given by

$$\begin{aligned}
\hat{\tau}_0 &= 1+2\sqrt{\frac{k^2\log\left(4n^2/\delta\right)}{N}}+2\frac{k\log\left(4n^2/\delta\right)}{N}+2\sqrt{\frac{\log\left(2n/\delta\right)}{N}}+2\frac{\log\left(2n/\delta\right)}{N} \\
\hat{\tau}_1 &= 1+\theta-2\left(1+\theta\right)\sqrt{\frac{\log\left(1/\delta\right)}{N}}.
\end{aligned}$$

Whenever $\hat{\tau}_1 > \hat{\tau}_0$, we take the threshold τ and define the following computationally efficient test

$$\hat{\psi}\left(\hat{\boldsymbol{\Sigma}}\right) = \mathbf{1}\left\{\mathrm{SDP}_k\left(\hat{\boldsymbol{\Sigma}}\right) > \tau\right\}.$$

It discriminates between $\mathcal{H}_0$ and $\mathcal{H}_1$ with probability $1-\delta$. It remains to find for which values of θ the condition $\hat{\tau}_1 > \hat{\tau}_0$ holds. It corresponds to our minimum detection level.

Theorem 10.12.2 (Berthet and Rigollet [514]). *Assume that n, N, k and δ are such that $\bar{\theta} \leq 1$, where*

$$\bar{\theta} = 2\sqrt{\frac{k^2 \log\left(4n^2/\delta\right)}{N}} + 2\frac{k \log\left(4n^2/\delta\right)}{N} + 2\sqrt{\frac{\log\left(2n/\delta\right)}{N}} + 4\sqrt{\frac{\log\left(1/\delta\right)}{N}}.$$

Then, for any $\theta > \bar{\theta}$, any $\tau \in [\hat{\tau}_0, \hat{\tau}_1]$, the test $\hat{\psi}\left(\hat{\mathbf{\Sigma}}\right) = \mathbf{1}\left\{\mathrm{SDP}_k\left(\hat{\mathbf{\Sigma}}\right) > \tau\right\}$ discriminates between $\mathcal{H}_0$ and $\mathcal{H}_1$ with probability $1 - \delta$.

By considering asymptotic regimes, for large n, N, k, taking $\delta = n^{-\beta}$ with $\beta > 0$, gives a sequence of tests $\hat{\psi}_N\left(\hat{\mathbf{\Sigma}}\right)$ that discriminates between $\mathcal{H}_0$ and $\mathcal{H}_1$ with probability converging to 1, for any fixed $\theta > 0$, as soon as

$$\frac{k^2 \log\left(n\right)}{N} \to 0.$$

Compared with Theorem 10.11.1, this price to pay for using this convex relaxation is to multiply the minimum detection level by a factor of $\sqrt{k}$. In most examples, k remains small so that this is not a very high price.

10.13 Sparse Vector Estimation

We follow [526]. The estimation of a sparse vector from noisy observations is a fundamental problem in signal processing and statistics, and lies at the heart of the growing field of compressive sensing. At its most basic level, we are interested in accurately estimating a vector $\mathbf{x} \in \mathbb{R}^n$ that has at most r non-zeros from a set of noisy linear measurements

$$\mathbf{y} = \mathbf{A}\mathbf{x} + \mathbf{z} \tag{10.35}$$

where $\mathbf{A} \in \mathbb{R}^{m \times n}$ and $\mathbf{z} \sim \mathcal{N}\left(\mathbf{0}, \sigma^2\mathbf{I}\right)$. We are often interested in the under-determined setting where m may be much smaller than n. In general, one would not expect to be able to accurately recover $\mathbf{x}$ when $m < n$ since there are more unknowns than observations. However it is by now well-known that by exploiting sparsity, it is possible to accurately estimate $\mathbf{x}$.

If we suppose that the entries of the matrix $\mathbf{A}$ are i.i.d. $\mathcal{N}(0, 1/n)$, then one can show that for any $\mathbf{x} \in \mathbb{B}_k := \{\mathbf{x} : \|\mathbf{x}\|_0 \leqslant k\}$, ℓ_1 minimization techniques such as the Lasso or the Dantzig selector produce a recovery $\hat{\mathbf{x}}$ such that

$$\frac{1}{n} \|\mathbf{x} - \hat{\mathbf{x}}\|_2^2 \leqslant C_0 \frac{k\sigma^2}{m} \log n \tag{10.36}$$

holds with high probability provided that $m = \Omega\left(k \log\left(n/k\right)\right)$ [527]. We consider the worst case error over all $\mathbf{x} \in \mathbb{B}_k$, i.e.,

$$E^{\star}(\mathbf{A}) = \inf_{\hat{\mathbf{x}}} \sup_{\mathbf{x}\in\mathbb{B}_k} \mathbb{E}\left[\frac{1}{n}\left\|\hat{\mathbf{x}}(\mathbf{y}) - \mathbf{x}\right\|_2^2\right]. \tag{10.37}$$

The following theorem gives a fundamental limit on the minimax risk which holds for any matrix $\mathbf{A}$ and any possible recovery algorithm.

Theorem 10.13.1 (Candès and Davenport [526]). *Suppose that we observe* $\mathbf{y} = \mathbf{A}\mathbf{x} + \mathbf{z}$ *where* $\mathbf{x}$ *is a* k*-sparse vector,* $\mathbf{A}$ *is an* $m \times n$ *matrix with* $m \geq k$*, and* $\mathbf{z} \sim \mathcal{N}(\mathbf{0}, \sigma^2\mathbf{I})$. *Then there exists a constant* $C_1 > 0$ *such that for all* $\mathbf{A}$,

$$E^{\star}(\mathbf{A}) \geqslant C_1 \frac{k\sigma^2}{\|\mathbf{A}\|_F^2} \log(n/k). \tag{10.38}$$

We also have that for all $\mathbf{A}$

$$E^{\star}(\mathbf{A}) \geqslant \frac{k\sigma^2}{\|\mathbf{A}\|_F^2}. \tag{10.39}$$

This theorem says that *there is no* $\mathbf{A}$ and *no recovery algorithm* that does fundamentally better than the Dantzig selector (10.36) up to a constant (say, 1/128); that is, ignoring the difference in the factors $\log n/k$and $\log n$. In this sense, the results of compressive sensing are, indeed, at the limit.

Corollary 10.13.2 (Candès and Davenport [526]). *Suppose that we observe* $\mathbf{y} = \mathbf{A}(\mathbf{x} + \mathbf{w})$ *where* $\mathbf{x}$ *is a* k*-sparse vector,* $\mathbf{A}$ *is an* $m \times n$ *matrix with* $k \leq m \leq n$*, and* $\mathbf{w} \sim \mathcal{N}(\mathbf{0}, \sigma^2\mathbf{I})$. *Then for all* $\mathbf{A}$

$$E^{\star}(\mathbf{A}) \geqslant C_1 \frac{k\sigma^2}{m} \log(n/k) \text{ and } E^{\star}(\mathbf{A}) \geqslant \frac{k\sigma^2}{m}. \tag{10.40}$$

The intuition behind this result is that when noise is added to the measurements, we can boost the SNR by rescaling $\mathbf{A}$ to have higher norm. When we instead add noise to the signal, the noise is also scaled by $\mathbf{A}$, and so no matter how $\mathbf{A}$ is designed there will always be a penalty of $1/m$.

The relevant work is [528] and [135]. We only sketch the proof ingredients. The proof of the lower bound (10.38) follows a similar course as in [135]. We will suppose that $\mathbf{x}$ is distributed uniformly on a finite set of points $\mathcal{X} \subset \mathbb{B}_k$, where $\mathcal{X}$ is constructed so that the elements of $\mathcal{X}$ are well separated. This allows us to show a lemma which follows from Fano's inequality combined with the convexity of the Kullback-Leibler (KL) divergence. The problem of constructing the packing set $\mathcal{X}$ exploits the matrix Bernstein inequality of Ahlswede and Winter.

10.14 Detection of High-Dimensional Vectors

We follow [529]. See also [530] for a relevant work. Detection of correlations is considered in [531, 532]. We emphasize the approach of the Kullback-Leibler divergence used in the proof of Theorem 10.14.2. Consider the hypothesis testing problem

$$\begin{aligned}\mathcal{H}_0 &: \mathbf{y} = \mathbf{z} \\ \mathcal{H}_1 &: \mathbf{y} = \mathbf{x} + \mathbf{z}\end{aligned} \tag{10.41}$$

where $\mathbf{x}, \mathbf{y}, \mathbf{z} \in \mathbb{R}^n$ and $\mathbf{x}$ is the unknown signal and $\mathbf{z}$ is additive noise. Here only one noisy observation is available per coordinate. The vector $\mathbf{x}$ is assumed to be sparse. Denote the scalar inner product of two column vectors $\mathbf{a}, \mathbf{b}$ by $\langle \mathbf{a}, \mathbf{b} \rangle = \mathbf{a}^T\mathbf{b}$. Now we have that

$$\begin{aligned}\mathcal{H}_0 &: y_i = \langle \mathbf{y}, \mathbf{a}_i \rangle = \langle \mathbf{z}, \mathbf{a}_i \rangle = z_i, \quad i = 1, \ldots, N \\ \mathcal{H}_1 &: y_i = \langle \mathbf{x}, \mathbf{a}_i \rangle + z_i, \quad i = 1, \ldots, N\end{aligned} \tag{10.42}$$

where the measurement vectors $\mathbf{a}_i$'s have Euclidean norm bounded by 1 and the noise z_i's are i.i.d. standard Gaussian, i.e., $\mathcal{N}(0,1)$. A test procedure based on N measurements of the form (10.42) is a binary function of the data, i.e., $T = T(\mathbf{a}_1, y_1, \ldots, \mathbf{a}_N, y_N)$, with $T = \varepsilon \in \{0,1\}$ indicating that T favors T_ϵ. The worst-case risk of a test T is defined as

$$\gamma(T) := \mathbb{P}_0(T=1) + \max_{\mathbf{x} \in \mathcal{X}} \mathbb{P}_{\mathbf{x}}(T=0),$$

where $\mathbb{P}_{\mathbf{x}}$ denotes the distribution of the data when $\mathbf{x}$ is the true underlying vector and the subset $\mathcal{X} \in \mathbb{R}^n \backslash \{0\}$. With a prior π on the set of alternatives $\mathcal{X}$, the corresponding average Bayes risk is expressed as

$$\gamma_\pi(T) := \mathbb{P}_0(T=1) + \mathbb{E}_\pi \mathbb{P}_{\mathbf{x}}(T=0),$$

where $\mathbb{E}_\pi$ denotes the expectation under π. For any prior π and any test procedure T, we have

$$\gamma(T) \geqslant \gamma_\pi(T). \tag{10.43}$$

For a vector $\mathbf{a} = (a_1, \ldots, a_m)^T$, we use the notation

$$\|\mathbf{a}\| = \left(\sum_{i=1}^m a_i^2 \right)^{1/2}, \quad |\mathbf{a}| = \sum_{i=1}^m |a_i|$$

to represent the Euclidea norm and ℓ_1-norm. For a matrix $\mathbf{A}$, the operator norm is defined as

$$\|\mathbf{A}\|_{op} = \sup_{\mathbf{x}\neq 0} \frac{\|\mathbf{A}\mathbf{x}\|}{\|\mathbf{x}\|}.$$

The $\mathbf{1}$ denotes the vector with all coordinates equal to 1.

Vectors with non-negative entries may be relevant to imaging processing.

Proposition 10.14.1 (Arias-Castro [529]). *Consider* $\mathbf{x} = \mathbf{1}/\sqrt{n}$. *Suppose we take N measurements of the form* (10.42) *for all i. Consider the test that rejects $\mathcal{H}_0$ when*

$$\sum_{i=1}^{N} y_i > \tau\sqrt{N},$$

where τ is some critical value. Its risk against $\mathbf{x}$ is equal to

$$1 - \Phi(\tau) + \Phi\left(\tau - \sqrt{N/n}\,|\mathbf{x}|\right),$$

where Φ is the standard normal distribution function. Hence, if $\tau = \tau_n \to \infty$, this test has vanishing risk against alternatives satisfying $\sqrt{N/n}\,|\mathbf{x}| - \tau_n \to \infty$.

We have used the result

$$\frac{1}{\sqrt{N}} \sum_{i=1}^{N} y_i \sim \mathcal{N}\left(\sqrt{N/n}\,|\mathbf{x}|, 1\right).$$

Let us present the main theorem of this section.

Theorem 10.14.2 (Theorem 1 of Arias-Castro [529]). *Let $\mathcal{X}(\mu, k)$ denote the set of vectors in $\mathbb{R}^n$ having exactly k non-zero entries all equal to $\mu > 0$. Based on N measurements of the form* (10.42), *possibly adaptive, any test for $\mathcal{H}_0 : \mathbf{x} = 0$ versus $\mathcal{H}_1 : \mathbf{x} \in \mathcal{X}(\mu, k)$ has risk at least $1 - \sqrt{N/(8n)}k\mu$.*

In particular, the risk against alternatives $\mathcal{H}_1 : \mathbf{x} \in \mathcal{X}(\mu, k)$ with $\sqrt{N/n}\,|\mathbf{x}| = \sqrt{N/n}k\mu \to 0$, goes to 1 uniformly over all procedures.

Proof. Since the approach is important, we follow [529] for a proof. We use the standard approach to deriving uniform lower bounds on the risk, by putting a prior on the set of alternatives and use (10.43). Here we simply choose the uniform prior on $\mathcal{X}(\mu, k)$, which is denoted by π. The hypothesis testing is now $\mathcal{H}_0 : \mathbf{x} = 0$ versus $\mathcal{H}_1 : \mathbf{x} \sim \pi$. By the Neyman-Pearson fundamental lemma, the likelihood ratio test is optimal. The likelihood ratio is defined as

$$L := \frac{\mathbb{P}_\pi(\mathbf{a}_1, y_1, \ldots, \mathbf{a}_N, y_N)}{\mathbb{P}_0(\mathbf{a}_1, y_1, \ldots, \mathbf{a}_N, y_N)} = \mathbb{E}_\pi \exp\left(\sum_{i=1}^{N} y_i\left(\mathbf{a}_i^T\mathbf{x}\right) - \left(\mathbf{a}_i^T\mathbf{x}\right)^2/2\right),$$

where $\mathbb{E}_\pi$ is the conditional expectation with respect to π, and the test is $T = \{L > 1\}$. It has the risk

$$\gamma_\pi(T) = 1 - \frac{1}{2}\|\mathbb{P}_\pi - \mathbb{P}_0\|_{\mathrm{TV}}, \tag{10.44}$$

where $\mathbb{P}_\pi = \mathbb{E}_\pi \mathbb{P}_{\mathbf{x}}$—the π-mixture of $\mathbb{P}_{\mathbf{x}}$— and $\|\cdot\|_{\mathrm{TV}}$ is the total variation distance [533, Theorem 2.2]. By Pinsker's inequality [533, Lemma 2.5]

$$\|\mathbb{P}_\pi - \mathbb{P}_0\|_{\mathrm{TV}} \leqslant \sqrt{K(\mathbb{P}_\pi, \mathbb{P}_0)/2}, \tag{10.45}$$

where $K(\mathbb{P}_\pi, \mathbb{P}_0)$ is the Kullback-Leibler divergence [533, Definition 2.5]. We have

$$\begin{aligned} K(\mathbb{P}_\pi, \mathbb{P}_0) &= -\mathbb{E}_0 \log L \\ &\leqslant \mathbb{E}_\pi \sum_{i=1}^{N} \mathbb{E}_0 \left[y_i \left(\mathbf{a}_i^T \mathbf{x}\right) - \left(\mathbf{a}_i^T \mathbf{x}\right)^2/2 \right] \\ &= \mathbb{E}_\pi \sum_{i=1}^{N} \mathbb{E}_0 \left[\left(\mathbf{a}_i^T \mathbf{x}\right)^2/2 \right] \\ &= \sum_{i=1}^{N} \mathbb{E}_0 \left[\mathbf{a}_i^T \mathbf{C} \mathbf{a}_i \right] \\ &\leqslant N \|\mathbf{C}\|_{\mathrm{op}} \end{aligned}$$

where $\mathbf{C} = \{c_{ij}\} := \mathbb{E}_\pi\left(\mathbf{x}\mathbf{x}^T\right)$. The first line is by definition; the second line follows from the definition of $\mathbb{P}_\pi/\mathbb{P}_0$, the use of Jensen's inequality justified by the convexity of $x \to -\log x$, and by Fubini's theorem; the third line follows from independence of $\mathbf{a}_i$, y_i and $\mathbf{x}$ (under $\mathbb{P}_0$) and the fact that $\mathbb{E}[y_i] = 0$; The fourth is by independence of $\mathbf{a}_i$, and $\mathbf{x}$ (under $\mathbb{P}_0$) and by Fubini's theorem; the fifth line follows since $\|\mathbf{a}_i\| \leqslant 1$ for all i.

Recall that under π the support of $\mathbf{x}$ is chosen *uniformly* at random among the subsets of size k. Then we have

$$c_{ii} = \mu^2 \mathbb{P}_\pi(x_i \neq 0) = \mu^2 \cdot \frac{k}{n}, \quad \forall i,$$

and

$$c_{ij} = \mu^2 \mathbb{P}_\pi(x_i \neq 0, x_j \neq 0) = \mu^2 \cdot \frac{k}{n} \cdot \frac{k-1}{n-1}, \quad i \neq j.$$

This simple matrix has operator norm $\|\mathbf{C}\|_{op} = \mu^2 k^2/n$.

Now going back to the Kullback-Leibler divergence, we thus have

$$K(\mathbb{P}_\pi, \mathbb{P}_0) \leqslant N \cdot \mu^2 k^2/n,$$

and returning (10.44) via (10.45), we bound the risk of the likelihood ratio test

$$\gamma(T) \geqslant 1 - \sqrt{K(\mathbb{P}_\pi, \mathbb{P}_0)/8} \geqslant 1 - \sqrt{N/(8n)} k\mu.$$

□

From Proposition 10.14.1 and Theorem 10.14.2, we conclude that the following is true in a minmax sense: Reliable detection of a nonnegative vector $\mathbf{x} \in \mathbb{R}^n$ from N noisy linear measurements is possible if $\sqrt{N/n}\,|\mathbf{x}| \to \infty$ and impossible if $\sqrt{N/n}\,|\mathbf{x}| \to 0$.

Theorem 10.14.3 (Theorem 2 of Arias-Castro [529]). *Let $\mathcal{X}(\pm\mu, k)$ denote the set of vectors in $\mathbb{R}^n$ having exactly k non-zero entries all equal to $\pm\mu$, with $\mu > 0$. Based on N measurements of the form* (10.42), *possibly adaptive, any test for $\mathcal{H}_0$: $\mathbf{x} = 0$ versus $\mathcal{H}_1 : \mathbf{x} \in \mathcal{X}(\pm\mu, k)$ has risk at least $1 - \sqrt{Nk/(8n)}\mu$.*

In particular, the risk against alternatives $\mathcal{H}_1 : \mathbf{x} \in \mathcal{X}(\pm\mu, k)$ with $N/n\|\mathbf{x}\|^2 = (N/n)\, k\mu^2 \to 0$, goes to 1 uniformly over all procedures.

We choose the uniform prior on $\mathcal{X}(\pm\mu, k)$. The proof is then completely parallel to that of Theorem 10.14.2, now with $\mathbf{C} = \mu^2 (k/n)\,\mathbf{I}$ since the signs of the nonzero entries of $\mathbf{x}$ are i.i.d. Rademacher. Thus $\|\mathbf{C}\|_{op} = \mu^2 (k/n)$.

10.15 High-Dimensional Matched Subspace Detection

The motivation of this section is to illustrate how concentration of measure plays a central role in the detection problem, freely taking material from [534]. See also the PhD dissertation [535]. The classical formulation of this problem is a binary hypothesis test of the following form

$$\begin{aligned}\mathcal{H}_0 &: \mathbf{y} = \mathbf{z} \\ \mathcal{H}_1 &: \mathbf{y} = \mathbf{x} + \mathbf{z}\end{aligned} \tag{10.46}$$

where $\mathbf{x} \in \mathbb{R}^n$ denotes a signal and $\mathbf{z} \in \mathbb{R}^n$ is a noise of known distribution. We are given a subspace $S \in \mathbb{R}^n$, and our task is to decide whether $\mathbf{x} \in S$ or $\mathbf{x} \notin S$, based on measurements $\mathbf{y}$. Tests are usually based on some measure of the energy of $\mathbf{y}$ in the subspace S, and these "matched subspace detectors" enjoy optimal properties [536,537]. See also for spectrum sensing in cognitive radio [156, 157, 159, 384, 385, 538].

Motivated by high-dimensional applications where it is prohibitive or impossible to measure $\mathbf{x}$ completely, we assume that only a small subset $\Omega \subset \{1, \ldots, n\}$ of the elements of $\mathbf{x}$ are observed with and without noise. Based on these observations, we test whether $\mathbf{x} \in S$ or $\mathbf{x} \notin S$. Given a subspace S of dimension $k \ll n$, how many elements of $\mathbf{x}$ must be observed so that we can reliably decide whether it belongs to S. The answer is that, under mild incoherence conditions, the number is $O(k \log k)$, such that reliable matched subspace detectors can be constructed from very few measurements, making them scalable and applicable to *large-scale* testing problems.

The main focus of this section is an estimator of the energy of $\mathbf{x}$ in S based on only observing the elements $x_i \in \Omega$. Let $\mathbf{x}_\Omega$ be the vector of dimension $|\Omega| \times 1$

composed of the elements $x_i, i \in \Omega$. We form the $n \times 1$ vector $\tilde{\mathbf{x}}$ with elements x_i if $x_i \in \Omega$ and zero if $i \notin \Omega$, for $i = 1, \ldots, n$. Filling missing elements with zero is a fairly common, albeit naive, approach to dealing with missing data.

Let $\mathbf{U}$ be an $n \times r$ matrix whose columns span the k-dimensional subspace S. For any $\mathbf{U}$, define $\mathbf{P}_S = \mathbf{U}\left(\mathbf{U}^T\mathbf{U}\right)^{-1}\mathbf{U}^T$. The energy of $\mathbf{x}$ in the subspace S is $\|\mathbf{P}_S\mathbf{x}\|_2^2$, where $\mathbf{P}_S$ is the projection operator onto S. Consider the case of partial measurement. Let $\mathbf{U}_\Omega$ denote the $|\Omega| \times r$ matrix, whose rows are the $|\Omega|$ rows of $\mathbf{U}$ indexed by the set Ω. Define the projection operator

$$\mathbf{P}_{S_\Omega} = \mathbf{U}_\Omega\left(\mathbf{U}_\Omega^T\mathbf{U}_\Omega\right)^{-1}\mathbf{U}_\Omega^T,$$

where the dagger $\dagger$ denotes the pseudoinverse. We have that if $\mathbf{x} \in S$, then $\|\mathbf{x} - \mathbf{P}_S\mathbf{x}\|_2^2 = 0$ and $\|\mathbf{x}_\Omega - \mathbf{P}_{S_\Omega}\mathbf{x}_\Omega\|_2^2 = 0$, whereas $\|\tilde{\mathbf{x}} - \mathbf{P}_S\tilde{\mathbf{x}}\|_2^2$ can significantly greater than zero. This property makes $\|\mathbf{P}_S\tilde{\mathbf{x}}\|_2^2$ a much better candidate estimator than $\|\mathbf{P}_{S_\Omega}\mathbf{x}_\Omega\|_2^2$. However, if $|\Omega| \leqslant r$, it is possible that $\|\mathbf{x}_\Omega - \mathbf{P}_{S_\Omega}\mathbf{x}_\Omega\|_2^2 = 0$, even if $\|\mathbf{x} - \mathbf{P}_S\mathbf{x}\|_2^2 > 0$. Our main result will show that if $|\Omega|$ is slightly greater than r, then with high probability $\|\mathbf{x}_\Omega - \mathbf{P}_{S_\Omega}\mathbf{x}_\Omega\|_2^2$ is very close to $\frac{|\Omega|}{n}\|\mathbf{x} - \mathbf{P}_S\mathbf{x}\|_2^2$.

Let $|\Omega|$ denote the cardinality of Ω. The coherence of the subspace S is

$$\mu(S) := \frac{n}{k}\max_i \|\mathbf{P}_S\mathbf{e}_i\|_2^2.$$

That is, $\mu(S)$ measures the maximum magnitude attainable by projecting a standard basis element onto S. We have $1 \leqslant \mu(S) \leqslant \frac{n}{k}$. For a vector $\mathbf{v}$, we let $\mu(\mathbf{v})$ denote the coherence of the subspace spanned by v. By plugging in the definition, we have

$$\mu(\mathbf{v}) = \frac{n\|\mathbf{v}\|_\infty^2}{\|\mathbf{v}\|_2^2}.$$

To state the main result, write

$$\mathbf{x} = \mathbf{y} + \mathbf{w}$$

where $\mathbf{y} \in S, \mathbf{w} \in S^\perp$. Again let Ω refer to the set of indices for observations of entries in $\mathbf{x}$, and denote $|\Omega| = m$. We split the quantity of interest into three terms and bound each with high probability. Consider

$$\|\mathbf{x}_\Omega - \mathbf{P}_{S_\Omega}\mathbf{x}_\Omega\|_2^2 = \|\mathbf{w}_\Omega - \mathbf{P}_{S_\Omega}\mathbf{w}_\Omega\|_2^2.$$

Let the k columns of $\mathbf{U}$ be an orthonormal basis for the subspace S. We want to show that

$$\|\mathbf{w}_\Omega - \mathbf{P}_{S_\Omega}\mathbf{w}_\Omega\|_2^2 = \|\mathbf{w}_\Omega\|_2^2 - \mathbf{w}_\Omega^T\mathbf{P}_{S_\Omega}\mathbf{w}_\Omega = \|\mathbf{w}_\Omega\|_2^2 - \mathbf{w}_\Omega^T\mathbf{U}_\Omega\left(\mathbf{U}_\Omega^T\mathbf{U}_\Omega\right)^{-1}\mathbf{U}_\Omega^T\mathbf{w}_\Omega \tag{10.47}$$

is nearly $\frac{m}{n}\|\mathbf{w}\|_2^2$ with high probability.

Theorem 10.15.1 (Theorem 1 of Balzano, Recht, and Nowak [534]). *Let* $\delta > 0$ *and* $m \geqslant \frac{8}{3} k\mu(S) \log\left(\frac{2k}{\delta}\right)$. *Then with probability at least* $1 - 4\delta$,

$$\frac{(1-\alpha)\, m - k\mu(S) \frac{(1+\beta)^2}{(1-\gamma)}}{n} \left\|\mathbf{x} - \mathbf{P}_S \mathbf{x}\right\|_2^2 \leqslant \left\|\mathbf{x}_\Omega - \mathbf{P}_{S_\Omega} \mathbf{x}_\Omega\right\|_2^2 \leqslant (1+\alpha)\frac{m}{n}\left\|\mathbf{x} - \mathbf{P}_S \mathbf{x}\right\|_2^2$$

where $\alpha = \sqrt{\frac{2\mu(\mathbf{w})^2}{m} \log(1/\delta)}, \beta = \sqrt{2\mu(\mathbf{w}) \log(1/\delta)}$, *and* $\gamma = \sqrt{\frac{8k\mu(S)^2}{3m} \log(2k/\delta)}$.

We need the following three equations [534] to bound three parts in (10.47). First,

$$(1-\alpha)\frac{m}{n}\left\|\mathbf{w}\right\|_2^2 \leqslant \left\|\mathbf{w}_\Omega\right\|_2^2 \leqslant (1+\alpha)\frac{m}{n}\left\|\mathbf{w}\right\|_2^2 \tag{10.48}$$

with probability at least $1 - 2\delta$. Second,

$$\left\|\mathbf{U}_\Omega^T \mathbf{w}_\Omega\right\|_2^2 \leqslant (1+\beta)^2 \frac{m}{n}\frac{k\mu(S)}{n}\left\|\mathbf{w}\right\|_2^2 \tag{10.49}$$

with probability at least $1 - \delta$. Third,

$$\left\|\left(\mathbf{U}_\Omega^T \mathbf{U}_\Omega\right)^{-1}\right\|_2 \leqslant \frac{n}{(1-\gamma)\, m} \tag{10.50}$$

with probability at least $1 - \delta$, provided that $\gamma < 1$. The proof tools for the three equations are McDiarmid's Inequality [539] and Noncommutative Bernstein Inequality (see elsewhere of this book).

10.16 Subspace Detection of High-Dimensional Vectors Using Compressive Sensing

We follow [540]. See also Example 5.7.6. We study the problem of detecting whether a high-dimensional vector $\mathbf{x} \in \mathbb{R}^n$ lies in a known low-dimensional subspace $\mathcal{S}$, given few compressive measurements of the vector. In high-dimensional settings, it is desirable to acquire only a small set of compressive measurements of the vector instead of measuring every coordinate. The objective is not to reconstruct the vector, but to detect whether the vector sensed using compressive measurements lies in a low-dimensional subspace or not.

One class of problems [541–543] is to consider a simple hypothesis test of whether the vector $\mathbf{x}$ is 0 (i.e. observed vector is purely noise) or a known signal vector $\mathbf{s}$:

$$\mathcal{H}_0 : \mathbf{x} = 0 \quad \text{vs.} \quad \mathcal{H}_1 : \mathbf{x} = \mathbf{s}. \tag{10.51}$$

Another class of problems [532, 534, 540] is to consider the subspace detection setting, where the subspace is known but the exact signal vector is unknown. This set up leads to the composite hypothesis test:

$$\mathcal{H}_0 : \mathbf{x} \in \mathcal{S} \quad \text{vs.} \quad \mathcal{H}_1 : \mathbf{x} \notin \mathcal{S}. \tag{10.52}$$

Equivalently, let $\mathbf{x}_\perp$ denote the component of $\mathbf{x}$ that does not lie in $\mathcal{S}$. Now we have the composite hypothesis test

$$\mathcal{H}_0 : \|\mathbf{x}_\perp\|_2 = 0 \quad \text{vs.} \quad \mathcal{H}_1 : \|\mathbf{x}_\perp\|_2 > 0. \tag{10.53}$$

The observation vector is modeled as

$$\mathbf{y} = \mathbf{A}\left(\mathbf{x} + \mathbf{w}\right) \tag{10.54}$$

where $\mathbf{x} \in \mathbb{R}^n$ is an unknown vector in $\mathcal{S}$, $\mathbf{A} \in \mathbb{R}^{m \times n}$ is a random matrix with i.i.d. $\mathcal{N}(0,1)$ entries, and $\mathbf{w} \sim \mathcal{N}\left(\mathbf{0}, \sigma^2 \mathbf{I}_{n \times n}\right)$ denotes noise with known variance σ^2 that is independent of $\mathbf{A}$. The noise model (10.54) is different from a more commonly studied case

$$\mathbf{y}' = \mathbf{A}\mathbf{x} + \mathbf{z} \tag{10.55}$$

where $\mathbf{z} \sim \mathcal{N}\left(\mathbf{0}, \sigma^2 \mathbf{I}_{m \times m}\right)$. For fixed $\mathbf{A}$, we have $\mathbf{y} \sim \mathcal{N}\left(\mathbf{A}\mathbf{x}, \sigma^2 \mathbf{A}\mathbf{A}^T\right)$ and $\mathbf{y}' \sim \mathcal{N}\left(\mathbf{A}\mathbf{x}, \sigma^2 \mathbf{I}_{m \times m}\right)$. Compressive linear measurements may be formed later on to optimize storage or data collection.

Define the projection operator $\mathbf{P}_\mathbf{U} = \mathbf{U}\mathbf{U}^T$. Then $\mathbf{x}_\perp = \left(\mathbf{I} - \mathbf{P}_\mathbf{U}\right)\mathbf{x}$, where $\mathbf{x}_\perp$ is the component of $\mathbf{x}$ that does not lie in $\mathcal{S}$, and $\mathbf{x} \in \mathcal{S}$ if and only if $\|\mathbf{x}_\perp\|_2^2 = 0$. Similar to [534], we define the test statistic $T = \left\|\left(\mathbf{I} - \mathbf{P}_{\mathbf{BU}}\right)\left(\mathbf{A}\mathbf{A}^T\right)^{-1/2}\mathbf{y}\right\|_2^2$ based on the observed vector $\mathbf{y}$ and study its properties, where $\mathbf{B} = \left(\mathbf{A}\mathbf{A}^T\right)^{-1/2}\mathbf{A}$. Here $\mathbf{P}_{\mathbf{BU}}$ is the projection operator onto the column space of $\mathbf{BU}$, specifically

$$\mathbf{P}_{\mathbf{BU}} = \mathbf{BU}\left[\left((\mathbf{BU})^T\mathbf{BU}\right)^{-1}(\mathbf{BU})^T\right]$$

if $(\mathbf{BU})^T\mathbf{BU}$ exists.

Now are ready to present the main result. For the sake of notational simplicity, we directly work with the matrix $\mathbf{B}$ and its marginal distribution. Writing $\mathbf{y} = \mathbf{B}\left(\mathbf{x} + \mathbf{w}\right)$, we have $T = \|\left(\mathbf{I} - \mathbf{P}_{\mathbf{BU}}\right)\mathbf{y}\|_2^2$. Since $\mathbf{A}$ is i.i.d. normal, the distribution of the row span of $\mathbf{A}$ (and hence $\mathbf{B}$) will be uniform over m-dimensional subspaces of $\mathbb{R}^n$ [544]. Furthermore, due to the $\left(\mathbf{A}\mathbf{A}^T\right)^{-1/2}$ term, the rows of $\mathbf{B}$ will be orthogonal (almost surely). First, we show that, in the absence of noise, the test statistic $T = \|\left(\mathbf{I} - \mathbf{P}_{\mathbf{BU}}\right)\mathbf{B}\mathbf{x}\|_2^2$ is close to $m\,\|\mathbf{x}_\perp\|_2^2/n$ with high probability.

Theorem 10.16.1 (Azizyan and Singh [540]). *Let* $0 < r < m < n$, $0 < \alpha_0 < 1$ *and* $\beta_0, \beta_1, \beta_2 > 1$. *With probability at least* $1 - \exp\left[(1-\alpha_0+\log\alpha_0)\, m/2\right] - \exp\left[(1-\beta_0+\log\beta_0)\, m/2\right] - \exp\left[(1-\beta_1+\log\beta_1)\, m/2\right] - \exp\left[(1-\beta_2+\log\beta_2)\, r/2\right]$

$$\left(\alpha_0\frac{m}{n} - \beta_1\beta_2\frac{r}{n}\right)\|\mathbf{x}_\perp\|_2^2 \leqslant \left\|(\mathbf{I}-\mathbf{P}_{\mathbf{BU}})\,\mathbf{Bx}\right\|_2^2 \leqslant \beta_0\frac{m}{n}\|\mathbf{x}_\perp\|_2^2 \qquad (10.56)$$

The proof of Theorem 10.16.1 follows from random projections and concentration of measure. Theorem 10.16.1 implies the following corollary.

Corollary 10.16.2 (Azizyan and Singh [540]). *If* $m \geqslant c_1 r\log m$, *then with probability at least* $1 - c_2\exp(-c_3 m)$,

$$d_1\frac{m}{n}\|\mathbf{x}_\perp\|_2^2 \leqslant \left\|(\mathbf{I}-\mathbf{P}_{\mathbf{BU}})\,\mathbf{Bx}\right\|_2^2 \leqslant d_2\frac{m}{n}\|\mathbf{x}_\perp\|_2^2$$

for some universal constants $c_1 > 0, c_2 > 0, c_3 \in (0,1), d_1 \in (0,1), d_2 > 1$.

Corollary 10.16.5 states that given just over r noiseless compressive measurements, we can estimate $\|\mathbf{x}_\perp\|_2^2$ accurately with high probability. In the presence of noise, it is natural to consider the hypothesis test:

$$T = \left\|(\mathbf{I}-\mathbf{P}_{\mathbf{BU}})\,\mathbf{y}\right\|_2^2 \underset{\mathcal{H}_1}{\overset{\mathcal{H}_0}{\lessgtr}} \eta \qquad (10.57)$$

The following result bounds the false alarm level and missed detection rate of this test (for appropriately chosen η) assuming a lower bound on $\|\mathbf{x}_\perp\|_2^2$ under $\mathcal{H}_1$.

Theorem 10.16.3 (Azizyan and Singh [540]). *If the assumptions of Corollary 10.16.5 are satisfied, and if for any* $\mathbf{x} \in \mathcal{H}_1$

$$\|\mathbf{x}_\perp\|_2^2 \geqslant \sigma^2\frac{4e+2}{d_1}\left(1-\frac{r}{m}\right)n,$$

then

$$\mathbb{P}\left(T \geqslant \eta|\mathcal{H}_0\right) \leqslant \exp\left[-c_4\,(m-r)\right]$$

and

$$\mathbb{P}\left(T \leqslant \eta|\mathcal{H}_1\right) \leqslant c_2\exp\left[-c_3 m\right] + \exp\left[-c_5\,(m-r)\right],$$

where $\eta = e\sigma^2\,(m-r)\,, c_4 = (e-2)\,/2, c_5 = (e+\log(2e+1))\,/2$, *and all other constants are as in Corollary 10.16.5.*

It is important to determine whether the performance of the test statistic we proposed can be improved further. The following theorem provides an information-theoretic lower bound on the probability of error of any test. A corollary of this theorem implies that the proposed test statistic is optimal, that is, every test with probability of missed detection and false alarm decreasing exponentially in the number of compressive samples m requires that the energy off the subspace scale as n.

Theorem 10.16.4 (Azizyan and Singh [540]). *Let P_0 be the joint distribution of $\mathbf{B}$ and $\mathbf{y}$ under the null hypothesis. Let P_1 be the joint distribution of $\mathbf{B}$ and $\mathbf{y}$ under the alternative hypothesis where $\mathbf{y} = \mathbf{B}(\mathbf{x} + \mathbf{w})$, for some fixed $\mathbf{x}$ such that $\mathbf{x} = \mathbf{x}_\perp$ and $\|\mathbf{x}\|_2 = M > 0$. If conditions of Corollary 10.16.5 are satisfied, then*

$$\inf_{\phi} \max_{i=0,1} P_i(\phi \neq i) \geqslant \frac{1}{8} \exp\left[-\frac{M^2}{2\sigma^2}\frac{m}{n}\right]$$

where the infimum is over all hypothesis tests ϕ.

Proof. Since the approach is interesting, we follow [540] to give a proof here. Let K be the Kullback-Leibler divergence. Then

$$\inf_{\phi} \max_{i=0,1} P_i(\phi \neq i) \geqslant \frac{1}{8} e^{-K(P_0,P_1)}$$

(see [533]). Let q be the density of $\mathbf{B}$ and $p(\mathbf{y}; \boldsymbol{\mu}, \boldsymbol{\Sigma})$ that of $\mathcal{N}(\boldsymbol{\mu}, \boldsymbol{\Sigma})$. Under P_0, $\mathbf{y} \sim \mathcal{N}(\mathbf{0}, \sigma^2 \mathbf{I}_{m\times m})$ since rows of $\mathbf{B}$ are orthonormal. So,

$$\begin{aligned} K(P_0, P_1) &= \mathbb{E}_{\mathbf{B}}\mathbb{E}_{\mathbf{y}} \log \tfrac{p(\mathbf{y};\mathbf{0},\sigma^2\mathbf{I}_{m\times m})q(\mathbf{B})}{p(\mathbf{y};\mathbf{B}\mathbf{x},\sigma^2\mathbf{I}_{m\times m})q(\mathbf{B})} \\ &= \tfrac{1}{2\sigma^2}\mathbb{E}_{\mathbf{B}} \|\mathbf{B}\mathbf{x}\|_2^2 \\ &= \tfrac{\|\mathbf{x}\|_2^2}{2\sigma^2}\tfrac{m}{n}. \end{aligned}$$

□

Corollary 10.16.5 (Azizyan and Singh [540]). *If there exists a hypothesis test ϕ based on $\mathbf{B}$ and $\mathbf{y}$ such that for all n and σ^2,*

$$\max_{i=0,1} P(\phi \neq i|\mathcal{H}_i) \leqslant C_0 \exp\left[-C_1(m-r)\right]$$

for some $C_0, C_1 > 0$, then there exists some $C > 0$ such that

$$\|\mathbf{x}_\perp\|_2^2 \geqslant C\sigma^2(1 - r/m)\, n$$

for any $\mathbf{x} \in \mathcal{H}_1$ and all n and σ^2.

Note that $r \leqslant \frac{m}{c_1 \log m}$, from Corollary 10.16.5.

10.17 Detection for Data Matrices

The problem of detection and localization of a small block of weak activation in a large matrix is considered by Balakrishnan, Kolar, Rinaldo and Singh [545]. Using information theoretic tools, they establish lower bounds on the minimum number of compressive measurements and the *weakest signal-to-noise ratio (SNR)* needed to detect the presence of an activated block of positive activation, as well as to localize the activated block, using both non-adaptive and adaptive measurements.

Let $\mathbf{A} \in \mathbb{R}^{n_1 \times n_2}$ be a signal matrix with unknown entries that we would like to recover. We consider the following observation model under which N noisy linear measurements of $\mathbf{A}$ are available

$$\mathbf{y}_i = \mathrm{Tr}\left(\mathbf{A}\mathbf{X}_i\right) + \mathbf{z}_i, \quad i = 1, \ldots, N \tag{10.58}$$

where $\mathbf{z}_1, \ldots, \mathbf{z}_N \overset{iid}{\sim} \mathcal{N}\left(0, \sigma^2\right)$, $\sigma > 0$ known, and the sensing matrices $\mathbf{X}_i$ satisfy either $\left\|\mathbf{X}_i\right\|_F \leqslant 1$ or $\mathbb{E}\left\|\mathbf{X}_i\right\|_F^2 = 1$. We are interested in two measurement schemes[2]: (1) adaptive or sequential, that is, the measurement matrices $\mathbf{X}_i$ is a (possibly randomized) function of $\left(\mathbf{y}_j, \mathbf{X}_j\right)_{j \in [i-1]}$; (2) the measurement matrices are chosen at once, that is, passively.

10.18 Two-Sample Test in High Dimensions

We follow [237] here. The use of concentration of measure for the standard quadratic form of a random matrix is the primary reason for this whole section. There are two independent sets of samples $\{\mathbf{x}_1, \ldots, \mathbf{x}_{n_1}\}$ and $\{\mathbf{y}_1, \ldots, \mathbf{y}_{n_2}\} \in \mathbb{R}^p$. They are generated in an i.i.d. manner from p-dimensional multivariate Gaussian distributions $\mathcal{N}\left(\boldsymbol{\mu}_1, \boldsymbol{\Sigma}\right)$ and $\mathcal{N}\left(\boldsymbol{\mu}_2, \boldsymbol{\Sigma}\right)$ respectively, where the mean vectors $\boldsymbol{\mu}_1$ and $\boldsymbol{\mu}_2$, and positive-definitive covariance matrix $\boldsymbol{\Sigma} > 0$, are all fixed and unknown. The hypothesis testing problem of interest here is

$$\mathcal{H}_0 : \boldsymbol{\mu}_1 = \boldsymbol{\mu}_2 \quad \text{versus} \quad \mathcal{H}_1 : \boldsymbol{\mu}_1 \neq \boldsymbol{\mu}_2. \tag{10.59}$$

The most well known test statistic for this problem is the Hotelling T^2 statistic, defined by

$$T^2 = \frac{n_1 n_2}{n_1 + n_2} \left(\bar{\mathbf{x}} - \bar{\mathbf{y}}\right)^T \hat{\boldsymbol{\Sigma}}^{-1} \left(\bar{\mathbf{x}} - \bar{\mathbf{y}}\right), \tag{10.60}$$

---

[2] We use $[n]$ to denote the set $=\{1, \ldots, n\}$.

where $\bar{\mathbf{x}} = \frac{1}{n_1}\sum_{i=1}^{n_1}\mathbf{x}_i$ and $\bar{\mathbf{y}} = \frac{1}{n_2}\sum_{i=1}^{n_2}\mathbf{y}_i$ are sample mean vectors, $\hat{\boldsymbol{\Sigma}}$ is the pooled sample covariance matrix, given by

$$\hat{\boldsymbol{\Sigma}} = \frac{1}{n}\sum_{i=1}^{n_1}(\mathbf{x}_i - \bar{\mathbf{x}})(\mathbf{x}_i - \bar{\mathbf{x}})^T + \frac{1}{n}\sum_{i=1}^{n_2}(\mathbf{y}_i - \bar{\mathbf{y}})(\mathbf{y}_i - \bar{\mathbf{y}})^T$$

and we define $n = n_1 + n_2 - 1$ for convenience.

When $p > n$, the matrix $\hat{\boldsymbol{\Sigma}}$ is singular, and the Hotelling test is not well defined. Even when $p \leq n$, the Hotelling test is known to perform poorly if p is nearly as large as n. It is well-known that $\hat{\boldsymbol{\Sigma}}$ is a degraded estimate of $\boldsymbol{\Sigma}$ in high dimensions, allowing for the data dimension p to exceed the sample size n.

The Hotelling T^2 test measures the separation of $\mathcal{H}_0$ and $\mathcal{H}_1$ in terms of the Kullback-Leibler (KL) divergence [546, p. 216] defined by

$$D_{KL}\left(\mathcal{N}(\boldsymbol{\mu}_1, \boldsymbol{\Sigma})\,|\mathcal{N}(\boldsymbol{\mu}_2, \boldsymbol{\Sigma})\right) = \frac{1}{2}\boldsymbol{\delta}^T\boldsymbol{\Sigma}^{-1}\boldsymbol{\delta},$$

with $\boldsymbol{\delta} = \boldsymbol{\mu}_1 - \boldsymbol{\mu}_2$. The relevant statistic distance is driven by the length of $\boldsymbol{\delta}$. This section is primarily motivated by the properties of random matrices. In particular, we are interested in the so-called random projection method [547]. If a projection matrix $\mathbf{P}_k^T \in \mathbb{R}^{k\times p}$ is used to project data from $\mathbb{R}^p$ to $\mathbb{R}^k$. After the projection, the classical Hotelling T^2 test, defined by (10.60), is thus used from the projected data. When this projection matrix is random—random projection method, this random projection matrix reduce the dimension and simultaneously preserve most of the length of $\boldsymbol{\delta}$.

We use the matrix $\mathbf{P}_k^T\hat{\boldsymbol{\Sigma}}\mathbf{P}_k$ as a surrogate for $\hat{\boldsymbol{\Sigma}}^{-1}$ in the high-dimensional setting. To eliminate the variability of a single random projection, we use the average of the matrix $\mathbf{P}_k\left(\mathbf{P}_k^T\hat{\boldsymbol{\Sigma}}\mathbf{P}_k\right)^{-1}\mathbf{P}_k^T$ over the ensemble $\mathbf{P}_k$, to any desired degree of precision. The resulting statistic is proportional to $\mathbb{E}_{\mathbf{P}_k}\left[\mathbf{P}_k\left(\mathbf{P}_k^T\hat{\boldsymbol{\Sigma}}\mathbf{P}_k\right)^{-1}\mathbf{P}_k^T\right]$.

Let $z_{1-\alpha}$ denote the $1-\alpha$ quartile of the standard normal distribution, and let $\Phi(\cdot)$ be its cumulative distribution function. Consider the Haar distribution on the set of matrices

$$\mathbf{P}_k^T\mathbf{P}_k = \mathbf{I}_{k\times k}, \quad \mathbf{P}_k \in \mathbb{R}^{p\times k}.$$

If $\mathbf{P}_k$ is drawn from the Haar distribution, independently of the data, then our random projection-based test statistic is defined by

$$\hat{T}_k^2 = \frac{n_1 n_2}{n_1 + n_2}(\bar{\mathbf{x}} - \bar{\mathbf{y}})^T\mathbb{E}_{\mathbf{P}_k}\left[\mathbf{P}_k\left(\mathbf{P}_k^T\hat{\boldsymbol{\Sigma}}\mathbf{P}_k\right)^{-1}\mathbf{P}_k^T\right](\bar{\mathbf{x}} - \bar{\mathbf{y}}).$$

For a desired nominal level $\alpha \in (0,1)$, our testing procedure rejects the null hypothesis $\mathcal{H}_0$ if and only if $\hat{T}_k^2 \geqslant t_\alpha$, where

$$t_\alpha \equiv \frac{y_n}{1-y_n}n + \sqrt{\frac{2y_n}{(1-y_n)^3}}\sqrt{n}z_{1-\alpha},$$

$y_n = k/n$ and $z_{1-\alpha}$ is the $1-\alpha$ quartile of the standard Gaussian distribution. $\hat{T}_k^2$ is asymptotically Gaussian.

To state a theorem, we make the condition $(\mathbf{A1})$.

A1 There is a constant $y \in (0,1)$ such that $y_n = y + o\left(\frac{1}{\sqrt{n}}\right)$.

We also need to define two parameters $\hat{\mu}_n = \frac{y_n}{1-y_n}n$, and $\hat{\sigma}_n = \sqrt{\frac{2y_n}{(1-y_n)^3}}\sqrt{n}$. $f(n) = o(g(n))$ means $f(n)/g(n) \to 0$ as $n \to 0$.

Theorem 10.18.1 (Lopes, Jacob and Wainwright [237]). *Assume that the null hypothesis $\mathcal{H}_0$ and the condition $(\mathbf{A1})$ hold. Then, as $(n,p) \to \infty$, we have the limit*

$$\frac{\hat{T}_k^2 - \hat{\mu}_n}{\hat{\mu}_n} \xrightarrow{d} \mathcal{N}(0,1), \tag{10.61}$$

(Here $\xrightarrow{d}$ stands for convergence in distribution) and as a result, the critical value t_α satisfies

$$\mathbb{P}\left(\hat{T}_k^2 \geqslant t_\alpha\right) = \alpha + o(1).$$

Proof. Following [237], we only give a sketch of the proof. Let $\tau = \frac{n_1+n_2}{n_1 n_2}$ and $\mathbf{z} \sim \mathcal{N}(0, \mathbf{I}_{p\times p})$. Under the null hypothesis that $\boldsymbol{\delta} = \mathbf{0}$, we have $\bar{\mathbf{x}} - \bar{\mathbf{y}} = \sqrt{\tau}\boldsymbol{\Sigma}^{1/2}\mathbf{z}$, and as a result,

$$\hat{T}_k^2 = \mathbf{z}^T \underbrace{\boldsymbol{\Sigma}^{1/2}\mathbb{E}_{\mathbf{P}_k}\left[\mathbf{P}_k\left(\mathbf{P}_k^T\hat{\boldsymbol{\Sigma}}\mathbf{P}_k\right)^{-1}\mathbf{P}_k^T\right]\boldsymbol{\Sigma}^{1/2}}_{\mathbf{A}}\mathbf{z}, \tag{10.62}$$

which gives us the standard quadratic form $\hat{T}_k^2 = \mathbf{z}^T\mathbf{A}\mathbf{z}$. The use of concentration of measure for the standard quadratic form is the primary reason for this whole section. Please refer to Sect. 4.15, in particular, Theorem 4.15.8.

Here, $\mathbf{A}$ is a random matrix. We may take $\bar{\mathbf{x}} - \bar{\mathbf{y}}$ and $\hat{\boldsymbol{\Sigma}}$ to be independent for Gaussian data [208]. As a result, we may assume that $\mathbf{z}$ and $\mathbf{A}$ are independent. Our overall plan is to work conditionally on $\mathbf{A}$, and use the representation

$$\mathbb{P}\left(\frac{\mathbf{z}^T\mathbf{A}\mathbf{z} - \hat{\mu}_n}{\hat{\sigma}_n} \leqslant x\right) = \mathbb{E}_{\mathbf{A}}\mathbb{P}_{\mathbf{z}}\left(\frac{\mathbf{z}^T\mathbf{A}\mathbf{z} - \hat{\mu}_n}{\hat{\sigma}_n} \leqslant x \,|\mathbf{A}\right),$$

where $x \in \mathbb{R}$.

Let $o_{\mathbb{P}_{\mathbf{A}}}(1)$ stand for a positive constant in probability under $\mathbb{P}_{\mathbf{A}}$. To demonstrate the asymptotic Gaussian distribution of $\mathbf{z}^T\mathbf{A}\mathbf{z}$, in Sect. B.4 of [237], it is shown that $\frac{\|\mathbf{A}\|_{op}}{\|\mathbf{A}\|_F} = o_{\mathbb{P}_{\mathbf{A}}}(1)$ where $\|\cdot\|_F$ denotes the Frobenius norm. This implies that the Lyupanov condition [58], which in turn implies the Lindeberg condition, and it then follows [87] that

$$\sup_{x\in\mathbb{R}}\left|\mathbb{P}_{\mathbf{z}}\left(\frac{\mathbf{z}^T\mathbf{A}\mathbf{z}-\mathrm{Tr}(\mathbf{A})}{\sqrt{2}\|\mathbf{A}\|_F}\leqslant x\,|\mathbf{A}\right)-\Phi(x)\right| = o_{\mathbb{P}_{\mathbf{A}}}(1). \tag{10.63}$$

The next step is to show that $\mathrm{Tr}(\mathbf{A})$ and $\|\mathbf{A}\|_F$ can be replaced with *deterministic* counterparts $\hat{\mu}_n = \frac{y_n}{1-y_n}n$, and $\hat{\sigma}_n = \sqrt{\frac{2y_n}{(1-y_n)^3}}\sqrt{n}$. More precisely

$$\mathrm{Tr}(\mathbf{A}) - \hat{\mu}_n = o_{\mathbb{P}_{\mathbf{A}}}(\sqrt{n}) \quad \text{and} \quad \|\mathbf{A}\|_F - \hat{\sigma}_n = o_{\mathbb{P}_{\mathbf{A}}}(\sqrt{n}). \tag{10.64}$$

Inserting (10.64) into (10.63), it follows that

$$\mathbb{P}\left(\frac{\mathbf{z}^T\mathbf{A}\mathbf{z}-\hat{\mu}_n}{\hat{\sigma}_n}\leqslant x\,|\mathbf{A}\right)-\Phi(x) = o_{\mathbb{P}_{\mathbf{A}}},$$

and the central limit theorem (10.61) follows from the dominated convergence theorem. □

To state another theorem, we need the following two conditions:

- **(A2)** There is a constant $b\in(0,1)$ such that $\frac{n_1}{n} = b + o\left(\frac{1}{\sqrt{n}}\right)$.
- **(A3)** (Local alternative) The shift vector and covariance matrix satisfy $\boldsymbol{\delta}^T\boldsymbol{\Sigma}^{-1}\boldsymbol{\delta} = o(1)$.

Theorem 10.18.2 (Lopes, Jacob and Wainwright [237]). *Assume that conditions (A1), (A2), and (A3) hold. Then, as $(n,p)\to\infty$, the power function satisfies*

$$\mathbb{P}\left(T_k^2 \geqslant t_\alpha\right) = \Phi\left(-z_{1-\alpha} + b(1-b)\cdot\sqrt{\frac{1-y}{2y}}\cdot\Delta_k\sqrt{n}\right) + o(1). \tag{10.65}$$

where

$$\Delta_k = \boldsymbol{\delta}^T\mathbb{E}_{\mathbf{P}_k}\left[\mathbf{P}_k\left(\mathbf{P}_k^T\hat{\boldsymbol{\Sigma}}\mathbf{P}_k\right)^{-1}\mathbf{P}_k^T\right]\boldsymbol{\delta}.$$

Proof. The heart of this proof is to use the conditional expectation and the concentration of quadratic form. We work under the alternative hypothesis. Let $\tau = \frac{n_1+n_2}{n_1n_2}$ and $\mathbf{z}\sim\mathcal{N}(0,\mathbf{I}_{p\times p})$. Consider the limiting value of the power function $\mathbb{P}\left(\frac{1}{n}\hat{T}_k^2\geqslant t_\alpha\right)$. Since the shift vector is nonzero $\boldsymbol{\delta}\neq 0$, We can break the test statistic into three parts. Recall from (10.62) that

$$\hat{T}_k^2 = \frac{n_1 n_2}{n_1 + n_2} (\bar{\mathbf{x}} - \bar{\mathbf{y}})^T \mathbb{E}_{\mathbf{P}_k} \left[\mathbf{P}_k \left(\mathbf{P}_k^T \hat{\boldsymbol{\Sigma}} \mathbf{P}_k \right)^{-1} \mathbf{P}_k^T \right] (\bar{\mathbf{x}} - \bar{\mathbf{y}}),$$

where

$$\bar{\mathbf{x}} - \bar{\mathbf{y}} = \sqrt{\tau} \boldsymbol{\Sigma}^{1/2} \mathbf{z} + \boldsymbol{\delta},$$

and $\mathbf{z}$ is a standard Gaussian p-vector. Expanding the definition of $\hat{T}_k^2$ and adjusting by a factor n, we have the decomposition

$$\frac{1}{n} \hat{T}_k^2 = I + II + III,$$

where

$$I = \frac{1}{n} \mathbf{z}^T \mathbf{A} \mathbf{z} \tag{10.66}$$

$$II = 2 \frac{1}{n\sqrt{\tau}} \mathbf{z}^T \mathbf{A} \boldsymbol{\Sigma}^{-1/2} \boldsymbol{\delta} \tag{10.67}$$

$$III = \frac{1}{n\tau} \boldsymbol{\delta}^T \boldsymbol{\Sigma}^{-1/2} \mathbf{A} \boldsymbol{\Sigma}^{-1/2} \boldsymbol{\delta}. \tag{10.68}$$

Recall that

$$\mathbf{A} = \boldsymbol{\Sigma}^{1/2} \mathbb{E}_{\mathbf{P}_k} \left[\mathbf{P}_k \left(\mathbf{P}_k^T \hat{\boldsymbol{\Sigma}} \mathbf{P}_k \right)^{-1} \mathbf{P}_k^T \right] \boldsymbol{\Sigma}^{1/2}.$$

We will work on the conditional expectation $\mathbb{E}_{\mathbf{A}}$ with the condition $\mathbf{A}$. Consider

$$\mathbb{P}\left(T_k^2 \geqslant t_\alpha\right) = \mathbb{E}_{\mathbf{A}} \mathbb{P}_{\mathbf{z}} \left(I \geqslant \frac{1}{n} t_\alpha - II - III \,|\mathbf{A} \right).$$

Working with $\mathrm{Tr}\left(\frac{1}{n}\mathbf{A}\right)$ and $\left\|\frac{1}{n}\mathbf{A}\right\|_F$, and multiplying the top and the bottom by $\sqrt{n}$, we have

$$\mathbb{P}\left(T_k^2 \geqslant t_\alpha\right) = \mathbb{E}_{\mathbf{A}} \mathbb{P}_{\mathbf{z}} \left(\frac{\mathbf{z}^T \left(\frac{1}{n}\mathbf{A}\right) \mathbf{z} - \mathrm{Tr}\left(\frac{1}{n}\mathbf{A}\right)}{\sqrt{2}\left\|\frac{1}{n}\mathbf{A}\right\|_F} \geqslant \frac{\sqrt{n}\left(\frac{1}{n} t_\alpha - \mathrm{Tr}\left(\frac{1}{n}\mathbf{A}\right) - II - III\right)}{\sqrt{n}\sqrt{2}\left\|\frac{1}{n}\mathbf{A}\right\|_F} \,|\mathbf{A} \right). \tag{10.69}$$

Recall the definition of the critical value

$$t_\alpha \equiv \frac{y_n}{1 - y_n} n + \sqrt{\frac{2y_n}{(1 - y_n)^3}} \sqrt{n} z_{1-\alpha}.$$

We also define the numerical sequence $l_n = \frac{1}{\tau n}\left(\frac{n}{n-k-1}\right)\Delta_k$. In Sects. C.1 and C.2 of [237], they establish the limits

$$\mathbb{P}_{\mathbf{z}}\left(\sqrt{n}\,|II| \geqslant \varepsilon\,|\mathbf{A}\right) = o_{\mathbb{P}_{\mathbf{A}}}(1), \quad \text{and} \quad \sqrt{n}\,(III - l_n) = o_{\mathbb{P}_{\mathbf{A}}}(1)$$

where $\epsilon > 0$. Inserting the limits (10.64) into (10.69), we have

$$\mathbb{P}\left(T_k^2 \geqslant t_\alpha\right) = \mathbb{E}_{\mathbf{A}}\mathbb{P}_{\mathbf{z}}\left(\frac{\mathbf{z}^T\mathbf{A}\mathbf{z} - \operatorname{Tr}(\mathbf{A})}{\sqrt{2}\|\mathbf{A}\|_F} \geqslant \frac{\sqrt{n}\left(\sqrt{\frac{2y_n}{n(1-y_n)^3}}\cdot z_{1-\alpha} - l_n - II\right)}{\sqrt{\frac{2y_n}{(1-y_n)^3}}} + o_{\mathbb{P}_{\mathbf{A}}}(1)\,|\mathbf{A}\right). \tag{10.70}$$

By the limit (10.63), we have

$$\begin{aligned}
&\mathbb{P}_{\mathbf{z}}\left(\frac{\mathbf{z}^T\mathbf{A}\mathbf{z} - \operatorname{Tr}(\mathbf{A})}{\sqrt{2}\|\mathbf{A}\|_F} z_{1-\alpha} - \sqrt{n}\frac{(1-y_n)^3}{2y_n}(l_n + II) + o_{\mathbb{P}_{\mathbf{A}}}(1)\,|\mathbf{A}\right) \\
&= \Phi\left(-z_{1-\alpha} + \sqrt{n}\sqrt{\frac{(1-y_n)^3}{2y_n}}l_n\right) + o_{\mathbb{P}_{\mathbf{A}}}(1),
\end{aligned}$$

where the error term $o_{\mathbb{P}_{\mathbf{A}}}(1)$ is bounded by 1. Integrating over $\mathbf{A}$ and applying the dominated convergence theorem, we obtain

$$\mathbb{P}\left(T_k^2 t_\alpha\right) = \Phi\left(-z_{1-\alpha} + \sqrt{n}\sqrt{\frac{(1-y_n)^3}{2y_n}}l_n\right) + o(1).$$

Using the assumptions $y_n = \frac{k}{n} = a + o\left(\frac{1}{\sqrt{n}}\right)$ and $\frac{n_1}{n} = b + o\left(\frac{1}{\sqrt{n}}\right)$, we conclude

$$\mathbb{P}\left(T_k^2 \geqslant t_\alpha\right) = \Phi\left(-z_{1-\alpha} + b(1-b)\cdot\sqrt{\frac{1-y}{2y}}\cdot\Delta_k\sqrt{n}\right) + o(1),$$

which is the same as (10.65). □

10.19 Connection with Hypothesis Detection of Noncommuntative Random Matrices

Consider the hypothesis detection of the problem

$$\begin{aligned}
&\mathcal{H}_0 : \mathbf{A} = \mathbf{R}_n \\
&\mathcal{H}_1 : \mathbf{B} = \mathbf{R}_x + \mathbf{R}_n
\end{aligned} \tag{10.71}$$

where $\mathbf{R}_x$ is the true covariance matrix of the unknown signal and $\mathbf{R}_n$ is the true covariance matrix of the noise. The optimal average probability of correct detection [5, p. 117] is

$$\frac{1}{2} + \frac{1}{4}\|\mathbf{A} - \mathbf{B}\|_1,$$

where the trace norm $\|\mathbf{X}\|_1 = \mathrm{Tr}\sqrt{\mathbf{X}\mathbf{X}^H}$ is the sum of the absolute eigenvalues. See also [548, 549] for the original derivation.

In practice, we need to use the sample covariance matrix to replace the true covariance matrix, so we have

$$\begin{aligned} \mathcal{H}_0 &: \hat{\mathbf{A}} = \hat{\mathbf{R}}_n \\ \mathcal{H}_1 &: \hat{\mathbf{B}} = \hat{\mathbf{R}}_x + \hat{\mathbf{R}}_n \end{aligned}$$

where $\hat{\mathbf{R}}_x$ is the true covariance matrix of the unknown signal and $\hat{\mathbf{R}}_n$ is the true covariance matrix of the noise. Using the triangle inequality of the norm, we have

$$\left\|\hat{\mathbf{A}} - \hat{\mathbf{B}}\right\|_1 \leqslant \left\|\hat{\mathbf{A}} - \mathbf{A}\right\|_1 + \left\|\mathbf{B} - \hat{\mathbf{B}}\right\|_1.$$

Concentration inequalities will connect the non-asymptotic convergence of the sample covariance matrix to its true value, via. See Chaps. 9 and 5 for covariance matrix estimation.

10.20 Further Notes

In [550], Sharpnack, Rinaldo, and Singh consider the basic but fundamental task of deciding whether a given graph, over which a noisy signal is observed, contains a cluster of anomalous or activated nodes comprising an induced connected subgraph.

Ramirez, Vita, Santamaria and Scharf [551] studies the existence of locally most powerful invariant tests for the problem of testing the covariance structure of a set of Gaussian random vectors. In practical scenarios the above test can provide better performance than the typically used generalized likelihood ratio test (GLRT).

Onatski, Moreira and Hallin [552] consider the problem of testing the null hypothesis of sphericity of a high-dimensional covariance matrix against an alternative of multiple symmetry-breaking directions (multispiked alternatives).

Chapter 11
Probability Constrained Optimization

In this chapter, we make the connection between concentration of measure and probability constrained optimization. It is the use of concentration inequality that makes the problem of probability constrained optimization *mathematically tractable*. Concentration inequalities are the enabling techniques that make possible probability constrained optimization.

11.1 The Problem

We follow Nemirovski [553] to set up the problem. We consider a probability constraint

$$\text{Prob}\left\{\boldsymbol{\xi} : \mathbf{A}\left(\mathbf{x}, \boldsymbol{\xi}\right) \in \mathbb{K}\right\} \geqslant 1 - \varepsilon \tag{11.1}$$

where $\mathbf{x}$ is the decision vector, $\mathbb{K}$ is a closed convex cone, and $\mathbf{A}\left(\mathbf{x}, \boldsymbol{\xi}\right)$ is defined as

$$\mathbf{A}\left(\mathbf{x}, \xi\right) = \mathbf{A}_0\left(\mathbf{x}\right) + \sigma \sum_{i=1}^{N} \xi_i \mathbf{A}_i\left(\mathbf{x}\right), \tag{11.2}$$

where

- $\mathbf{A}_i\left(\cdot\right)$ are affine mapping from $\mathbb{R}^n$ to finite-dimensional real vector space E;
- ξ_i are scalar random perturbations satisfying the relations

  1. ξ_i are mutually independent;
  2. $\mathbb{E}\left\{\xi_i\right\} = 0$;

$$\mathbb{E}\left\{\exp\left(\xi_i^2/4\right)\right\} \leqslant \sqrt{2}; \tag{11.3}$$

R. Qiu and M. Wicks, *Cognitive Networked Sensing and Big Data*,
DOI 10.1007/978-1-4614-4544-9_11,

- $\sigma > 0$ is the level of perturbations.
- $\mathbb{K}$ is a closed convex cone in E.

For ξ_i, we are primarily interested in the following cases:

- $\xi_i \sim \mathcal{N}(0,1)$ are Gaussian noise; the absolute constants in (11.3) comes exactly from the desire to use the standard Gaussian perturbations;
- $\mathbb{E}\{\xi_i\} = 0; \quad |\xi_i| \leqslant 1$ so ξ_i are bounded random noise.

For the vector space E and the closed pointed convex cone $\mathbb{K}$, we are interested in the cases: (1) if $E = \mathbb{R}$, (real), and $\mathbb{K} = \mathbb{R}_+$ (positive real); (11.2) is the special case for a scalar linear inequality. (2) If a real vector space is considered $E = \mathbb{R}^{m+1}$, and

$$\mathbb{K} = \left\{ \mathbf{x} \in \mathbb{R}^{m+1} : x_{m+1} \geqslant \sqrt{x_1^2 + \cdots + x_m^2} \right\};$$

here (11.2) is a randomly perturbed Conic Quadratic Inequality (CQI), where the data are affine in the perturbations. (3) $E = \mathbb{S}^m$ is the space of $m \times m$ symmetric matrices, $\mathbb{K} = \mathbb{S}_+^m$ is the cone of positive semidefinite matrices from $\mathbb{S}^m$; here (11.2) is a randomly perturbed Linear Matrix Inequality (LMI).

We are interested to describe $\mathbf{x}$'s which satisfy (11.2) with a given high *probability*, that is:

$$\text{Prob}\ \{\boldsymbol{\xi} : \mathbf{A}(\mathbf{x}, \boldsymbol{\xi}) \notin \mathbb{K}\} \leqslant \varepsilon, \tag{11.4}$$

for a given $\varepsilon \ll 1$. Our ultimate goal is to optimize over the resulting set, under the additive constraints on $\mathbf{x}$. A fundamental problem is that (11.4) is *computationally intractable*. The solution to the problem is connected with concentration of measure through large deviations of sums of random matrices, following Nemirovski [554].

Typically, the only way to estimate the probability for a probability constraint to be violated at a given point is to use Monte-Carlo simulations (so-called *scenario approach* [555–558]) with sample sizes of $\frac{1}{\varepsilon}$; this becomes too costly when ε is small such as 10^{-5} or less. A natural idea is to look for *tractable approximations* of the probability constraint, i.e., for efficiently verifiable *sufficient conditions* for its validity. The advantage of this approach is its generality, it imposes no restrictions on the distribution of $\boldsymbol{\xi}$ and on how the data enter the constraints.

An alternative to the scenario approximation is an approximation based on "closed form" upper bounding of the probability for the randomly perturbed constraints $\mathbf{A}(\mathbf{x}, \xi) \in \mathbb{K}$ to be violated. The advantage of the "closed form" approach as compared to the scenario one is that the resulting approximations are deterministic convex problems with sizes independent of the required value of ε, so that the approximations also remain practical in the case of very small values of ε. A new class of "closed form" approximations, referred to as Bernstein approximations, is proposed in [559].

Example 11.1.1 (Covariance Matrix Estimation). For independent random vectors $\mathbf{x}_k, k = 1, \ldots, K$, the sample covariance—a Hermitian, positive semidefinite, random matrix—defined as

$$\hat{\mathbf{R}} = \sum_{k=1}^{K} \mathbf{x}_k \mathbf{x}_k^T, \quad \mathbf{x}_k \in \mathbb{R}^n$$

Note that the n elements of the k-th vector $\mathbf{x}_k$ may be dependent random variables. Now, assume that there are $N+1$ observed sample covariances $\hat{\mathbf{R}}_0(\mathbf{x}), i = 0, 1, 2, \ldots, N$ such that

$$\hat{\mathbf{R}}(\mathbf{x}, \boldsymbol{\xi}) = \hat{\mathbf{R}}_0(\mathbf{x}) + \sigma \sum_{i=1}^{N} \xi_i \hat{\mathbf{R}}_i(\mathbf{x}), \tag{11.5}$$

Our task is to consider a probability constraint (11.1)

$$\text{Prob}\left\{\boldsymbol{\xi} : \hat{\mathbf{R}}(\mathbf{x}, \boldsymbol{\xi}) \geqslant 0\right\} \geqslant 1 - \varepsilon \tag{11.6}$$

where $\mathbf{x}$ is the decision vector. Thus, the covariance matrix estimation is recast in terms of an optimization. Later it will be shown that the optimization problem is convex and may be solved efficiently using the general-purpose solver that is widely available online. Once the covariance matrix is estimated, we can enter the second stage of detection process for extremely weak signal. This line of research seems novel. □

11.2 Sums of Random Symmetric Matrices

Let us follow Nemirovski [560] to explore the connection of sums of random symmetric matrices with the probability constrained optimization. Let $\|\mathbf{A}\|$ denote the standard spectral norm (the largest singular value) of an $m \times n$ matrix $\mathbf{A}$. We ask this question.

(Q1) Let $\mathbf{X}_i, 1 \leq i \leq N$, be independent $n \times n$ random symmetric matrices with zero mean and "light-tail" distributions, and let $\mathbf{S}_N = \sum_{i=1}^{N} \mathbf{X}_i$. Under what conditions is a "typical value" of $\|\mathbf{S}_N\|$ "of order 1" such that the probability for $\|\mathbf{S}_N\|$ to be $\geq t$ goes to 0 exponentially fast as $t > 1$ grows?

Let $\mathbf{B}_i$ be deterministic symmetric $n \times n$ matrices, and ξ_i be independent random scalars with zero mean and "of order of one" (e.g., $\xi_i \sim \mathcal{N}(0,1)$). We are interested in the conditions for the "typical norm" of the random matrix

$$\mathbf{S}_N = \xi_1 \mathbf{B}_1 + \cdots + \xi_N \mathbf{B}_N = \sum_{i=1}^{N} \xi_i \mathbf{B}_i$$

to be of order 1. An necessary condition is

$$\mathbb{E}\left[\mathbf{S}_N^2\right] \leqslant O(1)\mathbf{I}$$

which, translates to

$$\sum_{i=1}^{N} \mathbf{B}_i^2 \leqslant \mathbf{I}.$$

A natural conjecture is that the latter condition is sufficient as well. This answer is affirmative, as proven by So [122]. A relaxed version of this conjecture has been proven by Nemirovski [560]: Specifically, under the above condition, the typical norm of $\mathbf{S}_N$ is $\leqslant O(1)m^{1/6}$ with the probability

$$\text{Prob}\left\{\|\mathbf{S}_N\| > tm^{1/6}\right\} \leqslant O(1)\exp\left(-O(1)t^2\right)$$

for all $t > 0$.

We can ask the question.

(Q2) Let $\xi_1, \ldots, \xi_N$ be independent mean zero random variables, each of which is either (i) supported on [-1,1], or (ii) normally distributed with unit variance. Further, let $\mathbf{X}_1, \ldots, \mathbf{X}_N$ be arbitrary $m \times n$ matrices. Under what conditions on $t > 0$ and $\mathbf{X}_1, \ldots, \mathbf{X}_N$ will we have an exponential decay of the tail probability

$$\text{Prob}\left(\left\|\sum_{i=1}^{N} \xi_i \mathbf{X}_i\right\| \geqslant t\right)?$$

Example 11.2.1 (Randomly perturbed linear matrix inequalities). Consider a randomly perturbed Linear Matrix Inequalities

$$\mathbf{A}_0(\mathbf{x}) - \sum_{i=1}^{N} \xi_i \mathbf{A}_i(\mathbf{x}) \geqslant 0, \tag{11.7}$$

where $\mathbf{A}_1(\mathbf{x}), \ldots, \mathbf{A}_N(\mathbf{x})$ are affine functions of the decision vector $\mathbf{x}$ taking values in the space $\mathbb{S}^n$ of symmetric $n \times n$ matrices, and ξ_i are independent of each other random perturbations. Without loss of generality, ξ_i can be assumed to have zero means.

A natural idea is to consider the probability constraint

$$\text{Prob}\left\{\boldsymbol{\xi} = (\xi_1, \ldots, \xi_N) : \mathbf{A}_0(\mathbf{x}) - \sum_{i=1}^{N} \xi_i \mathbf{A}_i(\mathbf{x}) \geqslant 0\right\} \geqslant 1 - \epsilon, \tag{11.8}$$

where $\epsilon > 0$ is a small tolerance. The resulting probability constraint, however, typically is "heavily computationally intractable." The probability in the left hand side cannot be computed efficiently, its reliability estimation by Monte-Carlo simulations requires samples of order of $1/\epsilon$, which is prohibitively time-consuming when ϵ is small, like 10^{-6} or 10^{-8}. A natural way to overcome this difficulty is to replace "intractable" (11.7) with its "tractable approximation"—an explicit constraint on $\mathbf{x}$.

An necessary condition for $\mathbf{x}$ to be feasible for (11.7) is $\mathbf{A}_0(\mathbf{x}) \geq 0$; strengthening this necessary condition to be $\mathbf{A}_0(\mathbf{x}) > 0$, $\mathbf{x}$ is feasible for the probability constraint if and only if the sum of random matrices

$$\mathbf{S}_N = \sum_{i=1}^{N} \xi_i \underbrace{\mathbf{A}_0^{-1/2}(\mathbf{x})\,\mathbf{A}_i(\mathbf{x})\,\mathbf{A}_0^{1/2}(\mathbf{x})}_{\mathbf{Y}_i}$$

is $\leqslant \mathbf{I}_n$ with probability $\geq 1-\epsilon$. Assuming, as it is typically the case, that the distribution of ξ_i are symmetric,[1] this condition is essentially the same as the condition $\|\mathbf{S}_N\| \leqslant 1$ with probability $\geq 1-\epsilon$. If we know how to answer (**Q**), we could use this answer to build a "tractable" sufficient condition for $\|\mathbf{S}_N\|$ to be ≤ 1 with probability close to 1 and thus could build a tractable approximation of (11.8). □

Example 11.2.2 (Nonconvex quadratic optimization under orthogonality constraints). We take this example—the *Procrustes problem*—from [560]. In the Procrustes problem, we are given K matrices $\mathbf{A}[k], k = 1, \ldots, N$, of the same size $m \times n$. Our goal is to look for N orthogonal matrices $\mathbf{X}[k]$ of $n \times n$ minimizing the objective

$$\sum_{1 \leqslant k < k' \leqslant N} \|\mathbf{A}[k]\mathbf{X}[k] - \mathbf{A}[k']\mathbf{X}[k']\|_2^2$$

where $\|\mathbf{A}\|_2 = \sqrt{\mathrm{Tr}\left(\mathbf{A}\mathbf{A}^T\right)}$ is the Frobenius norm of a matrix. This problem is equivalent to the quadratic maximization problem

$$P = \max_{\mathbf{X}[1],\ldots,\mathbf{X}[N]} \left\{ 2\sum_{k<k'} \mathrm{Tr}\left(\mathbf{A}[k]\mathbf{X}[k]\mathbf{X}^T[k']\mathbf{A}^T[k']\right) : \mathbf{X}[k] \in \mathbb{R}^{n\times n}, \mathbf{X}[k]\,\mathbf{X}^T[k] = \mathbf{I}_n, k = 1, \ldots, N \right\}. \tag{11.9}$$

When $N > 2$, this problem is intractable. For $N = 2$, there is a closed form solution. Equation (11.29) allows for a straightforward semidefinite relaxation. Geometrically speaking, we are given N collections of points in $\mathbb{R}^n$ and are seeking for rotations which make these collections as close to each other as possible, the closeness being measured by the sum of squared distances.

---

[1]We say that, X and Y are *identically distributed*, or *similar*, or that Y is a *copy* of X, if $\mathbb{P}_X(A) = \mathbb{P}_Y(A)$, where $\mathbb{P}_X(A) = \mathbb{P}_X(X \in A)$ is a probability measure. A random element X in a measurable vector space is called *symmetric*, if X and $-X$ are identically distributed. If X is a symmetric random element, then its distribution is a symmetric measure.

Let $\mathbf{Y} = Y\left[\mathbf{X}[1], \ldots, \mathbf{X}[N]\right]$ be the symmetric matrix defined as follows: the rows and the columns in $\mathbf{Y}$ are indexed by triples (k, i, j), where k runs from 1 to K and i, j run from 1 to n; the entry $Y_{kij,k'i'j'}$ in $\mathbf{Y}$ is $x_{ij}[k]x_{i'j'}\left[k'\right]$. Note $\mathbf{Y}$ is a symmetric, positive semidefinite matrix of rank 1. In (11.29), the relation $\mathbf{X}\left[k\right]\mathbf{X}^T\left[k\right] = \mathbf{I}_n$ is equivalent to a certain system $\mathcal{S}_k$ of linear equations on the entries of $\mathbf{Y}$, while the relation $\mathbf{X}\left[k\right]^T\mathbf{X}\left[k\right] = \mathbf{I}_n$ is equivalent to a certain system $\mathcal{T}_k$ of linear equations on the entries of $\mathbf{Y}$. Finally, the objective in (11.29) is a linear function $\mathrm{Tr}\left(\mathbf{BY}\right)$ of $\mathbf{Y}$, where $\mathbf{B}$ be appropriate symmetric matrix of the size $Kn^2 \times Kn^2$. It is seen that (11.29) is equivalent to

$$\max_{\mathbf{Y}\in\mathbb{S}^{Kn^2}} \mathrm{Tr}\left(\mathbf{BY}\right) : \mathbf{Y} \geqslant 0, \mathbf{Y}\text{satisfies } \mathcal{S}_k, \mathcal{T}_k, k = 1, \ldots, K, \mathrm{Rank}\left(\mathbf{Y}\right) = 1;$$

removing the trouble-making constraint $\mathrm{Rank}\left(\mathbf{Y}\right) = 1$, we have an explicit semidefinite problem

$$SDP = \max_{\mathbf{Y}\in\mathbb{S}^{Kn^2}} \mathrm{Tr}\left(\mathbf{BY}\right) : \mathbf{Y} \geqslant 0, \mathbf{Y}\text{satisfies } \mathcal{S}_k, \mathcal{T}_k, k = 1, \ldots, K$$

which is a relaxation of (11.29), so that $\mathrm{Opt}\left(\mathrm{SDP}\right) \geqslant \mathrm{Opt}\left(P\right)$. In fact, we have

$$\mathrm{Opt}\left(\mathrm{SDP}\right) \leqslant O(1)\left(n^{1/3} + \ln K\right)\mathrm{Opt}\left(P\right)$$

and similarly for other problems of quadratic optimization under orthogonality constraints. □

Theorem 11.2.3 (Nemirovski [560]). *Let* $\mathbf{X}_1, \ldots, \mathbf{X}_N$ *be independent symmetric* $n \times n$ *matrices with zero mean such that*

$$\mathbb{E}\left[\exp\left(\left\|\mathbf{X}_i\right\|^2/\sigma_i^2\right)\right] \leqslant \exp\left(1\right), \quad i = 1, \ldots, N$$

where $\sigma_i > 0$ *are deterministic scalars factors. Then*

$$\mathrm{Prob}\left\{\left\|\mathbf{S}_N\right\| \geqslant t\sqrt{\sum_{i=1}^{N}\sigma_i^2}\right\} \leqslant O(1)\exp\left(-O(1)t^2/\ln n\right), \quad \forall t > 0, \quad (11.10)$$

with positive absolute constraints $O(1)$.

Theorem 11.2.4 (Nemirovski [560]). *Let* $\xi_1, \ldots, \xi_N$ *be independent random variables with zero mean and zero third moment taking values in* $[-1, 1]$, $\mathbf{B}_i, i = 1, \ldots, N$, *be deterministic symmetric* $m \times m$ *matrices, and* $\Theta > 0$ *be a real number such that*

$$\sum_{i=1}^{N}\mathbf{B}_i^2 \leqslant \Theta^2\mathbf{I}.$$

Then

$$t \geqslant 7m^{1/4} \Rightarrow \text{Prob}\left\{\left\|\sum_{i=1}^{N} \xi_i \mathbf{B}_i\right\| \geqslant t\Theta\right\} \leqslant \tfrac{5}{4}\exp\left(-t^2/32\right),$$

$$t \geqslant 7m^{1/6} \Rightarrow \text{Prob}\left\{\left\|\sum_{i=1}^{N} \xi_i \mathbf{B}_i\right\| \geqslant t\Theta\right\} \leqslant 22\exp\left(-t^2/32\right). \quad (11.11)$$

See [560] for a proof. Equation (11.11) is extended by Nemirovski [560] to hold for the case of independent, Gaussian, symmetric $n \times n$ random matrices $\mathbf{X}_1, \ldots, \mathbf{X}_N$ with zero means and $\sigma > 0$ such that

$$\sum_{i=1}^{N} \mathbb{E}\left[\mathbf{X}_i^2\right] \leqslant \sigma^2 \mathbf{I}_n. \quad (11.12)$$

Let us consider non-symmetric (and even non-square) random matrices $\mathbf{Y}_i, i = 1, \ldots, N$. Let $\mathbf{C}_i$ be deterministic $m \times n$ matrices such that

$$\sum_{i=1}^{N} \mathbf{C}_i \mathbf{C}_i^T \leqslant \Theta^2 \mathbf{I}_m, \qquad \sum_{i=1}^{N} \mathbf{C}_i^T \mathbf{C}_i \leqslant \Theta^2 \mathbf{I}_n, \quad (11.13)$$

and ξ_i be independent random scalars with zero mean and of order of 1. Then

$$t \geqslant O(1)\sqrt{\ln(m+n)} \Rightarrow \text{Prob}\left\{\boldsymbol{\xi} : \sum_{i=1}^{N} \xi_i \mathbf{C}_i \geqslant t\Theta\right\} \leqslant O(1)\exp\left(-O(1)t^2\right). \quad (11.14)$$

Using the deterministic symmetric $(m+n) \times (m+n)$ $\mathbf{B}_i$ defined as

$$\mathbf{B}_i = \begin{bmatrix} & \mathbf{C}_i^T \\ \mathbf{C}_i & \end{bmatrix},$$

the following theorem follows from Theorem 11.2.4 and (11.12).

Theorem 11.2.5 (Nemirovski [560]). *Let deterministic $m \times n$ matrices $\mathbf{C}_i$ satisfying* (11.13) *with $\Theta = 1.$, and let ξ_i be independent random scalars with zero first and third moment and such that either $|\xi_i| \leq 1$ for all $i \leq N$, or $\xi \sim \mathcal{N}(0,1)$ for all $i \leq N$. Then*

$$t \geqslant 7(m+n)^{1/4} \Rightarrow \text{Prob}\left\{\sum_{i=1}^{N} \xi_i \mathbf{C}_i \geqslant t\Theta\right\} \leqslant \tfrac{5}{4}\exp\left(-t^2/32\right).$$

$$t \geqslant 7(m+n)^{1/6} \Rightarrow \text{Prob}\left\{\sum_{i=1}^{N} \xi_i \mathbf{C}_i \geqslant t\Theta\right\} \leqslant 22\exp\left(-t^2/32\right). \quad (11.15)$$

We can make a simple additional statement: Let $\mathbf{C}_i, \xi_i$ be defined as Theorem 11.2.5, then

$$t \geqslant 4\sqrt{\min(m,n)} \Rightarrow \text{Prob}\left\{\sum_{i=1}^{N} \xi_i \mathbf{C}_i \geqslant t\Theta\right\} \leqslant \frac{4}{3}\exp\left(-t^2/16\right). \quad (11.16)$$

It is clearly desirable to equations such as (11.15) to hold for smaller values of t. Moreover, it is nice to remove the assumption that the random variables $\xi_1, \ldots, \xi_N$ have zero third moment.

Conjecture (Nemirovski [560]) Let $\xi_1, \ldots, \xi_N$ be independent mean zero random variables, each of which is either (i) supported on [-1,1], or (ii) normally distributed with unit variance. Further, let $\mathbf{X}_1, \ldots, \mathbf{X}_N$ be arbitrary $m \times n$ matrices satisfying (11.13) with $\Theta = 1$. Then, whenever $t \geq O(1)\sqrt{\ln(m+n)}$, one has

$$\text{Prob}\left(\left\|\sum_{i=1}^{N} \xi_i \mathbf{X}_i\right\| \geqslant t\right) \leqslant O(1) \cdot \exp\left(-O(1) \cdot t^2\right).$$

It is argued in [560] that the threshold $t = \Omega\left(\sqrt{\ln(m+n)}\right)$ is in some sense the best one could hope for. So [122] finds that the behavior of the random variable $\mathbf{S}_N \equiv \sum_{i=1}^{N} \xi_i \mathbf{X}_i$ has been extensively studied in the functional analysis and probability theory literature. One of the tools is the so-called Khintchine-type inequalities [121].

We say $\xi_1, \ldots, \xi_N$ are independent Bernoulli random variables when each ξ_i takes on the values ± 1 with equal probability.

Theorem 11.2.6 (So [122]). *Let $\xi_1, \ldots, \xi_N$ be independent mean zero random variables, each of which is either (i) supported on [-1,1], or (ii) Gaussian with variance one. Further, let $\mathbf{X}_1, \ldots, \mathbf{X}_N$ be arbitrary $m \times n$ matrices satisfying* $\max(m,n) \geqslant 2$ *and* (11.13) *with* $\Theta = 1$. *Then, for any $t \geq 1/2$, we have*

$$\text{Prob}\left(\left\|\sum_{i=1}^{N} \xi_i \mathbf{X}_i\right\| \geqslant \sqrt{2e(1+t)\ln\max\{m,n\}}\right) \leqslant (\max\{m,n\})^{-t}$$

if $\xi_1, \ldots, \xi_N$ are i.i.d. Bernoulli or standard normal random variables; and

$$\text{Prob}\left(\left\|\sum_{i=1}^{N} \xi_i \mathbf{X}_i\right\| \geqslant \sqrt{8e(1+t)\ln\max\{m,n\}}\right) \leqslant (\max\{m,n\})^{-t}$$

if $\xi_1, \ldots, \xi_N$ are independent mean zero random variables supported on [-1,1].

Proof. We follow So [122] for a proof. $\mathbf{X}_1, \ldots, \mathbf{X}_N$ are arbitrary $m \times n$ matrices satisfying (11.13) with $\Theta = 1$, so all the eigenvalues of $\sum_{i=1}^{N} \mathbf{X}_i \mathbf{X}_i^T$ and $\sum_{i=1}^{N} \mathbf{X}_i^T \mathbf{X}_i$ lie in $[0,1]$. Then, we have

$$\left\| \left(\sum_{i=1}^{N} \mathbf{X}_i \mathbf{X}_i^T \right)^{1/2} \right\|_{S_p} \leqslant m^{1/p}, \quad \left\| \left(\sum_{i=1}^{N} \mathbf{X}_i^T \mathbf{X}_i \right)^{1/2} \right\|_{S_p} \leqslant n^{1/p}.$$

Let $\xi_1, \ldots, \xi_N$ be i.i.d. Bernoulli random variables or standard Gaussian random variables. By Theorem 2.17.4 and discussions following it, it follows that

$$\mathbb{E}\left[\left\| \sum_{i=1}^{N} \xi_i \mathbf{X}_i \right\|_{\infty}^{p} \right] \leqslant \mathbb{E}\left[\left\| \sum_{i=1}^{N} \xi_i \mathbf{X}_i \right\|_{S_p}^{p} \right] \leqslant p^{p/2} \cdot \max\{m, n\}$$

for any $p \geq 2$. Note that $\|\mathbf{A}\|$ denotes the spectrum norm of matrix $\mathbf{A}$. Now, by Markov's inequality, for any $s > 0$ and $p \geq 2$, we have

$$\mathbb{P}\left(\left\| \sum_{i=1}^{N} \xi_i \mathbf{X}_i \right\|_{\infty} \geqslant s \right) \leqslant s^{-p} \cdot \mathbb{E}\left[\left\| \sum_{i=1}^{N} \xi_i \mathbf{X}_i \right\|_{\infty}^{p} \right] \leqslant \frac{p^{p/2} \cdot \max\{m, n\}}{s^p}.$$

By assumption $t \geqslant 1/2$ and $\max\{m, n\} \geq 2$, we set

$$s = \sqrt{2e(1+t)\ln\max\{m, n\}}, p = s^2/e > 2$$

through which we obtain

$$\mathbb{P}\left(\left\| \sum_{i=1}^{N} \xi_i \mathbf{X}_i \right\|_{\infty} \geqslant \sqrt{2e(1+t)\ln\max\{m, n\}} \right) \leqslant (\max\{m, n\})^{-t}$$

as desired.

Next, we consider the case where $\xi_1, \ldots, \xi_N$ are independent mean zero random variables supported on $[-1, 1]$. Let $\varepsilon_1, \ldots, \varepsilon_N$ be i.i.d. Bernoulli random variables; $\varepsilon_1, \ldots, \varepsilon_N$ are independent of the ξ_i's. A standard symmetrization argument (e.g., see Lemma 1.11.3 which is Lemma 6.3 in [27]), together with Fubini's theorem and Theorem 2.17.4 implies that

$$\begin{aligned}
\mathbb{E}\left\| \sum_{i=1}^{N} \xi_i \mathbf{X}_i \right\|_{S_p}^{p} &\leqslant 2^p \cdot \mathbb{E}_\xi \mathbb{E}_\varepsilon \left\| \sum_{i=1}^{N} \varepsilon_i \xi_i \mathbf{X}_i \right\|_{S_p}^{p} \\
&\leqslant 2^p \cdot p^{p/2} \cdot \mathbb{E}_\xi \left[\max\left\{ \left\| \left(\sum_{i=1}^{N} \xi_i^2 \mathbf{X}_i \mathbf{X}_i^T \right)^{1/2} \right\|_{S_p}^{p}, \left\| \left(\sum_{i=1}^{N} \xi_i^2 \mathbf{X}_i^T \mathbf{X}_i \right)^{1/2} \right\|_{S_p}^{p} \right\} \right] \\
&\leqslant 2^p \cdot p^{p/2} \cdot \max\{m, n\}.
\end{aligned}$$

□

Example 11.3.2 will use Theorem 11.2.6. A

11.3 Applications of Sums of Random Matrices

We are in a position to apply the sums of random matrices to the probability constraints.

Example 11.3.1 (Randomly perturbed linear matrix inequalities [560]—Continued). Consider a randomly perturbed Linear Matrix Inequalities

$$\mathbf{A}_0(\mathbf{x}) - \sum_{i=1}^{N} \xi_i \mathbf{A}_i(\mathbf{x}) \geqslant 0, \tag{11.17}$$

where $\mathbf{A}_1(\mathbf{x}), \ldots, \mathbf{A}_N(\mathbf{x})$ are affine functions of the decision vector $\mathbf{x}$ taking values in the space $\mathbb{S}^n$ of symmetric $n \times n$ matrices, and ξ_i are independent of each other's random perturbations. $\xi_i, i = 1, \ldots, N$ are random real perturbations which we assume to be independent with zero means "of order of 1" and with "light tails"—we will make the two assumptions precise below.

Here we are interested in the sufficient conditions for the decision vector $\mathbf{x}$ such that the random perturbed LMI (11.17) holds true with probability $\geq 1 - \epsilon$, where $\epsilon << 1$. Clearly, we have

$$\mathbf{A}_0(\mathbf{x}) \geqslant 0.$$

To simplify, we consider the strengthened condition $\mathbf{A}_0(\mathbf{x}) > 0$. For such decision vector $\mathbf{x}$, letting

$$\mathbf{B}_i(\mathbf{x}) = \mathbf{A}_0^{-1/2}(\mathbf{x})\, \mathbf{A}_i(\mathbf{x})\, \mathbf{A}_0^{-1/2}(\mathbf{x}),$$

the question becomes to describe those $\mathbf{x}$ such that

$$\text{Prob}\left\{ \sum_{i=1}^{N} \xi_i \mathbf{B}_i(\mathbf{x}) \geqslant 0 \right\} \geqslant 1 - \epsilon. \tag{11.18}$$

Precise description seems to be completely intractable. The trick is to let the *closed-form* probability inequalities for sums of random matrices "do the most of the job"! What we are about to do are verifiable *sufficient conditions* for (11.18) to hold true.

Using Theorem 11.2.3, we immediately obtain the following sufficient condition: Let $n \geq 2$, perturbations ξ_i be independent with zero means such that

$$\mathbb{E}\left[\exp\left({\xi_i}^2\right)\right] \leqslant \exp(1), \quad i = 1, \ldots, N.$$

Then the condition

$$\mathbf{A}_0(\mathbf{x}) > 0 \quad \& \quad \sum_{i=1}^{N} \left\| \mathbf{A}_0^{-1/2}(\mathbf{x})\, \mathbf{A}_i(\mathbf{x})\, \mathbf{A}_0^{-1/2}(\mathbf{x}) \right\|^2 \leqslant \frac{1}{450 \exp(1) \left(\ln \frac{3}{\varepsilon}\right)(\ln m)} \tag{11.19}$$

is *sufficient* for (11.17) to be valid with probability $\geq 1 - \epsilon$.

Although (11.19) is verifiable, it in general, defines a *nonconvex* set in the space of decision variables $\mathbf{x}$. The "problematic" part of the condition is the inequality

$$\sum_{i=1}^{N} \left\| \mathbf{A}_0^{-1/2}(\mathbf{x})\, \mathbf{A}_i(\mathbf{x})\, \mathbf{A}_0^{-1/2}(\mathbf{x}) \right\|^2 \leqslant \tau \tag{11.20}$$

on $\mathbf{x}, \tau$. Equation (11.20) can be represented by the system of *convex* inequalities

$$-\mathbf{A}_0(\mathbf{x}) \leqslant \mu_i \mathbf{A}_i(\mathbf{x}) \leqslant \mathbf{A}_0(\mathbf{x}), \quad \mu_i > 0, \quad i = 1, \ldots, N, \quad \sum_{i=1}^{N} \frac{1}{\mu_i^2} \leqslant \tau.$$

Consider another "good" case when $\mathbf{A}_0(\mathbf{x}) \equiv \mathbf{A}$ is constant. Equation (11.20) can be represented by the system of *convex* constraints

$$-\nu_i \mathbf{A} \leqslant \mathbf{A}_i(\mathbf{x}) \leqslant \nu_i \mathbf{A}, \qquad i = 1, \ldots, N, \quad \sum_{i-1}^{N} \nu_i{}^2 \leqslant \tau$$

in variables $\mathbf{x}, \nu_i, \tau$.

Using Theorem 11.2.3, we arrive at the following sufficient conditions: Let perturbations ξ_i be independent with zero mean and zero third moments such that $|\xi_i| \leqslant 1, i = 1, \ldots, N$, or such that $\xi_i \sim \mathcal{N}(0,1), i = 1, \ldots, N$. Let, further, $\epsilon \in (0,1)$ be such that one of the following two conditions is satisfied

$$\begin{aligned} &(a) \ln\left(\frac{5}{4\epsilon}\right) \geqslant \frac{49 m^{1/2}}{32} \\ &(b) \ln\left(\frac{22}{\epsilon}\right) \geqslant \frac{49 m^{1/3}}{32} \end{aligned} \tag{11.21}$$

Then the condition

$$\mathbf{A}_0(\mathbf{x}) > 0 \quad \& \quad \sum_{i=1}^{N} \left\| \mathbf{A}_0^{-1/2}(\mathbf{x})\, \mathbf{A}_i(\mathbf{x})\, \mathbf{A}_0^{-1/2}(\mathbf{x}) \right\|^2 \leqslant \begin{cases} \frac{1}{32 \ln\left(\frac{5}{4\epsilon}\right)}, & \text{case (a) of (11.21)} \\ \frac{1}{32 \ln\left(\frac{22}{\epsilon}\right)}, & \text{case(b) of (11.21)} \end{cases} \tag{11.22}$$

is *sufficient* for (11.17) to be valid with probability $\geq 1 - \epsilon$.

In contrast, (11.19) and (11.22) defines a *convex* domain in the space of design variables. Indeed, (11.22) is of the form

$$\mathbf{A}_0(\mathbf{x}) > 0 \quad \& \quad \begin{bmatrix} \mathbf{Y}_i & \mathbf{A}_i(\mathbf{x}) \\ \mathbf{A}_i(\mathbf{x}) & \mathbf{A}_0(\mathbf{x}) \end{bmatrix} \geqslant 0, \quad i = 1, \ldots, N, \quad \text{and} \quad \sum_{i=1}^{N} \mathbf{Y}_i \leqslant c(\epsilon) \mathbf{A}_0(\mathbf{x}) \tag{11.23}$$

in variables $\mathbf{x}, \mathbf{Y}_i$. □

Example 11.3.2 (Safe tractable approximation—So [122]). We demonstrate how Theorem 11.2.6 can be used. Let us consider a so-called safe tractable approximation of the following probability constrained optimization problem:

$$\begin{aligned} &\text{minimize} && \mathbf{c}^T\mathbf{x} \\ &\text{subject to} && \mathbf{F}(\mathbf{x}) \leqslant \mathbf{0} \\ & && \text{Prob}\left(\mathbf{A}_0(\mathbf{x}) - \sum_{i=1}^{N} \xi_i \mathbf{A}_i(\mathbf{x}) \geqslant 0\right) \geqslant 1-\epsilon, \quad (\dagger) \\ & && \mathbf{x} \in \mathbb{R}^n. \end{aligned} \tag{11.24}$$

Here, $\mathbf{c} \in \mathbb{R}^n$ is a given objective vector; $\mathbf{F} : \mathbb{R}^n \to \mathbb{R}^l$ is an efficiently computable vector-valued function with convex components; $\mathbf{A}_0, \ldots, \mathbf{A}_N : \mathbb{R}^n \to \varphi^m$ are affine functions in $\mathbf{x}$ with $\mathbf{A}_0(\mathbf{x}) > 0$ for all $\mathbf{x} \in \mathbb{R}^n$;$\xi_1, \ldots, \xi_N$ are independent (but not necessarily identically distributed) mean zero random variables; $\epsilon \in (0,1)$ is the error tolerance parameter. We assume $m \geq 2$ so that (11.24) is indeed a probability constraint linear matrix inequality.

Observe that

$$\text{Prob}\left(\mathbf{A}_0(\mathbf{x}) - \sum_{i=1}^{N} \xi_i \mathbf{A}_i(\mathbf{x}) \geqslant 0\right) = \text{Prob}\left(\sum_{i=1}^{N} \xi_i \tilde{\mathbf{A}}_i(\mathbf{x}) \leqslant \mathbf{I}\right), \tag{11.25}$$

where

$$\tilde{\mathbf{A}}_i(\mathbf{x}) = \mathbf{A}_0^{-1/2}(\mathbf{x})\,\mathbf{A}_i(\mathbf{x})\,\mathbf{A}_0^{-1/2}(\mathbf{x}).$$

Now suppose that we can choose $\gamma = \gamma(\epsilon) > 0$ such that whenever

$$\sum_{i=1}^{N} \tilde{\mathbf{A}}_i^2(\mathbf{x}) \leqslant \gamma^2 \mathbf{I} \tag{11.26}$$

holds, the constraint condition (†) in (11.24) is satisfied. Then, we say that (11.26) is a sufficient condition for (†) to hold. Using the Schur complement [560], (11.26) can be rewritten as a linear matrix inequality

$$\begin{bmatrix} \gamma\mathbf{A}_0(\mathbf{x}) & \mathbf{A}_1(\mathbf{x}) & \cdots & \mathbf{A}_N(\mathbf{x}) \\ \mathbf{A}_1(\mathbf{x}) & \gamma\mathbf{A}_0(\mathbf{x}) & & \\ \vdots & & \ddots & \\ \mathbf{A}_N(\mathbf{x}) & & & \gamma\mathbf{A}_0(\mathbf{x}) \end{bmatrix} \geqslant 0. \tag{11.27}$$

Thus, by replacing the constraint condition (†) with (11.27), the original problem (11.24) is tractable. Moreover, any solution $\mathbf{x} \in \mathbb{R}^n$ that satisfies $\mathbf{F}(\mathbf{x}) \leqslant \mathbf{0}$ and (11.27) will be feasible for the original probability constrained problem of (11.24).

Now we are in a position to demonstrate how Theorem 11.2.6 can be used for problem solving. If the random variables $\xi_1, \ldots, \xi_N$ satisfy the conditions of Theorem 11.2.6, and at the same time, if (11.26) holds for $\gamma \geqslant \gamma(\epsilon) \equiv \left(\sqrt{8e\ln(m/\epsilon)}\right)^{-1}$, then for any $\epsilon \in (0, 1/2]$, it follows that

$$\text{Prob}\left(\sum_{i=1}^{N} \xi_i \tilde{\mathbf{A}}_i(\mathbf{x}) \leqslant \mathbf{I}\right) = \text{Prob}\left(\left\|\sum_{i=1}^{N} \xi_i \tilde{\mathbf{A}}_i(\mathbf{x})\right\|_\infty \leqslant 1\right) > 1 - \epsilon. \quad (11.28)$$

We also observe that

$$\text{Prob}\left(\left\|\sum_{i=1}^{N} \xi_i \tilde{\mathbf{A}}_i(\mathbf{x})\right\|_\infty > 1\right) = \text{Prob}\left(\left\|\sum_{i=1}^{N} \xi_i \left(\frac{1}{\gamma}\tilde{\mathbf{A}}_i(\mathbf{x})\right)\right\|_\infty > \frac{1}{\gamma}\right).$$

Since $\sum_{i=1}^{N}\left(\frac{1}{\gamma}\tilde{\mathbf{A}}_i(\mathbf{x})\right)^2 \leqslant \mathbf{I}$, by Theorem 2.17.4 and Markov's inequality (see the above proof of Theorem 11.2.6), we have

$$\text{Prob}\left(\left\|\sum_{i=1}^{N} \xi_i \left(\frac{1}{\gamma}\tilde{\mathbf{A}}_i(\mathbf{x})\right)\right\|_\infty > \frac{1}{\gamma}\right) \leqslant m \cdot \exp\left(-1/\left(8e\gamma^2\right)\right) \leqslant \epsilon.$$

This gives (11.28). Finally, using (11.25), we have the following theorem.

Theorem 11.3.3. *Let $\xi_1, \ldots, \xi_N$ be independent mean zero random variables, each of which is either (i) supported on [-1,1], or Gaussian with variance one. Consider the probability constrained problem* (11.24). *Then, for any $\epsilon \in (0, 1/2]$, the positive semi-definite constraint with $\gamma \geqslant \gamma(\epsilon) \equiv \left(\sqrt{8e\ln(m/\epsilon)}\right)^{-1}$ is a safe tractable approximation of* (11.24).

This theorem improves upon Nemirovski's result in [560], which requires $\gamma = O\left(m^{1/6} + \sqrt{\ln(1/\epsilon)}\right)$ before one could claim that the constraint is a safe tractable approximation of (11.24). □

Example 11.3.4 (Nonconvex quadratic optimization under orthogonality constraints [560]—continued). Consider the following optimization problem

$$P = \max_{\mathbf{X} \in \mathbb{M}^{m\times n}} \left\{ \langle \mathbf{X}, \mathcal{A}\mathbf{X}\rangle : \begin{array}{lr} \langle \mathbf{X}, \mathcal{B}\mathbf{X}\rangle \leqslant 1 & (a) \\ \langle \mathbf{X}, \mathcal{B}_l\mathbf{X}\rangle \leqslant 1, \quad l = 1, \ldots, L & (b) \\ \mathcal{C}\mathbf{X} = 0, & (c) \\ \|\mathbf{X}\| \leqslant 1. & (d) \end{array} \right\}, \quad (11.29)$$

where

- $\mathbb{M}^{m\times n}$ is the space of $m \times n$ matrices equipped with the Frobenius inner product $\langle \mathbf{X}, \mathbf{Y}\rangle = \text{Tr}\left(\mathbf{X}\mathbf{Y}^T\right)$, and $\|\mathbf{X}\| = \max_{\mathbf{Y}}\{\|\mathbf{X}\mathbf{Y}\|_2 : \|\mathbf{Y}\| \leqslant 1\}$ is, as always, the spectral norm of $\mathbf{X} \in \mathbb{M}^{m\times n}$,
- The mappings $\mathcal{A}, \mathcal{B}, \mathcal{B}_l$ are symmetric linear mappings from $\mathbb{M}^{m\times n}$ into $\mathbb{M}^{m\times n}$,
- $\mathcal{B}$ is positive semidefinite,

- $\mathcal{B}_l, l = 1, \ldots, L$ are positive semidefinite,
- $\mathcal{C}$ is a linear mapping from $\mathbb{M}^{m\times n}$ into $\mathbb{R}^M$.

Equation (11.29) covers a number of problems of quadratic optimization under orthogonal constraints, including the Procrustes problem. For details, we refer to [560].

We must exploit the rich structure of (11.29). The homogeneous linear constraints (c) in (11.29) imply that $\mathbf{X}$ is a block-diagonal matrix

$$\begin{bmatrix} \mathbf{X}_1 & 0 & 0 \\ 0 & \ddots & 0 \\ 0 & 0 & \mathbf{X}_K \end{bmatrix}$$

with $m_k \times n_k$ diagonal blocks $\mathbf{X}_k, k = 1, \ldots, K$.

Let us consider a semidefinite relaxation of Problem (11.29), which in general, is NP-hard. Problem (11.29), however, admits a straightforward semidefinite relaxation as follows—following [560]. The linear mapping $\mathcal{A}$ in Problem (11.29) can be identified with a symmetric $mn \times mn$ matrix $\mathbf{A} = [A_{ij,kl}]$ with rows and columns indexed by pairs $(i,j), 1 \le i \le m, 1 \le j \le n$ satisfying the relation

$$[\mathcal{A}\mathbf{X}]_{ij} = \sum_{k=1}^{m}\sum_{l=1}^{n} A_{ij,kl} \cdot x_{kl}.$$

Similarly, $\mathcal{B}, \mathcal{B}_l$ can be identified with symmetric positive semidefinite $mn \times mn$ matrix $\mathbf{B}, \mathbf{B}_l$, with $\mathbf{B}$ of rank 1. Finally, $\mathcal{C}$ can be identified with a $M \times mn$ matrix $\mathbf{C} = [C]_{\mu,ij}$:

$$(\mathcal{C}\mathbf{X})_{\mu,ij} = \sum_{i=1}^{m}\sum_{j=1}^{n} C_{\mu,ij} \cdot x_{ij}. \tag{11.30}$$

Let $\mathbb{S}^{mn}$ stand for the $mn \times mn$ symmetric matrix, and $\mathbb{S}_+^{mn}$ stand for the $mn \times mn$ positive semidefinite matrix respectively. For $\mathbf{X} \in \mathbb{M}^{m\times n}$, let $\text{vec}(\mathbf{X})$ be the mn-dimensional vector obtained from the matrix $\mathbf{X}$ by arranging its columns into a single column, and let $\mathcal{X}(\mathbf{X}) \in \mathbb{S}_+^{mn}$ be the matrix $\text{vec}(\mathbf{X})\,\text{vec}^T(\mathbf{X})$, that is the $mn \times mn$ matrix $[x_{ij}x_{kl}]$

$$\mathcal{X}(\mathbf{X}) = \text{vec}(\mathbf{X})\,\text{vec}^T(\mathbf{X}).$$

Observe that

$$\mathcal{X}(\mathbf{X}) \geqslant 0,$$

and that $\sum\limits_{i=1}^{m}\sum\limits_{j=1}^{n} c_{ij} \cdot x_{ij} = 0$ if and only if

$$0=\left(\sum_{i=1}^{m}\sum_{j=1}^{n}c_{ij}\cdot x_{ij}\right)^2\equiv\sum_{i=1}^{m}\sum_{j=1}^{n}\sum_{k=1}^{m}\sum_{l=1}^{n}c_{ij}\cdot x_{ij}\cdot c_{kl}\cdot x_{kl}=\mathrm{Tr}\left(\mathcal{X}\left(\mathbf{C}\right)\mathcal{X}\left(\mathbf{X}\right)\right).$$

Further, we have that

$$\langle \mathbf{X}, \mathcal{A}\mathbf{X}\rangle=\sum_{i=1}^{m}\sum_{j=1}^{n}\sum_{k=1}^{m}\sum_{l=1}^{n}A_{ij,kl}\cdot x_{ij}\cdot x_{kl}=\mathrm{Tr}\left(\mathbf{A}\mathcal{X}\left(\mathbf{X}\right)\right),$$

and similarly

$$\langle \mathbf{X}, \mathcal{B}\mathbf{X}\rangle=\mathrm{Tr}\left(\mathbf{B}\mathcal{X}\left(\mathbf{X}\right)\right),\quad \langle \mathbf{X}, \mathcal{B}_l\mathbf{X}\rangle=\mathrm{Tr}\left(\mathbf{B}_l\mathcal{X}\left(\mathbf{X}\right)\right).$$

Finally, $\|\mathbf{X}\|\leqslant 1$ if and only if $\mathbf{X}\mathbf{X}^T\leqslant \mathbf{I_m}$, in other words,

$$\|\mathbf{X}\|\leqslant 1\Leftrightarrow \mathbf{X}\mathbf{X}^T\leqslant \mathbf{I}_m.$$

Since the entries in the matrix product $\mathbf{X}\mathbf{X}^T$ are linear combinations of the entries in $\mathcal{X}\left(\mathbf{X}\right)$, we have

$$\mathbf{X}\mathbf{X}^T\leqslant \mathbf{I}_m\Leftrightarrow \mathcal{S}\left(\mathcal{X}\left(\mathbf{X}\right)\right)\leqslant \mathbf{I}_m,$$

where $\mathcal{S}$ is an appropriate linear mapping from $\mathbb{S}^{mn}$ to $\mathbb{S}^{m}$. Similarly, $\|\mathbf{X}\|\leqslant 1$ if and only if $\mathbf{X}^T\mathbf{X}\leqslant \mathbf{I_n}$, which is a linear restriction on $\mathcal{X}\left(\mathbf{X}\right)$:

$$\mathbf{X}\mathbf{X}^T\leqslant \mathbf{I}_n\Leftrightarrow \mathcal{T}\left(\mathcal{X}\left(\mathbf{X}\right)\right)\leqslant \mathbf{I}_n,$$

where $\mathcal{T}$ is an appropriate linear mapping from $\mathbb{S}^{mn}$ to $\mathbb{S}^{n}$.

With the above observations, we can rewrite (11.29) as

$$\max_{\mathbf{X}\in\mathbb{M}^{m\times n}}\left\{\mathrm{Tr}\left(\mathbf{A}\mathcal{X}\left(\mathbf{X}\right)\right):\begin{array}{ll}\mathrm{Tr}\left(\mathbf{B}\mathcal{X}\left(\mathbf{X}\right)\right)\leqslant 1 & (a)\\ \mathrm{Tr}\left(\mathbf{B}_l\mathcal{X}\left(\mathbf{X}\right)\right)\leqslant 1, l=1,\ldots,L & (b)\\ \mathrm{Tr}\left(\mathbf{C}^{\mu}\mathcal{X}\left(\mathbf{X}\right)\right)=0, \mu=1,\ldots,M(\mathrm{c}) & (c)\\ \mathcal{S}\left(\mathcal{X}\left(\mathbf{X}\right)\right)\leqslant \mathbf{I}_m,\quad \mathcal{T}\left(\mathcal{X}\left(\mathbf{X}\right)\right)\leqslant \mathbf{I}_n. & (d)\end{array}\right\},$$

where $\mathbf{C}^{\mu}\in\mathbb{S}_+^{mn}$ is given by

$$C_{ij,kl}^{\mu}=C_{\mu,kl}\cdot C_{\mu,kl}.$$

Since $\mathcal{X}\left(\mathbf{X}\right)\geqslant 0$ for all $\mathbf{X}$, the problem

$$SDP=\max_{\mathbf{X}\in\mathbb{S}_+^{mn}}\left\{\mathrm{Tr}\left(\mathbf{A}\mathcal{X}\right):\begin{array}{ll}\mathrm{Tr}\left(\mathbf{B}\mathcal{X}\right)\leqslant 1 & (a)\\ \mathrm{Tr}\left(\mathbf{B}_l\mathcal{X}\right)\leqslant 1, l=1,\ldots,L & (b)\\ \mathrm{Tr}\left(\mathbf{C}^{\mu}\mathcal{X}\right)=0, \mu=1,\ldots,M & (c)\\ \mathcal{S}\left(\mathcal{X}\right)\leqslant \mathbf{I}_m,\quad \mathcal{T}\left(\mathcal{X}\left(\mathbf{X}\right)\right)\leqslant \mathbf{I}_n & (d)\\ \mathcal{X}\geqslant 0 & (e)\end{array}\right\}\qquad(11.31)$$

is a relaxation of (11.29), so that Opt $(P) \leqslant$ Opt (SDP). Equation (11.31) is a semidefinite problem and as such is computationally tractable.

The accuracy of the SDP (11.31) is given by the following result [560]: There exists $\tilde{\mathbf{X}} \in \mathbb{M}^{m\times n}$ such that

$$\begin{array}{ll} (*) \left\langle \tilde{\mathbf{X}}, \mathcal{A}\tilde{\mathbf{X}} \right\rangle = \mathrm{Opt}\,(SDP) & (a) \left\langle \tilde{\mathbf{X}}, \mathcal{B}\tilde{\mathbf{X}} \right\rangle \leqslant 1 \\ (b) \left\langle \mathbf{X}, \mathcal{B}_l \mathbf{X} \right\rangle \leqslant \Omega^2, \quad l = 1, \ldots, L & (c) \mathcal{C}\tilde{\mathbf{X}} = 0, \quad (d) \left\| \tilde{\mathbf{X}} \right\| \leqslant \Omega \end{array} \tag{11.32}$$

where

$$\Omega = \max\left(\max_{1\leqslant k\leqslant K} \mu_k + \sqrt{32\ln(132K)}, \sqrt{32\ln(12(L+1))} \right),$$

$$\mu_k = \min\left(7(m_k + n_k)^{1/6}, 4\sqrt{\min(m_k, n_k)} \right).$$

In particular, one has both the lower bound and the upper bound

$$\mathrm{Opt}\,(P) \leqslant \mathrm{Opt}\,(SDP) \leqslant \Omega^2\,\mathrm{Opt}\,(P). \tag{11.33}$$

The numerical simulations are given by Nemirovski [560]: The SDP solver mincx (LMI Toolbox for MATLAB) was used, which means at most 1,000–1,200 free entries in the decision matrix $\mathcal{X}(\mathbf{X})$. □

We see So [122] for data-driven distributionally robust stochastic programming. Convex approximations of chance constrained programs are studied in Shapiro and Nemirovski [559].

11.4 Chance-Constrained Linear Matrix Inequalities

Janson [561] extends a method by Hoeffding to obtain strong large deviation bounds for sums of dependent random variables with suitable dependency structure. So [122] and Cheung, So, and Wang [562]. The results of [562] generalizes the works of [122, 560, 563], which only deal with the case of where $\xi_1, \ldots, \xi_m$ are independent.

The starting point is that for $\mathbf{x} \in \mathbb{R}^n$

$$F(\mathbf{x}, \boldsymbol{\xi}) = \mathbf{A}_0(\mathbf{x}) + \sum_{i=1}^{N} \xi_i \mathbf{A}_i(\mathbf{x}) \tag{11.34}$$

where $\mathbf{A}_0(\mathbf{x}), \mathbf{A}_1(\mathbf{x}), \cdots, \mathbf{A}_N(\mathbf{x}) : \mathbb{R}^n \to \mathcal{S}^d$ are affine functions that take values in the space $\mathcal{S}^d$ of $d \times d$ real symmetric matrices. The key idea is to construct safe tractable approximations of the chance constraint

$$\mathop{\mathbb{P}}_{\boldsymbol{\xi}}\left(\mathbf{A}_0\left(\mathbf{x}\right)+\sum_{i=1}^{N}\xi_i\mathbf{A}_i\left(\mathbf{x}\right)\npreceq 0\right)$$

by means of the sums of random matrices $\sum_{i=1}^{N}\xi_i\mathbf{A}_i\left(\mathbf{x}\right)$. By concentration of measure, by using sums of random matrices, Cheung, So, and Wang [562] arrive at a safe tractable approximation of chance-constrained, *quadratically perturbed*, linear matrix inequalities

$$\mathop{\mathbb{P}}_{\boldsymbol{\xi}}\left(\mathbf{A}_0\left(\mathbf{x}\right)+\sum_{i=1}^{N}\xi_i\mathbf{A}_i\left(\mathbf{x}\right)+\sum_{1\leqslant j\leqslant k\leqslant N}\xi_j\xi_k\mathbf{B}_{jk}\leqslant 0\right)\geqslant 1-\epsilon, \tag{11.35}$$

where $\mathbf{B}_{jk}:\mathbb{R}^n\to\mathcal{S}^d$ are affine functions for $1\leqslant i\leqslant N$ and $1\leqslant j\leqslant k\leqslant N$ and $\xi_1,\ldots,\xi_N$ are i.i.d. real-valued mean-zero random variables with light tails. Dependable perturbations are allowed. Some dependence among the random variables $\xi_1,\ldots,\xi_N$ are allowed. Also, (11.35) does not assume precise knowledge of the covariance matrix.

11.5 Probabilistically Constrained Optimization Problem

We follow [236] closely for our exposition. The motivation is to demonstrate the use of concentration of measure in a probabilistically constrained optimization problem. Let $||\cdot||$ and $||\cdot||_F$ represent the vector Euclidean norm and matrix norm, respectively. We write $\mathbf{x}\sim\mathcal{CN}\left(\mathbf{0},\mathbf{C}\right)$ if $\mathbf{x}$ is a zero-mean, circular symmetric complex Gassian random vector with covarance matrix $\mathbf{C}\geq\mathbf{0}$.

Consider so-called multiuser multiple input single output (MISO) problem, where the base station, or the transmitter, sends parallel data streams to multiple users over the sample fading channel. The transmission is unicast, i.e., each data stream is exclusively sent for one user. The base station has N_t transmit antennas and the signaling strategy is beamforming. Let $\mathbf{x}(t)\in\mathbb{C}^{N_t}$ denote the multi-antenna transmit signal vector of the base station at time t. We have that

$$\mathbf{x}(t)=\sum_{k=1}^{K}\mathbf{w}_k s_k(t), \tag{11.36}$$

where $\mathbf{w}_k\in\mathbb{C}^{N_t}$ is the transmit beamforming vector for user k, K is the number of users, and $s_k(t)$ is the data stream of user k, which is assumed to have zero-mean and unit power $\mathbb{E}\left[|s_k(t)|^2\right]=1$. It is also assumed that $s_k(t)$ is statistically independent of one another. For user i, the received signal is

$$y_i(t)=\mathbf{h}_i^H\mathbf{x}(t)+n_i(t), \tag{11.37}$$

where $\mathbf{h}_i \in \mathbb{C}^{N_t}$ is the channel gain from the base station to user i, and $n_i(t)$ is an additive noise, which is assumed to have zero mean and variance $\sigma^2 > 0$.

A common assumption in transmit beamforming is that the base station has perfect knowledge of $\mathbf{h}_1, \ldots \mathbf{h}_K$, the so-called perfect channel state information (CSI). Here, we assume the CSI is not perfect and modeled as

$$\mathbf{h}_i = \bar{\mathbf{h}}_i + \mathbf{e}_i, \qquad i = 1, \ldots, K,$$

where $\mathbf{h}_i$ is the actual channel, $\bar{\mathbf{h}}_i \in \mathbb{C}^{N_t}$ is the presumed channel at the base station, and $\mathbf{e}_i \in \mathbb{C}^{N_t}$ is the respective *error* that is assumed to be random. We model the error vector using complex Gaussian CSI errors,

$$\mathbf{e}_i \sim \mathcal{CN}(\mathbf{0}, \mathbf{C}_i) \tag{11.38}$$

for some known error covariance $\mathbf{C}_i \geqslant 0, i = 1, .., K$.

The SINR of user $i, i = 1, \ldots, K$ is defined as

$$\mathrm{SINR}_i = \frac{\left|\mathbf{h}_i^H \mathbf{w}_i\right|^2}{\sum\limits_{k \neq i} \left|\mathbf{h}_i^H \mathbf{w}_k\right|^2 + \sigma_i^2}.$$

The goal here is to design beamforming vectors $\mathbf{w}_1, \ldots, \mathbf{w}_K \in \mathbb{C}^{N_t}$ such that the qualify of service (QoS) of each user satisfies a prescribed set of requirements under imperfect CSI of (11.38), while using the least amount of power to do so.

Probabilistic SINR constrained problem: Given minimum SINR requirements $\gamma_1, \ldots, \gamma_K > 0$ and maximum tolerable outage probabilities $\rho_1, \ldots, \rho_K \in (0, 1]$, solve

$$\underset{\mathbf{w}_1, \ldots, \mathbf{w}_K \in \mathbb{C}^{N_t}}{\text{minimize}} \quad \sum_{i=1}^{K} \left\|\mathbf{w}_i\right\|^2 \tag{11.39}$$

$$\text{subject to } \mathbb{P}\left(\mathrm{SINR}_i \geqslant \gamma_i\right) \geqslant 1 - \rho_i, \qquad i = 1, \ldots, K. \tag{11.40}$$

This is a chance-constrained optimization problem due to the presence of the probabilistic constrain (11.40).

Following [236], we introduce a novel relaxation-restriction approach in two steps: relaxation step and restriction step. First, we present the relaxation step. The motivation is that for each i, the inequality $\mathrm{SINR}_i \geqslant \gamma_i$ is nonconvex in $\mathbf{w}_1, \ldots, \mathbf{w}_K$; specifically, it is indefinite quadratic. This issue can be handled by semidefinite relaxation [564, 565]. Equation (11.40) is equivalently represented by

$$
\begin{aligned}
&\underset{\mathbf{W}_1,\ldots,\mathbf{W}_K\in\mathbb{H}^{N_t}}{\text{minimize}} && \sum_{i=1}^{K}\mathrm{Tr}(\mathbf{W}_i)\\
&\text{subject to} && \mathbb{P}\left((\bar{\mathbf{h}}_i+\mathbf{e}_i)^H\left(\frac{1}{\gamma_i}\mathbf{W}_i-\sum_{k\neq i}\mathbf{W}_k\right)(\bar{\mathbf{h}}_i+\mathbf{e}_i)\geqslant\sigma_i^2\right)\geqslant 1-\rho_i, \quad i=1,\ldots,K,\\
& && \mathbf{W}_1,\ldots,\mathbf{W}_K\geqslant 0\\
& && \mathrm{rank}\,(\mathbf{W}_i)=1,\quad i=1,\ldots,K,
\end{aligned}
\tag{11.41}
$$

where the connection between (11.40) and (11.41) lies in the feasible point equivalence

$$\mathbf{W}_i=\mathbf{w}_i\mathbf{w}_i^H,\quad i=1,\ldots,K.$$

The semidefinite relaxation (11.41) works by removing the nonconvex rank-one constraints on $\mathbf{W}_i$, i.e., to consider the relaxed problem

$$
\begin{aligned}
&\underset{\mathbf{W}_1,\ldots,\mathbf{W}_K\in\mathbb{H}^{N_t}}{\text{minimize}} && \sum_{i=1}^{K}\mathrm{Tr}(\mathbf{W}_i)\\
&\text{subject to} && \mathbb{P}\left((\bar{\mathbf{h}}_i+\mathbf{e}_i)^H\left(\frac{1}{\gamma_i}\mathbf{W}_i-\sum_{k\neq i}\mathbf{W}_k\right)(\bar{\mathbf{h}}_i+\mathbf{e}_i)\geqslant\sigma_i^2\right)\geqslant 1-\rho_i, \quad i=1,\ldots,K,\\
& && \mathbf{W}_1,\ldots,\mathbf{W}_K\geqslant 0.
\end{aligned}
\tag{11.42}
$$

where $\mathbb{H}^{N_t}$ denotes Hermitian matrix with size N_t by N_t. The benefit of this relaxation is that the inequalities inside the probability functions in (11.42) are *linear* in $\mathbf{W}_1,\ldots,\mathbf{W}_K$, which makes the probabilistic constraints in (11.42) more tractable. An issue that comes from the semidefinite relaxation is the solution rank: the removal of the rank constraints $\mathrm{rank}\,(\mathbf{W}_i)=1$ means that the solution $(\mathbf{W}_1,\ldots,\mathbf{W}_K)$ to problem (11.42) may have rank higher than one. A standard way of tacking this is to apply some rank-one approximation procedure to $(\mathbf{W}_1,\ldots,\mathbf{W}_K)$ to generate a feasible beamforming solution $(\mathbf{w}_1,\ldots,\mathbf{w}_K)$ to (11.39). See [565] for a review and references. An algorithm is given in [236], which in turn follows the spirit of [566].

Let us present the second step: restriction. The relaxation step alone above does not provide a convex approximation of the main problem (11.39). The semidefinite relation probabilistic constraints (11.42) remain intractable. This is the moment that the Bernstein-type concentration inequalities play a central role. Concentration of measure lies in the central stage behind the Bernstein-type concentration inequalities. The Bernstein-type inequality (4.56) is used here. Let $\mathbf{z}=\mathbf{e}$ and $\mathbf{y}=\mathbf{r}$ in Theorem 4.15.7. Denote $T(t)=\mathrm{Tr}\,(\mathbf{Q})-\sqrt{2t}\sqrt{\|\mathbf{Q}\|_F+2\|\mathbf{y}\|^2}-t\lambda_+(\mathbf{Q})$.

We reformulate the challenge as the following: Consider the chance constraint

$$\mathbb{P}\left(\mathbf{e}^H\mathbf{Q}\mathbf{e}+2\,\mathrm{Re}\left(\mathbf{e}^H\mathbf{r}\right)+s\geqslant 0\right)\geqslant 1-\rho, \tag{11.43}$$

where $\mathbf{e}$ is a standard complex Gaussian random vector, i.e., $\mathbf{z}_i \sim \mathcal{CN}(\mathbf{0}, \mathbf{I}_n)$, the 3-tuple $(\mathbf{Q}, \mathbf{r}, s) \in \mathbb{H}^{n\times n} \times \mathbb{C}^n \times \mathbb{R}$ is a set of (deterministic) optimization variables, and $\rho \in (0, 1]$ is fixed. Find an efficiently computable convex restriction of (11.43). Here $\mathbb{H}^{n\times n}$ stands for the set of Hermitian matrices of $n \times n$.

Indeed, for each constraint in (11.42), the following correspondence to (11.43) can be shown for $i = 1, \ldots, K$

$$\mathbf{Q} = \mathbf{C}_i^{1/2}\left(\frac{1}{\gamma_i}\mathbf{W}_i - \sum_{k\neq i}\mathbf{W}_k\right)\mathbf{C}_i^{1/2}, \quad \mathbf{r} = \mathbf{C}_i^{1/2}\left(\frac{1}{\gamma_i}\mathbf{W}_i - \sum_{k\neq i}\mathbf{W}_k\right)\bar{\mathbf{h}}_i,$$

$$s = \bar{\mathbf{h}}_i^H\left(\frac{1}{\gamma_i}\mathbf{W}_i - \sum_{k\neq i}\mathbf{W}_k\right)\bar{\mathbf{h}}_i - \sigma_i^2, \quad \rho = \rho_i. \tag{11.44}$$

Using the Bernstein-type inequality, we can use *closed-form upper bounds* on the violation probability to construct an *efficiently computable convex function* $f(\mathbf{Q}, \mathbf{r}, s, \mathbf{u})$, where $\mathbf{u}$ is an extra optimization variable, such that

$$\mathbb{P}\left(\mathbf{e}^H\mathbf{Q}\mathbf{e} + 2\,\mathrm{Re}\left(\mathbf{e}^H\mathbf{r}\right) + s \geqslant 0\right) \leqslant f(\mathbf{Q}, \mathbf{r}, s, \mathbf{u}). \tag{11.45}$$

Then, the constraint

$$f(\mathbf{Q}, \mathbf{r}, s, \mathbf{u}) \leq \rho \tag{11.46}$$

is, by construction, a *convex restriction* of (11.43).

Since $T(t)$ is monotonically decreasing, its inverse mapping is well defined. In particular, the Bernstein-type inequality (4.56) can be expressed as

$$\mathbb{P}\left(\mathbf{e}^H\mathbf{Q}\mathbf{e} + 2\,\mathrm{Re}\left(\mathbf{e}^H\mathbf{r}\right) + s \geqslant 0\right) \geqslant 1 - e^{-T^{-1}(-s)}.$$

As discussed in (11.45) and (11.46), the constraint $1 - e^{-T^{-1}(-s)} \leq \rho$, or equivalently,

$$\mathrm{Tr}(\mathbf{Q}) - \sqrt{-2\ln(\rho)}\sqrt{\|\mathbf{Q}\|_F + 2\|\mathbf{y}\|^2} + \ln(\rho)\cdot\lambda_+(\mathbf{Q}) + s \geqslant 0 \tag{11.47}$$

serves as a *sufficient* condition for achieving (11.43).

At this point, it is not obvious whether or not (11.47) is convex in $(\mathbf{Q}, \mathbf{r}, s)$, but we observe that (11.47) is equivalent to the following system of convex conic inequalities

$$\begin{aligned}
&\mathrm{Tr}(\mathbf{Q}) - \sqrt{-2\ln(\rho)}\cdot u_1 + \ln(\rho)\cdot u_2 + s \geqslant 0,\\
&\sqrt{\|\mathbf{Q}\|_F + 2\|\mathbf{y}\|^2} \leqslant u_1,\\
&u_2\mathbf{I}_n + \mathbf{Q} \geqslant 0,\\
&u_2 \geqslant 0,
\end{aligned} \tag{11.48}$$

where $u_1, u_2 \in \mathbb{R}$ are slack variables. Therefore, (11.48) is an efficiently computable convex restriction of (11.43).

Let us introduce another method: decomposition into independent random variables. The resulting formulation is solved more efficiently than the Bernstein-type inequality method (11.48). The idea is to first decompose the expression $\mathbf{e}^H\mathbf{Q}\mathbf{e} + 2\,\mathrm{Re}\left(\mathbf{e}^H\mathbf{r}\right) + s$ into several parts, each of which is a sum of independent random variables. Then, one obtains a closed-form upper bound on the violation probability [562]. Let us illustrate this approach briefly, following [236]. Let $\mathbf{Q} = \mathbf{U}\mathbf{\Lambda}\mathbf{V}^H$ be the spectral decomposition of $\mathbf{Q}$, where $\mathbf{\Lambda} = \mathrm{diag}\left(\lambda_1, \ldots, \lambda_n\right)$ and $\lambda_1, \ldots, \lambda_n$ are the eigenvalues of $\mathbf{Q}$. Since $\mathbf{e}_i \sim \mathcal{CN}\left(\mathbf{0}, \mathbf{I}_n\right)$, and $\mathbf{U}^H$ is unitary, we have $\tilde{\mathbf{e}} = \mathbf{U}^H\mathbf{e} \sim \mathcal{CN}\left(\mathbf{0}, \mathbf{I}_n\right)$. As a result, we have that

$$\psi = \mathbf{e}^H\mathbf{Q}\mathbf{e} + 2\,\mathrm{Re}\left(\mathbf{e}^H\mathbf{r}\right) = \tilde{\mathbf{e}}^H\mathbf{\Lambda}\tilde{\mathbf{e}} + 2\,\mathrm{Re}\left(\mathbf{e}^H\mathbf{r}\right) = \psi_q + \psi_l.$$

Now, let us decompose the $\tilde{\mathbf{e}}^H\mathbf{\Lambda}\tilde{\mathbf{e}}+2\,\mathrm{Re}\left(\mathbf{e}^H\mathbf{r}\right)$ into the sum of independent random variables. We have that

$$\psi_q = \tilde{\mathbf{e}}^H\mathbf{\Lambda}\tilde{\mathbf{e}} = \sum_{i=1}^{n} \lambda_i |e_i|^2, \quad \psi_l = 2\,\mathrm{Re}\left(\mathbf{e}^H\mathbf{r}\right) = 2\sum_{i=1}^{n} \left(\mathrm{Re}\left\{r_i\right\}\mathrm{Re}\left\{e_i\right\} + \mathrm{Im}\left\{r_i\right\}\mathrm{Im}\left\{e_i\right\}\right).$$

The advantage of the above decomposition approach is that the distribution of the random vector $\mathbf{e}$ may be non-Gaussian. In particular, $\mathbf{e} \in \mathbb{R}^n$ is a zero-mean random vector supported on $\left[-\sqrt{3}, \sqrt{3}\right]^n$ with independent components. For details, we refer to [236, 562].

11.6 Probabilistically Secured Joint Amplify-and-Forward Relay by Cooperative Jamming

11.6.1 Introduction

This section follows [567] and deals with probabilistically secured joint amplify-and-forward (AF) relay by cooperative jamming. AF relay with N relay nodes is the simple relay strategy for cooperative communication to extend communication range and improve communication quality. Due to the broadcast nature of wireless communication, the transmitted signal is easily intercepted. Hence, communication security is another important performance measure which should be enhanced. Artificial noise and cooperative jamming have been used for physical layer security [568, 569].

All relay nodes perform forwarding and jamming at the same time cooperatively. A numerical approach based on optimization theory is proposed to obtain forwarding complex weights and the general-rank cooperative jamming complex weight matrix simultaneously. SDR is the core of the numerical approach. Cooperative jamming with the general-rank complex weight matrix removes the non-convex rank-1 matrix constraint in SDR. Meanwhile, the general-rank cooperative jamming complex weight matrix can also facilitate the achievement of rank-1 matrix solution in SDR for forwarding complex weights.

In this section, physical layer security is considered in a probabilistic fashion. Specifically speaking, the probability that an eavesdropper's SINR is greater than or equal to its targeted SINR should be equal to the pre-determined violation probability. In order to achieve this goal, probabilistic-based optimization is applied.

Probabilistic-based optimization is more flexible than the well-studied robust optimization and stochastic optimization. However, the flexibility of probabilistic-based optimization will bring challenges to the corresponding solver. Advanced statistical signal processing and probability theory are needed to derive the safe tractable approximation algorithm.

In this way, the optimization problem in a probabilistic fashion can be converted to or approximated to the correspondingly deterministic optimization problem. "Safe" means approximation will not violate the probabilistic constraints and "tractable" means the deterministic optimization problem is convex or solvable. There are several approximation strategies mentioned in [236, 570], e.g., Bernstein-type inequality [235] and moment inequalities for sums of random matrices [122]. Bernstein-type inequality will be exploited explicitly here.

11.6.2 System Model

A two-hop half-duplex AF relay network is considered. A source Alice would like to send information to the destination Bob through N relay nodes. Meanwhile, there is an eavesdropper Eve to intercept wireless signal. All the nodes in the network are only equipped with a single antenna. Alice, Bob, and relay nodes are assumed to be synchronized. Perfect CSIs between Alice and relay nodes as well as between relay nodes and Bob are known. However, CSIs related to Eve are partially known.

There is a two-hop transmission between Alice and Bob. In the first hop, Alice transmits information I. In order to interfere with Eve, Bob generates artificial noise J. I and J are assumed to be independent real Gaussian random variables with zero mean and unit variance. The function diagram of the first hop is shown in Fig. 11.1.

For the nth relay node, the received signal plus artificial noise is,

$$y_{r_n1} = h_{sr_n1}g_{s1}I + h_{dr_n1}g_{d1}J + w_{r_n1}, n = 1, 2, \ldots, N \tag{11.49}$$

where g_{s1} is the transmitted complex weight for Alice and g_{d1} is the transmitted complex weight for Bob; h_{sr_n1} is CSI between Alice and the nth relay node; h_{dr_n1} is CSI between Bob and the nth relay node; w_{r_n1} is the background Gaussian noise with zero mean and $\sigma^2_{r_n1}$ variance for the nth relay node. Similarly, the received signal plus artificial noise for Eve is,

$$y_{e1} = h_{se1}g_{s1}I + h_{de1}g_{d1}J + w_{e1} \tag{11.50}$$

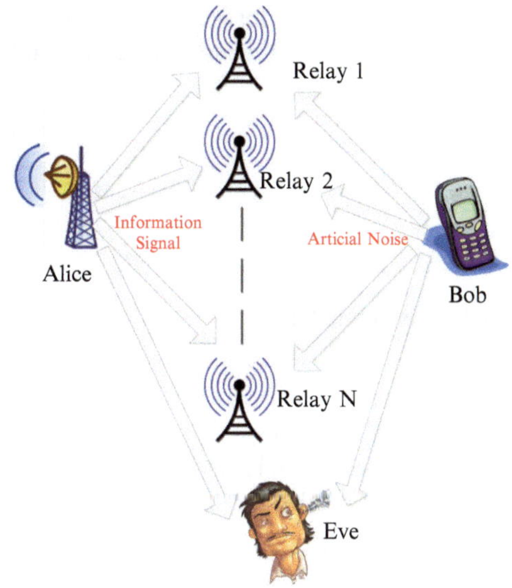

Fig. 11.1 The function diagram of the first hop

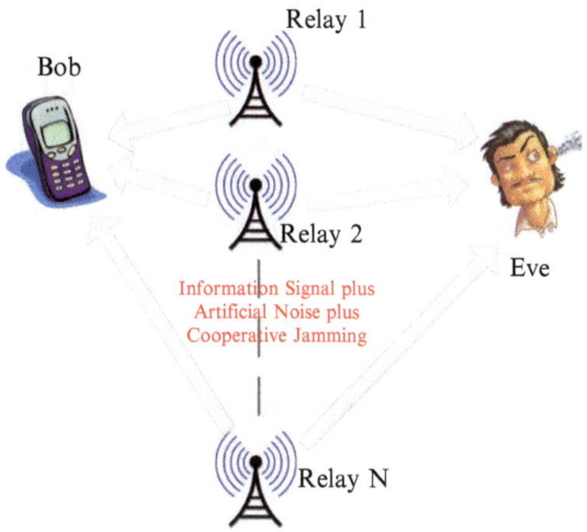

Fig. 11.2 The function diagram of the second hop

where h_{se1} is CSI between Alice and Eve; h_{de1} is CSI between Bob and Eve; w_{e1} is the background Gaussian noise with zero mean and σ_{e1}^2 variance for Eve. Thus, SINR for Eve in the first hop is,

$$\text{SINR}_{e1} = \frac{|h_{se1}g_{s1}|^2}{|h_{de1}g_{d1}|^2 + \sigma_{e1}^2} \tag{11.51}$$

In the second hop, N relay nodes perform joint AF relay and cooperative jamming simultaneously. The function diagram of the second hop is shown in Fig. 11.2.

The transmitted signal plus cooperative jamming for the nth relay node is

$$s_{r_n2} = g_{r_n2}y_{r_n1} + J_{r_n2} \tag{11.52}$$

where $g_{r_n 2}$ is the forwarding complex weight for the nth relay node and $J_{r_n 2}, n = 1, 2, \ldots, N$ constitutes cooperative jamming $\mathbf{j}_{r2}$,

$$\mathbf{j}_{r2} = \begin{bmatrix} J_{r_1 2} \\ J_{r_2 2} \\ \vdots \\ J_{r_N 2} \end{bmatrix} \tag{11.53}$$

and $\mathbf{j}_{r2}$ is defined as,

$$\mathbf{j}_{r2} = \mathbf{U}\mathbf{z} \tag{11.54}$$

where $\mathbf{U} \in C^{N \times r}, r \leq N$ is cooperative jamming complex weight matrix and $\mathbf{z}$, which follows $N(\mathbf{0}, \mathbf{I})$, is r-dimensional artificial noise vector. Hence, the transmitted power needed for the nth relay node is $|g_{r_n 2} h_{sr_n 1} g_{s1}|^2 + |g_{r_n 2} h_{dr_n 1} g_{d1}|^2 + |g_{r_n 2}|^2 \sigma^2_{r_n 1} + E\left\{|J_{r_n 2}|^2\right\}$ where $E\{\cdot\}$ denotes expectation operator and

$$E\left\{|\mathbf{j}_{r2}|^2\right\} = \text{diag}\left\{\mathbf{U}E\left\{\mathbf{z}\mathbf{z}^H\right\}\mathbf{U}^H\right\} \tag{11.55}$$

$$= \text{diag}\left\{\mathbf{U}\mathbf{U}^H\right\} \tag{11.56}$$

where $(\cdot)^H$ denotes Hermitian operator and diag $\{\cdot\}$ returns the main diagonal of matrix or puts vector on the main diagonal of matrix.

The received signal plus artificial noise by Bob is,

$$y_{d2} = \sum_{n=1}^{N} h_{r_n d2} \left(g_{r_n 2} y_{r_n 1} + J_{r_n 2}\right) + w_{d2} \tag{11.57}$$

where $h_{r_n d2}$ is CSI between the nth relay node and Bob; w_{d2} is the background Gaussian noise with zero mean and σ^2_{d2} variance for Bob. Similarly, the received signal plus artificial noise for Eve in the second hop is,

$$y_{e2} = \sum_{n=1}^{N} h_{r_n e2} \left(g_{r_n 2} y_{r_n 1} + J_{r_n 2}\right) + w_{e2} \tag{11.58}$$

where $h_{r_n e2}$ is CSI between the nth relay node and Eve; w_{e2} is the background Gaussian noise with zero mean and σ^2_{e2} variance for Eve.

In the second hop, SINR for Bob is,

$$\text{SINR}_{d2} = \frac{\left|\sum\limits_{n=1}^{N} h_{r_n d2} g_{r_n 2} h_{sr_n 1} g_{s1}\right|^2}{\left|\sum\limits_{n=1}^{N} h_{r_n d2} g_{r_n 2} h_{dr_n 1} g_{d1}\right|^2 + \sum\limits_{n=1}^{N} |h_{r_n d2} g_{r_n 2}|^2 \sigma^2_{r_n 1} + E\left\{\left|\sum\limits_{n=1}^{N} h_{r_n d2} J_{r_n 2}\right|^2\right\} + \sigma^2_{d2}} \tag{11.59}$$

Due to the known information about J, Bob can cancel artificial noise generated by itself. Thus, $\left|\sum_{n=1}^{N} h_{r_n d2} g_{r_n 2} h_{dr_n 1} g_{d1}\right|^2$ can be removed from the denominator in Eq. (11.59) which means SINR for Bob is

$$\mathrm{SINR}_{d2} = \frac{\left|\sum_{n=1}^{N} h_{r_n d2} g_{r_n 2} h_{sr_n 1} g_{s1}\right|^2}{\sum_{n=1}^{N} \left|h_{r_n d2} g_{r_n 2}\right|^2 \sigma_{r_n 1}^2 + E\left\{\left|\sum_{n=1}^{N} h_{r_n d2} J_{r_n 2}\right|^2\right\} + \sigma_{d2}^2} \tag{11.60}$$

SINR for Eve is,

$$\mathrm{SINR}_{e2} = \frac{\left|\sum_{n=1}^{N} h_{r_n e2} g_{r_n 2} h_{sr_n 1} g_{s1}\right|^2}{\left|\sum_{n=1}^{N} h_{r_n e2} g_{r_n 2} h_{dr_n 1} g_{d1}\right|^2 + \sum_{n=1}^{N} \left|h_{r_n e2} g_{r_n 2}\right|^2 \sigma_{r_n 1}^2 + E\left\{\left|\sum_{n=1}^{N} h_{r_n e2} J_{r_n 2}\right|^2\right\} + \sigma_{e2}^2} \tag{11.61}$$

Here, we assume there is no memory for Eve to store the received data in the first hop. Eve cannot do any type of combinations for the received data to decode information.

Let,

$$\mathbf{h}_{rd2} = \left[h_{r_1 d2}\ h_{r_2 d2} \cdots h_{r_N d2}\right] \tag{11.62}$$

Then,

$$E\left\{\left|\sum_{n=1}^{N} h_{r_n d2} J_{r_n 2}\right|^2\right\} = \mathbf{h}_{rd2} \mathbf{U} E\left\{\mathbf{z}\mathbf{z}^H\right\} \mathbf{U}^H \mathbf{h}_{rd2}^H \tag{11.63}$$

$$= \mathbf{h}_{rd2} \mathbf{U} \mathbf{U}^H \mathbf{h}_{rd2}^H \tag{11.64}$$

Similarly, let,

$$\mathbf{h}_{re2} = \left[h_{r_1 e2}\ h_{r_2 e2} \cdots h_{r_N e2}\right] \tag{11.65}$$

Then,

$$E\left\{\left|\sum_{n=1}^{N} h_{r_n e2} J_{r_n 2}\right|^2\right\} = \mathbf{h}_{re2} \mathbf{U} \mathbf{U}^H \mathbf{h}_{re2}^H \tag{11.66}$$

Based on the previous assumption, $h_{sr_n 1}$, $h_{dr_n 1}$, and $h_{r_n d2}$ are perfect known. While, h_{se1}, h_{de1}, and $h_{r_n e2}$ are partially known. Without loss of generality, h_{se1}, h_{de1}, and $h_{r_n e2}$ all follow independent complex Gaussian distribution with zero mean and unit variance (zero mean and $\frac{1}{2}$ variance for both real and imaginary parts). Due to the randomness of h_{se1}, h_{de1}, and h_{r_ne2}, SINR_{e1} and SINR_{e2} are also random variables.

Joint AF relay and cooperative jamming with probabilistic security consideration would like to solve the following optimization problems,

$$\begin{array}{l} \text{find} \\ g_{s1}, g_{d1}, g_{r_n2}, n = 1,2,\ldots,N, \mathbf{U} \\ \text{subject to} \\ |g_{r_n2}h_{sr_n1}g_{s1}|^2 + |g_{r_n2}h_{dr_n1}g_{d1}|^2 + |g_{r_n2}|^2\sigma_{r_n1}^2 + E\left\{|J_{r_n2}|^2\right\} \le P_{r_n2}, n = 1,2,\ldots,N \\ \Pr\left(\text{SINR}_{e1} \ge \gamma_e\right) \le \delta_e \\ \Pr\left(\text{SINR}_{e2} \ge \gamma_e\right) \le \delta_e \\ \text{SINR}_{d2} \ge \gamma_d \end{array} \tag{11.67}$$

where P_{r_n2} is the individual power constraint for the nth relay node; γ_e is the targeted SINR for Eve and δ_e is its violation probability; γ_d is the targeted SINR for Bob.

11.6.3 Proposed Approach

In order to make the optimization problem (11.67) solvable, the optimization problem in a probabilistic fashion need be converted to or approximated to the correspondingly deterministic optimization problem by the safe tractable approximation approach.

For SINR_{e1}, $|h_{se1}g_{s1}|^2$ and $|h_{de1}g_{d1}|^2$ are independent exponentially distributed random variables with means $|g_{s1}|^2$ and $|g_{d1}|^2$. $\Pr\left(\text{SINR}_{e1} \ge \gamma_e\right) \le \delta_e$ is equal to [571],

$$e^{\frac{-\gamma_e\sigma_{e1}^2}{|g_{s1}|^2}} \frac{|g_{s1}|^2}{|g_{s1}|^2 + \gamma_e |g_{d1}|^2} \le \delta_e \tag{11.68}$$

From inequality (11.68), given $|g_{s1}|^2$, we can easily get,

$$|g_{d1}|^2 \ge \frac{\left(e^{\frac{-\gamma_e\sigma_{e1}^2}{|g_{s1}|^2}} - \delta_e\right)|g_{s1}|^2}{\gamma_e\delta_e} \tag{11.69}$$

In this section, we are more interested in equally probabilistic constraints instead of too conservative or robust performance. In other words, $\Pr\left(\text{SINR}_{e1} \ge \gamma_e\right) = \delta_e$ and $\Pr\left(\text{SINR}_{e2} \ge \gamma_e\right) = \delta_e$ are applied. Hence, $|g_{d1}|^2$ is obtained by equality of inequality (11.69) to ensure the probabilistic security in the first hop with minimum power needed for Bob.

For SINR_{e2}, Bernstein-type inequality is explored [235, 236, 572].

Let

$$\mathbf{g}_{r2} = \begin{bmatrix} g_{r_1 2} \\ g_{r_2 2} \\ \vdots \\ g_{r_N 2} \end{bmatrix}, \tag{11.70}$$

$$\mathbf{h}_{sr1} = \left[h_{sr_1 1} \; h_{sr_2 1} \cdots h_{sr_N 1} \right] \tag{11.71}$$

and

$$\mathbf{h}_{dr1} = \left[h_{dr_1 1} \; h_{dr_2 1} \cdots h_{dr_N 1} \right] \tag{11.72}$$

Define $\mathbf{H}_{sr1} = \text{diag}\left\{\mathbf{h}_{sr1}\right\}$, $\mathbf{H}_{dr1} = \text{diag}\left\{\mathbf{h}_{dr1}\right\}$ and

$$\sigma_{r1}^2 = \begin{bmatrix} \sigma_{r_1 1}^2 & 0 & \cdots & 0 \\ 0 & \sigma_{r_2 1}^2 & \cdots & 0 \\ \vdots & \vdots & \vdots & \vdots \\ 0 & 0 & \cdots & \sigma_{r_N 1}^2 \end{bmatrix} \tag{11.73}$$

Based on SDR, define,

$$\mathbf{X} = \mathbf{g}_{r2}\mathbf{g}_{r2}^H \tag{11.74}$$

where $\mathbf{X}$ should be rank-1 semidefinite matrix and

$$\mathbf{Y} = \mathbf{U}\mathbf{U}^H \tag{11.75}$$

where $\mathbf{Y}$ is the semidefinite matrix and the rank of $\mathbf{Y}$ should be equal to or smaller than r.

$\Pr\left(\text{SINR}_{e2} \geq \gamma_e\right) \leq \delta_e$ can be simplified as,

$$\Pr\left(\text{h}_{re2}\mathbf{Q}\text{h}_{re2}^H \geq \sigma_{e2}^2\gamma_e\right) \leq \delta_e \tag{11.76}$$

where $\mathbf{Q}$ is equal to

$$\mathbf{Q} = \mathbf{H}_{sr1}\mathbf{X}\mathbf{H}_{sr1}^H\left|g_{s1}\right|^2 - \gamma_e\mathbf{H}_{dr1}\mathbf{X}\mathbf{H}_{dr1}^H\left|g_{d1}\right|^2 - \gamma_e\text{diag}\left\{\text{diag}\left\{\mathbf{X}\right\}\right\}\sigma_{r2}^2 - \gamma_e\mathbf{Y} \tag{11.77}$$

Then probabilistic constrain in (11.76) can be approximated as,

$$\begin{aligned} &\text{trace}\left(\mathbf{Q}\right) + \left(-2\log\left(\delta_e\right)\right)^{0.5} a - b\log\left(\delta_e\right) - \sigma_{e2}^2\gamma_e \leq 0 \\ &\left\|\mathbf{Q}\right\|_F \leq a \\ &b\mathbf{I} - Q \geq 0 \\ &b \geq 0 \end{aligned} \tag{11.78}$$

where trace$(\cdot)$ returns the sum of the diagonal elements of matrix; $\left\|\cdot\right\|_F$ return Frobenius norm of matrix.

Based on the definition of $\mathbf{X}$ and $\mathbf{Y}$, the individual power constraint for the nth relay node in the optimization problem (11.67) can be rewritten as,

$$(\mathbf{X})_{n,n}\left(|h_{sr_n1}g_{s1}|^2+|h_{dr_n1}g_{d1}|^2+\sigma_{r_n1}^2\right)+(\mathbf{Y})_{n,n}\le P_{r_n2}, n=1,2,\ldots,N \tag{11.79}$$

where $(\cdot)_{i,j}$ returns the entry of matrix with the ith row and jth column.

SINR_{d2} constraint $\mathrm{SINR}_{d2}\ge\gamma_d$ in the optimization problem (11.67) can be reformulated as,

$$\operatorname{trace}\left(\mathbf{a}^H\mathbf{a}\mathbf{X}\right)\ge\gamma_d\left(\operatorname{trace}\left(\mathbf{H}_{rd2}^H\mathbf{H}_{rd2}\sigma_{r1}^2\mathbf{X}\right)+\operatorname{trace}\left(\mathbf{h}_{rd2}{}^H\mathbf{h}_{rd2}\mathbf{Y}\right)+\sigma_{d2}^2\right) \tag{11.80}$$

where

$$\mathbf{H}_{rd2}=\operatorname{diag}\{\mathbf{h}_{rd2}\} \tag{11.81}$$

and

$$\mathbf{a}=\left(\operatorname{diag}\{\mathbf{H}_{sr1}\mathbf{H}_{rd2}g_{s1}\}\right)^T \tag{11.82}$$

where $(\cdot)^T$ denotes transpose operator

In this section, we mainly focus on the joint optimization $g_{r_n2}, n=1,2,\ldots,N$ and $\mathbf{U}$, i.e., $\mathbf{X}$ and $\mathbf{Y}$, in the second hop based on the given $|g_{s1}|^2$ and the calculated $|g_{d1}|^2$.

In order to minimize the total power needed by N relay nodes, the optimization problem (11.67) can be approximated as,

$$\begin{aligned}
&\text{minimize}\\
&\textstyle\sum_{n=1}^{N}((\mathbf{X})_{n,n}\left(|h_{sr_n1}g_{s1}|^2+|h_{dr_n1}g_{d1}|^2+\sigma_{r_n1}^2\right)+(\mathbf{Y})_{n,n})\\
&\text{subject to}\\
&(\mathbf{X})_{n,n}\left(|h_{sr_n1}g_{s1}|^2+|h_{dr_n1}g_{d1}|^2+\sigma_{r_n1}^2\right)+(\mathbf{Y})_{n,n}\le P_{r_n2}, n=1,2,\ldots,N\\
&\operatorname{trace}(\mathbf{Q})+(-2\log(\delta_e))^{0.5}a-b\log(\delta_e)-\sigma_{e2}^2\gamma_e\le 0\\
&\|\mathbf{Q}\|_F\le a\\
&b\mathbf{I}-Q\ge 0\\
&b\ge 0\\
&\operatorname{trace}\left(\mathbf{a}^H\mathbf{a}\mathbf{X}\right)\ge\gamma_d\left(\operatorname{trace}\left(\mathbf{H}_{rd2}^H\mathbf{H}_{rd2}\sigma_{r1}^2\mathbf{X}\right)+\operatorname{trace}\left(\mathbf{h}_{rd2}{}^H\mathbf{h}_{rd2}\mathbf{Y}\right)+\sigma_{d2}^2\right)\\
&\mathbf{X}\ge 0\\
&\operatorname{rank}(\mathbf{X})=1\\
&\mathbf{Y}\ge 0
\end{aligned} \tag{11.83}$$

Due to the non-convex rank constraint, the optimization problem (11.83) is an NP-hard problem. We have to remove rank constraint and the optimization problem (11.83) becomes an SDP problem which can be solved efficiently. However, the optimal solution to $\mathbf{X}$ cannot be guaranteed to be 1. Hence, the well-studied randomization procedure can be involved. In this section, we propose to use the sum of the least $N-1$ eigenvalues (truncated trace norm) minimization procedure to find feasible and near-optimal rank-1 solution to $\mathbf{X}$ [573]. This whole procedure will be presented as Algorithm 1.

In Algorithm 1, δ_e is given, then

Algorithm 1

1. Solve the optimization problem (11.83) without the consideration of rank constraint to get optimal solution $\mathbf{X}^*$, $\mathbf{Y}^*$, and minimum total power needed P^*; if $\mathbf{X}^*$ is the rank-1 matrix, then Algorithm 1 goes to step 3; otherwise Algorithm 1 goes to step 2;
2. Do eigen-decomposition to $\mathbf{X}^*$ to get the dominant eigen-vector $\mathbf{x}^*$ related to the maximum eigen-value of $\mathbf{X}^*$; solve the following optimization problem which is also an SDP problem,

$$\begin{array}{l}
\text{minimize} \\
\lambda(\text{trace}(\mathbf{X})-\text{trace}(\mathbf{X}\mathbf{x}^*(\mathbf{x}^*)^H))+((\sum_{n=1}^{N}((\mathbf{X})_{n,n}\left(|h_{sr_n1}g_{s1}|^2+|h_{dr_n1}g_{d1}|^2+\sigma_{r_n1}^2\right) \\
\qquad +(\mathbf{Y})_{n,n}))-P^*) \\
\text{subject to} \\
(\mathbf{X})_{n,n}\left(|h_{sr_n1}g_{s1}|^2+|h_{dr_n1}g_{d1}|^2+\sigma_{r_n1}^2\right)+(\mathbf{Y})_{n,n}\le P_{r_n2}, n=1,2,\ldots,N \\
\text{trace}(\mathbf{Q})+(-2\log(\delta_e))^{0.5}a-b\log(\delta_e)-\sigma_{e2}^2\gamma_e\le 0 \\
\|\mathbf{Q}\|_F\le a \\
b\mathbf{I}-Q\succeq 0 \\
b\ge 0 \\
\text{trace}(\mathbf{a}^H\mathbf{a}\mathbf{X})\geq\gamma_d\left(\text{trace}\left(\mathbf{H}_{rd2}^H\mathbf{H}_{rd2}\sigma_{r1}^2\mathbf{X}\right)+\text{trace}\left(\mathbf{h}_{rd2}{}^H\mathbf{h}_{rd2}\mathbf{Y}\right)+\sigma_{d2}^2\right) \\
\mathbf{X}\succeq 0 \\
\mathbf{Y}\succeq 0
\end{array} \tag{11.84}$$

   where λ is the design parameter; then optimal solution $\mathbf{X}^*$ and $\mathbf{Y}^*$ will be updated; if $\mathbf{X}^*$ is the rank-1 matrix, then Algorithm 1 goes to step 3; otherwise Algorithm 1 goes to step 2;
3. Get optimal solutions to $\mathbf{g}_{r2}$ and $\mathbf{U}$ by eigen-decompositions to $\mathbf{X}^*$ and $\mathbf{Y}^*$; Algorithm 1 is finished.

In the optimization problem (11.84), the minimization of trace($\mathbf{X}$) $-$ trace($\mathbf{X}\mathbf{x}^*(\mathbf{x}^*)^H$) tries to force $\mathbf{X}^*$ to be a rank-one matrix and the minimization of $(\sum_{n=1}^{N}((\mathbf{X})_{n,n}\left(|h_{sr_n1}g_{s1}|^2+|h_{dr_n1}g_{d1}|^2+\sigma_{r_n1}^2\right)+(\mathbf{Y})_{n,n}))-P^*$ tries to minimize the total transmitted power needed for N relay nodes.

As mentioned before, we are more interested in equally probabilistic constraints. If targeted violation probability is set to be $\delta_e^{\text{targeted}}$ which is also to be used as the parameter δ_e in the optimization problem (11.83), the violation probability in reality $\delta_e^{\text{reality}}$ by statistical validation procedure will be much smaller than $\delta_e^{\text{targeted}}$. Based on Bernstein-type inequalities, $\delta_e^{\text{reality}}$ will be a non-decreasing function of δ_e. Thus, we propose to exploit bi-section search to find suitable δ_e to make sure $\delta_{\rm e}^{\rm reality}$ is equal to $\delta_{\rm e}^{\rm targeted}$ [574, 575].

Overall, a novel numerical approach for joint AF relay and cooperative jamming with the consideration of probabilistic security will be proposed as Algorithm 2.

In Algorithm 2, $\delta_{\rm e}^l$ is set to be 0; $\delta_{\rm e}^u$ is set to be 1; and $\delta_{\rm e}^{\rm targeted}$ is given; then

Algorithm 2

1. Set δ_e to be $\delta_{\rm e}^{\rm targeted}$;
2. Invoke Algorithm 1 to get optimal solutions to $\mathbf{g}_{r2}$ and $\mathbf{U}$;
3. Perform Monte Carlo simulation to get $\delta_{\rm e}^{\rm reality}$; if $\delta_{\rm e}^{\rm reality} \geq \delta_{\rm e}^{\rm targeted}$, then $\delta_{\rm e}^u$ is set to be δ_e; otherwise $\delta_{\rm e}^l$ is set to be δ_e;
4. If $\delta_{\rm e}^u - \delta_{\rm e}^l \leq \xi$ where ξ is the design parameter, Algorithm 2 is finished; otherwise, δ_e is set to be $\frac{(\delta_{\rm e}^l + \delta_{\rm e}^u)}{2}$ and Algorithm 2 goes to step 2.

11.6.4 Simulation Results

In the simulation, $N = 10$; $\sigma_{r_n1}^2 = 0.15, n = 1, 2, \ldots, N$;$\sigma_{e1}^2 = 0.15$; $\sigma_{e2}^2 = 0.15$; $\sigma_{d2}^2 = 0.15$; $P_{r_n2} = 2, n = 1, 2, \ldots, N$; $\gamma_e = 1$. h_{sr_n1}, h_{dr_n1}, and $h_{r_nd2}, n = 1, 2, \ldots, N$ are randomly generated as complex zero-mean Gaussian random variables with unit covariance. CVX toolbox [488] is used to solve the presented SDPs.

In Fig. 11.3, we illustrate the relationship between total power needed by all relay nodes and the targeted SINR γ_d required by Bob. Meanwhile, different violation probabilities for Eve are considered. The smaller the pre-determined violation probability for Eve, the more power needed for N relay nodes. In other words, too conservative or robust performance for security requires a large amount of total transmitted power. Hence, we should balance the communication security requirement and total power budget through optimization theory.

In order to verify the correctness of the proposed numerical approach, 10,000 Monte Carlo simulations are run with randomly generated $h_{r_ne2}, n = 1, 2, \ldots, N$. $\delta_{\rm e}^{\rm targeted} = 0.08$. $\gamma_d = 20$. The histogram of the received SINRs for Eve is shown in Fig. 11.4. Similarly, if $\delta_{\rm e}^{\rm targeted} = 0.12$, the histogram of the received SINRs for Eve is shown in Fig. 11.5.

11.7 Further Comments

Probability constrained optimization is also used in [570]. Distributed robust optimization is studied by Yang et al. [576, 577] and Chen and Chiang [578] for communication networks. Randomized algorithms in robust control and smart grid are studied in [579–582].

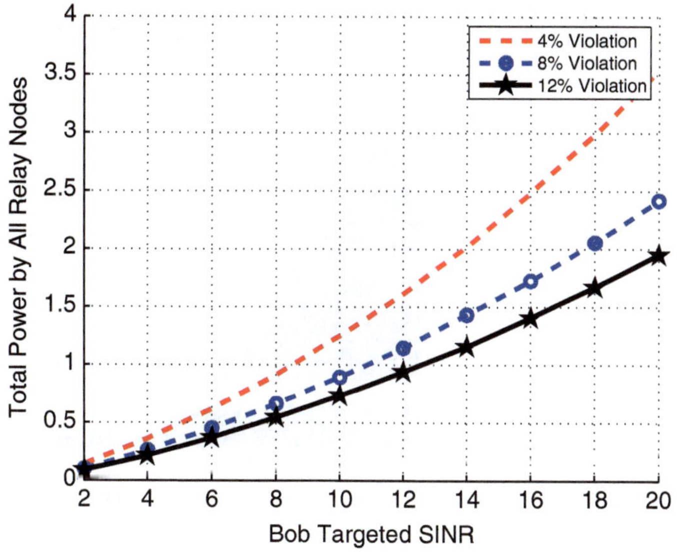

Fig. 11.3 Total power needed by all relay nodes

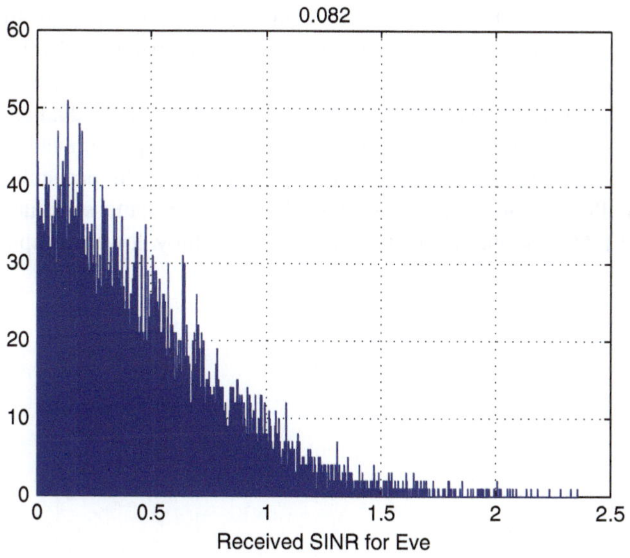

Fig. 11.4 The histogram of the received SINRs for Eve

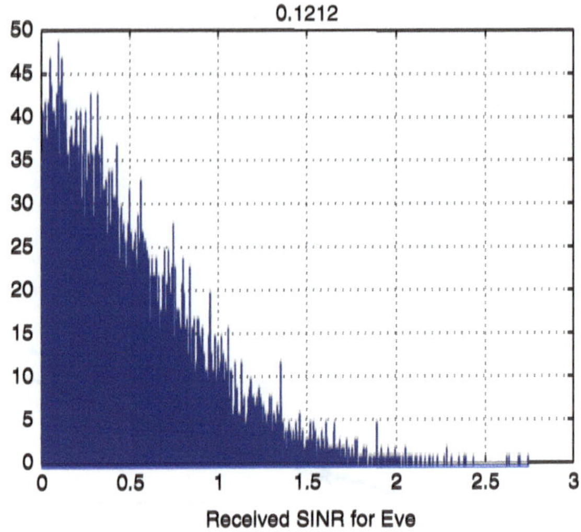

Fig. 11.5 The histogram of the received SINRs for Eve

In [583], distributionally robust slow adaptive orthogonal frequency division multiple access (OFDMA) with Soft QoS is studied via linear programming. Neither prediction of channel state information nor specification of channel fading distribution is needed for subcarrier allocation. As such, the algorithm is robust against any mismatch between actual channel state/distributional information and the one assumed. Besides, although the optimization problem arising from our proposed scheme is non-convex in general, based on recent advances in chance-constrained optimization, they show that it can be approximated by a certain linear program with provable performance guarantees. In particular, they only need to handle an optimization problem that has the same structure as the fast adaptive OFDMA problem, but they are able to enjoy lower computational and signaling costs.

Chapter 12
Database Friendly Data Processing

The goal of this chapter is to demonstrate how concentration of measure plays a central role in these modern randomized algorithms. There is a convergence of sensing, computing, networking and control. Data base is often neglected in traditional treatments in estimation, detection, etc.

Modern scientific computing demands efficient algorithms for dealing with large datasets—Big Data. Often these datasets can be fruitfully represented and manipulated as matrices; in this case, fast low-error methods for making basic linear algebra computations are key to efficient algorithms. Examples of such foundational computational tools are low-rank approximations, matrix sparsification, and randomized column subset selection.

12.1 Low Rank Matrix Approximation

Randomness can be turned to our advantage in the development of methods for dealing with these massive datasets [584, 585].

It is well known that a matrix $\mathbf{A}_k$ which minimizes both the Frobenious norm and the spectral norm error can be calculated via the singular value decomposition (SVD). The SVD takes cubic time, the computation cost of using it to form low-rank approximation can be prohibitive if the matrix is large.

For an integer $n = 2^p$, for $p = 1, 2, 3, \ldots$. the (non-normalized) $n \times n$ matrix of the Hadamard-Walsh transform is defined recursively as,

$$\mathbf{H}_n = \begin{bmatrix} \mathbf{H}_{n/2} & \mathbf{H}_{n/2} \\ \mathbf{H}_{n/2} & -\mathbf{H}_{n/2} \end{bmatrix}, \qquad \mathbf{H}_2 = \begin{bmatrix} +1 & +1 \\ +1 & -1 \end{bmatrix}.$$

The $n \times n$ matrix of the Hadamard-Walsh transform is equal to

$$\mathbf{H} = \frac{1}{\sqrt{n}} \mathbf{H}_n \in \mathbb{R}^{n \times n}. \tag{12.1}$$

R. Qiu and M. Wicks, *Cognitive Networked Sensing and Big Data*,
DOI 10.1007/978-1-4614-4544-9_12,

For integers r and $n = 2^p$ with $r < n$ and $p = 1, 2, 3, \ldots$. an subsampled randomized Hadamard transform matrix is an $r \times n$ matrix of the form

$$\mathbf{\Theta} = \sqrt{\frac{n}{r}} \cdot \mathbf{RHD}; \tag{12.2}$$

- $\mathbf{D} \in \mathbb{R}^{n \times n}$ is a random diagonal matrix whose entries are independent random signs, i.e., random variables uniformly distributed on $\{\pm 1\}$.
- $\mathbf{H} \in \mathbb{R}^{n \times n}$ is a normalized Walsh-Hadamard matrix.
- $\mathbf{R} \in \mathbb{R}^{r \times n}$ is a random matrix that restricts an n-dimensional vector to r coordinates, which are chosen uniformly at random and without replacement.

Theorem 12.1.1 (Subsampled randomized Hadamard transform [584]). *Let $\mathbf{A} \in \mathbb{R}^{m \times n}$ have rank ρ. For an integer k satisfying $0 < k \leqslant \rho$. Let $0 < \varepsilon < 1/3$ denote an accuracy parameter, $0 < \delta < 1$ be a failure probability, and $C \geq 1$ be a constant. Let $\mathbf{Y} = \mathbf{A}\mathbf{\Theta}^T$, $\mathbf{\Theta}$ is an $r \times n$ SRHT matrix with r satisfying*

$$6C^2\varepsilon^{-1}\left[\sqrt{k} + \sqrt{8\log(n/\delta)}\right]^2 \log(k/\delta) \leqslant r \leqslant n.$$

Then, with probability at least $1 - \delta^{C^2/24} - 7\delta$,

$$\left\|\mathbf{A} - \mathbf{Y}\mathbf{Y}^H\mathbf{A}\right\|_F \leqslant (1 + 50\varepsilon) \cdot \|\mathbf{A} - \mathbf{A}_k\|_F$$

and

$$\left\|\mathbf{A} - \mathbf{Y}\mathbf{Y}^H\mathbf{A}\right\|_2 \leqslant \left[6 + \sqrt{\varepsilon}\left(15 + \sqrt{\frac{\log(n/\delta)}{C^2\log(k/\delta)}}\right)\right] \cdot \|\mathbf{A} - \mathbf{A}_k\|_2 + \sqrt{\frac{\varepsilon}{8C^2\log(k/\delta)}} \cdot \|\mathbf{A} - \mathbf{A}_k\|_F.$$

The matrix $\mathbf{Y}$ can be constructed in $O(mn\log(r))$ time.

12.2 Row Sampling for Matrix Algorithms

We take material from Magdon-Ismail [586]. Let $\mathbf{e}_1, \ldots, \mathbf{e}_N$ the standard basis vectors in $\mathbb{R}^n$. Let $\mathbf{A} \in \mathbb{R}^{N \times n}$ denote an arbitrary matrix which represents N points in $\mathbb{R}^n$. In general, we represent a matrix such as $\mathbf{A}$ (bold, uppercase) by a set of vectors $\mathbf{a}_1, \ldots, \mathbf{a}_N \in \mathbb{R}^n$ (bold, lowercase), so that $\mathbf{A} = \left[\mathbf{a}_1\ \mathbf{a}_2 \cdots \mathbf{a}_N\right]^T$. Here $\mathbf{a}_i$ is the i-th row of $\mathbf{A}$, which we may also refer to by $\mathbf{A}_{(i)}$; similarly we refer to the i-th column as $\mathbf{A}^{(i)}$. Let $\|\cdot\|$ be the spectral norm and $\|\cdot\|_F$ the Frobenius norm. The numerical (stable) rank of $\mathbf{S}$ is defined as $\rho(\mathbf{S}) = \|\mathbf{S}\|_F^2 / \|\mathbf{S}\|^2$.

A *row-sampling matrix* samples k rows of $\mathbf{A}$ to form $\tilde{\mathbf{A}} = \mathbf{QA}$:

$$\mathbf{Q} = \begin{bmatrix} \mathbf{r}_1^T \\ \vdots \\ \mathbf{r}_k^T \end{bmatrix}, \qquad \tilde{\mathbf{A}} = \mathbf{QA} = \begin{bmatrix} \mathbf{r}_1^T\mathbf{A} \\ \vdots \\ \mathbf{r}_k^T\mathbf{A} \end{bmatrix},$$

where $\mathbf{r}_i^T\mathbf{A}$ samples the t_i-th row of $\mathbf{A}$ and rescales it. We are interested in random sampling matrices where each $\mathbf{r}_i$ is i.i.d. according to some distribution. Define a set of sampling probabilities $p_1, \ldots, p_N$, with $p_i > 0$ and $\sum\limits_{i=1}^{N} p_i = 1$; then $\mathbf{r}_i = \mathbf{e}_i/\sqrt{kp_t}$ with probability p_t. The scaling is also related to the sampling probabilities in all the algorithms we consider. We can rewrite $\mathbf{Q}^T\mathbf{Q}$ as the sum of k independently sampled matrices

$$\mathbf{Q}^T\mathbf{Q} = \frac{1}{k}\sum_{i=1}^{k} \mathbf{r}_i\mathbf{r}_i^T$$

where $\mathbf{r}_i\mathbf{r}_i^T$ is a diagonal matrix with only one non-zero entry; the t-th diagonal entry is equal to $1/p_t$ with probability p_t. Thus, by construction, for any set of non-zero sampling probabilities, $\mathbb{E}\left[\mathbf{r}_i\mathbf{r}_i^T\right] = \mathbf{I}_{N\times N}$. Since we are averaging k independent copies, it is reasonable to expect a concentration around the mean, with respect to k, and in some sense, $\mathbf{Q}^T\mathbf{Q}$ essentially behaves like the identity.

Theorem 12.2.1 (Symmetric Orthonormal Subspace Sampling [586]). *Let*

$$\mathbf{U} = \left[\mathbf{u}_1 \cdots \mathbf{u}_N\right]^T \in \mathbb{R}^{N\times n}$$

be orthonormal, and $\mathbf{D} \in \mathbb{R}^{n\times n}$ *be positive diagonal. Assume the row-sampling probabilities* p_t *satisfy*

$$p_t \geqslant \beta\frac{\mathbf{u}_t^T\mathbf{D}^2\mathbf{u}_t}{\mathrm{Tr}\left(\mathbf{D}^2\right)}.$$

Then, if $k \geqslant \left(4\rho\left(\mathbf{D}\right)/\beta\varepsilon^2\right)\ln\frac{2n}{\delta}$, *with probability at least* $1-\delta$,

$$\left\|\mathbf{D}^2 - \mathbf{D}\mathbf{U}^T\mathbf{Q}^T\mathbf{Q}\mathbf{U}\mathbf{D}\right\| \leqslant \varepsilon\|\mathbf{D}\|^2.$$

A linear regression is represented by a real data matrix $\mathbf{A} \in \mathbb{R}^{N\times n}$ which represents N points in $\mathbb{R}^n$, and a target vector $\mathbf{y} \in \mathbb{R}^N$. Traditionally, $N \gg n$. The goal is to find a regression vector $\mathbf{x}^\star \in \mathbb{R}^2$ which minimizes the ℓ_2 fit error (least squares regression)

$$\mathcal{E}\left(\mathbf{x}\right) = \|\mathbf{Ax} - \mathbf{y}\|_2^2 = \sum_{t=1}^{N}\left(\mathbf{a}_t^T\mathbf{x} - \mathbf{y}_t\right)^2.$$

This problem was formulated to use a non-commutative Bernstein bound. For details, see [587].

12.3 Approximate Matrix Multiplication

The matrix product $\mathbf{AB}^T$ for large dimensions is a challenging problem. Here we reformulate this standard linear algebra operation in terms of sums of random matrices. It can be viewed as non-uniform sampling of the columns of $\mathbf{A}$ and $\mathbf{B}$. We take material from Hsu, Kakade, and Zhang [113], converting to our notation. We make some comments and connections with other parts of the book at the appropriate points.

Let $\mathbf{A} = [\mathbf{a}_1 |\cdots| \mathbf{a}_n]$, and $\mathbf{B} = [\mathbf{b}_1 |\cdots| \mathbf{b}_n]$ be fixed matrices, each with n columns. Assume that $\mathbf{a}_j \neq 0$ and $\mathbf{b}_j \neq 0$ for all $j = 1, \ldots, n$. If n is very large, which is common in the age of Big Data, then the standard, straightforward computation of the product $\mathbf{AB}^T$ can be too computation-expensive. An alternative is to take a small (non-uniform) random samples of the columns of $\mathbf{A}$ and $\mathbf{B}$, say $\mathbf{a}_{j_1}, \mathbf{b}_{j_1}, \ldots, \mathbf{a}_{j_N}, \mathbf{b}_{j_N}$, or $\mathbf{a}_{j_i}, \mathbf{b}_{j_i}$ for $i = 1, 2, \ldots, N$. Then we compute a weighted sum of outer products[1]

$$\mathbf{AB}^T \cong \frac{1}{N} \sum_{i=1}^{N} \frac{1}{p_{j_i}} \mathbf{a}_{j_i} \mathbf{b}_{j_i}^T \tag{12.3}$$

where $p_{j_i} > 0$ is the a priori probability of choosing the column index $j_i \in \{1, \ldots, n\}$ from a collection of N columns. The "average" and randomness do most of the work, as observed by Donoho [132]: The regularity of having many "identical" dimensions over which one can "average" is a fundamental tool. The scheme of (12.3) was originally proposed and analyzed by Drinceas, Kannan, and Mahoney [313].

Let $\mathbf{X}_1, \ldots, \mathbf{X}_N$ be i.i.d. random matrices with the discrete distribution given by

$$\mathbb{P}\left(\mathbf{X}_i = \frac{1}{p_j} \begin{bmatrix} 0 & \mathbf{a}_j \mathbf{b}_j^T \\ \mathbf{b}_j^T \mathbf{a}_j & 0 \end{bmatrix} \right) = p_j$$

for all $j = 1, \ldots, n$, where

$$p_j = \frac{\|\mathbf{a}_j\|_2 \|\mathbf{b}_j\|_2}{Z}, \qquad \mathrm{Z} = \sum_{j=1}^{n} \|\mathbf{a}_j\|_2 \|\mathbf{b}_j\|_2.$$

Let

$$\hat{\mathbf{M}} = \frac{1}{N} \sum_{i=1}^{N} \mathbf{X}_i \quad \text{and} \quad \mathbf{M} = \begin{bmatrix} 0 & \mathbf{AB}^T \\ \mathbf{B}^T\mathbf{A} & 0 \end{bmatrix}.$$

---

[1] Outer products $\mathbf{xy}^T$ of two vectors $\mathbf{x}$ and $\mathbf{y}$ are rank-one matrices.

The spectral norm error $\left\|\hat{\mathbf{M}}-\mathbf{M}\right\|_2$ is used to describe the approximation of $\mathbf{AB}^T$ using the *average* of N outer products $\frac{1}{p_{i_j}}\mathbf{a}_{i_j}\mathbf{b}_{i_j}^T$, where the indices are such that

$$j_i = j \Leftrightarrow \mathbf{X}_i = \mathbf{a}_j\mathbf{b}_j^T/p_j$$

for all $i = 1,\ldots,N$. Again, the "average" plays a fundamental role. Our goal is to use Theorem 2.16.4 to bound this error. To apply this theorem, we must first check the conditions.

We have the following relations

$$\mathbb{E}[\mathbf{X}_i] = \sum_{j=1}^{n} p_j\left(\frac{1}{p_j}\begin{bmatrix} 0 & \mathbf{a}_j\mathbf{b}_j^T \\ \mathbf{b}_j^T\mathbf{a}_j & 0\end{bmatrix}\right) = \begin{bmatrix} 0 & \sum_{j=1}^{n}\mathbf{a}_j\mathbf{b}_j^T \\ \sum_{j=1}^{n}\mathbf{b}_j^T\mathbf{a}_j & 0\end{bmatrix} = \mathbf{M}$$

$$\mathrm{Tr}\left(\mathbb{E}\left[\mathbf{X}_i^2\right]\right) = \mathrm{Tr}\left(\sum_{j=1}^{n} p_j\left(\frac{1}{p_j^2}\begin{bmatrix} \mathbf{a}_j\mathbf{b}_j^T\mathbf{b}_j^T\mathbf{a}_j & 0 \\ 0 & \mathbf{b}_j^T\mathbf{a}_j\mathbf{a}_j\mathbf{b}_j^T\end{bmatrix}\right)\right) = \sum_{j=1}^{n}\frac{2}{p_j}\|\mathbf{a}_j\|_2^2\|\mathbf{b}_j\|_2^2 = 2Z^2$$

$$\mathrm{Tr}\left(\left(\mathbb{E}[\mathbf{X}_i]\right)^2\right) = \mathrm{Tr}\left(\mathbf{M}^2\right) = \begin{bmatrix} \mathbf{AB}^T\mathbf{B}^T\mathbf{A} & 0 \\ 0 & \mathbf{B}^T\mathbf{AAB}^T\end{bmatrix} = 2\mathrm{Tr}\left(\mathbf{AB}^T\mathbf{B}^T\mathbf{A}\right).$$

Let $\|\cdot\|_2$ stand for the spectral norm. The following norm inequalities can be obtained:

$$\begin{aligned}
\|\mathbf{X}_i\|_2 &\leqslant \max_{j=1,\ldots,n}\frac{1}{p_j}\left\|\begin{bmatrix} 0 & \mathbf{a}_j\mathbf{b}_j^T \\ \mathbf{b}_j^T\mathbf{a}_j & 0\end{bmatrix}\right\|_2 = \max_{j=1,\ldots,n}\frac{1}{p_j}\left\|\mathbf{a}_j\mathbf{b}_j^T\right\|_2 = Z \\
\|\mathbb{E}\mathbf{X}_i\|_2 &= \|\mathbf{M}\|_2 \leqslant \left\|\mathbf{AB}^T\right\|_2 \leqslant \|\mathbf{A}\|_2\|\mathbf{B}\|_2 \\
\left\|\mathbb{E}\left[\mathbf{X}_i^2\right]\right\|_2 &\leqslant \|\mathbf{A}\|_2\|\mathbf{B}\|_2 Z.
\end{aligned}$$

Using Theorem 2.16.4 and a union bound, finally we arrive at the following: For any $\varepsilon \in (0,1)$, and $\delta \in (0,1)$, if

$$N \geqslant \left(\frac{8}{3}+2\sqrt{\frac{5}{3}}\right)\frac{\left(1+\sqrt{r_A r_B}\right)\left(\log\left(4\sqrt{r_A r_B}\right)+\log(1/\delta)\right)}{\varepsilon^2},$$

then with probability at least $1-\delta$ over the random choice of column indices $j_1,\ldots,j_N$,

$$\left\|\frac{1}{N}\sum_{j=1}^{N}\frac{1}{p_{i_j}}\mathbf{a}_{i_j}\mathbf{b}_{i_j}^T - \mathbf{AB}^T\right\|_2 \leqslant \varepsilon\|\mathbf{A}\|_2\|\mathbf{B}\|_2,$$

where

$$r_A = \|\mathbf{A}\|_F^2 \, / \, \|\mathbf{A}\|_2^2 \in (1, \text{rank}\,(\mathbf{A})), \quad \text{and} \quad r_B = \|\mathbf{B}\|_F^2 \, / \, \|\mathbf{B}\|_2^2 \in (1, \text{rank}\,(\mathbf{B}))$$

are the numerical (or stable) rank. Here $||\cdot||_F$ stands for the Frobenius (or Hilbert-Schmidt) norm. For details, we refer to [113].

12.4 Matrix and Tensor Sparsification

In the age of Big Data, we have a data deluge. Data is expressed as matrices. One wonders what is the best way of efficiently generating "sketches" of matrices. Formally, we define the problem as follows: Given a matrix $\mathbf{A} \in \mathbb{R}^{n \times n}$, and an error parameter ε, construct a sketch $\tilde{\mathbf{A}} \in \mathbb{R}^{n \times n}$ of $\mathbf{A}$ such that

$$\left\| \mathbf{A} - \tilde{\mathbf{A}} \right\|_2 \leqslant \varepsilon$$

and the number of *non-zero entries* in $\tilde{\mathbf{A}}$ is *minimized.* Here $||\cdot||_2$ is the spectral norm of a matrix (the largest singular value), while $||\cdot||_F$ is the Frobenius form. See Sect. 1.4.5.

Algorithm 12.4.1 (Matrix sparsification [124]).

1. **Input**: matrix $\mathbf{A} \in \mathbb{R}^{n \times n}$, sampling parameter s.
2. **For all** $1 \le i, j \le n$ **do**

—**If** $A_{ij}^2 \leqslant \frac{\log^2 n}{n} \frac{\|\mathbf{A}\|_F^2}{s}$ **then** $\tilde{A}_{ij} = 0$,

—**elseIf** $A_{ij}^2 \geqslant \frac{\|\mathbf{A}\|_F^2}{s}$ **then** $\tilde{A}_{ij} = A_{ij}$,

—**Else** $\tilde{A}_{ij} = \begin{cases} \frac{A_{ij}}{p_{ij}}, & \text{with probability } p_{ij} = \frac{sA_{ij}^2}{\|\mathbf{A}\|_F^2} < 1 \\ 0, & \text{with probability } 1 - p_{ij} \end{cases}$

3. **Output**: matrix $\tilde{\mathbf{A}} \in \mathbb{R}^{n \times n}$.

An algorithm is shown in Algorithm 12.4.1. When $n \ge 300$, and

$$s = C \frac{n\,(\log n)\left(\log^2\left(n/\log^2 n\right)\right)}{\varepsilon^2},$$

then, with probability at least $1 - 1/n$, we have

$$\left\| \mathbf{A} - \tilde{\mathbf{A}} \right\|_2 \leqslant \varepsilon \|\mathbf{A}\|_F,$$

Theorem 2.17.7 has been used in [124] to analyze this algorithm.

Automatic generation of very large data sets enables the coming of Big Data age. Such data are often modeled as matrices. A generation of this framework permits the modeling of the data by higher-order arrays or tensors (e.g., arrays with more than two modes). A natural example is time-evolving data, where the third mode of the tensor represents time [588]. Concentration of measure has been used in [10] to design algorithms.

For any d-mode or order-d tensor $\mathcal{A} \in \mathbb{R}^{\overbrace{n \times n \times \cdots \times n}^{d \text{ times}}}$, its Frobenius norm $\|\mathcal{A}\|_F$ is defined as the square root of the sum of the squares of its elements. Now we define tensor-vector products: let $\mathbf{x}, y$ be vectors in $\mathbb{R}^n$. Then

$$\begin{aligned} \mathcal{A} \times_1 \mathbf{x} &= \sum_{i=1}^{n} \mathcal{A}_{ijk \cdots l} x_i, \\ \mathcal{A} \times_2 \mathbf{x} &= \sum_{i=1}^{n} \mathcal{A}_{ijk \cdots l} x_j, \\ \mathcal{A} \times_3 \mathbf{x} &= \sum_{i=1}^{n} \mathcal{A}_{ijk \cdots l} x_k, \quad \text{etc.} \end{aligned}$$

The outcome of the above operations is an order-$(d-1)$ tensor. The above definition may be extended to handle multiple tensor-vector products,

$$\mathcal{A} \times_1 \mathbf{x} \times_2 \mathbf{y} = \sum_{i=1}^{n} \mathcal{A}_{ijk \cdots l} x_i y_j,$$

which is an order-$(d - 2)$ tensor. Using this definition, the spectrum norm is defined as

$$\|\mathcal{A}\|_2 = \sup_{\mathbf{x}_1, \ldots, \mathbf{x}_d \in \mathbb{R}^n} |\mathcal{A} \times_1 \mathbf{x}_1 \cdots \times_d \mathbf{x}_d|,$$

where all the vectors $\mathbf{x}_1, \ldots, \mathbf{x}_d \in \mathbb{R}^n$ are unit vectors, i.e., $\|\mathbf{x}_i\|_2 = 1$, for all $i \in [d]$. The notation $[d]$ stands for the set $\{1, 2, \ldots, d\}$.

Given an order-d tensor $\mathcal{A} \in \mathbb{R}^{\overbrace{n \times n \times \cdots \times n}^{d \text{ times}}}$ and an error parameter $\epsilon > 0$, construct an order-d tensor sketch $\tilde{\mathcal{A}} \in \mathbb{R}^{\overbrace{n \times n \times \cdots \times n}^{d \text{ times}}}$ such that

$$\left\| \mathcal{A} - \tilde{\mathcal{A}} \right\|_2 \leqslant \varepsilon \left\| \tilde{\mathcal{A}} \right\|_2$$

and the number of non-zero entries in $\tilde{\mathcal{A}}$ is minimized.

Assume $n \geq 300$, and $2 \leqslant d \leqslant 0.5 \ln n$. If the sampling parameter s satisfies

$$s = \Omega \left(\frac{d^3 8^{2d} \mathbf{st}\left(\mathcal{A}\right) n^{d/2} \ln^3 n}{\epsilon^2} \right),$$

then, with probability at least $1 - 1/n$,

$$\left\| \mathcal{A} - \tilde{\mathcal{A}} \right\|_2 \leqslant \varepsilon \left\| \tilde{\mathcal{A}} \right\|_2.$$

Here, we use the stable rank $\mathbf{st}(\mathcal{A})$ of the tensor

$$\mathbf{st}(\mathcal{A}) = \frac{\|\mathcal{A}\|_F^2}{\|\mathcal{A}\|_2^2}.$$

Theorem 3.3.10 has been applied by Nguyen et al. [10] to obtain the above result.

12.5 Further Comments

See Mahoney [589], for randomized algorithms for matrices and data.

Chapter 13
From Network to Big Data

The main goal of this chapter is to put together all pieces treated in previous chapters. We treat the subject from a system engineering point of view. This chapter motivates the whole book. We only have space to see the problems from ten-thousand feet high.

13.1 Large Random Matrices for Big Data

Figure 13.1 illustrates the vision of big data that will be the foundation to understand cognitive networked sensing, cognitive radio network, cognitive radar and even smart grid. We will further develop this vision in the book on smart grid [6]. High dimensional statistics is the driver behind these subjects. Random matrices are natural building blocks to model big data. Concentration of measure phenomenon is of fundamental significance to modeling a large number of random matrices. Concentration of measure phenomenon is a phenomenon unique to high-dimensional spaces. The large data sets are conveniently expressed as a matrix

$$\mathbf{X} = \begin{bmatrix} X_{11} & X_{12} & \cdots & X_{1n} \\ X_{21} & X_{22} & \cdots & X_{2n} \\ \vdots & \vdots & \cdots & \vdots \\ X_{m1} & X_{m2} & \cdots & X_{mn} \end{bmatrix} \in \mathbb{C}^{m \times n}$$

where X_{ij} are random variables, e.g, sub-Gaussian random variables. Here m, n are finite and large. For example, $m = 100, n = 100$. The spectrum of a random matrix X tends to stabilize as the dimensions of $\mathbf{X}$ grows to infinity. In the last few years, local and non-asymptotic regimes, the dimensions of $\mathbf{X}$ are fixed rather than grow to infinity. Concentration of measure phenomenon naturally occurs. The eigenvalues $\lambda_i\left(\mathbf{X}^T\mathbf{X}\right), i = 1, \ldots, n$ are natural mathematical objects to study.

R. Qiu and M. Wicks, *Cognitive Networked Sensing and Big Data*,
DOI 10.1007/978-1-4614-4544-9_13,

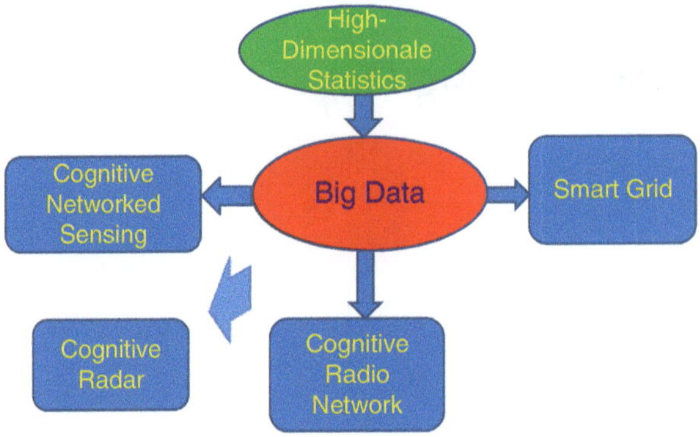

Fig. 13.1 Big data vision

The eigenvalues can be viewed as Lipschitz functions that can be handled by Talagrands concentration inequality. It expresses the insight: The sum of a large number of random variables is a constant with *high probability*. We can often treat both standard Gaussian and Bernoulli random variables in the unified framework of the sub-Gaussian family.

Theorem 13.1.1 (Talagrand's Concentration Inequality). *For every product probability $\mathbb{P}$ on $\{-1,1\}^n$, consider a convex and Lipschitz function $f:\mathbb{R}^n\to\mathbb{R}$ with Lipschitz constant L. Let $X_1,\ldots,X_n$ be independent random variables taking values $\{-1,1\}$. Let $Y=f(X_1,\ldots,X_n)$ and let $\mathbb{M}Y$ be a median of Y. Then For every $t>0$, we have*

$$\mathbb{P}\left(|Y-\mathbb{M}Y|\geqslant t\right)\leqslant 4e^{-t^2/16L^2}. \tag{13.1}$$

The random variable Y has the following property

$$\mathrm{Var}\left(Y\right)\leqslant 16L^2,\qquad \mathbb{E}\left[Y\right]-16L\leqslant\mathbb{M}\left[Y\right]\leqslant\mathbb{E}\left[Y\right]+16L. \tag{13.2}$$

For a random matrix $\mathbf{X}\in\mathbb{R}^{n\times n}$, the following functions are Lipschitz functions:

$$(1)\lambda_{\max}\left(\mathbf{X}\right);(2)\lambda_{min}\left(\mathbf{X}\right);(3)\ \mathrm{Tr}\left(\mathbf{X}\right);(4)\sum_{i=1}^{k}\lambda_i\left(\mathbf{X}\right);(5)\sum_{i=1}^{k}\lambda_{n-i+1}\left(\mathbf{X}\right)$$

where $\mathrm{Tr}\left(\mathbf{X}\right)$ has a Lipschitz constant of $L=1/n$, and $\lambda_i\left(\mathbf{X}\right),i=1,\ldots,n$ has a Lipschitz constant of $L=1/\sqrt{n}$. So the variance of $\mathrm{Tr}\left(\mathbf{X}\right)$ is upper bounded by $16/n^2$, while the variance of $\lambda_i\left(\mathbf{X}\right),i=1,\ldots,n$ by $16/n$. The variance of $\mathrm{Tr}\left(\mathbf{X}\right)$ is $1/n$ smaller than that of $\lambda_i\left(\mathbf{X}\right),i=1,\ldots,n$. For example, $n=100$, their ratio is 20 dB. The variance has a fundamental control over the hypothesis detection.

13.2 A Case Study for Hypothesis Detection in High Dimensions

Let

$$\mathbf{x} = \begin{bmatrix} X_1 \\ X_2 \\ \vdots \\ X_n \end{bmatrix}; \quad \mathbf{y} = \begin{bmatrix} Y_1 \\ Y_2 \\ \vdots \\ Y_n \end{bmatrix}; \quad \mathbf{s} = \begin{bmatrix} S_1 \\ S_2 \\ \vdots \\ S_n \end{bmatrix}; \quad \mathbf{x}, \mathbf{y}, \mathbf{s} \in \mathbb{C}^n.$$

The hypothesis detection is written as

$$\mathcal{H}_0 : \mathbf{y} = \mathbf{x}$$
$$\mathcal{H}_1 : \mathbf{y} = \mathbf{s} + \mathbf{x}.$$

The covariance matrices are defined as

$$\mathbf{R}_x = \mathbb{E}\left[\mathbf{x}\mathbf{x}^H\right], \mathbf{R}_y = \mathbb{E}\left[\mathbf{y}\mathbf{y}^H\right], \mathbf{R}_s = \mathbb{E}\left[\mathbf{s}\mathbf{s}^H\right], \quad \mathbf{R}_x, \mathbf{R}_y, \mathbf{R}_s \in \mathbb{R}^{n \times n}.$$

We have the equivalent form

$$\mathcal{H}_0 : \mathbf{R}_y = \mathbf{R}_x$$
$$\mathcal{H}_1 : \mathbf{R}_y = \mathbf{R}_s + \mathbf{R}_x = \text{Low rank matrix + Sparse matrix.}$$

where $\mathbf{R}_s$ is often of low rank. When the white Gaussian random vector is considered, we have

$$\mathbf{R}_x = \sigma^2 \mathbf{I}_{n \times n} = \sigma^2 \begin{bmatrix} 1 & 0 & \cdots & 0 \\ 0 & 1 & \cdots & \vdots \\ \vdots & \vdots & \ddots & 0 \\ 0 & \cdots & 0 & 1 \end{bmatrix},$$

which is sparse in that there are non-zero entries only along the diagonal line. Using matrix decomposition [590], we are able to separate the two matrices even when σ^2 is very small, say the signal to noise ratio $SNR = 10^{-6}$. Anti-jamming is one motivated example.

Unfortunately, when the sample size N of the random vector $\mathbf{x} \in \mathbb{R}^n$ is finite, the sparse matrix assumption of $\mathbf{R}_x$ is not satisfied. Let us study a simple example. Consider i.i.d. Gaussian random variables $X_i \sim \mathcal{N}(0,1), i = 1, \ldots, n$. In MATLAB code, we have the random data vector $\mathbf{x} = \text{randn}(n,1)$. So the true covariance matrix is $\mathbf{R}_x = \sigma^2 \mathbf{I}_{n \times n}$. For N dependent copies of $\mathbf{x}$, we define the sample covariance matrix

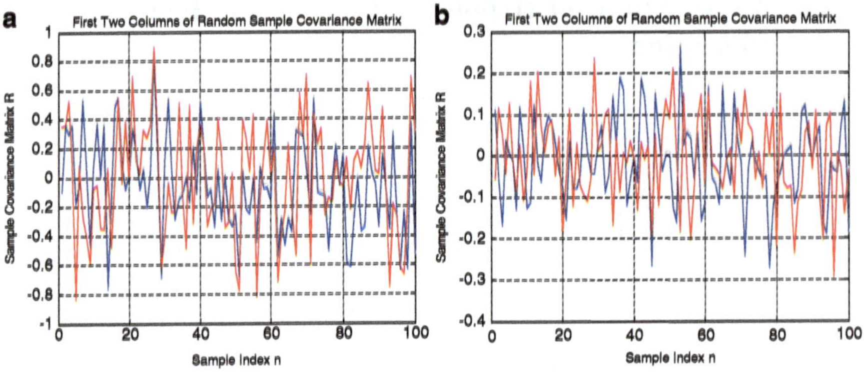

Fig. 13.2 Random sample covariance matrices of $n \times n$ with $n = 100$: (**a**) N = 10; (**b**) N = 100

$$\hat{\mathbf{R}}_x = \frac{1}{N}\sum_{i=1}^{N}\mathbf{x}_i\mathbf{x}_i^T = \frac{1}{N}\mathbf{X}\mathbf{X}^T \in \mathbb{R}^{n\times n},$$

where the data matrix is $\mathbf{X} = \left[\mathbf{x}_1\ \mathbf{x}_2 \cdots \mathbf{x}_N\right] \in \mathbb{R}^{n\times N}$. The first two columns of $\hat{\mathbf{R}}_x$ are shown in Fig. 13.2 for (a) $N = 10$ and (b) $N = 100$. The variance of case (b) is much smaller than that of case (a). The convergence is measured by $\hat{\mathbf{R}}_x - \mathbf{R}_x = \hat{\mathbf{R}}_x - \sigma^2\mathbf{I}_{n\times n}$. Consider the regime

$$\text{as} \quad n \to \infty, N \to \infty, \quad \frac{N}{n} \to c \in (0,1). \tag{13.3}$$

Under this regime, the fundamental question is

$$\hat{\mathbf{R}}_x \to \sigma^2\mathbf{I}_{n\times n}?$$

Under the regime of (13.3), we have the hypothesis detection

$$\mathcal{H}_0 : \hat{\mathbf{R}}_y = \hat{\mathbf{R}}_x$$
$$\mathcal{H}_1 : \hat{\mathbf{R}}_y = \hat{\mathbf{R}}_s + \hat{\mathbf{R}}_x$$

where $\hat{\mathbf{R}}_x$ is a random matrix which is not sparse. We are led to evaluate this following simple, intuitive test

$$\mathrm{Tr}\left(\hat{\mathbf{R}}_y\right) \geqslant \gamma_1 + \mathrm{Tr}\left(\hat{\mathbf{R}}_x\right), \quad \text{claim } \mathcal{H}_1.$$
$$\mathrm{Tr}\left(\hat{\mathbf{R}}_y\right) \leqslant \gamma_0 + \mathrm{Tr}\left(\hat{\mathbf{R}}_x\right), \quad \text{claim } \mathcal{H}_0.$$

Let us compare the classical likelihood ratio test (LRT).

13.3 Cognitive Radio Network Testbed

Let us use Fig. 13.3 to represent a cognitive radio network of N nodes. The network works for a TDMA manner. When one node, say, $i = 1$, transmits, the rest of the nodes $i = 2, \ldots, 80$ record the data. We can randomly (or deterministically) choose the next node to transmit.

13.4 Wireless Distributed Computing

For a segment of M samples, we can form a matrix of $N \times M$. For convenience, we collect the M data samples into a vector, denoted by $\mathbf{x} \in \mathbb{R}^M$ or $\mathbf{x} \in \mathbb{C}^M$. For N nodes, we have $\mathbf{x}_1, \ldots, \mathbf{x}_N$. We can collect the data into a matrix $\mathbf{X} = [\mathbf{x}_1, \ldots, \mathbf{x}_N]^T \in \mathbb{R}^{N \times M}$. The entries of the matrix $\mathbf{X}$ are (scalar) random variables. So X is a random matrix.

For statistics, we often start with the sample covariance matrix defined as

$$\hat{\mathbf{R}}_x = \frac{1}{N} \sum_{i=1}^{N} \mathbf{x}_i \mathbf{x}_i^T = \frac{1}{N} \mathbf{X}\mathbf{X}^T.$$

The true covariance matrix is $\mathbf{R}_x$. Of course one fundamental question is to answer how the sample size N affects the estimation accuracy $\left\| \mathbf{R}_x - \hat{\mathbf{R}}_x \right\|$ as a function of the dimension n.

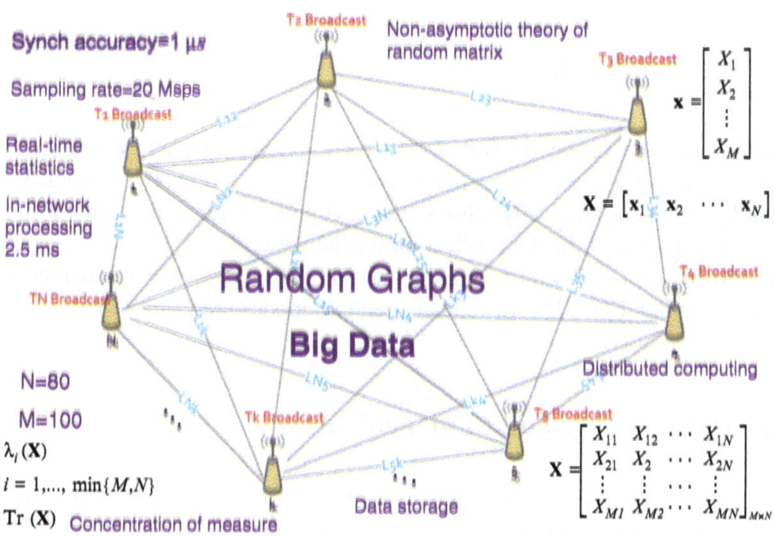

Fig. 13.3 Complex network model

When the sample size N is comparable to the dimension n, the so-called non-asymptotic random matrix theory must be used, rather than the asymptotic limits when N and n goes to infinity, i.e., $N \to \infty, n \to \infty$. Concentration of measure phenomenon is the fundamental tool to deal with this non-asymptotic regime. Lipschitz functions are the basic building blocks to investigate. Fortunately, most quantities of engineering interest belong to this class of functions. Examples include the trace, the largest eigenvalue, and the smallest eigenvalue of $\mathbf{X}$: $\mathrm{Tr}\left(\mathbf{X}\right), \lambda_{\max}\left(\mathbf{X}\right)$, and $\lambda_{\min}\left(\mathbf{X}\right)$.

Since the random matrix is viewed as the starting point for future statistical studies, it is important to understand the computing aspects, especially for some real-time applications. Practically, the sampling rate of the software-defined radio node is in the level of 20 Mega samples per second (Msps). For a data segment of $M = 100$ samples, the required sampling time is 5 μs for each node $i = 1, \ldots, N$.

At this stage, it is the right moment to talk about the network synchronization, which is critical. The Apollo testbed for cognitive radio network at Tennessee Technological University (TTU) [591–593] has achieved the synchronization error of less than 1 μs.

The raw data that are collected at each node $i = 1, \ldots, N$ are difficult to be moved around. It is feasible to disseminate the real-time statistics such as the parameters associated with the sample covariance matrix. To estimate the 100×100 sample covariance matrix for each node, for example we need collect a total of $M = 10,000$ sample points, which require the sampling time of 500 μs = 0.5 ms. It is remarkable to point out that the computing time is around 2.5 ms, which is five times larger than the required sampling time 0.5 ms. Parallel computing is needed to speed up the wireless distributed computing. Therefore, in-networking processing is critical.

13.5 Data Collection

As mentioned above, the network works for a TDMA manner. When one node, say, $i = 1$, transmits, the rest of the nodes $i = 2, \ldots, 80$ record the data. We can randomly (or deterministically) choose the next node to transmit. An algorithm can be designed to control the data collection for the network.

As shown in Fig. 13.4, communications and sensing modes are enabled in the Apollo testbed of TTU. The switch between two modes are fast.

13.6 Data Storage and Management

The large data sets are collected at each node $i = 1, \ldots, 80$. After the collection, some real-time in-network processing can be done to extract the statistical parameters that will be stored locally at each node. Of course, the raw data can be stored into the local database without any in-network processing.

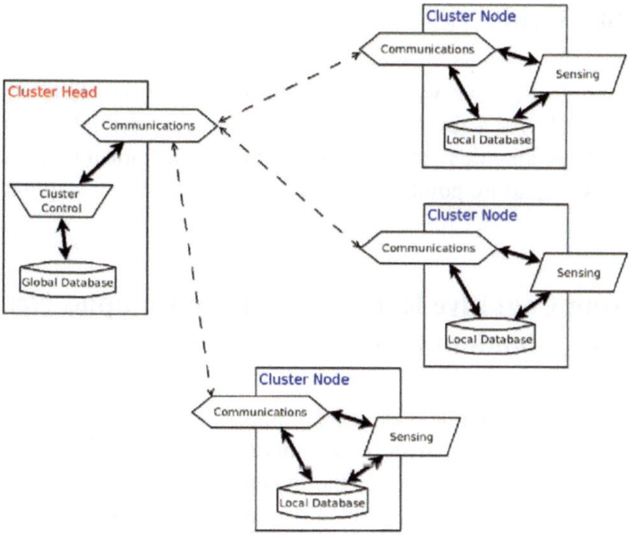

Fig. 13.4 Data collection and storage

It is very important to remark that the data sets are large and difficult to move around. To disseminate the information about these large data sets, we need rely some statistical tools to reduce the dimensionality, which is often the first thing to do with the data. Principal component analysis (PCA) is one such tool. PCA requires the sample covariance matrix as in the input to the algorithm.

Management of these large data sets is very challenging. For example, how to index these largest data sets.

13.7 Data Mining of Large Data Sets

At this stage, the data is assumed to be stored and managed property with effective indexing. We can mine the data for the information that is needed. The goals include: (1) higher spectrum efficiency; (2) higher energy efficiency; (3) enhanced security.

13.8 Mobility of Network Enabled by UAVs

Unmanned aerial vehicles (UAVs) enable the mobility of the wireless network. It is especially interesting to study the large data sets as a function of space and time. The Global Positioning System (GPS) is used to locate the 3-dimensional position together with time stamps.

13.9 Smart Grid

Smart Grid requires an two-way information flow [6]. Smart Grid can be viewed as a large network. As a result, Big Data aspects are essential. In another work, we systemically explore this viewpoint, using the mathematical tools of this current book as the new departure points. Figure 13.1 illustrates this viewpoint.

13.10 From Cognitive Radio Network to Complex Network and to Random Graph

The large size and dynamic nature of complex networks [594–598] enable a deep connection between statistic graph theory and the possibility of describing macroscopic phenomena in terms of the dynamic evolution of the basic elements of the system [599]. In our new book [5], a cognitive radio network is modeled as a complex network. (The use of a cognitive radio network in a smart grid is considered [592].) The book [600] also models the communication network as a random network.

So little is understood of (large) networks. The rather simple question "What is a robust network?" seems beyond the realm of present understanding [601]. Any complex network can be represented by a graph. Any graph can be represented by an adjacency matrix, from which other matrices such as the Laplacian are derived. One of the most beautiful aspects of linear algebra is the notion that, to each matrix, a set of eigenvalues can be associated with corresponding eigenvectors. As shown below, the most profound observation is that these eigenvalues for general random matrices are *strongly concentrated*. As a result, eigenvalues—which are certain scalar valued random variables—are natural metrics to describe the complex network (random graph). A close analogy is that the spectrum domain of the Fourier transform is the natural domain to study a random signal. A graph consists of a set of nodes connected by a set of links. Some properties of a graph in the topology domain is connected with the eigenvalues of random matrices.

13.11 Random Matrix Theory and Concentration of Measure

From a mathematical point of view, it is convenient to define a graph—and therefore a complex network—by means of the *adjacency matrix* $\mathbf{X} = \{x_{ij}\}$. This is a $N \times N$ matrix defined such that

$$x_{ij} = \begin{cases} 1, \text{ if } (i,j) \in \mathcal{E} \\ 0, \text{ if } (i,j) \notin \mathcal{E} \end{cases}. \tag{13.4}$$

For undirected graphs the adjacency matrix is symmetric,[1] $x_{ij} = x_{ji}$, and therefore contains redundant information. If the graph consists of N nodes and L links, then the $N \times N$ adjacency matrix can be written as

$$\mathbf{X} = \mathbf{U}\boldsymbol{\Lambda}\mathbf{U}^T \tag{13.5}$$

where the $N \times N$ matrix $\mathbf{U}$ contains as columns the eigenvectors $\mathbf{u}_1, \ldots, \mathbf{u}_N$ of $\mathbf{X}$ belonging to the real eigenvalues $\lambda_1 \geqslant \lambda_3 \geqslant \ldots \geqslant \lambda_N$ and where the diagonal matrix $\boldsymbol{\Lambda} = \text{diag}(\lambda_i)$. The basic relation (13.5) equates the topology domain, represented by the adjacency matrix, to the spectral domain of the graph, represented by the eigensystem in terms of the orthogonal matrix $\mathbf{U}$ of eigenvectors and diagonal matrix $\mathbf{U}\boldsymbol{\Lambda}$.

The topology of large complex networks makes the random graph model attractive. When a random graph—and therefore a complex network—is studied, the adjacency matrix $\mathbf{X}$ is a random matrix. Let G be an edge-independent random graph on the vertex set $[n] = \{1, 2, \ldots, n\}$; two vertices v_i and v_j are adjacent in G with probability p_{ij} independently. Here $\{p_{ij}\}_{i=1}^{n}$ are not assumed to be equal. Using the matrix notation, we have that

$$x_{ij} = \mathbb{P}(v_i \sim v_j) = p_{ij}. \tag{13.6}$$

As a result, the study of a cognitive radio network boils down to the study of random matrix $\mathbf{X}$.

When the elements of x_{ij} are random variables, the adjacency matrix $\mathbf{X}$ is a random matrix. So the so-called random matrix theory can be used as a powerful mathematical tool. The connection is very deep. Our book [5] has dedicated more than 230 pages to this topic. The most profound observation is that the spectra of the random matrix is highly concentrated. The surprisingly striking Talagrand's inequality lies at the heart of this observation. As a result, the statistical behavior of the graph spectra for complex networks is of interest [601].

Our useful approach is to investigate the eigenvalues (spectrum) of the random adjacency matrix and (normalized) Laplacian matrix. There is connection between the spectrum and the topology of a graph. The duality between topology and spectral domain is not new and has been studied in the field of mathematics called algebraic graph theory [602, 603]. What is new is the connection with complex networks [601].

The promising model is based on the concentration of the adjacency matrix and of the Laplacian in random graphs with independent edges along a line of research [604–608]. For example. see [604] for the model. We consider a random graph G such that each edge is determined by an independent random variable, where the probability of each edge is not assumed to be equal, i.e., $\mathbb{P}(v_i \sim v_j) = p_{ij}$. Each edge of G is independent of each other edge.

[1]Unless mentioned otherwise, we assume in this section that the graph is undirected and that $\mathbf{X}$ is symmetric.

For random graphs with such general distributions, several bounds for the spectrum of the corresponding adjacency matrix and (normalized) Laplacian matrix can be derived. Eigenvalues of the adjacency matrix has many applications in graph theory, such as describing certain topological features of a graph, such as connectivity and enumerating the occurrences of subgraphs [602, 609].

The data collection is viewed as a statistical inverse problem. Our results have broader applicability in data collection, e.g., problems in social networking, game theory, network security, and logistics [610].

Bibliography

1. N. Alon, M. Krivelevich, and V. Vu, "On the concentration of eigenvalues of random symmetric matrices," *Israel Journal of Mathematics*, vol. 131, no. 1, pp. 259–267, 2002. [33, 244]
2. V. Vu, "Concentration of non-lipschitz functions and applications," *Random Structures & Algorithms*, vol. 20, no. 3, pp. 262–316, 2002. [341]
3. V. Vu, "Spectral norm of random matrices," *Combinatorica*, vol. 27, no. 6, pp. 721–736, 2007. [238, 244, 245]
4. V. Mayer-Sch¨onberger and K. Cukier, *Big Data: A Revolution that will transform how we live, work and think.* Eamon Dolan Book and Houghton Mifflin Hardcourt, 2013. [xxi, 271]
5. R. Qiu, Z. Hu, H. Li, and M. Wicks, *Cognitiv Communications and Networking: Theory and Practice.* John Wiley and Sons, 2012. [xxi, 37, 85, 351, 352, 360, 441, 502, 526, 574, 575]
6. R. Qiu, *Introduction to Smart Grid.* John Wiley and Sons, 2014. [xxi, 567, 574]
7. G. Lugosi, "Concentration-of-measure inequalities," 2009. [5, 6, 16, 17, 121, 122]
8. A. Leon-Garcia, *Probability, Statistics, and Random Processing for Electrical Engineering.* Pearson-Prentice Hall, third edition ed., 2008. [9, 11, 85]
9. T. Tao, *Topics in Random Matrix Thoery.* American Mathematical Society, 2012. [9, 23, 24, 33, 54, 85, 97, 362, 363, 364, 473]
10. N. Nguyen, P. Drineas, and T. Tran, "Tensor sparsification via a bound on the spectral norm of random tensors," *arXiv preprint arXiv:1005.4732*, 2010. [12, 565, 566]
11. W. Hoeffding, "Probability inequalities for sums of bounded random variables," *Journal of the American Statistical Association*, pp. 13–30, 1963. [13, 14, 16]
12. G. Bennett, "Probability inequalities for the sum of independent random variables," *Journal of the American Statistical Association*, vol. 57, no. 297, pp. 33–45, 1962. [15, 389]
13. A. Van Der Vaart and J. Wellner, *Weak Convergence and Empirical Processes.* Springer-Verlag, 1996. [15, 283]
14. F. Lin, R. Qiu, Z. Hu, S. Hou, J. Browning, and M. Wicks, "Generalized fmd detection for spectrum sensing under low signal-to-noise ratio," *IEEE Communications Letters*, to appear. [18, 221, 226, 227]
15. E. Carlen, "Trace inequalities and quantum entropy: an introductory course," *Entropy and the quantum: Arizona School of Analysis with Applications, March 16–20, 2009, University of Arizona*, vol. 529, 2010. [18, 35, 39]
16. F. Zhang, *Matrix Theory.* Springer Ver, 1999. [18, 19, 96, 215, 311, 436, 472, 493, 498]
17. K. Abadir and J. Magnus, *Matrix Algebra.* Cambridge Press, 2005. [18, 19, 20, 445]
18. D. S. Bernstein, *Matrix Mathematics: Theory, Facts, and Formulas.* Princeton University Press, 2009. [20, 98, 204, 333]
19. J. A. Tropp, "User-friendly tail bounds for sums of random matrices." Preprint, 2011. [20]

R. Qiu and M. Wicks, *Cognitive Networked Sensing and Big Data*,
DOI 10.1007/978-1-4614-4544-9,

20. N. J. Higham, *Functions of Matrices: Theory and Computation*. Society for Industrial and Applied Mathematics, 2008. [20, 33, 86, 211, 212, 213, 264, 496]
21. L. Trefethen and M. Embree, *Spectra and Pseudospectra: The Behavior of Nonnormal Matrices and Operators*. Princeton University Press, 2005. [20]
22. R. Bhatia, *Positive Definite Matrices*. Princeton University Press, 2007. [20, 38, 44, 98]
23. R. Bhatia, *Matrix analysis*. Springer, 1997. [33, 39, 41, 45, 92, 98, 153, 199, 210, 214, 215, 264, 488, 500]
24. A. W. Marshall, I. Olkin, and B. C. Arnold, *Inequalities: Theory of Majorization and Its Applications*. Springer Verl, 2011. [20]
25. D. Watkins, *Fundamentals of Matrix Computations*. Wiley, third ed., 2010. [21]
26. J. A. Tropp, "On the conditioning of random subdictionaries," *Applied and Computational Harmonic Analysis*, vol. 25, no. 1, pp. 1–24, 2008. [27]
27. M. Ledoux and M. Talagrand, *Probability in Banach spaces*. Springer, 1991. [27, 28, 69, 71, 75, 76, 77, 163, 165, 166, 168, 182, 184, 185, 192, 198, 199, 307, 309, 389, 391, 394, 396, 404, 535]
28. J. A. Tropp, J. N. Laska, M. F. Duarte, J. K. Romberg, and R. G. Baraniuk, "Beyond nyquist: Efficient sampling of sparse bandlimited signals," *Information Theory, IEEE Transactions on*, vol. 56, no. 1, pp. 520–544, 2010. [27, 71, 384]
29. J. Nelson, "Johnson-lindenstrauss notes," tech. rep., Technical report, MIT-CSAIL, Available at web.mit.edu/minilek/www/jl_notes.pdf, 2010. [27, 70, 379, 380, 381, 384]
30. H. Rauhut, "Compressive sensing and structured random matrices," *Theoretical foundations and numerical methods for sparse recovery*, vol. 9, pp. 1–92, 2010. [28, 164, 379, 394]
31. F. Krahmer, S. Mendelson, and H. Rauhut, "Suprema of chaos processes and the restricted isometry property," *arXiv preprint arXiv:1207.0235*, 2012. [29, 65, 398, 399, 400, 401, 407, 408]
32. R. Latala, "On weak tail domination of random vectors," *Bull. Pol. Acad. Sci. Math.*, vol. 57, pp. 75–80, 2009. [29]
33. W. Bednorz and R. Latala, "On the suprema of bernoulli processes," *Comptes Rendus Mathematique*, 2013. [29, 75, 76]
34. B. Gnedenko and A. Kolmogorov, *Limit Distributions for Sums Independent Random Variables*. Addison-Wesley, 1954. [30]
35. R. Vershynin, "A note on sums of independent random matrices after ahlswede-winter." http://www-personal.umich.edu/~romanv/teaching/reading-group/ahlswede-winter.pdf. Seminar Notes. [31, 91]
36. R. Ahlswede and A. Winter, "Strong converse for identification via quantum channels," *Information Theory, IEEE Transactions on*, vol. 48, no. 3, pp. 569–579, 2002. [31, 85, 87, 94, 95, 96, 97, 106, 107, 108, 122, 123, 132]
37. T. Fine, *Probability and Probabilistic Reasoning for Electrical Engineering*. Pearson-Prentice Hall, 2006. [32]
38. R. Oliveira, "Sums of random hermitian matrices and an inequality by rudelson," *Elect. Comm. Probab*, vol. 15, pp. 203–212, 2010. [34, 94, 95, 108, 473]
39. T. Rockafellar, *Conjugative duality and optimization*. Philadephia: SIAM, 1974. [36]
40. D. Petz, "A suvery of trace inequalities." Functional Analysis and Operator Theory, 287–298, Banach Center Publications, 30 (Warszawa), 1994. www.renyi.hu/~petz/pdf/64.pdf. [38, 39]
41. R. Vershynin, "Golden-thompson inequality." www-personal.umich.edu/~romanv/teaching/reading-group/golden-thompson.pdf. Seminar Notes. [39]
42. I. Dhillon and J. Tropp, "Matrix nearness problems with bregman divergences," *SIAM Journal on Matrix Analysis and Applications*, vol. 29, no. 4, pp. 1120–1146, 2007. [40]
43. J. Tropp, "From joint convexity of quantum relative entropy to a concavity theorem of lieb," in *Proc. Amer. Math. Soc*, vol. 140, pp. 1757–1760, 2012. [41, 128]
44. G. Lindblad, "Expectations and entropy inequalities for finite quantum systems," *Communications in Mathematical Physics*, vol. 39, no. 2, pp. 111–119, 1974. [41]
45. E. G. Effros, "A matrix convexity approach to some celebrated quantum inequalities," vol. 106, pp. 1006–1008, National Acad Sciences, 2009. [41]

46. E. Carlen and E. Lieb, "A minkowski type trace inequality and strong subadditivity of quantum entropy ii: convexity and concavity," *Letters in Mathematical Physics*, vol. 83, no. 2, pp. 107–126, 2008. [41, 42]
47. T. Rockafellar, *Conjugate duality and optimization*. SIAM, 1974. Regional conference series in applied mathematics. [41]
48. S. Boyd and L. Vandenberghe, *Convex optimization*. Cambridge Univ Pr, 2004. [42, 48, 210, 217, 418, 489]
49. J. Tropp, "Freedman's inequality for matrix martingales," *Electron. Commun. Probab*, vol. 16, pp. 262–270, 2011. [42, 128, 137]
50. E. Lieb, "Convex trace functions and the wigner-yanase-dyson conjecture," *Advances in Mathematics*, vol. 11, no. 3, pp. 267–288, 1973. [42, 98, 129]
51. V. I. Paulsen, *Completely Bounded Maps and Operator Algebras*. Cambridge Press, 2002. [43]
52. T. M. Cover and J. A. Thomas, *Elements of Information Theory*. New York: John Wiley. [44]
53. J. Tropp, "User-friendly tail bounds for sums of random matrices," *Foundations of Computational Mathematics*, vol. 12, no. 4, pp. 389–434, 2011. [45, 47, 95, 107, 110, 111, 112, 115, 116, 121, 122, 127, 128, 131, 132, 140, 144, 493]
54. F. Hansen and G. Pedersen, "Jensen's operator inequality," *Bulletin of the London Mathematical Society*, vol. 35, no. 4, pp. 553–564, 2003. [45]
55. P. Halmos, *Finite-Dimensional Vector Spaces*. Springer, 1958. [46]
56. V. De la Peña and E. Giné, *Decoupling: from dependence to independence*. Springer Verlag, 1999. [47, 404]
57. R. Vershynin, "A simple decoupling inequality in probability theory," May 2011. [47, 51]
58. P. Billingsley, *Probability and Measure*. Wiley, 2008. [51, 523]
59. R. Dudley, *Real analysis and probability*, vol. 74. Cambridge University Press, 2002. [51, 230, 232]
60. M. A. Arcones and E. Giné, "On decoupling, series expansions, and tail behavior of chaos processes," *Journal of Theoretical Probability*, vol. 6, no. 1, pp. 101–122, 1993. [52]
61. D. L. Hanson and F. T. Wright, "A bound on tail probabilities for quadratic forms in independent random variables," *The Annals of Mathematical Statistics*, pp. 1079–1083, 1971. [53, 73, 380]
62. S. Boucheron, G. Lugosi, and P. Massart, "Concentration inequalities using the entropy method," *The Annals of Probability*, vol. 31, no. 3, pp. 1583–1614, 2003. [53, 198, 397, 404]
63. T. Tao, *Topics in Random Matrix Theory*. Amer Mathematical Society, 2012. [54, 55, 216, 238, 239, 241, 242, 355, 357, 358]
64. E. Wigner, "Distribution laws for the roots of a random hermitian matrix," *Statistical Theories of Spectra: Fluctuations*, pp. 446–461, 1965. [55]
65. M. Mehta, *Random matrices*, vol. 142. Academic press, 2004. [55]
66. D. Voiculescu, "Limit laws for random matrices and free products," *Inventiones mathematicae*, vol. 104, no. 1, pp. 201–220, 1991. [55, 236, 363]
67. J. Wishart, "The generalised product moment distribution in samples from a normal multivariate population," *Biometrika*, vol. 20, no. 1/2, pp. 32–52, 1928. [56]
68. P. Hsu, "On the distribution of roots of certain determinantal equations," *Annals of Human Genetics*, vol. 9, no. 3, pp. 250–258, 1939. [56, 137]
69. U. Haagerup and S. Thorbjørnsen, "Random matrices with complex gaussian entries," *Expositiones Mathematicae*, vol. 21, no. 4, pp. 293–337, 2003. [57, 58, 59, 202, 203, 214]
70. A. Erdelyi, W. Magnus, Oberhettinger, and F. Tricomi, eds., *Higher Transcendental Functions, Vol. 1–3*. McGraw-Hill, 1953. [57]
71. J. Harer and D. Zagier, "The euler characteristic of the moduli space of curves," *Inventiones Mathematicae*, vol. 85, no. 3, pp. 457–485, 1986. [58]
72. R. Vershynin, "Introduction to the non-asymptotic analysis of random matrices," *Arxiv preprint arXiv:1011.3027v5*, July 2011. [60, 66, 67, 68, 313, 314, 315, 316, 317, 318, 319, 320, 321, 324, 325, 349, 462, 504]

73. D. Garling, *Inequalities: a journey into linear analysis*. Cambridge University Press, 2007. [61]
74. V. Buldygin and S. Solntsev, *Asymptotic behaviour of linearly transformed sums of random variables*. Kluwer, 1997. [60, 61, 63, 76, 77]
75. J. Kahane, *Some random series of functions*. Cambridge Univ Press, 2nd ed., 1985. [60]
76. M. Rudelson, "Lecture notes on non-asymptotic theory of random matrices," *arXiv preprint arXiv:1301.2382*, 2013. [63, 64, 68, 336]
77. V. Yurinsky, *Sums and Gaussian vectors*. Springer-Verlag, 1995. [69]
78. U. Haagerup, "The best constants in the khintchine inequality," *Studia Math.*, vol. 70, pp. 231–283, 1981. [69]
79. R. Latala, P. Mankiewicz, K. Oleszkiewicz, and N. Tomczak-Jaegermann, "Banach-mazur distances and projections on random subgaussian polytopes," *Discrete & Computational Geometry*, vol. 38, no. 1, pp. 29–50, 2007. [72, 73, 295, 296, 297, 298]
80. E. D. Gluskin and S. Kwapien, "Tail and moment estimates for sums of independent random variable," *Studia Math.*, vol. 114, pp. 303–309, 1995. [75]
81. M. Talagrand, *The generic chaining: upper and lower bounds of stochastic processes*. Springer Verlag, 2005. [75, 163, 165, 192, 394, 396, 404, 405, 406, 407]
82. M. Talagrand, *Upper and Lower Bounds for Stochastic Processes, Modern Methods and Classical Problems*. Springer-Verlag, in press. Ergebnisse der Mathematik. [75, 199]
83. R. M. Dudley, "The sizes of compact subsets of hilbert space and continuity of gaussian processes," *J. Funct. Anal*, vol. 1, no. 3, pp. 290–330, 1967. [75, 406]
84. X. Fernique, "Régularité des trajectoires des fonctions aléatoires gaussiennes," *Ecole d'Eté de Probabilités de Saint-Flour IV-1974*, pp. 1–96, 1975. [75, 407]
85. M. Talagrand, "Regularity of gaussian processes," *Acta mathematica*, vol. 159, no. 1, pp. 99–149, 1987. [75, 407]
86. R. Bhattacharya and R. Rao, *Normal approximation and asymptotic expansions*, vol. 64. Society for Industrial & Applied, 1986. [76, 77, 78]
87. L. Chen, L. Goldstein, and Q. Shao, *Normal Approximation by Stein's Method*. Springer, 2010. [76, 221, 236, 347, 523]
88. A. Kirsch, *An introduction to the mathematical theory of inverse problems*, vol. 120. Springer Science+ Business Media, 2011. [79, 80, 289]
89. D. Porter and D. S. Stirling, *Integral equations: a practical treatment, from spectral theory to applications*, vol. 5. Cambridge University Press, 1990. [79, 80, 82, 288, 289, 314]
90. U. Grenander, *Probabilities on Algebraic Structures*. New York: Wiley, 1963. [85]
91. N. Harvey, "C&o 750: Randomized algorithms winter 2011 lecture 11 notes." http://www.math.uwaterloo.ca/~harvey/W11/, Winter 2011. [87, 95, 473]
92. N. Harvey, "Lecture 12 concentration for sums of random matrices and lecture 13 the ahlswede-winter inequality." http://www.cs.ubc.ca/~nickhar/W12/, Febuary 2012. Lecture Notes for UBC CPSC 536N: Randomized Algorithms. [87, 89, 90, 95]
93. M. Rudelson, "Random vectors in the isotropic position," *Journal of Functional Analysis*, vol. 164, no. 1, pp. 60–72, 1999. [90, 95, 277, 278, 279, 280, 281, 282, 283, 284, 292, 306, 307, 308, 441, 442]
94. A. Wigderson and D. Xiao, "Derandomizing the ahlswede-winter matrix-valued chernoff bound using pessimistic estimators, and applications," *Theory of Computing*, vol. 4, no. 1, pp. 53–76, 2008. [91, 95, 107]
95. D. Gross, Y. Liu, S. Flammia, S. Becker, and J. Eisert, "Quantum state tomography via compressed sensing," *Physical review letters*, vol. 105, no. 15, p. 150401, 2010. [93, 107]
96. D. DUBHASHI and D. PANCONESI, *Concentration of measure for the analysis of randomized algorithms*. Cambridge Univ Press, 2009. [97]
97. H. Ngo, "Cse 694: Probabilistic analysis and randomized algorithms." http://www.cse.buffalo.edu/~hungngo/classes/2011/Spring-694/lectures/l4.pdf, Spring 2011. SUNY at Buffalo. [97]
98. O. Bratteli and D. W. Robinson, *Operator Algebras amd Quantum Statistical Mechanics I*. Springer-Verlag, 1979. [97]

99. D. Voiculescu, K. Dykema, and A. Nica, *Free Random Variables*. American Mathematical Society, 1992. [97]
100. P. J. Schreiner and L. L. Scharf, *Statistical Signal Processing of Complex-Valued Data: The Theory of Improper and Noncircular Signals*. Ca, 2010. [99]
101. J. Lawson and Y. Lim, "The geometric mean, matrices, metrics, and more," *The American Mathematical Monthly*, vol. 108, no. 9, pp. 797–812, 2001. [99]
102. D. Gross, "Recovering low-rank matrices from few coefficients in any basis," *Information Theory, IEEE Transactions on*, vol. 57, no. 3, pp. 1548–1566, 2011. [106, 433, 443]
103. B. Recht, "A simpler approach to matrix completion," *Arxiv preprint arxiv:0910.0651*, 2009. [106, 107, 429]
104. B. Recht, "A simpler approach to matrix completion," *The Journal of Machine Learning Research*, vol. 7777777, pp. 3413–3430, 2011. [106, 431, 432, 433]
105. R. Ahlswede and A. Winter, "Addendum to strong converse for identification via quantum channels," *Information Theory, IEEE Transactions on*, vol. 49, no. 1, p. 346, 2003. [107]
106. R. Latala, "Some estimates of norms of random matrices," *AMERICAN MATHEMATICAL SOCIETY*, vol. 133, no. 5, pp. 1273–1282, 2005. [118]
107. Y. Seginer, "The expected norm of random matrices," *Combinatorics Probability and Computing*, vol. 9, no. 2, pp. 149–166, 2000. [118, 223, 330]
108. P. Massart, *Concentration Inequalities and Model Selection*. Springer, 2007. [120, 457]
109. M. Ledoux and M. Talagrand, *Probability in Banach Spaces: Isoperimetry and Processes*. Springer, 1991. [120]
110. R. Motwani and P. Raghavan, *Randomized Algorithms*. Cambridge Univ Press, 1995. [122]
111. A. Gittens and J. Tropp, "Tail bounds for all eigenvalues of a sum of random matrices," *Arxiv preprint arXiv:1104.4513*, 2011. [128, 129, 131, 144]
112. E. Lieb and R. Seiringer, "Stronger subadditivity of entropy," *Physical Review A*, vol. 71, no. 6, p. 062329, 2005. [128, 129]
113. D. Hsu, S. Kakade, and T. Zhang, "Tail inequalities for sums of random matrices that depend on the intrinsic dimension," 2011. [137, 138, 139, 267, 462, 463, 478, 562, 564]
114. B. Schoelkopf, A. Sola, and K. Mueller, *Kernl principal compnent analysis*, ch. Kernl principal compnent analysis, pp. 327–352. MIT Press, 1999. [137]
115. A. Magen and A. Zouzias, "Low rank matrix-valued chernoff bounds and approximate matrix multiplication," in *Proceedings of the Twenty-Second Annual ACM-SIAM Symposium on Discrete Algorithms*, pp. 1422–1436, SIAM, 2011. [137]
116. S. Minsker, "On some extensions of bernstein's inequality for self-adjoint operators," *Arxiv preprint arXiv:1112.5448*, 2011. [139, 140]
117. G. Peshkir and A. Shiryaev, "The khintchine inequalities and martingale expanding sphere of their action," *Russian Mathematical Surveys*, vol. 50, no. 5, pp. 849–904, 1995. [141]
118. N. Tomczak-Jaegermann, "The moduli of smoothness and convexity and the rademacher averages of trace classes," *Sp (1 [p¡.) Studia Math*, vol. 50, pp. 163–182, 1974. [141, 300]
119. F. Lust-Piquard, "Inégalités de khintchine dans c_p (1¡ p¡∞)," *CR Acad. Sci. Paris*, vol. 303, pp. 289–292, 1986. [142]
120. G. Pisier, "Non-commutative vector valued lp-spaces and completely p-summing maps," *Astérisque*, vol. 247, p. 131, 1998. [142]
121. A. Buchholz, "Operator khintchine inequality in non-commutative probability," *Mathematische Annalen*, vol. 319, no. 1, pp. 1–16, 2001. [142, 534]
122. A. So, "Moment inequalities for sums of random matrices and their applications in optimization," *Mathematical programming*, vol. 130, no. 1, pp. 125–151, 2011. [142, 530, 534, 537, 542, 548]
123. N. Nguyen, T. Do, and T. Tran, "A fast and efficient algorithm for low-rank approximation of a matrix," in *Proceedings of the 41st annual ACM symposium on Theory of computing*, pp. 215–224, ACM, 2009. [143, 156, 159]
124. N. Nguyen, P. Drineas, and T. Tran, "Matrix sparsification via the khintchine inequality," 2009. [143, 564]

125. M. de Carli Silva, N. Harvey, and C. Sato, "Sparse sums of positive semidefinite matrices," 2011. [144]
126. L. Mackey, M. Jordan, R. Chen, B. Farrell, and J. Tropp, "Matrix concentration inequalities via the method of exchangeable pairs," *Arxiv preprint arXiv:1201.6002*, 2012. [144]
127. L. Rosasco, M. Belkin, and E. D. Vito, "On learning with integral operators," *The Journal of Machine Learning Research*, vol. 11, pp. 905–934, 2010. [144, 289, 290]
128. L. Rosasco, M. Belkin, and E. De Vito, "A note on learning with integral operators," [144, 289]
129. P. Drineas and A. Zouzias, "A note on element-wise matrix sparsification via matrix-valued chernoff bounds," *Preprint*, 2010. [144]
130. R. CHEN, A. GITTENS, and J. TROPP, "The masked sample covariance estimator: An analysis via matrix concentration inequalities," *Information and Inference: A Journal of the IMA*, pp. 1–19, 2012. [144, 463, 466]
131. M. Ledoux, *The Concentration of Measure Pheonomenon*. American Mathematical Society, 2000. [145, 146]
132. D. Donoho *et al.*, "High-dimensional data analysis: The curses and blessings of dimensionality," *AMS Math Challenges Lecture*, pp. 1–32, 2000. [146, 267, 562]
133. S. Chatterjee, "Stein's method for concentration inequalities," *Probability theory and related fields*, vol. 138, no. 1, pp. 305–321, 2007. [146, 159]
134. B. Laurent and P. Massart, "Adaptive estimation of a quadratic functional by model selection," *The annals of Statistics*, vol. 28, no. 5, pp. 1302–1338, 2000. [147, 148, 485, 504]
135. G. Raskutti, M. Wainwright, and B. Yu, "Minimax rates of estimation for high-dimensional linear regression over; formula formulatype=," *Information Theory, IEEE Transactions on*, vol. 57, no. 10, pp. 6976–6994, 2011. [147, 166, 169, 170, 171, 424, 426, 510]
136. L. Birgé and P. Massart, "Minimum contrast estimators on sieves: exponential bounds and rates of convergence," *Bernoulli*, vol. 4, no. 3, pp. 329–375, 1998. [148, 254]
137. I. Johnstone, *State of the Art in Probability and Statastics*, vol. 31, ch. Chi-square oracle inequalities, pp. 399–418. Institute of Mathematical Statistics, ims lecture notes ed., 2001. [148]
138. M. Wainwright, "Sharp thresholds for high-dimensional and noisy sparsity recovery using; formula formulatype=," *Information Theory, IEEE Transactions on*, vol. 55, no. 5, pp. 2183–2202, 2009. [148, 175, 176, 182]
139. A. Amini and M. Wainwright, "High-dimensional analysis of semidefinite relaxations for sparse principal components," in *Information Theory, 2008. ISIT 2008. IEEE International Symposium on*, pp. 2454–2458, IEEE, 2008. [148, 178, 180]
140. W. Johnson and G. Schechtman, "Remarks on talagrand's deviation inequality for rademacher functions," *Functional Analysis*, pp. 72–77, 1991. [149]
141. M. Ledoux, *The concentration of measure phenomenon*, vol. 89. Amer Mathematical Society, 2001. [150, 151, 152, 153, 154, 155, 158, 163, 166, 169, 176, 177, 181, 182, 185, 188, 190, 194, 198, 232, 248, 258, 262, 313, 428]
142. M. Talagrand, "A new look at independence," *The Annals of probability*, vol. 24, no. 1, pp. 1–34, 1996. [152, 313]
143. S. Chatterjee, "Matrix estimation by universal singular value thresholding," *arXiv preprint arXiv:1212.1247*, 2012. [152]
144. R. Bhatia, C. Davis, and A. McIntosh, "Perturbation of spectral subspaces and solution of linear operator equations," *Linear Algebra and its Applications*, vol. 52, pp. 45–67, 1983. [153]
145. K. Davidson and S. Szarek, "Local operator theory, random matrices and banach spaces," *Handbook of the geometry of Banach spaces*, vol. 1, pp. 317–366, 2001. [153, 159, 160, 161, 173, 191, 282, 313, 325, 427, 486]
146. A. DasGupta, *Probability for Statistics and Machine Learning*. Springer, 2011. [155]
147. W. Hoeffding, "A combinatorial central limit theorem," *The Annals of Mathematical Statistics*, vol. 22, no. 4, pp. 558–566, 1951. [159]

148. M. Talagrand, "Concentration of measure and isoperimetric inequalities in product spaces," *Publications Mathematiques de l'IHES*, vol. 81, no. 1, pp. 73–205, 1995. [162, 216, 254, 313]
149. M. Ledoux, "Deviation inequalities on largest eigenvalues," *Geometric aspects of functional analysis*, pp. 167–219, 2007. [162, 190, 193, 325]
150. H. Rauhut, *Theoretical Foundations and Numerical Method for Sparse Recovery*, ch. Compressed Sensing and Structured Random Matrices, pp. 1–92. Berlin/New York: De Gruyter, 2010. [162]
151. J.-M. Azaïs and M. Wschebor, *Level sets and extrema of random processes and fields*. Wiley, 2009. [162, 163]
152. G. Pisier, *The volume of convex bodies and Banach space geometry*, vol. 94. Cambridge Univ Pr, 1999. [163, 202, 273, 292, 315, 386, 463]
153. Y. Gordon, A. Litvak, S. Mendelson, and A. Pajor, "Gaussian averages of interpolated bodies and applications to approximate reconstruction," *Journal of Approximation Theory*, vol. 149, no. 1, pp. 59–73, 2007. [166, 429]
154. J. Matousek, *Lectures on discrete geometry*, vol. 212. Springer, 2002. [172, 184, 425]
155. S. Negahban and M. Wainwright, "Estimation of (near) low-rank matrices with noise and high-dimensional scaling," *The Annals of Statistics*, vol. 39, no. 2, pp. 1069–1097, 2011. [174, 178, 181, 182, 185, 188, 416, 417, 418, 419, 421, 422]
156. P. Zhang and R. Qiu, "Glrt-based spectrum sensing with blindly learned feature under rank-1 assumption," *IEEE Trans. Communications*. to appear. [177, 233, 347, 514]
157. P. Zhang, R. Qiu, and N. Guo, "Demonstration of Spectrum Sensing with Blindly Learned Feature," *IEEE Communications Letters*, vol. 15, pp. 548–550, May 2011. [177, 233, 347, 491, 514]
158. S. Hou, R. Qiu, J. P. Browning, and Wick, "Spectrum sensing in cognitive radio with robust principal component analysis," in *IEEE Waveform Diversity and Design Conference 2012*, (Kauai, Hawaii), January 2012. [177, 491]
159. S. Hou, R. Qiu, J. P. Browning, and M. C. Wicks, "Spectrum sensing in cognitive radio with subspace matching," in *IEEE Waveform Diversity and Design Conference 2012*, (Kauai, Hawaii), January 2012. [233, 268, 491, 514]
160. S. Hou, R. Qiu, Z. Chen, and Z. Hu, "SVM and Dimensionality Reduction in Cognitive Radio with Experimental Validation," *Arxiv preprint arXiv:1106.2325, submitted to EURASIP Journal on Advances in Signal Processing*, 2011. [177]
161. P. Massart, "Concentration inequalities and model selection," 2007. [177, 254, 256]
162. L. Birgé, "An alternative point of view on lepski's method," *Lecture Notes-Monograph Series*, pp. 113–133, 2001. [179]
163. N. Karoui, "Operator norm consistent estimation of large-dimensional sparse covariance matrices," *The Annals of Statistics*, pp. 2717–2756, 2008. [179]
164. P. Loh and M. Wainwright, "High-dimensional regression with noisy and missing data: Provable guarantees with non-convexity," *arXiv preprint arXiv:1109.3714*, 2011. [186, 187, 188, 189]
165. E. Meckes, "Approximation of projections of random vectors," *Journal of Theoretical Probability*, vol. 25, no. 2, pp. 333–352, 2012. [195, 196]
166. A. Efimov, "Modulus of continuity," *Encyclopaedia of Mathematics. Springer*, 2001. [195]
167. V. Milman and G. Schechtman, *Asymptotic theory of finite dimensional normed spaces*, vol. 1200. Springer Verlag, 1986. [195, 258, 273, 280, 292]
168. E. Meckes, "Projections of probability distributions: A measure-theoretic dvoretzky theorem," *Geometric Aspects of Functional Analysis*, pp. 317–326, 2012. [196]
169. O. Bousquet, "A bennett concentration inequality and its application to suprema of empirical processes," *Comptes Rendus Mathematique*, vol. 334, no. 6, pp. 495–500, 2002. [198]
170. O. Bousquet, *Concentration inequalities and empirical processes theory applied to the analysis of learning algorithms*. PhD thesis, PhD thesis, Ecole Polytechnique, 2002. [198]
171. S. Boucheron, G. Lugosi, and O. Bousquet, "Concentration inequalities," *Advanced Lectures on Machine Learning*, pp. 208–240, 2004. [198]

172. M. Ledoux, "On talagrand's deviation inequalities for product measures," *ESAIM: Probability and statistics*, vol. 1, pp. 63–87, 1996. [198]
173. P. Massart, "About the constants in talagrand's concentration inequalities for empirical processes," *Annals of Probability*, pp. 863–884, 2000. [198]
174. E. Rio, "Inégalités de concentration pour les processus empiriques de classes de parties," *Probability Theory and Related Fields*, vol. 119, no. 2, pp. 163–175, 2001. [198]
175. S. Boucheron, O. Bousquet, G. Lugosi, and P. Massart, "Moment inequalities for functions of independent random variables," *The Annals of Probability*, vol. 33, no. 2, pp. 514–560, 2005. [198]
176. S. Boucheron, O. Bousquet, G. Lugosi, *et al.*, "Theory of classification: A survey of some recent advances," *ESAIM Probability and statistics*, vol. 9, pp. 323–375, 2005. []
177. S. Boucheron, G. Lugosi, P. Massart, *et al.*, "On concentration of self-bounding functions," *Electronic Journal of Probability*, vol. 14, no. 64, pp. 1884–1899, 2009. []
178. S. Boucheron, P. Massart, *et al.*, "A high-dimensional wilks phenomenon," *Probability theory and related fields*, vol. 150, no. 3, p. 405, 2011. [198]
179. A. Connes, "Classification of injective factors," *Ann. of Math*, vol. 104, no. 2, pp. 73–115, 1976. [203]
180. A. Guionnet and O. Zeitouni, "Concentration of the spectral measure for large matrices," *Electron. Comm. Probab*, vol. 5, pp. 119–136, 2000. [204, 218, 219, 227, 243, 244, 247, 268, 494]
181. I. N. Bronshtein, K. A. Semendiaev, and K. A. Hirsch, *Handbook of mathematics*. Van Nostrand Reinhold New York, NY, 5th ed., 2007. [204, 208]
182. A. Khajehnejad, S. Oymak, and B. Hassibi, "Subspace expanders and matrix rank minimization," *arXiv preprint arXiv:1102.3947*, 2011. [205]
183. M. Meckes, "Concentration of norms and eigenvalues of random matrices," *Journal of Functional Analysis*, vol. 211, no. 2, pp. 508–524, 2004. [206, 224]
184. C. Davis, "All convex invariant functions of hermitian matrices," *Archiv der Mathematik*, vol. 8, no. 4, pp. 276–278, 1957. [206]
185. L. Li, "Concentration of measure for random matrices." private communication, October 2012. Tenneessee Technological University. [207]
186. N. Berestycki and R. Nickl, "Concentration of measure," tech. rep., Technical report, University of Cambridge, 2009. [218, 219]
187. R. A. Horn and C. R. Johnson, *Matrix Analysis*. Cambridge University Press, 1994. [219, 465]
188. Y. Zeng and Y. Liang, "Maximum-minimum eigenvalue detection for cognitive radio," in *IEEE 18th International Symposium on Personal, Indoor and Mobile Radio Communications (PIMRC) 2007*, pp. 1–5, 2007. [221]
189. V. Petrov and A. Brown, *Sums of independent random variables*, vol. 197. Springer-Verlag Berlin, 1975. [221]
190. M. Meckes and S. Szarek, "Concentration for noncommutative polynomials in random matrices," in *Proc. Amer. Math. Soc*, vol. 140, pp. 1803–1813, 2012. [222, 223, 269]
191. G. W. Anderson, "Convergence of the largest singular value of a polynomial in independent wigner matrices," *arXiv preprint arXiv:1103.4825*, 2011. [222]
192. E. Meckes and M. Meckes, "Concentration and convergence rates for spectral measures of random matrices," *Probability Theory and Related Fields*, pp. 1–20, 2011. [222, 233, 234, 235]
193. R. Serfling, "Approximation theorems of mathematical statistics (wiley series in probability and statistics)," 1981. [223]
194. R. Latala, "Some estimates of norms of random matrices," *Proceedings of the American Mathematical Society*, pp. 1273–1282, 2005. [223, 330]
195. S. Lang, *Real and functional analysis*. 1993. [224]
196. A. Guntuboyina and H. Leeb, "Concentration of the spectral measure of large wishart matrices with dependent entries," *Electron. Commun. Probab*, vol. 14, pp. 334–342, 2009. [224, 225, 226]

197. M. Ledoux, "Concentration of measure and logarithmic sobolev inequalities," *Seminaire de probabilites XXXIII*, pp. 120–216, 1999. [224, 244]
198. Z. Bai, "Methodologies in spectral analysis of large-dimensional random matrices, a review," *Statist. Sinica*, vol. 9, no. 3, pp. 611–677, 1999. [225, 262, 502]
199. B. Delyon, "Concentration inequalities for the spectral measure of random matrices," *Electronic Communications in Probability*, pp. 549–562, 2010. [226, 227]
200. F. Lin, R. Qiu, Z. Hu, S. Hou, J. P. Browning, and M. C. Wicks, " Cognitive Radio Network as Sensors: Low Signal-to-Noise Ratio Collaborative Spectrum Sensing," in *IEEE Waveform Diversity and Design Conference*, 2012. Kauai, Hawaii. [226]
201. S. Chatterjee, "Fluctuations of eigenvalues and second order poincaré inequalities," *Probability Theory and Related Fields*, vol. 143, no. 1, pp. 1–40, 2009. [227, 228, 229, 230]
202. S. Chatterjee, "A new method of normal approximation," *The Annals of Probability*, vol. 36, no. 4, pp. 1584–1610, 2008. [229]
203. I. Johnstone, "High dimensional statistical inference and random matrices," *Arxiv preprint math/0611589*, 2006. [230, 490]
204. T. Jiang, "Approximation of haar distributed matrices and limiting distributions of eigenvalues of jacobi ensembles," *Probability theory and related fields*, vol. 144, no. 1, pp. 221–246, 2009. [230, 231]
205. R. Bhatia, L. Elsner, and G. Krause, "Bounds for the variation of the roots of a polynomial and the eigenvalues of a matrix," *Linear Algebra and Its Applications*, vol. 142, pp. 195–209, 1990. [231]
206. N. Gozlan, "A characterization of dimension free concentration in terms of transportation inequalities," *The Annals of Probability*, vol. 37, no. 6, pp. 2480–2498, 2009. [232]
207. P. Deift and D. Gioev, *Random matrix theory: invariant ensembles and universality*, vol. 18. Amer Mathematical Society, 2009. [232]
208. G. W. Anderson, A. Guionnet, and O. Zeitouni, *An Introduction to Random Matrices*. Cambridge University Press, 2010. [232, 488, 522]
209. R. Speicher, *Free probability theory and non-crossing partitions*. 39 Seminaire Lotharingien de Combinatoire, 1997. [234, 236]
210. V. Kargin, "A concentration inequality and a local law for the sum of two random matrices," *Probability Theory and Related Fields*, pp. 1–26, 2010. [235, 236]
211. H. Weyl, "Das asymptotische verteilungsgesetz der eigenwerte linearer partieller differentialgleichungen (mit einer anwendung auf die theorie der hohlraumstrahlung)," *Mathematische Annalen*, vol. 71, no. 4, pp. 441–479, 1912. [235]
212. A. Horn, "Eigenvalues of sums of hermitian matrices," *Pacific J. Math*, vol. 12, no. 1, 1962. [235]
213. S. Chatterjee, "Concentration of haar measures, with an application to random matrices," *Journal of Functional Analysis*, vol. 245, no. 2, pp. 379–389, 2007. [236]
214. S. Chatterjee and M. Ledoux, "An observation about submatrices," *Electronic Communications in Probability*, vol. 14, pp. 495–500, 2009. [237]
215. E. Wigner, "On the distribution of the roots of certain symmetric matrices," *The Annals of Mathematics*, vol. 67, no. 2, pp. 325–327, 1958. [238]
216. A. Soshnikov, "Universality at the edge of the spectrum of in wigner random matrices," *Communications in mathematical physics*, vol. 207, pp. 897–733, 1999. [243, 245]
217. A. Soshnikov, "Poisson statistics for the largest eigenvalues of wigner random matrices with heavy tails," *Electron. Comm. Probab*, vol. 9, pp. 82–91, 2004. [243]
218. A. Soshnikov, "A note on universality of the distribution of the largest eigenvalues in certain sample covariance matrices," *Journal of Statistical Physics*, vol. 108, no. 5, pp. 1033–1056, 2002. []
219. A. Soshnikov, "Level spacings distribution for large random matrices: Gaussian fluctuations," *Annals of mathematics*, pp. 573–617, 1998. []
220. A. Soshnikov and Y. Fyodorov, "On the largest singular values of random matrices with independent cauchy entries," *Journal of mathematical physics*, vol. 46, p. 033302, 2005. []

221. T. Tao and V. Vu, "Random covariance matrices: Universality of local statistics of eigenvalues," *Arxiv preprint arxiv:0912.0966*, 2009. [243]
222. Z. Füredi and J. Komlós, "The eigenvalues of random symmetric matrices," *Combinatorica*, vol. 1, no. 3, pp. 233–241, 1981. [245]
223. M. Krivelevich and V. Vu, "Approximating the independence number and the chromatic number in expected polynomial time," *Journal of combinatorial optimization*, vol. 6, no. 2, pp. 143–155, 2002. [245]
224. A. Guionnet, "Lecture notes, minneapolis," 2012. [247, 249]
225. C. Bordenave, P. Caputo, and D. Chafaï, "Spectrum of non-hermitian heavy tailed random matrices," *Communications in Mathematical Physics*, vol. 307, no. 2, pp. 513–560, 2011. [247]
226. A. Guionnet and B. Zegarlinski, "Lectures on logarithmic sobolev inequalities," *Séminaire de Probabilités, XXXVI*, vol. 1801, pp. 1–134, 1801. [248]
227. D. Ruelle, *Statistical mechanics: rigorous results*. Amsterdam: Benjamin, 1969. [248]
228. T. Tao and V. Vu, "Random matrices: Sharp concentration of eigenvalues," *Arxiv preprint arXiv:1201.4789*, 2012. [249]
229. N. El Karoui, "Concentration of measure and spectra of random matrices: applications to correlation matrices, elliptical distributions and beyond," *The Annals of Applied Probability*, vol. 19, no. 6, pp. 2362–2405, 2009. [250, 252, 253, 261, 262, 268]
230. N. El Karoui, "The spectrum of kernel random matrices," *The Annals of Statistics*, vol. 38, no. 1, pp. 1–50, 2010. [254, 268]
231. M. Talagrand, "New concentration inequalities in product spaces," *Inventiones Mathematicae*, vol. 126, no. 3, pp. 505–563, 1996. [254, 313, 404]
232. L. Birgé and P. Massart, "Gaussian model selection," *Journal of the European Mathematical Society*, vol. 3, no. 3, pp. 203–268, 2001. [254]
233. L. Birgé and P. Massart, "Minimal penalties for gaussian model selection," *Probability theory and related fields*, vol. 138, no. 1, pp. 33–73, 2007. [254]
234. P. Massart, "Some applications of concentration inequalities to statistics," in *Annales-Faculte des Sciences Toulouse Mathematiques*, vol. 9, pp. 245–303, Université Paul Sabatier, 2000. [254]
235. I. Bechar, "A bernstein-type inequality for stochastic processes of quadratic forms of gaussian variables," *arXiv preprint arXiv:0909.3595*, 2009. [254, 548, 552]
236. K. Wang, A. So, T. Chang, W. Ma, and C. Chi, "Outage constrained robust transmit optimization for multiuser miso downlinks: Tractable approximations by conic optimization," *arXiv preprint arXiv:1108.0982*, 2011. [255, 543, 544, 545, 547, 548, 552]
237. M. Lopes, L. Jacob, and M. Wainwright, "A more powerful two-sample test in high dimensions using random projection," *arXiv preprint arXiv:1108.2401*, 2011. [255, 256, 520, 522, 523, 525]
238. W. Beckner, "A generalized poincaré inequality for gaussian measures," *Proceedings of the American Mathematical Society*, pp. 397–400, 1989. [256]
239. M. Rudelson and R. Vershynin, "Invertibility of random matrices: unitary and orthogonal perturbations," *arXiv preprint arXiv:1206.5180*, June 2012. Version 1. [257, 334]
240. T. Tao and V. Vu, "Random matrices: Universality of local eigenvalue statistics," *Acta mathematica*, pp. 1–78, 2011. [259]
241. T. Tao and V. Vu, "Random matrices: The distribution of the smallest singular values," *Geometric And Functional Analysis*, vol. 20, no. 1, pp. 260–297, 2010. [259, 334, 499]
242. T. Tao and V. Vu, "On random$\pm$1 matrices: singularity and determinant," *Random Structures & Algorithms*, vol. 28, no. 1, pp. 1–23, 2006. [259, 332, 335, 499]
243. J. Silverstein and Z. Bai, "On the empirical distribution of eigenvalues of a class of large dimensional random matrices," *Journal of Multivariate analysis*, vol. 54, no. 2, pp. 175–192, 1995. [262]
244. J. Von Neumann, *Mathematische grundlagen der quantenmechanik*, vol. 38. Springer, 1995. [264]

245. J. Cadney, N. Linden, and A. Winter, "Infinitely many constrained inequalities for the von neumann entropy," *Information Theory, IEEE Transactions on*, vol. 58, no. 6, pp. 3657–3663, 2012. [265, 267]
246. H. Araki and E. Lieb, "Entropy inequalities," *Communications in Mathematical Physics*, vol. 18, no. 2, pp. 160–170, 1970. [266]
247. E. Lieb and M. Ruskai, "A fundamental property of quantum-mechanical entropy," *Physical Review Letters*, vol. 30, no. 10, pp. 434–436, 1973. [266]
248. E. Lieb and M. Ruskai, "Proof of the strong subadditivity of quantum-mechanical entropy," *Journal of Mathematical Physics*, vol. 14, pp. 1938–1941, 1973. [266]
249. N. Pippenger, "The inequalities of quantum information theory," *Information Theory, IEEE Transactions on*, vol. 49, no. 4, pp. 773–789, 2003. [267]
250. T. Chan, "Recent progresses in characterising information inequalities," *Entropy*, vol. 13, no. 2, pp. 379–401, 2011. []
251. T. Chan, D. Guo, and R. Yeung, "Entropy functions and determinant inequalities," in *Information Theory Proceedings (ISIT), 2012 IEEE International Symposium on*, pp. 1251–1255, IEEE, 2012. []
252. R. Yeung, "Facts of entropy," *IEEE Information Theory Society Newsletter*, pp. 6–15, December 2012. [267]
253. C. Williams and M. Seeger, "The effect of the input density distribution on kernel-based classifiers," in *Proceedings of the 17th International Conference on Machine Learning*, Citeseer, 2000. [268]
254. J. Shawe-Taylor, N. Cristianini, and J. Kandola, "On the concentration of spectral properties," *Advances in neural information processing systems*, vol. 1, pp. 511–518, 2002. [268]
255. J. Shawe-Taylor, C. Williams, N. Cristianini, and J. Kandola, "On the eigenspectrum of the gram matrix and the generalization error of kernel-pca," *Information Theory, IEEE Transactions on*, vol. 51, no. 7, pp. 2510–2522, 2005. [268]
256. Y. Do and V. Vu, "The spectrum of random kernel matrices," *arXiv preprint arXiv:1206.3763*, 2012. [268]
257. X. Cheng and A. Singer, "The spectrum of random inner-product kernel matrices," *arXiv:1202.3155v1 [math.PR]*, p. 40, 2012. [268]
258. N. Ross *et al.*, "Fundamentals of stein's method," *Probability Surveys*, vol. 8, pp. 210–293, 2011. [269, 347]
259. Z. Chen and J. Dongarra, "Condition numbers of gaussian random matrices," *Arxiv preprint arXiv:0810.0800*, 2008. [269]
260. M. Junge and Q. Zeng, "Noncommutative bennett and rosenthal inequalities," *Arxiv preprint arXiv:1111.1027*, 2011. [269]
261. A. Giannopoulos, "Notes on isotropic convex bodies," *Warsaw University Notes*, 2003. [272]
262. Y. D. Burago, V. A. Zalgaller, and A. Sossinsky, *Geometric inequalities*, vol. 1988. Springer Berlin, 1988. [273]
263. R. Schneider, *Convex bodies: the Brunn-Minkowski theory*, vol. 44. Cambridge Univ Pr, 1993. [273, 292]
264. V. Milman and A. Pajor, "Isotropic position and inertia ellipsoids and zonoids of the unit ball of a normed n-dimensional space," *Geometric aspects of functional analysis*, pp. 64–104, 1989. [274, 277, 291]
265. C. Borell, "The brunn-minkowski inequality in gauss space," *Inventiones Mathematicae*, vol. 30, no. 2, pp. 207–216, 1975. [274]
266. R. Kannan, L. Lovász, and M. Simonovits, "Random walks and an o*(n5) volume algorithm for convex bodies," *Random structures and algorithms*, vol. 11, no. 1, pp. 1–50, 1997. [274, 275, 277, 280, 282, 442]
267. O. Guédon and M. Rudelson, "Lp-moments of random vectors via majorizing measures," *Advances in Mathematics*, vol. 208, no. 2, pp. 798–823, 2007. [275, 276, 299, 301, 303]
268. C. Borell, "Convex measures on locally convex spaces," *Arkiv för Matematik*, vol. 12, no. 1, pp. 239–252, 1974. [276]

269. C. Borell, "Convex set functions ind-space," *Periodica Mathematica Hungarica*, vol. 6, no. 2, pp. 111–136, 1975. [276]
270. A. Prékopa, "Logarithmic concave measures with application to stochastic programming," *Acta Sci. Math.(Szeged)*, vol. 32, no. 197, pp. 301–3, 1971. [276]
271. S. Vempala, "Recent progress and open problems in algorithmic convex geometry," in *=IARCS Annual Conference on Foundations of Software Technology and Theoretical Computer Science (FSTTCS 2010)*, Schloss Dagstuhl–Leibniz-Zentrum fuer Informatik, 2010. [276]
272. G. Paouris, "Concentration of mass on convex bodies," *Geometric and Functional Analysis*, vol. 16, no. 5, pp. 1021–1049, 2006. [277, 290, 291, 292, 295]
273. J. Bourgain, "Random points in isotropic convex sets," *Convex geometric analysis, Berkeley, CA*, pp. 53–58, 1996. [277, 278, 282]
274. S. Mendelson and A. Pajor, "On singular values of matrices with independent rows," *Bernoulli*, vol. 12, no. 5, pp. 761–773, 2006. [277, 281, 282, 283, 285, 287, 288, 289]
275. S. Alesker, "Phi 2-estimate for the euclidean norm on a convex body in isotropic position," *Operator theory*, vol. 77, pp. 1–4, 1995. [280, 290]
276. F. Lust-Piquard and G. Pisier, "Non commutative khintchine and paley inequalities," *Arkiv för Matematik*, vol. 29, no. 1, pp. 241–260, 1991. [284]
277. F. Cucker and D. X. Zhou, *Learning theory: an approximation theory viewpoint*, vol. 24. Cambridge University Press, 2007. [288]
278. V. Koltchinskii and E. Giné, "Random matrix approximation of spectra of integral operators," *Bernoulli*, vol. 6, no. 1, pp. 113–167, 2000. [288, 289]
279. S. Mendelson, "On the performance of kernel classes," *The Journal of Machine Learning Research*, vol. 4, pp. 759–771, 2003. [289]
280. R. Kress, *Linear Integral Equations*. Berlin: Springer-Verlag, 1989. [289]
281. G. W. Hanson and A. B. Yakovlev, *Operator theory for electromagnetics: an introduction*. Springer Verlag, 2002. [289]
282. R. C. Qiu, Z. Hu, M. Wicks, L. Li, S. J. Hou, and L. Gary, "Wireless Tomography, Part II: A System Engineering Approach," in *5th International Waveform Diversity & Design Conference*, (Niagara Falls, Canada), August 2010. [290]
283. R. C. Qiu, M. C. Wicks, L. Li, Z. Hu, and S. J. Hou, "Wireless Tomography, Part I: A NovelApproach to Remote Sensing," in *5th International Waveform Diversity & Design Conference*, (Niagara Falls, Canada), August 2010. [290]
284. E. De Vito, L. Rosasco, A. Caponnetto, U. De Giovannini, and F. Odone, "Learning from examples as an inverse problem," *Journal of Machine Learning Research*, vol. 6, no. 1, p. 883, 2006. [290]
285. E. De Vito, L. Rosasco, and A. Toigo, "Learning sets with separating kernels," *arXiv preprint arXiv:1204.3573*, 2012. []
286. E. De Vito, V. Umanità, and S. Villa, "An extension of mercer theorem to matrix-valued measurable kernels," *Applied and Computational Harmonic Analysis*, 2012. []
287. E. De Vito, L. Rosasco, and A. Toigo, "A universally consistent spectral estimator for the support of a distribution," [290]
288. R. Latala, "Weak and strong moments of random vectors," *Marcinkiewicz Centenary Volume, Banach Center Publ.*, vol. 95, pp. 115–121, 2011. [290, 303]
289. R. Latala, "Order statistics and concentration of lr norms for log-concave vectors," *Journal of Functional Analysis*, 2011. [292, 293]
290. R. Adamczak, R. Latala, A. E. Litvak, A. Pajor, and N. Tomczak-Jaegermann, "Geometry of log-concave ensembles of random matrices and approximate reconstruction," *Comptes Rendus Mathematique*, vol. 349, no. 13, pp. 783–786, 2011. []
291. R. Adamczak, R. Latala, A. E. Litvak, K. Oleszkiewicz, A. Pajor, and N. Tomczak-Jaegermann, "A short proof of paouris' inequality," *arXiv preprint arXiv:1205.2515*, 2012. [290, 291]
292. J. Bourgain, "On the distribution of polynomials on high dimensional convex sets," *Geometric aspects of functional analysis*, pp. 127–137, 1991. [290]

293. B. Klartag, "An isomorphic version of the slicing problem," *Journal of Functional Analysis*, vol. 218, no. 2, pp. 372–394, 2005. [290]
294. S. Bobkov and F. Nazarov, "On convex bodies and log-concave probability measures with unconditional basis," *Geometric aspects of functional analysis*, pp. 53–69, 2003. [290]
295. S. Bobkov and F. Nazarov, "Large deviations of typical linear functionals on a convex body with unconditional basis," *Progress in Probability*, pp. 3–14, 2003. [290]
296. A. Giannopoulos, M. Hartzoulaki, and A. Tsolomitis, "Random points in isotropic unconditional convex bodies," *Journal of the London Mathematical Society*, vol. 72, no. 3, pp. 779–798, 2005. [292]
297. G. Aubrun, "Sampling convex bodies: a random matrix approach," *Proceedings of the American Mathematical Society*, vol. 135, no. 5, pp. 1293–1304, 2007. [292, 295]
298. R. Adamczak, "A tail inequality for suprema of unbounded empirical processes with applications to markov chains," *Electron. J. Probab*, vol. 13, pp. 1000–1034, 2008. [293]
299. R. Adamczak, A. Litvak, A. Pajor, and N. Tomczak-Jaegermann, "Sharp bounds on the rate of convergence of the empirical covariance matrix," *Comptes Rendus Mathematique*, 2011. [294]
300. R. Adamczak, R. Latala, A. E. Litvak, A. Pajor, and N. Tomczak-Jaegermann, "Tail estimates for norms of sums of log concave random vectors," *arXiv preprint arXiv:1107.4070*, 2011. [293]
301. R. Adamczak, A. E. Litvak, A. Pajor, and N. Tomczak-Jaegermann, "Quantitative estimates of the convergence of the empirical covariance matrix in log-concave ensembles," *Journal of the American Mathematical Society*, vol. 23, no. 2, p. 535, 2010. [294, 315]
302. Z. Bai and Y. Yin, "Limit of the smallest eigenvalue of a large dimensional sample covariance matrix," *The annals of Probability*, pp. 1275–1294, 1993. [294, 324]
303. R. Kannan, L. Lovász, and M. Simonovits, "Isoperimetric problems for convex bodies and a localization lemma," *Discrete & Computational Geometry*, vol. 13, no. 1, pp. 541–559, 1995. [295]
304. G. Paouris, "Small ball probability estimates for log-concave measures," *Trans. Amer. Math. Soc*, vol. 364, pp. 287–308, 2012. [295, 297, 305]
305. J. A. Clarkson, "Uniformly convex spaces," *Transactions of the American Mathematical Society*, vol. 40, no. 3, pp. 396–414, 1936. [300]
306. Y. Gordon, "Gaussian processes and almost spherical sections of convex bodies," *The Annals of Probability*, pp. 180–188, 1988. [302]
307. Y. Gordon, *On Milman's inequality and random subspaces which escape through a mesh in* $\mathbb{R}^n$. Springer, 1988. Geometric aspects of functional analysis (1986/87), 84–106, Lecture Notes in Math., 1317. [302]
308. R. Adamczak, O. Guedon, R. Latala, A. E. Litvak, K. Oleszkiewicz, A. Pajor, and N. Tomczak-Jaegermann, "Moment estimates for convex measures," *arXiv preprint arXiv:1207.6618*, 2012. [303, 304, 305]
309. A. Pajor and N. Tomczak-Jaegermann, "Chevet type inequality and norms of sub-matrices," [303]
310. N. Srivastava and R. Vershynin, "Covariance estimation for distributions with 2+\ epsilon moments," *Arxiv preprint arXiv:1106.2775*, 2011. [304, 305, 322, 475, 476]
311. M. Rudelson and R. Vershynin, "Sampling from large matrices: An approach through geometric functional analysis," *Journal of the ACM (JACM)*, vol. 54, no. 4, p. 21, 2007. [306, 307, 310]
312. A. Frieze, R. Kannan, and S. Vempala, "Fast monte-carlo algorithms for finding low-rank approximations," *Journal of the ACM (JACM)*, vol. 51, no. 6, pp. 1025–1041, 2004. [310]
313. P. Drineas, R. Kannan, and M. Mahoney, "Fast monte carlo algorithms for matrices i: Approximating matrix multiplication," *SIAM Journal on Computing*, vol. 36, no. 1, p. 132, 2006. [562]
314. P. Drineas, R. Kannan, and M. Mahoney, "Fast monte carlo algorithms for matrices ii: Computing a low-rank approximation to a matrix," *SIAM Journal on Computing*, vol. 36, no. 1, p. 158, 2006. [311]

315. P. Drineas, R. Kannan, M. Mahoney, *et al.*, "Fast monte carlo algorithms for matrices iii: Computing a compressed approximate matrix decomposition," *SIAM Journal on Computing*, vol. 36, no. 1, p. 184, 2006. [311]
316. P. Drineas and R. Kannan, "Pass efficient algorithms for approximating large matrices," in *Proceedings of the fourteenth annual ACM-SIAM symposium on Discrete algorithms*, pp. 223–232, Society for Industrial and Applied Mathematics, 2003. [310, 311]
317. P. Drineas, A. Frieze, R. Kannan, S. Vempala, and V. Vinay, "Clustering large graphs via the singular value decomposition," *Machine Learning*, vol. 56, no. 1, pp. 9–33, 2004. [313]
318. V. Mil'man, "New proof of the theorem of a. dvoretzky on intersections of convex bodies," *Functional Analysis and its Applications*, vol. 5, no. 4, pp. 288–295, 1971. [315]
319. K. Ball, "An elementary introduction to modern convex geometry," *Flavors of geometry*, vol. 31, pp. 1–58, 1997. [315]
320. R. Vershynin, "Approximating the moments of marginals of high-dimensional distributions," *The Annals of Probability*, vol. 39, no. 4, pp. 1591–1606, 2011. [322]
321. P. Youssef, "Estimating the covariance of random matrices," *arXiv preprint arXiv:1301.6607*, 2013. [322, 323]
322. M. Rudelson and R. Vershynin, "Smallest singular value of a random rectangular matrix," *Communications on Pure and Applied Mathematics*, vol. 62, no. 12, pp. 1707–1739, 2009. [324, 327]
323. R. Vershynin, "Spectral norm of products of random and deterministic matrices," *Probability theory and related fields*, vol. 150, no. 3, p. 471, 2011. [324, 327, 328, 329, 331]
324. O. Feldheim and S. Sodin, "A universality result for the smallest eigenvalues of certain sample covariance matrices," *Geometric And Functional Analysis*, vol. 20, no. 1, pp. 88–123, 2010. [324, 326, 328]
325. M. Lai and Y. Liu, "The probabilistic estimates on the largest and smallest q-singular values of pre-gaussian random matrices," *submitted to Advances in Mathematics*, 2010. []
326. T. Tao and V. Vu, "Random matrices: The distribution of the smallest singular values," *Geometric And Functional Analysis*, vol. 20, no. 1, pp. 260–297, 2010. [328, 338, 339, 340, 341]
327. D. CHAFAÏ, "Singular values of random matrices," *Lecture Notes*, November 2009. Université Paris-Est Marne-la-Vallée. []
328. M. LAI and Y. LIU, "A study on the largest and smallest q-singular values of random matrices," []
329. A. Litvak and O. Rivasplata, "Smallest singular value of sparse random matrices," *Arxiv preprint arXiv:1106.0938*, 2011. []
330. R. Vershynin, "Invertibility of symmetric random matrices," *Arxiv preprint arXiv:1102.0300*, 2011. []
331. H. Nguyen and V. Vu, "Optimal inverse littlewood-offord theorems," *Advances in Mathematics*, 2011. []
332. Y. Eliseeva, F. Götze, and A. Zaitsev, "Estimates for the concentration functions in the littlewood–offord problem," *Arxiv preprint arXiv:1203.6763*, 2012. []
333. S. Mendelson and G. Paouris, "On the singular values of random matrices," *Preprint*, 2012. [324]
334. J. Silverstein, "The smallest eigenvalue of a large dimensional wishart matrix," *The Annals of Probability*, vol. 13, no. 4, pp. 1364–1368, 1985. [324]
335. M. Rudelson and R. Vershynin, "Non-asymptotic theory of random matrices: extreme singular values," *Arxiv preprint arXiv:1003.2990*, 2010. [325]
336. G. Aubrun, "A sharp small deviation inequality for the largest eigenvalue of a random matrix," *Séminaire de Probabilités XXXVIII*, pp. 320–337, 2005. [325]
337. G. Bennett, L. Dor, V. Goodman, W. Johnson, and C. Newman, "On uncomplemented subspaces of lp, 1¡ p¡ 2," *Israel Journal of Mathematics*, vol. 26, no. 2, pp. 178–187, 1977. [326]
338. A. Litvak, A. Pajor, M. Rudelson, and N. Tomczak-Jaegermann, "Smallest singular value of random matrices and geometry of random polytopes," *Advances in Mathematics*, vol. 195, no. 2, pp. 491–523, 2005. [326, 462]

339. S. Artstein-Avidan, O. Friedland, V. Milman, and S. Sodin, "Polynomial bounds for large bernoulli sections of l1 n," *Israel Journal of Mathematics*, vol. 156, no. 1, pp. 141–155, 2006. [326]
340. M. Rudelson, "Lower estimates for the singular values of random matrices," *Comptes Rendus Mathematique*, vol. 342, no. 4, pp. 247–252, 2006. [326]
341. M. Rudelson and R. Vershynin, "The littlewood–offord problem and invertibility of random matrices," *Advances in Mathematics*, vol. 218, no. 2, pp. 600–633, 2008. [327, 334, 336, 338, 342, 344, 345]
342. R. Vershynin, "Some problems in asymptotic convex geometry and random matrices motivated by numerical algorithms," *Arxiv preprint cs/0703093*, 2007. [327]
343. M. Rudelson and O. Zeitouni, "Singular values of gaussian matrices and permanent estimators," *arXiv preprint arXiv:1301.6268*, 2013. [329]
344. Y. Yin, Z. Bai, and P. Krishnaiah, "On the limit of the largest eigenvalue of the large dimensional sample covariance matrix," *Probability Theory and Related Fields*, vol. 78, no. 4, pp. 509–521, 1988. [330, 502]
345. Z. Bai and J. Silverstein, "No eigenvalues outside the support of the limiting spectral distribution of large-dimensional sample covariance matrices," *The Annals of Probability*, vol. 26, no. 1, pp. 316–345, 1998. [331]
346. Z. Bai and J. Silverstein, "Exact separation of eigenvalues of large dimensional sample covariance matrices," *The Annals of Probability*, vol. 27, no. 3, pp. 1536–1555, 1999. [331]
347. H. Nguyen and V. Vu, "Random matrices: Law of the determinant," *Arxiv preprint arXiv:1112.0752*, 2011. [331, 332]
348. A. Rouault, "Asymptotic behavior of random determinants in the laguerre, gram and jacobi ensembles," *Latin American Journal of Probability and Mathematical Statistics (ALEA),*, vol. 3, pp. 181–230, 2007. [331]
349. N. Goodman, "The distribution of the determinant of a complex wishart distributed matrix," *Annals of Mathematical Statistics*, pp. 178–180, 1963. [332]
350. O. Friedland and O. Giladi, "A simple observation on random matrices with continuous diagonal entries," *arXiv preprint arXiv:1302.0388*, 2013. [334, 335]
351. J. Bourgain, V. H. Vu, and P. M. Wood, "On the singularity probability of discrete random matrices," *Journal of Functional Analysis*, vol. 258, no. 2, pp. 559–603, 2010. [334]
352. T. Tao and V. Vu, "From the littlewood-offord problem to the circular law: universality of the spectral distribution of random matrices," *Bulletin of the American Mathematical Society*, vol. 46, no. 3, p. 377, 2009. [334]
353. R. Adamczak, O. Guédon, A. Litvak, A. Pajor, and N. Tomczak-Jaegermann, "Smallest singular value of random matrices with independent columns," *Comptes Rendus Mathematique*, vol. 346, no. 15, pp. 853–856, 2008. [334]
354. R. law Adamczak, O. Guédon, A. Litvak, A. Pajor, and N. Tomczak-Jaegermann, "Condition number of a square matrix with iid columns drawn from a convex body," *Proc. Amer. Math. Soc.*, vol. 140, pp. 987–998, 2012. [334]
355. L. Erdős, B. Schlein, and H.-T. Yau, "Wegner estimate and level repulsion for wigner random matrices," *International Mathematics Research Notices*, vol. 2010, no. 3, pp. 436–479, 2010. [334]
356. B. Farrell and R. Vershynin, "Smoothed analysis of symmetric random matrices with continuous distributions," *arXiv preprint arXiv:1212.3531*, 2012. [335]
357. A. Maltsev and B. Schlein, "A wegner estimate for wigner matrices," *Entropy and the Quantum II*, vol. 552, p. 145, 2011. []
358. H. H. Nguyen, "Inverse littlewood–offord problems and the singularity of random symmetric matrices," *Duke Mathematical Journal*, vol. 161, no. 4, pp. 545–586, 2012. [335]
359. H. H. Nguyen, "On the least singular value of random symmetric matrices," *Electron. J. Probab.*, vol. 17, pp. 1–19, 2012. [335]
360. R. Vershynin, "Invertibility of symmetric random matrices," *Random Structures & Algorithms*, 2012. [334, 335]

361. A. Sankar, D. Spielman, and S. Teng, "Smoothed analysis of the condition numbers and growth factors of matrices," *SIAM J. Matrix Anal. Appl.*, vol. 2, pp. 446–476, 2006. [335]
362. T. Tao and V. Vu, "Smooth analysis of the condition number and the least singular value," *Mathematics of Computation*, vol. 79, no. 272, pp. 2333–2352, 2010. [335, 338, 343, 344]
363. K. P. Costello and V. Vu, "Concentration of random determinants and permanent estimators," *SIAM Journal on Discrete Mathematics*, vol. 23, no. 3, pp. 1356–1371, 2009. [335]
364. T. Tao and V. Vu, "On the permanent of random bernoulli matrices," *Advances in Mathematics*, vol. 220, no. 3, pp. 657–669, 2009. [335]
365. A. H. Taub, "John von neumann: Collected works, volume v: Design of computers, theory of automata and numerical analysis," 1963. [336]
366. S. Smale, "On the efficiency of algorithms of analysis," *Bull. Amer. Math. Soc.(NS)*, vol. 13, 1985. [336, 337, 342]
367. A. Edelman, "Eigenvalues and condition numbers of random matrices," *SIAM Journal on Matrix Analysis and Applications*, vol. 9, no. 4, pp. 543–560, 1988. [336]
368. S. J. Szarek, "Condition numbers of random matrices," *J. Complexity*, vol. 7, no. 2, pp. 131–149, 1991. [336]
369. D. Spielman and S. Teng, "Smoothed analysis of algorithms: Why the simplex algorithm usually takes polynomial time," *Journal of the ACM (JACM)*, vol. 51, no. 3, pp. 385–463, 2004. [336, 338, 339]
370. L. Erdos, "Universality for random matrices and log-gases," *arXiv preprint arXiv:1212.0839*, 2012. [337, 347]
371. J. Von Neumann and H. Goldstine, "Numerical inverting of matrices of high order," *Bull. Amer. Math. Soc*, vol. 53, no. 11, pp. 1021–1099, 1947. [337]
372. A. Edelman, *Eigenvalues and condition numbers of random matrices.* PhD thesis, Massachusetts Institute of Technology, 1989. [337, 338, 341, 342]
373. P. Forrester, "The spectrum edge of random matrix ensembles," *Nuclear Physics B*, vol. 402, no. 3, pp. 709–728, 1993. [338]
374. M. Rudelson and R. Vershynin, "The least singular value of a random square matrix is $o(n^{-1/2})$," *Comptes Rendus Mathematique*, vol. 346, no. 15, pp. 893–896, 2008. [338]
375. V. Vu and T. Tao, "The condition number of a randomly perturbed matrix," in *Proceedings of the thirty-ninth annual ACM symposium on Theory of computing*, pp. 248–255, ACM, 2007. [338]
376. J. Lindeberg, "Eine neue herleitung des exponentialgesetzes in der wahrscheinlichkeitsrechnung," *Mathematische Zeitschrift*, vol. 15, no. 1, pp. 211–225, 1922. [339]
377. A. Edelman and B. Sutton, "Tails of condition number distributions," *simulation*, vol. 1, p. 2, 2008. [341]
378. T. Sarlos, "Improved approximation algorithms for large matrices via random projections," in *Foundations of Computer Science, 2006. FOCS'06. 47th Annual IEEE Symposium on*, pp. 143–152, IEEE, 2006. [342]
379. T. Tao and V. Vu, "Random matrices: the circular law," *Arxiv preprint arXiv:0708.2895*, 2007. [342, 343]
380. N. Pillai and J. Yin, "Edge universality of correlation matrices," *arXiv preprint arXiv:1112.2381*, 2011. [345, 346, 347]
381. N. Pillai and J. Yin, "Universality of covariance matrices," *arXiv preprint arXiv:1110.2501*, 2011. [345]
382. Z. Bao, G. Pan, and W. Zhou, "Tracy-widom law for the extreme eigenvalues of sample correlation matrices," 2011. [345]
383. I. Johnstone, "On the distribution of the largest eigenvalue in principal components analysis," *The Annals of statistics*, vol. 29, no. 2, pp. 295–327, 2001. [347, 461, 502]
384. S. Hou, R. C. Qiu, J. P. Browning, and M. C. Wicks, "Spectrum Sensing in Cognitive Radio with Subspace Matching," in *IEEE Waveform Diversity and Design Conference*, January 2012. [347, 514]
385. S. Hou and R. C. Qiu, "Spectrum sensing for cognitive radio using kernel-based learning," *arXiv preprint arXiv:1105.2978*, 2011. [347, 491, 514]

386. S. Dallaporta, "Eigenvalue variance bounds for wigner and covariance random matrices," *Random Matrices: Theory and Applications*, vol. 1, no. 03, 2011. [348]
387. Ø. Ryan, A. Masucci, S. Yang, and M. Debbah, "Finite dimensional statistical inference," *Information Theory, IEEE Transactions on*, vol. 57, no. 4, pp. 2457–2473, 2011. [360]
388. R. Couillet and M. Debbah, *Random Matrix Methods for Wireless Communications*. Cambridge University Press, 2011. [360, 502]
389. A. Tulino and S. Verdu, *Random matrix theory and wireless communications*. now Publishers Inc., 2004. [360]
390. R. Müller, "Applications of large random matrices in communications engineering," in *Proc. Int. Conf. on Advances Internet, Process., Syst., Interdisciplinary Research (IPSI), Sveti Stefan, Montenegro*, 2003. [365, 367]
391. G. E. Pfander, H. Rauhut, and J. A. Tropp, "The restricted isometry property for time–frequency structured random matrices," *Probability Theory and Related Fields*, pp. 1–31, 2011. [373, 374, 400]
392. T. T. Cai, L. Wang, and G. Xu, "Shifting inequality and recovery of sparse signals," *Signal Processing, IEEE Transactions on*, vol. 58, no. 3, pp. 1300–1308, 2010. [373]
393. E. Candes, "The Restricted Isometry Property and Its Implications for Compressed Sensing," *Comptes rendus-Mathematique*, vol. 346, no. 9–10, pp. 589–592, 2008. []
394. E. Candes, J. Romberg, and T. Tao, "Stable Signal Recovery from Incomplete and Inaccurate Measurements," *Comm. Pure Appl. Math*, vol. 59, no. 8, pp. 1207–1223, 2006. []
395. S. Foucart, "A note on guaranteed sparse recovery via ℓ_1-minimization," *Applied and Computational Harmonic Analysis*, vol. 29, no. 1, pp. 97–103, 2010. [373]
396. S. Foucart, "Sparse recovery algorithms: sufficient conditions in terms of restricted isometry constants," *Approximation Theory XIII: San Antonio 2010*, pp. 65–77, 2012. [373]
397. D. Needell and J. A. Tropp, "Cosamp: Iterative signal recovery from incomplete and inaccurate samples," *Applied and Computational Harmonic Analysis*, vol. 26, no. 3, pp. 301–321, 2009. [373]
398. T. Blumensath and M. E. Davies, "Iterative hard thresholding for compressed sensing," *Applied and Computational Harmonic Analysis*, vol. 27, no. 3, pp. 265–274, 2009. [373]
399. S. Foucart, "Hard thresholding pursuit: an algorithm for compressive sensing," *SIAM Journal on Numerical Analysis*, vol. 49, no. 6, pp. 2543–2563, 2011. [373]
400. R. Baraniuk, M. Davenport, R. DeVore, and M. Wakin, "A Simple Proof of the Restricted Isometry Property for Random Matrices." Submitted for publication, January 2007. [373, 374, 375, 376, 378, 385]
401. D. L. Donoho and J. Tanner, "Counting faces of randomly-projected polytopes when the projection radically lowers dimension," *J. Amer. Math. Soc*, vol. 22, no. 1, pp. 1–53, 2009. []
402. M. Rudelson and R. Vershynin, "On sparse reconstruction from fourier and gaussian measurements," *Communications on Pure and Applied Mathematics*, vol. 61, no. 8, pp. 1025–1045, 2008. [374, 394]
403. H. Rauhut, J. Romberg, and J. A. Tropp, "Restricted isometries for partial random circulant matrices," *Applied and Computational Harmonic Analysis*, vol. 32, no. 2, pp. 242–254, 2012. [374, 393, 394, 395, 396, 397, 398, 404]
404. G. E. Pfander and H. Rauhut, "Sparsity in time-frequency representations," *Journal of Fourier Analysis and Applications*, vol. 16, no. 2, pp. 233–260, 2010. [374, 400, 401, 402, 403, 404]
405. G. Pfander, H. Rauhut, and J. Tanner, "Identification of Matrices having a Sparse Representation," in *Preprint*, 2007. [374, 400]
406. E. Candes and T. Tao, "Near-Optimal Signal Recovery From Random Projections: Universal Encoding Strategies?," *IEEE Transactions on Information Theory*, vol. 52, no. 12, pp. 5406–5425, 2006. [373, 384]
407. S. Mendelson, A. Pajor, and N. Tomczak-Jaegermann, "Uniform uncertainty principle for bernoulli and subgaussian ensembles," *Constructive Approximation*, vol. 28, no. 3, pp. 277–289, 2008. [373, 386]

408. A. Cohen, W. Dahmen, and R. DeVore, "Compressed Sensing and Best k-Term Approximation," in *Submitted for publication, July*, 2006. [373]
409. S. Foucart, A. Pajor, H. Rauhut, and T. Ullrich, "The gelfand widths of lp-balls for $0 < p < 1$," *Journal of Complexity*, vol. 26, no. 6, pp. 629–640, 2010. []
410. A. Y. Garnaev and E. D. Gluskin, "The widths of a euclidean ball," in *Dokl. Akad. Nauk SSSR*, vol. 277, pp. 1048–1052, 1984. [373]
411. W. B. Johnson and J. Lindenstrauss, "Extensions of lipschitz mappings into a hilbert space," *Contemporary mathematics*, vol. 26, no. 189–206, p. 1, 1984. [374, 375, 379]
412. F. Krahmer and R. Ward, "New and improved johnson-lindenstrauss embeddings via the restricted isometry property," *SIAM Journal on Mathematical Analysis*, vol. 43, no. 3, pp. 1269–1281, 2011. [374, 375, 379, 398]
413. N. Alon, "Problems and results in extremal combinatorics-i," *Discrete Mathematics*, vol. 273, no. 1, pp. 31–53, 2003. [375]
414. N. Ailon and E. Liberty, "An almost optimal unrestricted fast johnson-lindenstrauss transform," in *Proceedings of the Twenty-Second Annual ACM-SIAM Symposium on Discrete Algorithms*, pp. 185–191, SIAM, 2011. [375]
415. D. Achlioptas, "Database-friendly random projections," in *Proceedings of the twentieth ACM SIGMOD-SIGACT-SIGART symposium on Principles of database systems*, pp. 274–281, ACM, 2001. [375, 376, 408, 410]
416. D. Achlioptas, "Database-friendly random projections: Johnson-lindenstrauss with binary coins," *Journal of computer and System Sciences*, vol. 66, no. 4, pp. 671–687, 2003. [375, 379, 385, 414]
417. N. Ailon and B. Chazelle, "Approximate nearest neighbors and the fast johnson-lindenstrauss transform," in *Proceedings of the thirty-eighth annual ACM symposium on Theory of computing*, pp. 557–563, ACM, 2006. [375]
418. N. Ailon and B. Chazelle, "The fast johnson-lindenstrauss transform and approximate nearest neighbors," *SIAM Journal on Computing*, vol. 39, no. 1, pp. 302–322, 2009. [375]
419. G. Lorentz, M. Golitschek, and Y. Makovoz, "Constructive approximation, volume 304 of grundlehren math. wiss," 1996. [377]
420. R. I. Arriaga and S. Vempala, "An algorithmic theory of learning: Robust concepts and random projection," *Machine Learning*, vol. 63, no. 2, pp. 161–182, 2006. [379]
421. K. L. Clarkson and D. P. Woodruff, "Numerical linear algebra in the streaming model," in *Proceedings of the 41st annual ACM symposium on Theory of computing*, pp. 205–214, ACM, 2009. []
422. S. Dasgupta and A. Gupta, "An elementary proof of a theorem of johnson and lindenstrauss," *Random Structures & Algorithms*, vol. 22, no. 1, pp. 60–65, 2002. []
423. P. Frankl and H. Maehara, "The johnson-lindenstrauss lemma and the sphericity of some graphs," *Journal of Combinatorial Theory, Series B*, vol. 44, no. 3, pp. 355–362, 1988. []
424. P. Indyk and R. Motwani, "Approximate nearest neighbors: towards removing the curse of dimensionality," in *Proceedings of the thirtieth annual ACM symposium on Theory of computing*, pp. 604–613, ACM, 1998. []
425. D. M. Kane and J. Nelson, "A derandomized sparse johnson-lindenstrauss transform," *arXiv preprint arXiv:1006.3585*, 2010. []
426. J. Matoušek, "On variants of the johnson–lindenstrauss lemma," *Random Structures & Algorithms*, vol. 33, no. 2, pp. 142–156, 2008. [379]
427. M. Rudelson and R. Vershynin, "Geometric approach to error-correcting codes and reconstruction of signals," *International Mathematics Research Notices*, vol. 2005, no. 64, p. 4019, 2005. [384]
428. J. Vybíral, "A variant of the johnson–lindenstrauss lemma for circulant matrices," *Journal of Functional Analysis*, vol. 260, no. 4, pp. 1096–1105, 2011. [384, 398]
429. A. Hinrichs and J. Vybíral, "Johnson-lindenstrauss lemma for circulant matrices," *Random Structures & Algorithms*, vol. 39, no. 3, pp. 391–398, 2011. [384, 398]

430. H. Rauhut, K. Schnass, and P. Vandergheynst, "Compressed Sensing and Redundant Dictionaries," *IEEE Transactions on Information Theory*, vol. 54, no. 5, pp. 2210–2219, 2008. [384, 385, 387, 388, 389, 391, 392]
431. W. U. Bajwa, J. Haupt, A. M. Sayeed, and R. Nowak, "Compressed channel sensing: A new approach to estimating sparse multipath channels," *Proceedings of the IEEE*, vol. 98, no. 6, pp. 1058–1076, 2010. [400]
432. J. Chiu and L. Demanet, "Matrix probing and its conditioning," *SIAM Journal on Numerical Analysis*, vol. 50, no. 1, pp. 171–193, 2012. [400]
433. G. E. Pfander, "Gabor frames in finite dimensions," *Finite Frames*, pp. 193–239, 2012. [400]
434. J. Haupt, W. U. Bajwa, G. Raz, and R. Nowak, "Toeplitz compressed sensing matrices with applications to sparse channel estimation," *Information Theory, IEEE Transactions on*, vol. 56, no. 11, pp. 5862–5875, 2010. [400]
435. K. Gröchenig, *Foundations of time-frequency analysis*. Birkhäuser Boston, 2000. [401]
436. F. Krahmer, G. E. Pfander, and P. Rashkov, "Uncertainty in time–frequency representations on finite abelian groups and applications," *Applied and Computational Harmonic Analysis*, vol. 25, no. 2, pp. 209–225, 2008. [401]
437. J. Lawrence, G. E. Pfander, and D. Walnut, "Linear independence of gabor systems in finite dimensional vector spaces," *Journal of Fourier Analysis and Applications*, vol. 11, no. 6, pp. 715–726, 2005. [401]
438. B. M. Sanandaji, T. L. Vincent, and M. B. Wakin, "Concentration of measure inequalities for compressive toeplitz matrices with applications to detection and system identification," in *Decision and Control (CDC), 2010 49th IEEE Conference on*, pp. 2922–2929, IEEE, 2010. [408]
439. B. M. Sanandaji, T. L. Vincent, and M. B. Wakin, "Concentration of measure inequalities for toeplitz matrices with applications," *arXiv preprint arXiv:1112.1968*, 2011. [409]
440. B. M. Sanandaji, M. B. Wakin, and T. L. Vincent, "Observability with random observations," *arXiv preprint arXiv:1211.4077*, 2012. [408]
441. M. Meckes, "On the spectral norm of a random toeplitz matrix," *Electron. Comm. Probab*, vol. 12, pp. 315–325, 2007. [410]
442. R. Adamczak, "A few remarks on the operator norm of random toeplitz matrices," *Journal of Theoretical Probability*, vol. 23, no. 1, pp. 85–108, 2010. [410]
443. R. Calderbank, S. Howard, and S. Jafarpour, "Construction of a large class of deterministic sensing matrices that satisfy a statistical isometry property," *Selected Topics in Signal Processing, IEEE Journal of*, vol. 4, no. 2, pp. 358–374, 2010. [410]
444. Y. Plan, *Compressed sensing, sparse approximation, and low-rank matrix estimation*. PhD thesis, California Institute of Technology, 2011. [411, 414, 415]
445. R. Vershynin, "Math 280 lecture notes," 2007. [413]
446. B. Recht, M. Fazel, and P. A. Parrilo, "Guaranteed minimum-rank solutions of linear matrix equations via nuclear norm minimization," *SIAM review*, vol. 52, no. 3, pp. 471–501, 2010. [414]
447. R. Vershynin, "On large random almost euclidean bases," *Acta Math. Univ. Comenianae*, vol. 69, no. 2, pp. 137–144, 2000. [414]
448. E. L. Lehmann and G. Casella, *Theory of point estimation*, vol. 31. Springer, 1998. [414]
449. S. Negahban, P. Ravikumar, M. Wainwright, and B. Yu, "A unified framework for high-dimensional analysis of m-estimators with decomposable regularizers," *arXiv preprint arXiv:1010.2731*, 2010. [416]
450. A. Agarwal, S. Negahban, and M. Wainwright, "Fast global convergence of gradient methods for high-dimensional statistical recovery," *arXiv preprint arXiv:1104.4824*, 2011. [416]
451. B. Recht, M. Fazel, and P. Parrilo, "Guaranteed Minimum-Rank Solutions of Linear Matrix Equations via Nuclear Norm Minimization," in *Arxiv preprint arXiv:0706.4138*, 2007. [417]
452. H. Lütkepohl, "New introduction to multiple time series analysis," 2005. [422]
453. A. Agarwal, S. Negahban, and M. Wainwright, "Noisy matrix decomposition via convex relaxation: Optimal rates in high dimensions," *arXiv preprint arXiv:1102.4807*, 2011. [427, 428, 429, 485, 486]

454. C. Meyer, *Matrix analysis and applied linear algebra*. SIAM, 2000. [430]
455. M. McCoy and J. Tropp, "Sharp recovery bounds for convex deconvolution, with applications," *Arxiv preprint arXiv:1205.1580*, 2012. [433, 434]
456. V. Chandrasekaran, S. Sanghavi, P. A. Parrilo, and A. S. Willsky, "Rank-sparsity incoherence for matrix decomposition," *SIAM Journal on Optimization*, vol. 21, no. 2, pp. 572–596, 2011. [434]
457. V. Koltchinskii, "Von neumann entropy penalization and low-rank matrix estimation," *The Annals of Statistics*, vol. 39, no. 6, pp. 2936–2973, 2012. [434, 438, 439, 447]
458. M. A. Nielsen and I. L. Chuang, *Quantum Computation and Quantum Information*. Cambridge Press, 10th edition ed., 2010. [438]
459. S. Sra, S. Nowozin, and S. Wright, eds., *Optimization for machine learning*. MIT Press, 2012. Chapter 4 (Bertsekas) Incremental Gradient, Subgradient, and Proximal Method for Convex Optimization: A Survey. [440, 441]
460. M. Rudelson, "Contact points of convex bodies," *Israel Journal of Mathematics*, vol. 101, no. 1, pp. 93–124, 1997. [441]
461. D. Blatt, A. Hero, and H. Gauchman, "A convergent incremental gradient method with a constant step size," *SIAM Journal on Optimization*, vol. 18, no. 1, pp. 29–51, 2007. [443]
462. M. Rabbat and R. Nowak, "Quantized incremental algorithms for distributed optimization," *Selected Areas in Communications, IEEE Journal on*, vol. 23, no. 4, pp. 798–808, 2005. [443]
463. E. Candes, Y. Eldar, T. Strohmer, and V. Voroninski, "Phase retrieval via matrix completion," *Arxiv preprint arXiv:1109.0573*, 2011. [443, 444, 446, 447, 456]
464. E. Candes, T. Strohmer, and V. Voroninski, "Phaselift: Exact and stable signal recovery from magnitude measurements via convex programming," *Arxiv preprint arXiv:1109.4499*, 2011. [443, 444, 456]
465. E. Candès and B. Recht, "Exact matrix completion via convex optimization," *Foundations of Computational Mathematics*, vol. 9, no. 6, pp. 717–772, 2009. [443]
466. J. Cai, E. Candes, and Z. Shen, "A singular value thresholding algorithm for matrix completion," *Arxiv preprint Arxiv:0810.3286*, 2008. [446, 449]
467. E. Candes and T. Tao, "The power of convex relaxation: Near-optimal matrix completion," *Information Theory, IEEE Transactions on*, vol. 56, no. 5, pp. 2053–2080, 2010. []
468. A. Alfakih, A. Khandani, and H. Wolkowicz, "Solving euclidean distance matrix completion problems via semidefinite programming," *Computational optimization and applications*, vol. 12, no. 1, pp. 13–30, 1999. []
469. M. Fukuda, M. Kojima, K. Murota, K. Nakata, *et al.*, "Exploiting sparsity in semidefinite programming via matrix completion i: General framework," *SIAM Journal on Optimization*, vol. 11, no. 3, pp. 647–674, 2001. []
470. R. Keshavan, A. Montanari, and S. Oh, "Matrix completion from a few entries," *Information Theory, IEEE Transactions on*, vol. 56, no. 6, pp. 2980–2998, 2010. []
471. C. Johnson, "Matrix completion problems: a survey," in *Proceedings of Symposia in Applied Mathematics*, vol. 40, pp. 171–198, 1990. []
472. E. Candes and Y. Plan, "Matrix completion with noise," *Proceedings of the IEEE*, vol. 98, no. 6, pp. 925–936, 2010. [443]
473. A. Chai, M. Moscoso, and G. Papanicolaou, "Array imaging using intensity-only measurements," *Inverse Problems*, vol. 27, p. 015005, 2011. [443, 446, 456]
474. L. Tian, J. Lee, S. Oh, and G. Barbastathis, "Experimental compressive phase space tomography," *Optics Express*, vol. 20, no. 8, pp. 8296–8308, 2012. [443, 446, 447, 448, 449, 456]
475. Y. Lu and M. Vetterli, "Sparse spectral factorization: Unicity and reconstruction algorithms," in *Acoustics, Speech and Signal Processing (ICASSP), 2011 IEEE International Conference on*, pp. 5976–5979, IEEE, 2011. [444]
476. J. Fienup, "Phase retrieval algorithms: a comparison," *Applied Optics*, vol. 21, no. 15, pp. 2758–2769, 1982. [444]
477. A. Sayed and T. Kailath, "A survey of spectral factorization methods," *Numerical linear algebra with applications*, vol. 8, no. 6–7, pp. 467–496, 2001. [444]

478. C. Beck and R. D'Andrea, "Computational study and comparisons of lft reducibility methods," in *American Control Conference, 1998. Proceedings of the 1998*, vol. 2, pp. 1013–1017, IEEE, 1998. [446]
479. M. Mesbahi and G. Papavassilopoulos, "On the rank minimization problem over a positive semidefinite linear matrix inequality," *Automatic Control, IEEE Transactions on*, vol. 42, no. 2, pp. 239–243, 1997. [446]
480. K. Toh, M. Todd, and R. Tütüncü, "Sdpt3-a matlab software package for semidefinite programming, version 1.3," *Optimization Methods and Software*, vol. 11, no. 1–4, pp. 545–581, 1999. [446]
481. M. Grant and S. Boyd, "Cvx: Matlab software for disciplined convex programming," *Available httpstanford edu boydcvx*, 2008. [446]
482. S. Becker, E. Candès, and M. Grant, "Templates for convex cone problems with applications to sparse signal recovery," *Mathematical Programming Computation*, pp. 1–54, 2011. [446]
483. E. Candes, M. Wakin, and S. Boyd, "Enhancing sparsity by reweighted l_1 minimization," *Journal of Fourier Analysis and Applications*, vol. 14, no. 5, pp. 877–905, 2008. [446]
484. M. Fazel, H. Hindi, and S. Boyd, "Log-det heuristic for matrix rank minimization with applications to hankel and euclidean distance matrices," in *American Control Conference, 2003. Proceedings of the 2003*, vol. 3, pp. 2156–2162, Ieee, 2003. [446]
485. M. Fazel, *Matrix rank minimization with applications*. PhD thesis, PhD thesis, Stanford University, 2002. [446]
486. L. Mandel and E. Wolf, *Optical Coherence and Quantum Optics*. Cambridge University Press, 1995. [447, 448]
487. Z. Hu, R. Qiu, J. Browning, and M. Wicks, "A novel single-step approach for self-coherent tomography using semidefinite relaxation," *IEEE Geoscience and Remote Sensing Letters*. to appear. [449]
488. M. Grant and S. Boyd, "Cvx: Matlab software for disciplined convex programming, version 1.21." http://cvxr.com/cvx, 2010. [453, 556]
489. H. Ohlsson, A. Y. Yang, R. Dong, and S. S. Sastry, "Compressive phase retrieval from squared output measurements via semidefinite programming," *arXiv preprint arXiv:1111.6323*, 2012. [454]
490. A. Devaney, E. Marengo, and F. Gruber, "Time-reversal-based imaging and inverse scattering of multiply scattering point targets," *The Journal of the Acoustical Society of America*, vol. 118, pp. 3129–3138, 2005. [454]
491. L. Lo Monte, D. Erricolo, F. Soldovieri, and M. C. Wicks, "Radio frequency tomography for tunnel detection," *Geoscience and Remote Sensing, IEEE Transactions on*, vol. 48, no. 3, pp. 1128–1137, 2010. [454]
492. O. Klopp, "Noisy low-rank matrix completion with general sampling distribution," *Arxiv preprint arXiv:1203.0108*, 2012. [456]
493. R. Foygel, R. Salakhutdinov, O. Shamir, and N. Srebro, "Learning with the weighted trace-norm under arbitrary sampling distributions," *Arxiv preprint arXiv:1106.4251*, 2011. [456]
494. R. Foygel and N. Srebro, "Concentration-based guarantees for low-rank matrix reconstruction," *Arxiv preprint arXiv:1102.3923*, 2011. [456]
495. V. Koltchinskii and P. Rangel, "Low rank estimation of similarities on graphs," *Arxiv preprint arXiv:1205.1868*, 2012. [456]
496. E. Richard, P. Savalle, and N. Vayatis, "Estimation of simultaneously sparse and low rank matrices," in *Proceeding of 29th Annual International Conference on Machine Learning*, 2012. [456]
497. H. Ohlsson, A. Yang, R. Dong, and S. Sastry, "Compressive phase retrieval from squared output measurements via semidefinite programming," *Arxiv preprint arXiv:1111.6323*, 2011. [456]
498. A. Fannjiang and W. Liao, "Compressed sensing phase retrieval," in *Proceedings of IEEE Asilomar Conference on Signals, Systems and Computers, Pacific Grove, CA*, 2011. []
499. A. Fannjiang, "Absolute uniqueness of phase retrieval with random illumination," *Arxiv preprint arXiv:1110.5097*, 2011. []

500. S. Oymak and B. Hassibi, "Recovering jointly sparse signals via joint basis pursuit," *Arxiv preprint arXiv:1202.3531*, 2012. []
501. Z. Wen, C. Yang, X. Liu, and S. Marchesini, "Alternating direction methods for classical and ptychographic phase retrieval," []
502. H. Ohlsson, A. Yang, R. Dong, and S. Sastry, "Compressive phase retrieval via lifting," [456]
503. K. Jaganathan, S. Oymak, and B. Hassibi, "On robust phase retrieval for sparse signals," in *Communication, Control, and Computing (Allerton), 2012 50th Annual Allerton Conference on*, pp. 794–799, IEEE, 2012. [456]
504. Y. Chen, A. Wiesel, and A. Hero, "Robust shrinkage estimation of high-dimensional covariance matrices," *Signal Processing, IEEE Transactions on*, vol. 59, no. 9, pp. 4097–4107, 2011. [460]
505. E. Levina and R. Vershynin, "Partial estimation of covariance matrices," *Probability Theory and Related Fields*, pp. 1–15, 2011. [463, 476, 477]
506. S. Marple, *Digital Spectral Analysis with Applications*. Prentice-Hall, 1987. [468, 469]
507. C. W. Therrien, *Discrete Random Signals and Statistical Signal Processing*. Prentice-Hall, 1992. [471]
508. H. Xiao and W. Wu, "Covariance matrix estimation for stationary time series," *Arxiv preprint arXiv:1105.4563*, 2011. [474]
509. A. Rohde, "Accuracy of empirical projections of high-dimensional gaussian matrices," *arXiv preprint arXiv:1107.5481*, 2011. [479, 481, 484, 485]
510. K. Jurczak, "A universal expectation bound on empirical projections of deformed random matrices," *arXiv preprint arXiv:1209.5943*, 2012. [479, 483]
511. A. Amini, "High-dimensional principal component analysis," 2011. [488]
512. L. Vandenberghe and S. Boyd, "Semidefinite programming," *SIAM review*, vol. 38, no. 1, pp. 49–95, 1996. [489]
513. D. Paul and I. M. Johnstone, "Augmented sparse principal component analysis for high dimensional data," *arXiv preprint arXiv:1202.1242*, 2012. [490]
514. Q. Berthet and P. Rigollet, "Optimal detection of sparse principal components in high dimension," *Arxiv preprint arXiv:1202.5070*, 2012. [490, 503, 505, 506, 508, 509]
515. A. d'Aspremont, F. Bach, and L. Ghaoui, "Approximation bounds for sparse principal component analysis," *Arxiv preprint arXiv:1205.0121*, 2012. [490]
516. A. d'Aspremont, L. El Ghaoui, M. Jordan, and G. Lanckriet, *A direct formulation for sparse PCA using semidefinite programming*, vol. 49. SIAM Review, 2007. [501, 506]
517. H. Zou, T. Hastie, and R. Tibshirani, "Sparse principal component analysis," *Journal of computational and graphical statistics*, vol. 15, no. 2, pp. 265–286, 2006. [501]
518. P. Bickel and E. Levina, "Covariance regularization by thresholding," *The Annals of Statistics*, vol. 36, no. 6, pp. 2577–2604, 2008. [502]
519. T. Cai, C. Zhang, and H. Zhou, "Optimal rates of convergence for covariance matrix estimation," *The Annals of Statistics*, vol. 38, no. 4, pp. 2118–2144, 2010. []
520. N. El Karoui, "Spectrum estimation for large dimensional covariance matrices using random matrix theory," *The Annals of Statistics*, vol. 36, no. 6, pp. 2757–2790, 2008. [502]
521. S. Geman, "The spectral radius of large random matrices," *The Annals of Probability*, vol. 14, no. 4, pp. 1318–1328, 1986. [502]
522. J. Baik, G. Ben Arous, and S. Péché, "Phase transition of the largest eigenvalue for nonnull complex sample covariance matrices," *The Annals of Probability*, vol. 33, no. 5, pp. 1643–1697, 2005. [503]
523. T. Tao, "Outliers in the spectrum of iid matrices with bounded rank perturbations," *Probability Theory and Related Fields*, pp. 1–33, 2011. [503]
524. F. Benaych-Georges, A. Guionnet, M. Maida, *et al.*, "Fluctuations of the extreme eigenvalues of finite rank deformations of random matrices," *Electronic Journal of Probability*, vol. 16, pp. 1621–1662, 2010. [503]
525. F. Bach, S. Ahipasaoglu, and A. d'Aspremont, "Convex relaxations for subset selection," *Arxiv preprint ArXiv:1006.3601*, 2010. [507]

526. E. J. Candès and M. A. Davenport, "How well can we estimate a sparse vector?," *Applied and Computational Harmonic Analysis*, 2012. [509, 510]
527. E. Candes and T. Tao, "The Dantzig Selector: Statistical Estimation When p is much larger than n," *Annals of Statistics*, vol. 35, no. 6, pp. 2313–2351, 2007. [509]
528. F. Ye and C.-H. Zhang, "Rate minimaxity of the lasso and dantzig selector for the lq loss in lr balls," *The Journal of Machine Learning Research*, vol. 9999, pp. 3519–3540, 2010. [510]
529. E. Arias-Castro, "Detecting a vector based on linear measurements," *Electronic Journal of Statistics*, vol. 6, pp. 547–558, 2012. [511, 512, 514]
530. E. Arias-Castro, E. Candes, and M. Davenport, "On the fundamental limits of adaptive sensing," *arXiv preprint arXiv:1111.4646*, 2011. [511]
531. E. Arias-Castro, S. Bubeck, and G. Lugosi, "Detection of correlations," *The Annals of Statistics*, vol. 40, no. 1, pp. 412–435, 2012. [511]
532. E. Arias-Castro, S. Bubeck, and G. Lugosi, "Detecting positive correlations in a multivariate sample," 2012. [511, 517]
533. A. B. Tsybakov, *Introduction to nonparametric estimation*. Springer, 2008. [513, 519]
534. L. Balzano, B. Recht, and R. Nowak, "High-dimensional matched subspace detection when data are missing," in *Information Theory Proceedings (ISIT), 2010 IEEE International Symposium on*, pp. 1638–1642, IEEE, 2010. [514, 516, 517]
535. L. K. Balzano, *Handling missing data in high-dimensional subspace modeling*. PhD thesis, UNIVERSITY OF WISCONSIN, 2012. [514]
536. L. L. Scharf, *Statistical Signal Processing*. Addison-Wesley, 1991. [514]
537. L. L. Scharf and B. Friedlander, "Matched subspace detectors," *Signal Processing, IEEE Transactions on*, vol. 42, no. 8, pp. 2146–2157, 1994. [514]
538. S. Hou, R. C. Qiu, M. Bryant, and M. C. Wicks, "Spectrum Sensing in Cognitive Radio with Robust Principal Component Analysis," in *IEEE Waveform Diversity and Design Conference*, January 2012. [514]
539. C. McDiarmid, "On the method of bounded differences," *Surveys in combinatorics*, vol. 141, no. 1, pp. 148–188, 1989. [516]
540. M. Azizyan and A. Singh, "Subspace detection of high-dimensional vectors using compressive sampling," in *IEEE SSP*, 2012. [516, 517, 518, 519]
541. M. Davenport, M. Wakin, and R. Baraniuk, "Detection and estimation with compressive measurements," tech. rep., Tech. Rep. TREE0610, Rice University ECE Department, 2006, 2006. [516]
542. J. Paredes, Z. Wang, G. Arce, and B. Sadler, "Compressive matched subspace detection," in *Proc. 17th European Signal Processing Conference, Glasgow, Scotland*, pp. 120–124, Citeseer, 2009. []
543. J. Haupt and R. Nowak, "Compressive sampling for signal detection," in *Acoustics, Speech and Signal Processing, 2007. ICASSP 2007. IEEE International Conference on*, vol. 3, pp. III–1509, IEEE, 2007. [516]
544. A. James, "Distributions of matrix variates and latent roots derived from normal samples," *The Annals of Mathematical Statistics*, pp. 475–501, 1964. [517]
545. S. Balakrishnan, M. Kolar, A. Rinaldo, and A. Singh, "Recovering block-structured activations using compressive measurements," *arXiv preprint arXiv:1209.3431*, 2012. [520]
546. R. Muirhead, *Aspects of Mutivariate Statistical Theory*. Wiley, 2005. [521]
547. S. Vempala, *The random projection method*, vol. 65. Amer Mathematical Society, 2005. [521]
548. C. Helstrom, "Quantum detection and estimation theory," *Journal of Statistical Physics*, vol. 1, no. 2, pp. 231–252, 1969. [526]
549. C. Helstrom, "Detection theory and quantum mechanics," *Information and Control*, vol. 10, no. 3, pp. 254–291, 1967. [526]
550. J. Sharpnack, A. Rinaldo, and A. Singh, "Changepoint detection over graphs with the spectral scan statistic," *arXiv preprint arXiv:1206.0773*, 2012. [526]
551. D. Ramírez, J. Vía, I. Santamaría, and L. Scharf, "Locally most powerful invariant tests for correlation and sphericity of gaussian vectors," *arXiv preprint arXiv:1204.5635*, 2012. [526]

552. A. Onatski, M. Moreira, and M. Hallin, "Signal detection in high dimension: The multispiked case," *arXiv preprint arXiv:1210.5663*, 2012. [526]
553. A. Nemirovski, "On tractable approximations of randomly perturbed convex constraints," in *Decision and Control, 2003. Proceedings. 42nd IEEE Conference on*, vol. 3, pp. 2419–2422, IEEE, 2003. [527]
554. A. Nemirovski, "Regular banach spaces and large deviations of random sums," *Paper in progress, E-print: http://www2.isye.gatech.edu/nemirovs*, 2004. [528]
555. D. De Farias and B. Van Roy, "On constraint sampling in the linear programming approach to approximate dynamic programming," *Mathematics of operations research*, vol. 29, no. 3, pp. 462–478, 2004. [528]
556. G. Calafiore and M. Campi, "Uncertain convex programs: randomized solutions and confidence levels," *Mathematical Programming*, vol. 102, no. 1, pp. 25–46, 2005. []
557. E. Erdoğan and G. Iyengar, "Ambiguous chance constrained problems and robust optimization," *Mathematical Programming*, vol. 107, no. 1, pp. 37–61, 2006. []
558. M. Campi and S. Garatti, "The exact feasibility of randomized solutions of uncertain convex programs," *SIAM Journal on Optimization*, vol. 19, no. 3, pp. 1211–1230, 2008. [528]
559. A. Nemirovski and A. Shapiro, "Convex approximations of chance constrained programs," *SIAM Journal on Optimization*, vol. 17, no. 4, pp. 969–996, 2006. [528, 542]
560. A. Nemirovski, "Sums of random symmetric matrices and quadratic optimization under orthogonality constraints," *Mathematical programming*, vol. 109, no. 2, pp. 283–317, 2007. [529, 530, 531, 532, 533, 534, 536, 538, 539, 540, 542]
561. S. Janson, "Large deviations for sums of partly dependent random variables," *Random Structures & Algorithms*, vol. 24, no. 3, pp. 234–248, 2004. [542]
562. S. Cheung, A. So, and K. Wang, "Chance-constrained linear matrix inequalities with dependent perturbations: A safe tractable approximation approach," *Preprint*, 2011. [542, 543, 547]
563. A. Ben-Tal and A. Nemirovski, "On safe tractable approximations of chance-constrained linear matrix inequalities," *Mathematics of Operations Research*, vol. 34, no. 1, pp. 1–25, 2009. [542]
564. A. Gershman, N. Sidiropoulos, S. Shahbazpanahi, M. Bengtsson, and B. Ottersten, "Convex optimization-based beamforming," *Signal Processing Magazine, IEEE*, vol. 27, no. 3, pp. 62–75, 2010. [544]
565. Z. Luo, W. Ma, A. So, Y. Ye, and S. Zhang, "Semidefinite relaxation of quadratic optimization problems," *Signal Processing Magazine, IEEE*, vol. 27, no. 3, pp. 20–34, 2010. [544, 545]
566. E. Karipidis, N. Sidiropoulos, and Z. Luo, "Quality of service and max-min fair transmit beamforming to multiple cochannel multicast groups," *Signal Processing, IEEE Transactions on*, vol. 56, no. 3, pp. 1268–1279, 2008. [545]
567. Z. Hu, R. Qiu, and J. P. Browning, "Joint Amplify-and-Forward Relay and Cooperative Jamming with Probabilistic Security Consideration," 2013. submitted to IEEE COMMUNICATIONS LETTERS. [547]
568. H.-M. Wang, M. Luo, X.-G. Xia, and Q. Yin, "Joint cooperative beamforming and jamming to secure af relay systems with individual power constraint and no eavesdropper's csi," *Signal Processing Letters, IEEE*, vol. 20, no. 1, pp. 39–42, 2013. [547]
569. Y. Yang, Q. Li, W.-K. Ma, J. Ge, and P. Ching, "Cooperative secure beamforming for af relay networks with multiple eavesdroppers," *Signal Processing Letters, IEEE*, vol. 20, no. 1, pp. 35–38, 2013. [547]
570. D. Ponukumati, F. Gao, and C. Xing, "Robust peer-to-peer relay beamforming: A probabilistic approach," 2013. [548, 556]
571. S. Kandukuri and S. Boyd, "Optimal power control in interference-limited fading wireless channels with outage-probability specifications," *Wireless Communications, IEEE Transactions on*, vol. 1, no. 1, pp. 46–55, 2002. [552]
572. S. Ma and D. Sun, "Chance constrained robust beamforming in cognitive radio networks," 2013. [552]

573. D. Zhang, Y. Hu, J. Ye, X. Li, and X. He, "Matrix completion by truncated nuclear norm regularization," in *Computer Vision and Pattern Recognition (CVPR), 2012 IEEE Conference on*, pp. 2192–2199, IEEE, 2012. [555]
574. K.-Y. Wang, T.-H. Chang, W.-K. Ma, A.-C. So, and C.-Y. Chi, "Probabilistic sinr constrained robust transmit beamforming: A bernstein-type inequality based conservative approach," in *Acoustics, Speech and Signal Processing (ICASSP), 2011 IEEE International Conference on*, pp. 3080–3083, IEEE, 2011. [556]
575. K.-Y. Wang, T.-H. Chang, W.-K. Ma, and C.-Y. Chi, "A semidefinite relaxation based conservative approach to robust transmit beamforming with probabilistic sinr constraints," in *Proc. EUSIPCO*, pp. 23–27, 2010. [556]
576. K. Yang, J. Huang, Y. Wu, X. Wang, and M. Chiang, "Distributed robust optimization (dro) part i: Framework and example," 2008. [556]
577. K. Yang, Y. Wu, J. Huang, X. Wang, and S. Verdú, "Distributed robust optimization for communication networks," in *INFOCOM 2008. The 27th Conference on Computer Communications. IEEE*, pp. 1157–1165, IEEE, 2008. [556]
578. M. Chen and M. Chiang, "Distributed optimization in networking: Recent advances in combinatorial and robust formulations," *Modeling and Optimization: Theory and Applications*, pp. 25–52, 2012. [556]
579. G. Calafiore, F. Dabbene, and R. Tempo, "Randomized algorithms in robust control," in *Decision and Control, 2003. Proceedings. 42nd IEEE Conference on*, vol. 2, pp. 1908–1913, IEEE, 2003. [556]
580. R. Tempo, G. Calafiore, and F. Dabbene, *Randomized algorithms for analysis and control of uncertain systems*. Springer, 2004. []
581. X. Chen, J. Aravena, and K. Zhou, "Risk analysis in robust control-making the case for probabilistic robust control," in *American Control Conference, 2005. Proceedings of the 2005*, pp. 1533–1538, IEEE, 2005. []
582. Z. Zhou and R. Cogill, "An algorithm for state constrained stochastic linear-quadratic control," in *American Control Conference (ACC), 2011*, pp. 1476–1481, IEEE, 2011. [556]
583. A. M.-C. So and Y. J. A. Zhang, "Distributionally robust slow adaptive ofdma with soft qos via linear programming," *in IEEE J. Sel. Areas Commun.[Online]. Available: http://www1.se.cuhk.edu.hk/~manchoso*. [558]
584. C. Boutsidis and A. Gittens, "Im@articlegittens2012var, title=Var (Xjk), author=GITTENS, A., year=2012 ," *arXiv preprint arXiv:1204.0062*, 2012. [559, 560]
585. A. GITTENS, "Var (xjk)," 2012. [559]
586. M. Magdon-Ismail, "Using a non-commutative bernstein bound to approximate some matrix algorithms in the spectral norm," *Arxiv preprint arXiv:1103.5453*, 2011. [560, 561]
587. M. Magdon-Ismail, "Row sampling for matrix algorithms via a non-commutative bernstein bound," *Arxiv preprint arXiv:1008.0587*, 2010. [561]
588. C. Faloutsos, T. Kolda, and J. Sun, "Mining large time-evolving data using matrix and tensor tools," in *ICDM Conference*, 2007. [565]
589. M. Mahoney, "Randomized algorithms for matrices and data," *Arxiv preprint arXiv:1104.5557*, 2011. [566]
590. R. Ranganathan, R. Qiu, Z. Hu, S. Hou, Z. Chen, M. Pazos-Revilla, and N. Guo, *Communication and Networking in Smart Grids*, ch. Cognitive Radio Network for Smart Grid. Auerbach Publications, Taylor & Francis Group, CRC, 2013. [569]
591. R. C. Qiu, Z. Chen, N. Guo, Y. Song, P. Zhang, H. Li, and L. Lai, "Towards a real-time cognitive radio network testbed: architecture, hardware platform, and application to smart grid," in *Networking Technologies for Software Defined Radio (SDR) Networks, 2010 Fifth IEEE Workshop on*, pp. 1–6, IEEE, 2010. [572]
592. R. Qiu, Z. Hu, Z. Chen, N. Guo, R. Ranganathan, S. Hou, and G. Zheng, "Cognitive radio network for the smart grid: Experimental system architecture, control algorithms, security, and microgrid testbed," *Smart Grid, IEEE Transactions on*, no. 99, pp. 1–18, 2011. [574]

593. R. Qiu, C. Zhang, Z. Hu, , and M. Wicks, "Towards a large-scale cognitive radio network testbed: Spectrum sensing, system architecture, and distributed sensing," *Journal of Communications*, vol. 7, pp. 552–566, July 2012. [572]
594. E. Estrada, *The structure of complex networks: theory and applications*. Oxford University Press, 2012. [574]
595. M. Newman, *Network: An Introduction*. Oxford University Press, 2010. []
596. M. Newman, "The structure and function of complex networks," *SIAM review*, vol. 45, no. 2, pp. 167–256, 2003. []
597. M. Newman and M. Girvan, "Finding and evaluating community structure in networks," *Physical review E*, vol. 69, no. 2, p. 026113, 2004. []
598. S. Strogatz, "Exploring complex networks," *Nature*, vol. 410, no. 6825, pp. 268–276, 2001. [574]
599. A. Barrat, M. Barthelemy, and A. Vespignani, *Dynamical processes on complex networks*. Cambridge University Press, 2008. [574]
600. M. Franceschetti and R. Meester, *Random networks for communication: from statistical physics to information systems*. Cambridge University Press, 2007. [574]
601. P. Van Mieghem, *Graph spectra for complex networks*. Cambridge University Press, 2011. [574, 575]
602. D. Cvekovic, M. Doob, and H. Sachs, *Spectra of graphs: theory and applications*. Academic Press, 1980. [575, 576]
603. C. Godsil and G. Royle, *Algebraic graph theory*. Springer, 2001. [575]
604. F. Chung and M. Radcliffe, "On the spectra of general random graphs," *the electronic journal of combinatorics*, vol. 18, no. P215, p. 1, 2011. [575]
605. R. Oliveira, "Concentration of the adjacency matrix and of the laplacian in random graphs with independent edges," *arXiv preprint arXiv:0911.0600*, 2009. []
606. R. Oliveira, "The spectrum of random k-lifts of large graphs (with possibly large k)," *arXiv preprint arXiv:0911.4741*, 2009. []
607. A. Gundert and U. Wagner, "On laplacians of random complexes," in *Proceedings of the 2012 symposuim on Computational Geometry*, pp. 151–160, ACM, 2012. []
608. L. Lu and X. Peng, "Loose laplacian spectra of random hypergraphs," *Random Structures & Algorithms*, 2012. [575]
609. F. Chung, *Speatral graph theory*. American Mathematical Society, 1997. [576]
610. B. Osting, C. Brune, and S. Osher, "Optimal data collection for improved rankings expose well-connected graphs," *arXiv preprint arXiv:1207.6430*, 2012. [576]

Index

R. Qiu and M. Wicks, *Cognitive Networked Sensing and Big Data*,
DOI 10.1007/978-1-4614-4544-9,

D